BOILERS

A PRACTICAL REFERENCE

BOILERS

A PRACTICAL REFERENCE

KUMAR RAYAPROLU

CRC Press is an imprint of the
Taylor & Francis Group, an **informa** business

CRC Press
Taylor & Francis Group
6000 Broken Sound Parkway NW, Suite 300
Boca Raton, FL 33487-2742

First issued in paperback 2017

Version Date: 20121011

ISBN 13: 978-1-4665-0053-2 (hbk)
ISBN 13: 978-1-138-07327-2 (pbk)

Visit the Taylor & Francis Web site at
http://www.taylorandfrancis.com

and the CRC Press Web site at
http://www.crcpress.com

To my late parents for all that they had done

To my wife Usha

and daughters Ramya and Amulya

without whose tacit support and help this book would not have been possible

Contents

List of Figures

List of Tables

Preface

In his long association with boilers, the author has always found that there have been very few books on the subject of boilers for practising engineers. Books from the academic community tend to have more stress on theory, which does not quite sustain the interest of practising engineers, as the issues in the actual field are very different. There are a few books by working professionals but they are more on operational aspects.

To contribute to this acute shortage of literature on boilers, the author compiled his first book, *Boilers for Power and Process,* which was published in the USA in early 2009 by CRC Press. This was a structured exposition of the subject covering the fundamentals, engineering and major types of boilers based on different combustion techniques. Since then, there were several requests that another book be written, which should be even more user friendly, where one can get to the subject matter straightaway without having to search in the usual manner. This book, written in the format of an encyclopaedia, is the result of such a suggestion.

The word 'boiler' means different things to different people. From a small device to heat domestic water to the gigantic generators producing steam at supercritical conditions in large power stations, all of them pass for boilers. However, to provide focus and bring out the personal expertise, this book restricts itself to water tube boilers as found in the process industries and power plants. This includes fired and unfired process waste heat boilers and those behind gas turbines. Specifically excluded are fire tube (except for a passing mention), marine and miniature boilers as well as the boilers for nuclear power plants.

There are around 550 key boiler words in the book which are elaborated along with nearly the same number of illustrations in support. It was a revelation of a kind to learn that in these few words, almost the entire boiler technology can be covered.

The book explains, broadly, the following topics:

- Almost the whole range of boilers and main auxiliaries, along with steam and gas turbines
- Traditional firing techniques like grates, oil/gas and PF and modern systems like FBC, Hrsg and so on
- Industrial, utility, waste heat, MSW and bio-fuel-fired boilers, including supercritical boilers
- The underlying scientific fundamentals of combustion, heat transfer, fluid flow and so on as relevant
- Basics of fuels, water, ash, high-temperature steels, structurals, refractory, insulation and so on
- Other engineering topics like boiler instruments, controls, welding, corrosion, wear and so on
- Air pollution, its abatement techniques and their effect on the design of boilers and auxiliaries
- Emerging technologies like carbon capture, oxy-fuel combustion, PFBC and so on

Any effort of this type is sure to fall short of expectations simply because of the vastness of the subject. Also, there is no 'syllabus'. However, in the experience of the author, almost all topics needed by boiler engineers in process and power plants are covered here.

For want of a more appropriate word, the book is titled as a 'reference'. It would be more appropriate to name this work as an encyclopaedia but it would fall short in size and therefore the expectations of the readers. By naming it as 'Practical Reference', the author hopes to convey that the book is a reference manual directed more towards practising engineers.

By its size and focus, it is a regular professional book, to be used by boiler engineers of all walks as a desk book for constant reference. Design, project, operation, consulting engineers connected with boilers should find the book of good use. Students in power plant and heat power engineering, seeking a secondary reference, would also find the volume useful. It is strong on fundamentals and design aspects, besides being elaborate on the practical side. The numerous pictures should greatly aid in enhancing understanding, as they are very carefully chosen to add to the explanation and not serve as repetition.

Finally, it is hoped that this unique book of reference, the like of which is not there in the market, besides adding to the sparsely filled gallery of practical books on boilers, serves the needs of serious readers well.

Kumar Rayaprolu

Acknowledgements

Both active and passive support from various quarters is involved in compiling a volume of this size and breadth. The least that a grateful author could do is to acknowledge their contribution for wider appreciation. Accordingly, I would like to thank the following organisations for lending their support by allowing their pictures to be incorporated, which greatly enhances the value of the book. A list of such illustrations is attached in picture credits.

Boiler Makers:

The Babcock and Wilcox Company, USA
Foster Wheeler Corporation, USA
Riley Power Inc., USA
Innovative Steam Technologies, Canada
Hitachi Power Europe GmbH, Germany

Fuel Feeding Equipment Makers:

Maschinenfabrik Besta & Meyer GmbH & Co. KG M.A., Germany

Fan Makers:

TLT-Turbo GmbH, Germany

I should also place on record the help offered by Mr. Y.M. Joshi in the secretarial area and Mr. Pawan Umarji in providing many pictures.

Author

Kumar Rayaprolu has been associated with boilers of various types in different capacities since his graduation in 1969, starting as a postgraduate trainee with the erstwhile ACC-Vickers-Babcock Limited (AVB). This was a subsidiary of the former Babcock and Wilcox Limited of the UK, established in the mid-1960s in India for making utility and industrial boilers. After a year on the shop floor, he was selected for a two-year post-graduate training with the parent company at Renfrew (Scotland) and London and also various construction sites, where he could get to learn boilers and firing equipment design and boiler commissioning. B&W Ltd. at that time was Europe's largest boiler maker and naturally could provide some excellent learning opportunities.

After returning to India, he spent the next five years in commissioning and proposals of utility and industrial boilers with AVB. He moved on to work with a boiler company making industrial boilers (mainly stoker-, oil- and bagasse-fired boilers) to the designs of Foster Wheeler (Canada) in 1978 and then with a consulting engineering company, Engineers India Ltd., which was originally founded as Bechtel India. Each assignment lasted for two to three years.

FBC boilers made their first entry into India in the early 1980s and the author moved to a JV formed for making ignifluid boilers with the former Fives-Cail-Babcock of France, to head their Engineering and later both Engineering and Projects. After six years, there was an opportunity to head the marketing function of the new JV–Thermax Babcock Limited, formed for a variety of industrial boilers with Babcock and Wilcox of the USA. After three years, the author received the opportunity to head the newly formed boiler division at Krupp Industries Limited in 1990, which obtained the licence to make CFBC boilers from the former Deutsche Babcock of Germany.

The 10-year period that followed was very creative as it provided a pioneering opportunity of building CFBC boilers in India, besides establishing a new business unit and nurturing it to health and growth. He moved in 2000 to India's largest engineering company, Larsen and Toubro Limited, to head their Captive and Cogeneration Power and Hrsg divisions for five years. Finally, he spent the last three years as the president of utility boilers trying to establish a new business line.

The author considers himself rather fortunate in having an early mover advantage in the 1970s when boiler manufacturing had started gathering momentum in India. Naturally, he enjoyed the benefits of extensive interaction with European and American boiler makers, who were the technology partners those days. This helped in an accelerated learning and vast exposure. As part of the technology transfer process, the author could visit the manufacturing and engineering facilities of many leaders in the industry such as Babcock of UK, USA, Canada and Germany, Foster Wheeler of Canada and the USA, LLB of Germany and so on. The author has gathered a well-rounded and overall experience of design, development, engineering, project execution, estimation, proposals, marketing, manufacturing and commissioning. He was a business head for more than 15 years but managed to retain engineering function all through his career. The author has had an uninterrupted association with boilers for over four decades and has naturally developed some deep technical and business insights.

The author's first book on boilers, *Boilers for Power and Process*, was published in April 2009 by CRC Press in the USA and was well received. This is the author's second book.

Structure of the Book

The structure of the book is quite simple. Words are all explained alphabetically. At the same time, all words related to any particular topic are grouped together so that the entire subject matter stays intact at one place. The reader is suitably directed. This way, the reader gets the explanation for the word he is searching for and also sees all the related words and their explanations, getting to learn about the whole topic if so interested. In addition, repetition of description is avoided as the full topic is elaborated at one place and not fragmented.

Also, all related words in a description are bold to indicate that they are elaborated elsewhere. The reader can learn the explanation of the word as well as the entire subject when the related words are gone through.

In my long association with boilers, I have worked with quite a few organisations which were in competition in some areas. This has helped me to develop an understanding which is technology and company neutral. This is reflected throughout in the book.

The book is extensively illustrated to supplement the explanation. In most cases, the pictures describe more than what is contained in the text and the readers are requested to note this fact and get benefitted.

In a book of this type, each chapter would contain descriptions of several diverse topics. It was, therefore, felt that it will be appropriate for all the references and further reading material to be grouped and placed at the end of the book, listed topic wise.

A conscious effort is made to keep explanations as direct and brief as possible. Further, to make the text short for ease of reading, chemical symbols and other standard abbreviations are used extensively. These are listed at the rear of the book and the reader is urged to familiarise himself with them.

A Table of Combustion constants and a Periodic Table of Elements are included as Appendices which are needed for reference many times. Also attached there is a write up on the Coals of the World.

SI units are used in the book and in most places equivalent mks and Imperial units are also provided for the convenience of readers. It may be noted that conversions are rounded off suitably. Conversions of commonly used units are given at the beginning of the book in the front matter.

As indicated earlier, certain related words in many paragraphs are underlined to denote that they are elaborated elsewhere in the book. This cross referencing should improve the readability and utility of the book.

Finally, it is hoped that the book meets its primary object of making the learning of a big subject like boilers a little easier for the reader community.

Conversion of Units

Length

1 inch	2.54 cm
	25.4 mm
1 foot	12″
	30.48 cm
	0.333 yards
1 cm	0.3937″
1 mm	0.03937″
	1000 microns
1 m	3.281′
	39.37″
1 mil or thou	1/1000″
	0.0254 mm
	25.4 μm
1 micron	10^{-6} m
	10^{-3} mm

Area

1 in²	6.45 cm²
	645 mm²
1 ft²	144 in²
	929 cm²
	0.0929 m²
1 cm²	100 mm²
	0.155 in²
1 m²	10,000 cm²
	10.76 ft²
1 acre	4047 m²
	4840 yards²
	43545 ft²
	0.405 ha
1 ha	10,000 m²
	2.47 acres

Volume

1 in³	16.39 cm³/mL
1 cc/mL	0.061 in³
1 ft³	1728 in³
	28.32 L
	0.0283 m³
1 L	61 in³
	1000 cc
	0.22 Imperial gallon
	0.2642 US gallon
1 m³	1000 L
	35.31 ft³
	220 Imperial gallons
	264.2 US gallons
	6.29 barrels
1 oz	28.13 cc
1 Imperial gallon	277.4 in³
	4.55 L
1 US gallon	231 in³
	3.875 L
	0.833 Imperial gallon
1 barrel (oil)	42 US gallons
	34.97 Imperial gallons
	159 L
	0.159 m³

Weight

1 US long ton (t)	2240 lb
	1016 kg
	1.016 te
1 metric ton (Mt) or 1 tonne (te)	2205 lb
	1000 kg
	0.984 long tons
	1.102 short tons
1 short ton	2000 tons
	907 kg
1 lb	16 ounces (oz)
	7000 grains
	454 g
	0.454 kg
1 kg	2.205 lb
1 gram (g)	15.43 grains
1 ounce (oz)	28.35 g
	437.5 gr
1 grain (gr)	0.0648 g
	64.8 mg
	0.0023 oz

Force

1 lbf	4.45 Newton
1 kgf	9.807 N
1 Newton	0.248 lbf
	0.102 kgf
1 tonf	9.964 kN
1 tef	9.807 kN

Pressure

1 atmosphere (physical)	760 mm Hg
	10 m of water
	14.696 psi
	1.033 kg/cm²
	1013 mbar
1 atm (metric) or 1 kg/cm²	736 mm Hg
	9.684 m of water
	14.22 psi
	981 mbar
1 bar	1000 mbar
	10^5 Pa
	100 kPa
	10^6 dynes/cm²
	750 mm Hg
	1.02 kg/cm²
	14.5 psi
1 mbar	0.4″ wg
	10.02 mm wg
	100 Pa
1 Pa	1 N/m²
	1 kg/m-s²
	0.102 mm wg
	0.01 mbar
1 kPa	1000 Pa
	10 mbar
	102 mm wg
1 lb/ft²	0.1922″ wg
	4.88 kg/m²
	47.88 Pa
1 psi	2.036″ Hg
	2.307′ water
	.0703 kg/cm²
	703 mm wg
	0.0695 bar
	68.947 mbar

1 kg/m^2	1 mm wg
	9.81 Pa
	0.2048 lb/ft^2
1 kg/mm^2	0.635 tons/in^2
	0.645 te/mm^2
1″ wg	5.2 lb/ft^2
	0.036 psi
	249 Pa
	2.49 mbar
1′ water head	0.434 psi
	28.89 mbar
1 m water head	0.1 kg/cm^2
	9807 Pa
	73.6 mm Hg
	1.42 psi
	39.4 in wg
	2.9 in Hg
1 mm water head	9.807 Pa
	0.98 kPa
	0.0394 in wg
1″ Hg	0.491 psi
	3386.4 Pa
	33.864 mbar
1 m Hg	1.36 kg/cm^2
	1333 mbar
	1000 torr
1 ton/in^2	1.575 kg/cm^2
	15.44 MPa
1 MPa or 1 N/mm^2	145 psi
	10.2 kg/cm^2
	10 bar

Velocity

1 fps	0.3048 m/s
1 fpm	0.0051 m/s
	0.3048 m/m
	18.29 m/h
1 mps	3.28 fps
	196.8 fpm
1 m/min	0.0547 fps
	3.28 fpm

Mass Velocity

1 lb/ft^2 h	4.8824 kg/m^2 h
	1.355×10^{-3} kg/m^2 s
1 kg/m^2 h	0.2048 lb/ft^2 h
1 kg/m^2 s	738 lbs/f^{2}t h

Volume Flow

1 ft^3/s	0.0283 m^3/h
1 ft^3/min	0.472×10^{-3} m^3/s 0.0283 m^3/min
	1.699 m^3/h
1 m^3/h	0.589 ft^3/min
	3.675 I.gpm
	4.41 US gpm

Density

1 ft^3/lb	0.0624 m^3/kg
1 lb/ft^3	16.02 kg/m^3
1 gr/ft^3	2.288 g/m^3
1 m^3/kg	16.02 ft^3/lb
1 kg/m^3	.0624 lb/ft^3

Concentration

1 gr/ft^3	2.288 g/m^3
1 g/m^3	0.437 gr/ft^3
	0. 0703 gr/Igpm
1 gr/I.gallon	1426 gr/m^3
	14.26 mg/L
1 g/L	70.2 gr/I. gallon

Power and Heat

1 Therm	10,000 Btu
1 Btu	778 ft lb
	107.6 kg m
	0.2520 kcal
	1.055 kJ
1 kcal	3088 ft lb
	427 kg m
	3.968 Btu
	4.1868 kJ
Watt (power/heat flow)	1 J/s
	1 Nm/s
	1 kg m^2/s^2
	738 ft lb/s
Kw (power/heat flow)	102 kg m/s
	1.341 hp
	1.360 metric hp (PS)
	3413 Btu/h
	860 kcal/h
1 MW thermal	3.413×10^6 Btu/h
	860×10^3 cal/h
1 horse power (hp)	33,000 ft lb/min
	550 ft lb/s
	76 kg m/s
	1.014 PS
	0.746 kw
1 metric hp 1 PS (pferde starke)	32550 ft lb/min
	542 ft lb/s
	0,986 hp
	0.735 kw
kWh (energy/work)	3413 btu
	860 kcal
	3.6 MJ
1 Btu/h	0.293 W
1 joule	10^7 ergs
	0.7375 ft lb
	0.000948 Btu
	1 W s
	1 kg m^2/s^2
1 Btu/lb (calorific value)	0.556 kcal/kg
	2.326 kJ/kg
1 kcal/kg	1.8 Btu/lb
	4.1868 kJ/kg
1 kJ/kg	0.423 Btu/lb
	0.2352 kcal/kg
1 MJ/kg	423 Btu/lb
	235.2 kcal/kg
1 Btu/ft^3/h (volumetric release)	8.90 kcal/m^3/h
	10.3497 W/m^3
1 W/m^3	0.86 kcal/m^3 h
	0.0966 Btu/ft^3 h
1 kW/m^3	860 kcals/m^3 h
	96.6 Btu/ft^3 h
1 kcal/m^3/h	0.1124 Btu/ft^3/h
1 Btu/ft^2/h (heat flux)	2.712 kcal/m^2/h
	3.154 W/m^2
1 kcal/m^2/h	0.3687 Btu/ft^2/h

	0.2846 kcal/m^2/h
	4.1868 kJ/m^2
1 MW/m^2	317193 Btu/ft^2
	860,000 kcal/m^2/h
1 Btu/lb °F (specific entropy)	4.1868 kJ/kg K
1 kJ/kg K	0.2388 Btu/lb °F
1 Btu/ft^{2}h · °F (heat transfer coefficient)	4.88 kcal/m^{2}h °C
	5.678 W/m^2 °K
1 kcal/m^{2}h °C	0.2048 Btu/ft^2 h°F
	1.163 W/m^2 °K
1 W/m^2 °K	0.1761 Btu/ft^2 h°F
	0.860 kcal/m^2 h°C
1 Btu/ft^2 °hF/ft (thermal conductivity)	1.488 kcal/m^{2}h C/m
	1.731 W/m-k
1 kcal/m^2 °hC/m	0.672 Btu/ft^2 °hF/ft
	8.06 Btu/ft^2 h°F/in
1 Btu/ft^2 °hF/in	0.124 kcal/m^{2}h C/m
	0.144 W/m-k
kJ/kWh (Heat rate)	0.948 Btu/kWh
	0.144 W/m °K
1 kcal/kWh	3.968 Btu/kWh
	4.1868 kJ/kWh
1 Btu/kWh	0.252 kcal/kWh
	1.055kJ/kWh
Water at 16.7°C (62°F)	
1 ft^3	62.3 lb
1 lb	0.01604 ft^3
1 I. gallon	10 lb
Water at 4°C (39.2°F)	
1 ft^3	62.4 lb
1 lb	0.01602 ft^3
1 m^3	1000 kg
1 L	1 kg
1 kg/m^3	1 part/1000 or 1 g/L
1 g/m^3	1 ppm or 1 mg/L
Viscosity—Dynamic	
1 Poise (P)	0.1 N s/m^2
	0.1 kg/m s
	0.1 Pa s
	0.00209 lb s/ft^2
1 centi Poise (cP)	0.01 P
	1 m Pa s
Viscosity—Absolute/Kinematic	
1 Stoke (st)	1 cm^2/s
	100 mm^2/s
	10.76×10^{-4} ft^2 s
1 centistoke (cSt)	1 mm^2/s
	10.76×10^{-6} ft^2 s
1 ft^2/s	92,903 mm^2/s
Emission: NO_x	
1 ppm dv (dry volume)	1.912 mg/DSCM
	2.05 mg/Nm^3
	1/1196 gr/DSCF
1 mg/Nm^3	0.932 mg/DSCM
	0.487 ppm
	1/2455 gr/DSCF
1 lb/M Btu on GCV @ 3% O2 (±2% accuracy)	760 ppm on NG
	800 ppm on RG
	725 ppm on FO
1 lb/M Btu on GCV @ 6% O2 (±2% accuracy)	598 ppm coal
	1230 mg/DSCM
Emission: SO_2	
1 ppm dv (dry volume)	2.66 mg/DSCM
	2.857 mg/Nm^3
	1/860 gr/DSCF
1 mg/Nm^3	0.932 mg/DSCM
	0.35 ppm
	1/2455 gr/DSCF
Emission: CO	
1 ppm dv (dry volume)	1.164 mg/DSCM
	1.25 mg/Nm^3
	1/1965 gr/DSCF
1 mg/Nm^3	0.932 mg/DSCM
	1.250 ppm
	1/2455 gr/DSCF
Emission: VOC	
1 ppm dv (dry volume)	0.666 mg/DSCM
	0.714 mg/Nm^3
	1/3440 gr/DSCF
1 mg/Nm^3	0.932 mg/DSCM
	1.40 ppm
	1/2455 gr/DSCF

Picture Credits

S. No.	Firm	Country	S. No.	Figure No.	Description
1	B&W	USA	1	1.5	Permissible minimum metal temperature versus sulphur in fuel for rotary AHs
			2	1.14	Permissible carbon steel tube temperature versus S in fuel for tubular AH
			3	2.5	Lower furnace of a BFBC boiler with open bottom arrangement
			4	8.27	Typical pendant SH in a large utility boiler
			5	8.35	Combined cycle power generation using GT, Hrsg and ST—isometric and schematic presentations
			6	9.32	Typical single drum high pressure non-reheat IR-CFB boiler for CPP
			7	9.33	Cross section of U-beams
			8	15.11	A small single burner oil and gas fired FM boiler
			9	16.21	A large ring roll pulveriser
			10	20.3	A modified tower type boiler with drainable eco in second pass
			11	20.12	Furnace tube arrangement in UP boiler
			12	20.22	Typical start up schematic of spiral walled SC boiler with variable operation
2	FWC	USA	1	3.5	Water- or steam-cooled cyclone
			2	3.8	'Intrex' coil arrangement in compact boilers of FW
			3	3.15	'Compact'-type utility range reheat CFBC boiler of FW
			4	16.27	Combustion flexibility of CFBC compared to PF firing
			5	16.33	Long flames in the down shot firing
			6	20.7	W-type down shot fired boiler with divided back pass
			7	20.24	460 MWe SC RH boiler of compact design in operation in Poland
			8	22.1	Solid wastes arranged by their heating value and ease of burning
3	Riley	USA	1	16.19	Roller- and bowl-type MPS mill
			2	16.23	High-speed horizontal Atrita mills
			3	16.24	Milling plant in a boiler
4	IST	Canada	1	15.17	Size comparison of Otsg and Hrsg behind LM 6000 turbine
5	HPE	Germany	1	6.43	Gravimetric belt feeder
			2	12.5	Beater mill for medium moisture in fuel—isometric and cross-sectional views
			3	12.6	Beater mill for high moisture in fuel—isometric and cross-sectional views
			4	12.10	A large tower type SC brown coal-fired boiler with eight mills and three levels of firing
			5	16.20	Low speed horizontal ball mills—isometric and cross-sectional views
			6	18.18	Isometric view of the boiler structure for a large top supported utility boiler
			7	20.1	Tower-type SC boiler with opposed firing SOFA and SCR arrangements
6	TLT	Germany	1	1.25	Cross-sectional views of a twin stage axial fan supported on anti vibration mountings
			2	1.26	Vertical axial flow fans
			3	1.27	Various arrangements of axial fans
			4	1.28	A comparison of characteristic curves of centrifugal and axial flow fan
7	Besta	Germany	1	6.41	Drag chain volumetric feeder
			2	6.42	Drag chain feeder in assembled condition
			3	12.2	Plate belt feeder with twin fuel inlets
			4	12.3	Isometric view of a plate belt feeder

1

A

1.1 Abrasion and Abrasion Index (*see* Wear)

1.2 Absolute or Dynamic Viscosity (μ) (*see* Viscosity in Fluid Characteristics)

1.3 Acid Cleaning (*see* Commissioning)

1.4 Acid Rain (*see also* Air Pollution Emissions and Controls *and* Gas Cleaning)

Acid rain is a broad term that refers to a *mixture of wet and dry acidic deposition from the atmosphere* containing higher than normal amounts of HNO_3 and H_2SO_4. Primarily, the emissions of SO_2 and NO_x resulting from fossil fuel combustion, mainly from automobiles and power plants, are responsible. Acid rain occurs when these gases react in the atmosphere with H_2O, O_2 and other chemicals to form various acidic compounds and precipitate along with rain, fog, snow and dew. The pH value is often as low as 4, which is highly detrimental to plants, aquatic animals and the infrastructure. The deposition need not be always wet. In tropical countries, it is in dry form, carried by the airborne particulate matter (PM). SO_2 and NO_x released from power plants and other sources are blown by the prevailing winds sometimes over hundreds of kilometres. Figure 1.1 taken from EPA publication illustrates the concept of acid rain very lucidly.

Acid rain is a serious matter as it contributes to, mainly,

- Acidification of lakes and streams
- Damage of trees at high elevations (e.g., red spruce trees above 600 m)
- Damage of many sensitive forest soils
- Decay of building materials and paints, including statues and sculptures as well as cars
- Visibility degradation
- Harm to public health in general

A serious effort from all governments, world over, since the 1970s has mitigated the problem to a great extent, but a lot more still needs to be done.

As the main offenders are **SO_x** and **NO_x gases cleaning** in boilers is invariably undertaken to limit the emissions to the accepted levels. These measures are described at various places in the book.

1.4.1 Continuous Emission Monitoring Systems

In the 1970s and 1980s, continuous emission monitoring systems (CEMS) were installed primarily to monitor boiler flue gas for O_2, CO_2 and CO, mainly for the optimisation of **combustion** control. However, with *air pollution* gaining a lot of attention CEMS are currently used as online means to comply with air emission standards such as the EPA's Acid Rain Program. CEMS today continuously collect, record and report the emissions data as required by the local regulations.

Standard CEMS consist of a sample probe, a filter, a sample line, a gas conditioning and a gas calibration

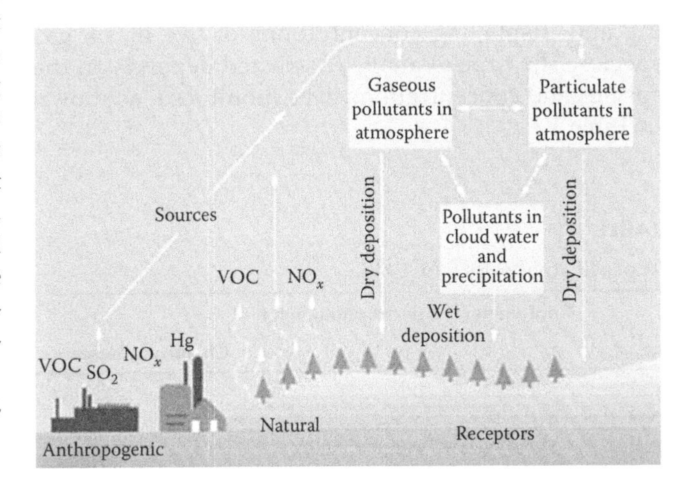

FIGURE 1.1
Acid rain phenomenon.

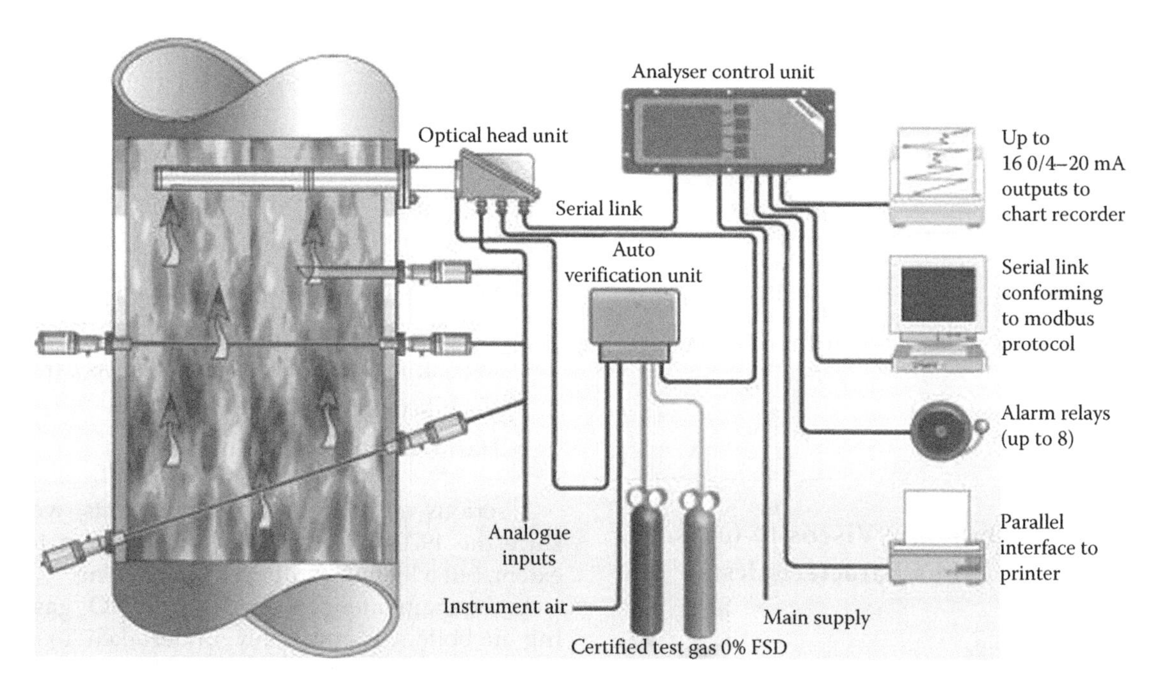

FIGURE 1.2
In situ integrated continuous stack monitoring system.

equipment, and a series of gas analysers for the parameters observed. Typical monitored emissions include SO_2, NO_x, CO, CO_2, O_2, HCl, Hg, volatile organic compounds/hydrocarbons (VOC/HC) and PM. CEMS can also measure flue gas flow, opacity and moisture (M). A typical *in situ* integrated stack monitoring system is shown in Figure 1.2. Besides the gases, PM and O_2 are also frequently measured.

In the past, each analyser was designed to monitor a single gas species, thereby needing a series of analysers, if multiple gas analysis was required. The modern CEMS is capable of simultaneously monitoring and displaying concentrations of five or six gas species. The type of analyser selected depends on the species and concentrations to be monitored, as shown in Table 1.1.

TABLE 1.1
Types of Analysers for CEMS

	Photometer		Spectrophotometer		
Species	IR	UV	IR	UV	Chemiluminescence
CO	X		X		
NO	X	X	X	X	X
SO_2	X	X	X	X	
NO_2	X	X	X	X	X
CO_2	X		X		
H_2O	X		X	X	

1.5 Acid Sludge (*see* Refuse Fuels from Refinery in Liquid Fuels)

1.6 Acid Smuts (*see* Oil Ash)

1.7 Acoustic Soot Blowers (*see* Sonic Horns)

1.8 Acoustic Enclosure (*see* Noise Control)

1.9 Acoustic Leak Detection System

Boiler tube leakage is a major cause of boiler outage in most coal-fired plants. An early detection can

- Simplify the repair
- Reduce downtime
- Effect big savings

mainly because considerable secondary damage to pressure parts (PPs) and refractory which follows tube leaks is avoided. It is the secondary damage which can prolong the repair work and make it expensive. In the case of black liquor recovery (BLR) boilers, the much-dreaded explosion on account of smelt water reaction can be averted.

A tube leakage produces a characteristic hissing noise which the plant operators can detect. There are also changes in the furnace pressure, increase of feed water (fw) flow to compensate for the leak flow, increased makeup water flow to maintain the de-aerator level, and so on, which an alert operating team would notice. But it may take some time for the leak to grow big enough to show the changes. Sometimes it takes as many as 72 h for a small leak to grow big enough for detection.

Acoustic leak detection monitoring systems employ sensors and microphones to detect high-frequency noises emitted from leaking boiler tubes. Piezoelectric sensors mounted on the boiler structure transform the sound waves into electrical voltage, which are then amplified, filtered, and processed to indicate the leakage size and exact location with the aid of an online distributed control system (**DCS**).

1.10 Adiabatic Flame Temperature (*see* Combustion)

1.11 Aeroderivative (*see* Types of GTs in Turbines, Gas)

1.12 Ageing of Boiler Components

Ageing or degradation is the steady decrease in the strength of materials with time. Ageing of boiler components can be attributed to four main factors, namely, *corrosion, erosion, fatigue* and overheating.

1.12.1 Corrosion

Corrosion can affect the PPs both on the inside and outside. (***See*** the topic of Corrosion for a broader understanding.)

a. Internal corrosion
   - This is mainly due to the deficiencies in **water conditioning**. (*See* Water.)
   - It can also be due to improper **chemical cleaning** or **storage** when the chemicals may not have been fully rinsed out. (*See* the topic on Acid Cleaning under Commissioning or start-up of boilers.)

b. External corrosion

   This is mostly due to the attack of ash and flue gas or reducing conditions. (*See* Ash Corrosion in Ash.)
   - *High-temperature corrosion* of the PPs, mainly superheat (SH) and reheat (RH), is due to the attack of low-melting metallic compounds in ash.
   - *Low-temperature corrosion* is due to the S compounds turning into sulphurous gases on cooling to temperatures <200°C and corroding the PP as well as non-PPs like AH, ESP and flues.
   - *Reducing conditions* in furnace, mainly due to the staging of combustion air, are conducive to corrosion attack unless the tubes are fully protected by a refractory layer.

1.12.2 Erosion

Erosion is due to both ash and coal. PPs in the path of flue gases are prone to external erosion due to the current of the ash-borne gas stream at high velocities. Lowering ash, reducing gas velocities and external shielding of areas prone to erosion are ways to minimise the problem. (*See* Ash Erosion in Ash.)

Coal erosion affects the mill internals and coal pipes when the entrained sand, shale, pyrites and other impurities are high. *Basalt* or ceramic lining can minimise erosion due to coal. (*See* Wear in Boilers in Wear.)

1.12.3 Fatigue

Fatigue is severe in high-temperature components which are thick and stiff, when they are subject to frequent stress reversals as in cyclic loading. Repeated starting and stopping of the boiler also causes thermal stresses, leading to cracks in welds and parent materials. Fatigue cracks are normally noticed in the tube to header, pipe to drum, fitting to tube, support attachment welds and other areas of stress concentration.

1.12.4 Overheating

Overheating occurs mainly in tubes, either due to higher heat or lower cooling. Improper water conditioning leading to **scale** buildup in the furnace tubes and their rupture is a common phenomenon. Usually, problems of overheating of surfaces occur in the early stages of plant operation. Unless there are changes to fuel or operating conditions, overheating should not occur after plant stabilisation.

1.13 Agro-Fuels and Firing

Termed variously as agro-, bio-, or vegetable fuels, these are essentially organic wastes generated by various crops, forest and purpose-grown trees. They can also be specially manufactured fuels like ethanol, which is produced on a large scale as replacement of petrol for vehicles. But what are usually burnt in boilers are mainly the agro-wastes and purpose-grown trees.

Bagasse and **wood** are also agro-fuels, but due to their abundance and regular use in boilers for several decades, boilers using them are extensively standardised and are treated somewhat separately.

Along with power from wind and solar energy, generating power from agro-fuels is part of the overall renewable energy or green power scheme. There are limits to the use of agro-fuels as explained here.

- They are characterised by their limited and seasonal availability. They are fluffy and bulky, making long-distance transportation unviable. Usually, their ash is soft and ash content is low but ash fusion temperatures are low and ash is both fouling and corrosive. Agro-fuels need bulky fuel handling but light ash handling equipments.
- Agro-fuels come in a wide variety with differing compositions which vary with season and geographic location. The types of tropical fuels are very different from those of temperate climates.
- Besides being seasonal, fuel availability is limited due to issues of production, harvest and collection.
- Being cellulosic, agro-fuels degrade fairly quickly and hence cannot be stored for a long time. Sizing of fuels can be a problem particularly with stringy fuels.
- For maintaining steam generation round the year, support of a fossil fuel is required during the periods of non-availability or shortage of agro-fuels.

With these limitations, only small power stations of generally 5–30 MWe can be built using purely agro-fuels, which are brought from nearby areas that are within short distances. The power plant also has to be close to the consuming centres.

These limitations are favourable for creating a distributed generation network, which is gaining increasing acceptance world over.

Agro-fuels lately are being used more and more in large coal-fired power stations in **co-firing** or **co-combustion** mode with a view to reduce the overall greenhouse gas (**GHG**) emissions and also to lower NO_x production. In many old boilers where steam generation has been forced to be curtailed by the new air pollution norms, burning agro-wastes has come in very handy to produce more steam without overstepping the pollution limits.

Agro-fuels, in general, have high dry ash free (daf) volatile matter (VM) ranging from 70% to 90% and low fixed carbon (FC). M is generally ~10%.

Gross calorific value (Gcv) on daf basis ranges from 19 to 23 MJ/kg, 4500 to 5500 kcal/kg or 8000 to 10,000 Btu/lb. The properties of typical agro-fuels of Northern Europe are given in Table 1.2. It can be seen that the properties exhibit a wide variation.

Ignition and **combustion** are quite easy and rapid due to high **VM**, low **FC** and practically no **ash**. **Unburnt losses** are low. On the whole, combustion η is quite good even with simpler firing equipment like **grates**. As the steam generation is modest due to limited availability of fuel, grate firing is most widely employed. Grates enjoy the benefits of

- Simplicity of construction
- Reduced operation and maintenance (O&M) costs
- Multi-fuel firing capability
- Low auxiliary power consumption
- ~1:4 seamless turndown of capacity

but the combustion η is lower by 2–4% points in comparison with bubbling fluidised bed combustion (BFBC), if it is possible.

BFBC is also employed when ash in fuel is reasonably high, such as in **rice husk**. BFBC enjoys advantages of

- Having few moving parts
- Delivering slightly higher combustion η of 2–4% over grate firing

but the auxiliary power consumption is slightly higher and the **turndown** is lower.

Various agro-fuels elaborated here are

1. Bagasse and its firing
2. Bark and its firing

TABLE 1.2

Properties of Typical Agro-Fuels of Northern Europe

Fuel and Property	Peat	Wood without Bark	Bark	Forest Residues (Coniferous Tree with Needles)	Willow	Straw	Reed Canary Grass (Spring Harvested)	Olive Residues
Ash%	4–7	0.4–0.5	2–3	1–3	1.1–4.0	5	6.2–7.5	2–7
M%	40–55	5–60	45–65	50–60	50–60	17–25	15–20	60–70
Ncv J/Mkg	20.9–21.3	18.5–20.0	18.5–23	18.5–20	18.4–19.2	17.4	17.1–17.5	17.5–19
C%	52–56	48–52	48–52	48–52	47–51	45–47	45.5–46.1	48–50
H_2%	5.0–6.5	6.2–6.4	5.7–6.8	6.0–6.2	5.8–6.7	5.8–6.0	5.7–5.8	5.5–6.5
N_2%	1–3	0.1–0.5	0.3–0.8	0.3–0.5	0.2–0.8	0.4–0.6	0.65–1.04	0.5–1.5
O_2%	30–40	38–42	24.3–40.2	40–44	40–46	40–46	44	34
S%	<0.05–0.3	<0.05	<0.05	<0.05	0.02–0.10	0.05–0.2	0.08–0.13	0.07–0.17
Cl%	0.02–0.06	0.01–0.03	0.01–0.03	0.01–0.04	0.01–0.05	0.14–0.97	0.09	0.1
K%	0.8–5.8	0.02–0.05	0.1–0.4	0.1–0.4	0.2–0.5	0.69–1.3	0.3–0.5	30
Ca%	0.05–0.1	0.1–1.5	0.02–0.08	0.2–0.9	0.2–0.7	0.1–0.6	9	

3. Rice husk/hull and its firing
4. Wood and its firing
5. Hog fuels

1.13.1 Bagasse and Its Firing (see *also* Bagasse-Fired Boilers *and* Grates)

The word 'bagasse' comes from the Spanish word *bagazo* which means trash or waste. In cane sugar mills in the tropical countries, waste generated after cane crushing and juice extraction in the milling section of boiler plant is called bagasse.

Bagasse forms about a quarter (24–30%) of the weight of sugarcane. Bagasse is a seasonal fuel. A season lasts from 6 to 10 months; longer seasons are with farms close to the equator.

Even though there are some advanced processes of turning bagasse into newsprint or chemicals, as of now, most of the bagasse ends up being the prime fuel to the boilers in the sugar factories.

Steam is generated from bagasse for

- Heating cane juice in boiling house section
- Driving steam turbines (STs)
- Producing electrical power

With distributed bio-power gaining greater acceptance, increasingly there is more power generated from bagasse in high-pressure (HP) cogeneration power plants.

Properties of bagasse

Bagasse is an extremely consistent fuel and the properties stay within ±2% across the globe.

Bagasse has

- High M varying from 45% to 52% depending on the design and operation of mills.
- Low ash of 1–2% depending on the type of cane cutting (manual or mechanical).
- High VM of 42–44%.
- Low FC between 5% and 6%.
- Daf VM is 85–87% comparable with any fuel gas.
- Once the M is removed, the bagasse ignites and burns extremely rapidly.
- High M keeps the flame temperature low, thereby producing very low NO_x.
- Bagasse has extremely low bulk ρ of 120 and 200 kg/cum (7.5–12.5 lb/cu ft) in loose and stacked conditions and hence is very voluminous.
- In-plant transportation is easy but the equipment sizes are large.
- Long-distance transportation is unviable.

The other salient features are

- The fibres of bagasse are small and uniform and hence are quite easy to transport, distribute and spread evenly inside the boilers and burn uniformly.
- Bagasse is reasonably soft and compressible, making it possible to bale for storage or transport.
- With only 1–2% ash in fuel, ash produced from bagasse boilers is very nominal. The ash disposal plant therefore is quite simple.
- Ash fusion temperatures of bagasse are quite low. In properly dimensioned and fully

water-cooled furnaces, slagging and fouling are rare problems.

- Inherent ash of sugar cane is soft and non-abrasive but the soil and dirt, particularly with mechanical harvesting, carried along with fuel contributes to the abrasive nature of ash. With conservative gas velocities and tube spacing, erosion can be avoided substantially.
- The low ash fusion temperature can create a problem in case of multi-fuel firing, such as bagasse and oil or coal. Ash from two fuels can form eutectic whose melting point is lower than the melting temperatures of ash of either fuel. With careful dimensioning of furnace, this problem can be overcome.

1.13.2 Bark and Its Firing

The classical definition of bark is that it is *the outermost layer of stems and roots of woody plants*. Plants with bark include trees, woody vines and shrubs. The main function of bark in trees is to protect the delicate inner layers of the tree from loss of water, temperature extremes, intense sunlight, and disease organisms. Being porous, bark helps the tree to breathe.

In the industry bark is a pulp mill waste resulting from the debarking operation of tree trunks. Bark, constituting about 10% volume of the wood as it is peeled off the trunks in long strips, is difficult to handle as fuel because of both size and M (which can be as high as 80%).

Cellulose and hemicelluloses content is lower in bark than in wood and the properties of lignin are also different. Variations also exist in physical and mechanical properties—ρ, hygroscopicity, dimensional stability (shrinkage and swelling) and so on. However, thermal properties and heating value are nearly same as that of wood. Composition and properties of bark are highly variable depending on the type of tree and the season.

Burning of bark is similar to the burning of **wood** in boilers. **Grate firing** of chipped bark is most popular and both travelling and stationary grates are employed in the pulp and paper industry where bark is mostly generated. As the quantity of bark is small compared with wood, it is often fired along with other fuels in multi-fuel-fired boilers.

1.13.3 Rice Husk/Hull and Its Firing

Rice husk or rice hull is a seasonal waste product from the rice mills in tropical countries such as India, Thailand, Malaysia and so on. Rice husk is the protective outer shell of the seed which is removed in the de-husking/de-hulling operation.

Rice husk forms 20–25% of the weight of the dried paddy. It has a uniform size of <3 mm needing no further preparation.

Rice husk is mainly used for producing captive steam and power. A greater value addition lies in producing anhydrous (as opposed to crystalline) SiO_2 from rice husk, which forms around 20% of total SiO_2.

Rice husk is characterised by its high ash of 15–18% of which SiO_2 forms nearly 90%. This makes the ash extremely abrasive. A typical analysis of rice husk is given in Table 1.3 from which it can be inferred that the combustion properties of this fuel are very good.

The bulk ρ of husk is 100–300 kg/cum (6.25–18.75 lb/cft) with particle ρ of about 600 kg/cum (37.5 lb/cft). Being bulky, husk requires a large storage space. Unlike bagasse, husk is very dry, with about 5–10% M and not compressible. This fuel is not amenable to baling. Husk ash, on account of its high SiO_2, has very high initial ash deformation temperature (IADT) of >1400°C and fusion temperature of >1600°C.

Rice husk demands a fairly high **ignition temperature** and good **residence time** in furnace and grate/bed as the FC is high. Rice husk is suitable for combustion both in grate-fired and BFBC boilers.

The advantage of **stoker firing** is its simplicity of construction and operation, low fan power and seamless 1:4 turndown but the combustion η, compared with BFBC, is slightly lower by 2–4%.

With 15–20% ash, **BFBC** also makes a good alternative, as there is enough ash in fuel to replenish the loss of bed material. Fan power in BFBC is higher to compensate for its better η. **Overbed** firing is better than **underbed** firing as the ash is very erosive.

Rice husk is fired in specially pulverised fuel (**PF**) **boilers** only when the aim is to capture amorphous SiO_2. Fuel is ground in specially designed pulverisers and fired in conservatively designed furnaces, with temperatures not exceeding 860°C so that the SiO_2 in ash does not turn crystalline.

TABLE 1.3

Range of Proximate Analysis of Rice Husk

	Air-Dried Range (%)	Daf Range (%)
A	15–18	—
M	5–10	—
VM	65–70	80
FC	12–15	20
Gcv MJ/kg	13.8–15.1	20.1–20.9
kcal/kg	3300–3600	4800–5000
Btu/lb	5940–6480	8640–9000

1.13.4 Wood and Its Firing

Wood is a complex vegetable tissue composed mainly of carbohydrates. Wood has relatively lower calorific value (cv) in comparison with fossil fuels. Resins, gums and other substances along with M contribute to the lowering of Gcv of wood. Without this burden, the cv of the woods would be nearly the same as bagasse, at ~20.2 MJ/kg, 4830 kcal/kg or 8700 Btu/lb on daf basis.

There are two types of woods, namely *hard wood* and *soft wood*. This classification is not based on the hardness but on botanical terminology. *Hard woods* belong to trees with broad leaves. They have less resin and hence burn slowly for a long time. *Soft woods* belong to trees with scale or needle-like leaves. Soft woods burn quickly.

Seasoning refers to the drying time before combustion. Wood needs a seasoning time of a minimum 4–6 months. Freshly cut wood would have M ranging from 30% to 50%. After a year of seasoning, it would reduce to 18–25% along with some loss of fuel value.

Woods with <50% M burn well.

At more than 65% M, combustion of wood is not self-sustaining, needing a support fuel.

Such high M is because of rain-, snow-, or water-based transportation.

M in wood logs is ~40% while it is 15–25% in saw dust and wood chips.

M depends largely on the type, handling, storage and age of the wood.

Ash in wood is rarely higher than 2.5%.

Ash fusion temperatures are low, needing larger furnaces with lower furnace exit gas temperature (FEGT). Wood has practically no S or N. Hence, there is no fear of SO_2 or NO_x formation and there is no need for deSO_x or deNO_x plants for gas cleaning. Like with the other bio-fuels, it is not economical to transport wood beyond say, 100 km, because of relatively low cv and large volume.

Steam and power generation from wood has been practiced from as early as the nineteenth century until coal and then oil became very popular. Wood firing is still done regularly in lumber and paper mills in the Northern hemisphere. Its popularity is on the rise again due to its low air polluting characteristic, such as

- Large boiler plants can be built on wood.
- It is not so seasonal like many other agro-fuels.
- Wood firing helps to reduce *GHG*.
- Wood firing is environmentally friendly with no SO_x and NO_x generation.
- Co-firing of wood is easier.

Wood firing is relatively modest in the tropical countries as hard wood has more value-added uses like in the making of furniture, paper rayon and so on, and hence not much soft wood is grown.

In the Northern hemisphere countries of Scandinavia, Canada, the United States and certain South American countries, where soft wood is abundantly available, it is an important fuel and for this reason big power plants are planned and built.

1.13.5 Hog Fuels

Hog fuels are wood residues processed through a chipper or mill to produce coarse chips and those that are normally used for fuel are bark, sawdust, planer shavings, wood chunks, dirt and fines. A pulp mill normally produces one or the other type of hog fuel.

Hog fuels, being solid, are easier to mix with the main fuel or can be burnt on their own when quantities are sufficient. Sludge also can be burnt with hog fuel.

1.14 Air Ducts (*see* Draught Plant)

1.15 Air Flow Measurement (*see* Instruments for Measurement)

1.16 Airheater (*see also* Backend Equipment in Heating Surfaces)

AH is almost always the last heat trap in the boiler. Only in some exceptional cases, part of *economiser* may be placed after AH.

AHs in boiler plants operate in the gas temperature range of 450–120°C, heating the incoming combustion air from ~100°C to 450°C.

- The extent of flue gas cooling in AH is dictated by the *low temperature or dew-point corrosion* of the AH surfaces.
- The maximum hot air temperature at the AH outlet depends on the ability of the downstream equipment (grates, burners and mills) to withstand heat, based on either the material

construction or process limitation. For example, it is

- 150–200°C for grates
- ~400°C for coal mills
- ~450°C for burners

Because of the low temperatures operating in AHs, the entire construction is normally of *carbon steel* (CS). At the cold end, where the low-temperature **corrosion** is to be combated, it is not unusual to employ weather-resistant steel like **COR-TEN** or enamelled parts.

This topic is sub-divided into the following sub-topics:

AH classification

a. Rotary airheater (RAH)
   1. Ljungstrom (moving rotor) design
   2. Rothemuhle (moving/rotating hood) design
b. Tubular airheater (TAH)
c. Plate-type AH
d. Heat pipe AH
e. Steam coil air preheater (Scaph)

AH classification

AHs can be classified as (a) recuperative and (b) regenerative.

a. *Recuperative* AHs are equipments where heat is directly transferred from flue gas to air.
   - TAHs where gas flows through tubes and air over them, or vice versa, fall in this category.
   - *Plate-type AHs*, which are less popular, are also recuperative. The gas and air flow alternately through a set of parallel plates.
b. *Regenerative* AH, as the name suggests, employs an intermediate media for transferring the heat.
   - RAH employs a large rotor/stator containing baskets of corrugated metal sheets, which are heated by the hot gases, only to transfer the heat to the cold air.

   Either (a) the rotor can be moving with connecting ducts remaining stationary as in Ljungstrom design or (b) the stator can be stationary with the connecting ducts rotating as in Rothemuhle design.
   - *Heat pipe AHs*, described later, are also regenerative AHs.

Both types of AHs, TAH and RAH, are popular, with each design having its specific advantages. They are described under their respective headings. Table 1.4 provides a comparison of the TAH and RAH.

TABLE 1.4
Comparison of Tubular and Rotary AHs

Item	TAH	RAH
Construction	Static Simple Less parts	Rotating More parts Specialised
Space required	More. AH is large. Tube withdrawal needs a lot of space	Less. AH is compact
Erection	More site assembly	More shop assembly
Air leakage into gas space	None	Yes
		Fans have to be large
Fire hazard	None	Yes. Particularly in oil-fired boilers at low loads
Repair of corroded elements	Easy	Not possible
	Short tubes to be inserted	Baskets to be replaced

With the growing manufacturing of RAHs worldwide, TAHs are slowly giving way to the RAH, on account of their compactness and ease of layout. However, in fluidised bed combustion (FBC) boilers, TAHs are preferred because of excessive pressure difference between air and gas sides which puts a lot of strain on the AH seals.

AHs cannot be employed in certain types of boilers. In BLR boilers, the severe fouling nature of black liquor (*BL*) prohibits use of AH. In heat recovery steam generators (**Hrsgs**), no air is supplied to the boiler, making AH irrelevant.

1.16.1 Airheater, Rotary

The *rotor* or *stator* is the heart of any recuperative RAH, which absorbs heat from flue gases and transfers to cold air. The heat transfer from hot gases to cold air is through the medium of rotor/stator, which is packed with corrugated, undulated or flat sheets of 18–26 **SWG** (0.46–1.22 mm) spaced at 5–10 mm apart for the passage of gas and air. It is a turbulent flow through the sheets for efficient heat transfer. Typically, the baskets, packed between 300–350 m^2/m^3 of heating surface (HS; 90–105 ft^2/ft^3), are filled in the rotor/stator. The thin sheets are of usually CS. At times, the cold section is packed with either **COR-TEN** or enamelled sheets. Heat transfer rates are much higher at 17–57 W/m^2 K or 3–10 Btu/ft^2 h°F compared with TAHs where the heat transfer rate is ~17 W/m^2 K or 3 Btu/ft^2 h°F, which makes them very compact.

1.16.1.1 Ljungstrom (Moving Rotor) Design

In *Ljungstrom (moving rotor) design*, the rotor rotates at speeds varying from 1 to 3 rpm. The connecting ducts are stationary. There are seals provided on the top and

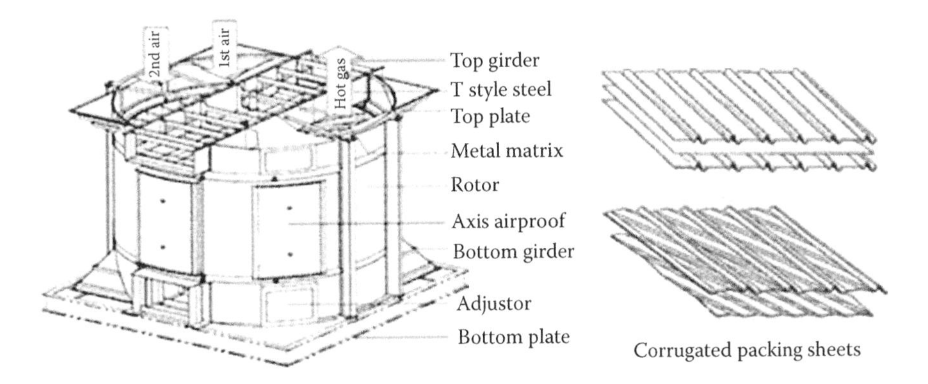

FIGURE 1.3
Vertical tri-sector Ljungstrom (moving rotor) AH.

bottom of the rotor to prevent the escape of HP air into the low-pressure (LP) gas zone. Mass-type SBs are installed on the gas inlet side for on load cleaning of elements, which are prone to clogging by ash. Soot blowing also tends to blow away the unburnt fuel deposits, which can become a fire hazard. In fact, to prevent this, the start-up logic is so set that an AH cannot start unless the SBs operate for a predetermined length of time.

AHs can be *vertical* or *horizontal* (based on the orientation of the drive shaft) to suit plant layouts. Usually, horizontal AHs are popular with package boilers and vertical AHs are the mainstay of utility boilers.

A *bisector AH* is popular in oil- and gas-fired boilers where there is only a single stream of hot air.

A *vertical tri-sector AH* along with the corrugated packing sheets, which is shown in Figure 1.3, is employed in PF boilers which have two streams of hot air, as needed at different pressure and temperature (p & ts) for primary air (PA) and secondary air (SA). The air and gas bypass ducts are located at the corners (which are not shown in the picture).

In place of a tri-sector AH, there can be two bisector AHs for PA and SA separately depending on the layout considerations.

1.16.1.2 Rothemuhle (Moving/Rotating Hood) Design

In *Rothemuhle (moving/rotating hood)* design, shown in Figure 1.4, the stator, which is packed with HS, remains stationary and connecting ducts/hoods rotate slowly. All features of the various alternative arrangements of the Ljungstrom design are available here.

The permissible gas exit temperatures for RAHs are nearly the same or marginally lower than TAHs. A graph indicating the minimum permissible metal temperature versus S in fuel for long corrosion free operation is given in Figure 1.5, which is taken from The Babcock and Wilcox Company (B&W).

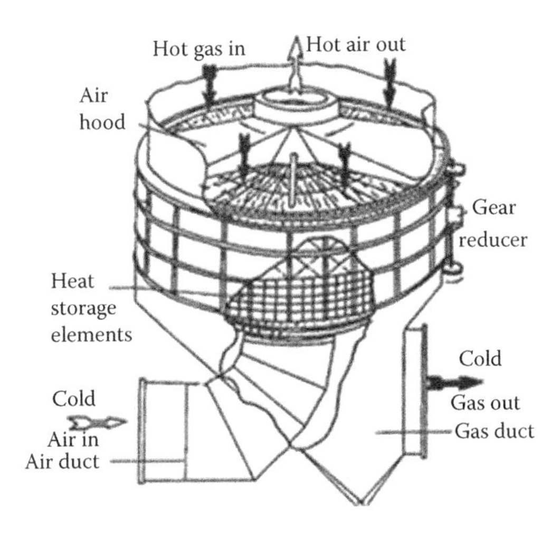

FIGURE 1.4
Vertical Rothemuhle (moving hood) AH.

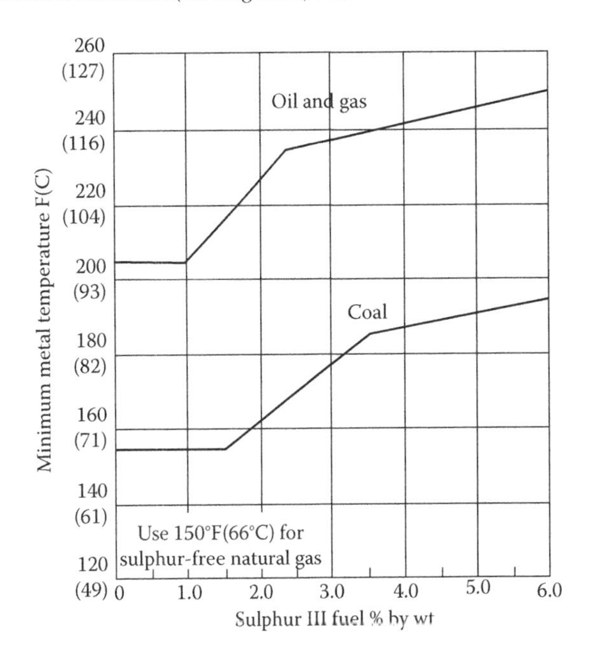

FIGURE 1.5
Permissible minimum metal temperature versus sulphur in fuel for RAHs. (From The Babcock and Wilcox Company, USA, with permission.)

Fires in RAHs

Fires in AHs, particularly RAHs, are not frequent but also not uncommon. They are more common in heavy fuel oil (FO) than in PF-fired boilers. When they occur, they are usually severe enough to destroy the entire AH. Fires are mostly initiated during the starting of boilers at the cold-end side in the fouled AH, where the accumulated unburnt fuel gets to agitate and ignite. Leaking AH bearings are another source of ignition.

Fires are detected by noticing the large variation in AH air and gas temperature profiles. Special detection devices are also employed. SBs are installed on either end; frequent blowing keeps the AH clean. Proper tuning of the firing equipment and frequent soot blowing are the only precautions needed to avert fires in TAH.

Water washing arrangement along with draining is also provided in TAHs. On detecting fire, the boiler is tripped to stop the air flow and full water washing is turned on, while the rotor/hoods are kept under rotation until the fire is completely extinguished.

1.16.2 Airheater, Tubular

Because of the simplicity of construction and lower cost, TAHs are very popular with industrial and lower end of utility boilers of up to ~200 MWe. The essential differences between the tubular and RAHs are given in Table 1.4.

TAHs can be in *vertical* or *horizontal* arrangement, based on the tube orientation, to suit the boiler layout. With air flowing through the tubes, they are smaller and therefore enjoy slightly superior heat transfer rates. This makes the horizontal AHs little more compact than the vertical TAHs. The differences between the two designs can be observed in Table 1.5.

TAHs are normally made of CS tubes and plates. Many times, the cold end is made out of weather-resistant **COR-TEN** steel for better corrosion resistance and life. The tubes are usually expanded in the tube plates. Sometimes, they are even welded but the welding is done with thin wire to avoid bending of the tube plates due to excessive weld heat. During the boiler operation, the tubes expand and there are suitable arrangements for absorbing the expansion. The tube lengths are rarely in excess of 5.5 m in both the designs. The tube thickness is normally 2.03/2.3 mm (14 or 13 SWG) for all fuels except for spreader stoker (**SS**) firing and coal firing where 2.3 or even 2.95 mm (11 SWG) tube is used. In exceptional circumstances, tubes as thin as 1.24 mm are known to have been used. The supporting tube plates are 25 or 30 mm thick with appropriate stiffening. The baffle plates, located in the tube bank for keeping the tubes straight and preventing vibration, are usually made of 16 or 20 mm thickness.

There are several ways of arranging the tube bundles. Typical vertical and horizontal TAHs are shown in Figures 1.6 and 1.7.

Methods to prevent corrosion of AH internals

Cold-end corrosion takes place in the zone of contact of exit gas and inlet air, which happens to be the coldest part of any AH. It is essential that an AH sizing should be done in a way that the gas exit temperature is not too cold to lower the metal temperature below safe limits. When the boiler is operated at less than the full load, this dip in temperature is inevitable and the following remedies are built into the scheme to prevent corrosion:

1. **Scaph** can raise the inlet temperature of combustion air and prevent corrosion. (*See* Scaph in Section 1.16.5.)
2. AH air bypassing avoids the gas cooling to the same extent and improves the metal temperatures.
3. AH gas bypassing achieves the same result.

TABLE 1.5

Design Features of Vertical and Horizontal TAHs

Parameter	Units	VAH	HAH
Flow through tube		Gas	Air
Minimum tube temperature			Slightly higher
Tube sizes OD	mm	63.5, 70	38.1, 50.8
Usual tube thickness	mm	2 or 2.3	2 or 2.3
Usual tube arrangement		Staggered	In-line
Max tube lengths	m	5.5 bottom support	5.5
Heat transfer rate	Btu/ft² h°F	~3	3–10
	W/m² °K	~17	17–57

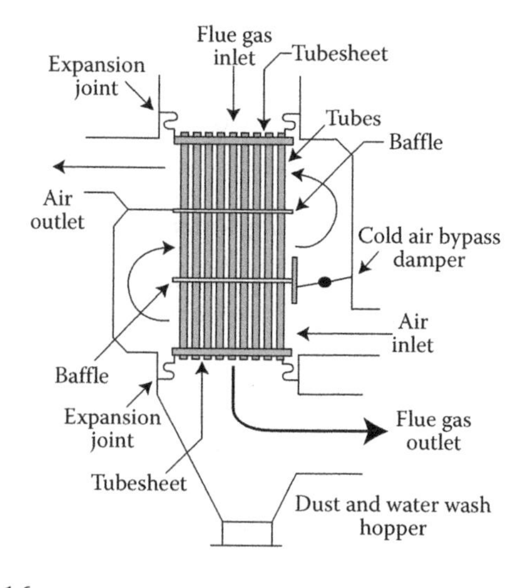

FIGURE 1.6
Typical vertical TAH.

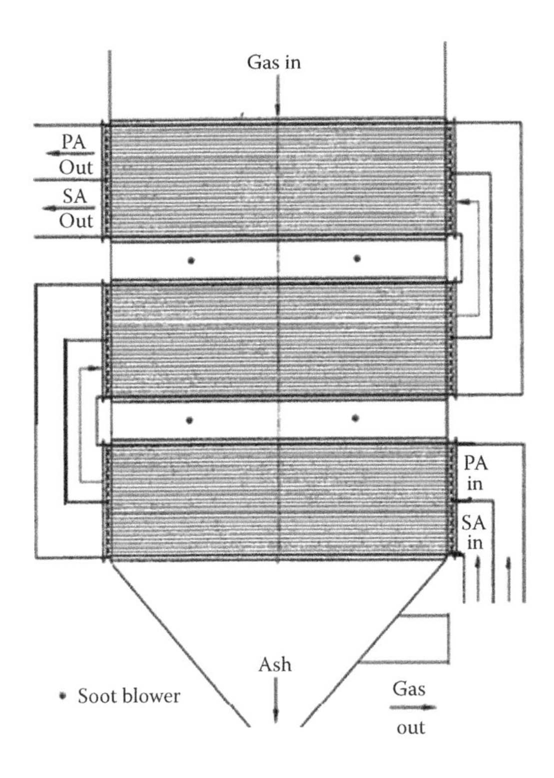

FIGURE 1.7
Typical horizontal TAH.

4. Hot air recirculation to the fan suction raises the air inlet temperature.
5. Providing thicker tubes, enamelled baskets, or weather-resistant alloy tubes or internals prolong the life of the cold-end section.

Each has its advantages and disadvantages and has to be examined on the merits of the case and layout considerations.

1.16.3 Plate-Type AHs

These are less expensive than TAHs and are employed in industrial boilers by some boiler makers but are not as popular as TAHs. Air and gas flow in alternate channels and heat transfer takes place through the thin sheets. A typical arrangement is shown in Figure 1.8. A variety of configurations are possible. They can be a bit more compact then TAHs. Air-to-gas leakage is also negligible.

1.16.4 Heat Pipe AHs

Heat pipe AHs comprise of numerous sealed CS tubes which contain a heat-transfer liquid. Usually, either toluene or naphthalene is used for operating temperature limits of 290°C and 430°C, respectively, at an internal pressure of ~28 bar. The liquid inside

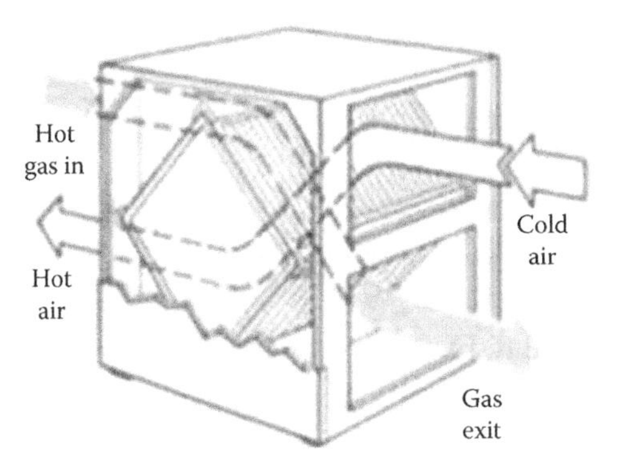

FIGURE 1.8
Typical plate-type airheater.

the tubes is vapourised by the heat absorbed from flue gases on one side and condensed by outside cold combustion air being drawn into the furnace on the other side. Heat is transferred from gas to air by this repeated evaporation and condensation of the heat transfer liquid. It is a regenerative type of AH as the heat is indirectly transferred with the help of another medium.

The heat pipe is tilted upwards at the condenser end (air side) at an angle of 4–10°, thereby providing gravity assistance to the flow of the working fluid back to the evaporator (flue gas) side (as shown in Figure 1.9). Tubes are finned over the entire tube length both to increase the surface area and enhance heat transfer for making the AH compact. 40–240 fins/m (~1–6 fins/in.) are used to make the AH suitable for both clean and dirty fuels. Tubes of 12 m and longer are employed to make sizable AHs.

A steel plate separates the combustion air and flue gas streams into parallel ducts. Individual heat pipe tubes transport thermal energy from one gas stream to the other with no measurable leakage, ensuring the highest η. A typical AH with air and gas flowing horizontally is shown in Figure 1.10. The orientation can be different to suit the layout.

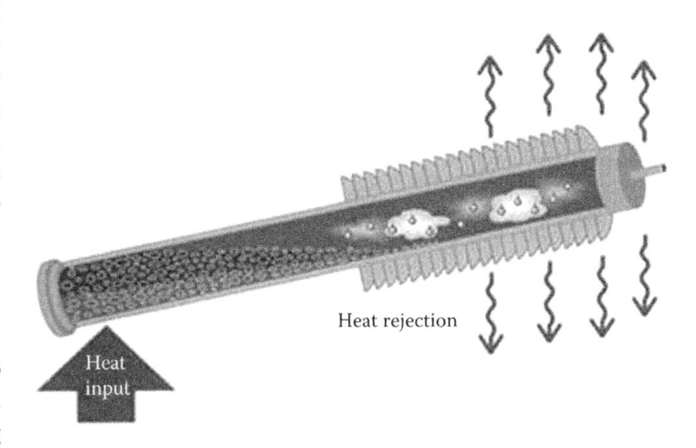

FIGURE 1.9
Typical individual heat pipe.

The benefits claimed for heat pipe AH are as follows:

- AH assembly works out more compact with less weight.
- Pressure losses are lower.
- With no leakage the fan volumes can be marginally smaller.
- Overall reduction in power.
- No moving parts and practically no maintenance.
- Minimum cold-end corrosion.
- Slightly higher costs of heat pipe AH usually make it economically viable mainly if there are space constraints. They are used more in industrial boilers.

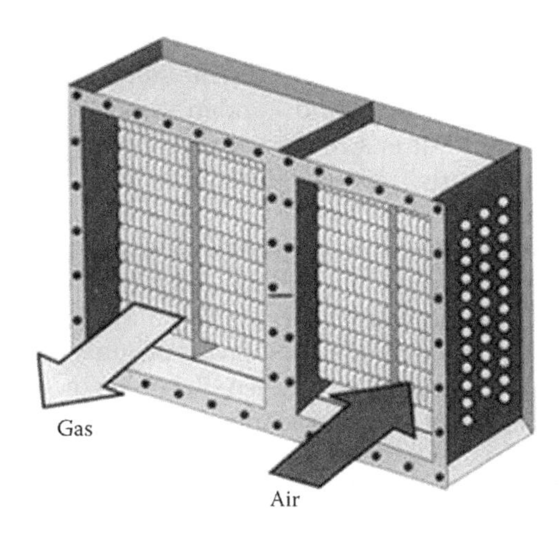

FIGURE 1.10
Horizontal gas–air AH using heat pipes.

1.16.5 Steam Coil Air Preheater

Cold-end metal temperature in an AH is approximately the arithmetic mean of the incoming air and outgoing gas temperatures. In winter months when the ambient temperatures drop precipitously, the cold-end metal temperature can be lowered to levels when dew-point corrosion can set in, which can be avoided if the air inlet temperature can be raised. Scaph is a finned tube AH inserted at the discharge of forced draught (FD) fan, which heats the combustion air with the help of steam taken from the boiler at some suitable point. As the fins are not subjected to high air temperatures, tension-wound fins are adequate for the duty. Usually, Al fins are wound over CS tubes.

Scaph is sometimes designed for continuous use. AH exit gas temperature can be cooled more and better fuel η can be obtained. But this has to be examined carefully as the heating of combustion air is after all derived from the heat of steam.

In *BLR boilers* where hot air is always needed for the combustion of high-M fuel and no conventional AH can be installed as it would get severely fouled due to dust in flue gases, large Scaphs are required to be provided to supply hot air on a continuous basis.

Scaph consists of a frame which houses one or more steam coils. The frame is installed in the air ducting and each steam coil is individually removable from the frame so that they can be cleaned, repaired, or replaced without disturbing the ducting. In applications with a large heat load, a bank of several steam coils is often a better solution than one large steam coil because the large steam coils are susceptible to performance problems due to uneven steam distribution and they are difficult to handle during installation and removal. Figure 1.11 displays the schematic of a large Scaph.

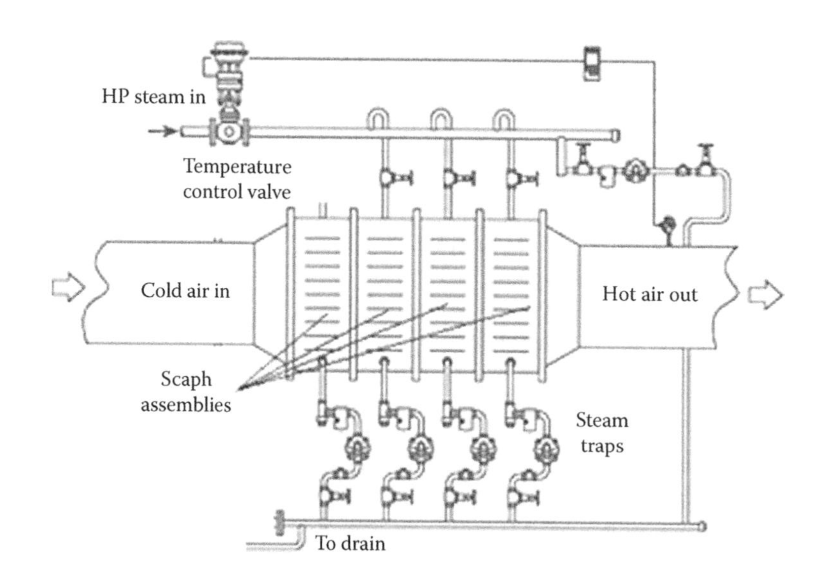

FIGURE 1.11
Specific volume variation of air with altitude.

1.17 Air Infiltration or Air Ingress in Boiler Setting (*see* Combustion)

1.18 Air Nozzles (in FBC Boilers) (*see* Fluidised Bed Combustion)

1.19 Air Pollution (*see also* Acid Rain, Emissions and Controls Gas Cleaning *and* Green House Gases)

Air pollution is the *introduction of PM and hazardous chemicals into the atmosphere* by natural phenomena like volcanic activity, emissions from the industrial or transportation activity. Pollutants are substances that can cause harm to humans, infrastructure or the environment. They can be solid particles, liquid droplets or gaseous matter.

Next to transportation, power plants are the highest polluters. In the power plants, it is the boilers that emit the maximum pollutants as fuels are burnt mainly in boilers. Combustion of natural gas (NG) is practically without harmful emissions while burning of coal in all forms is highly offending, requiring many modes of gas cleaning efforts.

The topic of air pollution is elaborated under the following sub-topics:

a. PM in flue gas
b. Hazardous gases
   1. Carbon monoxide (CO)
   2. Nitrogen oxides (NO_x)
   3. Sulphur oxides (SO_x)
   4. Ozone (O_3) at ground level

1.19.1 Particulate Matter in Flue Gas

PM is a general term used for describing a heterogeneous mixture of both solid particles (namely dust, dirt, soot, smoke) and liquid droplets found in air or gas. PM can be suspended for long periods of time. Large or dark particles are seen as soot or smoke while the small ones cannot be detected by the naked eye.

PM can be either primary or secondary pollutant.

- Primary PM in air, such as dust and soot, is directly emitted from a variety of sources such as vehicular traffic, factories, construction sites, tilled lands, unpaved roads and crushing and burning activities.
- Secondary PM is formed indirectly from the chemical change of primary PM when gases from burning fuels react with sunlight and water vapour.

PM in flue gas is essentially the fly ash and unburnt C particles elutriated from the combustion chamber. They cause dust nuisance in the surroundings, harming the infrastructure as well as humans. They are captured by appropriate **dust collecting equipment**.

1.19.2 Hazardous Gases

1.19.2.1 Carbon Monoxide

CO is a colourless, odourless and highly toxic gas formed when C in fuel is not burnt completely. Motor vehicle exhaust contributes maximum to CO emissions in air. The others are mainly industrial processes such as metal processing and chemical manufacturing, residential wood burning and natural sources such as forest fires.

CO can cause harmful health effects by reducing O_2 delivery to the body's organs such as the heart, brain and tissues. At high levels, CO can cause death. In boiler plants, CO in combustion flue gas has to be minimised for reducing unburnt C loss and improving boiler η. Increasing excess air for the reduction of CO is counterproductive as it can be seen in Figure 5.6. Traces of CO in flue gas in ppm level, rather than zero, is what is aimed.

NO_x-limiting measures like low NO_x burners (**LNB**), staged combustion, flue gas recirculation (**FGR**) and so on tend to increase CO formation due to lower combustion temperature.

The best way to limit CO is to optimise the combustion parameters and adopt prudent combustion techniques that provide the **3Ts** (time, temperature and turbulence) in adequate measure.

*1.19.2.2 Nitrogen Oxides (*see also *Nitrogen Oxides in Flue Gas)*

NO_x is the generic term used to describe the sum of NO, NO_2 and other monoatomic oxides of N_2, many of them being colourless and odourless. Only NO_2, however, along with particles in the air can often be seen as a reddish-brown layer over many urban areas. Diatomic nitrous oxide (N_2O), commonly known as laughing gas, is not a part of NO_x even though it contributes to the destruction of stratospheric O_3.

NO_x is a group of highly reactive gases that play a major role in the formation of O_3. All fossil-fuel combustion generates some level of NO_x due to high temperature and the presence of N_2 and O_2, both in fuel and in air. However, it is only at high temperatures that NO_x generation becomes objectionably high. The primary sources of NO_x are motor vehicles, electric utilities and other industrial, commercial and residential sources that burn fuels.

Fuel NO_x and *thermal* NO_x are the two types produced in boilers at high temperatures from N_2 in fuel and N_2 in air, respectively. NO_x generation is reduced by lowering the flame temperature through one of the several techniques available. Post-combustion it can be trapped by subjecting the flue gases to reduction reactions of selective catalytic reaction (**SCR**) or selective non-catalytic reduction (**SNCR**).

1.19.2.3 Sulphur Oxides

In power plants and industries, colourless SO_x gases are formed when any fuel containing S, such as **coal** and oil, is burnt in normal ways. They are also produced when gasoline is extracted from oil or metals are extracted from ore.

SO_2 is the criteria pollutant that is the indicator of SO_x concentrations. SO_2 dissolves in water vapour in air or gas to form acid, and interacts with other gases and particles to form SO_4 and other products that can be harmful to humans and their environment.

Over 65% of SO_2 released to the air comes from electric utilities that burn coal. Other sources of SO_2 are industrial facilities that derive their products from raw materials like metallic ore, coal and crude oil, or that burn coal or oil to produce process heat. Examples are petroleum refineries, cement manufacturing and metal processing facilities.

SO_x in flue gases is detrimental as they cause **low-temperature** or **dew-point corrosion** and attack AH and low-temperature flues. Flue gas cooling in boilers is limited by this factor.

Flue gas desulphurisation (**FGD**) or deSO_x plants are installed near the boilers, before the stack, to precipitate SO_x out of the flue gases. Alternately **FBC**, with its in-furnace desulphurisation reaction with lime stone addition, can prevent generation of SO_x altogether.

1.19.2.4 Ozone (O_3) at Ground Level

O_3 present in the stratosphere protects life on earth against damage by UV radiation. It is a colourless compound that has an electric-discharge-type odour. It is exclusively a secondary pollutant, that is, it is not emitted directly into the air, but is created at ground level by a chemical reaction between NO_x and VOCs in the presence of heat and sunlight. The formation of ground-level O_3 takes place in the lower atmosphere, especially during hot weather.

At the ground level, however, the presence of O_3 is highly undesirable.

- O_3 is one of the main ingredients of smog.
- When inhaled, even in very low concentrations, O_3 can cause respiratory problems and asthma attack.
- O_3 in high concentrations in rural areas may result in reduced agricultural production.

1.20 Air Registers (*see* Burners)

1.21 Alkali Boil Out (*see* Commissioning)

1.22 Allowable Stresses (*see* Pressure Part Design)

1.23 Alloy Steels (*see* Metals)

1.24 All Volatile Treatment (*see* Water)

1.25 Alternate Fuel (*see* Fuel Firing Modes in Fuels)

1.26 Altitude

Altitude of a site is its *height above mean sea level (MSL)*. The higher the altitude, the rarer is the ambient air and the lower the barometric pressure, as shown in Table 1.6.

TABLE 1.6

Variation of Barometric Pressure with Altitude

Altitude (m)	0	75	150	250	300	450	600	750	900	1000	1200	1350	1500	1800	2100	2400
Barometer (mmHg)	760	753	746	739	733	719	706	693	681	668	656	644	632	609	586	564

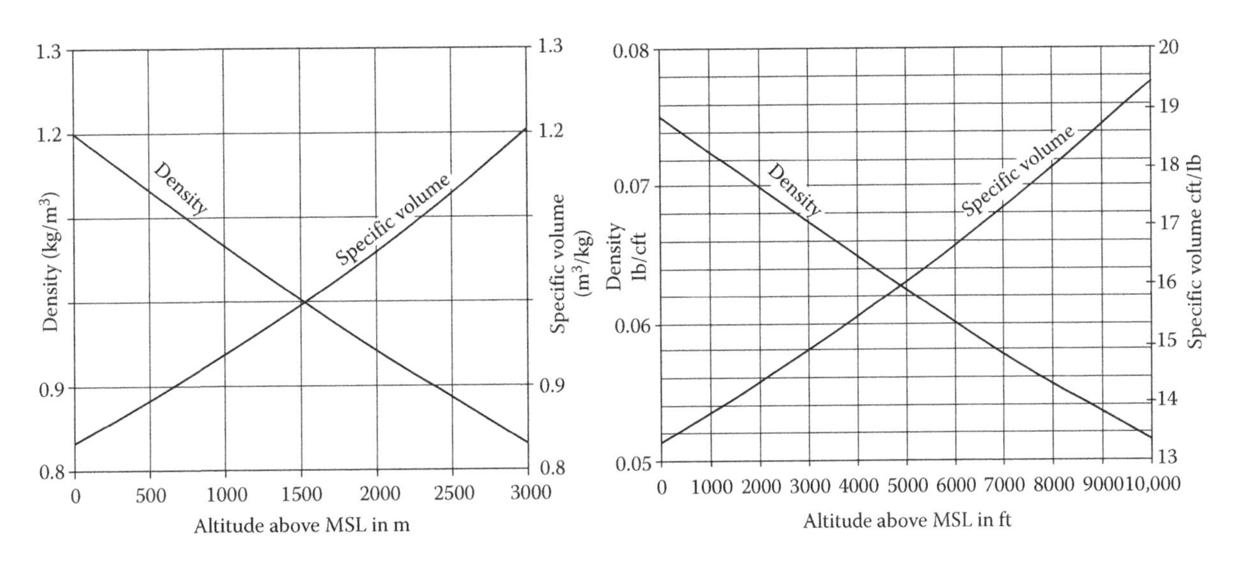

FIGURE 1.12
Schematic arrangement of a large Scaph.

TABLE 1.7

Altitude Correction Factor for Discharge Volume of Machines

Altitude (m)	0.0	480	990	1500	1980	2460	2970
Altitude (ft)	0.0	1600	3300	5000	6600	8200	9900
Volume correction factor	1.00	1.05	1.11	1.17	1.24	1.31	1.39

As the pressure is lower, the specific volume of air is correspondingly higher and ρ lower at higher altitudes. This is shown in Figure 1.12 in SI and British units.

For the **volumetric machines** like **fans** and gas turbines (*GTs*), this means a lowered air weight and reduced throughput at higher altitudes. To achieve the rated output, the machines need to be larger by a factor given in Table 1.7. Auxiliary power consumption will be higher. In GTs, output reduces between 3% and 5% for every 300 m rise from MSL.

1.26.1 Volumetric Machines

Many machines like the compressors, GTs and fans have finite capability of how much of air they can aspirate depending on their internal construction. These are known as volumetric machines.

Their performance is, therefore, directly volume and not weight related. Depending upon the ambient conditions, the weight of air entering the machine varies in proportion to the ambient specific volume. The output of the machine in kg/s and kg/cm^2 is not constant but highly ambient and altitude dependent.

1.27 Ambient Conditions

Prevailing conditions at the plant site are called ambient, namely, the surrounding

- Temperature (dry bulb reading)
- Pressure as denoted by the altitude
- Humidity (as given by the combination of dry and wet bulb readings)

Ambient condition affects the performance of the plant quite seriously and hence is considered one of the most important of the design factors. The effect of humidity is rather small but those of ambient temperature and altitude are really significant.

- Thermal η figure of any boiler is incomplete without the ambient being specified. For each 5°C rise in the ambient temperature, the gross thermal η of boiler would be lower by ~0.25%.

- The performance of all volumetric machines, like fans and GTs, is highly ambient dependent. Ambient temperature varies from –5°C to 50°C in tropics and –20°C to 35°C in temperate climates. This contributes to a ~20% change in specific volume of air. This requires that the fans and the drives are correspondingly bigger. The auxiliary power consumption also escalates.
- The motor withstands temperatures and the insulation selected should be adequate for the maximum ambient temperature.
- The thickness of boiler insulation nearly doubles if the temperature of the lagging has to be maintained at, say 60°C, at all ambient conditions.
- In GTs, the ambient temperature plays a very vital role. As it rises, the air weight (for the same volume) reduces and the resultant output power drops. GT exhaust gas temperature rises and the GT η lowers. Correspondingly, the **heat rate** rises.

The reverse happens when the ambient temperature reduces. Typically, the output of a 100 MWe ISO rated GT varies from ~110 to 75 MWe when the ambient changes from 0°C to 40°C. For each 1°C variation of ambient temperature, the output of GT varies by 0.5–0.9%.

1.28 Angle of Repose (*see* Properties of Substances)

1.29 Amines (*see* Water)

1.30 Anthracite

Anthracites are the most mature and oldest of *coals*. They are fully formed coals in the process of *coalification*. See Figures 3.23 and 18.7. They have low VM (2–14%) and very high FC (86% or more) both on daf basis.

Graphite is the only form of coal more fully formed than anthracite. Anthracites are placed between bituminous coals and graphite. Anthracite, semi-anthracite and meta-anthracite are the three types of anthracites. Meta-anthracite approaches graphite in structural composition and burns with short smokeless blue flame. Its ignition temperature is high and it is slow to ignite and difficult to burn. It has practically no commercial importance.

Most anthracites and semi-anthracites have lower Gcv than the highest grade of bituminous coal. They have low **VM**, high **FC** and consequently high **ignition temperatures**. They are not easy to burn. Yet they are burnt in commercial scale in both industrial and utility boilers boilers.

Anthracite has greater hardness, higher relative ρ (of 1.3–1.4), and semi-metallic lustre. It is free from the included soft or fibrous notches and does not soil the fingers when rubbed. The M content of freshly mined anthracite is generally less than 15%. The Gcv ranges from 26 to 33 MJ/kg, ~6110 to 7760 kcal/kg or 11,000 to 14,000 Btu/lb) on daf basis. Anthracites need about 600°C to ignite. They burn with short smokeless blue flames. They can be burnt on gravity-fed chain grate/travelling grate (CG/TG) but not on SS, as there is very little VM to burn in suspension to support the combustion of FC on the grate below.

In PF mode, they are burnt in **vertical/arch-mounted burners** in **downshot firing** employing high PA temperatures and higher fuel fineness (more than 85% through 70 μ mesh). FBC has also demonstrated its ability to burn anthracite very efficiently in a clean manner due to its low temperature of combustion.

Anthracite is the least plentiful form of coal but widely spread in the world. There is some anthracite in Russia, Ukraine, China, Korea, South Africa, Australia, Western Canada, some parts of Western Europe (Spain and Wales), and the eastern United States. The main uses are in domestic heating, gas production and power generation.

1.31 Anti-Segregation Chute (*see* Grates)

1.32 API Gravity (*see* Liquid fuels)

1.33 Approach Temperature (*see also* (1) Steaming and Non-Steaming Economiser in Economiser Classification *and* (2) Thermal and Mechanical Design Aspects of HRSG)

In any heat exchanger (HX), as the hot and cold fluids come into contact, the hot fluid loses its heat and turns cold while the cold fluid gains heat to turn hot. When there are two HXs in series, the difference in temperatures between the cold end of hot fluid of the first HX and the hot end of the cold fluid of the second HX is called the approach temperature.

1.34 Arch-Fired Boilers (*see* Downshot *or* Arch-Fired Boilers in Utility Boilers)

1.35 Arch Firing (*see* Arch Mounted/Downshot/Vertical Burners in Pulverised Fuel Burners)

1.36 Ash (*see also* Coal Ash *and* Oil Ash)

Ash is the residual mineral matter (MM) of fuel after combustion. (See Proximate Analysis in Coal.) It is the non-combustible residue left behind after completion of fuel combustion and is usually slightly less than the MM originally present. However, in coals containing high amounts of Ca, ash can be higher than the MM due to retention of oxides of S. Ash is not present in natural or many synthetic gases. But it is there in gases like blast furnace gas (BFG), COG, CO gas and so on.

In most liquid fuels, ash is present, albeit in small quantities (<0.2%) but not in light distillates. The origin of this ash is the plant and animal matter in the oil wells. Ash in the fuel oils is generally low melting. Residuals concentrate all ash with its metallic compounds and S.

A significant amount of ash is associated with solid fuels. At the both the ends of the band of solid fuel band, that is, peat and anthracite, ash is low (2–10%). It is quite high in bituminous and sub-bituminous coals (at times exceeding 40%) and also lignites of some fields. Coal ash comes, to an extent, from the MM of the original plant material and, to a large extent, from soil, rocks and stony material of the coal seams. When high-ash coals are washed, the rejects can contain ash in excess of 65%. Unlike the oil ash, coal ash is characterised by high melting points (see ash fusibility).

A fuel is only as good as its ash, affecting it both by quality and quantity. Basically, ash is undesirable in fuels because

- It retards the combustion speed and hence demands larger furnace volume.
- Excess air may be needed leading to a possible slight drop in thermal η.
- Serious problems of slagging, fouling, erosion and corrosion are created.
- There is a wasteful expenditure in the transportation of ash along with fuel.
- Ash collection and disposal systems are necessary.
- A soot blowing system is required for maintaining cleanliness of heat transfer surfaces.

Ash can be either inherent or extraneous. *Inherent/fixed ash* is one that forms a structural part of fuel and is inseparable by mechanical means. *Extraneous ash* is one that gets added during mining, transportation and so on and can be separated by mechanical action. This topic of Ash is covered in the following sub-topics:

1. Coal ash analysis
2. Ash: bottom and fly
3. Clinker
4. Ash fusibility
5. Ash corrosion or gas-side corrosion
6. Ash-side erosion
7. Slagging and Fouling
8. Base–acid (B/A) ratio

1.36.1 Coal Ash Analysis

Ash analysis has to be understood both in chemical and mineralogical terms to gain a complete picture. Coal ash analysis is different from pure ash analysis of, say, bed or fly ash.

Ash is a mixture of various mineralogical compounds such as calcite ($CaCO_3$), dolomite ($CaCO_3.MgCO_3$), siderite ($FeCO_3$), pyrite (FeS_2), gypsum ($CaSO_4.2H_2O$), quartz (SiO_2), haematite (Fe_2O_3), magnetite (Fe_3O_4), rutile (TiO_2) and so on.

Chemical analysis on the other hand gives ash analysis in terms of

a. Acidic oxides, namely, SiO_2, Al_2O_3 and TiO_2
b. Basic oxides, namely, Fe_2O_3, Na_2O, K_2O, CO and MgO

ASTM standards provide guidelines for testing a sample of powdered coal. This is required to be burnt to completion at 700–750°C in a ceramic crucible during the test.

It is observed that

- Ash high in acidic oxides has a high softening temperature
- Basic oxides in ash tend to reduce the softening temperature.

- Iron compounds such as FeS_2 form corrosive gases like SO_2 and SO_3 in furnace at high temperature.
- S combines with Na and K to form low-melting compounds that cause corrosion and slagging.

From the point of ash fusibility, the various compounds of ash can be classified into three categories:

1. High-melting compounds such as SiO_2, Al_2O_3, Fe_2O_3, CaO and MgO are pure oxides with melting points ranging from 1600°C to 2800°C. They do not melt during combustion and end up in fly ash with their original structure. Being crystalline, they cause **erosion** but not **slagging** or **fouling**.
2. Medium-melting compounds such as Na_2SiO_3, K_2SO_4 and FeS_2 have fusibility in the range of 900–1100°C. They are the cause for slagging as they form a sticky base layer on water walls, wing walls, division walls and platens.
3. Low-melting compounds, mainly chlorides and sulphates of alkali metals, such as NaCl, Na_2SO_4, $CaCl_2$ and $MgCl_2$, have a fusibility range of 700–850°C and form the sticky base layer on SH and RH tubes causing fouling.

1.36.2 Ash: Bottom and Fly

In all types of combustion of solid fuels, lighter ash is elutriated from furnace and heavier ash is rejected at the bottom. The amounts of fly and bottom ash depend on the type of fuel, its preparation and firing equipment. In coal firing, it is typical to have fly ash of 10–20% in gravity-fed grates, 20–30% in SS, 20–30% in PF firing and >90% in CFBC.

The higher the fly ash, the greater the tendency for ash/outside erosion of the tubes and the refractory. Ash handling is more elaborate and expensive. Fly ash cools down along with the flue gases and heat losses are lower as compared with bottom ash losses. With more bottom ash the handling is relatively easier but heat loss to the ash pit is higher.

1.36.3 Clinker

In grate firing, during the process of combustion, fuel ash melts and turns jellylike, which attaches coal and ash particles and also fines turning into lumps called clinker. Towards the end of combustion, such small clinker formation is quite natural. When this process starts in the middle of combustion, the clinker does not stop at small lumps but grows to form big stone-like masses disturbing the bed and combustion. In CGs/TGs, it gets carried to the ash pit, but in DGs and RGs, it stays at the same place. Clinker formation is a nightmare to any stoker operator.

The root cause is the softening of ash particles, which can be traced to the following:

- Ash having low IADT
- Overloading of grate leading to excessive bed temperature
- Coals having high **swelling index**
- Formation of reducing conditions in fuel bed depressing the ash melting points

Usually, a combination of these factors leads to clinker formation. Reducing conditions in the bed develop when

- Coal bed is uneven on grate and not of uniform thickness.
- Segregation of large and small pieces exists in the bed.
- Air flow is not uniform over the entire bed surface.
- Air leakages result in improper flow through bed.
- There are excessive fines in coal.

By correcting these underlying factors, in most cases, the clinker problem can be overcome. If the coal ash deformation temperatures are lower or the swelling index is higher than design, changing the coal is the solution.

1.36.4 Ash Fusibility

Ash fusibility is the *gradual softening and melting of fuel ash with increase in ash temperature* due to the melting of constituents and chemical reactions.

Ash fusibility studies help in predicting the behaviour of ash in boiler based on which the sizing of furnace can be done. Also, when there are slagging and fouling problems in furnace and SH/RH, respectively, fusibility studies help in understanding the phenomenon and take corrective steps.

Even though the ash fusibility figures are not entirely reliable or adequate for design and analysis, they are regularly used for want of any better data, with full knowledge of their limitations.

The procedure for ash fusibility study is given in ASTM D1857. An ash pyramid of 19 mm height and 6.35 mm equilateral base is heated in a gas or in furnace in both oxidising and reducing atmospheres and the deformation temperatures are noted.

Ash fusibility consists of four stages of deformation as depicted in Figure 1.13.

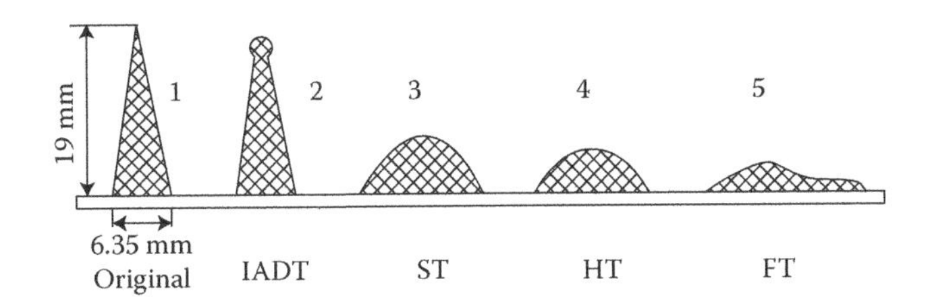

FIGURE 1.13
Ash fusibility stages.

1. IADT when the top of the cone begins to take a round shape
2. *Softening temperature* (ST) or *Spherical temperature* when the shape of the cone is lost and height and breadth become equal
3. *Hemispherical temperature* (HT) when the cone reduces to a hemisphere with the base twice the height
4. *Fluid temperature* (FT) when the specimen has turned flat (height 1.6 mm) and molten ash begins to flow

IADT is the most important temperature for boiler furnace design. Furnace area and volume are dimensioned in such a way that the resultant FEGT is at least 55°C lower than IADT.

ST is also important as the stickiness of ash increases from here on. Normally, ST is higher by about 100°C over IADT. All the four temperatures are measured in oxidising and reducing conditions. Temperatures are lower in reducing conditions. For furnace design, it is the readings under reducing conditions which are considered.

1.36.5 Ash Corrosion or Gas Side Corrosion

Ash corrosion is the same as *gas-side* or *external corrosion* for the PPs. Corrosion is caused on the external surfaces of the PPs by ash and on the internal parts by water. Ash or gas-side corrosion is mainly of two types:

a. High-temperature or hot corrosion of SH and RH surfaces.
b. Low-temperature or cold or dew-point corrosion of economiser and AH.

1.36.5.1 High-Temperature or Hot Corrosion

High-temperature or hot corrosion is mainly due to the attack of alkali (Na and K) metals or V when the tube metal temperatures are usually >620°C. It is found to occur when the gas and metal temperatures both attain high levels. The other aspects of hot corrosion are

- Coals with >3.5% S and 0.25% Cl attack SH and RH tubes.
- Corrosion attack is normally the upstream of the tube, facing the gas flow, rather than the downstream side.
- Ferritic low AS like 2¼% Cr 1% Mo are more susceptible than stainless-steel (ss) tubes with 18Cr 8Ni tubes.
- Outlet tubes of SH and RH platens facing SBs experience the highest corrosion rates.

1.36.5.2 Low-Temperature, Cold Corrosion or Dew Point

Low-temperature, cold corrosion or dew point is on account of S in the fuel oxidising to SO_2 and SO_3 and attacking the tube metals when cooled. They turn to sulphurous and sulphuric acids and the gas temperatures cool down to <200°C when the acids precipitate. Then, corrosion starts to occur. Cold corrosion is combated by raising the metal temperatures at the cold end from where water or air enters.

- In economisers, the principle is to raise the inlet feed temperature to a safe limit by
  - Increasing the regenerative fw heating
  - Adopting drum feed heating
  - Recirculation of the hot fw
- In AHs also, the lowest metal temperature at inlet of air needs to be raised by
  - Raising the air temperature by Scaph
  - Bypassing air or gas in the AH

Figure 1.14 shows the relationship between the onset of corrosion in CS tubes and S in fuel. This is an experience-based graph and not derived from dew-point calculation. The graph suggests the minimum metal temperature to be maintained with CS tubes to prevent dew-point corrosion in both economisers and TAHs.

1.36.6 Ash-Side Erosion

External tube erosion in boilers is entirely due to ash in fuel. Oil- and gas-fired boilers are free of erosion but coal- and lignite-fired boilers suffer from erosion problems in one or the other manner. Containing erosion is more practical than elimination.

Erosion is mainly in the PPs and equipments in contact with coal and ash, like feeders, mills, coal pipes, ash pipes and so on. The hard constituents of ash, namely, SiO_2 and Al_2O_3, are very abrasive in nature and when they move at high velocity, wear is inevitable. Fe_2O_3 is also an abrasive compound which contributes to

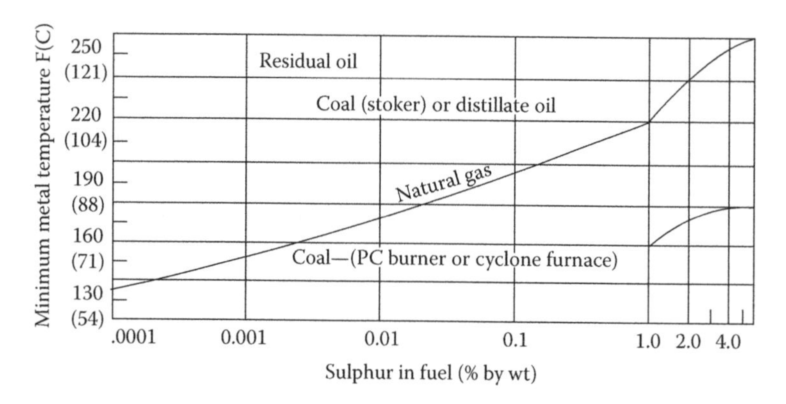

FIGURE 1.14
Permissible CS tube temperature versus sulphur in fuel for TAH and economiser. (From The Babcock and Wilcox Company, with permission.)

erosion. The higher the ash content in coal, the greater is the susceptibility to erosion.

Coal ash (60–80%) turns into fly ash in modern solid fuel-fired boilers. Erosion is best prevented by limiting the gas velocities in boiler passes.

Gas-side erosion is influenced mainly by the dust load in flue gas and its velocity. In fact, erosion is proportional to gas velocity$^{3.5}$, which shows how important it is to reduce the gas velocity which, in turn, makes the boiler larger. The other factors are

- Non-uniformity of dust loading across the cross section.
- Density of tubes in a bank.
- Type of tube bank—inline or staggered construction.
- Gas turns that segregate ash due to centrifugal action.
- Angle of ash impingement.

At right angles or parallel to the surfaces, erosion is negligible whereas at 20–30° of inclination to the surface, the erosion is maximum.

Erosion can be minimised by adopting

a. Design stage precautions before the boiler is built
b. Providing sacrificial materials at the erosion locations

The design stages precautions are

- Limiting the gas velocities to safe levels
- Adopting inline tubes for tube banks in gas paths
- Providing a larger furnace with proper secondary air (SA) and tertiary air (TA) nozzles, which can help in lowering the ash carryover
- Adopt tower-type boiler for high-ash coals

Sacrificial erosion protective measures adopted are

- Wear liners on tubes
- Studding on tubes
- Square or omega tubes for wing wall or platen SH construction in CFBC boilers
- Wear hoods for the protection of tube bends, particularly after the gas turns and enters the second pass

In coal firing, it has been experienced that it is the economisers with higher gas velocities and closer pitching that suffer erosion more than SH and RH.

1.36.7 Slagging and Fouling

Both slagging and fouling are high-temperature ash deposit-forming mechanisms.

Slagging is the formation of *molten or partially fused ash deposits on radiant surfaces* like the furnace, division, wing walls or convection surfaces like platens exposed to radiant heat.

Fouling is the formation of bonded, cemented or sintered ash deposits on the convection surfaces of SH and RH. Areas of slagging and fouling in a PF-fired boiler are shown in Figure 1.15.

Slag is formed when molten or semi-molten ash particles come in contact with cooler tube surfaces and stick to the tube walls on cooling down partly. Wall blowers for furnace and long retractable **SBs** for platens are needed to dislodge the slag. Ash viscosity determines the ease of deslagging.

Fouling is usually caused by the vapourisation of volatile inorganic elements of ash during combustion. Firmly adherent ash layer is formed on convection tube nest on which further layers of ash attach themselves to form bonded deposits, as the outer layer of the original deposit is at higher temperature with glue-like consistency.

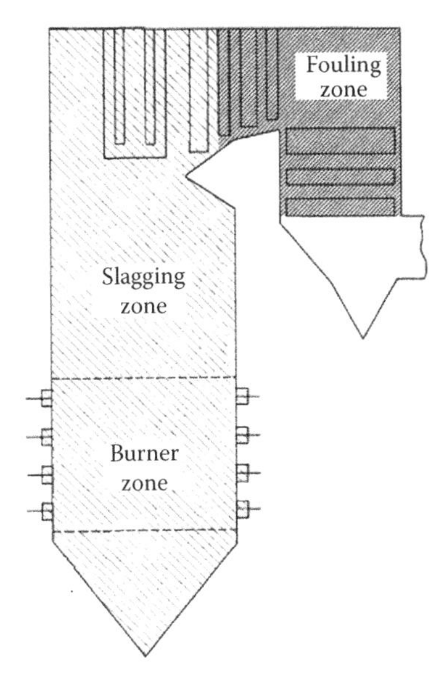

FIGURE 1.15
Slagging and fouling zones in a PF-fired boiler.

Long retractable SBs between tube banks are provided to remove these sintered deposits.

Slagging and fouling are interlinked. Slagging hinders the heat absorption in furnace and raises the gas temperature in convection SH/RH zone, which promotes more fouling.

Slagging is minimised by adequately dimensioned furnace (with FEGT < IADT of coal by 50–100°C). Another method is to have FGR where flue gas from economiser exit is introduced at the bottom of the furnace, to reduce the furnace gas temperatures.

Fouling is minimised by increasing the tube spacing of high-temperature SH and RH and reducing the depth of banks. Figure 1.16 captures these points.

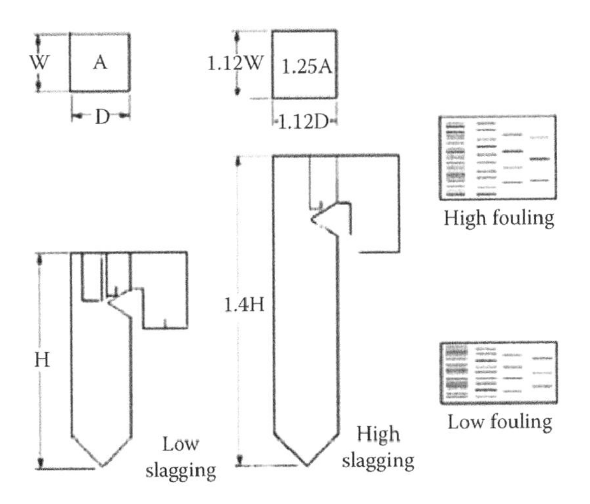

FIGURE 1.16
Effect of slagging and fouling potential on boiler sizing.

Both slagging and fouling can be controlled to a certain extent by positioning strategically adequate number of SBs of the right type.

Boiler sizes and therefore the costs go up sharply for burning highly slagging and fouling fuels. In other words, boilers designed for normal fuels have to be sharply derated when high slagging and fouling fuels have to be burnt.

1.36.8 Base–Acid Ratio

B/A ratio in ash is very useful to estimate the slagging potential of coals.

Chemical analysis of coal ash distributes the ash constituents as

- Acidic oxides, namely, SiO_2, Al_2O_3 and TiO_2
- Basic oxides, namely, Fe_2O_3, Na_2O, K_2O, CaO and MgO

Acidic oxides, the sum of which can range from 20% to 90%, increase the ash fusion temperature and slag viscosities. On the other hand, the basic oxides, which can add up from 5% to 80%, generally lower the ash fusion temperatures except at the extremes. Low ash fusion temperatures occur when the basic oxides sum up to 30–40%.

B/A ratio is $(Fe_2O_3 + CaO + MgO + Na_2O + K_2O)/SiO_2 + Al_2O_3 + TiO_2$, which gives the slagging potential. A ratio of 0.4 indicates a high slagging tendency.

1.37 ASME Boiler and Pressure Vessel Code (*see* Boiler Codes)

1.38 Ash Recirculation or Recycling or Re-Injection

This refers to the act of recirculating ash from the second or subsequent boiler passes into the combustion chamber with a view to improve the C burn up η by burning the elutriated char (FC). In the firing of solid fuels, despite providing generous furnace volume and HP SA to act as a curtain, carryover of char is inevitable.

There is more char in fly ash if the FC in fuel is high or the fuel is very light.

- **Sub-bituminous coals** and *lignites* have such quick and complete burning that ash recycling may be of little advantage. In fact it is often a disadvantage because of the dust problem.

- **Agro-fuels** have low FC but being light they get carried away in suspension firing. Fly ash reinjection is almost always needed to improve C burn up.
- **Mass burning** may not need ash recirculation as the fuel hugs the grate.
- Ash recirculation in **FBC** boilers, particularly **CFBC**, is beneficial for bituminous and higher-grade coals.
- Ash recirculation is not usually done in **PF firing** as the fuel is well ground and fired in the first place so that combustion is almost complete.

Char needs more time and temperature to burn. The char-laden fly ash can be reheated to furnace temperature by returning it to the bottom of the furnace where it will remain for sometime; this will help in burning it further. Usually, an additional 0.5% point of improvement in unburnt losses can be realised.

At the same time, the dust inside the **furnace** also increases making **PPs** susceptible to erosion. The components of the fly ash reinjection system are also subject to erosion risk.

Ash recirculation is, thus, a balance between gain in **thermal** η and downtime due to gas-side **erosion**.

1.39 Asphalt (*see* Liquid Fuels)

1.40 Assisted Circulation or Full Load Recirculation (*see* Circulation)

1.41 Atmospheric Fluidised Bed Combustion

FBC is of two types, namely, atmospheric (**AFBC**) and pressurised (**PFBC**), depending upon the operating pressure of the combustor. AFBC was developed nearly four decades ago and fully commercialised with mainly three different types—**BFBC, CFBC** and expanded or turbulent bed combustion. **PFBC** has also been under development for nearly a couple of decades now and some more experience from the reference and pilot plants has to be gathered to bring it to a commercial level.

Essential differences between AFBC and PFBC are as follows:

- While AFBC is for producing steam in **Rankine cycle** mode, PFBC is for generating power in combined cycle **CC** mode.
- Because of pressurised combustion in the PFBC, the combustor and steam generator are extremely compact.

Both AFBC and PFBC are described at length in Chapter 6.

1.42 Atomisation

Atomisation is the act of splitting a stream into fine particles. For efficient combustion of oil, it must be in a fine mist for instant ignition and complete combustion. The residence time in furnace is minimised by making the spray as fine as possible. In other words, within the confines of the same furnace, combustion is more complete when the particles are small.

1.43 Atomiser

Atomiser is the equipment which converts a stream of oil into a fine spray as needed for complete combustion by

1. Dispersion in a high-speed rotary cup (centrifugal atomisation)
2. Squeezing HP oil stream through an orifice (pressure atomisation) or
3. Shear the oil film with another low-viscosity medium like steam or air (steam or air atomisation)

Methods 1 and 2 are both classified as mechanical atomisation while method 3 as steam/air atomisation. Heating of oil is a prerequisite for proper atomisation in any of the above methods to lower the viscosity to appropriate level unless it is a light oil like kerosene, naphtha and so on. For

1. Mechanical atomisation, **viscosity** should be between 80 and 120 **SSU**
2. Steam/air atomisation, the viscosity can be higher between 200 and 250 SSU

For a wider turndown, the oil has to be heated further to attain 100–150 SSU.

In other words, light FOs are to be heated to ~60–70°C and heavy FOs to 95–105°C. Figure 12.12 shows the variation of viscosity with temperature and the desired ranges.

1.43.1 Mechanical Atomisation

Mechanical atomisation produces coarser oil spray resulting in slightly higher unburnt C loss, but the main advantage is that it needs no atomising fluid like steam or air. But the FO has to be heated to a higher temperature. It is suited for

- Smaller applications where simplicity rather than η is the consideration
- Start-up and low load support applications of the boiler plants

There are two types of mechanical atomisers.

1.43.1.1 Rotary Cup Atomiser

This works on the principle of *centrifugal dispersion*. A cast aluminium cup is set into high-speed rotation and heated FO is introduced at the centre. The oil spreads outwards by the action of centrifugal force and disperses into a fine mist and burns.

- These atomisers are simple, rugged and reliable.
- They are available for capacities as large as 4000 kg/h.
- The turndown is limited.
- Maintenance is somewhat heavier as there are several moving parts.

Rotary cup burners are used in heating furnaces, marine boilers and small package boilers that require short flames and are not quite popular for regular land boiler applications. A typical burner with rotary cup atomiser is shown in Figure 1.17. It is fully self-contained with its own fan. Large burners can be built with the capability of burning as high as ~2500 kg/h of oil with ~30 MW_{th} output.

1.43.1.2 Pressure Atomisation

This involves raising the FO pressure to a high level of 40–70 ata and forcing it through the fine holes of the tip of the atomiser gun, thereby creating a mist of oil. Here also the FO has to be heated to a higher degree to reach the required viscosity. There are two types of atomisation processes, namely

a. *Variable pressure atomisation:* Here, only the required amount of FO is admitted into the oil guns. Oil pressure and consequently the flow is regulated by an oil CV located upstream of the

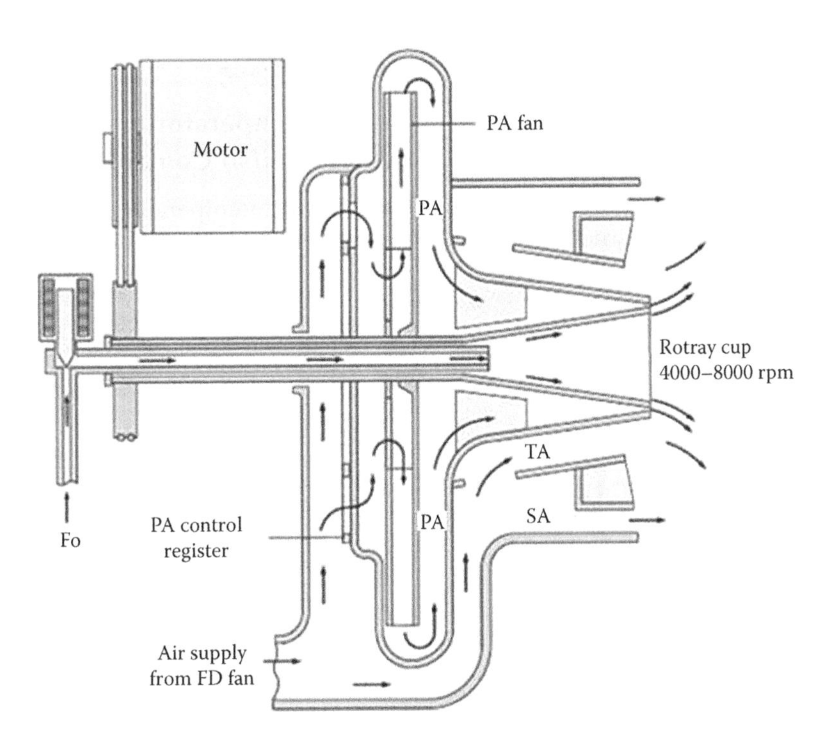

FIGURE 1.17
Burner with rotary cup atomiser.

burners. The advantage of the system is that there is no hot oil recirculating back to the tank. But the disadvantage is that the atomisation turns poorer with falling oil pressures at low loads.

b. *Fixed pressure return flow atomisation:* Here, the burners receive full oil pressure with feed control valve (FCV) regulating the return flow instead of the main flow. The atomisation quality is superior as oil pressure does not fall with load. But there is a constant flow of hot oil back to the oil tank. This system is more popular. A 10:1 turndown is possible with oil pressure of 70 bar. A burner turndown of 1:3 is achievable with windbox pressure of 200–300 mm water gauge (wg).

Pressure atomisation is quite popular in power plants for start-up duties as no other atomising medium is required. The system is simple and convenient. The limitations are

- The power consumption is high as oil has to be compressed to HP.
- The burner turndown is limited to 1:3 or 1:4 at best.
- Wearing of atomiser tips is quite fast leading to a deterioration of spray quality.

1.43.2 Steam/Air Atomisation

Steam or air is used for shearing of FO film to produce a fine mist in this type of atomisation. The main features are

- Oil pressure required is only 20 bar (~300 psig) for a turndown as wide as 20:1.

  Most systems operate at 7 or 10 bar (~100/150 psig) of oil pressure giving an atomiser turndown of 6:1
- Steam/air for atomisation is required at 5, 10, or 15 bar (~75, 150 and 200 psig) with 10 bar (~150 psig) level being quite popular.
- The usual steam consumption is ~10% of oil consumption or ~1.5% of steam generation.
- In most systems, steam pressure is held constant and FO pressure is varied with load.

1.43.3 Atomiser Construction

Both pressure and steam atomisers are of 'pipe in pipe' construction with the diameter of the outer pipe at 50–80 mm. Inside pipes carry return oil in the case of return flow atomiser and steam in the case of steam jet atomiser. Atomiser tips are the places from where the atomised spray emanates into the furnace.

In steam jets, as the steam expands into the outlet nozzle, the pressure drops temporarily at the vena contrata to vacuum levels. This induces oil to enter the steam space and get sheared into fine droplets. This happens irrespective of the supply pressure of the steam whether it is higher or lower than oil pressure. Construction of the discharge end of a typical steam jet atomiser is depicted in Figure 1.18. A comparison of steam/air and pressure atomiser is given in Table 1.8.

1.44 Atrita Mills (*see* Horizontal Mills *under* Pulverisers)

1.45 Attemperator and Desuperheater (*see also* Control Valves in Valves)

DeSH and attemperator are equipments for reducing the SH of steam and maintaining it at a constant level

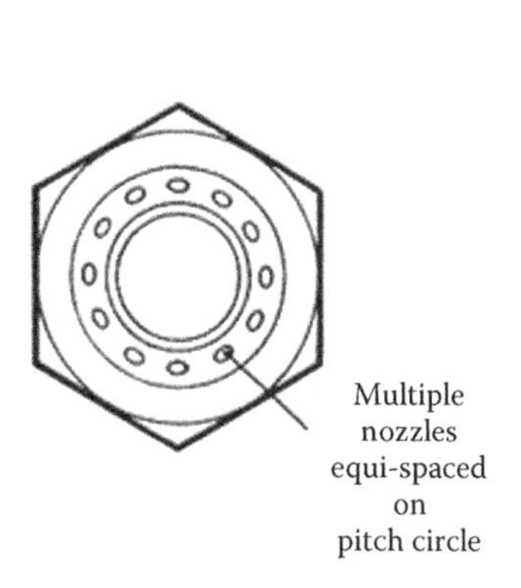

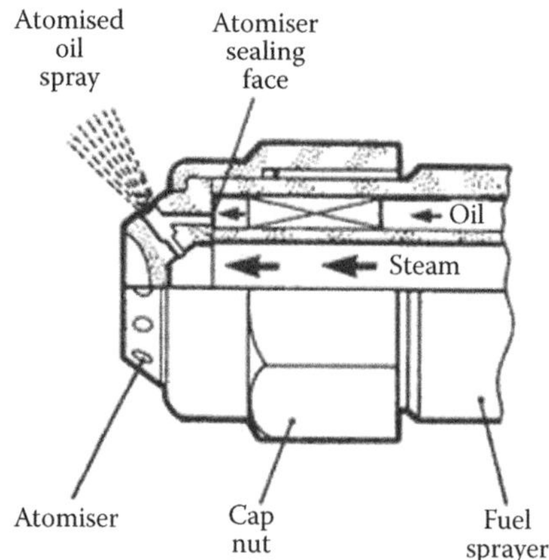

FIGURE 1.18
Cross-sectional view of a typical steam jet atomiser.

TABLE 1.8

Pressure and Steam/Air Atomisation Comparison

S. No.	Parameter	Pressure Atomisation	Steam/Air Atomisation
1	Action	Pressure	Shear
2	Atomisation	Coarse	Fine
3	Viscosity required SSU	80–120	100–150
4	Steam/air required	—	10–15%
5	Steam/air pressure	—	5–15 bar saturated
6	Oil pressure required	High 70–100 bar (~1000–1500 psig)	Low 7–20 bar (~100–300 psig)
7	Oil temperature required	Higher	Lower
8	Pumping power	Higher	Lower
9	Pumping temperature	Higher	Lower
10	Atomiser turndown	Lower 10:1	Higher 20:1
11	Burner turndown	Lower 3:1	Higher 5:1
12	Suitability	Intermittent duty	Continuous duty
		Small amount of oil	Larger amount of oil
		Start-up	Load carrying

as required by the process or within the control range of the boiler, respectively. In other words, it is the deSH in process plants and attemperator in boiler plant that is needed for steam temperature regulation. As it *tempers* the steam temperature, it is an attemperator. It is also a *steam conditioner* as it is sometimes called.

Desuperheating action can be either

- Direct as in spray type where spray water is injected into the steam or
- Indirect as in the case of surface (HX), deSH, or a drum attemperator

In their action and construction, deSH and attemperator bear close resemblance but there are some finer points of difference between the two.

- DeSH is for LP and medium-pressure (MP)/temperature application as required in process plants while attemperator covers the entire range.
- DeSH can tolerate water of moderate quality in most cases. On the contrary, in an attemperator, very high quality water is mostly needed as the final steam has to be fed into an ST.
- The temperature regulation in an attemperator is usually more sophisticated than in a deSH, as it is required to vary with boiler load.
- Being integrated with the boiler, reliability expectation of attemperator is much higher than deSH.
- In other words, an attemperator is a sophisticated and reliable deSH in construction, control and operation. Also, attemperators are usually employed to perform an additional task of limiting the tube metal temperatures by placing them between or ahead of SH or RH banks.

1.45.1 Spray Attemperator

The response of a spray system is much faster and steam pressure drop is lower. Spray-type equipment is also compact. The final steam temperature can be regulated within a very narrow band. However, the spray water quality required is of very high order (typically 20 ppb SiO_2 content) as the water mixes with the steam directly and goes to the ST in most cases. Figure 1.19 shows a typical spray attemperator of a large boiler. Attemperator nozzle receives spray water from a CV to mix with the high-temperature steam in the form of a fine spray. This is for a quick absorption of the mist into steam so that its temperature reduces to the required level within a short distance. A thermal sleeve is usually provided to protect the steam line from receiving the water spray, which can cause a severe shock leading to cracks in pipe. A venturi shape of the thermal sleeve helps to accelerate steam while lowering its pressure, so that water gets sucked into steam and evaporates very quickly. An inter-stage attemperator provided between two steam headers has a rather short distance to absorb the entire water

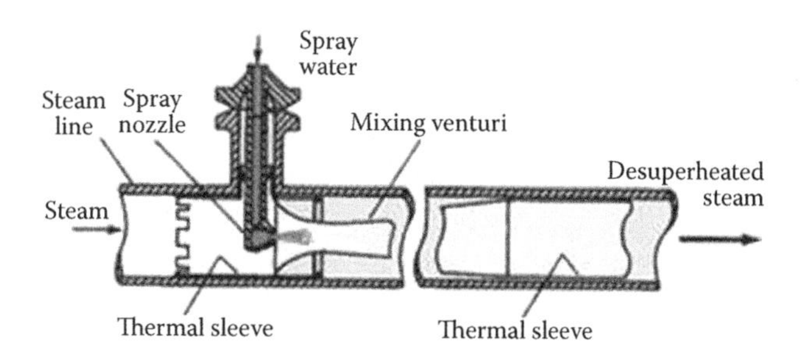

FIGURE 1.19
Spray attemperator of a large boiler.

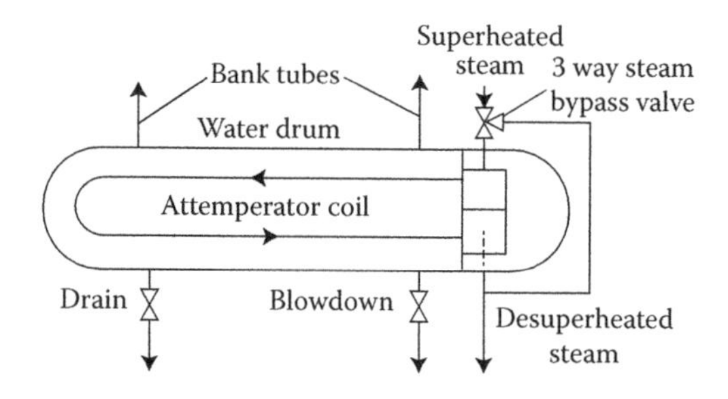

FIGURE 1.21
Schematic arrangement of a drum-type attemperator.

sprayed. At lower pressures, thermal sleeves are not installed. Venturi shape is also design dependent and it can be substituted by a straight pipe.

In many boilers these days, the spray CV and spray nozzle are combined into an integral unit called combined spray and desuperheating (CSDH) unit, which can be simply mounted on the steam line. See Figure 21.11 for constructional details. Such units offer multiple water nozzles which open progressively, keeping the spray uniformly fine over the whole range, which is not possible with a conventional single nozzle. This permits higher rangeability besides making the arrangement more compact. Liners inside the pipe can also be avoided. Figure 1.20 shows the control scheme of the conventional and CSDH arrangements. Refer to CV section under the heading of valves for additional details.

1.45.2 Surface Attemperator

In an indirect system or surface attemperator, the

- Water quality is not of much concern as there is no mixing of steam and spray water.
- Steam pressure drop, including that in the inlet and outlet pipes, is quite high.
- Overall response is sluggish.
- Range of the final steam temperature is quite wide, generally around ±10–15°C in a drum attemperator.
- Equipment is also bigger.

Figure 1.21 shows a schematic of a drum-type attemperator which is common in industrial boilers. Steam from SH header is taken into the inlet box of the attemperator coil through a three-way valve which regulates the amount of steam to be desuperheated. Attemperator coil is a tube bundle in which steam passes through the tubes with drum water surrounding it. Steam is cooled during its passage through the bundle. Instead of drum, there can be an external HX with fw as the coolant.

Attemperator types and suitability

Spray-type attemperator is employed in HP boilers as high-quality fw is available for spraying. Surface attemperators are preferred when the water is not of spray quality as is the case with many MP and LP boilers. Another option is to condense the drum water and use it for spray if the control range is limited. This is because the drum water at saturation temperature has limited cooling ability and the spray quantity requirement is large.

A schematic explains both arrangements of water and condensate spray systems in inter-stage attemperators in Figure 1.22.

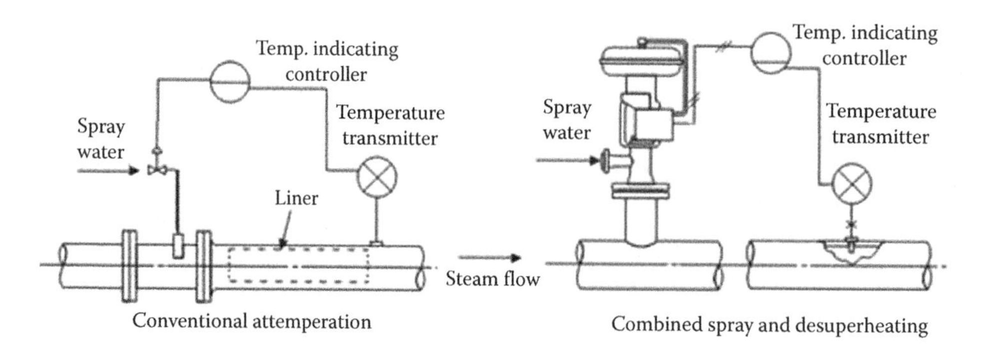

FIGURE 1.20
Spray attemperators—conventional and CSDH.

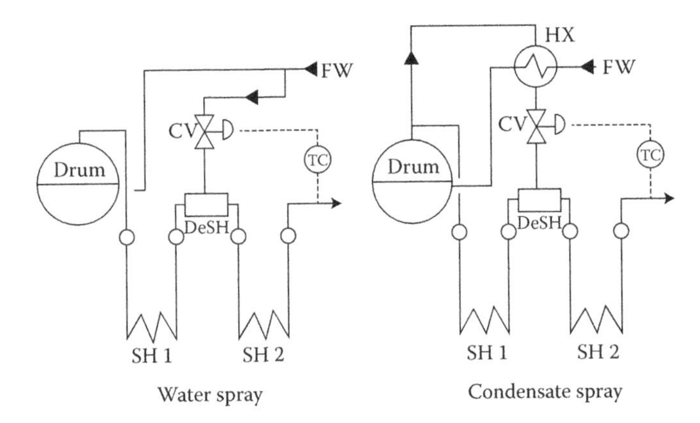

FIGURE 1.22
Schematics of attemperators with water and condensate sprays.

RH attemperation

Spray-type attemperators for RH application are usually located on the outlet side and are meant only for emergency use as the main RH steam temperature control is done by

- Flame control in the furnace
- FGR
- Gas bypassing in the second pass

Location of attemperator

An attemperator is placed

- At the outlet of SH for low final steam temperature of <400°C (outlet attemperator) when SH metal temperature is low. SH construction is simple.
- Between the SHs for higher temperatures (interstage attemperator). The attemperator here limits the metal temperature and often helps to lower tube metallurgy.

Number

For temperatures <500°C, it is usual to have one attemperator and for higher temperatures two attemperators. Besides a better control and rangeability, the metallurgy of tubes can be beneficially lowered.

Outlet temperature

No attemperators should be designed for steam outlet temperatures of less than 5°C over saturation temperature. If the setting is at saturation temperature, the spray CV would not know when to stop and therefore wet steam would result.

1.46 Austenitic Steels (*see* Stainless Steels in Metals)

1.47 Automatic Combustion Control (*see* Control Range *and* Boiler Control)

1.48 Auxiliaries

A boiler is mainly combustion equipment followed by heat transfer equipment (with no combustion in unfired boilers). In a way, both are substantially static in nature, requiring some additional equipment to make them perform which are called the auxiliaries.

There are broadly three sets of auxiliary equipments in most boilers, namely

- Draught plant equipment consisting
  - *Fans* for setting air and gas flows into motion in boiler setting (*see* Fans)
  - *Dampers* in flues and ducts/breechings for diverting and regulating flows (*see* Dampers)
  - *Dust collecting equipment* for appropriately dedusting the flue gases before letting into the atmosphere (*see* Dust Collection Equipment)
- Equipment in steam and water circuit consisting
  - *Feed and circulating pumps* for pumping fw into boiler at proper pressure and circulating boiler water (bw) as needed
  - *Mountings and fittings* for safety of boiler which are mandated by Boiler Codes, namely
    1. *Main steam stop valve (MSSV)* (*see* Valves)
    2. Non-return valve *(NRV) on steam line* (*see* Valves)
    3. *SVs* for relieving over pressure of steam for PP safety (*see* Valves)
    4. *Blowdown valves* for regulation of bw quality (*see* Valves)
    5. *High and low water alarms* on steam drum for drum water regulation
    6. *WLIs* (*see* Water Level Indicators)
    7. *PGs* on drum, SH and RH headers
  - *SBs* for removal of ash deposits from HSs for optimum heat transfer (*see* Sootblowers)

1.49 Auxiliary Fuel (*see* Fuel Firing Modes in Fuels)

1.50 Axial Fan (*see also* Fans)

The nomenclature of fans is derived from the direction of fluid flow with respect to the fan shaft–axial fan for flow parallel to the shaft and **centrifugal fan** for flow normal to the shaft.

A good example for axial fan is the common table fan. Axial fans are good for high volumes and moderate heads. *High η and self-limiting characteristics are the main advantages* of the axial flow fans.

Blades of an axial fan have generally aerofoil cross-section and can either

a. Be fixed in position or
b. Rotate around their longitudinal axis

Blade pitch, which is the angle of the blade to the airflow, can be fixed or adjustable. Changing the blade angle, or pitch, is one of the major advantages of an axial fan. Small blade pitch produces lower flows while increasing the pitch produces higher flow. The flow and pressure are both regulated by varying the blade pitches/angles. The axial fans are more expensive because of the high-quality blade movement mechanism and superior manufacturing techniques/materials required. A typical blade control hub is shown in Figure 1.23. The main advantage is that the axial fans can maintain higher efficiencies at part loads than the centrifugal fans with their inlet vane controls (IVCs).

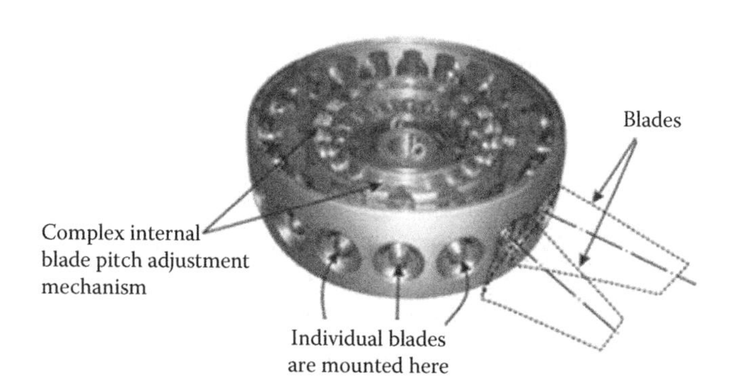

FIGURE 1.23
Blade control hub of an axial fan.

Variable pitch axial fans are capable of changing the blade angles during the running of the fan, which is made possible by the sophisticated hub that carries the blades, shown in Figure 1.23. Each blade is connected to a spindle, which is rotated by a lever and a servo-controlled hydraulic cylinder that moves all the levers simultaneously while the fan impeller is rotating. This varies the output of the fan.

The blade/hub assembly is mounted on a shaft, referred to as the rotor, which rotates in the fan casing. The casing may have an open inlet, but more commonly it will have a right-angled bend to allow the motor to sit outside the ductwork. The discharge casing gently expands to slow down the air or gas flow and convert kinetic energy into useful static pressure.

Axial fans for boilers come in two types: single stage and two stages (as shown in Figure 1.24). Single-stage fans are usually capable of generating a maximum head of ~1200 mm wg while two-stage fans ~2500 mm wg. Both can handle a maximum of ~1300 cum/s of air. Very large fans with 5.3 m diameters have been built with input motor ratings of 13500 kw so far.

A cross-sectional view of a typical two-stage large axial flow fan supported on anti-vibration mountings is shown in Figure 1.25. Some of the main constructional features of an axial fan are described below.

- The blades are adjusted simultaneously by the actuating mechanism provided on the outlet side of the impeller.
- The rotor, consisting of the impellers, the main bearing assembly and the blade adjustment mechanism, can be installed and removed as a complete subassembly on both single-stage and dual-stage fan models.
- The fan housing with its removable top portion is connected to the diffuser and inlet box via a quickly removable non-metallic bandage held down by a steel strap.
- With this design the rotor replacement on the induced draught (ID) fan of a 600 MWe boiler can be done in three shifts.
- The fan is powered by a constant-speed electric motor normally arranged outside the fan itself.

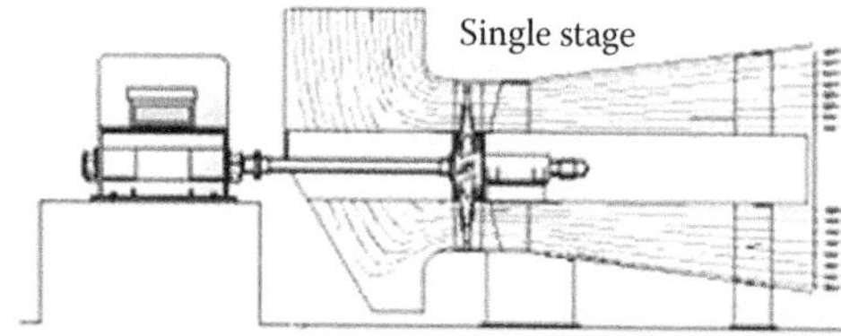
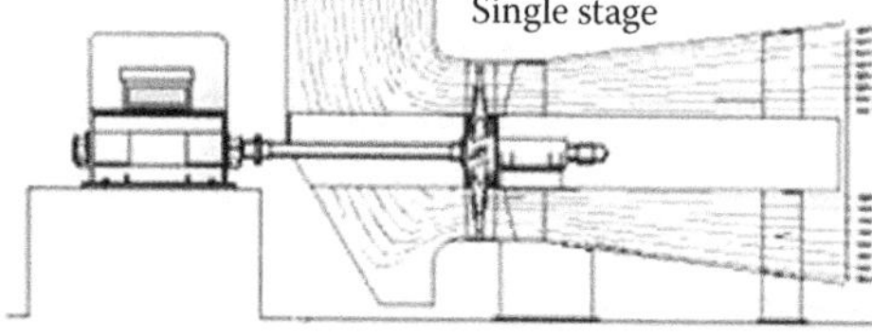

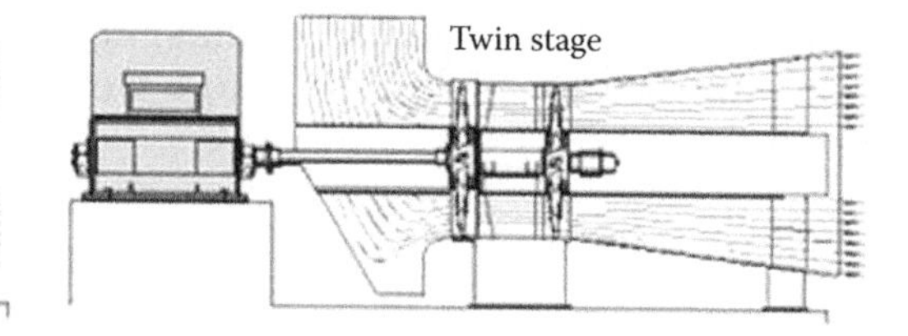

FIGURE 1.24
Single- and twin-stage axial flow fans.

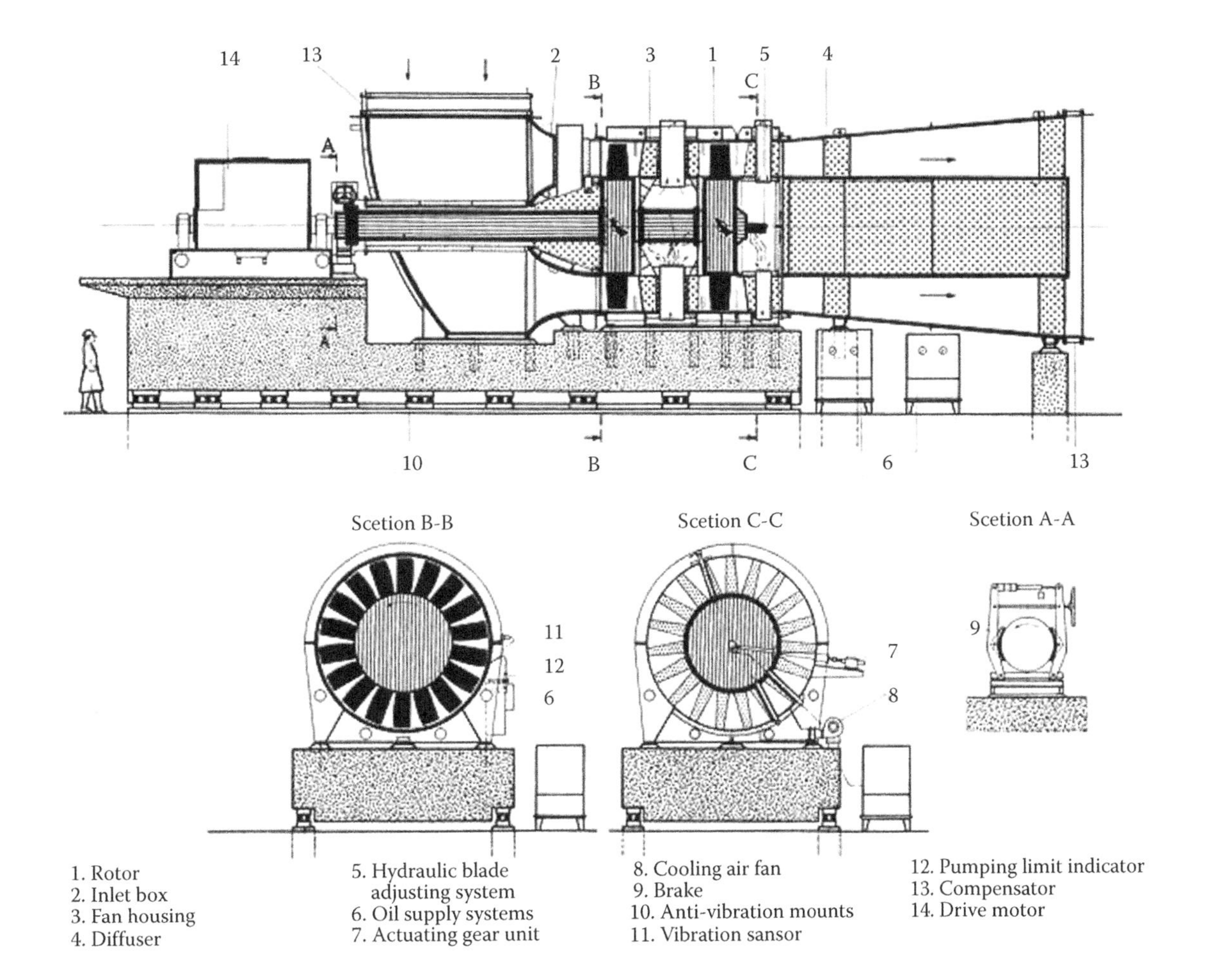

FIGURE 1.25
Cross-sectional views of a twin-stage axial fan supported on anti-vibration mountings. (From TLT-Turbo GmbH, Germany, with permission.)

The motor is connected to the rotor via a hollow shaft with a torsionally flexible curved-tooth or multiple spring disc coupling.

- Owing to the temperature loads acting on ID fans, the interior of the hub is thermally insulated in order to protect the rotating components.
- Cooling air is supplied into the hub through the hollow bracing and blades by a set of separate external fans. It is important that the cooling air-carrying ducts are insulated to prevent temperatures below the dew point.
- If the rotor is supported in sliding bearings, a brake is fitted on the drive-side coupling to protect the bearings against running in mixed-friction conditions and to prevent rotor spinning once the motor has been de-energised.
- Lubricating oil for the main bearing and hydraulic oil for the hydraulic actuating mechanism are supplied by oil supply units mounted outside the fan.

Axial fans have a unique advantage of vertical arrangement, shown in Figure 1.26, which helps in placing them at the bottom of a stack or on steel structure saving space.

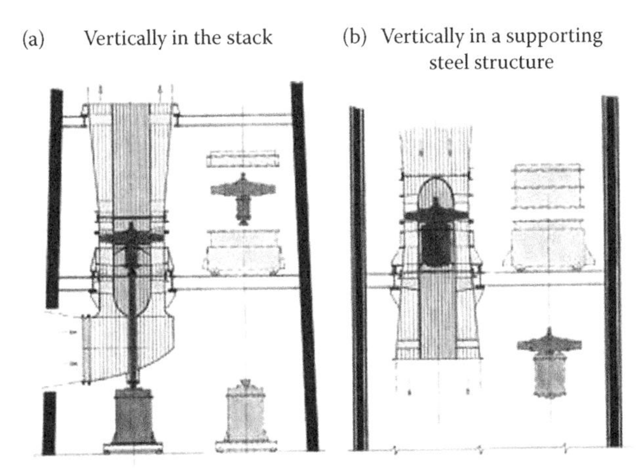

FIGURE 1.26
Method of placing vertical axial flow fans (a) inside a stack or (b) on steel structure. (From TLT-Turbo GmbH, Germany, with permission.)

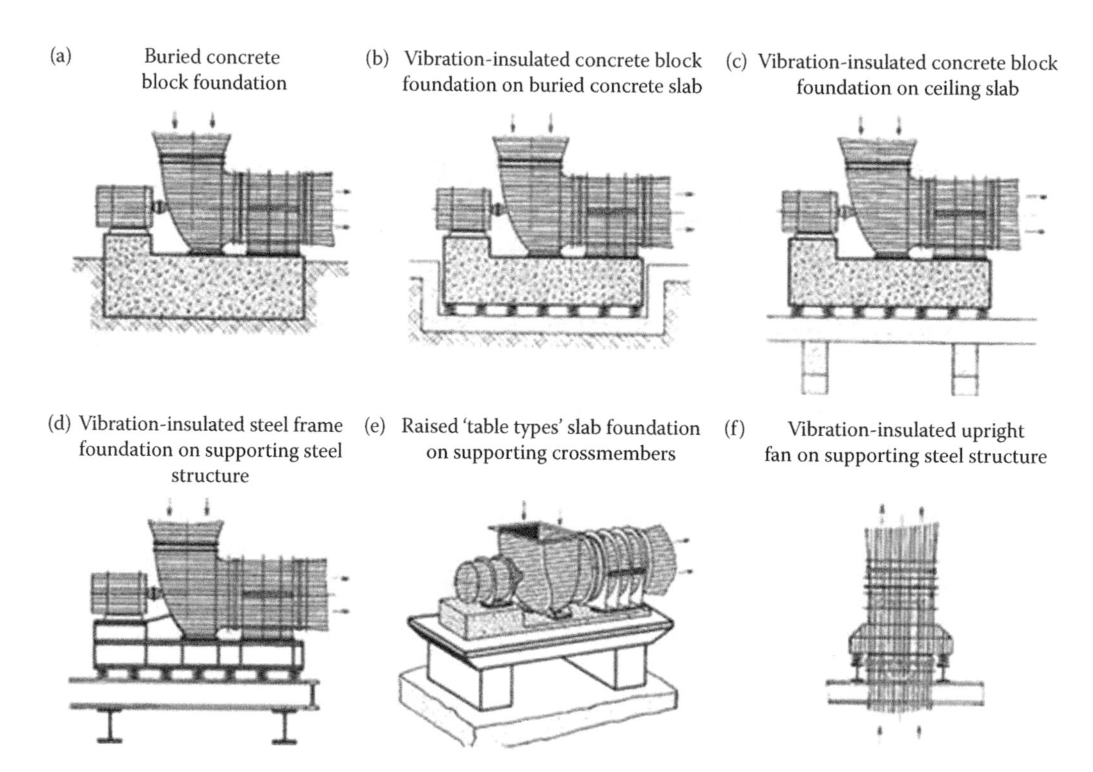

FIGURE 1.27
Types of foundations for horizontal (a–e) and vertical (f) axial fans. (From TLT-Turbo GmbH, Germany, with permission.)

There are several ways of placing axial fans on their foundations (as shown in Figure 1.27).

There are two important points of difference between axial and centrifugal fans in their performance, as shown in Figure 1.28, which provides a comparison of the two characteristic curves.

- The iso-η curves of variable-pitch axial flow run approximately parallel to the system resistance graph, which gives good efficiencies throughout a broad operating range. On the contrary, for centrifugal fans with variable inlet vanes, the iso-η curves intersect the system resistance

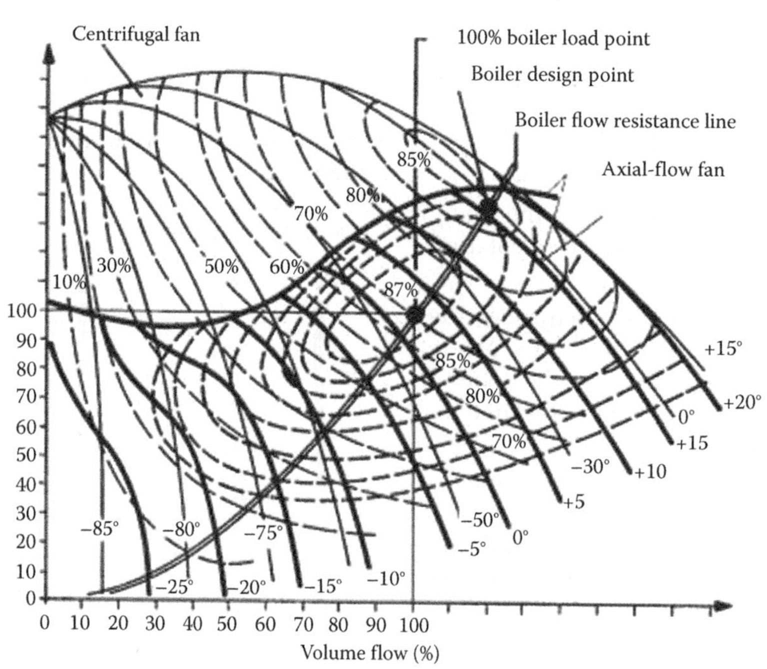

FIGURE 1.28
Comparison of characteristic curves of centrifugal and axial flow fan. (From TLT-Turbo GmbH, Germany, with permission.)

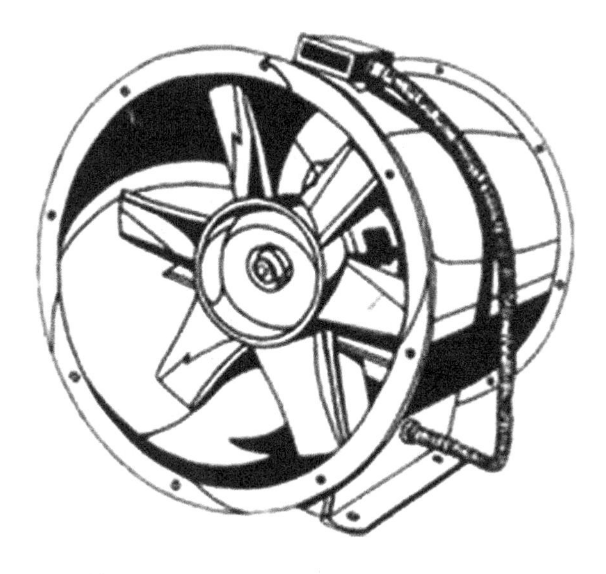

FIGURE 1.29
Typical propeller fan.

curves, meaning that their η under part-load condition is automatically lower than with axial-flow units.

- An axial-flow fan can be selected in such a way that the boiler design point is located above the maximum η range so that the operating points of maximum interest can be in the highest η spectrum.

Escalating fuel and power costs are steadily strengthening the case for axial fans—at least for FD fan applications in utility boilers. However, the ID application is weighed down with the possibility of fan erosion, even with slightly higher or aggressive ash in fuel. Erosion of blades is the main concern as they are made of Al or Mg alloys to keep the weights down, which reduces the hub strength and bearing loads. The blades in many cases are protected by hardened steel strips but the axial fans are still not considered suitable for ID fan duties for either high-ash or abrasive ash applications. Another important aspect is that the performance degradation on erosion is greater for axial fans than for centrifugals.

In comparison with centrifugal fans the axial fans are

- More expensive
- More prone to erosion
- More maintenance-prone due to more moving parts
- Better in η and hence lower in power consumption
- Higher η over a wider range of boiler load

Propeller fans

Propeller fans are axial fans where air flow is parallel to the fan axis. They move large amounts of air by striking at an angle and developing small heads, typically ~300 N/m^2. The housing, which may or may not be present, plays little part. The flow control is minimal. Household ceiling and table fans are examples of propeller fans. Boiler ventilation room exhaust and ID cooling tower fans fall in this category. Figure 1.29 carries a picture of a typical propeller fan which has housing.

2

B

2.1 Back-End Equipment (*see* Heating Surfaces)

2.2 Bag Filter/Cloth Filter/Fabric Filter (*see* Dust Collection Equipment)

2.3 Bagasse and Firing (*see* Agro-Fuels)

2.4 Bagasse-Fired Boilers (*see also* Bagasse in Agro-Fuels *and* Grates)

A bagasse-fired boiler has always been treated as one of the equipments forming part of a sugar plant, as its input and output are closely enmeshed with the plant operations. Decades ago, when sugar factories were located in the remote countryside, a sugar plant and its boiler could survive in splendid isolation as the bagasse produced was more than adequate to supply all the steam needed for process and power for the plant as well as its neighbourhood. As excess bagasse would pose problems of disposal, the boilers were not efficient. Simple boilers that could be operated by unskilled operators were in demand. **Pile burning** in stationary step grate or **horseshoe type** of boilers with low capacities and low p & t were very popular.

As the sugar factories grew larger in size and plant operations improved, more steam and power became necessary and bagasse shortages were experienced. Efficient boilers were needed working at better p & t. Stationary inclined water-cooled grate (**IWCG**), pinhole grate (**PHG**), dumping grates (**DG**) and **TG** replaced the earlier grates. Boilers equipped with *spreader stoker* (SS) became the industry standard, with large boilers built up to 200 tph with ~60 bar and ~450°C.

Over the last couple of decade, as green power from renewable energy sources along with distributed power generation gained greater attention, bagasse firing for cogen became very popular all over the world. High pressure (HP) boilers up to ~200 tph up to 130 bar and 540°C have been built in the last decade. Unit sizes could not be increased upwards owing to the limitation of grate size.

The salient features of bagasse-fired boilers are listed below.

- Type

  Boilers up to ~60 tph are generally bottom supported.

  Bi-drum boilers are popular for pressures up to ~60 bar.
- Firing system

  DGs can be employed up to ~100 tph. TG and PHG are popular from 60 tph onwards.

  DGs with their inherent intermittent ash discharge (IAD) are less efficient by ~1–2% when compared with TGs.
- Auxiliary fuel for off season

  Coal, oil or gas is the most popular off-season fuel for maintaining steam generation.

  Off season steam is needed only if it is a boiler for cogen. Then the auxiliary fuel is to be designed for full steam generation.

  If there is no cogen it is adequate to size the auxiliary fuel firing for ~25% of full load as needed for start-up duties.
- Multi-fuel firing

  DG boilers are suitable only for oil/gas as auxiliary fuels. If coal is to be used it should be low ash with <15%.

  PHGs and IWGs can only admit oil/gas. No coal firing is possible.

  TGs are the most versatile as any auxiliary fuel namely oil/gas/coal can be burnt at full capacity.
- Grate rating

  Heat release rates (HRRs) for TGs and PHGs are nearly the same. For DGs it is ~20% lower as ash is not continuously discharged. Table 2.1 gives the maximum permissible rates with hot combustion air.

TABLE 2.1

Maximum Permissible Grate Rating for Burning Bagasse

Grate	MW/m²	kcal/m²/h	Btu/ft² h
TG or PHG/IWG	3.47	2.98×10^6	1.1×10^6
DG	2.85	2.44×10^6	0.9×10^6

Bagasse with high amount of volatile matter (VM) is an easy fuel to burn. 25–30 and excess air should be sufficient for a good burn out with PA of 70–80%.

- Furnace sizing

Unlike some of the other bio-fuels bagasse does not have high amount of alkalis in its ash. Ash deformation temperature is reasonably high permitting a fairly high furnace exit gas temperature (FEGT). High SOTs are therefore possible without slagging and fouling problems. The furnace sizing parameters are shown in Table 2.2.

Furnace gas velocity should be limited to 5.5 m/s to limit carryover of dust.

- Hot air temperature

Hot air improves the combustion η and speed and it is almost mandatory for burning high M fuel-like bagasse. The temperature withstanding ability of grate limits hot air temperature. For TG it is ~230°C (~450°F), DG ~230°C (~450°F) and PHG/IWCG ~315°C (~600°F)

- Grit re-firing (GR)/ash re-circulation

As bagasse is an extremely light fuel there is a tendency for it to get carried away with flue gases. This results in a reasonable amount of char to make it worthwhile to return the fly ash to the furnace before it gets too cool and subject it to the heat of the flames once more. GR from the first set of hoppers is usually carried out to get an additional improvement in η to the extent of ~0.5%.

But opinion on GR is divided. It is felt by some that both char and furnace are not hot enough to give the desired incremental η while on ash recirculation increases the gas side erosion problem.

- PA, SA and TA

LP PA and HP SA are needed.

SA of 20–30% of total air at pressure of ~750 mm (~30″) wg is normal for which a separate SA fan is required.

Hot versus cold SA is a much debated issue. The advantage of hot SA is that it does not quench the flame and reduce combustion efficiency. At the same time it does not provide good penetration like cold SA. It is the reverse with cold SA.

A part of SA is diverted as TA with nozzles provided at ~2 m below the nose in some of the installations of late, to improve the combustion η and also reduce carryover of dust.

- SH

High SH is possible with platen-type SH hung in the furnace or wing wall SH arranged in the front wall. SH bank placed immediately after the furnace exit should have a spacing of no <200 mm (8″) and subsequent banks 100 mm (4″) to avoid ash deposits.

Owing to the high M in fuel and large gas volume, HS required for SH is comparatively small in bagasse firing. When full steam temperature is required on auxiliary fuel SH area needs to be enlarged significantly.

This would result in a larger attemperator too.

- Boiler bank (BB)

When spreader firing is employed BBs are usually of single-pass baffleless construction to avoid erosion, at least in larger sizes of boilers. Two-pass longitudinal gas flow is popular in smaller range. When coal firing is needed to fully load for off-season duty, single-pass BB is desirable for all sizes, unless coal has low ash with no erosion tendency.

- Back-end equipment

In olden boilers it was sufficient to provide only AH at the back end as η was not as great a consideration as the cost and simplicity. Also hot air temperatures were required to be high for stationary horseshoe furnaces.

Modern boilers are invariably equipped with both Eco and AH and flue gases are cooled to 160–170°C (~320–340°F). Economisers are plain tube with in-line arrangement to withstand gas side erosion. TAHs are chosen in preference to RAH as the latter tend to get fouled with fluffy dust in flue gas. Also the char in dust poses a fire hazard.

- Gas velocities

As dust carryover is low and ash is less, permissible gas velocities are higher than in coal firing. The normal limits are 15–18 m/s (50–60 fps) when ash is very low ~2%. However, bagasse is a mildly abrasive fuel and all care for erosion needs

TABLE 2.2

Factors for Sizing Furnace in Bagasse Firing

Volumetric HRR			Residence Time	FEGT		SOT	
MW/m^3	kcal/m^3	Btu/ft^3	s	°C	°F	°C	°F
<0.31	<2,65,000	<30,000	~2.5	~1000	1830	~540	~1000

to be taken, including limiting average velocities to <12 m/s for long life of pressure parts.

- SBs

  Wall blowers are almost never required as there is no slagging because of low amounts of ash and low furnace temperatures because of high fuel M. But high-temperature SH, immediately at the furnace exit, is susceptible to fouling and retractable SBs are usually provided. Rotary SBs are needed in other areas like SH, boiler and Eco banks.

- Dust collection

  Depending on the firing technique, ash content and boiler configuration, dust loading at AH exit can be in the range 2.5–8.0 g/Nm^3 (~1.0–3.5 grains/cft).

  - Lightness of bagasse ash is the real problem making MDCs unsatisfactory.
  - ESPs, bag filters and wet scrubbers are in use with various degrees of success.
  - High ash resistivity comes in the way of good results from ESPs.
  - Collection in bag filter is satisfactory but there is always a danger of char particles burning the bags. It is necessary to have a knock out baffle to give a change to the gas direction so that the char and ash can precipitate out of the gas stream before entering the bag filter.
  - Wet scrubbers are also used but the power costs are high. Air pollution problem is transferred to water pollution issue and unless there is a proper solution to that wet scrubbers cannot be adopted.

- Fans

  Compared with an equivalent boiler on coal firing, FD fans are nearly of the same size whereas SA fans are bigger because of both higher volume and head. ID fans are really larger as the gas volumes are significantly more. This is because the high M of fuel (~50% of fuel weight) is transferred to flue gas.

Shown in Figure 2.1 is a top-supported DG-fired boiler with two stages of SH, two-pass longitudinal flow BB, two stages of Eco and two stage vertical TAH. It is a medium-size boiler with no coal firing. Such units would have a typical gas exit temperature of ~180°C (~355°F) and operate at ~68% gross η on Gcv.

In Figure 2.2 a large top-supported bagasse-fired boiler equipped with continuous ash discharge (CAD) TG is depicted. This boiler is an indoor installation as there is a boiler house around the unit. The BB is short with single pass with no baffles that can withstand erosion better.

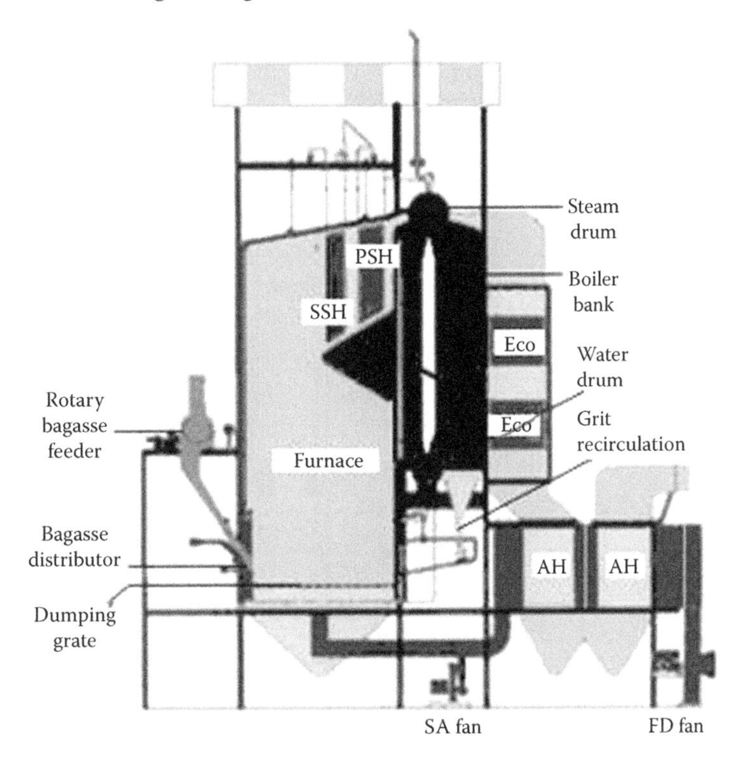

FIGURE 2.1
Top-supported DG spreader stoker bagasse-fired boiler with two-pass BB.

2.5 Ball and Racer Mill (*see* Pulverisers)

2.6 Ball and Tube Mill (*see* Pulverisers)

2.7 Ballast or Burden in Fuel (*see* Proximate Analysis in Coal)

2.8 Banking of Boiler (*see* Operation and Maintenance Topics)

2.9 Bark (*see* Agro-Fuels)

2.10 Basalt

Basalt is a hard, black volcanic rock with <52% SiO_2. Because of this low SiO_2 content, basalt has a low viscosity (thickness).

Cast basalt is produced by melting basalt rock and shaping into tiles or cylinders. Special heat treatment produces a fine crystalline structure which makes it resist sliding abrasion. It is used for enhancing the life of various process equipments in power, steel and cement plants that suffer abrasion.

Basalt lining also possesses an ability to withstand external impact strength. Even if broken the liner does not shatter but stays in place. It has extremely high resistance to sliding abrasion. The lining gets polished with time because of the flow of material and the friction reduces. Basalt lining has a limitation in its operating temperature which should be <~350°C.

Many times PF lines from mills to boiler are lined with basalt to improve its life. Typical properties are given in Table 2.3.

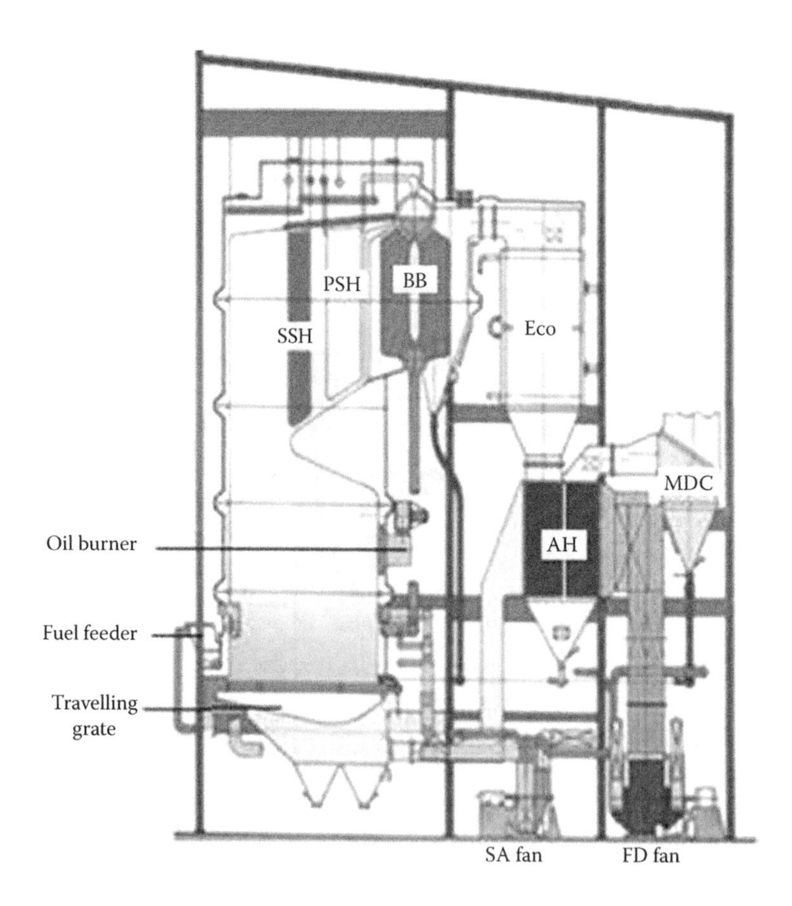

FIGURE 2.2
Top-supported TG spreader stoker bagasse-fired boiler with single-pass BB.

TABLE 2.3
Properties of Basalt Lining

Property	Value
Specific gravity	2.9–3.1 te/cum
Hardness	7–8 moh scale
Compressive strength	4500 kg/cm^2
Abrasive resistance	0.06–0.08 cm
Bending strength	300 kg/cm^2
Water absorption	Nil

2.11 Base–Acid Ratio in Coal Ash (*see* Ash)

2.12 Base Fuel (*see* Fuel Firing Modes)

2.13 Beater Wheel Mills (*see* Lignite)

2.14 Belt Feeders in Pulverised Firing (*see* Fuel Feeders)

2.15 Benson Boiler (*see* Once-Through Boiler)

2.16 Bernoulli's Equation (*see* Fluid Flow)

2.17 Bi-Colour Gauges (*see* Direct Water Level Indicators)

2.18 Bi-Drum Boilers (*see* Industrial Boilers)

2.19 0-Fuels (*see* Agro-Fuels)

2.20 Biomass

Biomass is a renewable energy resource from the carbonaceous waste generated from all activities. It is the biological material derived from living, or recently living organisms. It is the organic matter formed directly or indirectly by virtue of photosynthesis.

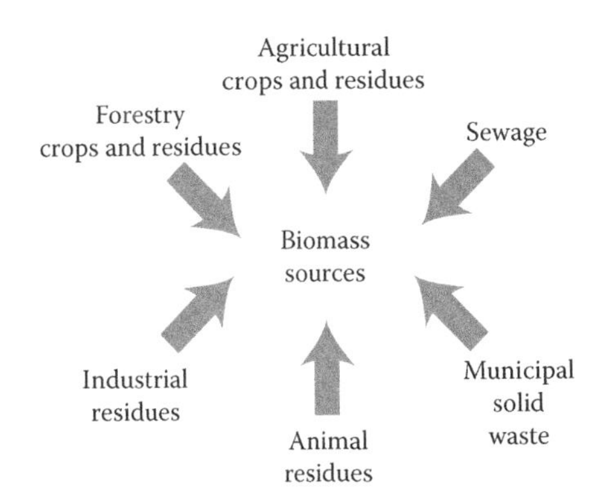

FIGURE 2.3
Sources of biomass.

Biomass is a comprehensive expression covering wide sources. Biomass includes:

- Agro-/bio-fuels such as crops, forest/agro industrial residues, purpose grown trees and so on
- Aquatic plants
- Municipal wastes
- Animal wastes (poultry litter)
- All bio-degradable waste materials

This is shown in Figure 2.3.

Even though organic and originating from ancient plant tissues, **fossil fuels** are not considered biomass because the C in them is out of the C cycle for millennia.

In burning of biomass no additional CO_2 is added to the atmosphere since it absorbs the same amount of C in growing as it releases when burnt. This is the main attraction today in turning to biomass as fuel, despite its several limitations.

2.21 Bi-Metal Tubes (*see* Boiler Tubes *under* Materials of Construction)

2.22 Birmingham Wire Gauge (BWG) (*see* Wire Gauge)

2.23 Bitumen (*see* Liquid Fuels)

2.24 Bituminous Coal

This is the most abundant form of coal with the broadest range of commercial uses. These coals have long been used for steam generation in electric power plants and industrial boiler plants. They are also used for gas production.

Bituminous coal is also called *black coal* or *hard coal* (Stein Kohle in German). It derives its name from the tar or bitumen-like substance that it contains which makes it somewhat soft.

It is dark brown to black and has well-defined bright and dull bands. The C content of bituminous coal is ~60–80% on daf basis and the rest is composed of water, air, H_2 and S, which have not been driven off from the S-bearing mineral called maceral. It has a relatively high Gcv varying between 24 and 34 MJ/kg (5650–8000 kcal/kg or 10,150–14,400 Btu/lb) on daf basis.

It is relatively easy to handle and burn this coal, as its burden (A + M) is not high. Bituminous coals burn with smoky yellow flame like **bitumen** and their distillation product is coal tar which is of a bituminous nature. S content varies from ~0.3% to 4.0% in bituminous coals needing many times a suitable **desulphurisation** for the flue gases.

True ρ of coal is ~1300 kg/m^3 and bulk ρ is ~800 kg/m^3 (81 and 50 lb/cft, respectively)

Certain varieties, which have good **swelling properties**, are also used to make **coke**, a hard substance of almost pure C that is important for smelting iron ore.

2.25 Bituminous Sands (*see* Liquid Fuels)

2.26 Black Liquor (*see* Recovery Boiler *and* Kraft Process)

2.27 Black Liquor Recovery Boiler (*see* Recovery Boiler)

2.28 Black Body (*see* Heat Transfer)

2.29 Blast Furnace Gas (*see* Gaseous Fuels)

2.30 Blast Furnace Gas-Fired Boilers (*see* Waste Heat Boilers)

2.31 Blowdown (*see also* Boiler Blowdown in Water *and* Safety Valves *under* Valves)

The term blowdown is used in two different contexts—in boilers and in SVs.

In boilers by blowdown is meant the disposal of bw continuously Continuous blowdown (CBD) for limiting the **carryover** or intermittent blowdown (IBD) for removal of **sludge**.

In SVs blowdown is the difference between the set pressure and the disc-reseating pressure expressed as percentage of the set pressure.

2.32 Blowdown Tanks (*see also* Boiler Blowdown *under* the Topic of Water)

In any boiler, there are two types of blowdown tanks.

1. *IBD tank or dirty drain vessel:* This is to collect and dispose the water of the IBD carried out periodically to drain the **sludge** of the bw. Because the discharge is intermittent, no heat recovery is possible. The tank is vented to atmosphere to a suitable safe location. An overflow drain from tank discharges the dirty water to the drains.
2. *CBD tank or clean drain vessel:* CBD is taken from the steam drum on a continuous basis to regulate the Total dissolved solids (**TDS**) at LP and **dissolved SiO_2** at HP. Usually the CBD is ~2% of maximum continuous rating (MCR). As the flow is continuous heat recovery is possible. Usually the flash steam from CBD tank is taken to the **deaerator**. The water from the CBD tank is further flashed at LP to recover some more steam, if there is need, before the drain water is let off.

Figure 2.4 displays a typical vertical IBD tank. The tank can also be made horizontal to save on the height.

2.33 Blowdown Steam Recovery

Heat recovery from IBD is not practised as the blowdown is intermittent and also dirty. It contains a lot of **sludge**. CBD is just the opposite and is eminently suited for heat recovery when the boilers are sufficiently large or there are multiple units. It is normal to lead the flash

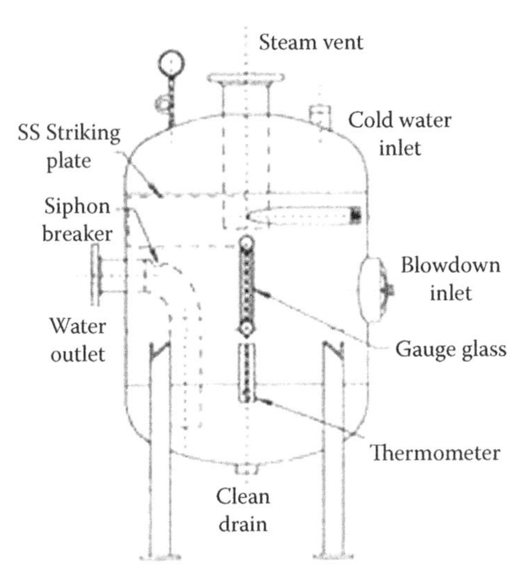

FIGURE 2.4
Vertical IBD tank.

steam to deaerator from CBD tank. The condensed drain water, which is at the same pressure as CBD tank, is led to the IBD tank for further disposal. Heat recovery at this stage is possible by installing a suitable HX which can transfer heat, typically to the fw. This is shown schematically in Figure 2.5.

Heat recovery is regularly practised in Hrsgs with multiple pressure levels as the blowdown is from several drums.

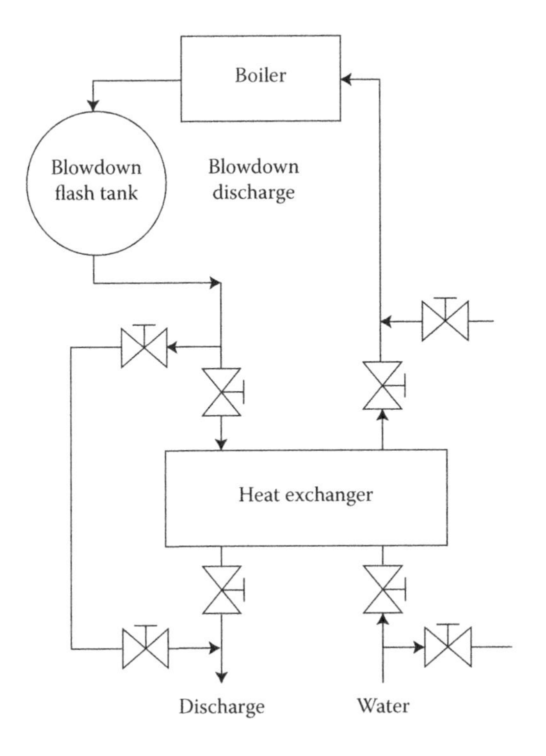

FIGURE 2.5
Heat recovery scheme from boiler blowdown.

2.34 Blowing, Steam Line (*see* Commissioning)

2.35 Blow Off (*see also* Blowdown Tanks *and see* Boiler Blowdown *under* the Topic of Water)

Blow off is an American term for IBD.

2.36 Boil Out (*see* Commissioning)

2.37 Boiler (or Steam Generator) (*see also* Boiler Classification)

Boiler and steam generator for all practical purposes mean the same. Boiler is an English term and somewhat colloquial. Steam generator is European and slightly more scientific sounding. Also a boiler can generate hot water when it is termed as hot water boiler.

Mainly, the boilers convert *water into steam* for a variety of downstream uses. Boiler is primarily *an energy converter*, converting the *chemical energy of fuels or heat energy of waste gas streams into heat energy of steam.*

Boilers are used in numerous applications depending on where the steam is employed. The main areas of use are in process steam and power generation, with the latter being overwhelmingly bigger.

Boilers are made in a variety of sizes, using diverse *fuels* to produce steam at various ps & ts, using assorted firing techniques and employing different types of circulation systems. In WH boilers there is no fuel and hence no fuel firing equipment at all and the boilers have to perform the task of transferring heat from hot gas steam to water. Boiler designs have been undergoing progressively significant changes because of the ever tightening **air pollution** norms.

Any fired boiler is a combination of firing equipment followed by HX. Thermodynamically, it is a blend of **combustion** and **heat transfer**. How effectively the two tasks are performed decides the η of a boiler. In WH or unfired boilers firing equipment is conspicuous by its absence and the boiler there is a sophisticated HX.

For firing **liquid** and **gaseous fuels** burner firing is employed which is relatively simpler. It is the **solid fuels** that need a variety of firing techniques depending on the fuel and its quantity. There are a variety of **grates, pile burning** methods, **FBC** systems, **PF firing, cyclone**

furnaces and so on. Prime fuels namely coal, lignite, oil and gas are the major fuels for burning in boilers both in industry and power stations. Oil and gas are used in combined cycle mode than steam cycle except where process steam alone is desired. **Bio-fuels** are increasingly consumed either alone or in cofiring mode in existing or new boilers with a view to bringing down the **GHG** emissions. **Waste fuels** and **WH** are used employed in the process industry where recovery of energy makes a big difference to the economics of the particular plant. These boilers come in a variety of configurations such as medium to small. Fuel flexibility is another important feature of the modern boilers recognising the inescapable fact that the fuels are getting expensive, scarce and new fuels are forced on the market.

Steam is generated at a variety of ps & ts. Process steam in industry is rarely required at >32 bar and 400°C for heating and other process applications. As cogen is usually adopted in the industry, to maximise the fuel η by generating both power and steam/heat, steam may be generated up to ~140 bar and ~530°C. For pure power generation, however, highest possible steam conditions are employed to produce power at the least cost. In drum boilers, employing natural or assisted circulation, steam is generated up to ~190 bar and 540/540°C. In supercritical (**SC**) and Ultra SC (**USC**) **boilers** steam conditions of ~250 bar/565/590°C and 350 bar/600/620°C are, respectively, employed.

Industrial and **utility boilers** are the two major divisions in boilers addressing the needs of process/cogen and utility power, respectively. Industrial boilers are usually <400 tph, natural circulation and non-RH boilers operating at <140 bar and 540°C. Flexibility of multi-fuel firing is usually desired. Utility boilers are larger sub-critical (sub-c) or SC pressure RH boilers running on a prime fuel with no particular fuel flexibility needed and with the sole purpose of producing steam at the best η for power generation.

Natural circulation boilers constitute over 90% of all the boiler population being easier to build and operate and needing relatively simpler Water treatment (WT). For large utility boilers >500 MW OT **forced circulation** boilers are increasingly preferred because of

- Higher cycle efficiency
- Variable pressure operation
- Reduced GHG emissions

The significant environmental factors affecting boiler design and construction are mainly the limits set for NO_x and SO_x. The rapid acceptance of FBC boilers is principally because of their environmental friendliness. Introduction of low NO_x burners (**LNB**), separated overfire air (**SOFA**) systems and **SCR** and **SNCR** in PF boilers are all meant to lower NO_x. Scrubbers for SO_x are also needed as *in situ* desulphurisation is not possible. In many instances PF boilers have become a more expensive yielding ground to CFBs.

Water tube boilers are made from as small a size as ~5 tph for process applications to as large as ~4000 tph with double RH for power generation, with even larger units being planned in the near future.

2.38 Boiler Bank or Convection Bank or Evaporator Bank (*see* Heating Surfaces)

2.39 Boiler Bank Tubes (*see* Pressure Parts)

2.40 Boiler Blowdown, Continuous and Intermittent (*see* Water)

2.41 Boiler Classification

Boilers come in a bewildering variety of shapes and sizes for a variety of end uses in different constructions. The following is a brief classification.

The *most important* classification of boilers is based on what flows through the boiler tubes:

- *Fire tube or flue tube or shell type* when flue gases are inside and water is outside.
- *Water tube* when water is inside the tubes and flue gases are outside.
- *Combination type (combo boiler)* when flue gas and water flow both outside and inside the tubes in a boiler that contains an external furnace and shell-type boiler in a sequence.

By the *end use* of the boilers they can be classified as

- Industrial boilers
- Utility boilers
- Marine boilers
- Nuclear boilers

Based on the *type of firing* the boilers can be grouped and further sub-divided as

- Mass/pile burning boilers
- Stoker-fired boilers—CG/TG, DG, SS, RG and so on

- Burner-fired boilers
- BFBC—under-bed and over-bed
- CFBC—hot cyclone, cold cyclone, U-beam and compact designs
- Pulverised fuel-fired boilers—wall fired and corner fired
- Liquor-fired boilers—BLR boilers
- WH boilers—Process waste heat (WH) recovery boilers—waste heat recovery boilers (Whrb)
- Heat recovery steam generators behind GTs–heat recovery steam generator (Hrsg)—vertical and horizontal

On the basis of *operating pressure* the boilers are made out into:

- Sub-critical (sub-c) boilers
- Super-critical (SC) boilers

From *circulation* considerations the boilers are separated as:

- Natural circulation or drum-type boilers
- Forced circulation boilers
- Once-through (OT) or no drum boilers

From the *draught* considerations they are classified as:

- Natural or balanced draught boilers
- FD or pressurised fired boilers

Based on *construction* they are divided as:

- Package boilers
- Field erected boilers

Based on the *type of support* they fall into:

- Top-supported boilers
- Bottom-supported boilers
- Middle-/girdle-supported boilers

Based on the *furnace construction* they are classified as:

- Two-pass boilers
- 1½ pass boilers
- Single or tower-type boilers
- Down shot or Arch fired boilers

2.42 Boiler Codes

Boiler codes are essentially concerned with the safety aspects of the boilers to prevent the occurrence of disastrous explosions that marked the early period of boiler history. They are basically extensions of pressure vessel (PV) codes customised for fired equipment. They are not concerned with the sizing or thermal design of the boilers or the firing equipment.

Many leading nations have their own codes but ASME boiler and pressure vessel (BPV) codes are the most widely accepted and used worldwide.

These codes deal with the safety aspects in all areas of boiler making and operation and thus cover

- Mechanical design
- Manufacturing and workmanship
- Materials and stresses
- Inspection and testing
- Documentation and marking
- Care and operation
- Welder qualifications

2.42.1 ASME Boiler and Pressure Vessel Code

ASME BPV codes establish rules of safety governing the design, fabrication and inspection of boilers and PVs and nuclear power plant components during construction. The objective of the rules is to provide a margin for deterioration in service. Advancements in design and material and the evidence of experience are constantly added by Addenda. The code is updated every 3 years, the last being in the year 2010.

The Code consists of 12 sections and contains over 15 divisions and subsections. The Code Sections are:

I. Power Boilers
II. Materials
III. Rules for Construction of Nuclear Facility Components
IV. Heating Boilers
V. Non-Destructive Examination (NDE)
VI. Recommended Rules for the Care and Operation of Heating Boilers
VII. Recommended Guidelines for the Care of Power Boilers
VIII. Pressure Vessels
IX. Welding and Brazing Qualifications
X. Fibre-Reinforced Plastic Pressure Vessels

XI. Rules for In-Service Inspection of Nuclear Power Plant Components
XII. Rules for Construction and Continued Service of Transport Tanks

Of these, Sections I, II, IV, V, VII, and IX are of relevance to water tube boilers. These are briefly explained below.

I. Power Boilers

This section provides requirements for all methods of construction of power, electric and miniature boilers; high-temperature water boilers used in stationary service; and power boilers used in locomotive, portable, and traction service in which:

- Steam or other vapour is generated at >15 psig
- High-temperature water boilers intended for operation at >160 psig and or >250°F

SHs, **Ecos**, and other **PPs** connected directly to the boiler without intervening **valves** are considered as part of the scope of Section 2.1.

II. Materials

There are four parts to this volume divided as:

- Part A—Ferrous Material Specifications
- Part B—Non-Ferrous Material Specifications
- Part C—Specifications for Welding Rods, Electrodes and Filler Metals
- Part D—Properties

IV. Heating Boilers

This subsection provides requirements for design, fabrication, installation and inspection of steam-generating boilers, and hot water boilers intended for LP service that are directly fired by oil, gas, electricity or coal. It contains appendices which cover approval of new material, methods of checking SV and safety relief valve capacity, examples of methods of checking SV and safety relief valve capacity, examples of methods of calculation and computation, definitions relating to boiler design and welding and quality control systems.

V. Non-Destructive Examination

This covers the requirements and methods for **NDE** which are referenced and required by other code sections. It also includes manufacturer's examination responsibilities, duties of authorised inspectors and requirements for qualification of personnel, inspection and examination. Examination methods are intended to detect surface and internal discontinuities in materials, **welds** and fabricated parts and components.

VII. Recommended Guidelines for the Care of Power Boilers

Guidelines are provided to promote safety in the use of stationary, portable and traction-type heating boilers. Emphasis is placed on industrial boilers because of their extensive use.

The contents include fuels for routine operation; operating and maintaining boiler appliances; inspection; prevention of direct causes of boiler failure; design of installation; operation of boiler auxiliaries; control of internal chemical conditions.

IX. Welding and Brazing Qualifications

Rules are provided relating to the qualification of welding and brazing procedures as required by other Code Sections for component manufacture. It also covers rules relating to the qualification and re-qualification of welders and brazers, in order that they may perform welding or brazing as required by other Code Sections in the manufacture of components.

2.43 Boiler Control

A boiler is meant to convert fw into steam at certain parameters of p & t by burning the specified fuels for use in process or ST over a range of loads. The operating parameters of the boiler should be continuously monitored and regulated such that it is fuel efficient and safe for personnel and equipment. Control systems are needed for performing this task supported by various measuring and control instruments.

If combustion and other associated systems are considered apart, the essential boiler control involves balancing of main inputs namely fw, fuel and air in step with the steam demand. For this, the only parameters required to be kept constant at their set points are, the:

1. Steam header pressure
2. Furnace **draught**

To achieve this, the controls required are

1. Steam pressure
2. Combustion
3. Furnace draught
4. Drum level/fw

In addition, the following controls are also needed, namely:

5. Steam temperature
6. O_2 trim

The essential control systems used in drum-type boilers are described below after familiarising the readers with symbols employed which are as per Scientific Apparatus Makers Association (SAMA). Figure 2.6 depicts the enclosure symbols and Figure 2.7 depicts the signal processing symbols.

1. Steam Pressure Control

   Steam pressure is the most important boiler variable because it indicates the balance between the supply and demand for steam. When steam supply exceeds demand the header pressure rises and when demand exceeds supply the pressure falls in a standalone single boiler.

   Figure 2.8 shows the *plant master control* with a single-loop control diagram that varies the firing rate to control steam pressure at the desired set point.

   When several boilers in parallel supply a common steam header, usual with industrial boilers, optimal load sharing is needed based on the total load and the performance of the individual boilers. Some boilers may be shutdown, some may be base loaded with constant firing and the remaining are allowed to swing with the load with variable firing rate. Figure 2.9 shows a *boiler master control* that provides these adjustments. Each boiler master provides a bias adjustment and an auto/manual transfer switch. In manual, the operator can reduce the firing rate to a low fire condition for shutdown or hold the firing rate at any appropriate base-loading condition. In auto, the boiler master follows the master firing rate demand signal except as altered by the bias adjustment. The operator can adjust the boiler master bias up or down to increase or decrease its share of the load.

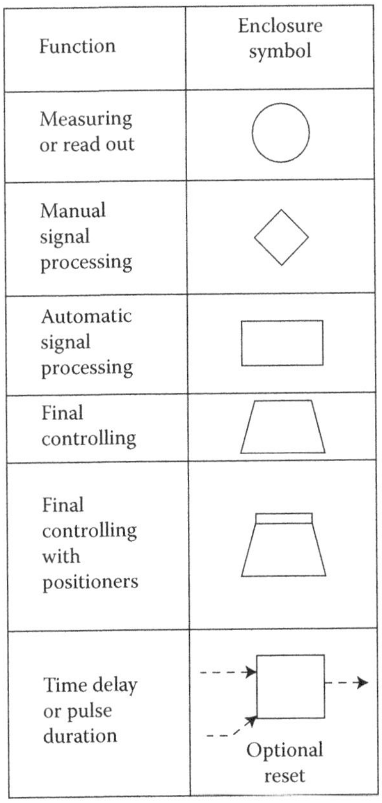

Function	Enclosure symbol
Measuring or read out	
Manual signal processing	
Automatic signal processing	
Final controlling	
Final controlling with positioners	
Time delay or pulse duration	Optional reset

FIGURE 2.6
Instrumentation symbols as per SAMA.

2. Combustion Control

   Combustion control delivers air and fuel to the firing equipment to match the firing rate demand from plant/boiler master at an air/fuel ratio which provides safe and efficient combustion.

   Less air flow not only leads to incomplete combustion and waste of fuel but can also cause an accumulation of unburnt fuel or combustible gases that can be ignited explosively by hot spots in the furnace.

   At the same time too much air flow also wastes fuel by excessive **stack loss.**

   Combustion controls achieve the optimum air/fuel ratio as well as safe firing.

   Single-point positioning or jack shaft control is the simplest combustion control system which can be applied to very small oil/gas-fired boilers which uses a mechanical linkage to manipulate the properly aligned fuel CV and combustion air flow damper in a fixed relationship. Figure 2.10 shows such a scheme.

   Parallel positioning control system uses two outputs in parallel to control fuel CV and air damper. It is commonly used on package boilers and is shown in Figure 2.11.

   Full metered lead-lag (cross limited) control system is the standard control arrangement in larger oil/gas-fired boilers. Fuel and combustion air flows are measured and are used to improve the control of the air-to-fuel ratio which compensates for fuel and combustion air flow variations and places active safety constraints to prevent hazardous conditions

   Three measurements namely, steam header pressure, the fuel flow and the air flows are used to balance the fuel/air mixture (as shown in Figure 2.12). Combustion controls consist of

Function	Signal Processing Symbol
Summing	Σ or +
Averaging	Σ/n
Difference	Δ or –
Proportional	K or P
Integral	∫ or |
Derivative	d/dt or D
Multiplying	×
Dividing	÷
Root extraction	$\sqrt[n]{}$
Exponential	x^n
Non-linear function	f(x)
Tri-state signal (raise, hold, lower)	↕
Integrate/Totalise	Q
High selecting	>
Low selecting	<
High limiting	$\not>$
Low limiting	$\not<$
Reverse proportional	–K or –P
Velocity limiting	V$\not>$
Bias	±
Time function	f(t)
Variable signal generator	A
Transfer	T
Signal monitor	H/, H/L, /L

Function		Signal Processing Symbol
Logical signal generator		B
Logical AND		AND
Logical OR		OR
Qualified logical OR	> n	GTn
	< n	LTn
n = integer	= n	EQn
Logical NOT		NOT
Set memory		S, SO
Reset memory		R, RO
Pulse duration		PD
Pulse duration of the lesser time		LT
Time delay of initiation		DI or GT
Time delay of termination		DT
Input/output	Analog	A
	Digital	D
Signal converter	Voltage	E
	Frequency	F
Examples D/A I/P	Hydraulic	H
	Current	I
	Electro magnetic	O
	Pneumatic	P
	Resistance	R

FIGURE 2.7
Signal processing symbols.

fuel flow and air flow control loops that are driven by the firing rate demand signal. The cross-limiting (or lead-lag) circuit assures an air-rich mixture since the air flow set point will always lead the fuel on an increasing load and lag when the load is decreasing.

Stoker-fired boiler combustion control is depicted in Figure 2.13. It is a fuel flow–air flow control system.

Combustion control for a typical small coal-based PF-fired boiler with two feeders and fans is shown in Figure 2.14. There is a cross limiting/led lag system to maintain proper air/fuel ratio. There is a derivative control on the output of the drum pressure. There is a speed control and damper control on the FD fan.

3. O_2 Trim Control

This control for trimming air/fuel ratio is more common with oil/gas-fired boilers which do not have the problem of stratification of flue gases across the boiler due to vigorous combustion, which often happens in solid fuel-fired boilers.

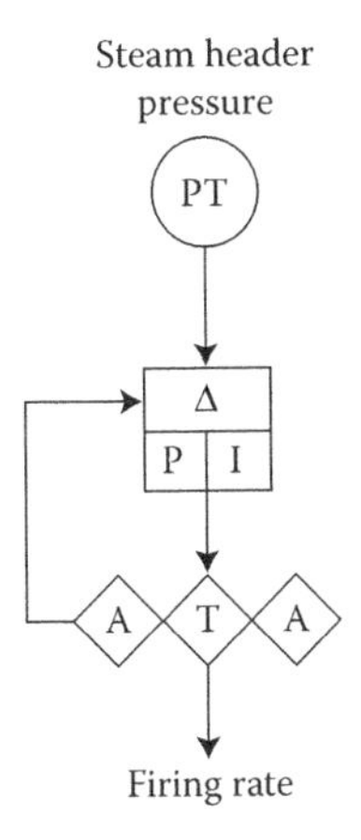

FIGURE 2.8
Typical plant master control.

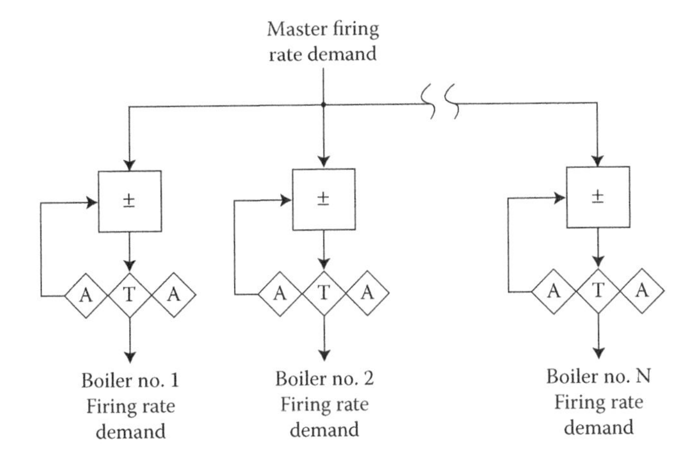

FIGURE 2.9
Boiler master control for multiple boilers running in parallel.

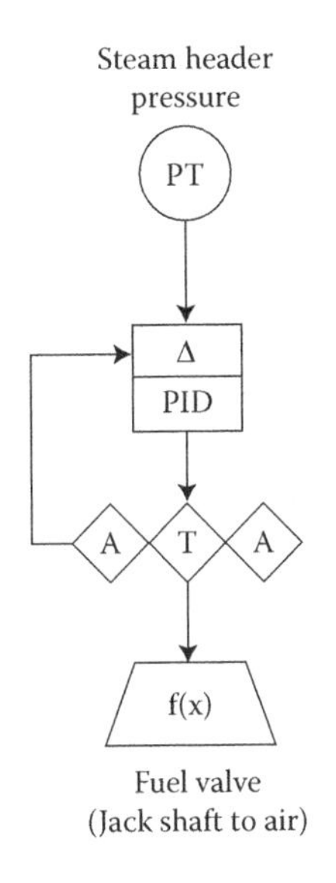

FIGURE 2.10
Single positioning or jack shaft combustion control.

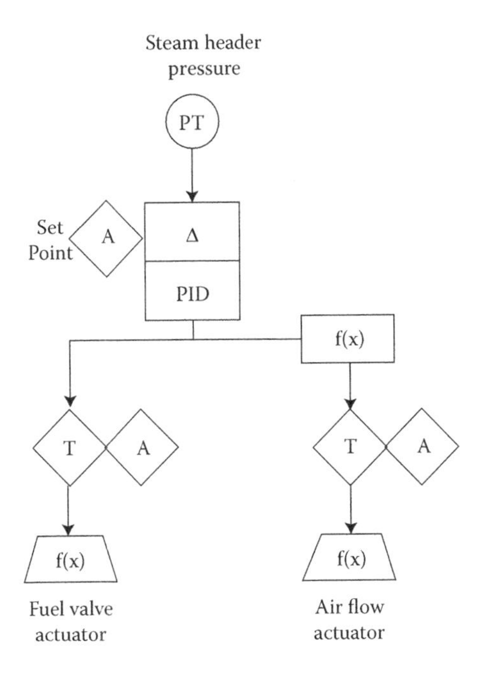

FIGURE 2.11
Parallel positioning combustion control.

Automatic air/fuel ratio adjustment is generally based on the percentage of excess O_2 in the flue gas with feedback taken from an O_2 analyser. Figure 2.15 shows one method of trimming the air/fuel ratio. The optimum percentage of O_2 in the flue gas depends on the type of fuel and varies with load.

4. Furnace Draught Control

This is needed in balanced draft systems that are equipped with both FD and ID fans. The furnace operates at slightly negative pressures to prevent flue gas leakage to the surroundings; not too low to avoid air leakage into the furnace or furnace implosion in the extreme.

As shown in Figure 2.16, FD fan damper is generally manipulated by the air flow controller while the ID fan damper by the furnace pressure controller. When the air flow controller manipulates the flow into the furnace, the pressure will be disturbed unless there is a corresponding change to the flow out of the furnace. An impulse feed forward connection couples the two dampers to minimise the furnace pressure disturbance on a change in air flow. As the impulse decays, external reset feedback to the furnace pressure controller drives the integral component to maintain the new steady-state ID damper position. The furnace pressure controller trims the feed forward compensation as required to control the pressure at set point.

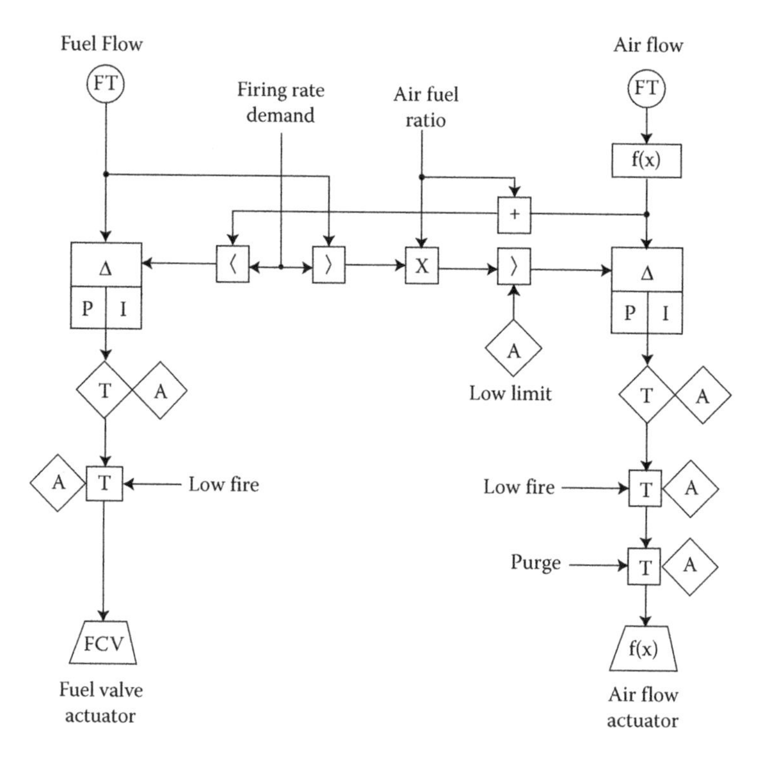

FIGURE 2.12
Combustion control using lead–lag/cross-limiting circuit.

5. Drum Level (fw) Control

   Maintaining the correct drum level is critical for the safe operation of a boiler as:

   - Low water level may uncover the water tubes exposing them to heat stress and damage and
   - High level may carry over water droplets exposing steam turbine to damage.

The level control problem is complicated by transients known as *shrink* and ***swell***. This is a decreased or an increased drum level signal

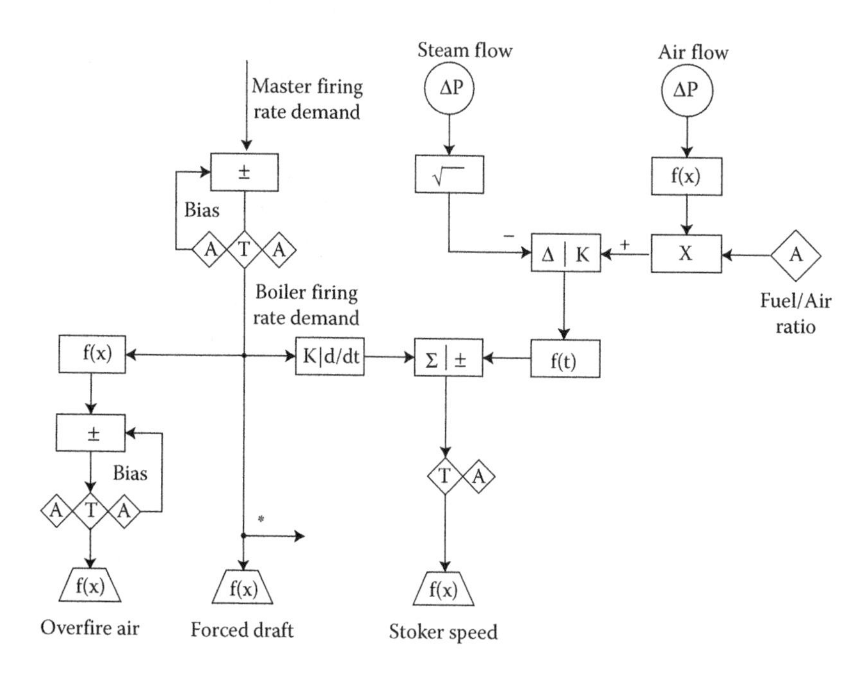

FIGURE 2.13
Combustion control for stoker-fired boilers.

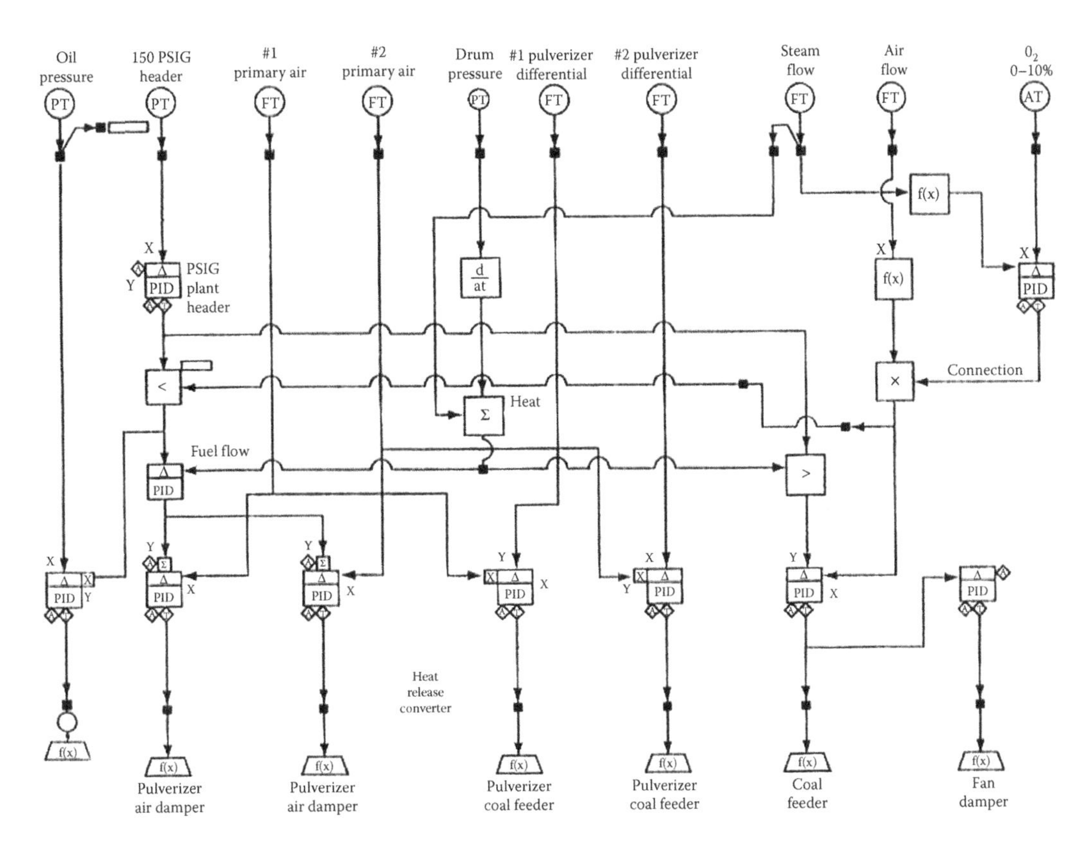

FIGURE 2.14
PF-fired boiler control system.

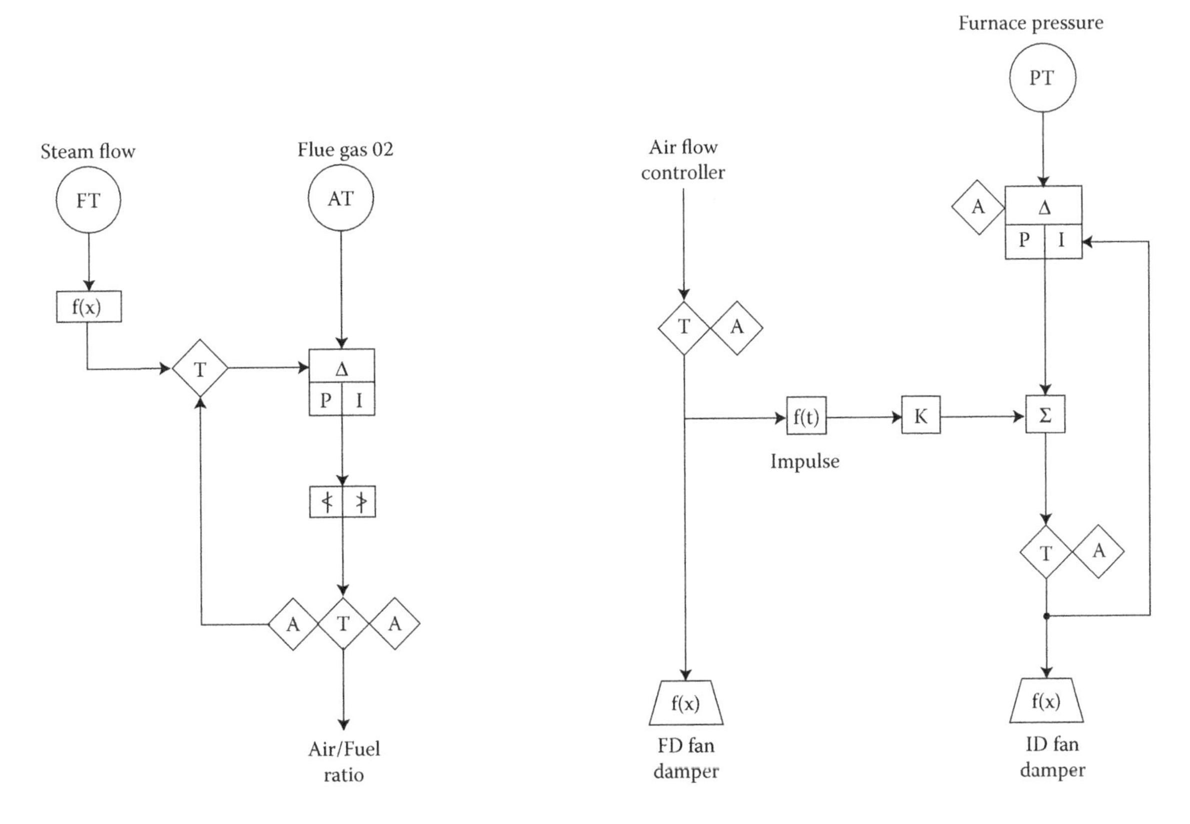

FIGURE 2.15
Oxygen trim control.

FIGURE 2.16
Furnace draught control.

because of the formation of less or more vapour bubbles in the drum water while there is actually no change in the amount of water in the drum. When there is a sudden increase in load the steam pressure falls signalling for more feeding of water. But the bubbles formed make the drum level swell and a signal is erroneously given for reduction of water supply. The reverse happens when load reduces and pressure increases. This way level changes happen in the opposite direction of what is expected with a particular load variation causing, albeit temporarily, severe control system overshoot or undershoot.

Single-element drum level control is the simplest type used for controlling fire tube and small packaged water tube boilers. Control is based only on boiler drum level and hence has no compensation of any shrink or swell. It is suitable only for small boilers with slow load changes.

Two-element drum level control measures steam flow along with boiler drum level. The steam flow signal is used in a feed forward control loop to anticipate the need for an increase in fw to maintain a constant drum level. This requires the Δp across the fw CV to remain constant, as well as the CV signal versus flow profile. Small- and medium-sized boilers with moderate load changes can be controlled this way.

Three-element drum level control further adds an fw flow signal to those used in two-element fw control. The drum level controller manipulates the fw flow set point in conjunction with feed forward from the steam flow measurement. The feed forward component keeps the fw supply in balance with the steam demand. The drum level controller trims the fw flow set point to compensate for errors in the flow measurements or any other unmeasured load disturbances (e.g. blowdown) that may affect the drum level. Three-element control is used in boilers that experience wide, fast load changes and is the most widely used form of fw and drum level control.

All the three types of control—single, two, and three element controls—are depicted in Figure 2.17.

1. During startup or low-load operation the flow measurements used in the three-element control may fall well below the rangeability limits of the flow metres making the control erratic. The level control switches automatically change from three- to single-loop (one-element) control strategy. Separate level controllers, not shown here, are used to allow different controller tuning for one- versus three-element control and to provide appropriate controller tracking strategies to provide for bump-less transfer between the two modes.
2. At design steam pressure >35 bar pressure compensation of the drum level signal is recommended to provide accurate drum level signal over the full operating range,

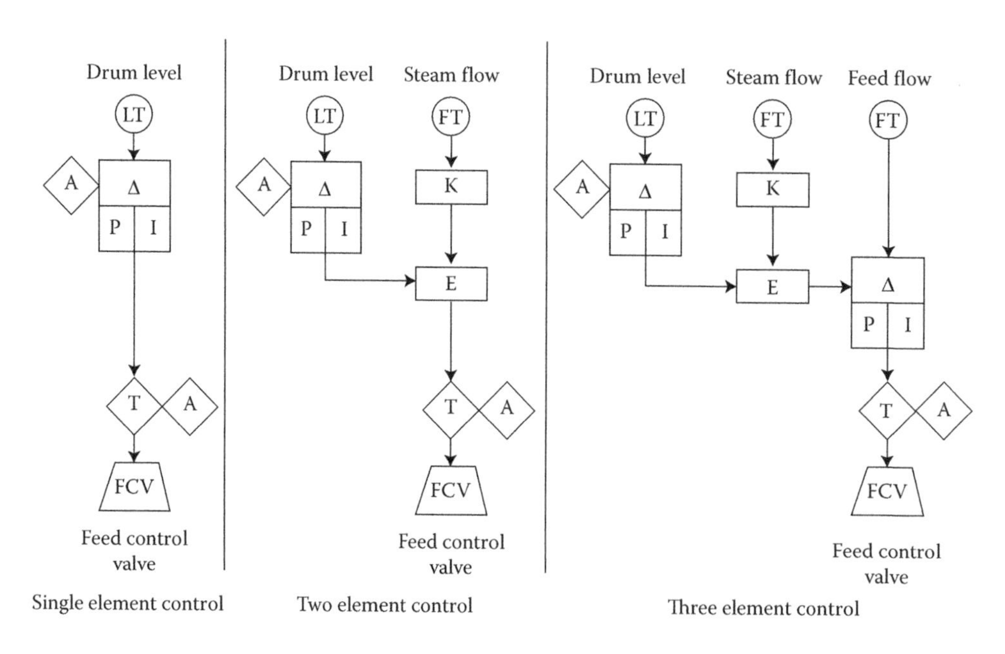

FIGURE 2.17
Different drum level controls—single, two, and three element controls.

from ambient conditions up to normal operating p & t. The level measurement is a hydrostatic measurement based on the Δp developed between top and bottom pressure taps in the drum.

6. Steam temperature control is usually a single-loop control as depicted under an attemperator for single-stage attemperation in Figure 1.20.

2.44 Boiler Evaporation (*see* Boiler Steam Flow)

2.45 Boiler Heat Duty

Boiler heat duty is the net heat output from a boiler expressed in MW_{th}, kcal/h, or M Btu/h.

It is the total heat pick up in the boiler by:

1. Fw in turning to steam
2. Cold RH steam turning to hot RH steam
3. Any other heat duty like external SH

See Figure 8.3.

Heat duty of a non-RH boiler with no other heating = Ws (Hs – Hw) (2.1)

Heat duty of an RH boiler with no other heating = Ws (Hs – Hw) + Wsrh (Hro – Hri) (2.2)

where

Ws = main steam flow
Wsrh = RH flow
Hs = enthalpy of SH outlet steam
Hw = enthalpy of fw at inlet
Hro and Hri = RH steam enthalpies at RH outlet and inlet, respectively

2.46 Boiler Horsepower

Boiler hp was developed in the nineteenth century to measure the capacity of boilers to deliver steam to steam engines. It is relevant only for very small boilers of earlier times. Boiler hp is obsolete today and is replaced by **boiler heat duty**. Boiler hp is only of academic and historical interest.

One boiler hp is equal to a boiler thermal output needed to produce 15.65 kg/h (34.5 lbs/h) of steam at a p & t of 0 barg (0 psig) and 100°C (212°F) with fw at the same condition. In other words, it is for producing steam from and at 100°C at room pressure. One boiler hp is ~9810 W, 8430 kcal/h, 33,479 Btu/h.

2.47 Boiler Island

The word boiler island has two different meanings:

1. In a utility power plant, the total equipments are divided among the three islands, namely boiler, ST and balance of plant (BOP). Here, boiler island means the main boiler, in-plant fuel feeding and ash discharge systems, dust collection equipment, $deNO_x$ and $deSO_x$ plants, chemical dosing equipment and some HP piping (fw, main steam and RH).
2. In large industrial plants such as refineries and petrochemicals, boiler island means a certain designated area allocated for boiler. All equipments coming within this area are known as boiler island. Usually fair amounts of piping on pipe rack would also form part of the scope in addition to various auxiliary plants listed above.

2.48 Boiler Layup (*see* Operation and Maintenance Topics)

2.49 Boiler Mountings and Fittings (*see* Auxiliaries)

2.50 Boiler Operating Modes

There are several modes in which boilers are operated:

- *Base-loaded plant*: When boilers are loaded steadily without much variation of loads it is base load operation. Typically, earlier coal-fired boilers or boilers at pit head are in base load mode so that optimum η is derived from the base-loaded boilers. These boilers are subjected to *cold starting* only occasionally as they are rarely shut down.

- *Weekend mode*: Here the boilers work during the week and are shut over the weekend. Boilers are required to be cold started at the beginning of each week. In case of CFBC boilers it is possible to box up the boilers over the weekend and have a *warm start* to bring the boiler on range.
- *Cycling mode*: Here the plant operates at full load during the daytime and at minimum load of usually 20% during nighttime. It may or may not be shutdown during the weekends.
- *Two shift mode*: Here the plants are run only during daytime and closed during the nights. Boilers are returned to service every morning by having a warm start.

In a power plant load variations are taken up by more efficient units. GT-based CCPPs are used for cycling. If coal fired units are to meet all conditions the units are required to be designed especially for cycling duty and or on-off duties.

2.51 Boiler Preservation (*see* Operation and Maintenance Topics)

2.52 Boiler Pressures

There are several pressures in a boiler which are listed below:

Superheater outlet pressure (SOP or SHOP)

This is the pressure at which steam is available from final SH header of boiler for further use.

This is usually ~5% higher than the pressure required in the process or by the ST to account for line losses from the boiler to the consuming point.

In fixed pressure boilers, SOP is held constant while in variable pressure boilers, SOP varies with TG load.

Drum pressure

Drum pressure is the pressure inside the drum. It is usually ~10% higher than SOP to account for losses in SH and attemperator.

Circulation of steam and water takes place at the drum pressure in evaporator circuits.

Design pressure

Design pressure of the boiler is usually the highest safety valve lifting pressure (SVLP).

Design pressure limits the maximum operation pressure in the boiler.

For RH, the design pressure is usually the highest SVLP on the inlet piping.

Operating pressure

Operating pressure is the pressure at which a component is under operation normally at the boiler MCR.

Maximum operating pressure is the pressure at which the component can be operated continuously without exceeding the permissible stress levels.

Maximum operating pressure is same as the allowable or maximum allowable pressure.

Allowable pressure

On the basis of the thickness of the PP we can perform a calculation to find what maximum pressure can be safely withstood at the operating temperature. This is called the allowable maximum operating pressure of the part.

The allowable pressure for the boiler is of interest, as the difference between the allowable and the design pressures indicates the margin of safety. This may be utilised in case the boiler pressure needs a certain upgradation for any eventuality.

Calculation pressure

Calculation pressure of the various components is the design pressure adjusted for factors like static head.

RH inlet and outlet pressures

Unlike SOP, RH outlet pressure (RHOP) is not controlled.

RH inlet pressures and temperatures are ST dependent.

2.53 Boiler Quality Plates (*see* Materials of Construction)

2.54 Boiler Rating (*see also* Boiler Steam Flow)

Boiler rating is nearly the same as its *steam flow* or *evaporation,* at least in small boilers. This figure is incomplete unless it is qualified with the p & t at which steam is generated along with fwt. There are several aspects to boiler rating which are described below.

All the ratings have to be further defined as *gross* or *net*.

Gross rating is the total evaporation without considering the steam consumption of the auxiliaries.

Net rating deducts from the gross rating such internal uses and indicates the steam flow.

Internal uses may include steam for various drives (fans and pumps), deaerator and so on.

2.54.1 Boiler Maximum Continuous

- BMCR/MCR is the ability of a boiler to generate and supply the declared nameplate rating of steam at specified p & t from fw made available at certain temperature, continuously and effortlessly with no short fall or side effects (such as tube over heating or furnace slagging or auxiliary overloading and so on).
- In actual fact, it is the minimum guaranteed steam flow under specified conditions and fuels.
- Usually well-designed boilers with adequate margins in all the auxiliaries can provide between 5% and 10% higher evaporation when the boilers are new. With ageing, a substantial part of this excess capacity is consumed mainly because of the fouling of surfaces and wearing of parts.
- In power plants it is usual to select the BMCR to match or exceed the valve wide open (VWO) condition of the ST.

2.54.2 Boiler Normal Continuous Rating

NCR is a rating, slightly lower than MCR, usually corresponding to the MCR of ST (TMCR) in a power plant or a steady steam requirement in a process plant. This is the rating at which the boiler would be operating most of the time. Naturally it is at NCR that the boiler η is optimised.

2.54.3 Boiler Peak Rating

- Boiler peak rating is the sporadic over capacity that a boiler is required to demonstrate for short durations of 2–4 h in a day to meet the enhanced process or power requirements.
- If the peak rating is <110% for 4 h in a day, the duty is met without any increase in the boiler size but building adequate margins in the firing equipment and the auxiliaries.
- If the peak duty is higher, the boiler, firing equipment and auxiliaries are all designed for the enhanced duty with boiler η optimised for MCR condition and peak rating at off-best efficiency.
- Running the boiler continuously at peak duty is detrimental to the life of the boiler as all the surfaces would be running a little hotter and gas speeds would be higher.
- In Hrsgs, peak rating does not apply, as a GT is already designed to run at the maximum temperature at MCR condition. Any further over firing would reduce the life of hot parts significantly.

2.54.4 Valve Wide Open Condition

VWO condition refers to the ST. At full-load running of ST (TMCR) the inlet CVs or governor valves are nearly fully open to admit almost the entire steam. There is, however, some more valve travel available to reach the VWO condition. This margin is kept intentionally to permit additional steam flow, usually to the extent of 5–7% or even more on smaller STs.

The purpose of this margin is to make up for the ageing of the ST. With time, the performance and output of the ST are bound to fall off marginally, mainly on account of the increased clearances between parts and scaling of the blades. Then the ST would demand more steam flow to maintain the output. VWO condition will tackle this problem suitably.

It is normal to select the MCR flow of the boiler to match or slightly exceed the VWO condition. Naturally, in relatively new STs it is possible to get some additional output of ~5–10% under VWO conditions.

2.55 Boiler Scope

There are mainly four important inputs and outputs each for any conventional boiler. If we add an RH inlet and outlet, they become five each. Scope of a conventional boiler plant lies within the TPs of these inputs and outputs.

The inputs/outputs and the TPs can be typically in the following manner, as shown in Table 2.4.

Figure 2.18 describes these for a typical utility boiler.

Many times, for the industrial boilers, the **deaerators** and **BFPs** are made part of the boiler scope. But in utility boilers, these form part of the ST island as they are physically located in the turbine hall. Likewise, the fuel hoppers form part of industrial boilers in most occasions but are part of fuel handling plant package in utilities.

2.56 Boiler Specifications (*see* Boiler Steaming Conditions)

TABLE 2.4

Inputs and Outputs for a Conventional Boiler

Sl No	Inputs	TPs for Inputs	Outputs	TPs for Outputs
1.	Water	NRV upstream of Eco or drum	Steam	MSSV or NRV
2.	Air	FD and PA/SA fan inlets	Gas	ID fan outlet
3.	Fuels	Fuel Feeder, fuel pipes	Ash	Hopper bottom
4.	Chemicals	NRVs in dosing lines	Blowdowns and drains	Dirty or clean blowdown vessels
5.	Cold RH steam	Cold RH header	Hot RH steam	Hot RH header

2.57 Boiler Start up (*see* Commissioning)

2.58 Boiler Steam Flow

- Steam flow or evaporation in kg/s, kg/h, lbs/h is the amount of steam generated at the specified **SOP** and **SOT** from fw supplied at certain temperature.
- When **RH** is present, the concept of evaporation must be expanded to include RH flow as well.
- In smaller boilers, steam flow can be treated as boiler rating. As the boilers grow in size, fw temperatures greatly vary and RH flow has to be accounted. Heat duty is better at expressing the boiler rating than steam flow.

2.59 Boiler Steaming Conditions/Parameters

Boilers are always specified by four important attributes besides the type of fuel and firing equipment. These are:

1. Steam flow or evaporation
2. SOP
3. SOT
4. fwt

In case of RH boilers, the RH inlet and outlet p & t conditions and flow are also to be stated.

Type of fuel such as coal, oil, gas and so on and type of firing, that is, grate, burner or FBC are also to be indicated to give an overall picture.

2.60 Boiler Steam Temperatures

There are several temperatures in a boiler which are given below.

2.60.1 Superheater Outlet Temperature

This is the temperature to which the steam is finally heated in the boiler. Usually it is ~5°C higher than required at ST or process inlet to account for steam temperature loss in steam lines.

For <350°C SOT is not usually controlled, being at low temperature.

Up to 400°C usually an attemperator is located at the outlet of SH as the SH tube metal temperatures are quite low.

Up to 500°C a single attemperator located between the stages of SH (**inter-stage attemperator**) is generally adequate.

For >500°C usually a two-stage attemperator is provided for faster regulation as well as limiting the SH metal temperatures. Also the SH metal temperatures remain under control.

Normally SOT is held constant between 70% and 100% MCR which is called the *steam temperature control range* (STC).

2.60.2 Reheater Outlet Temperature

This is the temperature at which the steam leaves the RH outlet header and it is usually held constant by:

- Varying the centre of heat input in the furnace by burner tilt in tangential firing or by cutting in and out the circular burners in wall firing in and out.
- Controlling the flue gas bypass in the boiler second pass where RH is located or by FGR.

RHOT is not regulated on a continuous basis by spray attemperators which are provided only for emergency purpose mainly to limit the tube metal temperatures.

2.60.3 Feed Water Temperature

This is the temperature at which the fw enters the Eco or drum after being heated in deaerator and FWHs.

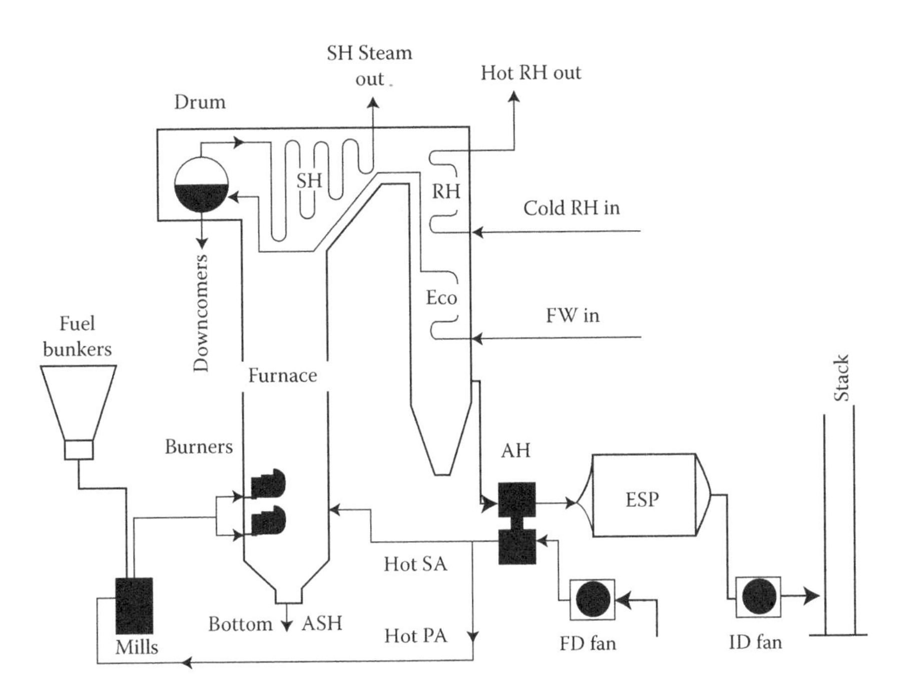

FIGURE 2.18
Schematic of a typical PF-fired boiler showing TPs.

The FWT can be as low as 85°C for small process boilers with no **deaeration** and as high as >300°C for very large utility boilers with several stages of **fw heating**.

In Hrsgs where there is no deaeration and regenerative feed heating, FWT can be as low as 30–40°C. In fact it is the condensate from the hotwell of the condenser and it enters preheater section of the Hrsg.

2.61 Boiler Storage (*see* Operation and Maintenance Topics)

2.62 Boiler Supporting (*see* Structure)

2.63 Boiler Tubes (*see* Materials of Construction)

2.64 Boiler Types (*see* Boiler Classification)

2.65 Boiling (*see* Properties of Steam and Water)

2.66 Bomb Calorimeter (*see* Measurement of Fuel cv by Bomb Calorimeter *under* Instruments *and* Measurements)

2.67 Bottom Ash (*see* Ash: Bottom *and* Fly)

2.68 Bottom-Supported Boiler (*see* Structure)

2.69 Bourdon Gauge (*see* Pressure Measurement *under* Instruments *and* Measurements)

2.70 Brayton Cycle or Joule Cycle (*see* Thermodynamic Cycles)

2.71 Breechings (*see* Flues and Ducts in Draught Plant)

Breechings is another word commonly used for flues and ducts together.

2.72 BRIL (*see also* Refractory, Insulation *and* Lagging)

BRIL, which is an acronym for Bricks–*Refractory*–Insulation–*Lagging*, essentially consists of all the non-ferrous parts of the boiler. The assembly surrounding all the hot parts of the boiler, namely the boiler setting or enclosure, is substantially made of BRIL so that:

- The heat is contained within
- Heat loss is restricted to a minimum
- Personnel safety is ensured

Boilers of the earlier part of last century were small and employed refractory bricks to a large extent to achieve the above objectives and also support the boiler PPs in many cases. In the course of boiler development as the boiler sizes and the extent of water wall cooling increased, the importance of BRIL has considerably altered. Greater awareness for aesthetics and environment has increased the use of insulation and lagging in boilers now. At the same time, with greater usage of **membrane panels** and the progressive discontinuance of many mass burning practices, the refractory bricks have been rendered practically obsolete along with reduced use of refractories in general. *I* and *L* have grown at the expense of B and R.

2.73 Brinnel Hardness Test (*see* Testing of Materials [Metals])

2.74 Brittleness (*see* Metal Properties)

2.75 Brown Coal (*see* Lignite)

2.76 Bubble Flow (*see* Flow of Liquids)

2.77 Bubbling Fluidised-Bed Combustion (*see also* Fluidisation, Fluidised-Bed Combustion *and* Circulating Fluidised-Bed Combustion)

Combustion taking place in the lower end of the fluidisation regime is called bubbling fluidised-bed combustion. For coals crushed to <10 mm in size BFBC is in the velocity range ~1–3.5 m/s.

This topic is elaborated under the following sub-topics:

1. The fluidised bed
2. Bed thickness—shallow and deep
3. Fuel feeding
4. Bed combustion
5. Bed temperature
6. Bed regulation at part loads
7. Bed construction
8. Air nozzles
9. Freeboard in BFBC boilers
10. Bed coil in BFBC boilers
11. BFBC versus CFBC and stoker firing

2.77.1 The Fluidised Bed

The bed essentially consists of ~95–97% inert bed material and only ~3–5% active fuel. The coarse bed material in a size range 0.5–1.5 mm is made up of mostly:

- Crushed and sized refractory
- Rounded river sand with no quartz content having sharp edges (sea sand is unsuitable as it is too alkaline and contains chlorides) or
- Bed ash from stoker-fired boilers properly screened and sized

Crushed refractory is more expensive but a better choice. If the fuel has >15% ash, bed material filling up is needed only at the start of the boiler and no further replenishment is usually required, as the ash left behind by the burning fuel is enough to provide for the loss in the system.

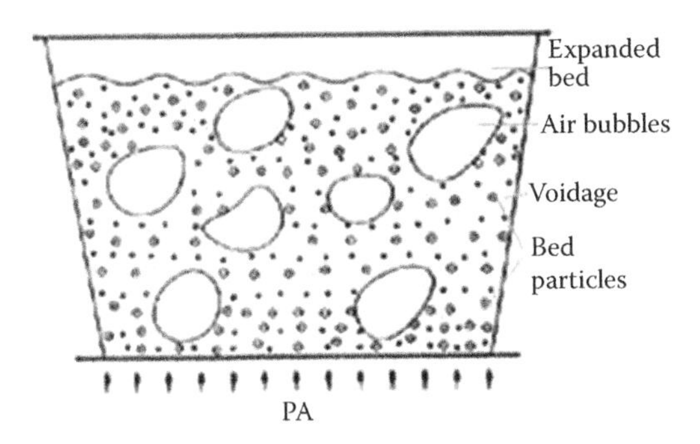

FIGURE 2.19
Bubbling fluidised bed.

When HP PA is supplied in this velocity range of ~1–3.5 m/s, the *slumped bed* becomes an *expanded bed* and the ratio is called the *bed expansion factor* which in most cases is ~2. The bed particles are set into an active swirling motion by the fluidising air. The expanded bed naturally has lot of *voidage* and air finds its way through to the top in the form of bubbles. Enclosed by the four walls and located at the bottom of the furnace with almost all the combustion air supplied from the bottom through air nozzles and the air bubbling through the bed the fluid motion bears a close resemblance to boiling water in a pan. Hence the name 'bubbling bed'. Figure 2.19 describes the phenomenon very clearly

Bed ρ depends on the voidage. Typical slumped bed ρ for various coals is ~14.5 kg/m^3 (~90 lbs/cft) and that of expanded bed with ~65% voidage is ~9.3 kg/m^3 (~58 lbs/cft).

Figure 2.20 depicts the lower part of a BFBC boiler furnace with the over-bed fuel feeding, under-bed air feeding through air nozzles, in-bed coils, ash removal and so on.

2.77.2 Bed Thickness—Shallow and Deep

Bubbling beds are of two types based on their bed thickness

a. *Shallow bed*: The slumped bed height is between 300 and 400 mm and expanded bed height is about twice that. Shallow bed BFBC enjoys the benefit of lower pressure drop through the bed and slightly lower thermal inertia.
b. *Deep bed*: The slumped bed height is ~500–600 mm and when expanded it is about twice. The pressure drop required here is higher. A deep bed provides adequate residence time for helping higher C burn out and good turbulence for better desulphurisation reaction.

Deep bed is suitable for slow burning fuels with high FC and fuels with large S content while shallow bed is

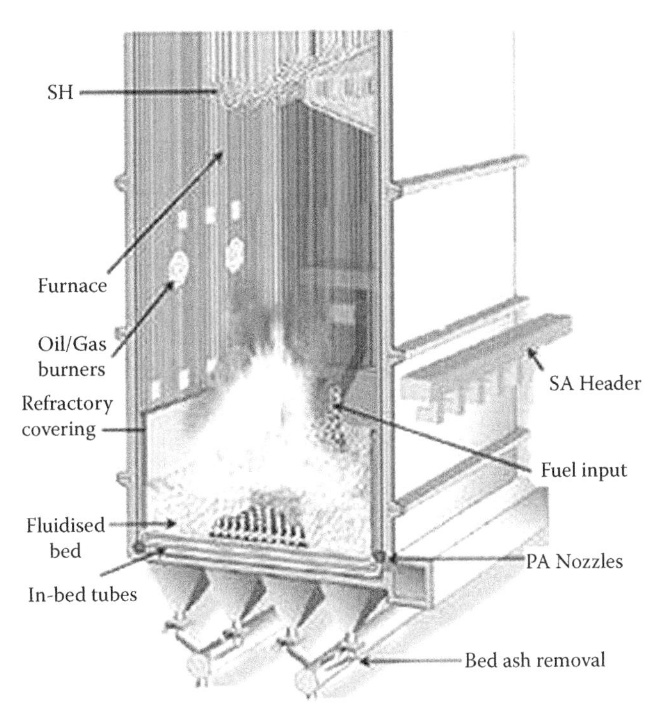

FIGURE 2.20
Lower furnace of a BFBC boiler.

adequate for quick burning fuels like sub-bituminous coals with low S content. As the bed area grows larger, distributing air uniformly in shallow bed becomes difficult resulting in improper fluidisation and clinker formation.

2.77.3 Fuel Feeding

There are two types of fuel feeding techniques in the BFBC boilers namely:

Under-bed feeding
Over-bed feeding

Under-bed feeding is classical feeding in the beginning of the development of the BFBC boilers mainly for coal. Over-bed feeding is a concept adapted from grate-fired boilers because of convenience. Today, as BFBC boilers are shifting from coal firing to bio-fuel firing, over-bed feeding is getting more popular.

The essential difference between the two techniques is that

Fuel is pushed into the bed from below in under-bed feeding, as shown in Figure 2.21, while
Fuel is spread on to the bed from above in over-bed feeding, as shown in Figure 2.22

2.77.3.1 Under-Bed Feeding

Under-bed feeding in which HP air carries the fuel through numerous fuel pipes and distributes it fairly

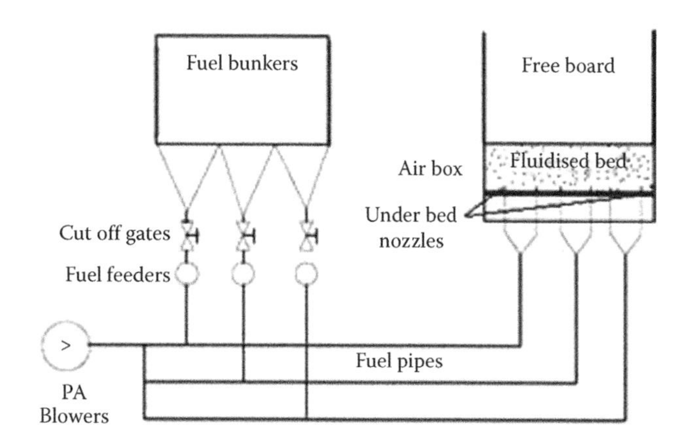

FIGURE 2.21
Under-bed fuel feeding scheme in a BFBC boiler.

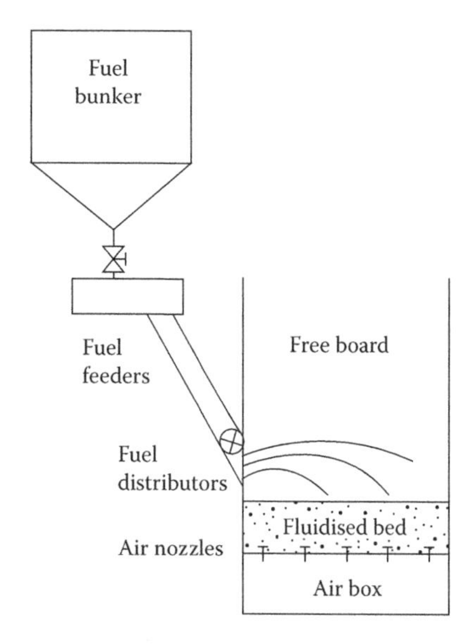

FIGURE 2.22
Over-bed fuel feeding scheme in BFBC boiler.

uniformly across the bed. This can be done only when the fuels are hard, dry and nutty, like various coals. However, in over-bed feeding, the limitations of M and sizing are more relaxed as the fuel is only thrown on to the bed either by mechanical rotors or by pneumatic spreaders. Fuel crushing is less and permissible M is higher in over-bed feeding.

Under-bed feeding enjoys the following advantages because very less fuel burns in suspension:

- Carryover is less
  - As more fines burn in bed than above, the bed temperature is slightly higher. Part of VM also burns in the bed.
  - Residence time is fractionally higher as the fuel travels from bottom to top.
  - With higher residence time, more fines burn. Together with slightly higher bed temperature combustion η is better by ~0.5–1.0% for coals.
- S capture is better and Ca/S ratio is also smaller.
- NO_x generation is lower because of slightly lowered freeboard temperature.
- Fines as high as 40% <1 mm can be burnt.

The furnace can be made narrow and deep as there is greater freedom in distributing the fuel. The reduced width of the boiler can bring down the overall boiler cost despite more equipment.

The limitations of under-bed design are:

- Multiple feeding points and long erosion-prone fuel pipes make the arrangement quite clumsy.
- With surface M >6%, the choking of fuel pipes is a recurrent problem. The remedial measures such as construction of covered fuel storage are expensive and space consuming. Using hot air for fuel conveying is also a mild concern.
- Crushing costs are higher for sizing the fuel to <6 mm.
- Multiple outlets of the bunkers and corresponding fuel feeders is also a messy arrangement.
- Fuel feeding is power intensive as HP hot air from Rootes blowers is required at 3500–5000 mm wg for overcoming line losses and feeding against the bed pressure.
- Control of numerous coal pipes and feeders adds to the cost of instrumentation and control (I&C).

2.77.3.2 Over-Bed Feeding

Over-bed feeding is simpler and more elegant but slightly less efficient. Furnace depth is restricted to ~7 m because of the limitation of fuel throw of the feeder, which has to be compensated by making the furnace wider. Generally, the boiler tends to be more expensive despite less equipment in the fuel feeding system.

Its advantages are

- Better layout with fewer bunker outlets and **fuel feeders**.
- No coal pipes as there is no under-bed coal transportation involved.
- Less crushing of fuel as it needs to be spread in furnace and not transported by pipes. Typically fuel is to be crushed to <20 mm for coal and <30 mm for lignites.
- Higher tolerance to surface M—as high as 20%.

- Easy regulation of lesser number of fuel feeders making I&C simpler.
- Less number of equipments and hence less O&M costs.

The disadvantages are that the boilers

- Tend to be wider and heavier
- Are slightly less efficient
- Have lesser tolerance to fines

On the whole, over-bed design is gaining greater acceptance now that the BFBC boilers are employed more and more for bio-fuels and sludge. They are also favoured for quick burning fuels like lignite, low S fuels requiring little desulphurisation or high ash coals which are also erosive. Small η gain of the under-bed design is often sacrificed for the simplicity of arrangement and operational convenience of the over-bed designs.

2.77.4 Bed Combustion

When combustion takes place in the bubbling mode, there is no air staging as practically all the combustion air is passed through the **air nozzles** at the bottom. In order to keep the combustion:

a. Cool, the bed coils are submerged in the expanded bed, which otherwise would heat up the bed to higher temperatures. Water, steam or even air is passed through the tubes such that the temperature of the bed remains between 800 and 900°C.
b. Clean, suitably sized **limestone** (<2 mm) is added to the bed for desulphurisation. Limestone can be added through separate nozzles at the bottom in under-bed firing or along with the fuel in over-bed firing.

Drying, ignition and burning of char are the three classical successive modes through which any combustion goes through. A bubbling bed consists almost entirely of the inert bed material swirling violently at temperatures of 800–900°C. When fresh fuel is admitted, either through the bottom or at the top of the bed, it almost instantly loses its M and VM. VM burns in the *freeboard* above the bed with the heat for ignition supplied from the hot bed below. The char or FC is left behind in the bed where it attains the bed temperature and burns out in no time. The swirling action of bed particles provides adequate scrubbing to remove the ash layer formed on top of the fuel particle and exposing fresh surface. Thus the entire burning process is very vigorous and fast.

When the particle attains a size of ~0.4 mm, it gets lifted off into the freeboard and carried away by the flue gases. It is trapped in the hoppers or cyclone dust collectors. Hopper dust is usually returned to the combustion chamber for re-burning while the material in the dust collector is discarded.

Time, temperature and turbulence are the three essential requirements for any **combustion**. The expanded bed provides ~0.5 s of residence time in the bed while the freeboard above provides ~2.5 s in suspension. For low volatile and difficult to burn fuels like **anthracite** and **petcoke**, the freeboard is made some 20% larger with a residence time of ~3 s. There is ample turbulence in the bed. With 20 to 30 times the weight of incoming fuel the inert bed also provides high temperature as well as a good flywheel effect to burn out most fuels satisfactorily. The net result of all this is that the combustion efficiencies are quite high at ~92–96% for coals and ~98–99% for **agro-fuels**.

2.77.5 Bed Temperature

Bed temperature is one of the most important parameters in controlling the operation of any FBC boiler. It represents a practical compromise between the two divergent needs of clean/cool and efficient combustion. At lower temperatures, NO_x generated is low and the lime–sulphur reaction is also good, but the combustion η is low with high unburnt C loss. The reverse is the case with high temperature.

Approximately 700–750°C can be considered as the minimum acceptable bed temperature when the unburnt loss increases sharply while 900–950°C can be considered as the maximum when NO_x levels and limestone consumption escalate severly, depending on the fuel properties. Accepted bed temperatures are broadly as follows for various fuels:

- ~800°C (1470°F) for high VM and low FC fuels like lignites
- ~850°C (1560°F) for fuels having good amount of S as lime–sulphur reaction is optimum at this level
- ~900°C (1650°F) for low VM and high FC fuels like anthracite and petcoke

High bed temperatures are also recommended for high ash and low S coals.

2.77.6 Bed Regulation at Part Loads

Part load operation of BFBC boilers is not as simple as in the other boilers. At low loads when fuel and air are reduced the bed temperature also reduces because of the constant amount of cooling by the bed coil. Naturally the combustion η also reduces because of the increased unburnt C loss. To overcome this η loss three methods of load regulation are employed for conventional BFBC boilers with bed coils:

a. *Velocity regulation*: Air and fuel are proportionately reduced by up to 70–75% of load when the bed attains the minimum acceptable bed temperature of 700–750°C. This type of control is the simplest but less efficient and with limited turn down. Small boilers employ this.
b. *Bed height regulation*: If the bed height is lowered in line with the load demand, a part of the bed coil tube would get uncovered. This reduces the cooling of the bed and can help in maintaining a constant bed temperature. There is almost no reduction in η but the practical aspect of collecting the hot bed material and raising it to a silo at some reasonable height for returning to bed along with the controls is quite expensive and complicated.
c. *Bed slumping*: If the air box at the bottom of the bed is compartmentalised, it is possible to cut out an air compartment, de-fluidise a part of the bed and maintain the bed temperature at low load. This is simpler than bed height variation, though not as efficient.

The normal practice is to adopt velocity regulation between 100% and ~75% load and slump the beds one after the other to provide boiler turn down of 40–50% of boiler MCR. However, this type of control is not seamless.

In boilers burning high M fuels like sludge and having no bed coils velocity regulation is adopted like in any conventional boiler.

2.77.7 Bed Construction

A bubbling bed also needs some type of bottom plate for resting of the bed. In smaller boilers of <50 tph, the bottom plate can be a refractory covered CS plate of 10–20 mm thickness, suitably stiffened and supported. In larger boilers, it is normal to bend one of the side membrane walls to form a fully water-cooled bottom for the bed. This bottom membrane panel is turned into a wind box so that the differential expansion problems associated with the attachment of a separate fabricated wind box are eliminated.

Air nozzles, described later, are welded to the bottom plate which also carries ash pipes of 150–200 mm NB placed in the ratio of one for every 15–20 m^2 of bed area.

TCs for the measurement of bed temperature and Δp measurement for bed height are also attached to the bottom plate. Bed material is filled till ~300–600 mm over the top of air nozzles.

Fuel nozzles, in case of under-bed feeding, are also inserted through the bottom plate. These are 150–200 mm NB and spaced at one for every 2–3 m^2 of bed.

When fluidised, the bed divides into a *static layer* of ~100 mm around the air nozzles and a *fluidised layer* of 600–1200 mm above the nozzles depending on whether it is a shallow or deep bed operation.

Ash removal is initiated whenever the under-bed pressure reaches pre-detemined level. Bottom ash contains pieces of shale and rock received with coal and unburnt C is rarely >1%.

The furnace walls around the expanded bed are required to be lined with Al_2O_3 or SiC castable refractory to prevent erosion because of the continuously impinging bed particles.

2.77.8 Air Nozzles

Air nozzles are required to provide HP PA at the bottom of the furnace to fluidise the bed material. For proper fluidisation, it is essential that the air is provided uniformly throughout the bed.

The design of air nozzle should be such that there is no back flow of bed material into the wind box during load changes or boiler shutdowns. Several changes were made before each manufacturer developed his proprietary design overcoming this menace.

A typical air nozzle is shown in Figure 2.23. It is normal to have one, two or three rows of drilled holes of ~1.5 mm Ø spaced equally around the periphery. The holes are inclined downwards generally at ~15°. The small holes and downward inclination together often help in preventing backflow. In some cases, a top shroud is also provided.

The spacing of nozzles is usually ~100–125 mm (4″–5″) on approximately square pitch with ~60–80 nozzles/m^2. But the spacing can also be increased to ~200–250 mm

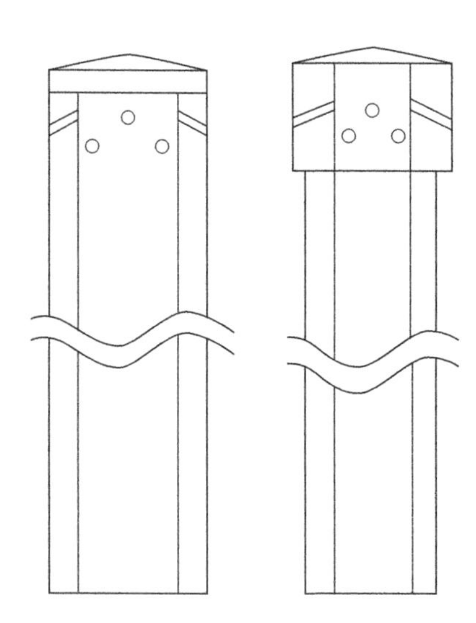

FIGURE 2.23
Air nozzles.

with ~20–30 nozzles/m². The nozzles then are made of appropriately larger diameter. Basically the nozzles are designed for an air pressure loss of ~300 mm wg.

Typical air nozzle for bubbling bed boilers has a bore of 20 mm with a thick wall of ~6 mm. On a flat bottom bed, 160–180 mm long air nozzles are needed as the static part of the bed itself is ~100 mm in thickness. When the bottom floor is inclined for circulation considerations the nozzles will be very long. Longer the nozzle the thicker is its stem for proper rigidity.

ss is the usual material of construction for the nozzles mainly because the nozzles attain a temperature of 800–900°C. This is not during the operation as combustion air keeps them cool but during the slumped condition when hot bed material inundates the nozzles. Also during start-ups, if the PA is heated by either hot gas generator or duct burner, air attains temperatures of ~900°C. Both these conditions demand ss material. During the running, however, the PA is limited to ~300°C. For very small BFBC boilers starting on charcoal, it is not unusual to find nozzles made of high Cr castings.

The nozzle body has to be sufficiently thick to not only withstand the erosive forces of the churning bed but also to withstand certain amount of abuse when clinker forms for some reason in the bed and has to be removed by chipping.

2.77.9 Freeboard in Bubbling Fluidised-Bed Combustion Boilers

The freeboard above the bed provides residence time of ~2.5 s for normal fuels and 3 s for slow burning fuels. It is usually 4–6 m in height for most types of coal application. The freeboard is not lined with refractory and heat transfer takes place like in any conventional furnace. SA nozzles are usually provided to burn the VM and reduce the carryover. SA is limited to ~10%. The FEGT is usually higher than the bed temperature by ~30–40°C.

Large amount of SA is not to be supplied as the bed cannot be allowed to go into reduced conditions to prevent bed coil corrosion. But in case of bio-fuel firing, where a large amount of VM is to be burnt, high SA (30–40%) is provided in free board. But there is no bed coil to worry about. The bottom plate is replaced by several large PA pipes to which the air nozzles are attached. This so called *open bottom* arrangement is described in Section 2.78.1.

The level of ash carryover from the freeboard in coal firing is somewhat similar to the dust levels in PF burning. It is necessary, therefore, to have in-line tube arrangement for all HSs. Single-pass baffleless cross flow or longitudinal flow BBs are recommended from erosion consideration. The gas velocities are also kept quite low, similar to those in PF firing.

2.77.10 Bed Coil in Bubbling Fluidised-Bed Combustion Boilers

Bed coil is a set of tubes immersed in the fuel bed of a bubbling bed boiler for extracting heat from the bed and maintaining its temperature at the desired value. Usually, it is the evaporator tubes which are placed in the bed coil. However they can also be cooled by steam and at times, even air. Bed coil surface is peculiar to only BFBC boilers and not to be found in any other type of boiler.

Open bottom BFBC boilers burning high M fuels do not need bed coils as the fuel M is sufficient to keep the bed temperature under control.

The following are some of the design aspects related to bed coils.

2.77.10.1 Fully or Partly Immersed Bed Coils

Bed height can be kept constant with bed coil fully immersed all the time and the bed temperatures can be allowed to vary with load, as described earlier.

Alternatively, bed height can be varied exposing part of the bed coil and bed temperature can be held constant under all loads. This system is a little more complicated as the bed level has to be continuously raised and lowered needing additional equipment and control.

In most boilers, therefore, the bed coil is kept fully immersed.

2.77.10.2 Sizing of Bed Coils

In a bed coil, there is no control on the flow of coolant. Therefore, the sizing of the fully immersed bed coil has to be done very carefully to attain the desired bed temperature correctly. A larger surface would depress the bed temperature, lower the turn down, reduce combustion efficiency, increase the CO and make the SH underperform. A smaller bed surface would raise the bed temperature and all the above factors would have reverse effects.

For sizing of bed coils heat transfer rate of ~255 W/m² °C, 220 kcal/m² h °C or 45 Btu/ft² h °F is taken which is 5–8 times more than the heat transfer rate in any other HS.

2.77.10.3 Flat or Sloping Tubes

Bed coil tubes immersed in highly turbulent bed of solids is subject to continuous bombardment from the bed particles. The tubes have to be rigidly supported and adequately protected against erosion. Bed coils placed horizontally experience minimum erosion as the impinging solids do not travel along the tube as they would, in case of slightly inclined tubes. But all horizontal tubes would call for external pumping arrangement which is not required if the tubes are gently inclined at ~5°. A practical solution is to adopt inclined tubes with protective measure for the tubes such as studding or weld overlays.

2.77.10.4 *Tube Materials*

ss is not a preferred material for the bed tubes as the Cl in ash can cause corrosion. C or C-Mo steels for tubes carrying water and low AS (T22 or T91) for tubes carrying steam are used. Ribbed tubes are also used in HP boilers when DNB conditions are anticipated. It is usual to provide 1 mm of sacrificial thickness in bed tubes for erosion protection.

2.77.10.5 *Bed Coil SH*

For superheating to >480°C, many times, a bed coil SH is employed where the finishing stage of SH is arranged inside the bubbling bed. It is not a preferred arrangement as a lot of care during start up is needed to protect the tubes from overheating. Moreover, the complicated steam piping arrangements make the plant layout a little untidy. Also slumping of bed containing SH coils is not possible as the final steam temperature would be short of requirement.

2.77.10.6 *Access*

Adequate space for access between the air nozzles and bed coils is necessary for maintenance purpose. Likewise, access to all tubes from above is important. The spacing of tubes and depth of coil is governed by these factors.

2.77.10.7 *Bed Coil Erosion*

Bed coil erosion is a serious problem particularly for coals with high ash content. Over the years the bed velocity has been reduced from 3.0 to 2.5 m/s and even 2.0 m/s in some exceptional cases. This would imply a larger cross section of boiler and hence a more expensive plant.

In addition, bed tubes are protected by studding or any other suitable erosion shields. As erosion is unpredictable many plant owners keep a spare bed coil and replace the old one as part of preventive maintenance which seems to be more economical.

2.77.11 Bubbling Fluidised-Bed Combustion versus Circulating Fluidised-Bed Boilers Combustion and Stoker Firing

Table 6.4 along with the write up under the topic of FBC covers the subject adequately.

2.78 Bubbling Fluidised-Bed Combustion Boilers

Commercialisation of the BFBC boilers dates back to the mid-1970s. During the late 1970s and early 1980s, a lot of euphoria was generated because of the early success of BFBC boilers in areas of cool, clean and efficient combustion. It was felt that environmental friendliness was high and fuel flexibility was good. Utility sized BFBC boilers were also designed and even constructed up to a capacity of 160 MWe on coal firing. Soon it was realised that BFBC had definite limitations for coal firing by way of erosion of the bed coils, boiler turndown requiring bed slumping and complicated controls and operation.

Simultaneously, the developments with CFBC showed that in every aspect CFBC was better than BFBC and made a mark as utility grade technology. Soon BFBC came to be accepted as an industrial boiler for process and cogen needs, largely eclipsing SS firing for almost all normal fuels. Today, BFBC is employed for burning coal and other bio-fuels such as rice husk, having reasonable FC or firing high VM and high M fuels like sludge. The former employs design with bottom plate while the latter uses design with open bottom.

BFBC boilers for coal and other conventional fuels are made in the range ~5–150 tph. Both under-bed and over-bed feeding systems are used, with under-bed being marginally more efficient and economical. Figure 2.24 shows a typical BFBC boiler with a double feeding arrangement. Under-bed feeding operates for most part of the year and the over-bed feeding is brought into picture during very wet conditions. Salient features of boilers with bottom plate design are briefly given here –

- With low HRR rates of ~1.5 MW/m^2 BFBC occupies more or less the same plan area as an equivalent **SS boiler**. Naturally, a lot of conversion of boilers from SS to BFBC has been done.
- The **residence time** is ~2.5 s for normal fuels and 3.0 s for fuels with high FC. A ~0.5 s residence time is provided at the bed. The freeboard height rarely exceeds ~6.0 m even in larger units.
- Both shallow and deep bed boilers are in operation with shallow beds for quick burning sub-bituminous coals with very little S. Shallow bed boilers enjoy the benefit of reduced auxiliary fan power over deep beds.
- Bed tubes are very effective and produce ~60–70% steam in a BFBC boiler. Consequently, the boilers need less convective HS. Boilers can be of **single-drum design** or can be **bi-drum boilers** with very small BB containing less tubes.
- Dust levels in the flue gas at freeboard exit are ~300 g/N m^3 with high ash coals. This is higher than the dust loading in PF boilers burning similar fuels. All the convection surfaces like BB, Eco and AH are to be designed with low **gas velocities** ranging from 6 to 10 m/s to avoid erosion.

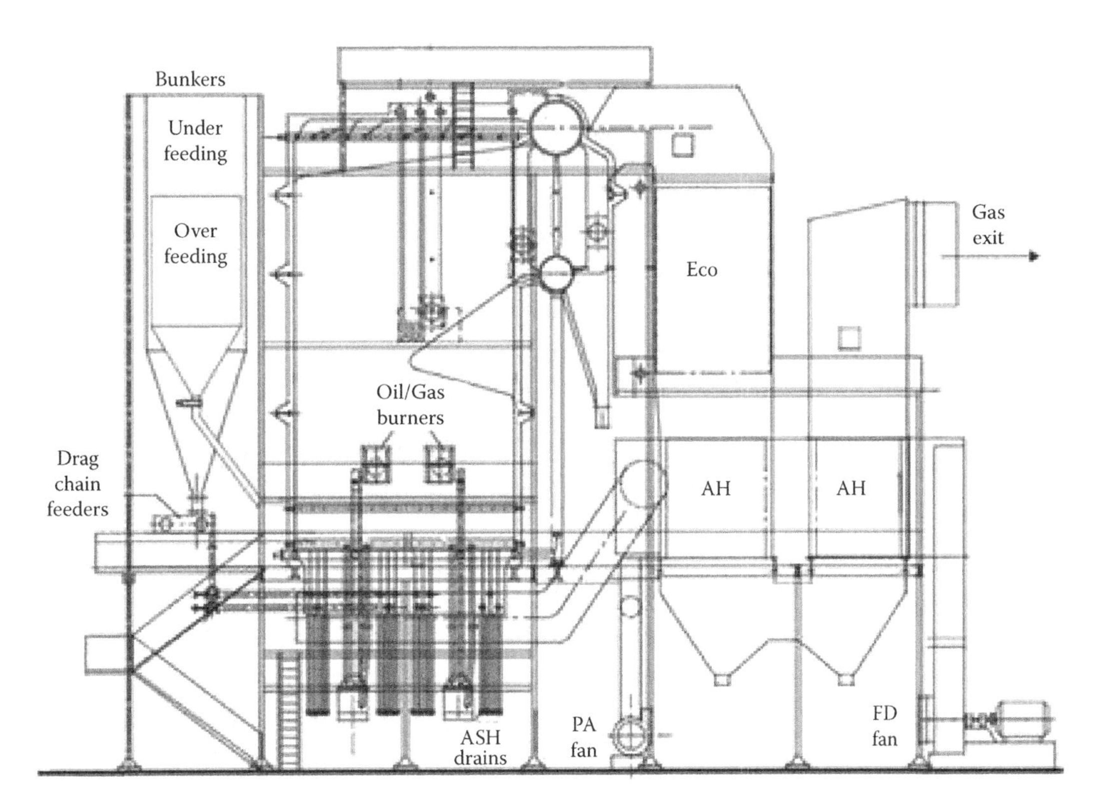

FIGURE 2.24
Multi-fuel-fired BFBC boiler with both over and under feeding systems.

- **Erosion** protection is also desired both for bed coils and all the first tube rows in various convection banks.
- Bed velocities at full load have been reduced from 3.0 to 2.5 m/s over the years. For high ash coals, they are kept even lower between 1.8 and 2.2 m/s, to reduce bed coil erosion.
- **Ash recirculation** is done between 2.0 and 2.5 times the coal feeding rate taking the ash from the Eco exit for improving carbon burn up for bituminous coals.
- Boilers of 60 tph and below for coal and lignite are usually made **bottom supported**.

  For pressures >60 bar, single-drum designs are normally employed.
- Excepting the design and arrangement of the bed and freeboard, the rest of the boiler, with various convection sections, is designed like a conventional SS-fired boiler.

2.78.1 Bubbling Fluidised-Bed Combustion Boilers with Open Bottom

Fuels like **sludge** from sewage plants or pulp mills, wood wastes, husk from coffee grounds and so on are difficult to burn because of their low cv and high M. But they can be burnt in BFBC boilers very well because the hot bed of inert solids provides a large heat sink and helps to evaporate the M. Thereafter, the combustion of dried fuel is relatively easy. A conventional BFBC boiler with in-bed coil of tubes is not the solution as the M in fuel will further depress the bed temperatures too heavily to enable an efficient combustion. The required bed temperature can be maintained by the M in fuel alone and the bed coils can be eliminated. This type of design is called an open bottom.

Combustion air is spread across the bed by means of air nozzles fitted over large air pipes with small gaps among them (as shown in Figure 2.25). The cross-sectional and isometric views of the lower furnace explain the construction of open bottom bubbling bed and also bring out the difference with the conventional bed coil construction. Bed material is filled from the bottom ash doors upwards. Fuel is always fed from the top in an overthrow fashion. Ash settles down at the bottom and slowly travels the ash doors in due course.

2.79 Buckstays (*see* Furnace Pressure in Heating Surfaces)

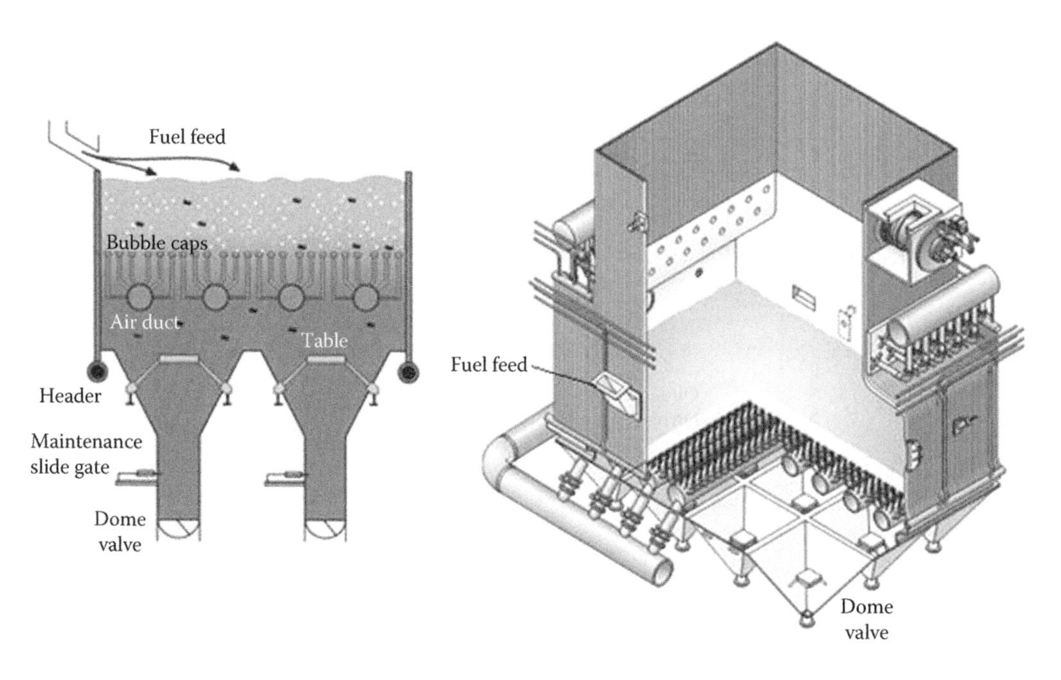

FIGURE 2.25
Lower furnace of a BFBC boiler with open bottom arrangement. (From Babcock & Wilcox Company, USA, with permission.)

2.80 Burden in Fuel (*see* Ballast in Proximate Analysis in Coal)

2.81 Burners (*see* 1. Oil and Gas Burners, 2. Pulverised Fuel Burners, 3. Lignite and Brown Coal Firing *and* 4. Low NO_x Burners)

Burner is the combustion equipment for burning fluids, such as various types of oils, gases, PF and biomass wastes in air suspension. Constructional features of burners vary for different fuels. However, there are certain common aspects in all burners which are listed below:

- Mounted on furnace, a burner basically combines air, fuel and ignition energy in the right proportion and admits the fuel–air mixture into the furnace, where the combustion reaction takes place.
- A burner does not control the input quantities but only mixes them in the most optimum manner.
- Fuel and air are controlled from outside the burner with their respective regulating valves/systems and dampers.
- The external air–fuel regulation is done in such a way that a burner can be cut in or cut out independent of the other burners.
- A burner is normally fitted to the windbox, which in turn, is attached to the furnace.

Various types of burners used in water tube boilers are described in the specified sections listed above.

2.82 Burner Management System (*see* Operation and Maintenance Topics)

2.83 Bypass Stack (*see* Draught Plant)

3

C

3.1 Calcination Reaction (*see* Desulphurisation)

3.2 Calorific Value of Fuels (*see* Thermodynamic Properties)

3.3 Calorimeter (*see* Steam Quality Measurement *under* Instruments *and* Measurements)

3.4 Calorising

Calorising is a high-temperature diffusion process using a blend of ferro aluminium, Al_2O_3, NH_4Cl and a number of other alloys. The process creates an Al-rich outer layer which increases the surface hardness of the component (600–650 Brinnel hardness number, BHN) significantly resulting in greatly improved wear resistance. At elevated temperatures, the Al_2O_3 surface that forms to a depth of 100–150 microns, provides an excellent high-temperature abrasion resistance. The inter-metallic layer seals the base metal improving resistance to oxidation.

Diffusion processing involves the creation of an inter-metallic layer on the surface of the metal components, providing protection against acid corrosion, high-temperature corrosion and wear. The process is often referred to as pack cementation and is accomplished by introducing metal(s) into a substrate alloy through high-temperature chemical reactions.

The diffusion process of calorising extends the life of industrial components exposed to aggressive operating environments. Calorising has significantly improved the life of burner nozzles, coal feed pipes, corrosion resistant fasteners, soot blower (SB) lances, SB hangers, Eco supports and many other components that suffer extreme wear and/or heat degradation.

Calorising is a very cost-effective way of imparting high-temperature hardness in many applications without adopting expensive alloys. Calorising is adopted in boiler field mainly for small components for standing up to temperature of ~900°C.

3.5 Capillarity or Capillary Action (*see* Fluid Properties *under* Fluid Characteristics)

3.6 Carbon Monoxide (*see* Air Pollution)

3.7 Carbon Monoxide Fuel Gas (*see* Gaseous Fuels)

3.8 Carbon Residue (*see* Fuel Properties *under* Liquid Fuels)

3.9 Carbon Steels (*see* Metals)

3.10 Carbon Sequestration and Storage

The process of *capturing* CO_2 *(from flue gases) and storing it safely in underground reservoirs to avoid increasing GHG effect* is called C capture and sequestration (CCS).

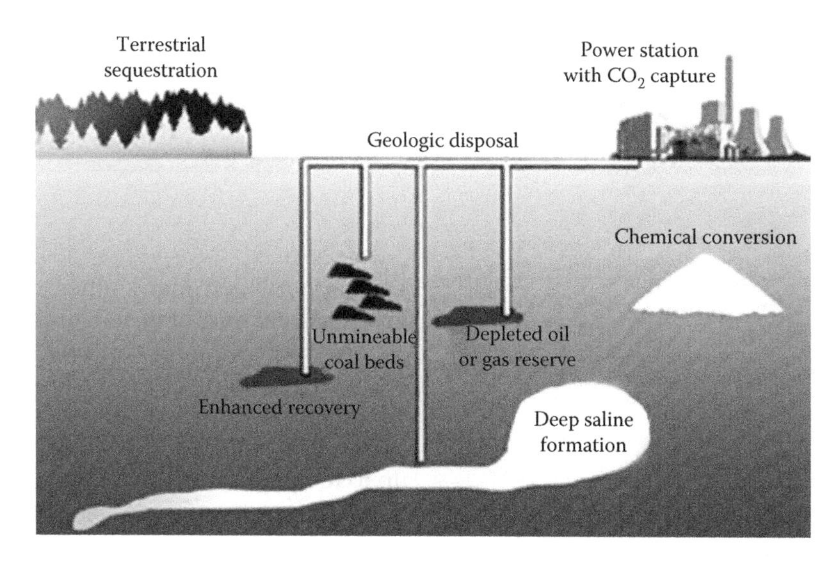

FIGURE 3.1
Carbon sequestration process.

Storage options include using subsurface saline aquifers, reservoirs, ocean water, ageing oilfields or other C sinks (as shown in Figure 3.1).

A lot of serious research is still needed to get a breakthrough in almost all stages of the activity namely, capturing, liquefaction, transportation, injection and storing of CO_2 competitively.

This matter is of great interest because over two-thirds of electrical power is generated from coal across many countries and coal firing is the single largest contributor to the GHG production. Unless CO_2 is captured from flue gases and disposed off safely, the long-term view for coal-based power generation remains cloudy, particularly in view of resurgence of nuclear power of late. Currently, it is estimated that the cost of power from coal would go up by at least ~60% with C sequestration and storage, making it uncompetitive and too expensive for customers. Gas-based power in many places can become cheaper.

3.11 Carnot Cycle (*see* Thermodynamic Cycles)

3.12 Carry over (*see* Steam Purity)

3.13 Cast Iron (*see* Metals)

3.14 Castings (*see* Metals)

3.15 Caustic Embrittlement or Caustic Stress Corrosion (*see* Water)

3.16 Cavitation (*see also* Control Valves *under* Valves)

Cavitation is the *formation of vapour bubbles in a moving body of liquid when the pressure falls lower than the* **vapour pressure** of liquid at **vena contracta**. Subsequently when there is a pressure recovery, the vapour bubbles collapse and produce noise, vibration and erosion.

Cavitation occurs in high-speed hydraulic machines such as water turbines, **centrifugal pumps** and in certain portions of pipeline and also in **CVs** of the boiler fw line.

3.17 Ceramic Fibre Lining (*see* Insulating Materials)

3.18 Chang Cycle or Steam Injected Gas Turbine Cycle (*see* GT Performance *under* Turbines, Gas)

3.19 Char or Fixed Carbon (*see* Coal Analyses *under* Coal)

3.20 Charpy Test (*see* Impact Testing of Metals [Charpy and Izod] *under* Testing of Materials)

3.21 Chemical Dosing of Boilers (*see* Water Treatment)

3.22 Chemical Cleaning of Boilers (*see* Commissioning)

3.23 Chimney (*see* Stack)

3.24 Chimney Effect (*see* Stack Effect)

3.25 Choked Flow Condition (*see* Flow Types in Vertical and Horizontal Tubes)

3.26 Chordal Thermocouple (*see* Temperature Measurement *under* Instruments for Measurements)

3.27 Chromising (*see* Boiler Tubes in Materials of Construction)

3.28 Circulating Fluidised-Bed Combustion (*see also* Fluidisation, Fluidised-Bed Combustion *and* Bubbling Fluidised-Bed Combustion)

CFBC has its origins dating to late 1970s and early 1980s in Europe. Ahlstrom of Finland and Lurgi of Germany have done pioneering work during that period culminating in the commercialisation of CFBC boilers by mid 1980s. CFBC has steadily progressed since then to rival the established pulverised fuel (PF) technology to a large extent in certain areas in the last 25 years.

CFBC operates at the higher end of the **fluidisation curve**, just below the transport regime by adopting velocities in the range 4–7 m/s. There is no defined bed, like in **bubbling** or **expanded-bed** combustors, but only a cloud of dust (bed material) filling the entire chamber. The cloud is denser at the bottom and thinner at the top as can be seen in Figures 6.28 and 6.30. This high-fluidisation velocity results in a rather slim and tall combustor when compared with bubbling fluidised-bed combustion (BFBC) but slightly stouter than PF. The heat release rate (HRR) in plan area of ~7 MW/m^2 is fairly close to the **PF firing**. These boilers are called classical, fast circulation or full-circulation CFBC boilers to differentiate them from expanded or turbulent bed CFBC boilers developed in the next stage.

CFBC is characterised by

- Staged combustion
- Ash re-circulation from cyclone

for accomplishing cool, clean and efficient combustion.

The combustor temperatures are maintained at the same levels of bed temperatures as in BFBC boiler. The considerations for the temperature are also the same.

The unique feature of CFBC is the cyclone which collects the elutriated dust particles from the top of the 1st pass and returns them to the bottom of combustor.

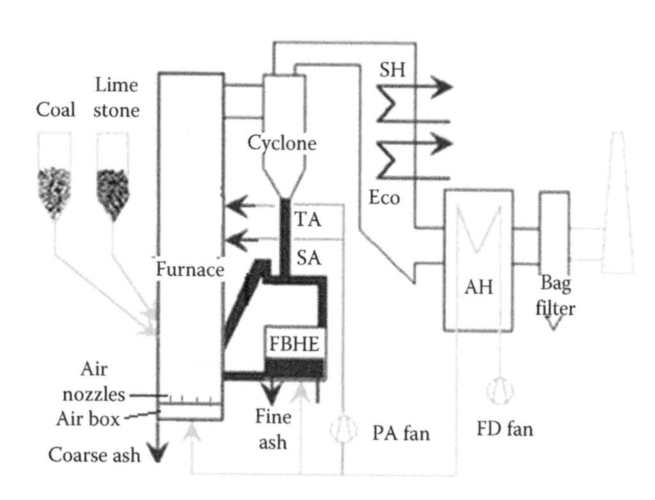

FIGURE 3.2
Schematic of CFBC process.

This continuous circulation of bed particles (along with the burning C) several times from combustor to cyclone provides ample residence time for near completeness of combustion. CFBC has 99+% of C burn-up, almost the same as PF firing. The high-ash re-circulation also helps to make a uniform temperature profile in the combustor, free board and cyclones, all of which are maintained at ~850°C (~1560°F). A schematic representation of the classical CFBC boiler is shown in Figure 3.2.

As the cyclone operates at such high-temperature levels it is called *hot cyclone* as opposed to the cyclone in an expanded-bed design that operates between 450 and 500°C and hence termed as *cold cyclone.* The boilers are also called *hot and cold cyclone boilers.*

The much greater acceptance of CFBC has been because of the following advantages:

- Cool and clean combustion and yet very efficient
- Very high fuel flexibility
- Very good level of simultaneous multi-fuel firing capability
- High efficiency, remaining practically the same over the entire load range
- Very superior boiler dynamics, comparable to PF

It is exceptionally versatile, efficient and environment friendly combustion.

The topic of CFBC is covered in the following sub-topics.

1. CFBC versus BFBC
2. CFBC versus PF
3. Types of CFBC
4. Combustor
5. Freeboard
6. Cyclones
7. Loop Seal/Seal Pot/L valve/J valve/Siphon
8. FBHE
9. Bottom ash
    a. Screw coolers
    b. Rotary coolers
    c. Fluidised-bed ash coolers (FBAC)
10. Heating the bed for start up

3.28.1 Circulating Fluidised-Bed Combustion versus Bubbling Fluidised-Bed Combustion (see *also* Fluidisation)

Both CFBC and BFBC are fluidised-bed boilers. While BFBC is at lower end of fluidisation phenomenon CFBC is at the higher end. This makes the two entirely different. Table 6.4 in the topic of fluidisation provides a good comparison.

3.28.2 Circulating Fluidised-Bed Combustion versus Pulverised Fuel (see *also* Pulverised Fuel Firing)

CFBC, with its set of unique advantages, has practically edged out PF firing from industrial and also small utility boilers. In future it is expected to provide even more challenge to the PF firing in larger utility boilers as well. A detailed write-up and a comparison (Table 16.8) of the two technologies are provided under the topic of PF firing.

3.28.3 Types of Circulating Fluidised-Bed Combustion

The CFBC employing gas velocity close to ~7 m/s at full load is the *classical CFBC* which is also called *fast circulation CFB.* For all practical purposes, the velocity range over the years has been reduced to 4–6 m/s to reduce the incidence of refractory and tube erosion.

In the course of development, another variant with reduced full-load gas velocities of ~4 m/s has been developed, combining the advantages of

a. The simplicity of BFBC
b. Efficient operation of CFBC

This is called *expanded or turbulent bed CFBC.*

As explained earlier the classical CFB boilers are also called the *hot cyclone* boilers.

- In the expanded-bed design of the former Deutsche Babcock, gases are cooled to 450–500°C before the cyclone by having HS in the first pass itself. Such boilers are also called *cold cyclone* boilers. It was patented as *Circofluid* boilers.

- In another design by B&W of the USA the cyclones are replaced by inertial separators with a trade name of U-beams. U-beams are placed inside the furnace, just before exit, in such a way that ash precipitates directly into the furnace dispensing with the loop seal arrangement. This way an internal circulation is set up. Such designs are called *U-beam* or internal recirculation (*IR*) boilers.

As the bed velocities are lower, the combustor size of the expanded-bed CFBC is ~1.5 times larger than in full-circulation boilers. The combustor has a definite bed at the bottom, with a fair amount of elutriation of particles to the cyclone, but in much lesser proportion than in the classical CFBC. Even though the ash re-circulation is lower, because of the presence of an active bed where the fuel particles get to reside and burn out substantially, it makes this system practically as efficient as full-circulation CFB. On the contrary, the erosion issues are lower and control easier. However, as the sizes progress towards large utility range, the full-circulation CFB becomes more cost-effective because of smaller cross section of combustor. The two technologies are compared in Table 3.1.

3.28.4 Combustor

Combustor to CFBC is like bed to BFBC. It is the space where there is a turbulent mixing of air and fuel leading to combustion. Combustor starts from the bottom plate and extends up to TA nozzles with PA nozzles at the bottom and SA nozzles in between. Owing to staged combustion, reduced conditions prevail in the combustor zone needing the enclosure walls to be protected by suitable layer of refractory.

TABLE 3.1

Comparison of BFBC and CFBC Boiler Systems

S. No.	Parameter	Expanded-Bed CFBC (1)	Full-Circulation CFBC (2)	Remarks
1	Fluidising velocity m/s and fps	3.5–4.5 11.5–15	4.5–7 15–23	~50% higher velocity in (2)
2	Auxiliary power KW/MW th	~15	~26	>50% higher power in (2)
3	Ash recirculation	10–12	10–100	Times fuel input
4	Total refractory	Less	Heavy	4–5 Times more in (2)
5	Start-up time from cold (h)	3–4	10–12	
6	Start-up oil consumption (te)	5–6	~20–25	
7	fw topping up pump	Nor required	Required	

Owing to staged combustion only ~60% of the combustion air is provided at the bottom of the combustor through uniformly spaced air nozzles. The combustor bottom tends to be much smaller than the bottom of bed in BFBC due to two or three times higher the bed velocities and nearly half the amount of PA. The remaining 40% air is distributed nearly equally as SA and TA in the free board.

Combustor sides made of membrane walls are lined usually with castable SiC refractory to protect the tubes from

a. Corrosion due to the reduced conditions prevailing in combustor
b. Erosion due to the impingement of bed particles

SiC is used as it is the hardest refractory material to withstand the duty and also because it is a good conductor of heat.

In the classical CFB, the bed is expanded to a height of 25–30 m. The bed is made of finer particles of 0.1–0.3 mm. The expansion results in a cloud of particles filling up the combustor up to its ceiling. The lower part of the cloud is denser than the upper part.

Substantial amount of dust returns along the combustor walls while some escape to the cyclones where it is captured and returned to the bottom of combustor. This re-circulated ash helps to cool the bed and improve the C burn up.

Like in the fast circulating BFBC, the bed material forms 95–97% while the active fuel forms 3–5%.

In the expanded-bed CFB the bed material is coarser at 0.75 mm. For nearly the same air pressure at the bottom, the bed expands to a height of ~1.2–1.4 m with elutriation of particles of ~0.4 mm in size. The dust load in the cyclone is lower. Consequently, a conventional cyclone is replaced by an inertial cyclone in IR-CFB design. The material re-circulating within the combustor is also less.

A typical picture of the CFBC combustor and cyclone is shown in Figure 3.3.

Bed material, as explained in the BFBC, can be crushed refractory, river sand or sized stoker ash. The loss of bed material is to be continuously replenished unless the fuel has >~15% ash.

Fuel and limestone (where necessary) are added to the combustor from the top. Fuels are crushed to nominal size of 10, 8 and 6 mm for lignite coal and low VM coals/anthracites for expanded-bed CFB. For classical CFB, 1 or 2 mm lower sizing may be needed.

Fines in fuel <1 mm can be as high as 40% which is a great advantage. BFBC is more sensitive to fines with a limit of <20%.

The surface M can be as high as 15%.

The limestone is required to be crushed to <1 mm depending on purity and reactivity.

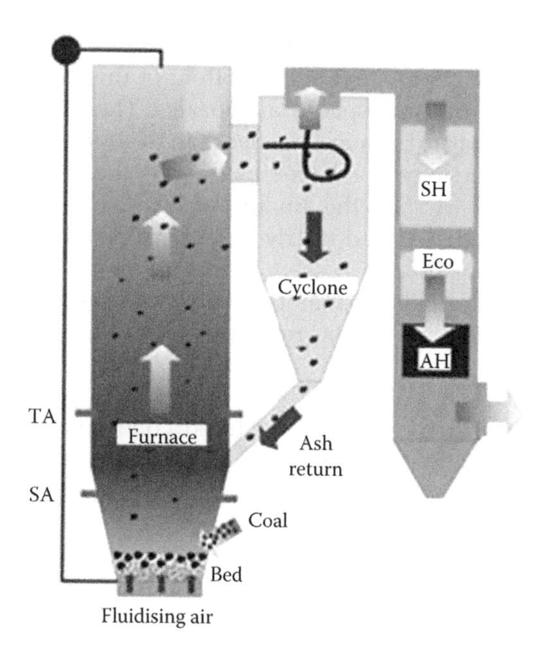

FIGURE 3.3
CFBC combustor and cyclone system.

In course of operation, as the combustor bottom accumulates shale and stony material the Δp of air increases. Periodic discharge of bed material from the bottom is necessary to restore the Δp. Bed material usually contains <1% unburnt C and constitutes <10% of the total ash even with very high ash coals.

In full-circulation CFB boilers, fly ash is removed from the ESP or bag filter for disposal. Ash re-circulation from the select hoppers of ESP is usually done to enhance the burning η with high-grade coals such as bituminous coals and anthracites. For quick burning coals, ash re-circulation may not be necessary as combustion is completed in the combustor itself. In case of expanded-bed CFB, fly ash is also discharged from the cyclones in addition to the ESP/bag filter.

3.28.5 Freeboard

Freeboard is the space reckoned from the top of air nozzles to the exit of the first pass in classical CFBC boilers. For expanded-bed boilers, it is taken from the top of the expanded bed up to beginning of the convection surfaces or inertial separators. Freeboard, thus, contains part of the combustor and the entire open space above.

Freeboard is the area where fuel gets time to complete the combustion. Freeboard volume is usually expressed in seconds of the residence time. It is the freeboard volume divided by the total gas quantity at the freeboard temperature which is usually 850 ± 50°C. The residence times are usually between 4 and 5.5 s with 4.5 secs for most coals. A range of 5–5.5 s is required for slow burning and high FC fuels such as pet coke and anthracites.

Above the TA nozzles, the freeboard is not required to be lined with refractory. In a full circulation, CFBC boiler a layer of ~200 mm bed material continuously slides down along the walls after being lifted to full height by the air jets from the PA nozzles below. The walls of the freeboard, therefore, receive their heat more from the conduction of the hot ash air than by radiation unlike in conventional boilers.

For steam temperatures >480°C, it is a normal practice to provide a part of SH surface in the upper freeboard usually as **wing wall panels**. The wing wall panels are studded and refractory covered in the lower part to withstand the severe erosion of the dust-laden flue gases of the first pass. They can also be placed as **horizontal platens** constructed of **omega tubes** for erosion protection. In cold cyclone boilers of tower type construction, substantial HS consisting of SH, Evap and Eco are placed in the first pass above the freeboard. **Division walls** are also be placed in the free board when more cooling of flue gas is needed.

3.28.6 Cyclones

A cyclone in CFB a boiler is perhaps the most important component which has a strong bearing on the overall performance. Static and bulky, a typical cyclone has an unimpressive look but its performance and reliability are very vital to realise the optimum thermal performance of a CFBC boiler.

The basic duty of a cyclone is to separate the dust particles from the flue gas to send them down to the combustor for re-circulation. Collection η should be typically ~99% for particles >100 μm and should be maximised as the dust-escaping cyclone can cause erosion to the downstream tube banks. If the erosion in the rear pass is to be avoided, the gas velocities have to be reduced by keeping the tube banks wider apart which increases the HS and boiler cost. At the same time, at reduced collection η the unburnt loss will also be increased fractionally.

Typically the dust-laden gases are accelerated to a velocity of 25–28 m/s in the inlet scroll section in a cyclone (see Figure 3.4). As the gases rotate in the cylindrical part at such high velocities the dust gets separated by centrifugal action, travels to the periphery and gets collected along the walls. The sloping walls of the cyclone help to move the dust spirally downwards and free the gases to exit through the central pipe. What sets aside the cyclones for CFB boiler is the combination of a very large amount of dust in the flue gases which is typically between 10 and 20 kg/Nm^3 (0.6–1.25 lb/Nft^3) and the high temperature of ~850°C. To withstand the dust impingement and provide sufficient insulation the refractory thickness required is ~500–600 mm. This poses a real challenge to construct such thick layers which can stay in position for months

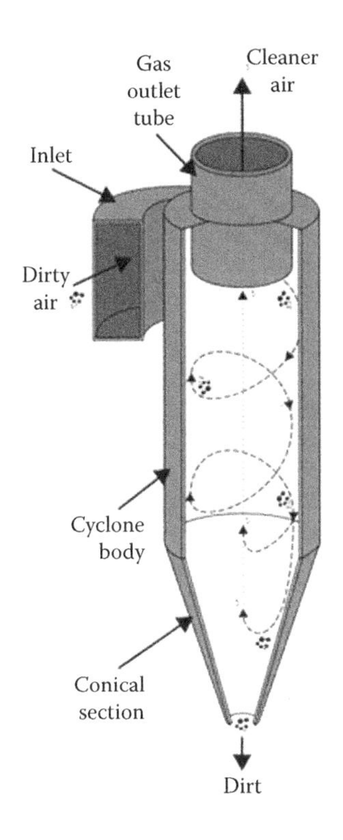

FIGURE 3.4
Action of cyclone in a CFBC boiler.

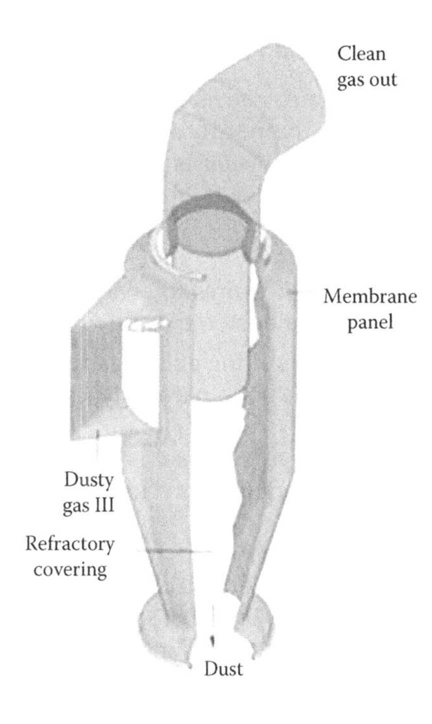

FIGURE 3.5
Water- or steam-cooled cyclone. (From Foster Wheeler Corporation, USA, with permission.)

at a stretch. Typical dense **refractory bricks** or **castable refractory** layers are employed which have also to withstand erosion.

Hot cyclones are made in sizes of ~3—8 m Ø. The massive refractory in cyclone not only increases the difficulty in supporting the cyclones from the structure but also makes the boiler start-up and cooling unduly long. Cold start-up times can be 8–10 h requiring large amount of FO for start-up purposes. Also, the cooling down time is quite long, normally 2–3 days. Further, an auxiliary boiler feed pump (BFP) is required to top up the water in steam drum during boiler shutdown. This is to compensate for the evaporation of water from boiler because of the residual heat in the refractory.

To overcome these problems cyclones are sometimes either *water cooled* or *steam cooled*. Cyclones are made of membrane panels and lined with a thin castable refractory of ~50 mm applied on the inside of for erosion protection (as shown in Figure 3.5). Such cyclones are naturally more expensive but are easy on maintenance. Start up and cooling down times are significantly lower. Water- or steam-cooled cyclones are felt to be not so ideal for slow burning fuels like anthracite.

Conventional cyclones with their long inlet and outlet pipes are replaced by somewhat octagonal or square cyclones made of membrane panels practically attached to the furnace walls which are called the *compact design* patented by feed water control (FWC).

A cold cyclone boiler operates at lower temperatures of 400°C and 500°C. Inlet dust loading is also lower between 1 and 2 kg/Nm^3. Because of low temperature and low-dust loading, steel cyclones with ~50 mm of castable refractory are adequate. Naturally, these cyclones are also much smaller in diameter ranging from 2 to 4 m because of smaller gas volumes.

In IR designs conventional cyclones are altogether replaced by inertial separators called U-beams by B&W (as shown in Figure 3.6). U-beams enjoy the advantage of lower Δp of ~25–30 mm wg but are also less efficient with ~90% collection efficiency. Two rows of U-beams are usually suspended just inside the first pass at the exit in such a way that the dust-collected (~75%) cascades into the furnace along the rear wall. U-beams are made with high grade ss typically SS 309 H, SS 310 H or RA 253 MA to withstand high gas temperatures ~850 ± 50°C.

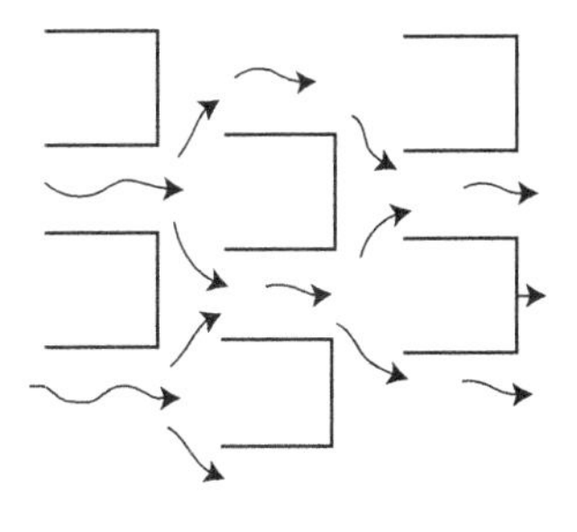

FIGURE 3.6
Inertial dust separators termed as U-beams by B&W.

3.28.7 Loop Seal/Seal Pot/L Valve/J Valve/Siphon

All these are various names for the proprietary ash valves installed in the return ash leg from cyclone to combustor. As the combustor is at a higher pressure of nearly 200 mm wg than the exit of cyclone, an ash pump is necessary to

- Equivalise the pressures to enable cyclone ash flow into the combustor and
- Prevent short circuiting of flue gases from combustor to cyclone bottom directly

Proper functioning of loop seal is very vital to create the re-circulation of ash. Each cyclone has an independent ash valve.

Ash valve is a fluidised-bed chamber with the inlet pipe connected to the bottom of the cyclone and the outlet to the combustor terminated at a suitable height above the PA nozzles. An independent HP blower is connected to the ash valve to pump air to fluidise the cyclone ash and make it flow over ash valve's side wall and slide into the combustor. Speed variation of the blower regulates the flow of re-circulated cyclone ash.

Owing to high temperature, the entire return ash circuit consisting of the ash valve and its upstream and downstream pipes are all internally lined with thick refractory of ~200 mm.

In the cold cyclone no refractory lining is required as ash is at a relatively low temperature of ~450°C.

In the current **IR CFB** boilers the entire loop seal arrangement is avoided which can be noted in Figure 9.30.

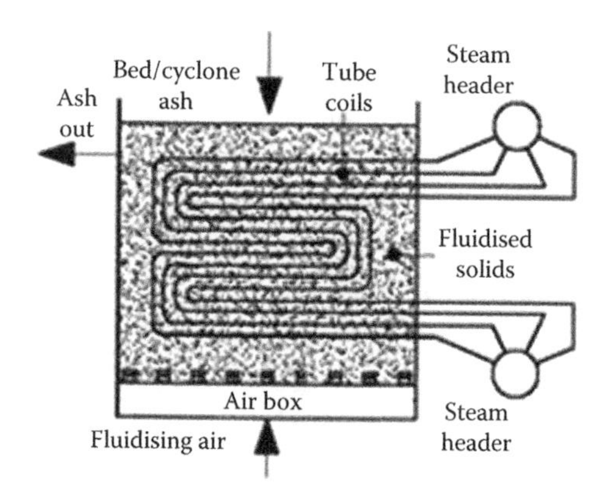

FIGURE 3.7
Concept of FBHE.

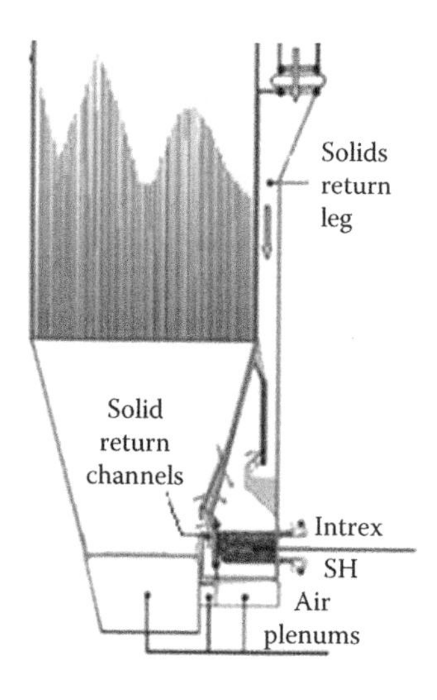

FIGURE 3.8
'Intrex' coil arrangement in compact boilers of FW. (From Foster Wheeler Corporation, USA, with permission.)

3.28.8 Fluid Bed Heat Exchanger

In boilers >250 tph when

a. High SOT >~480°C is desired or
b. Steam is to be re-heated

FBHE is employed in Lurgi or FW type of hot cyclone/compact designs. Classical FBHE is a heavy refractory chamber (600 mm thick) which has a bubbling-type fluidised-bed heater. FBHE draws a part of ash after the ash valve to heat the steam to the final temperature. Steam coils are placed in FBHE and hot ash entering the chamber is fluidised. Heat is transferred to the steam and the cooled ash is returned to the combustor. The concept is explained in Figure 3.7.

By this arrangement a certain amount of heat duty from the combustor is transferred to FBHE. The SH and RH coils in FBHE are very compact as heat transfer rates in the bubbling bed are some 5–10 times higher at ~300–600 W/m^2°C, ~250–500 kcal/m^2°F h or ~50–100 Btu/ft^2 h°F than in convection banks.

Moreover, the hot ash which is now cooled by the FBHE, when returned to the combustor, helps to stabilise the combustor temperature very effectively.

Although there are advantages in FBHE of having a compact external HS for SH and RH and obtaining a more uniform combustor temperature, there are usual disadvantages associated with a bubbling bed. There are distinct possibilities of tube side erosion and corrosion as well as refractory maintenance. In FWC design the FBHE is termed by their patented name INTREX where the refractory is practically eliminated (as shown in Figure 3.8).

All boiler makers do not favour FBHE. **Platen SH/RH** in furnace made with **omega tubes** to withstand erosion is preferred by some. Also **wing wall construction** is another option to combat erosion.

3.28.9 Bottom Ash

The accumulated bottom ash has to be periodically removed from any CFB boiler as the PA pressure starts increasing. Heavier particles namely, stones, shale and oversized fuel pieces gravitate to the bottom. As their quantity increases, the air pressure required for fluidisation also increases.

The bottom discharge system is usually designed for ~10% of total ash even though it rarely reaches that magnitude. Such large capacity is useful in case of speedy draining of bed during emergencies, such as a tube leakage. The unburnt C is also scarcely higher than 1%. The temperature of the ash discharged is nearly the same as the bed temperature which has to be further cooled to 80–150°C depending on the downstream discharge system and the local environmental requirements.

No ash coolers are needed if the pollution norms are relaxed enough to permit discharge of hot ash into a water impounded hopper from where the slurry is taken to a pond. This is strictly possible with fuels having little S needing no lime addition to the furnace. With increased environmental awareness such arrangements of ash disposal are rare.

Ash coolers are installed between the ash discharge pipes from the bed and the ash discharge system. There are three types of ash coolers namely

a. Screw coolers
b. Rotary coolers
c. FBAC

of which the FBACs are placed by the side of the boiler while the others are placed directly underneath.

3.28.9.a Screw Coolers

A typical ash screw cooler is shown in Figure 3.9. The casing, shaft and the flights are all cooled with water so that the incoming ash of ~850°C is cooled to the desired temperature during its travel from inlet to the outlet.

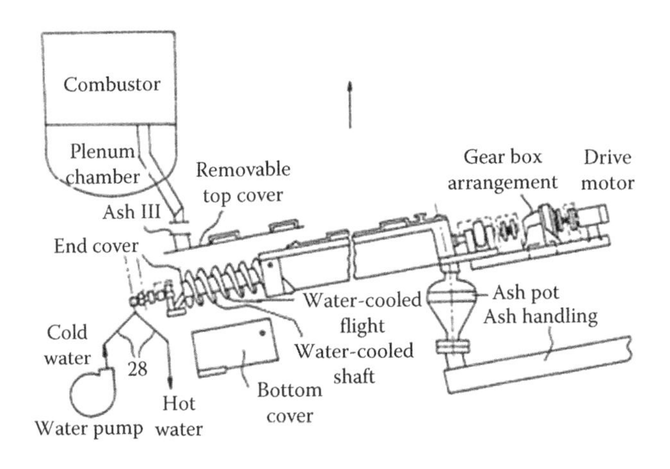

FIGURE 3.9
Ash screw cooler arrangement together with ash disposal system.

The initial part of the cooler is made of ss to withstand high ash temperature. Properly designed, screw coolers can cool ~6 tph of ash in ~6–7 m length from ~850°C to ~80°C. Hence, multiple ash coolers are required in a boiler. They are arranged at proper inclination so that direct loading either on the trucks or into the ash vessels of the discharge system is possible at the outlet end. Screw coolers are suitable for low and medium ash installations. There is no heat recovery. The treated cooling water (CW) which gets hot is cooled again by plant service water and the heat is lost.

3.28.9.b Rotary Coolers

These are bigger equipment and placed on the ground horizontally. The cooled ash is discharged into scraper ash extractor, placed below the ground level which carries the ash for further disposal. Water is passed through the flights attached to the inside of the drum and the annular space around the drum. Here also, there is no heat recovery. A typical rotary ash cooler is shown in Figure 3.10.

3.28.9.c Fluidised-Bed Ash Coolers

These are popular with boilers with high ash coal as heat recovery is possible. FBAC is a refractory box with fluidising nozzles at the bottom to create a bubbling bed with cooling coil below. Ash overflows from the bed and enters the cooler where it is fluidised by air drawn from PA stream. Water passes through the cooling coil, abstracts heat and cools the ash which is periodically drained away. The air from cooler which is now hot is introduced into the furnace without any loss of heat. A typical FBAC is shown in Figure 3.11.

3.28.10 Heating the Bed for Start Up (*see also* Hot Gas Generators *and* Duct Burners)

For start up of CFB boilers the bed has to be heated to a temperature of 400–600°C depending on the ignition temperature of the fuel. For heating the bed hot gas is

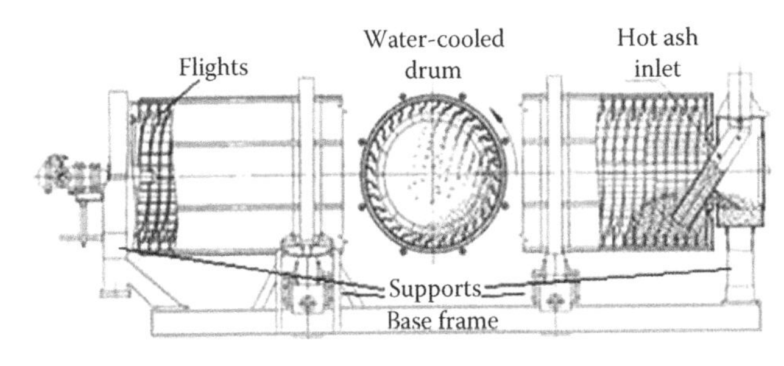

FIGURE 3.10
Rotary ash cooler.

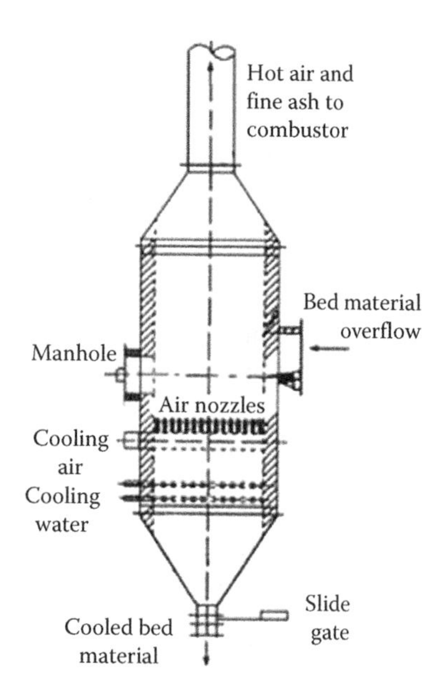

FIGURE 3.11
Fluidised-bed ash cooler.

generated using either a hot gas generators (**HGG**) or **duct burner** depending on the type of start up fuel. HGGs are used for all FOs while the duct burners strictly for NG or light FO. Alternatively, retractable start up oil guns installed at some suitable height in the boiler walls can also be used to heat the top of the bed. Heating the bed by radiation this way is slower and less positive than heating from below.

A typical HGG is shown in Figure 3.12. Oil is burnt with high excess air of ~70% using a circular turbulent burner which is mounted on a cylindrical drum with double casing with refractory lining on the inside. The remaining process air is passed through the annular gap between the inner and outer casings and mixed with combustion gases either progressively or at the end, depending on the design, such that the flue gases exit HGG at temperatures of 600–1100°C.

Multiple HGGs are attached to the windbox in such a way that the flue gases are spread all across the underside of the bed. The hot gases go through the fluidising air nozzles of the bed and heat the bed uniformly to the required temperature.

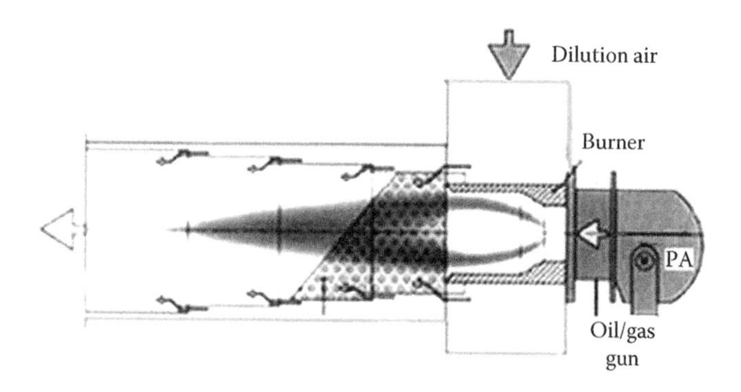

FIGURE 3.12
Hot gas generator (HGG).

3.29 Circulating Fluidised-Bed Combustion Boilers (*see also* Utility Circulating Fluidised-Bed Combustion Boilers)

Starting modestly in the middle of 1980s the CFBC boilers have grown in size, complexity and diversity to spread in most parts of the world where there are solid fuels to be burnt. The largest boiler which was commissioned in 2009 is a supercritical (SC) boiler for generating 462 MWe after the successful operation of several sub-critical (sub-c) boilers in 250 and 300 MWe range.

CFBC boilers have substantially replaced the PF boilers in the industrial range of up to 150 MWe mainly on account of the following features:

- Nearly the same thermal η
- Very good fuel flexibility
- Better environmental compliance with cooler combustion and in-furnace desulphurisation
- Stepless turndown
- Better operator friendliness
- Simpler O&M
- Minimal oil requirement for start ups and low load stabilisation
- No fear of boiler explosion

Despite marginally more fan power and slightly bigger-sized boiler, the advantages stated above, together with more competitive pricing, have been weaning the customers away from PF in this segment. This is because the modern PF boiler

- Tends to be somewhat oversized with low NO_x burners and larger furnaces
- Needs additional deNO_x and deSO_x units for meeting pollution norms

making the boiler plant more expensive.

In the higher utility range PF continues to be more competitive, now that the prices of the gas cleaning equipment have become cheaper, unless

- There is a need to burn very difficult fuels like anthracite or petcoke

- Very stringent pollution norms with high S in fuel or
- Some additional fuel flexibility is desired

CFBC boilers in the higher utility range have been only the full-circulation type either with conventional hot cyclone or the compact design. The high fluidising velocities make the units slimmer and more competitive than the expanded-bed CFBC boilers. It is also because the full-circulation boilers have the early mover advantage and built references.

In the industrial range, however, both types of CFB boilers compete. The expanded-bed CFBs are found to enjoy the advantages of

- Lower power consumption
- Lesser maintenance because of lower refractory
- Faster start up and shutdown because of lesser refractory
- Very little consumption of support oil
- Simpler O&M

CFBC boilers will rival PF in years to come as there are designs being readied for the next steps of 600 and 800 MWe. The three concerns are

- Marginally higher fan power
- Fractionally lower thermal efficiency
- Higher amount of refractory

3.29.1 Hot Cyclone Circulating Fluidised-Bed Combustion Boilers

A typical modern large top supported full-circulation conventional twin hot cyclone RH CFBC boiler for high p and t is shown in Figure 3.13. It is a classical location for the cyclone which is placed between the combustor/first pass and second pass. First pass contains the combustor and free board. Hot PA for combustion, taken from the AH, is given from the bottom as fluidising air. **Duct burners** heat the air initially to ~400–600°C for bed heating before the introduction of main fuel. The entire RH is placed in the second pass above the Eco and TAH while the SH is split between the first and the second passes. The second pass enclosure is made of membrane panels. Wing wall Evap is also located in the first pass in addition to horizontal platen SH and wing wall vertical SH. All surfaces are drainable.

Figure 3.14 shows a typical small hot cyclone boiler of ~100 tph. Here all HS is placed in the second pass. Cyclone is not in its classical location but is placed at the front to

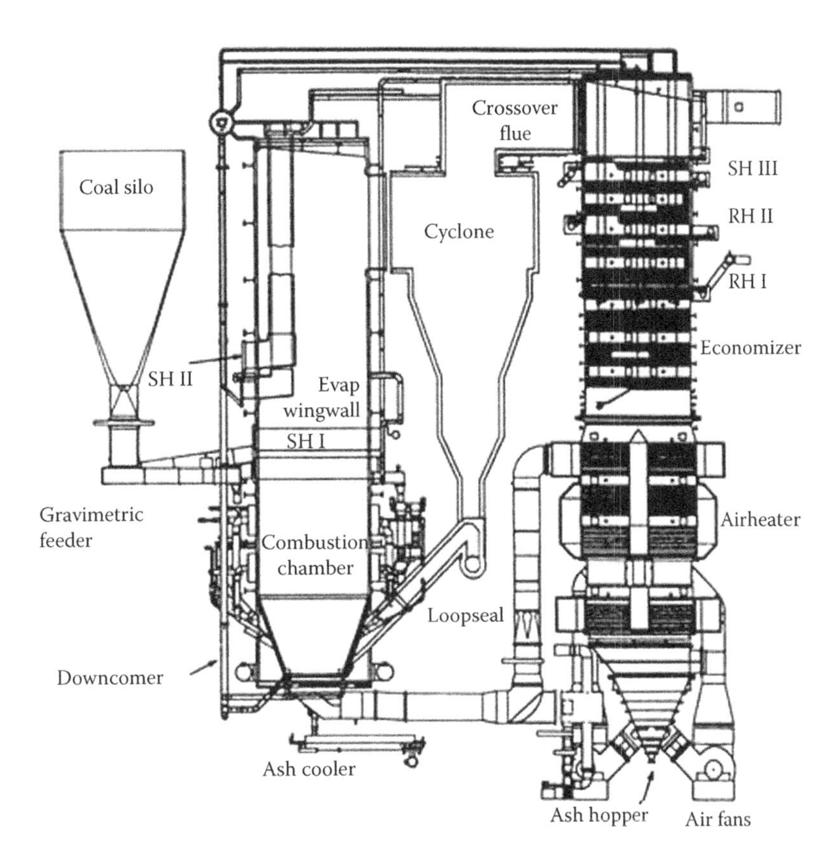

FIGURE 3.13
Large conventional hot cyclone utility range CFBC re-heat boiler.

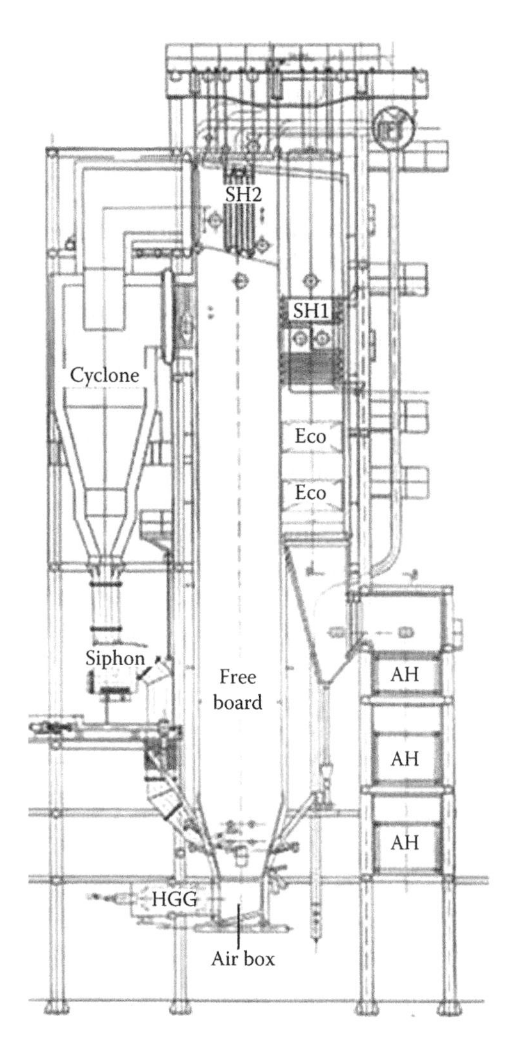

FIGURE 3.14
Industrial range conventional hot cyclone CFBC boiler.

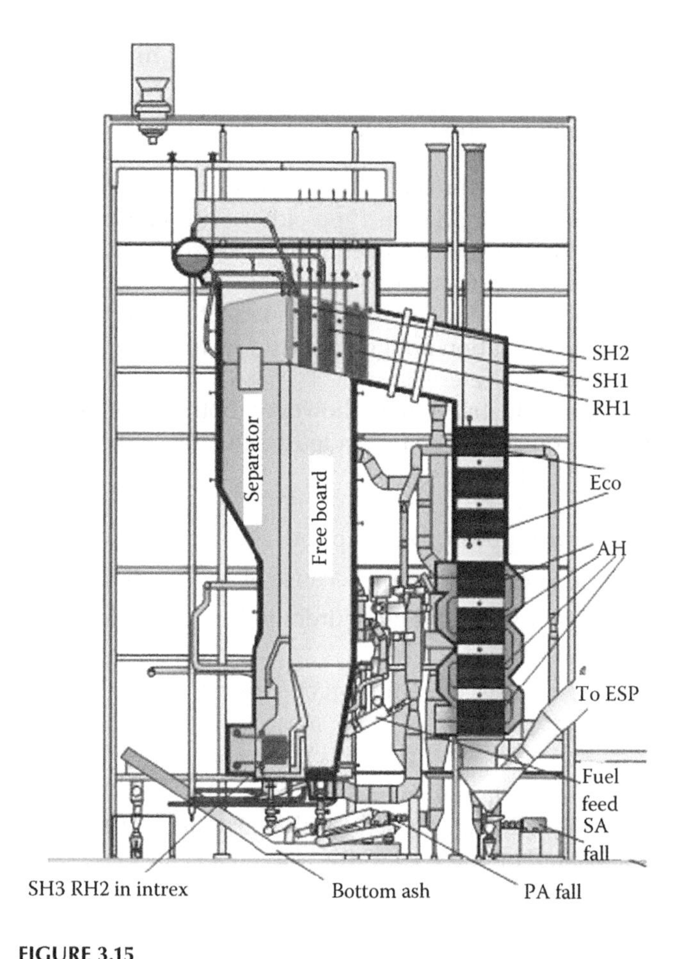

FIGURE 3.15
'Compact'-type utility range re-heat CFBC boiler of FW. (From Foster Wheeler Corporation, USA, with permission.)

enable the first and the second passes to share rear wall and make a compact arrangement for cost effectiveness.

3.29.2 Compact Boilers

Compact design is the patented CFBC boiler of FWC in which the cyclone is practically attached to the first pass with very short inlet and outlet pipes and loop seal arrangement. The cyclones are water cooled and made of octagonal shape with very little refractory. Such boiler arrangement is shown in Figure 3.15. In the return leg, it is possible to accommodate FBHE for SH or RH.

3.29.3 Cold Cyclone Boilers (see Circofluid or Cold Circulating Fluidised-Bed Combustion System)

3.29.4 Internal Re-Circulation Circulating Fluidised-Bed Combustion Boilers (see Internal Re-Circulation Circulating Fluidised-Bed System)

3.30 Circofluid or Cold Circulating Fluidised-Bed Combustion System (*see also* Fluidised-Bed Combustion, Circulating)

This patented technology was developed in the early 1990s by the former Deutsche Babcock of Germany. This is an expanded-/turbulent-bed CFB system with full-load bed velocities ranging from 3.5 to 4.5 m/s and falling between the bubbling and classical/full-circulation CFBC systems.

This boiler is essentially a *tower type boiler* with a CFB combustor attached at its bottom instead of a PF furnace. Because of the presence of HSs in the first pass itself, directly above the combustor, the flue gases cool down to temperatures between 400°C and 500°C at the end of the first pass. They are de-dusted in cyclones and further cooled in the second pass to ~140°C. The cyclones are naturally smaller because of the lower gas temperature. The casing of the second pass is made of simple steel enclosure requiring no membrane wall construction.

The first pass HSs receive heat both from hot gases and circulating bed material making them very effective somewhat like the HSs of FBHE, though not as effective since the HSs are suspended in dusty gas stream and not immersed in a bubbling ash bed. The lower fluidising velocities also carry lesser amount of dust into the cyclone, anywhere from 1 to 2 kg/Nm^3 as against 10–20 kg/Nm^3 in the full-circulation CFBC boilers. The cyclones require a thin 50 mm castable refractory as against 500–600 mm brick construction of the hot cyclone, greatly reducing the total refractory.

These features are listed in Table 3.2. As the HSs are lower the boilers are cheaper. They are also easy on O&M. The specific advantages as compared with classical CFBC boilers are

- Lower fan power because of lower bed velocities
- Much smaller refractory of ~20% of classical CFBC
- Very short start-up times of 4–5 h as against 10–12 h in full-circulation CFBCs
- Second pass made of only steel casing instead of membrane walls
- Bed temperature extremely steady because of the re-circulation of cooler ash

As the HSs are in the first pass, there are chances of erosion unless the preventive protection measures are taken fully. Presently boilers with 2 and 4 cyclone arrangements are built with 200 and 400 tph capacities on coal. Higher capacities up to 800 tph should be possible with 6–8 cyclones. As the bed velocities are lower the furnace cross-section is larger by 150% which makes the conventional CFBCs cheaper in the higher capacity ranges.

TABLE 3.2

Design Parameters of Cold Cyclone Boilers

Parameter	Value
Fluidising velocity at full load	3.5–4.5 m/s 11.5–14.7 fps
Bed temperature	800–900°C 1470–1650°F
Excess air at full load	18–25%
Expanded-bed height	1.2–1.4 m 4–4.5 ft
Freeboard height	15–17 m 50–55 ft
Residence time	3.5–5.0 s
Dust loading in gas	1–2 kg/Nm^3
Gas velocity in first pass	<5 m/s 16 fps
Cyclone entry temperature	400–500°C 750–930°F
Start-up method	By HGG or duct burner below bed
Second pass enclosure	Plain CS

Presently, cold cyclone boilers are not made for lack of demand in Europe but are designed and manufactured in India and China and have earned sizeable market share in India. Figure 3.16 shows the arrangement of a typical two-cyclone boiler for coal/lignite firing with Eco and horizontal TAH as backend equipment.

Figure 3.17 shows the schematic arrangement of a typical cold cyclone boiler.

- The first pass contains the entire SH and part/full Eco.
- The screen tubes are fully gun studded, as shown in Figure 16.6, to withstand erosion from dusty gases and provide protection to the all the tube banks arranged directly above by suitably deflecting the dusty gases.
- The combustor has three levels of air—PA or fluidising air at the bottom and SA and TA at suitable heights in the combustor.
- Furnace walls are lined with SiC castable refractory up to a little higher than TA level to protect the tubes from corrosion because of the prevailing reducing conditions in the combustor created by the staged air supply. Additionally, the dense SiC protects the tube walls from erosive forces of the turbulent bed at the bottom and high-velocity dust laden gases above. SiC is especially chosen for helping in transmitting heat to water.
- Cyclones are smaller in diameter as the flue gases are colder and are long for good collection efficiency.
- Ash is returned to the combustor, mixed with the fresh fuel, through the inlet fuel and ash pipes. Fuel feeding is by gravity as the furnace in nominally under suction.
- Ash pump, called the siphon, moves the ash from LP region of cyclone to the HP level of furnace.
- Fresh fuel is withdrawn from the bunker by the fuel feeder and transported to the inlet pipes downstream of siphons, in tune with the boiler fuel requirement.
- Limestone addition, if needed, is done by withdrawing from the limestone bunkers and discharging into the fuel conveyor.
- PA and SA fans feed the fresh air as required for combustion, at different pressure levels.
- Combustion air is often heated while cooling the flue gas to the desired level in TAHs. Although this is desirable, it is not always necessary to heat the air like in PF boilers where hot air is absolutely necessary to dry the coal while grinding.

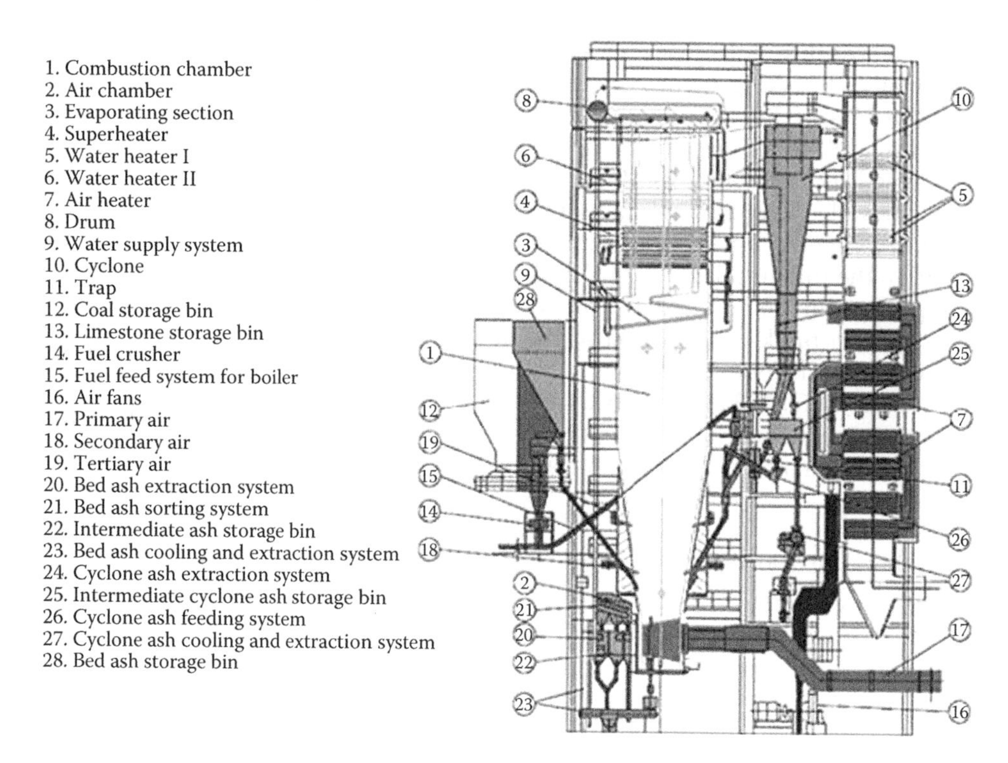

FIGURE 3.16
Small cold cyclone boiler with two cyclones.

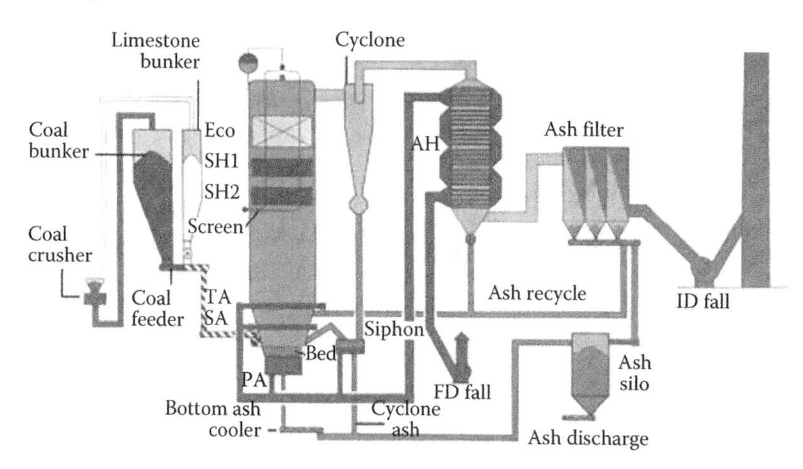

FIGURE 3.17
Schematic arrangement of a cold cyclone CFBC boiler system.

If feed water temperature (fwt) is sufficiently low AH can be dispensed with in favour of large Eco.

- Bed heating is usually by HGGs and hot gas is admitted from the bottom for positive and quick heating. Duct burners are employed when the start up fuel is light oil or NG.
- Bed level control is by intermittently draining the bed ash whenever the bed pressure exceeds the limit.
- Bed temperature control is by continuously varying the recycled ash by regulating the ash discharge from siphons.
- Ash re-cycling is usually found necessary for fuels with high amount of FC to improve the combustion efficiency. Ash from Eco, AH and first field ESP hoppers are returned to the combustor for re-circulation.

3.31 Circulation

Circulation can be defined as the act of water–steam mixture moving around in Evap circuits before getting converted completely to steam. Circulation is required to prevent overheating of the heat-absorbing surfaces. This may be achieved

a. Naturally by gravitational forces as in *natural circulation*
b. Mechanically by pumps as in *forced circulation* or
c. By a combination of both methods as in *assisted circulation* or combined circulation

The topic of circulation is covered in the following sub-topics.

Circulation ratio:

1. Natural circulation
2. Assisted circulation or full-load re-circulation
3. Forced circulation

Aim of good circulation:

1. Exit quality
2. Minimum velocity
3. SWH

3.31.1 Circulation Ratio

Circulation ratio is the number of times a unit quantity of water circulates in a circuit before fully turning to steam. It is the amount of steam and water circulating in a circuit for each unit of steam produced. For the entire boiler, it is the amount of steam–water mixture in circulation in the boiler divided by the steam flow.

A boiler consists of several circuits. A circuit is a group of similarly patterned and heated generating tubes with upward flow of steam–water mixture with its own feeder and riser pipes. Circulation ratio in each circuit will be different depending on the heat incident on the circuit, tube geometry and dimensions and so on.

Of all the circuits the *critical circuit is the one which has the lowest circulation ratio or closest to departure from nucleate boiling (DNB) regime.* Making the circuit safer involves increasing the inlet water velocity to at least 1 m/s by increasing the feeder and/or riser areas. Sometimes, for HP boilers >150 bar, ribbed or rifled tubes may have to employed in areas of high heat flux like the burner zone.

Circulation ratio of a boiler is the average ratio of all circuits. This can be obtained by calculating the total amount of steam–water flow and dividing by the gross steam flow.

3.31.1.1 Natural Circulation

Circulation of steam and water in boiler circuits because of natural force like gravity, without any external aid such as pumps, is natural circulation. It is because of the thermo-syphonic head set up between the lighter water in the heated tubes of furnace and the relatively denser and cooler water in the downcomers.

The circulating head is proportional to

$$H \times (\rho_d - \rho_r)$$

where

ρ is the density of water and

d and r are the downcomer and riser tube, respectively

H is the height of the circuit, as shown in Figure 3.18

From this it follows that for good circulation

- Higher the location of the steam drum the better it is for circulation as it increases H. Taller boilers and higher pressures go together generally.
- **Downcomers** should have steam-free water under all conditions for enhancing ρ_d. For this reason the **steam–water separation** in drum should be very good.
- Water velocities at the entry to the large bore downcomers should be low enough to prevent steam bubbles being drawn into them which would lower ρ_d. Vortex breakers are therefore inserted at the inlets of large bore downcomers.
- ρ_d can also be enhanced by using unheated downcomers.
- It is for this reason that Eco outlet water is taken only to the steam drum and not to the bottom drum.

Natural circulation is the most commonly adopted type of circulation for sub-c boilers, having been employed for as high drum pressures as 211 bar (3000 psia) where the ρ differential between water and steam are as low as 2.5. At higher pressures than this it becomes very expensive as unduly large downcomers and risers are needed.

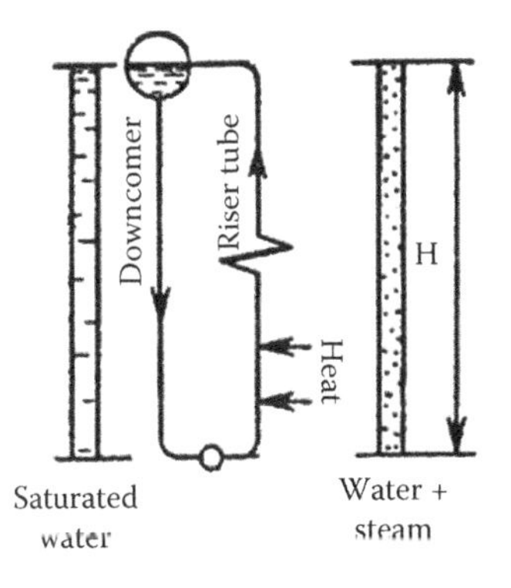

FIGURE 3.18
Process of natural circulation.

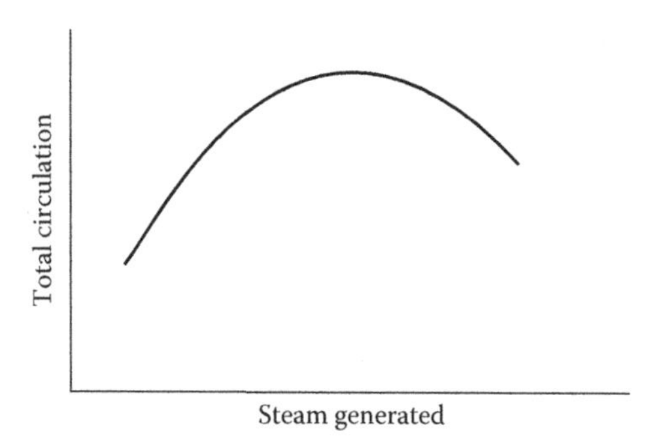

FIGURE 3.19
Self-limiting feature of natural circulation boilers.

The main advantages are

- Simplicity
- Safety due to self-limiting characteristic

Self-limiting feature of a circuit is the ability to induce more circulation and keep the tube metal temperature under check when there is more heat input. To explain further, when there is more heat on the furnace tubes there is greater steam generation and lowering of ρ of steam water mixture. With the ρ differential increasing there is a tendency for more circulation in risers which is easily supported by cooler water rushing in from the downcomers. This balancing is natural and automatic because there is no pump in between and the circuits operate on the rising part of the curve depicted in Figure 3.19. Beyond a point, however, steam generated in the risers is so large that the friction losses will be more than the thermo-syphonic head available resulting in **choking of flow** in riser tubes causing tube failures due to overheating. Natural circulation boilers always operate on the rising part of the circulation curve.

3.31.1.2 Assisted Circulation or Full-Load Re-Circulation

Assisted circulation or full-load re-circulation, depicted in Figure 3.20 is more suitable for pressures of 150 bar and above for large utility boilers for fixed pressure operation, where it is found beneficial to supplement the thermo-syphonic head by boiler circulating pump (BCP).

At ~200 bar this arrangement is said to provide good security against possible circulation lapses. Several large boilers are operating satisfactorily on this principle.

The advantages are that

- Smaller and thinner tubes can be employed for furnace
- Downcomer and riser tubes are of lower sizes and reduced lengths

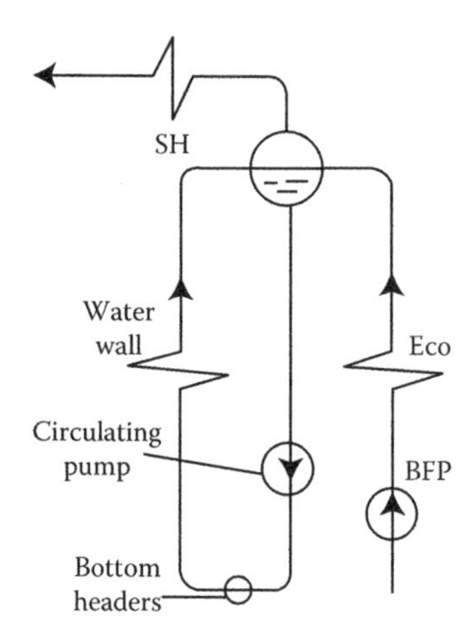

FIGURE 3.20
Concept of assisted circulation or full-load re-circulation.

- As the flow can be controlled in tubes a greater flexibility in the disposition of surfaces is possible

These benefits offset the additional cost of **BCPs**, their drives and flow balancing ferrules at the entrance of riser tubes.

The additional pumping power of the BCP is substantially recovered in the small heating of circulating water due to friction in the ferrules.

3.31.1.3 Forced Circulation

Forced circulation in a boiler is the flow of water and vapour set into motion by a circulation pump, as opposed to natural circulation where there is thermo-syphonic head (combination of gravity head and convection currents) at work.

Forced circulation is adopted both for sub-c and SC pressures, unlike natural circulation which is suited only for sub-c application.

Forced circulation is popular in sub-c pressure usually with waste heat (WH) boilers where

- There is low heat flux in furnace. Usually there is little or no radiation in furnace.
- Generating tubes are horizontally disposed.
- Load fluctuations are rapid.

Figure 3.21 depicts the use of forced circulation for sub-c boilers.

SC boilers have no evaporators and water directly turns into steam. They have to work on forced circulation only (see Figure 3.22).

3.31.2 Aims of Good Circulation

The main aim of circulation is *to maintain wet wall flow or nucleate boiling under all load conditions firing all fuels.* For

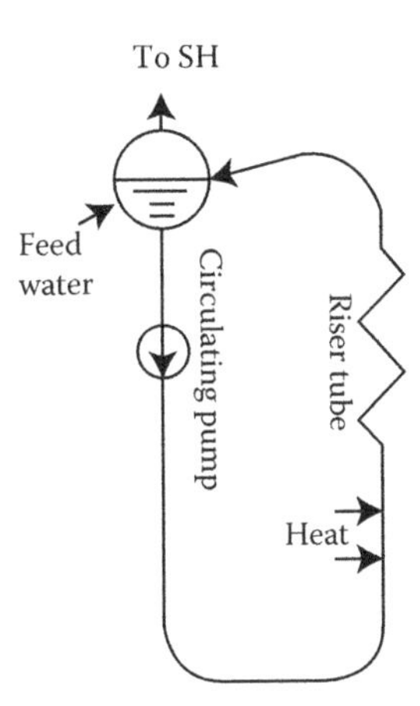

FIGURE 3.21
Forced circulation in sub-c boilers.

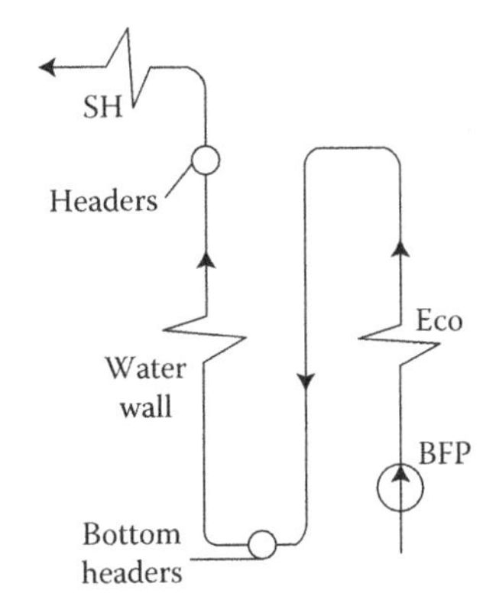

FIGURE 3.22
Forced circulation in SC boilers.

achieving it the following three conditions are required to be satisfied for each water wall circuit.

A *circuit* is defined as a set of

- Heated water wall tubes
- With upward flow of water
- Having similar shape and heat input

with independent supply of saturated water from the bottom and evacuation of steam and water mixture at the top.

3.31.2.1 Exit Quality

At the top of any circuit steam by weight (SBW) should be less than the specified limit to prevent **film boiling**. This depends on the drum pressure and location of burners—whether at the top or bottom.

For drum pressures up to 100 bar it is the SBW at the top of a circuit that governs the circulation, that is, if the SBW is within limits at the top of the circuit, safe SBW conditions would result at all the lower levels as well. But at higher pressures such assumption is not safe and a more detailed analysis of the heat flux at all levels would be required.

TABLE 3.3

Permissible Steam by Volume from Risers at Various Drum Pressures

Drum pressure in bar	20	40	70	100	140	180
SBV-steam by volume %	80	75	70	67	65	55

The exit steam water quality is limited to ~55–80% SBV. SBV to SBW conversion is given in properties of steam and water. Steam occupies more volume at lower pressure. Hence SBV increases at lower pressures and decreases at higher pressures, as shown in Table 3.3.

3.31.2.2 Minimum Velocity

Water velocity at the entry to a circuit should exceed the specified limit depending on the inclination of the tubes. The speed of water should be high enough to prevent the steam bubbles from adhering to the tube walls which can cause overheating of tubes and sludge accumulation. Lower the inclination higher is the limit for minimum velocity (as shown in Table 3.4).

3.31.2.3 Saturated Water Head

SWH, the ratio of pressure loss (including static head) to the pressure produced by a column of saturated water of the same height, is required to be a certain specified minimum to prevent flow reversal.

The minimum values for % SWH at various pressures are established from experience. When % SWH falls below these levels flow reversals can occur between tubes of the same circuit leading to stagnation and eventual tube failures.

The usual remedy for meeting this requirement is to increase the water flow to the defaulting circuit.

The typical minimum % SWH ranges from ~50% to 70% increasing with the pressure as given by Table 3.5.

TABLE 3.4

Minimum Water Velocities at the Inlet of Circuits

	Minimum Velocity m/s at Pressure	
Type of Wall	**<100 bar**	**>100 bar**
Furnace water wall with or without slope	—	0.6
Furnace water wall with slope >30°	0.3	—
Furnace water wall with slope <30°	1.0	1.0
Furnace water wall with slope <30° with heat on top	1.7	1.7
Vertical boiler tubes	0.15	0.15
Water-cooled burner throats	0.15	0.15

TABLE 3.5
Percentage of Minimum-Saturated Water Head Required at Various Pressures

Drum pressure in bar	35	70	105	140	175	200
Percentage of SWH	51	56	60	63	65	66

Circulation ratio is higher at lower pressures than at higher pressures.

3.32 Clad Tubes (*see* Boiler Tubes *under* Materials of Construction)

3.33 Cladding (*see* Lagging)

3.34 Clean Coal Combustion

Coal is perhaps the dirtiest of all fuels as it pollutes environment in several ways. Yet its use for generation of power is inevitable as it is the cheapest prime fuel and most widely distributed geographically available. Together with lignite it contributes to over two-thirds of power generated and is likely to stay that way for the foreseeable future.

Environmental pollution occurs in several ways when coal is burnt. **Fly ash** contributes to dust in air. **GHG** like CO_2 add to the general global warming. **Sulphurous gases** cause acid rain. **NO_x** destroys protective O_3 layer in the atmosphere and also cause respiratory problems to humans. **Ash** pollutes the ground and water.

Clean combustion is the burning of coal without polluting the atmosphere adversely. Several steps are to be implemented simultaneously at various stages in coal chain to clean up combustion.

- Washing of coal removes ash and minerals significantly at the mine mouth. Burning washed coal, with reduced **burden**, produces less ash.
- **Dust collecting equipments** such as ESP, bag filters or wet scrubbers remove almost all the fly ash.
- Sulphurous gases are removed from flue gases post-combustion by **FGD** or deSO_x plants.
- Denitrification (**deNO_x**) can be done in several ways either during combustion stage or post-combustion by catalytic or non-catalytic reduction. **LNBs**, separated overfire air (**SOFA**), **re-burning, FGR** are the methods for NO_x abatement at combustion stage.
- **FBC** is a process where clean combustion results because of reduced combustion temperature and in-furnace desulphurisation.
- SC boilers reduce the GHG by using lesser amount of coal due to better cycle η.
- Big reduction in CO_2 emission into atmosphere is achieved by **CCS** which is a future technology.

Most of these processes are explained at appropriate places in the book.

3.35 Clinker (*see* Ash)

3.36 Cloth Filter (*see* Dust Collection Equipment)

3.37 Coal

Coal is a complex heterogeneous mixture composed of both organic and inorganic materials containing as many as 65 chemical elements and geologically formed over several millennia from the ancient vegetation by the combination of time, pressure and heat of the earth. Depending on how long it has been under these conditions the resulting coal assumes several properties. The most ancient coals under higher pressure would have converted practically all the vegetable matter into FC while the youngest coals with lesser pressure are still substantially woody matter with much more VM than FC. All the coals stack up between these two extremes. Ash is the sand that got trapped along with wood in the formation of coal.

This topic of coal is elaborated under the following sub-topics.

1. Coalification
2. Coal classification
3. Coal analyses
   a. Proximate analysis
   b. UA
4. Analysis-reporting methods
5. Coal cv
6. Coal grindability index

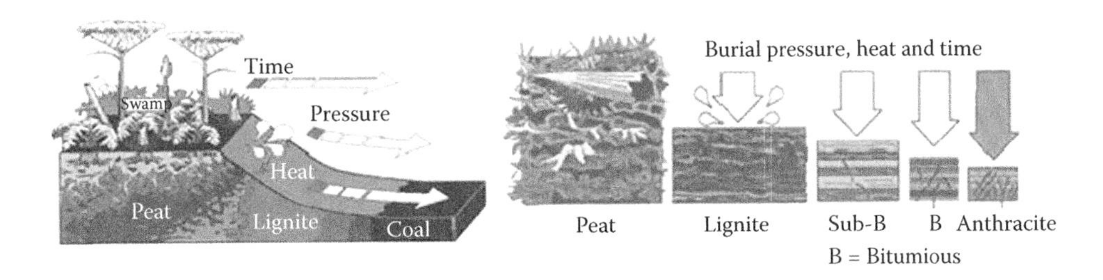

FIGURE 3.23
Coalification process.

7. Caking or agglomerating coals
8. Coking coals
9. Coal erosion

3.37.1 Coalification

Coalification is the process by which vegetable matter is transformed into coal by the compression and hardening over long periods of time (as shown in Figure 3.23).

Coal can also be defined as a black or brownish black combustible substance that is formed by the decomposition of vegetation over several millennia in absence of air under pressure and heat of earth.

3.37.2 Coal Classification

Coals are classified in slightly different ways in various countries.

ASTM D388 classifies coal by rank, which denotes the degree of coalification and is quite popular. Table 3.6 presents this ASTM classification of coals.

In Europe, the classification is different and they are separate for hard and **brown coals**.

Certain interesting points about coals to be noted are

- With increasing coalification the rank of coal is higher.
- In higher rank FC and Gcv increase while M and VM decrease.
- The ratio of FC to VM is called the *fuel ratio*, which naturally increases with the rank.

Higher the fuel ratio higher is the Gcv. But with anthracites the Gcv, in fact, drops, as there is not enough VM to contribute to the heat value.

- Lower the rank higher is the VM and O_2 content which aids ignition and enhances combustibility and also the flame stability.
- Low-rank coals produce *self-pulverisation* during combustion, as the inherent M locked in the pores of coal, gets heated and expands rapidly (volume expansion of 1:1600!) to fragment the fuel particles.

3.37.3 Coal Analyses

There are two types of analyses. Here the word coal is used synonymously with solid fuels. Almost all solids are analysed in these ways.

TABLE 3.6
Classification of Coals by ASTM

Class	Group	FC% (dmmf)	VM% (dmmf)	Gcv (Moist mmf) MJ/kg	kcal/kg	Btu/lb	Remarks
i. Anthracitic	Meta-anthracite	>98%	<2%				
	Anthracite	92–98%	2–8%				
	Semi-anthracite	86–92%	8–14%				Non-agglomerating
ii. Bituminous	Low-volatile coal	78–86%					Commonly agglomerating
	Medium-volatile coal	69–78%					Commonly agglomerating
	High-volatile coal A, B, and C	<69%		24.4–32.6	5830–7780	10,500–14,000	Commonly agglomerating
iii. Sub-bituminous	A, B, and C coals			19.3–24.4	4610–5830	8300 –10,500	
iv. Lignitic	Lignite A			14.66–19.3	3500–4610	6300–8300	
	Lignite B			<14.66	<3500	<6300	Also called Brown coals

3.37.3.1 Proximate Analysis

PA gives a good indication of how fuel burns.

The four items that comprise the proximate analysis of fuel, namely

- Moisture (M)
- Ash (A)
- VM
- Fixed carbon (FC) or char

are depicted in Figure 3.24. Air and gas weights and further combustion calculations are performed by graphical methods.

Burden or ballast
M + A collectively is known as *burden or ballast* as they have no contribution to the cv of fuel. Combustion is rendered more difficult and slow with increasing ballast.

1. Moisture (M)
M is classified as follows:

- *Inherent M* is the M trapped inside the fuel. On heating, it is progressively released along with VM up to 700°C.
- *Surface M* is the M picked up by coal because of its hygroscopic nature from the surroundings. It is from the
  - Humidity in air
  - Rain or snow during transportation
  - Washery operations
  - Mining operations

M content of coal increases as the rank deceases 15%, 30%, and >40% M is possible with bituminous, sub-bituminous coals and lignites. Normal PF systems cannot handle >40% M due to ignition difficulty.

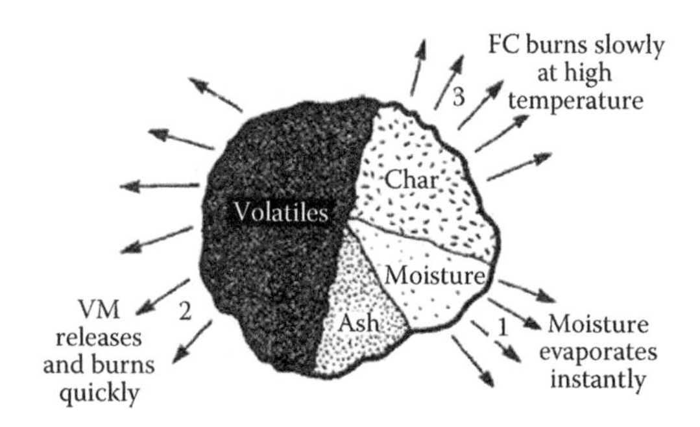

FIGURE 3.24
Pictorial presentation of proximate analysis of coal.

2. Coal ash (A) *see also* Ash
Coal ash is the incombustible MM left behind when coal is burnt.

In the laboratory it is the residue left on complete combustion of coal sample performed in a carefully controlled and specified manner in a muffle furnace at 700–750°C.

Ash is of two types: (1) adventitious and (2) inherent.

- *Adventitious* or *extraneous* ash is the MM derived from shale, clay, pyrites and dirt from earthy or stony bands in the coal seam while
- *Inherent* ash is the mineral substance derived from the plant vegetable matter which is organically combined with coal during coalification process. It forms <2% of the total ash

Coal ash is composed of compounds of Si, Al, Fe and Ca and, to a lesser extent, compounds of Mg, Ti, Na and K. Although reported as oxides in analysis they occur in ash as a mixture of silicates, oxides and sulphates.

Mineral Matter
Ash forms nearly 90% of the MM. The remaining accounts for the combined M and CO_2 expelled from clays and other minerals by heat.

3. Volatile matter
This is a complex mixture of organic materials which starts to distils off at ~300°C and burns in suspension in furnace. The procedure for determining VM is laid out in ASTM method D 3175. Approximately 1 g of fuel is heated in a crucible for 7 min at 950 ± 20°C with no contact with air under standard conditions. Weight of VM is the loss of weight less the weight of expelled M which is noted separately.

Notable aspects of VM are briefly listed below:

- Higher VM is desirable in fuel as it stabilises the flame and accelerates burning of FC.
- VM is a complex mixture of organic materials which contains the following:
  - Water from the combustion of fuel
  - Complex mixture of H_2, O_2, CO, CH_4, C_2H_6 and other HCs
  - Tar, a mixture of HCs and organic compounds especially phenols
- These organic materials volatilise on heating to ~300°C and burn in suspension in furnace.
- The higher the VM, the lower the ignition temperature and correspondingly greater the

combustion speed. So it is easier to ignite and burn fuel if VM is higher. Anthracites and low-volatile bituminous coals are difficult to ignite and take longer to burn needing special furnaces and burners.

- VM contributes to higher proportions the total cv as the coal rank decreases. It is ~10% in anthracites, 40% in high-volatile bituminous and sub-bituminous coals and >50% for lignites.
- Coals having less than ~19% VM on daf basis are termed as low-volatile coals and are difficult to ignite and burn.
- High VM renders coals more environmentally friendly as the NO_x generation is less because of lower FC and correspondingly lower flame temperatures.
- cv of VM from high-rank coals such as the high-volatile bituminous coals can be twice as high as from low-rank coals, as the former is rich in HCs while the latter contains a lot of M and CO.
- The high VM in fuels lowers the ignition temperature making it easy to ignite and burn.

4. Fixed carbon or char
This is the carbonaceous residue left in the crucible after the distillation of VM. The following are worth noting. FC is not

- C of UA, neither is it the total C in fuel as some C forms part of VM.
- Pure C as there are traces of H_2, O_2, and N_2 and about half S in it.
- Directly measured but calculated by difference by subtracting % of M, A and VM from 100.

Together with VM, FC forms the burnable portion of fuel; the balance being the unburnable ballast or burden.

FC or char takes a long time, between 1 and 2 s, to burn unlike VM. FC burns on the grate in grate-fired boilers, in the bed in FBC or in furnace in PF firing. The speed of combustion of char depends on factors like porosity, size, surrounding heat, O_2 availability and so on. Char of higher rank coals is dense and rock such as while char of lower rank coals is lighter and more fibrous. Char reactivity, therefore, lessens as the rank increases.

3.37.4 Ultimate Analysis

UA is the elemental analysis of coal. C, H_2, N_2, S and O_2 (by difference) are given. UA is amenable to directly calculate the air and gas weights and the combustion calculations thereafter by using appropriate equations.

- C is the sixth most abundantly available element in the universe. Naturally occurring C exists in three forms, namely amorphous, graphite and diamond.

  Elemental C in UA is the sum of all C present in both FC and VM of PA.

  C is the principal element to contribute to the cv of coal. The others are H_2 and S but their contributions are small.

  During combustion, C turns into CO_2, which is one of the principal GHGs contributing to the global warming, and hence becoming objectionable.
- H_2 in coal is small and usually under 5% on daf basis. Most H_2 turns into H_2O vapour on combustion.
- S is always present in coal and it can be in both *inorganic* and *organic* forms.

The inorganic coal S is mostly in the form of pyritic S in pyrites (FeS_2) or marcasite crystals as FeS_x and as sulphate S in $CaSO_4$ or $FeSO_4$. The sulphate S is usually low (<0.1%) unless the pyrite has been oxidised. Larger pyritic S can be removed by physical cleaning.

The forms of organic S are less well established. Organic S cannot be removed by physical means and existing chemical process are generally of high cost.

S in coal broadly reduces with rank though there are lignites with high amount of S. Usually it varies from ~0.2% to 4% by weight. Coals with S of 8% is less common but are present in certain areas.

S turns into mostly SO_2 and also partly to SO_3 on combustion. Even though the absolute amounts of these gases may be low they cause the maximum damage by way of low-temperature corrosion. S is therefore the most undesirable element in coal.

- N_2 being an inert material it does not participate in combustion. At high temperature it combines with O_2 in air and produces traces of fuel NO_x. Whereas it has no effect on boiler the environment is seriously polluted. Low temperature combustion and denitrification are the ways to control the problem.
- The higher the O_2 the lower is the rank and Gcv of fuel.

O_2 content in the ultimate analysis (UA) is always determined by difference, that is $O_2 = 100 - C + H_2 + N_2 + S$.

O_2 reduces the air requirement for combustion.

3.37.5 Analysis Reporting Method

Coal analysis is reported in several ways

a. *As received basis*: This is the analysis as received in plant and sent to the laboratory directly.
b. *Air dried basis*: This is the usual way of reporting the analysis. The coal is ground and sieved through 0.2 mm screen and powder is spread thinly and air dried when the M attains equilibrium with the ambient humidity. With surface M gone what is reported in this analysis is the inherent/interstitial M.
c. *Dry or M-free basis*: This is the analysis without M and is obtained by multiplying VM, FC and ash by

$(100/100 - M)$ for analysis on as received basis or

$$\frac{100}{100\text{-inherent moisture}}$$ for analysis on air dried basis.

d. daf basis:

This is the VM% and FC% without the burden of M and A. This signifies the pure burning portion of the fuel and is obtained by multiplying VM and FC by $(100/100 - M - A)$.

The bar graph of PA in Figure 3.25 brings out these aspects very clearly. The UA corresponding to the daf fuel is also indicated alongside. The difference between the C in UA and FC can be noticed—C is FC + carbonaceous content of VM.

3.37.6 Coal Calorific Value (see *also* Thermodynamic Properties)

cv of coal is determined in the laboratory by **bomb calorimeter** in a specified manner. For a quick and close estimate it is also possible to calculate the Gcv.

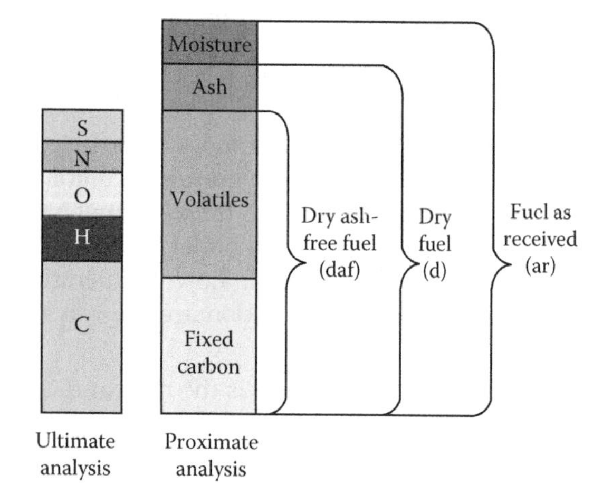

FIGURE 3.25
Various methods of reporting of coals analysis.

For anthracite and bituminous coals (not for lignites) Gcv can be calculated using Dulong formula from the UA which is accurate within 2–3% of the bomb calorimeter's results. The heating value (hv) of coal is due to C, H_2, O_2 and S and also because of heats of dissociation and other phenomena which cannot be captured correctly leading to the inaccuracy of the result of the formula.

$$\text{Gcv} = 80.8\,\text{C} + 344.9\,(\text{H–O}/8) + 22.5\,\text{S in kcal/kg} \quad (3.1)$$

$$= 145.44\,\text{C} + 620.28\,(\text{H–O}/8) + 40.5\,\text{S in Btu/lb} \quad (3.2)$$

where H, O and S are % in UA.

Heat produced on combustion depends largely on the C, H_2 and O_2 present in the coal and, to a lesser extent, on the S and these components vary by coal rank.

- C, by far the major component of coal, is the principal source of heat, generating ~33.83 MJ/kg (8080 kcal/kg or 14,450 Btu/lb). Typical C on daf basis ranges from >60% for lignite to >80% for anthracite.
- Although H_2 generates large amount of heat at ~144.3 MJ/kg (~34,490 kcal/kg or 62,030 Btu/lb) it accounts for <5% on daf basis and not all of this is available for heat as most of the H_2 combines with O_2 to form H_2O vapour.
- The higher the O_2 the lower is its cv. This inverse relationship occurs because O_2 in coal is bound to the C and has, therefore, already partially oxidised C, decreasing its ability to generate heat.
- Heat contributed by S is small, because the cv of S is only 9.42 MJ/kg (~2250 kcal/kg or 4,050 Btu/lb) and S generally averages 1– 2% by weight.

This explains why Dulong formula has its limitations.

3.37.7 Coal Grindability Index (see Hardgrove Index Hardgrove Index)

3.37.8 Caking or Agglomerating Coals (see *also* Coking Coals)

Certain bituminous coals swell on heating, become plastic and fuse together to form cakes. Such coals are called *caking coals*.

When caking coals form large masses, they disrupt the combustion on grate or bubbling bed. Such coals need extreme care or blending when used in stoker and BFB firing. In PF and CFBCs, the caking properties do not hinder the combustion because the fuel is in flight and does not rest on a grate or bed.

Caking is determined by a *free swelling index test* in which coal is heated on a flame in specified manner till M and VM are driven off. If the heated sample does not remain as powder but fuses together to form coke button, the coal is said to possess caking properties. Depending on the cross-sectional area of coke button, its free-swelling index is decided in a scale of 1–9 by comparison. Figure 3.26 shows the coke profiles from standard swelling test.

Free burning coals are the opposite of caking coals. Unlike caking coals that become plastic and fuse together to form cakes free burning coals do not fuse together to form buttons

Caking coals are different from *coking coals* which are bituminous coals high in FC and produce metallurgical coke.

3.37.9 Coking Coals

Certain varieties of bituminous coals, when heated out of contact with air, lose their VM to form hard dense coke consisting entirely of FC and ash of the original coal. Such coals make good coke required for metallurgical purposes and are known as coking coals.

Because of high FC, coking coals produce high temperatures during their combustion.

All coking coals exhibit small tendencies of caking, that is, swelling on heating but all coking coals do not necessarily make good caking coals.

Sub-bituminous coals, lignites and anthracites are non-coking coals.

Table 3.7 presents a list of relevant ASTM standards for coals.

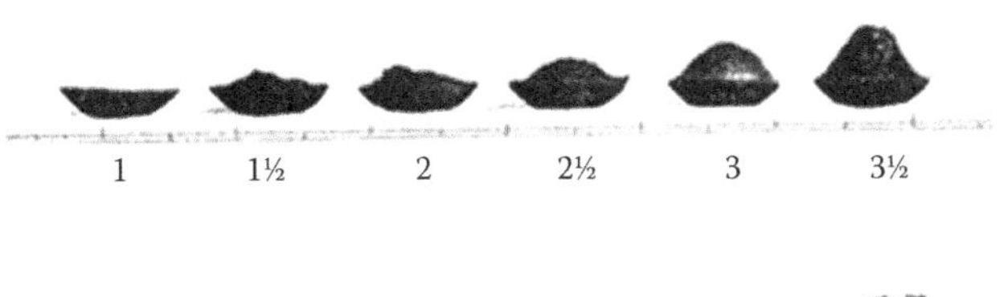

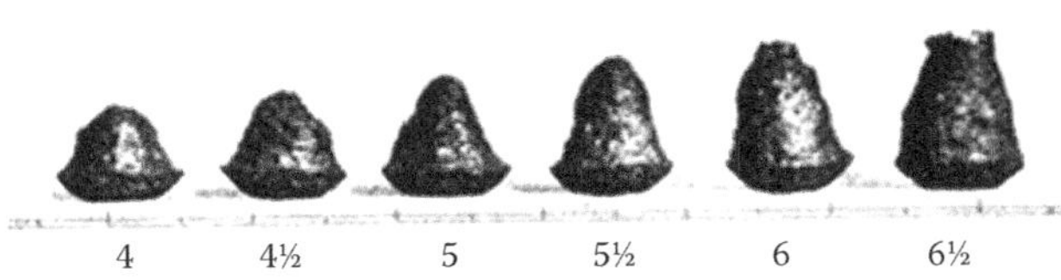

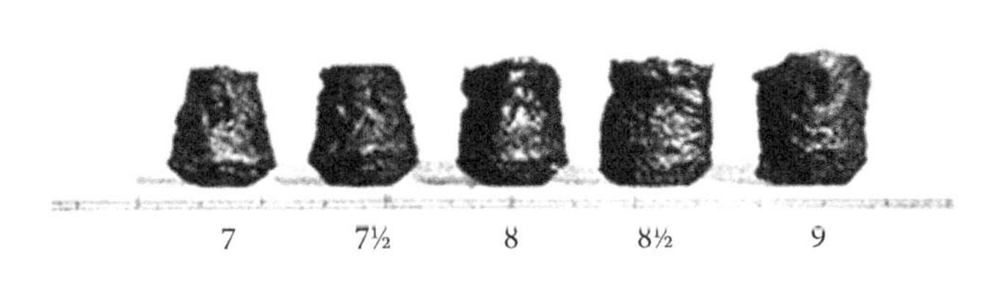

FIGURE 3.26
Coke profiles from swelling test.

TABLE 3.7
List of ASTM Standards for Coals

D121	Coal and coke
D197	Sampling and fineness test of powdered coal
D271	Sampling and analysis of coal and coke
D291	Cubic foot weight of crushed bituminous coal
D310	Size of anthracite
D311	Sieve analysis of crushed bituminous coal
D323	Perforated-plate sieves for testing purpose
D388	Classification of coals by rank
D407	Gross cv and net cv of solid and liquid fuels
D409	Test method for grindability of coal by the Hardgrove machine method
D410	Screen analysis of coal
D431	Size of coal designated from its screen analysis
D440	Drop shatter test for coal
D441	Tumbler test for coal
D492	Sampling coals classified according to ash content
D547	Dustiness index of coal and coke
D720	Test method for free-swelling index of coal
D1412	Test method for equilibrium M of coal at 96%
D1756	Carbon dioxide in coal
D1757	Sulphur in coal ash
D1812	Plastic properties of coal by the Gieseler plastometer
D1857	Fusibility of coal ash
D2013	Method of preparing coal samples for analysis
D2014	Expansion or contraction of coal by the sole-heated oven
D2015	Test method for gross cv of coal and coke by the adiabatic bomb calorimeter
D2234	Test methods for collection of a gross sample of coal
D2361	Test method for chlorine in coal
D2492	Sulphur forms in coal
D2639	Plastic properties of coal by the automatic Gieseler plastometer
D2796	Lithological classes and physical components of coal
D3173	Test method for M in the analysis sample of coal and coke
D3174	Test method for ash in the analysis sample of coal and coke
D3175	Test method for VM in the analysis sample of coal and coke
D3176	Practice for UA of coal and coke
D3177	Test methods for total sulphur in the analysis sample of coal and coke
D3302	Test method for total M in coal
D3286	Test method for gross cv of coal and coke by the isoperibol bomb calorimeter
D3682	Test method for major and minor elements in coal and coke ash by atomic absorption
D4239	Test methods for sulphur in the analysis sample of coal and coke using high-temperature tube furnace
D4208	Test method for total chlorine in coal by the oxygen bomb combustion/ion-selective electrode method
D4326	Test method for major and minor elements in coal and coke ash by x-ray fluorescence
D4371	Test method for determining the washability characteristics of coal
D4749	Test method for performing the sieve analysis of coal and designating coal size
D5142	Test methods for proximate analysis of the analysis sample of coal and coke by instrumental procedures
E121	Wire cloth sieves for testing purpose
E323	Perforated-plate sieves for testing purpose

3.37.10 Coal Erosion (*see* Wear)

3.38 Coal Ash (*see* Ash, Oil Ash)

3.39 Coal Bed Methane Coal Bed Methane (*see* Gaseous Fuels)

3.40 Coal Pipes (*see* Pulverised Fuel Piping)

3.41 Coal Tar (*see* Liquid Fuels)

3.42 Co-Firing/Co-Combustion

Co-firing or co-combustion is the burning of two or more different fuels at the same time in a boiler. It is basically the same as multi-fuel firing. But in today's context of reducing pollutants it has come to mean mainly biomass being burned with a fossil fuel, the most common fossil fuel being coal. In other words, co-firing is supplementing coal with biomass in a coal-fired boiler.

Co-firing can be direct and indirect. Direct co-firing involves firing the fuels simultaneously in the same boiler while in indirect co-firing the supplementary fuel is burnt separately.

- *Direct co-firing* is done two ways, namely
  - Blending biomass and coal in the fuel handling system and feeding the blend to the boiler.
  - Having separate fuel handling systems along with separate burners for biomass. This way there is no contact with the main coal delivery and firing system.
- *Indirect co-firing* also can be done in two ways, namely
  - Thermal conversion of biomass to gas and co-firing the converted fuel with the main fuel.
  - Separate firing of the supplementary fuels and guide the resulting gases to mix with the main flue gases in the boiler.

Co-firing, is beneficial in a number of ways.

- CO_2 reduction is a big benefit when a fossil fuel and bio-fuel are fired together.
- NO_x reduction takes place as NO_x from bio-fuels is low.
- SO_x emission is lowered because of very low sulphur content in biomass.
- Overall fuel costs are lowered as bio-fuels are cheaper to source and burn.
- It is a cost-effective way of using renewable fuels and conserving fossil fuels.
- Combustion quality is improved as the bio-fuels have greater amounts of VM and lower FC which is just the opposite of coal.

Co-firing is always not easy to adopt as there are a number of factors which have to be carefully assessed regarding the compatibility of the dissimilar fuels and the benefits that are likely to accrue.

- Problems are usually related to fuel preparation, storage, delivery, quality and quantity.
- Fouling of the boiler HSs due to low-melting substances in bio-fuels is a serious matter.
- Increased ash deposition on HSs of a boiler calls for more frequent soot blowing.
- Increased corrosion rate of high-temperature components is another likely problem.
- Negative impact on flue gas cleaning (SCR) if any has to be carefully evaluated.

Co-firing needs time to smoothen out problems associated with O&M before it can be made successful.

Co-firing is gradually gaining acceptance world wide as a way of reducing harmful pollutants, increasing the use of renewable fuels and reducing the cost of operations all put together.

3.43 Carbon Monoxide Gas (*see* Gaseous Fuels)

3.44 Carbon Monoxide Boilers (*see* Waste Heat Boilers)

3.45 Cogen Cycle (*see* Thermodynamic Cycles)

3.46 Coke

Coke is made only from coking coals in coke ovens. *Coke is the devolatilised coking coal produced by heating coal in absence of air.* VM and M are driven out leaving behind the FC, A and S to form a hard substance called coke.

The process of decomposing coal into the gaseous and solid fractions is known as the *destructive distillation or carbonisation*. Only some coals with certain resinous substances can become plastic and fuse together to form coke. The quality of coke varies based on the coal properties, ps & ts employed and the type of coke oven.

High-temperature distillation is carried out between 900°C and 1100°C in by-product ovens while low-temperature carbonisation is done at 500–700°C.

At higher temperatures there is a greater evolution of gases producing metallurgical and gas making coke. Owing to the high FC the coke produces a smokeless high-temperature burning which is required in mostly blast furnaces and to a much smaller extent in pig iron cupolas.

Low-temperature carbonisation yields coke more suitable for domestic use whose production and consumption has declined over years.

Table 3.8 shows coke from high- and low-temperature carbonisation.

TABLE 3.8

Coke from High- and Low-Temperature Carbonisation

		High-Temperature Carbonisation By-Product Oven	Low-Temperature Carbonisation
Temperature range	C	900–1100	500–700
Type of coke		Hard	Not so hard
Coke yield/100 kg coal	kg	65–75	75–80
VM in coke	%	1–3	5–15
Gas/te of coke	M^3	280–370	110–250
LFO and tar/te of coke	l	40–65	70–190
Light fuel oil $((NH_4))_2SO_4$/te of coke	kg	11–12	5–7
Gcv	kcal/m^3	4700–5300	6000–9800

3.47 Coke Breeze

Approximately 5% of coke is undersized at <16 mm which is unsuitable for metallurgical industry and is used for mainly steam generation. Coke breeze is very abrasive as it contains higher ash than the rest of coke. As it has practically no VM the ignition requires a high temperature.

Coke breeze is best fired by MB on gravity fed CG/TG along with high-volatile coal in a sandwich manner. Erosion of firing equipments is limited as there are less moving parts and erosion of PP is the minimum as ash carryover in flue gas is low.

Table 3.9 provides typical PA of coke breeze.

TABLE 3.9

Typical Proximate Analysis of Coke Breeze

	Typical%
M	7.3
Ash	11.0
VM	2.3
FC	79.4
Total	100.0
Gcv MJ/kg	~27.1
Gcv kcal/kg	~6480
Gcv Btu/lb	~11,660

3.48 Coke Oven Gas (*see* Gaseous Fuels)

3.49 Coke Oven Gas-Fired Boilers (*see* Waste Heat Recovery Boilers)

3.50 Coking Coal (*see* Coal)

3.51 Cold Drawn Seamless Cold Drawn Seamless (CDS) Tubes (*see* Materials of Construction)

3.52 Combination (Combo) Boilers (*see* Hybrid Boilers in Smoke Tube Boilers)

3.53 Combined Cycle (*see* Thermodynamic Cycles)

3.54 Combined Cycle Power Plant or Combined Cycle Gas Turbine

Modern high η gas-based power generation is the result of a combination of GT and steam cycles—Brayton cycle superimposed on Rankine cycle. GT generates primary power and ST generates additional power from steam produced from GT exhaust gases. Figure 3.27 depicts a typical CCPP.

This is thermodynamically the most efficient form of power generation with cycle efficiencies touching nearly 60% on Ncv basis with large GTs operating at high temperatures and compression ratios.

Figure 8.35 shows the components and schematic arrangement of a CCPP. Typical layout of a CCPP is also shown here in Figure 3.28 taken from a recent catalogue of Siemens. Heat recovery steam generator (Hrsg) shown in this picture is a horizontal unit. Both GT and ST drive a single Gen. There is no bypass stack.

Figure 3.28b depicts the approximate corresponding heat balance. GT is typically a *9 FA* unit with firing temperatures of ~1300°C and gas exhaust temperature of ~600°C (Table 19.5). GT produces ~31% power (~220 Mwe) while ST produces ~17% (~120 MWe). Together they generate ~340 Mwe depending upon the ambient

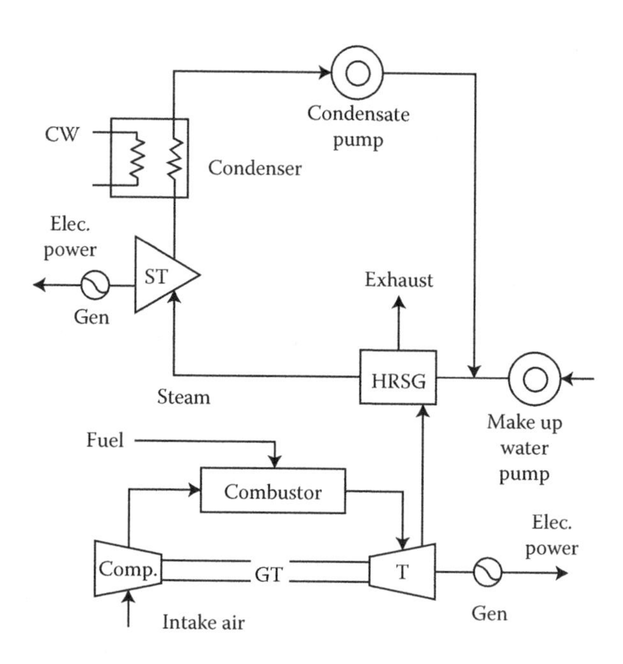

FIGURE 3.27
Schematic of a CC power plant.

conditions. Hrsg here operates between ~600°C and ~100°C recovering 47% of total energy in 3P+ RH configuration. Even in a CC plant the condenser losses can be as high as ~29% and exhaust gas loss ~20% of the total heat input. Compared to a conventional plant there is reduction of ~40% in cooling water for condenser. The overall η of the CCPP is ~48%.

3.55 Combined Heat Transfer (*see* Heat Transfer)

3.56 Combustion

Combustion is a

- Self-sustaining
- Rapid
- Exothermic (heat producing)
- Chemical reaction

To make itself sustaining, the reaction temperature should exceed the ignition temperature of the fuel.

The sub-topics under the main topic of combustion are listed below.

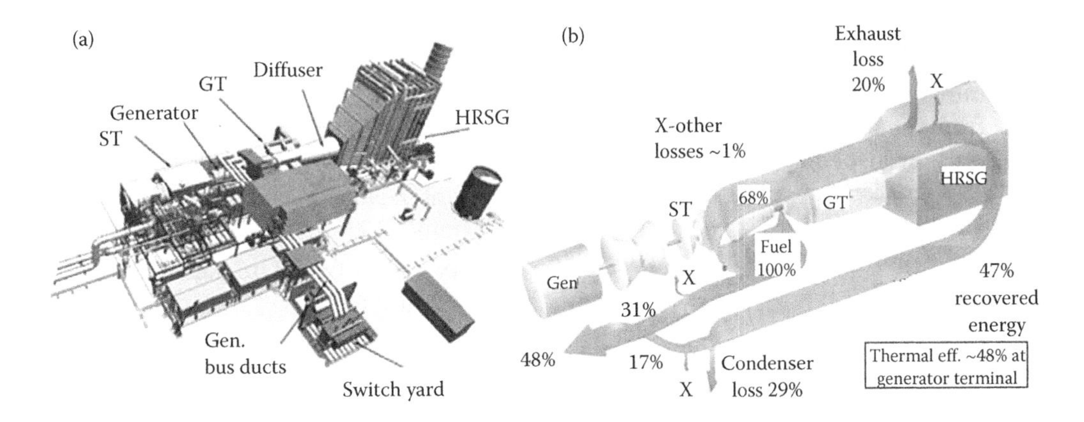

FIGURE 3.28
Large CCPP with in line GT, ST and common TG (a) isometric view (b) heat balance.

1. Combustible
2. Ignition temperature
3. Adiabatic flame temperature
4. Stages of combustion
5. Combustion air
6. Air infiltration or air ingress in boiler setting
7. Primary, secondary and tertiary airs (PA, SA and TA)
8. Combustion calculations
9. Heat of combustion
10. Three Ts
11. Combustion reaction
12. Dissociation
13. Spontaneous combustion

3.56.1 Combustible

Combustible is a substance that *has great affinity for* O_2. A fuel is made substantially of various combustibles. When combustibles react with O_2 (usually from air), combustion results and heat is produced.

The strength of affinity of a combustible towards O_2 determines the speed of combustion.

This oxidation speed can vary enormously—very slow as in the case of rust formation or instantaneous as in the case of gas firing.

Combustion is complete only when all combustibles have reacted with O_2 to the full extent to which they are capable of reacting. This is important to note, as C can react to form both CO and CO_2 but the combustion is complete only when all C is converted to CO_2.

There are only three combustibles in fossil fuels, namely, C, H_2 and S of which S is of minor significance as the heat produced is small in comparison, as shown in Table 3.10. Full list of combustibles appears in Table 3.11.

TABLE 3.10
Selected Combustibles in Fossil Fuels and Heats of Combustion

	Heat Produced			
Combustible	**MJ/kg**	**kcal/kg**	**Btu/lb**	**Product**
C	32.8	7836	14093	CO_2
H	120.0	28666	51558	H_2O
S	9.265	2213	3980	SO_2

- C and H are beneficial constituents producing good amount of heat.
- S is the most undesirable because it not only produces the least amount of heat but also causes the maximum harm to the heating surfaces by the way of **low-temperature corrosion**.

3.56.2 Ignition Temperature

Ignition temperature is *the minimum temperature required to be attained by combustible to sustain the combustion reaction*. Then the heat produced by the chemical reaction

TABLE 3.11
Combustibles and Their Ignition Temperatures

Combustible	Ignition Temperature (°C)
FC Bit coal	408
FC Semi-Bit coal	465
FC anthracite	450–600
S	243
Kerosene	255–295
Gasolene	260–430
Acetylene C_2H_2	304–440
Ethylene C_2H_4	480–550
Ethane C_2H_6	470–575
H_2	575–590
CO	610–655
CH_4	630–750

should be more than the heat lost to the surroundings to ensure the continuation of combustion.

Ignition temperatures of gases of VM are higher than FC. They distil off from the fuel at relatively much lower temperatures and burn in suspension in the open space of furnace. Hence the ignition temperature of a fuel is considered same as that of its FC.

Table 3.8 gives the ignition temperatures of various combustibles.

3.56.3 Adiabatic Flame Temperature

Adiabatic flame temperature is the theoretical temperature that would be attained by the products of combustion if the

- Entire chemical energy
- Sensible heat content of the fuel above the datum temperature

were transferred to the products on combustion. This assumes that

- There is no heat loss to surroundings
- There is no **dissociation**

3.56.4 Stages of Combustion

Every combustion proceeds in three distinct but somewhat overlapping stages

- *Drying* where the M leaves the fuel instantly due to the heat of furnace and hot air (flash drying).
- *Burning of VM* in the furnace chamber drawing heat from furnace radiation and the hot gases. The various constituents of VM distil from the fuel progressively and burn in the furnace rapidly.
- *Burning of FC* takes place on the grate surface or fluidised bed or furnace in case of PF boilers. The dry devolatilised FC burns slowly, releasing maximum heat.

3.56.5 Combustion Air

Combustion air is the *air that reacts with fuel in combustion reaction*. It is different from *tramp air or leakage air* which enters the combustion air as the ingress but does not participate in combustion.

Dry air is treated as a mixture (not a compound) of mainly N_2 and O_2 in the ratio of

- 79:21 by volume
- 77:23 by weight

Small amounts of CO_2 and inert gases such as Ar are treated as part of N_2.

Dry air has

- An apparent MW of 29 kg/kg mol
- ρ of 1.29 kg/m^3 or 0.0805 lb/cft at NTP conditions (760 mmHg, 0°C)
- sp. vol. of 0.7728 m^3/kg or 12.38 cft/lb at NTP conditions
- Relative ρ (with respect to H_2 of 1) of 14.38

Dry air/unit O_2 is 4.32 kg/kg.

All combustion calculations are performed on dry air basis and humidity is added separately to arrive at the wet air.

3.56.6 Air Infiltration or Air Ingress in Boiler Setting

Whenever a boiler operates in **balanced draught** mode, that is, with furnace at slight negative pressure, tramp air infiltration into the downstream setting is a strong possibility. Membrane panel construction can prevent the air leakage substantially. But it is in expansion joints that leakage is difficult to prevent, at least after some time of operation. Closer to the ID fans the tendency for the air leakage is more due to heavy suction.

Tramp air leakage is detrimental as it does not participate in combustion but contributes to the stack loss. The thermal η is lowered and also the ID fans are required to be made larger, with corresponding increase in the auxiliary fan power.

Another source is the tramp air leakage around the combustion equipment such as the grates.

Over the years the air infiltration has been brought down by adopting better wall construction techniques. **Refractory, spaced, tangent** and **membrane walls,** which were progressively developed over decades, are all superior to each other and have reduced the furnace infiltration to the minimum. Likewise, the superior constructional features of all **dampers, expansion joints, duct** work and **casing** have contributed to lowering of tramp air.

From furnace to ID fan it was not unusual to have 15–20% air leakage in the old boilers with spaced tube construction. This is reduced to ~5% in very large modern boilers with full membrane walls and welded ductwork.

Air infiltration does not take place when boilers operate in pressurised mode with only FD and PA fans and no ID fans.

3.56.7 Primary, Secondary and Tertiary Airs

Fuel and air have to mix intimately for proper combustion. Also air has to be provided where combustion occurs. All fuels are composed of VM and FC. VM burns quickly and in openness of furnace while FC needs long time and burns on the grate, fluid bed or in the inside swirls of burner flames. Combustion air is apportioned suitably by splitting into PA and SA broadly for burning FC and VM, respectively. Usually SA is at HP varying from 500 to 750 mm wg (20–30″ wg) in grate and FBC firing for penetrating the furnace. TA is nothing but upper SA provided for capturing and burning the escaping fuel and also to provide a curtain of air to reduce dust carry over.

Staged combustion is another reason for air split. In staged combustion by providing air in stages the combustion, starved of full air, is not complete and the attendant combustion temperature is low. This is a technique to lower the NO_x generation. Additionally, in CFBC in-bed desulphurisation can also be achieved by further lowering the temperature to ~850°C with further aid of ash re-circulation.

SA is also called overfire air (OFA).

SOFA is a pre-combustion stage technique adopted in boilers to attain lower levels of NO_x. Keeping the same amount of total air a part of the SA/OFA is separated and provided at an even higher level than SA and TA, lowering the flame temperatures and protracting the combustion even more. This is discussed separately under **NO_x** in flue gas.

3.56.8 Combustion Calculations

Combustion calculations are always based on

- *Dry air* which is calculated purely from the O_2 requirements of all the combustibles in the fuel but
- *Wet air* is what enters combustion reaction as the ambient air is never free of *humidity*

So wet air = dry air + humidity in air

Psychrometric chart gives the humidity in air at various temperatures and other related properties.

Stoichiometric air (with 0% excess air) is the theoretical air required for combustion of fuel. This can be calculated from the fuel composition.

However in actual practice combustion reaction needs slightly more air which is called the *excess air.*

Excess air depends on the fuel, firing equipment, combustion temperature and so on, and is indicated as a % of the theoretical air.

Table 3.12 gives the typical range of excess air required for various fuels.

Excess air control is very important. Higher than required excess air increases the stack losses while lower excess air increases the unburnt loss. Both ways the boiler η is reduced. The optimum excess air lies somewhere within the limits set in Table 3.12 which has to be discovered by trial and error during the actual boiler operation. This is shown in Figure 5.6.

TABLE 3.12

Excess Air Required for Various Fuels and Firing Equipments

Fuel	Firing	Excess Air by Weight at Full Load (%)
PF	Water-cooled furnace	15–20
Coal	Stoker	30–45
FO	Register burner	3–15
NG, RG,COG	Register burner	3–10
BFG	Scroll burner	15–20
Bagasse	All grates	25–35
BL	Recovery furnace	5–7

Excess air, as % of part load air, increases. This is because the leakage areas and consequently the tramp air ingress practically remain the same while the combustion air is reduced. As % of air at part load it increases as compared to full load.

Although higher excess air lowers the boiler η it is very helpful in increasing the steam temperatures of SH and RH. In many designs part load excess air is increased to meet the temperature control range instead of adding the HS. Excess air control is thus a double edged weapon.

3.56.9 Heat of Combustion

This is the amount of heat produced on combustion of a combustible.

Table 3.13 gives the heats of combustion for select combustibles on Gcv basis. These hvs are measured by **bomb calorimeter** for solid and liquid fuels in which the combustible substances are burned to completion in a constant volume of O_2. Calorimeters of continuous or constant flow are used for gaseous fuels.

A comprehensive table of combustion constants is provided in Appendix 1.

3.56.10 Three Ts

The three classical factors contributing to the completeness of combustion are as shown in Figure 3.29.

1. *Time*: Combustion reaction needs adequate time for completion which is to be provided between furnace and firing equipment.
2. *Temperature*: Fuel should first reach the ignition temperature for the initiation of combustion. Thereafter, there has to be enough supply of

TABLE 3.13

Selected Combustibles and Heats of Combustion

Combustible	Symbol	Molecular Weight	Combustion Reaction	Heat (GCV) MJ/kg	kcal/kg	Btu/lb
Carbon	C	12	$C + O_2 = CO_2$	32.79	7833	14,100
Hydrogen	H	2	$H_2 + O = H_2O$	142.12	33,945	61,100
Sulphur	S	32	$S + O_2 = SO_2$	9.26	2212	3983
Hydrogen sulphide	H_2S	34	$H_2S + 1.5O_2 = SO_2 + H_2O$	16.51	3944	7100
Methane	CH_4	16	$CH_4 + 2O_2 = CO_2 + 2H_2O$	55.54	13,266	23,879
Ethane	C_2H_6	30	$C_2H_6 + 3O_2 = 2CO_2 + 3H_2O$	51.91	12,400	22,320
Propane	C_3H_8	44	$C_3H_8 + 4H_2O = 3CO_2 + 4H_2O$	50.38	12,033	21,660
Butane	C_4H_{10}	58	$C_4H_{10} + 5H_2O = 4CO_2 + 5H_2O$	49.56	11,838	21,308
Pentane	C_5H_{12}	72	$C_5H_{12} + 6O_2 = 5CO_2 + 6H_2O$	49.06	11,717	21,091

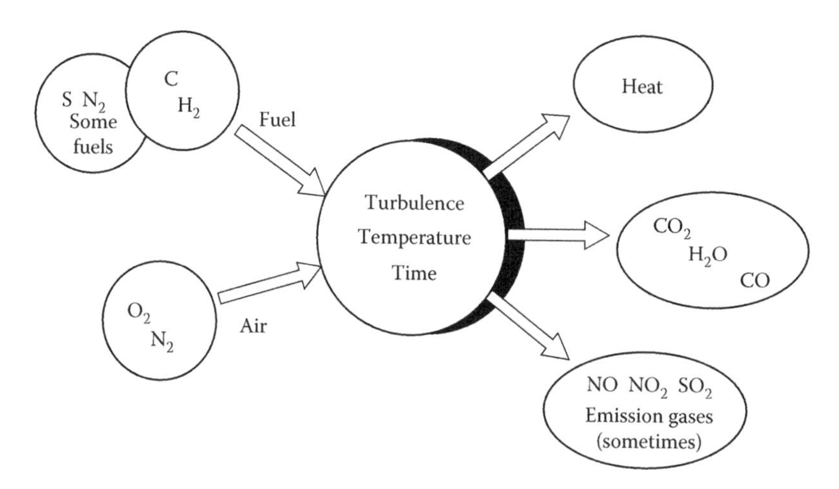

FIGURE 3.29
Combustion process of fossil fuel.

hot air and radiation from furnace to speed up the combustion reaction rate and generate high-flame temperature. This in turn accelerates the combustion further.

3. *Turbulence*: The particles of air and fuel should intimately mix with each other for any combustion to rapidly progress. Thereafter the ash particles residing on the fuel must be scrubbed away to expose fresh surface of fuel for the continuation of reaction. Both are possible when there is an intense turbulence in the flows.

3.56.11 Combustion Reaction

Combustion occurs when fuels react with O_2 in the air to produce heat and combustion products. Fossil fuels are essentially HC, meaning they are composed primarily of C and H. When fossil fuels are burnt CO_2 and H_2O are the principal chemical products, formed from the reactants C and H in the fuel and O_2 in the air. The simplest example of HC fuel combustion is the reaction of CH_4, the largest component of NG, with O_2 in the air. When this reaction is balanced, or stoichiometric, each molecule of CH_4 reacts with two molecules of O_2 producing one molecule of CO_2 and two molecules of H_2O. When this occurs, energy is released as heat.

$$CH_4 + 2O_2 \rightarrow CO_2 + 2H_2O \quad (3.3)$$

$$\text{Reactants} \rightarrow \text{Products} + \text{Heat}$$

Heat of combustion for various combustibles is listed in Table 3.13.

In actual combustion processes, other products are also often formed. A typical example of an actual combustion process is shown in Figure 3.29.

The combining of O_2 in the air and C in the fuel to form CO_2 and generate heat is a complex process, requiring

- The right mixing *turbulence*
- Sufficient activation *temperature*

- Enough *time* for the reactants to come into contact and combine

as it has been explained earlier.

Unless combustion is properly controlled, high concentrations of undesirable products can form. CO and soot, for example, result from poor mixing of fuel and air or too little air. Other undesirable products, such as **nitrogen oxides** (NO, NO_2), form in excessive amounts when the flame temperature is too high. If a fuel contains S, SO_2 gas is formed. For solid fuels, such as coal and wood, ash forms from incombustible materials in the fuel.

3.56.12 Dissociation

Dissociation is the breakdown of CO_2 and H_2O into their elements at temperatures >1650°C when they are present in excess of ~10% and 3%, respectively.

It is a kind of reverse combustion, with endothermic (heat consuming) reactions consuming 2414 and 33,945 kcal/kg of heat respectively, which lower the flame temperature.

Once the temperatures are sufficiently low, again the elements reunite and generate heat.

There is no loss of heat but a lowering of flame temperature in dissociation.

3.56.13 Spontaneous Combustion (*see* Spontaneous Combustion)

3.57 Combustion Turbines (*see* Gas Turbines)

3.58 Combustor (in Fluidised-Bed Boilers) (*see* Circulating Fluidised-Bed Combustion)

3.59 Commissioning of Boilers

After the erection of boiler is complete, it has to be commissioned to make it suitable for commercial operation. Commissioning involves a set of systematic procedures of cleaning the boiler PPs drying refractory and conducting trial runs of various auxiliaries involving definite mile stones so that the boiler and its auxiliaries are made available for safe, smooth and reliable operation.

By completion of erection it is meant that substantial installation of mechanical, electrical, instrumentation, civil and refractory works is finished. Mechanical completion forms the bulk of the task and it is marked by the completion of mainly

1. HT of PPs
2. Air leak test of flues and ducts
3. Alignment tests of auxiliaries

Commissioning activity starts thereafter while the tail end erection jobs would go on for a few more weeks. Insulation and refractory applications cannot start until the HT of PPs is complete. Insulation and lagging is almost the last erection task to be finished which would go on along with commissioning till end.

Boiler commissioning involves the following important activities:

1. Commissioning all auxiliaries and firing equipment
2. Alkali boil out of PPs
3. Chemical cleaning of PPs
4. Steam blowing of SH, RH and steam piping
5. SV setting

Main activity revolves around cleaning the internals of PPs as uncontaminated steam can only come from very clean internal surfaces. Improper steam quality injures both boiler and ST.

- Alkali boil out is to remove oily materials and dirt.
- Acid cleaning is to clean up mill and other strongly adhering scales.
- Steam blowing is to clean SH, RH and piping of dirt and mill scales.

3.59.1 Alkali Boil Out

Alkali boil out is primarily for

- Removing oil, grease and loose dirt from the internal surfaces
- Depositing a protective layer of magnetite Fe_3O_4 on the cleaned surfaces

This is the first activity on boiler PPs and is taken up after the HT of the boiler is completed and the oil burners are commissioned. This will also be the first lighting up of a new boiler.

Boiler is first of all inspected from inside thoroughly and all loose erection materials removed prior to the light up. Minimum number of auxiliaries, instruments and safeties as required for boil out are commissioned. The drum internals are placed inside the drum to let them get cleaned and get the magnetite deposit. They are not fitted in place to help the inside of the drum to get cleaned and coated unless they are simple baffle type separators which do not hinder the inside of the inside of the drum from getting the magnetite coating. The magnetite layer should cover the entire surface with no break lest the corrosive products should attack the steel surface and initiate corrosion.

Chemicals for boil out are Na-based and can vary, as shown in Table 3.14, depending on the plant practice and the recommendations of the boiler maker. Additionally, detergent is also added at the rate of 0.25 kg/te of water.

The boil out procedures may vary in some detail depending on the type of boiler and the practice of the boiler manufacturer. Usually the alkaline boil out lasts for 4 days.

- Boil out chemicals are required to be first mixed in water to make proper solution which is then pumped into the steam drum by chemical feed pumps. They should not be added in solid form.
- In smaller boilers the chemical solution can be fed directly into the drum where the water level is at least 150–200 mm.
- Slow firing is started after filling the drum with water till its normal level. Extreme care is needed during this period as it is the first firing of the unit.
- When steam comes out freely the vents are closed and boiler pressure is slowly raised to ~15% of the operating pressure in ~8 h. After ~4 h of soaking at this pressure the fire is put out and the boiler is allowed to cool overnight to precipitate the sludge to the bottom.
- On the following day the boiler is first drained through the bottom drains till the water level in the steam drum reaches the bottom of the gauge glass.
- Then the pressure is built up to ~30% and all operations are repeated like on the first day.
- On the third day the pressure is raised to 50% all operations are repeated like on the first day.
- On the fourth day when the water cools to ~90°C final draining is done to wash away all the chemicals and sludge and draining until all traces of oil disappear from water. The man holes of the steam drum are opened and the boiler is rinsed with clear water and flushed for a few times until the drained water is clear.
- During the boil out samples are taken at regular intervals to check for the drop in chemical concentration in water. When it drops to less than half further chemicals are added to make the up concentration.

TABLE 3.14

Chemicals Used in Chemical Boil Out Process

Mixture	Chemicals	Kilogram of Chemicals/1000 kg Water
1	NaOH	3
	Na_2CO_3	3
2	$Na_3PO_4.12H_2O$	2.5
	NaOH	2.5
3	$Na_3PO_4.12H_2O$	1.5
	NaOH	1.5
	Na_2CO_3	1.5
4	NaOH	3

After completion of rinsing the internal surfaces are examined for the cleanliness of surfaces and deposition of magnetite layer. In the unlikely event of either of them being unsatisfactory the procedure is repeated.

During the boil out the internals of the **SVs** are required to be protected by means of hydrostatic plugs. After the boil out the **gauge glasses** need a thorough cleaning or even change of glass parts.

3.59.2 Acid Cleaning

Internal surfaces of modern boilers can only be cleaned chemically unlike the very old boilers which were amenable to mechanical wire brushing because of the straight tube construction. Chemical cleaning embraces both acid- and alkali-based cleaning. Alkali boil out to remove dirt, mill scale and oily matter is compulsorily carried out in all new boilers as part of start up procedure. *Acid cleaning is primarily for the removal of scales and sludge and also for dirt and mill scale.* Scales are formed because of the precipitation of salts on the inside of tubes. Mill scale, dirt and oily stuff are inherent to the tube making and tube manipulation processes in the tube mills and boiler making shops.

Acid cleaning is done for *new as well as working boiler.* Removal of scales and sludge from the tube surface is of great value as the heat transfer is restored and η is improved. As a new boiler is yet to accumulate any scales and the other impurities are effectively removed by alkali boil out many boiler makers do not recommend acid cleaning of new boilers. It is strongly felt that there is always a danger of the acid dissolving the parent tube material due to the slightest oversight as there is practically no separating scale between the acid and metal. Also there is a fear of acid being trapped in some of the bends and horizontal tubes due to the difficulty of adequate flushing. At any rate acid cleaning is the activity of an expert with great deal of experience which a boiler maker does not possess.

Acid cleaning of a working boiler is relatively safer. The timing of cleaning is best decided by the owner and carried out by an expert. The opinion on acid cleaning of a running boiler is also divided even as it is getting more acceptable.

Sometimes acid cleaning is performed on a new boiler after it has worked for 3–6 weeks at a slightly reduced load of ~75%. During this first operation a lot of mill scale and rust dislodged from condenser and fw system accumulates in the boiler. In the same period because of upsets in water conditions and frequent load changes deposits are formed on the inside of boiler tubes, despite adhering to very good water treatment (WT) procedures.

Acid cleaning enjoys the advantage of

- Speed: It needs normally 3 days and rarely more than 4 days of outage
- Cost effectiveness
- Good cleaning of even the most inaccessible locations

There are four steps in cleaning.

1. Washing of HSs with inhibitor-laden acid solvent to dissolve deposits
2. Flushing the loose deposits and salts with clean water which removes the acid solvent as well
3. Boil out with soda ash solution to neutralise the remnant acid
4. Final rinsing with clean water

Two methods, each with its own advantages, are adopted for chemical cleaning.

a. Circulation method
b. Soaking method

In circulation method the solvent is re-circulated through all circuits until its strength reaches an equilibrium indicating an end to the reaction with deposits when it is drained and bw rinsed. In the soaking method, the solvent is filled and left to soak for 4–8 h depending on the deposits. It is drained away after reaching equilibrium. The former method is considered to be safer, cheaper and more appropriate for cleaning Ecos and SHs where the solvent can reach because of circulation. The latter method has the advantages of simpler piping, less need for supervision and assurance that all parts are reached by the solvent. Figure 3.30 shows the more popular circulation method of acid cleaning.

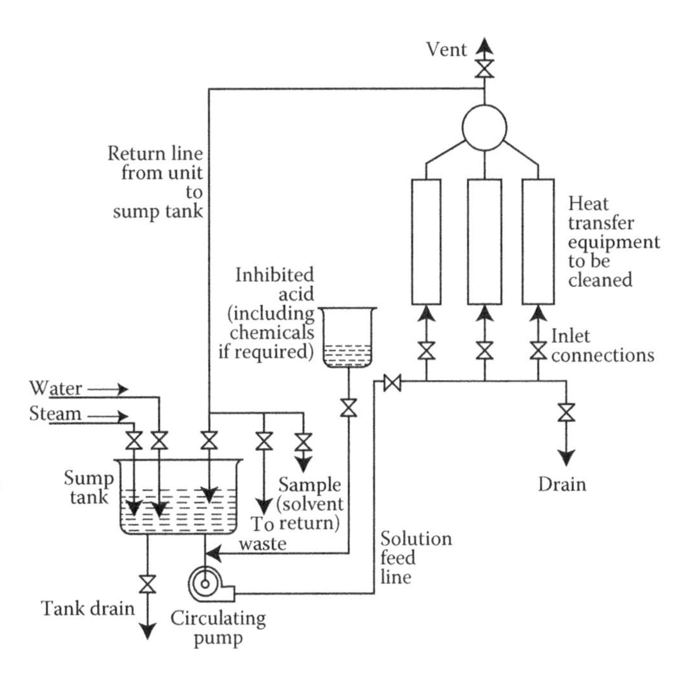

FIGURE 3.30
Circulation method of acid cleaning.

3.59.3 Steam Blowing

The purpose of steam blowing is to *clear the SH and RH coils and the steam lines of any foreign matter remaining inside*. These are the parts usually not cleaned during the alkaline boil out. The impurities are dirt, mill scale, welding waste and corrosion products accumulated during storage. Severe damage to ST can result if such foreign matter is allowed to enter the equipment.

In the running units major PP repairs where the possibility of introduction of foreign material into the system exists, need to be followed by steam blowing.

In order to carryout steam blowing the main firing equipment is required to be substantially completed. In alkaline boil out availability of oil/gas burners was adequate as only slow firing was required. Simultaneously most instruments and controls are also required to be available.

The principle behind the blowing of steam lines is *to loosen the mill scale by the repeated expansion and contraction of steam pipes followed by washing at high velocities*, all with the aid of steam. It is desirable to have nearly the same flow conditions as normal during steam blowing also. As this is not possible when exhausting to the atmosphere a lower pressure of 40–55 bar is used with satisfactory results. Steam blowing should be done before insulating the pipe lines to maximise the effect of thermal cycling. With insulated pipes greater time is needed between blows to allow the system to cool.

Steam blowing of a utility boiler is done in four stages in order not to pile up debris from one line to the other.

Stage I: SHs and main steam lines up to HP stop cum CV at ST inlet.

Stage II: SHs, main steam lines and cold RH lines. During the last few blows the HP bypass lines to condenser are included in parallel. If necessary, HP bypass lines are blown separately.

Stage III: SH, main steam lines, Cold RH lines, RH, Hot RH lines upstream of LP bypass valve.

Stage IV: Extraction lines and miscellaneous auxiliary steam lines.

A temporary steam piping to exhaust steam into the atmosphere is required to be connected at the end of main steam piping for performing steam blowing of stage I. For the stage II the temporary pipe is shifted to the end of cold RH pipe and another temporary connection has to be introduced between main steam and cold RH pipes. For stages III and IV the temporary pipe is shifted to the end of hot RH pipe. A careful engineering of temporary piping is needed by experienced engineers along with suitable protection for over pressure and over temperature. Temporary piping is often equipped with shut off valves which are used also as blowing valves.

There are two methods of blowing the steam lines, namely the continuous and intermittent/puffing methods. Both are followed.

1. *Continuous method:* The procedure here is to raise the boiler pressure to just below the permissible pressure of the temporary atmospheric discharge piping and blow the line while steadily the increasing the firing rate and maintaining the pressure during the blow. When the pressure starts dropping, despite full firing, the stop valve is gradually closed. Several successive blows are made with adequate cooling periods between each blow. This procedure has the advantage of long and steady blowing periods. The inside of steam piping experiences a continuously high-steam velocity because of the high-steam volume (due to LP) which helps in vigorous scraping. When the p & t change, the pipe expands and contracts which helps to loosen the mill scale.

   Usually the waste steam discharged to atmosphere is dark in colour when blowing starts and becomes clear as the steam blowing is complete and the foreign matter is expelled from the system.

2. *Puffing method*: Here the boiler pressure is raised to ~40 ata or just below the permissible pressure in the temporary atmospheric discharge piping and steam blowing is done shutting off the firing. Steam pressure is allowed to decay to 10–15 ata and stop valve is closed. The procedure is repeated until the lines are cleaned up. This method has the advantage that the sharp decay in steam pressure produces a rapid change in saturated steam temperature causing a thermal shock which helps to loosen the adhered scale from the inside of tubes and piping. Also steam expands simultaneously as the pressure drops causing an increase in the steam velocity. The scales are then removed by the high-velocity steam. However all the PPs, including the drum, are subjected to certain levels of thermal shock. To ensure a thorough cleaning a number of suc cessive blows are made.

   In both methods water level in the steam drum fluctuates severely often disappearing from the gauge glasses. A lot of care is needed to handle the boiler with no damage.

Experienced observation decides the adequacy of blowing. A representative of the ST company usually sets the required standards of acceptable indentation on target plates. Clean metal target plates, usually of Al in the early stages, are rigidly secured in the path of the discharging steam at a suitable distance at the end of temporary pipe. After each blow the target plates are inspected for any denting by the debris. ss plates having mirror finish are used in final blows. Steam blowing is suspended only when the target plates are clean and free of any marks. It is an accepted practice to limit the number of blows to ~6 or 8 per day at intervals of 2 h between the consecutive blows and with overnight cooling.

Temporary piping is disconnected and normal routing is restored after steam blowing is complete making the unit ready for the next stage of commissioning, which is the SV floating.

3.59.4 Safety Valve Floating

Readiness for commercial operation for a new boiler is marked on completing the mile stone of floating of SVs. This is witnessed by an independent third-party inspector from the insurance company or the appropriate government agency. The details are recorded and a formal authorisation to steam the boiler commercially is also accorded.

In running boilers this is done routinely every year or so as laid down by the statute.

SVs are type tested for each model for relieving capacity and individually set for every valve for **lifting and re-seating pressures**. The adjusting rings are locked in position by the valve maker before despatch.

At site the valves are floated once again in position to confirm that their lifting and re-seating pressures are the same as marked on the valve and shown in certificates. Also the sharpness of lifting and re-seating of valves are checked. **Chattering of valve**, if any, is also inspected for. Minor irregularities like these may be induced during transportation and erection. Suitable adjustments are made to eliminate such defects during the course of SV setting. No capacity tests are done.

As the first step all the hydrostatic plugs introduced in the SVs for their protection during the boil out and steam-blowing operations are removed. Pressure gauges in the vicinity of the valves which are used for noting the readings are re-calibrated.

Even though the main firing equipment is in readiness, having performed the steam blowing operation, auxiliary oil/gas guns may be sufficient for SV floating as only pressure testing is done. The usual sequence of setting the SVs is SH, drum and RH. Boiler pressure is raised and the SV with the lowest setting is first blown. During the blowing the firing rate is increased. Lifting and re-seating pressures are noted and compared with the original settings. Adjustments to the **blowdown rings** is made if needed until the desired settings are reached without any chatter. The valve is gagged and the next valve at higher setting is taken up.

RH SV floating is also carried out by admitting steam into RH section with the help of temporary connection made at the time of steam blowing between main steam and cold RH pipes at the ST end.

Floating of SVs formally marks the end of the commissioning of boiler paving the way for the final step of handing over the plant. It is also a major milestone for commercial transaction.

3.60 Compensators (*see* Expansion Joints *under* Draught Plant)

3.61 Composite Tubes (*see* Boiler Tubes *under* Materials of Construction)

3.62 Compressibility (*see* Fluid Properties *under* Fluid Characteristics)

3.63 Compressible Fluid (*see* Types of Fluid *under* Fluid Characteristics)

3.64 Condensate (*see* Water Types in Water)

3.65 Conduction (*see* Heat Transfer)

3.66 Conductivity (for Total Dissolved Solids in Measurement Steam and Water)

Conductivity is the *inverse of resistivity* and is defined as *the ability to conduct or transmit heat, electricity or sound*. Its units are Siemens per meter (S/m) in SI and millimhos per centimeter (mmho/cm) in customary units. Its symbol is k or s. It may mean in the context of boilers

- Electricalconductivity, a measure of a material's ability to conduct electric current
- Electrolytic conductivity, a measurement of an electrolytic solution, such as water
- Ionicconductivity, a measure of the conductivity through ionic charge carriers
- **Thermal conductivity**, the intensive property of a material that indicates its ability to conduct heat

Pure water is not a good conductor of electricity. Because the electrical current is transported by the ions in solution, the conductivity increases as the concentration of ions increases. Typical conductivity of waters is

Ultra pure water 5.5×10^{-6} S/m
Drinking water 0.005–0.05 S/m
Sea water 5 S/m

TDS is a measure of the total ions in solution. Electrolytic conductivity (EC) is actually a measure of the ionic activity of a solution in terms of its capacity to transmit current. In dilute solution, TDS and EC are reasonably comparable. The TDS of a water sample based on the measured EC value can be calculated using the following equation:

$$TDS\ (mg/l\ or\ ppm) = 0.5 \times EC\ (dS/m\ or\ mmho/cm)\ or\ = 0.5 \times 1000 \times EC\ (mS/cm) \quad (3.4)$$

The above relationship can also be used to check the acceptability of water chemical analyses. It does not apply to waste water. As the solution becomes more concentrated (TDS > 1000 mg/l, EC > 2000 ms/cm/mmho/cm), the proximity of the solution ions to each other depresses their activity and consequently their ability to transmit current, although the physical amount of dissolved solids is not affected. At high TDS values, the ratio TDS/EC increases and the relationship tends towards TDS = 0.9 × EC.

3.67 Contact Corrosion of Steel Materials (*see* Corrosion)

3.68 Continuous Blowdown (*see* Boiler Blowdown in Water)

3.69 Continuous Emissions Monitoring Systems (*see* Acid Rain)

3.70 Control Range (*see also* Boiler Control)

There are two ranges for automatic control in boilers, namely

1. Automatic combustion control (ACC)
2. Automatic steam temperature control (STC)

Usually boilers operate in a range 40–100% maximum continuous rating (MCR) in auto-combustion mode. It means that there is no manual intervention needed in this steaming range. This depends on the type of firing equipment and the level of sophistication of I&C. A seamless ACC is possible in **grate** and **CFBC** boilers while it is a stepped control in **PF** and **BFBC** boilers. In PF firing mills are required to be cut in and out while in BFBC bed slumping is involved. A higher turndown of 30–100% is also easily possible in grate and CFB while 40–100% is quite difficult in PF and BFBC boilers. It needs more I&C expense to increase the range.

Approximately 70–100% STC range is generally adequate. Sometimes a range 50–100% is demanded in certain process boilers. Larger **SH**, more **attemperation** and superior metallurgy are the penalties to pay for wider range of STC.

In many **Whrbs** and all **Hrsgs** no ACC or STC ranges are possible as these boilers are only slave units of the upstream process plants or GTs.

3.71 Convection (*see* Heat Transfer)

3.72 Convection Bank (*see* Heating Surfaces)

3.73 Convective Superheater (*see* Superheater Classification in Heating Surfaces)

3.74 Corner Firing (*see* Pulverised Fuel Burner Arrangement)

3.75 Corrosion (*see also* Ash Corrosion)

Corrosion is the *destructive chemical or electrochemical action* of the aggressive chemicals in the surrounding environment over the unprotected metal or alloy surface. Corrosion *results in loss of metal* surface slowly or rapidly, progressively or abruptly. The aggressors can be

O_2, CO_2, SO_2 in gases or liquids or even humidity in air, chemically reacting in a destructive manner to produce salts, oxides, hydroxides and chlorides. *Corrosion happens primarily because the metals and alloys in their refined form are inherently unstable and would want to revert to their stable mineral form,* for which the surrounding aggressors help their cause.

There are several types of corrosion of which only a few are relevant for boilers, which are described here.

Corrosion in boilers is principally categorised as

a. External or ash side corrosion or
b. Internal or water/steam corrosion

The most common type in boilers is the external corrosion because of ash causing both **high- and low-temperature corrosion**, which is covered in 'ash corrosion'. Water-side corrosion is mainly by dissolved O_2 which is explained in 'Pitting' below. The others are described below.

- *Contact corrosion* between the assembled components occurs when different metals or alloys are in contact because of the formation of electrochemical cells between them.
- *Crevice corrosion* occurs under the heads of bolts, nuts and so on because of uneven ventilation. This localised corrosion frequently occurs with small amounts of stagnant solution left by holes, lap joints, gaskets and so on.
- *Erosion corrosion* is a formation of groves, channels, ridges, valleys, and rounded holes caused by the movement of corrosive fluid over metal, where erosion and corrosion act together. Erosion is the mechanical destruction because of the presence of hard particles in movement and the action is purely mechanical, unlike corrosion whose action is chemical or electrochemical. Erosion corrosion is minimised by the
    - Application of coatings such as hard facing, weld overlay or suitable repair
    - Design change involving change of shape, geometry and selection of materials
    - Use of better materials which can withstand this phenomenon
    - Change of process and addition inhibitors
- *Selective corrosion* is said to take place when the constituents are selectively attacked in any alloy, such as zinc in brass.
- *Stress corrosion* is an abrupt failure of a component because of the corrosion when materials are not properly stress relieved or annealed. Cracks are formed underneath the metal with the surface remaining intact. The stress produces abrupt inter-crystalline or trans-crystalline separation without the formation of any corrosion products.
- *Caustic embrittlement or caustic corrosion* in PPs (explained separately) is an example of stress corrosion cracking.
- ss SHs, characteristically used for high p & t, are highly susceptible because of chlorides and O_2 in flue gases. It is also called *chloride stress corrosion*.
- *Surface or uniform corrosion* is the chemical or electrochemical corrosion that takes place over the whole exposed surface. It can be prevented or minimised by appropriate materials selection, protective coating, inhibitor application, and cathodic protection.

Corrosion loss is measured in mm/year or mils/year (mils is 1/1000 of an inch). There are six levels of corrosion attack ranging from 10 μm/year (high resistance) to 10 mm/year (non-resistant).

3.75.1 Pitting or Pitting Corrosion

Pitting corrosion is a localised form of corrosion by which cavities or 'holes' are produced in the material (as shown in Figure 3.31). Pitting is considered more dangerous than uniform corrosion damage because it is more difficult to detect and predict. Corrosion products often cover the pits. A small, narrow pit with minimal overall metal loss can lead to the failure of an entire engineering system.

Ecos in all boiler plants are susceptible to pitting. O_2 pitting, caused by the presence of O_2 even in traces and

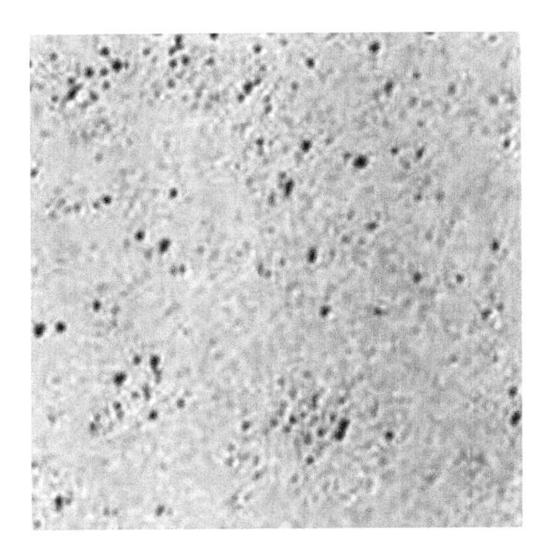

FIGURE 3.31
Surface pitting of metals.

increase in temperature, is a major problem in economisers calling for maintaining essentially O_2-free water. Hence there is the need for **O_2 scavenging** treatment, of fw in addition to **deaeration**.

The inlet is subject to severe pitting because it is often the first area after the deaerator to be exposed to increased heat. Whenever possible, tubes in this area along with inlet header should be inspected closely for evidence of corrosion.

3.76 COR-TEN Steel

COR-TEN is the registered trade name of US Steel Corporation for high-tensile weathering steel which exhibits increased resistance to atmospheric corrosion compared with unalloyed steels.

It is low AS of Cu, Cr, and Ni and its resistance is due to the formation of a rust coloured protective layer on its surface, which develops and regenerates continuously under the influence of the weather. In cases of particular air pollution by aggressive agents, contact with water for long periods, permanently exposed to M or use in the vicinity of the sea conventional surface protection coating is absolutely necessary.

COR-TEN is used for various types of welded, bolted and riveted constructions, for example, steel structures, bridges, tanks and containers, exhaust systems, vehicles, and equipment constructions. In boilers it is used mainly for low-temperature tubes of TAH, baskets of RAH, parts of steel stacks and flues and ducts.

COR-TEN A applies to plates up to 12.5 mm and COR-TEN B up to 60 mm in thickness. Tables 3.15 and 3.16 provide chemical and mechanical properties of this steel.

3.77 Counter Flow of Fluids (*see* Fluid Flow Patterns in Fluid Flow)

3.78 Cracking (of Petroleum) (*see also* Crude Oil *or* Petroleum in Major Liquid Fuels *under* Liquid Fuels)

Cracking is a process of thermal decomposition of heated oil. It is the breaking down of heavier HC molecules with high boiling points into the simpler HCs under heat and pressure. Cracking of crude oil yields light products like petrol and diesel. Cracking can be divided into

- Catalytic cracking
- Thermal cracking
- Hydrocracking

Catalytic cracking is employed for converting heavy HC fractions obtained by vacuum distillation into a mixture of light products such as petrol, light FOs, liquefied petroleum gas (LPG) and so on by breaking down the feed stock under controlled heat (450–500°C) and pressure in the presence of a catalyst.

Thermal cracking employs heat to break down the feed stock from vacuum distillation to produce light FOs and petrol.

Hydrocracking is catalytic cracking in the present of H_2 to increase the yield of petrol and light FOs.

3.79 Creep (*see* Metal Properties)

3.80 Crevice Corrosion (*see* in Corrosion)

3.81 Critical Flow and Pressure (for Nozzles)

The flow of gas or vapour is limited by the nozzle area, upstream pressure and the sonic velocity. The

TABLE 3.15

Chemical Composition (Heat Analysis, %)

Grade	C	Si	Mn	P	S	Cr	Cu	V	Ni
COR-TEN A	0.12	0.25–0.75	0.20–0.50	0.07–0.15	0.030	0.50–1.25	0.25–0.55		0.65
COR-TEN B	0.16	0.30–0.50	0.80–1.25	0.030	0.030	0.40–0.65	0.25–0.40	0.02–0.10	0.40

TABLE 3.16

Mechanical Properties at Room Temperature, in the State of Delivery Condition

Grade	Material Thickness	Minimum YP (Re H Mpa)	UTA Rm MPa	Elongation min. A (Lo = 5.65√So)%
COR-TEN A	≥3 mm	355	470–630	20
COR-TEN B	≤16 mm	355	470–630	20
	>16 ≤ 50	345		

maximum flow through a nozzle is determined by the downstream critical pressure which is governed by the critical pressure ratio. When the gas expands its velocity reaches sonic velocity at some stage. The pressure at that point is called critical pressure. Even though the pressure may fall further flow cannot increase as its velocity has already reached the maximum.

Critical flow nozzles are also called *sonic chokes*. By establishing a shock wave the sonic chokes establish a fixed flow rate unaffected by the Δp or any fluctuations or changes in downstream pressure. A sonic choke may provide a simple way to regulate a gas flow. A typical convergent–divergent nozzle is shown in Figure 3.32.

The ratio between the critical pressure and the initial pressure for a nozzle can be expressed as

$$p_c/p_1 = (2/(n+1))^{n/(n-1)} \quad (3.5)$$

where

p_c = critical pressure (in Pa)

p_1 = inlet pressure (in Pa)

n = index of isentropic expansion or compression—or polytropic constant.

For a perfect gas undergoing an adiabatic expansion the index—n—is the ratio of specific heats—$k = c_p/c_v$. There is no unique value for—n. Values of n for some common gases are

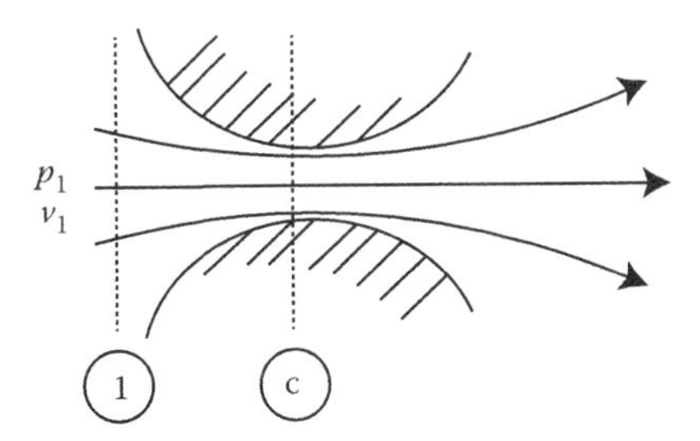

FIGURE 3.32
Convergent–divergent nozzle.

- Steam saturated and wet: 1.135
- Steam superheated: 1.30
- Air: 1.4

Critical pressure ratio for an air nozzle can be calculated as

$$p_c/p_1 = (2/(1.4+1))^{1.4/(1.4-1)} = 0.528 \quad (3.6)$$

Mass flow through a nozzle with sonic flow where the downstream pressure equals the critical pressure is

$$m_c = A_c\,(n\,p_1\,\rho_1)^{1/2} \times (2/(n+1))^{(n+1)/2(n-1)} \quad (3.6)$$

where

m_c = mass flow at sonic flow (kg/s)

A_c = nozzle area (m^2)

ρ_1 = initial ρ (kg/m^3)

3.82 Critical Pressure Point of Steam and Water (*see also* Properties of Steam and Water)

When heat is added to sub-cooled water its temperature raises until it reaches its *saturation temperature* corresponding to its pressure and starts *boiling*. Further heating adds to the heat content of water without raising its temperature and turns it progressively to more and more steam. At the end of boiling, all water would have turned into steam. In other words the *wet steam*, containing partly water and partly steam would have turned into *dry steam*, with no water content. Further heating takes the steam beyond the saturation temperature into the *superheated steam*.

However, this process of steam formation involving boiling is relevant at pressures lower than *critical pressure*, that is, *sub-c pressures*.

Critical point is just the point at which no change of state takes place when the steam pressure is increased or heat is added. At the critical point, water and steam cannot be distinguished. Water turns into steam directly without the intermediate stage of evaporation.

This critical point is reached when the steam pressure attains 221.2 bar/225.5 kg/cm^2 absolute or 3206.2 psia with the corresponding steam temperature of 374.1°C or 705.4°F. These values are based on ASME tables and can be fractionally different from the other steam tables. This is called *critical pressure*. Steam pressure higher than critical pressure is termed as *super-critical pressure*.

In general pressures above the critical point are called *super-critical pressure* while those below are *sub-c pressure*.

The T–s graph in Figure 3.33 depicts this phenomenon clearly.

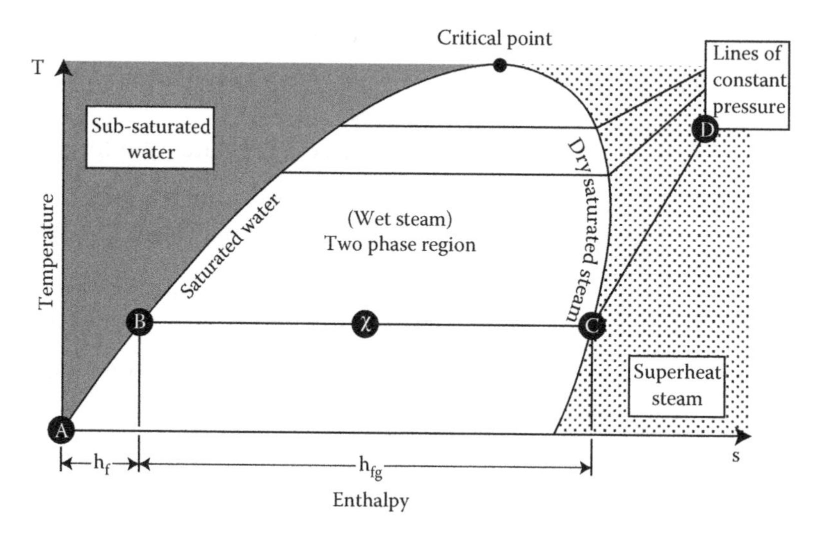

FIGURE 3.33
Temperature–enthalpy (T–s) diagram of steam.

3.83 Critical Pressure (*see* Critical Pressure Point of Steam and Water)

3.84 Critical Temperature (*see* Critical Pressure Point of Steam and Water)

3.85 Critical Heat Flux (*see* Boiling *under* Properties of Steam and Water)

3.86 Cross Flow of Fluids (*see* Fluid Flow Patterns in Fluid Flow)

3.87 Crude Oil (*see* Liquid Fuels)

3.88 Cycle Efficiency (*see also* Thermodynamic Cycles)

Any heat engine takes heat from a hot body, converts a part of it into work and rejects the rest to a cold body. Thermodynamic cycle η is defined as the ratio of the net external work done by the engine during one cycle to the heat absorbed from the source during that cycle.

The work W obtained in each cycle is called the output while the heat Q_1 extracted from the source in one cycle is called the input. If Q_2 is the heat rejected $W = Q_1 - Q_2$.

$$\eta = \frac{\text{Output}}{\text{Input}} = \frac{W}{Q_1} = \frac{Q_1 - Q_2}{Q_1} = 1 - \frac{Q_2}{Q_1} \quad (3.7)$$

3.89 Cycling Mode of Operation (*see* Boiler-Operating Modes)

3.90 Cyclone Furnace and Firing

Cyclone-firing was developed in the United States in 1940s basically for firing *poor grade coals with a low ash fusion temperature and high M content* which are difficult to use in PF boilers. It is suitable for coals with

- VM > 15% (dry basis) as radiant heat needed for high slag temperature inside the cyclone has to be supplied by the burning of volatiles.
- Minimum ash content of 6% and 4% for bituminous and sub-bituminous coals, respectively, for developing the slag belt.
- Maximum ash limited to 25% as higher ash increases the ash losses quite sharply.
- Ash having particular slag viscosity characteristics. Basically the molten ash has to flow along the horizontal surface which is possible at 250 cP. Ash with higher viscosity or cyclone temperature lower than attaining 250 cP is unsuitable. The maximum temperature at which the slag has a viscosity of 250 cP is 1340°C for bituminous and 1260°C for sub-bituminous coals. Ash slag behaviour is critical to satisfactory operation of cyclone.
- M content <20% for bituminous and 30% for sub-bituminous coals.

A variety of coals have been burnt in cyclone furnaces starting from low-volatile bituminous coals to high-volatile lignites. In co-firing mode fuels like wood bark, chips, saw dust, RDF and so on have been burnt.

The main advantages claimed for cyclone firing are

1. Less fuel preparation—only crushing and no grinding. Milling equipment is totally dispensed with
2. Smaller furnace and boiler as combustion takes place in cyclones and relatively clean gases go into furnace and tube banks
3. Easy ash handling as bulk of ash is removed at the bottom as slag and fly ash is therefore small

Cyclone furnace can be explained as follows:

- It is a water-cooled horizontal cylinder where crushed coal is burnt to release heat at extremely high rates of 4.65–8.3 MW/cum, ~4.0 × 10^6–7.1 × 10^6 kcal/cum or ~450,000–800,000 Btu/cft.
- It is of 1.8–3 m (6–10 ft) in diameter constructed with tangent tube wall which is protected by a refractory layer held by pin studding.
- Cyclone furnaces are attached to the main boiler furnace on one or both sides. The cyclone furnace chambers have a narrow (or tapered) base, together with an arrangement for slag removal.

Concept of a typical small cyclone-fired boiler is shown in Figure 3.34.

- Crushed coal of <95% through four meshes (4.75 mm) is introduced tangentially into the burner at the front end of the cyclone (as shown in Figure 3.35).
- Excess air employed is quite low at 10–13% as higher air cools the furnace temperature.
- ~15% of the combustion air, used as PA, carries the fuel along with it while entering the burner tangentially. Owing to the centrifugal action the relatively large coal/char fuel particles are thrown to the periphery and retained in the cyclone while the air passes through, promoting intense scrubbing action.

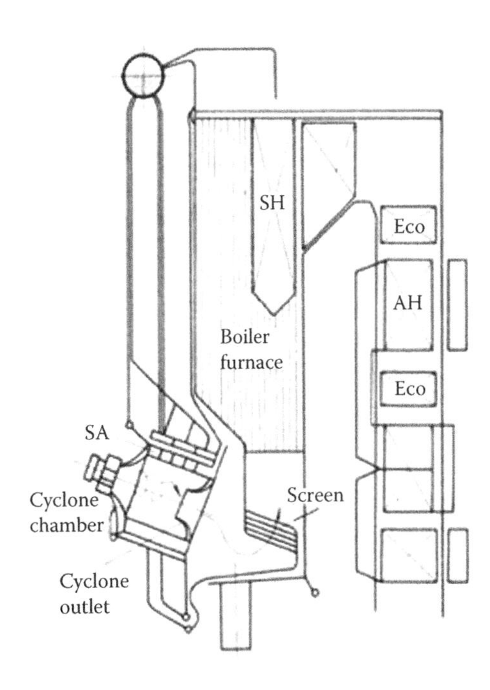

FIGURE 3.34
Small cyclone-fired boiler.

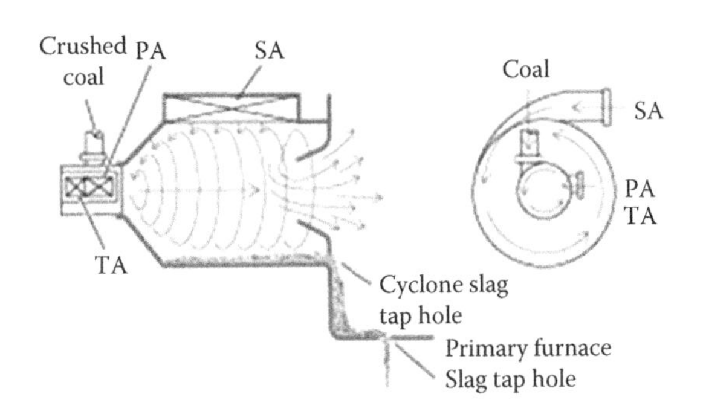

FIGURE 3.35
Cyclone furnace.

- Another ~15% of air is introduced into the burner as TA at the centre of the burner, along the cyclone axis directly into the central vortex. It is used to control the vortex vacuum, and hence the position of the main combustion zone which is the primary source of radiant heat. An increase in TA moves that zone towards the furnace exit and the main boiler.
- The whirling, or centrifugal, action on the fuel is further increased by the tangential admission of ~70% high-velocity SA into the cyclone.
- In the passage of the fuel mass through the cyclone the larger particles are trapped in the molten and sticky layer which covers the entire surface of the cyclone interior, except for the area in front of the air inlets. The coal is fired under conditions of intense heat input, and while the finest particles may pass through the vortex in the centre, the larger ones are thrown towards the walls and are re-circulated to achieve adequate burnout.

Combustion: Combustion inside the cyclone is extremely vigorous. VM leaves the fuel and burns almost immediately in the middle of the cyclone. As the fuel mass rotates inside the cylindrical portion of the cyclone the FC or char gets time to burn. To withstand the erosive forces of the fuel and air the cyclone wall made of tangent tube construction is refractory protected. This acts as insulation layer and prevents high heat absorption by cyclone promoting high-combustion temperatures which escalate to the levels of 1650–2000°C (3000–3600°F). Ash in coal turns into a flowing slag further adding to the protection of the cyclone walls. The char and ash particles from the incoming fuel get attached to the slag layer. The high temperature of the slag and heavy air scrubbing burns the char and melts the ash and increases its own thickness. As a result of the intense combustion conditions, NO_x formation tends to be considerably higher than in PF firing, typical values being 1050–3900 mg/Nm^3.

Ash: Molten slag with ~70–80% ash leaves the cyclone from the bottom to enter the slag pit. Molten ash flows by gravity from the base of the cyclone furnaces, and is removed from the system at the bottom of the boiler. It drops out into a quench tank, thus losing a substantial amount of heat. Precautions against gas build-up and explosions are essential in and around the slag quench tank.

Flue gas with ~20–30% ash is discharged through a water-cooled re-entrant throat at the rear of the cyclone into the boiler furnace. Flue gases go through the rest of the boiler which is like any conventional PF boiler. As the dust loading in gas is less the tube spacing can be closer in all tube banks making the boiler much smaller than a comparable PF boiler.

Cyclone versus PF: The essential difference between cyclone furnace and PF furnace is the manner in which combustion takes place. In PF firing the fuel is suspended in a moving gas stream and hence relatively large furnace is required to complete the combustion of. On the contrary, in cyclone firing, the coal is held by the slag in the cyclone and the air is passed over the fuel, thereby burning the fuel particle very fast. Thus, large quantities of fuel can be fired very quickly and combustion completed in a relatively small volume. The boiler furnace is used thereafter to cool the products of combustion to levels acceptable to convection surfaces.

Limitations of cyclone firing: Cyclone furnaces are not without limitations and disadvantages.

- The coal used should have a relatively low S content in order for most of the ash to melt for collection.
- High-power fans are required to move the larger coal pieces and air forcefully through the furnace.
- More NO_x pollutants are produced compared with pulverised coal (PC) combustion.
- Burner requires almost yearly replacement of liners because of erosion by the high velocity of the coal.

Experience: There are over 100 units, mainly in the USA (operating on lower rank coals with low-ash fusion temperature) and in Germany. Virtually none has been commissioned for the last couple of decades because of the high-combustion temperatures which result in high NO_x emission.

As with PF boilers, cyclone-fired units can be of almost any size. Existing units range in capacity from 33 to 1150 MWe, and there are more than 20 units sized between 400 and 900 MWe.

As the units which were built mainly during the 1960s and 1970s, sub-c steam cycles are used, and the efficiencies of these units were in the range 35–38%, similar to equivalent size PF units.

Flue gas cleaning/emissions: The principles for flue gas cleaning are similar to those for **PF** boilers. In addition to the use of **SCR** for NO_x reduction various other combustion methods can be used, such as two-stage combustion, and a re-burn concept (in which 15–35% of the coal is pulverised, and added into the main boiler chamber bypassing the cyclone furnace).

Shown in Figure 3.36 is a utility-sized cyclone-fired two-pass boiler equipped with two rows of cyclone furnaces.

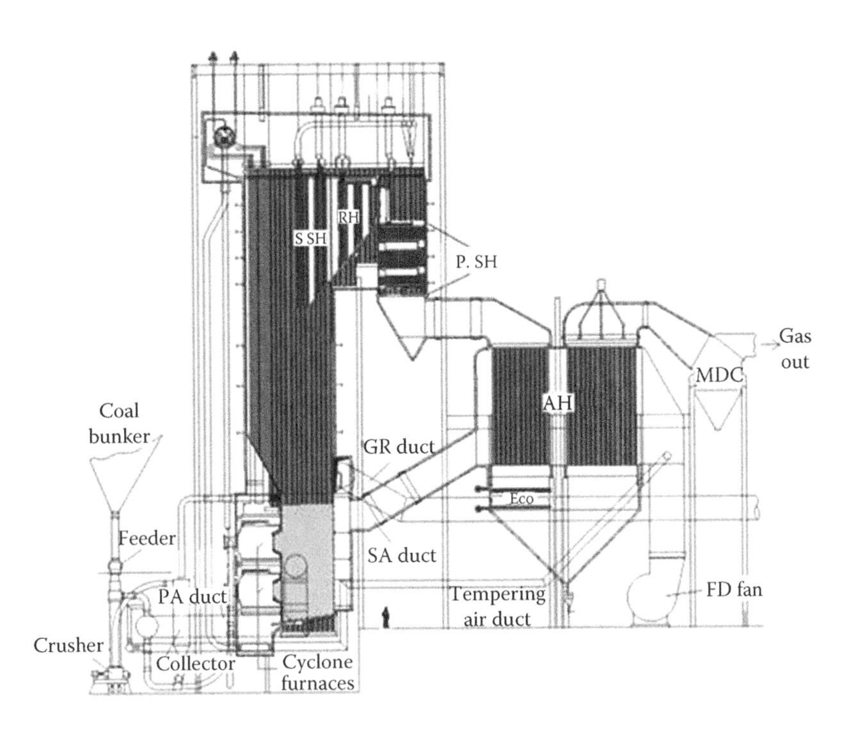

FIGURE 3.36
Utility-sized cyclone-fired re-heat boiler with two rows of cyclones.

3.91 Cyclones, Hot and Cold (*see also* Circulating Fluidised-Bed Combustion)

In CFBC boilers dust escaping the free board in the first pass is captured by large cyclones and returned to the combustor for bed temperature control and better C burn up.

Gas temperatures exiting the first pass can be at ~800–900°C as in full-circulation CFBC boilers or at ~400–500°C as in expanded-bed CFBC boilers. The cyclones accordingly operate at these temperatures and are known as hot or cold, respectively. This is discussed in greater detail under CFBC.

3.91.1 Cyclones Steam or Water Cooled

When hot cyclones are sometimes cooled by water or steam-cooled membrane walls for reducing the refractory and imparting longer life they are also called steam or water-cooled cyclones.

They are also at times called cold cyclones because of the cooling effect provided, although it is not strictly right as the incoming gas is at high temperature.

3.92 Cyclone Separators

A cyclone is a centrifugal separator for separating fluids and solids based on their widely differing densities. The mixture of fluids is accelerated in the inlet chamber of the cyclone and given a high-speed swirl in the circular body. Centrifugal force separates the solid particles to the periphery where they slide down the cone. The lighter fluid, now free of dust, leaves from the central exit pipe.

There are mainly three areas in boilers where cyclones are employed, namely, for

- Steam separation from water inside the steam drum in vertical or horizontal steam separators
- Dust separation from flue gas in FBC boilers in hot or cold cyclones
- Fly ash precipitation from flue gas exit in single or multiple cyclones

3.93 Cyclone Steam Separators (*see* Drum Internals)

4

D

4.1 Dampers (*see* Draught Plant)

4.2 Darcy–Weisbach Formula (*see* Fluid Flow)

4.3 Deaeration/Deaerator (*see* Water)

4.4 Dealkalisation (*see* Water)

4.5 Deflection Baffles in Steam Drum (*see* Drum Internals)

4.6 Degrees Engler (*see* Liquid Fuels *and* Instruments for Measurement)

4.7 Deionisation (*see* Demineralisation of Water in Water)

4.8 Demineralisation (*see* Water)

4.9 Denitrification or $DeNO_x$ (*see* NO_x Reduction)

4.10 Density of Air and Gas (*see* Properties of Air and Flue Gas)

4.11 Density of Solids (*see* Bulk and True Densities of Solids in Properties of Substances)

4.12 Density (ρ) of Steam and Water (*see* Properties of Steam and Water)

4.13 Deoxidation of Steels (*see* Metallurgy)

4.14 Departure from Nucleate of Boiling (*see* Properties of Steam and Water)

4.15 Design Pressure of Boiler (*see* Boiler Pressure)

4.16 Desulphurisation or DeSO$_x$ Reaction

Desulphurisation is the *removal of S oxides* so that the flue gas is free of SO_2 and SO_3 which are harmful to boiler HS, humans and environment. High levels of sulphurous gases produce respiratory problems in humans and damage to vegetation and buildings because of the effects of **acid rain**.

Removal of S from the fuel itself would be the most desirable but it has been so far very expensive. Desulphurisation can be either before or after combustion.

- *In situ* or in-furnace desulphurisation, possible in FBC boilers, is only some 30 years old and removes SO_2 formed in the bed itself by sulphation reaction.
- FGD, a post-combustion process, captures SO_2 and SO_3 produced by combustion reaction and has been practised since the 1930s.

4.16.1 *In Situ* Desulphurisation

This is realised by adding limestone ($CaCO_3$) to the fuel and heating to the right temperatures of ~700–900°C when S reacts with $CaCO_3$ and $CaSO_4$ is formed where S is captured.

Lime–S reaction taking place within the confines of the fluidised bed is called *in-bed desulphurisation*. Together with the combustion reaction taking place and the resultant gypsum coming out smoothly, along with ash in FBC boilers, has really been a breakthrough.

Desulphurisation is not a single reaction but three reactions in sequence, namely

1. *Calcination*
2. *Combustion*
3. *Sulphation reactions*

a. *Calcination reaction* is an endothermic (heat absorbing) reaction where CaO, quick lime, is produced by the decomposition of limestone ($CaCO_3$) when CO_2 is driven off on heating.

$$CaCO_3 + 425.5 \text{ kcal/kg (766 btu/lb)} = CaO + CO_2\text{... on heating} \quad (4.1)$$

b. *Combustion* reaction

$$S + O_2 = SO_2 \text{ on combustion} \quad (4.2)$$

c. *Sulphation reaction:* Lime–S reaction is basically turning the SO_2 produced (from the oxidation of S on combustion) to gypsum ($CaSO_4$) by its reaction with CaO, which is an exothermic (heat producing) reaction

$$SO_2 + CaO + 1/2O_2 = CaSO_4 + 3740.5 \text{ kcal/kg (6733 btu/lb)... on heating} \quad (4.3)$$

Both sulphation and calcination reactions start by ~700°C (~1300°F) and reach their optimum at 840–850°C (~1545–1565°F). The lime consumption is the least in this temperature range. Even then the theoretical requirement of 1 mol of Ca per mol of S (Ca/s ratio of 1) cannot be reached because

- The sulphation takes place on the surface of the lime particle in the bed while the core of the particle fails to participate.
- Some S in fuel which is inorganically bound does not oxidise to SO_2.
- Some SO_2 escapes when sorbent is less or along with VM of fuel.
- Limestone purity is lower than optimum-typically ~92%.

Bed temperature range of 800–900°C (1470–1650°F) is most important for desulphurisation because

- Calcination is not complete at temperatures <800°C.
- Sulphation reaction falls off rapidly beyond 850°C because $CaSO_4$ formed on the surface of CaO particle melts because of high temperature and forms a coating inhibiting further exposure to reaction.

Limestone consumption becomes excessive outside of this range. The graph in Figure 4.1 shows the manner in which the sulphation reaction peaks ~850°C and how the limestone demand escalates for greater S removal in BFBC.

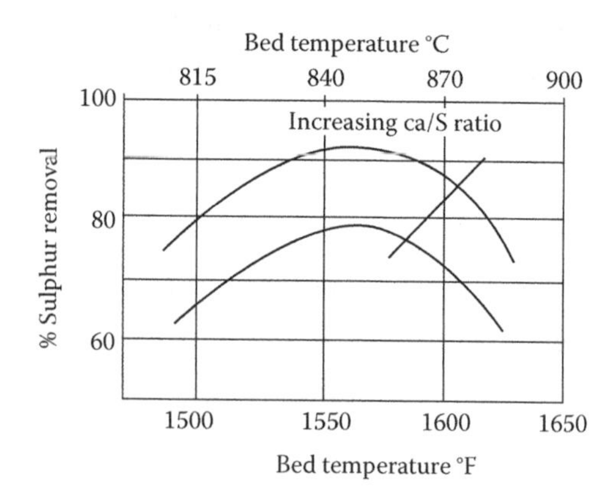

FIGURE 4.1
Effect of bed temperature and Ca/S ratio on sulphur removal in BFBC.

A practical limit for desulphurisation for BFBC can be considered as 85% with Ca/S ratio of *2.5–3.0*. But with CFBC

- Having a longer residence time of about 5 secs (2.5 in BFBC)
- Smaller fuel and limestone particle size

This limit can be considered as 95% with Ca/S ratio of *1.8–2.5* depending on

- The S content of the fuel
- Purity and reactivity of limestone

At higher levels of ash of >40% and S >5% Ca/S ratios escalate rather sharply to meet the limits.

4.16.2 Flue Gas Desulphurisation

Most flue gas desulphurisation (FGD) processes remove the acidic SO_2 from flue gases by reaction with a suitable alkaline substance such as

- Limestone ($CaCO_3$)
- Quicklime (CaO)
- Hydrated lime ($Ca(OH)_2$)

$CaCO_3$ is an abundant and therefore relatively cheap material and both CaO and $Ca(OH)_2$ are produced from $CaCO_3$ by heating. Other alkalis sometimes used include Na_2CO_3, $MgCO_3$ and NH_3.

The alkali reacts with SO_2 to produce a mixture of SO_3 and SO_4 salts (of Ca, Na, Mg or NH_4, depending on the alkali used) in proportions determined by the process conditions; in some processes, all the SO_3 is converted to SO_4.

The reaction between the SO_2 and the alkali can take place either

- In bulk solution ('wet' FGD processes) where alkali is in slurry form or
- At the wetted surface of the solid alkali ('dry' and 'semi-dry' FGD processes)

Wet System

In wet FGD systems, the slurry and flue gas contact each other in a spray tower. The SO_2 in the flue gas dissolves in the water to form a dilute solution of acid which reacts with and is neutralised by the dissolved alkali. The SO_3 and SO_4 salts produced precipitate out of solution, depending on the relative solubility of the different salts present. $CaSO_4$, for example, is relatively insoluble and readily precipitates out. Na_2SO_4 and $(NH_4)_2SO_4$ are more highly soluble.

Dry and Semi-Dry Systems

In dry and semi-dry systems, the solid alkali 'sorbent' is brought into contact with the flue gas, either by injecting or spraying the alkali into the gas stream or by passing the flue gas through a bed of alkali. In either case, the SO_2 reacts directly with the solid to form the corresponding SO_3 and SO_4. For this to be effective, the solid has to be quite porous and/or finely divided. In semi-dry systems, water is added to the flue gas to form a liquid film on the particles in which the SO_2 dissolves, promoting the reaction with the solid.

Limestone Gypsum Wet Scrubber

Selection of the most appropriate FGD process for a particular application will normally be made on economic grounds, that is, the process with the lowest overall through-life cost. The limestone gypsum wet scrubbing process, producing $CaSO_4 \cdot 2H_2O$ (gypsum) is the most common FGD process now being installed worldwide and has evolved over almost 30 years which is illustrated in Figure 4.2. Nowadays, a plant would normally be designed to achieve a high-quality gypsum product, which is suitable for wallboard manufacture.

The arrangement consists of a spray tower with a rotary regenerative RH, which is a gas–gas AH. The limestone–gypsum plant is located downstream of the ID fan, so that most of the fly ash from combustion (>99.5% for coal firing) is already removed by ESP or bag filter before the flue gas reaches the FGD plant.

Hot flue gas from the ESP passes through an ID and/or booster fan and enters the gas/gas RH. Here, the gas is cooled by the exhaust gas from absorber. The cooled gas from the RH enters the absorber and mixes with the process liquor. Some of the water is evaporated, and the gas is further cooled. The gas is scrubbed with the recirculating limestone slurry to remove the required amount of SO_2 which is >95%. This process also removes almost 100% of HCl if any in the flue gas.

At the top of the absorber, the gas passes through demisters to remove suspended water droplets. After leaving the absorber, the exhaust gas is passed through the RH again, as mentioned earlier to raise its temperature before being exhausted to the stack. Absorber outlet temperatures are typically 50–70°C, depending mainly on the type of fuel burnt. As the slurry falls down the tower it contacts the rising flue gas. The SO_2 is dissolved in the water, neutralised and thus removed from the flue gas. $CaCO_3$ from the limestone reacts with the SO_2 and O_2 from air, ultimately to produce gypsum, which precipitates from solution in the sump. HCl is also dissolved in the water and neutralised to produce $CaCl_2$ solution. Fresh limestone slurry is pumped into the sump to maintain the required pH.

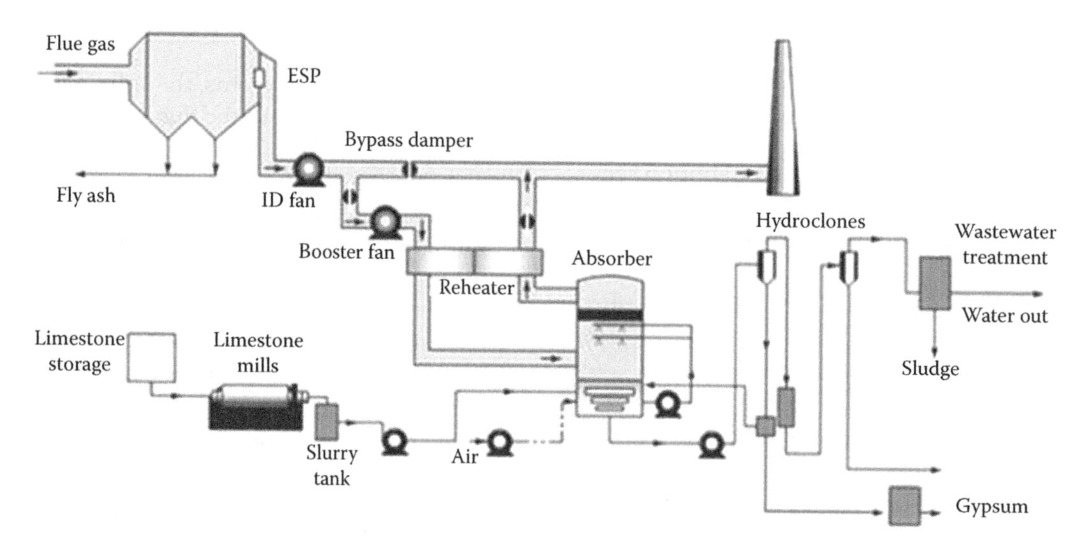

FIGURE 4.2
Limestone–gypsum wet scrubbing FGD process.

The reactions involved are as under.

The reaction taking place in wet scrubbing using limestone slurry

$$CaCO_3 \text{ (solid)} + SO_2 \text{ (gas)} \rightarrow CaSO_3 \text{ (solid)} + CO_2 \text{ (gas)} \quad (4.4)$$

When wet scrubbing with a **lime** slurry

$$Ca(OH)_2 \text{ (solid)} + SO_2 \text{ (gas)} \rightarrow CaSO_3 \text{ (solid)} + H_2O \text{ (liquid)} \quad (4.5)$$

When wet scrubbing with $Mg(OH)_2$ slurry

$$Mg(OH)_2 \text{ (solid)} + SO_2 \text{ (gas)} \rightarrow MgSO_3 \text{ (solid)} + H_2O \text{ (liquid)} \quad (4.6)$$

To partially offset the cost of the FGD installation, in some designs, the $CaSO_3$ is further oxidised to produce marketable **gypsum**. This technique is also known as forced oxidation

$$CaSO_3 \text{ (solid)} + H_2O \text{ (liquid)} + 1/2O_2 \text{ (gas)} \rightarrow CaSO_4 \text{ (solid)} + H_2O \quad (4.7)$$

A natural alkaline material to absorb SO_2 is seawater. SO_2 is absorbed in the water, and when O_2 is added it reacts to form sulphate ions SO_4^- and free H^+. The surplus of H^+ is offset by the carbonates in seawater pushing the carbonate equilibrium to release CO_2 gas:

$$SO_2 \text{ (gas)} + H_2O + 1/2O_2 \text{ (gas)} \rightarrow SO_4^{2-} \text{ (solid)} + 2H^+ \quad (4.8)$$

$$HCO_3^- + H^+ \rightarrow H_2O + CO_2 \text{ (gas)} \quad (4.9)$$

4.17 Desuperheater (*see* Attemperator)

4.18 Deslaggers (*see* Wall Blowers in Sootblowers)

4.19 Dew Point Corrosion (*see* Ash Corrosion)

4.20 Dew Point or Dew Point Temperature (T_{dp})

The dew point or the dew point temperature is the *temperature at which air or gas can no longer hold all of its watervapour and some of it must condense into liquid phase.* If the air/gas temperature cools to the dew point or if the dew point rises to equal the air/gas temperature, then dew, fog or clouds begin to form. At this point where the dew point temperature equals the air/gas temperature, the relative humidity is 100%.

Dew point of flue gas is dependent on a number of factors such as the temperature, pressure, fuel type and S content. In boilers the flue gases should exit above the dew point to avoid condensation of water vapour from flue which leads to low-temperature corrosion.

The presence of SO_2 will increase the dew point very dramatically. For instance, flue gas with 5% water vapour and no S oxides has a dew point of ~32°C (~90°F). The same flue gas with just 0.01% SO_3 added has a dew point of ~120°C (~248°F)! If this gas cools to 120°C, H_2SO_4 acid will begin to condense out of the gas. It is to be noted that

- Adding the same amount of SO_3 presents much less problem as the dew point change is not nearly as severe.
- H_2SO_3 is much less corrosive than H_2SO_4.

The dew point can be calculated based on the SO_2 and the M content of the flue gases. This dew point should not be confused with the acid dew point related to the formation of H_2SO_4.

S compounds such as H_2S, CH_4S (methyl mercaptan) as well as S burn in the presence of O_2 to produce SO_2 and SO_3 in the flue gas which always contains substantial water vapour. When the gases cool to dew point the water vapour condenses out and S oxides dissolve to form H_2SO_3 and H_2SO_4 which attack the steel. Neither of the S oxides presents much of a problem with normal materials of construction unless the flue gas temperature falls below the gas dew point temperature.

4.21 Diameter, Equivalent (d_e) and Hydraulic (d_h)

d_e of a rectangular duct or pipe is the *diameter of a circular duct or pipe of the same length which gives the same pressure loss*.

d_e of a rectangular tube or duct can be calculated as

$$d_e = 1.30 \times ((a \times b)^{0.625}/(a + b)^{0.25}) \quad (4.10)$$

d_e is not the same as d_h.

Hydraulic diameter is used for non-circular pipes and ducts to calculate pressure loss and other parameters just like in circular profiles. d_h is the ratio between the four times cross-sectional area and the wetted perimeter of the duct or tube.

For a circular pipe of diameter d flowing full of fluid

$$d_h = \frac{4 \times \text{area}}{\text{perimeter}} = \frac{(4\pi d^2/4)}{\pi d} = d \quad (4.11)$$

and for a square duct of sides a and flowing full

$$d_e = \frac{4a^2}{4a} = a \quad (4.12)$$

4.22 Dimensionless Numbers in Heat Transfer (*see* Heat Transfer)

4.23 Direct Water Level Indicators (*see also* Water Level Indicators *and* Remote Water Level Indicators)

Direct water level indicator (DWLI) or gauge glasses are directly attached to the steam drum. Depending on the drum pressure different constructions of gauge glasses apply which are

1. Tubular
2. Reflex
3. Transparent
4. Bi-colour
5. Port-type bi-colour

1. *Tubular gauge glasses* shown in Figure 4.3 are employed for low pressures up to ~15 barg. These are the oldest and simplest. At higher pressures and consequently higher saturation temperatures the affinity of alkaline drum water to dissolve glass increases and does not permit the use of plain glass.
2. *Reflex type gauge glasses*, shown in Figure 4.4, are suitable up to a drum pressure of ~32 barg. The borosilicate flat glass has serrations on the inside which is in contact with water. Owing to the difference in the refractive indices of steam and water, the incident light rays slanted at 45° suffer refraction in the steam space but not in the water space. As a result, the steam space appears silver white on the transparent water

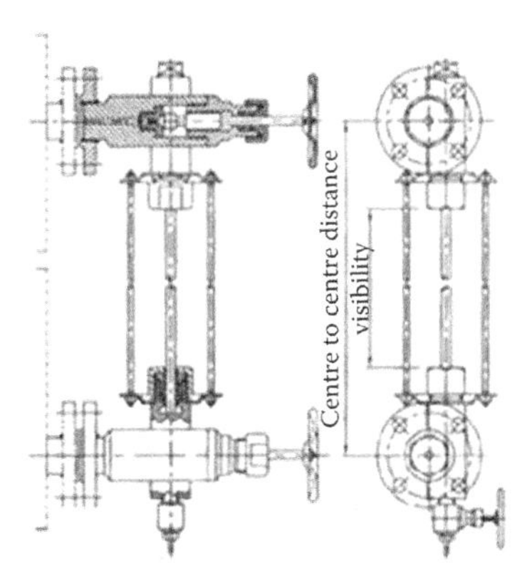

FIGURE 4.3
Tubular gauge glasses.

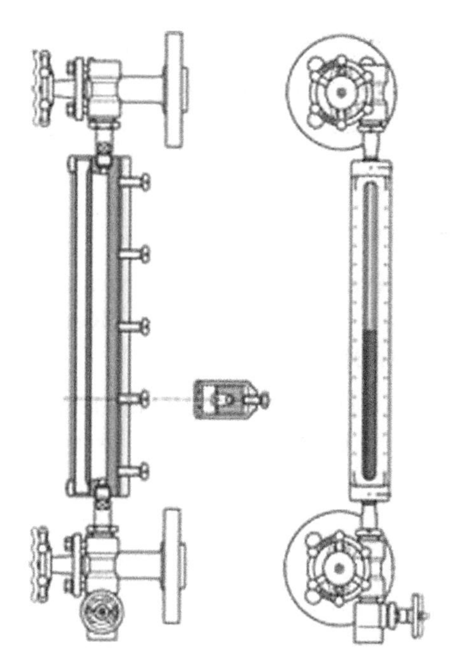

FIGURE 4.4
Reflex gauge glasses.

space. Borosilicate glass can withstand the corrosive attack of alkaline high-temperature water far better than plain glass.

3. *Transparent water gauges*, shown in Figure 4.5, are used up to 120 barg with a view in distance of ~20 m at a certain angle. Two flat mica lined glass sheets trapping the steam and water between them are illuminated at an angle from the rear. The light rays impinging on the water meniscus are refracted and the viewer sees the water level in the illuminated gauge. It is the mica lining of glass that helps to separate the glass from water and helps it withstand higher temperature.

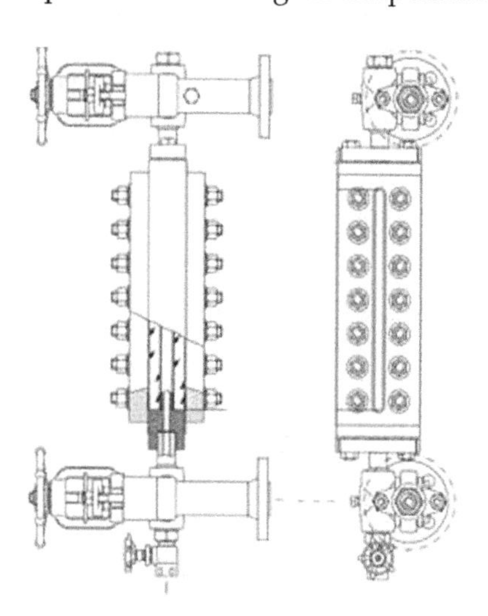

FIGURE 4.5
Transparent gauges.

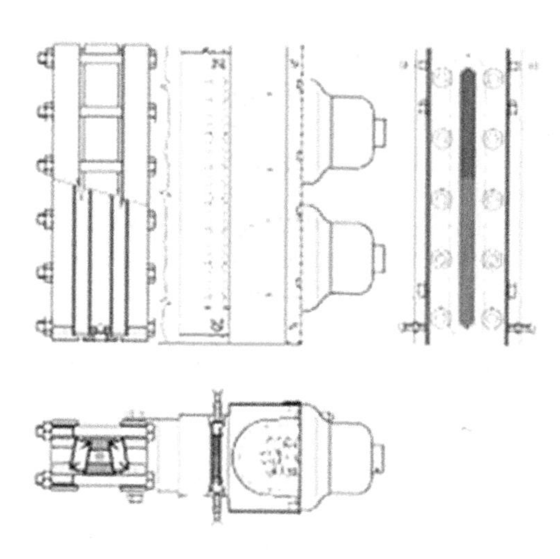

FIGURE 4.6
Bi-colour gauges.

4. *Bi-colour gauges*, shown in Figure 4.6, can be used up to ~180 barg. They have wedge-shaped glasses to provide differential refraction to light in steam and water space. Otherwise they are similar to the transparent gauges. An illuminator is located at the rear of the gauge with red and green colour slides fixed in place. It provides red colour for steam and green colour for water as the light gets deflected sideways.
5. *Port-type bi-colour gauges* also known as *Bull's eye gauges* employ the bi-colour action with port-type glasses. This design is suitable up to ~220 barg, the highest pressure for which steam drums are employed (Figure 4.7).

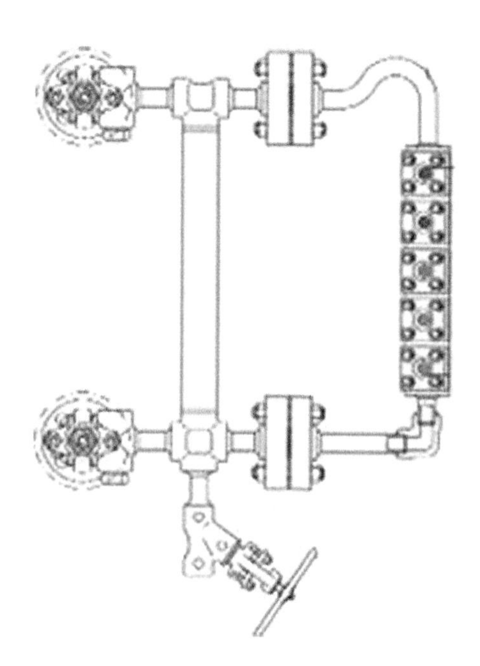

FIGURE 4.7
Port-type gauge.

The construction differs significantly in port-type gauge glasses where long glasses are replaced by small glass ports held rigidly by steel body. As the glass exposed to steam at high p & t is small in size and it is further held in place by steel plates, the life of the gauge glass is sufficiently long. The borosilicate glass ports are backed by thin transparent mica sheets to prevent contact with the highly corrosive steam and water.

4.24 Dissociation of Flue Gas (*see* Combustion)

4.25 Distillation (*see also* Major Liquid Fuels: Crude Oil or Petroleum)

Distillation is a *unit operation which is used for separating mixtures based on their volatilities by boiling a liquid mixture. It is not a chemical reaction.*

Distillation is the *most common separation technique*. It is energy intensive.

Various fractions of crude oil are separated by distillation. Because crude oil is a mixture of HCs with different boiling temperatures, it can be separated by distillation into groups of HCs that boil between two specified boiling points. Two types of distillation are performed:

Atmospheric distillation
Vacuum distillation

4.26 Distributed Control Systems

Over the last couple of decades distributed control systems (DCS) have become very popular both in large and small industrial and process power plants mainly, as they can handle and control large-scale process systems very efficiently and effectively. DCS have far outpaced the programmable logic controllers (PLCs) and personal computer (PC)-based systems because of their

a. Size—ability to process thousands of signals
b. Centralised administration—total control of all distributed units from a single node on the network
c. Data management—ability to store thousands of data points in real time

A typical DCS has a bi-directional dual redundant network capable of transferring a large amount of data between the node points (as shown in Figure 4.8).

Connected to the network are four different types of devices with very diverse functions. Listed in the order of hierarchy these are

1. *Engineering work station* provides a total control over the DCS. Programming the DCS, adding/deleting the spare I/O capacity to the network are the main functions.
2. *Data management work station* manages all the process database containing all the process data points or tags on DCS.
3. *Operator work station* is for controlling the plant with the help of mimics and data displayed

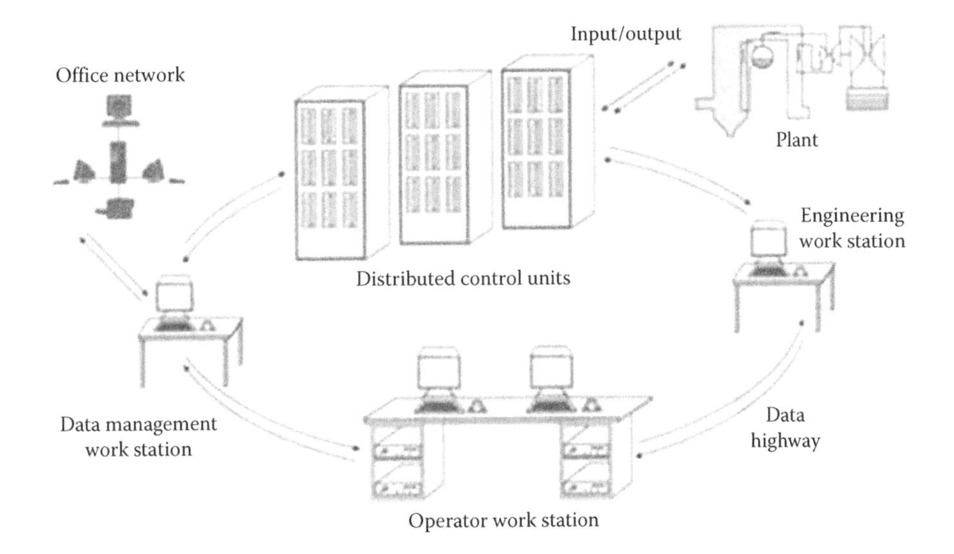

FIGURE 4.8
Typical DCS.

on the screens. This is where there is a man-machine interface.

4. *Distributed control units* actually implement plant control. Input/output signals are directly exchanged between the plant and distributed control units and hence can operate independent of the rest of the system.

4.27 Division Wall (in Furnace) (*see* Pressure Parts)

4.28 Dosing of Chemicals (*see* Water Treatment)

4.29 Downcomers or Supplies (*see* Pressure Parts)

4.30 Down Shot Burners (*see* PF Burners)

4.31 Down Shot-Fired Boilers (*see* Utility Boilers)

4.32 Drain Piping (*see* Integral Piping in Pressure Parts)

4.33 Draught (*see also* Furnace Draught *under* Fans)

Draught is defined as *current of air/gas*, especially the one intruding into an enclosed space. In boiler draught signifies the pressure difference below that of the atmosphere. Flow of air and gas inside boiler is also generally termed as boiler draught.

Early boilers operated under *induced/natural draught*, which relied entirely on the stack without the aid of fans, to provide air and gas flow through the boiler, purely by the ρ difference of air between the bottom and top of the stack.

Modern boilers with so many tube banks inside and HP SA/TA can scarcely operate on the meagre stack effect and need fans for creating adequate draught.

Balanced and FD are the two modes of boiler operation depending on whether the gas pressure in furnace is neutral (furnace static pressure = atmospheric pressure) or positive. Both types of operations are depicted in Figure 4.9 which shows the profile of air and gas pressures across a typical boiler.

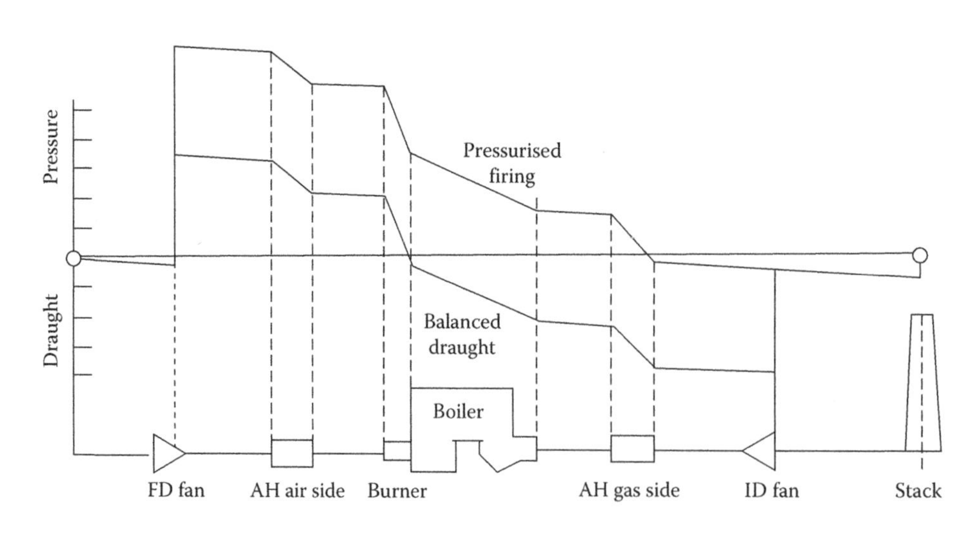

FIGURE 4.9
Draught profiles in forced and balanced draught firing.

4.34 Draught Plant

This is an all inclusive term to describe all parts in the air and gas flow circuit outside the main heat transfer surfaces. Draught plant usually consists of

a. Equipments like fans, dust collection equipment and so on
b. Fabricated items such as flues and ducts, expansion joints, dampers and bypass stack

Equipments in (a) are covered separately while items in (b) are elaborated here under the headings of

a. Flues and ducts
b. Expansion joints or compensators
   1. Metallic expansion joints
   2. Non-metallic/cloth/fabric joints
c. Dampers
   1. Isolation dampers
      a. Multi-flap or louver dampers
      b. Guillotine dampers
   2. Control dampers
   3. Diverter dampers
   4. Weather/stack dampers
d. Bypass stack

4.34.a Flues and Ducts

Ducts that carry flue gases and combustion air are termed, respectively, as flues and ducts in a boiler.

Construction wise there is no difference but by convention *ducts carry air* and *flues carry gas.*

They are purely fabricated items employing sheet steel of 4 and 5 mm for ducts and 5 and 6 mm for flues. It is not usually economical to fabricate them in the boiler shops. They are normally subcontracted or even site fabricated.

Compact layout and optimum sizing of the flues and ducts is essential to contain the overall costs of boiler. In most cases, their size is decided by the equipments they are connected to. There are prescribed velocity limits for optimal sizing of various ducts to avoid excessive pressure loss, noise, erosion and vibration. Adequate stiffening and supporting is also essential besides proper sizing.

Normal range of permissible velocities is as given in Table 4.1.

It is usual to internally brace ducts >2 m. Square or rectangular ducts are most common, but it is the round ducts which are most economical and are used extensively for diameters up to 1.5 m as the permissible air/gas velocities can be higher and stiffening minimum.

CS is most common constructional material but when the temperatures are >450°C, low AS or ss are used. At times brick or insulation lining on the inside is adopted. The former is needed for abrasive gases such as those from coal while the latter is employed for non-abrasive flue gases from oil/gas firing. In extreme high temperatures, like the downstream of duct burners, lining with C fibre mattresses is employed if the gases are clean.

Layout engineering of flues and ducts is very important as it contribute in a big way for the pressure losses and the fan power. Generous sizing, no doubt, reduces losses but increases the auxiliary power cost and vice versa. A good layout is a fine balancing act.

A typical air and gas duct diagram of PF boiler is shown in Figure 4.10 along with places where gas analysis is usually taken.

TABLE 4.1
Permissible Air and Gas Velocities in Flues and Ducts

	Location	Velocity		Remarks
		m/s	fpm	
Gas Ducts	Up to ID fan—coal firing	12.5–15	2500– 3000	<12.5 ash builds up and >15 erosion in flues
	Up to ID fan—oil/gas firing	14	2800	Also for other non-erosive fuels
	Beyond ID fan—coal firing	9–14	1800–2800	
	Beyond ID fan—oil/gas firing	14	2800	Also for other non-erosive fuels
	Bypass and cross over ducts	14	2800	
Air Ducts	FD suction	9–11	1800–2200	Maximum 14 m/s (2800 fpm) if space is less
	FD discharge—cold	11	2200	
	FD discharge—hot	15	3000	
	Tempering air	14 and 18	2800 and 3600	For rectangular and circular ducts
	Mill inlet	18	3600	High velocity required for air flow measuring
	Seal air	20 and 25	3600 and 5000	For rectangular and circular ducts

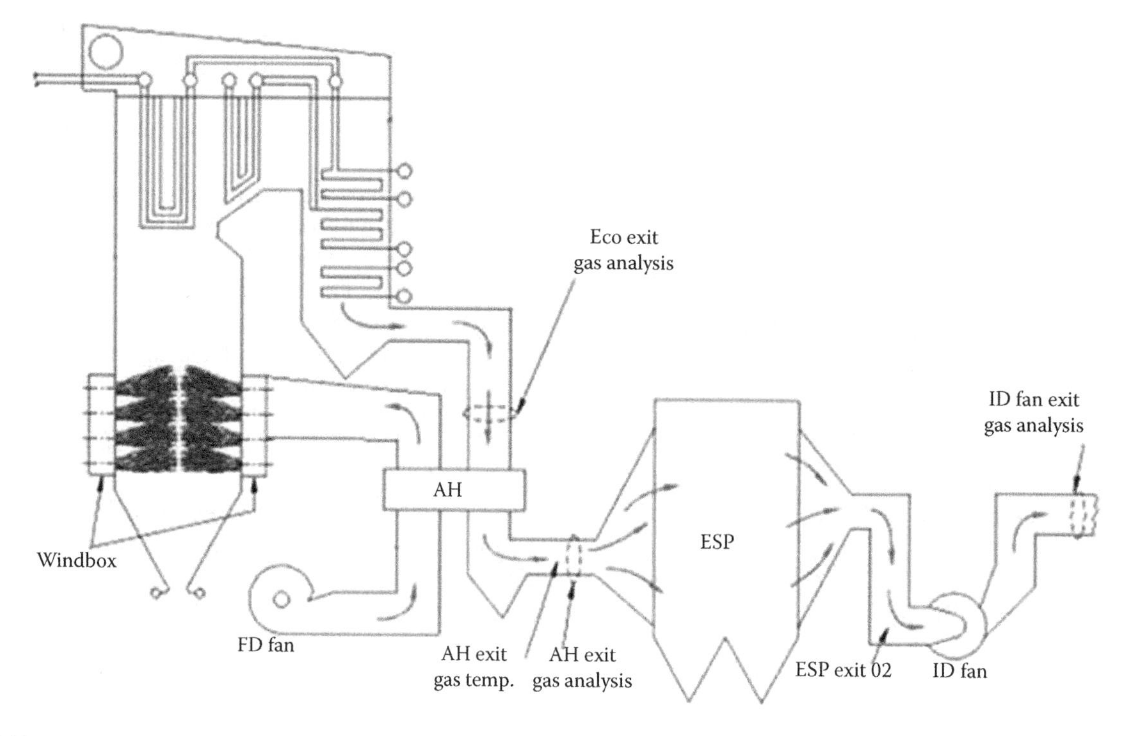

FIGURE 4.10
Flues and ducts in a typical PF-fired boiler.

4.34.b Expansion Joints or Compensators

The main purpose of expansion joints in any duct work is to *absorb the differential expansion* arising out of duct expansion between the hot and cold conditions. Small expansion joints, called V breathers, are invariably provided at the discharge of fans to prevent vibrations from the rotating machinery to be transferred to stationary duct work.

Expansion joints are of two types:

a. Metallic expansion joints
b. Non-metallic/cloth/fabric expansion joints

FIGURE 4.11
Rectangular metallic expansion joint.

Metallic expansion joints are the oldest and most proven. The expansion leaves, held in place by the flanges, are pressed from very thin ss sheets with deep drawing qualities. Metallic expansion joints are quite rugged and have a long life. They can withstand erosive gases quite easily as they are made of metal. Once installed, they stay there permanently, needing practically no further attention. However, compared with the fabric joints the metallic joints are much poorer in accommodating large misalignment in the connecting ducts, demanding a better quality of erection, which also consumes more time.

Figure 4.11 illustrates a typical metallic expansion joint with bolted flanges.

Non-metallic/cloth/fabric joints are of recent origin. Properly selected, they are also very rugged, long lasting and capable of withstanding high temperatures. Additionally they can accommodate angular movement because of misalignment between the two ducts which is possible for metallic joints only to a very limited extent. They have also demonstrated that they can withstand the erosive gases just as well as the metallic expansion joints. Another big advantage is their low weight.

Typical fabric expansion joint is shown in Figure 4.12. The flanges shown are of welded type here.

Naturally the metallic joints are progressively yielding place to the fabric joints.

4.34.c Dampers

Dampers are to ducts what valves are to piping. Dampers are used in boiler application for

FIGURE 4.12
Circular fabric or cloth expansion joint.

- Isolation
- Control
- Diversion
- Prevention of back flow

1. *Isolation Dampers*

Isolation dampers are of two types to suit the permissible leakage levels.

a. Multi-flap or louver dampers

These are the most common variety of dampers. They are routinely used at the discharge of fans and for flow isolation in ducts. Motorised multi-flap type of dampers shown in Figure 4.13 are light, occupy less duct space, can be inserted in any position and are mechanically simple. The pressure drop and leakage are both a little higher than in guillotine dampers. The more flaps in the damper, the less space the damper needs and more uniform the downstream flow. Besides on-off duty, they can also be used for modulating the flows.

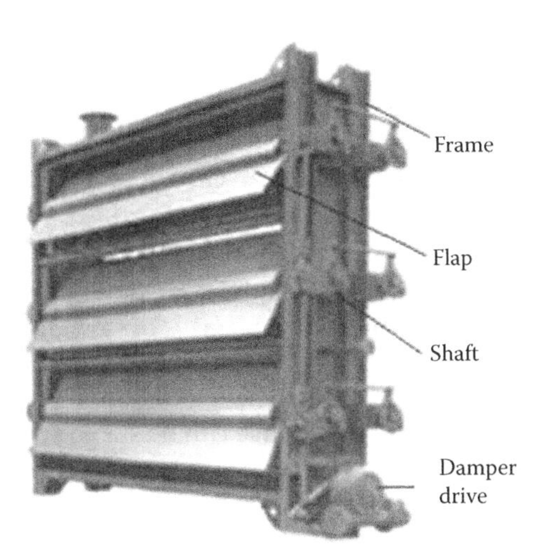

FIGURE 4.13
Multi-flap motorised louver damper.

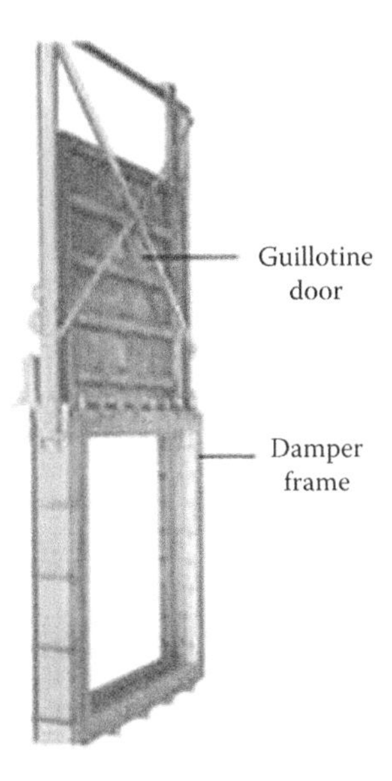

FIGURE 4.14
Guillotine damper.

b. Guillotine dampers

These are employed for isolation duty where better leak tightness is desired. Isolating FG ducts and ESPs are some of the examples. Guillotine dampers employ a single stiffened plate which has seals all around (as shown in Figure 4.14). Usually they are motor operated because of the size. For extreme leak tightness like 99.9% the seals are inflated by means of air pressure. Guillotine dampers provide the highest leak tightness and offer the least pressure drop. Even though they need short space in the ducts they require a considerable space outside for the retraction of the door and locating the air compressor station. They are also not suitable for modulating duties.

2. *Control Dampers*

These are essentially multi-flap dampers with flaps made to aerofoil shapes so that the flow pattern is more uniform and regulated. The flow can be calibrated against the angle of rotation of shaft.

3. *Diverter Dampers*

The most common diverter damper is the one placed between **GT** and **Hrsg**. The hot exhaust gases can be diverted either to Hrsg or to the bypass stack. This damper shown in Figure 4.15 is primarily a large flap

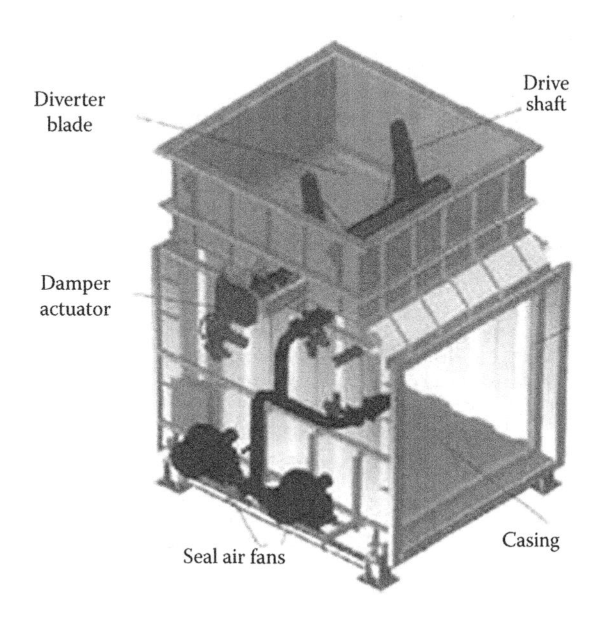

FIGURE 4.15
Diverter damper.

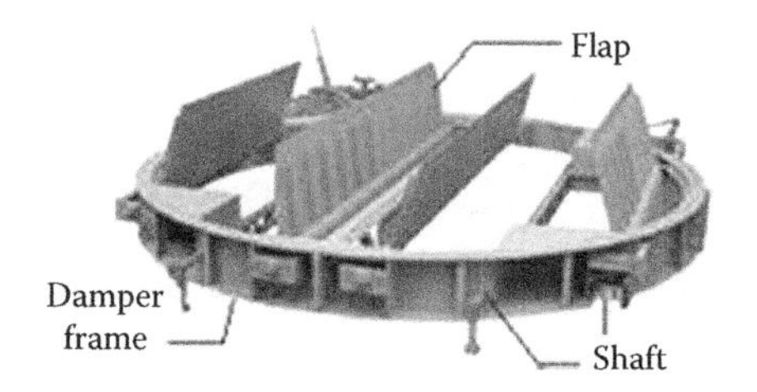

FIGURE 4.16
Multi-flap weather damper.

which is placed in one or the other position to stop the flow through one channel and divert it entirely to the other. The seal system is made quite elaborate so that the gases do not leak. Diverter damper in GT application is subject to very harsh conditions. The gas temperatures vary from 50°C to 650°C and gas pressures from nil to 400 mm WG. The dimensions are also very large.

4. *Weather/Stack Dampers*

There is hardly any requirement of preventing back flow of air or gas by means of dampers in conventional boilers. There is a rare requirement in case of Hrsgs where rain or snow has to be prevented from entering the unit through its large stack during idle conditions. A special type of non-return damper, called the weather/stack damper is usually installed at the top of the stack.

Weather damper consists of two or more flaps which are supported eccentrically on their shafts. During the running of Hrsg the flue gas pressure forces the damper plates to remain open. As soon as the GT stops and the pressure decays, the flaps close by themselves due to the eccentric positioning of the shaft. Motorisation of the damper is done to help the initial lifting of the flaps. To reduce the total space requirement the number of flaps is increased, as shown in Figure 4.16.

4.34.d Bypass Stack

Bypass stack is the stack assembly, including its silencer, which located above the two way diverter damper that is placed between the exhaust duct of the GT and inlet duct of the Hrsg. The purpose of the diverter damper is to guide the hot GT exhaust gases to the atmosphere when not required in the Hrsg located at the downstream. The high-velocity hot exhaust gases have also to be silenced. See Figure 4.17 which shows a complete CCPP with diverter damper and bypass stack assembly placed in a classical in-line arrangement.

Diverter damper and bypass duct are always together and are treated as single unit. As they are subject to high temperatures and high gas velocities hence are large, robustly built and expensive.

Most of the time there is no flow through the bypass stack in a CCPP.

If the plant is intended to operate in either open cycle or **CC** mode for the entire life time no bypass stack is required. Diverter damper and bypass stack assembly is provided only some times in CC mode to provide for the following special flexibility:

- GT can be initially installed in open cycle and Hrsg and ST are added a few years later to convert to CC. If a bypass stack is already installed the addition of ST and Hrsg can take place with no stoppage of GT.
- In a new CC plant the GT, whose delivery is fast, is installed first. The GT then works on open cycle and starts generating revenues until Hrsg and ST are installed later on, which may take nearly a year. Open cycle (in GT) (OC) to CC conversion can take place with no stoppage of GT if a diverter damper were already installed.
- Without a bypass stack the Hrsg is subject to the fury of the hot discharge gases of GT whenever there is a start up. This calls for more flexibility in its construction and up gradation to superior metallurgy. This expense can be saved with the bypass stack.
- The bypass stack separates the GT and ST operation thereby permitting the GT to operate in open cycle even when there is any problem with ST or Hrsg.

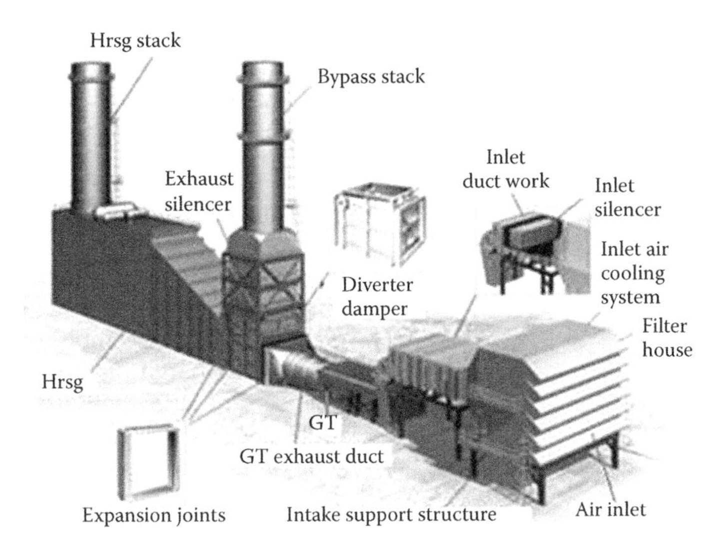

FIGURE 4.17
Bypass stack assembly in CCPP along with other main equipment.

The decision to include a diverter damper has to be very carefully evaluated. Against the advantages of flexibility in operation there are the following serious drawbacks as well:

- The damper assembly is quite expensive and complicated. This is because
  - Due to high-temperature operation the damper and stack both have to be of ss construction.
  - The gas pressure experienced is also high calling for a heavy duty execution.
  - As gas leakage through the idle stack contributes to loss of heat/η the diverter. Damper has to be thoroughly leak proof.
- This equipment is large and occupies sizable space along with its ducting.
- There is an element of gas pressure drop involved in damper and ducting, which together with small gas leakage in the bypass stack, contributes to loss of power which is sizable when reckoned over the life of the plant.

The present trend is to avoid the bypass stack but it is also a matter of individual plant philosophy.

4.35 Drums (*see* Pressure Parts)

4.36 Drum Anchoring (*see* Drum Supporting)

4.37 Drum Internals (*see also* Steam Purification)

Drum internals is often mistaken for merely steam separators. On the contrary, it is an all inclusive term referring to all the internals placed inside the steam drum which includes

- Steam purifiers consisting of
  - Steam separators
  - Steam scrubbers/driers
  - Dry box
- CBD pipe
- Chemical dosing pipe
- fw distribution pipe

Proper design, fitment and operation of all the drum internals are very important for satisfactory performance of the boiler.

Diameter of the drum is decided, particularly at HPs, on the ease of fitting the steam purifier equipment.

Figure 4.18 depicts all the drum internals. Horizontal steam separators are shown here.

For steam purification, the following are installed in the drum for steam–water mixture to pass serially in

1. Steam separators for separating small amount of steam from bulk of water
2. Steam scrubbers or driers for separating traces of water from bulk of steam

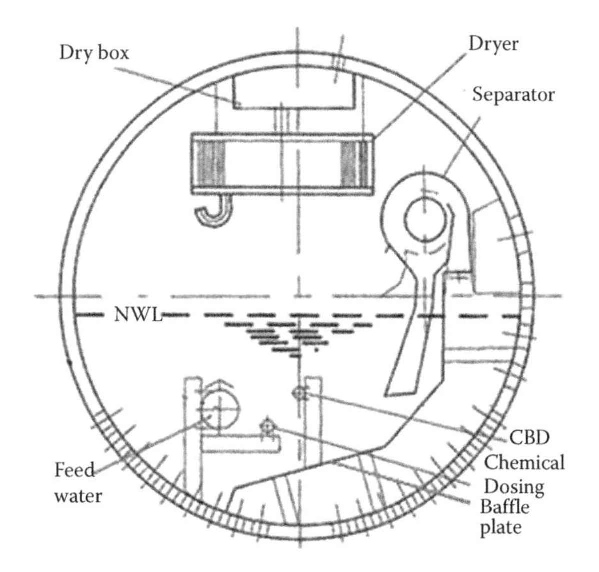

FIGURE 4.18
Cross-section of steam drum showing the drum internals.

Steam Separators

Steam separators are of two types.

1. Deflection baffles and inertial separators
2. Circular or cyclone separators

1. *Deflection baffles and inertial separators* are simple in construction; have less pressure drop and easy to fit. They are separators where the mixture is subjected to change of direction and are used up to ~130 bar drum pressure. A typical inertial separator is shown in Figure 4.19.
2. *Circular or cyclone separators* are more complicated in construction, consume more pressure drop and are more difficult to install. They work on the principle of separation by centrifugal action. They can be used from very low to very high drum pressures—up to 210 bar. Between the
   a. Vertical cyclone separators
   b. Horizontal cyclone separators

the latter are more compact. Figure 4.20 shows the vigorous separation action in a vertical cyclone while Figure 4.21 shows the assembly of steam separating devices, namely the baffle plate, vertical cyclone and scrubber. Intense separation activity takes place inside the steam separators as the mixture passing through the steam separators is between 5 and 50 times the steam output, depending on the boiler pressure.

Figure 4.22 shows the internals of a large-boiler drum where there are two rows of cyclones on each side.

Centrifugal separation of steam from water can employ both horizontal and vertical separators. Figure 4.16 depicting the assembly of drum internals shows horizontal separators.

Steam Scrubbers

Steam scrubbers or driers work on the principle of adsorption to abstract the last traces of steam in the water. Water has great affinity for steel. Scrubbers can be in the form of *corrugated plates or demister*. Steam passes through them very slowly to stick to the plate or mesh. Scrubbers should have smooth and corrosion-resistant surfaces and hence are made of ss 304 or equivalent material. Corrugated scrubbers are bolted together to form packs and are placed above vertical cyclones or just ahead of dry box in horizontal separators in such a way that all steam necessarily passes through them for drying.

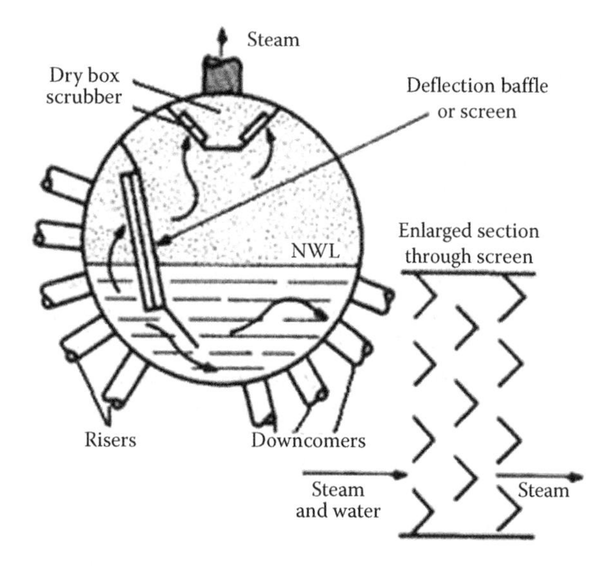

FIGURE 4.19
Inertial separation of steam and water.

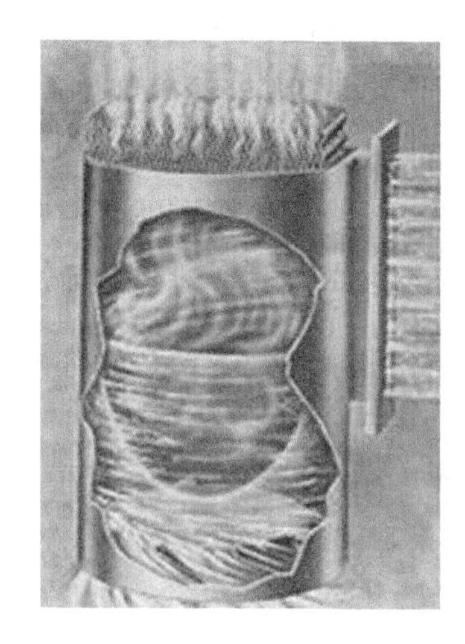

FIGURE 4.20
Vigorous swirling motion of steam and water inside a vertical cyclone separator.

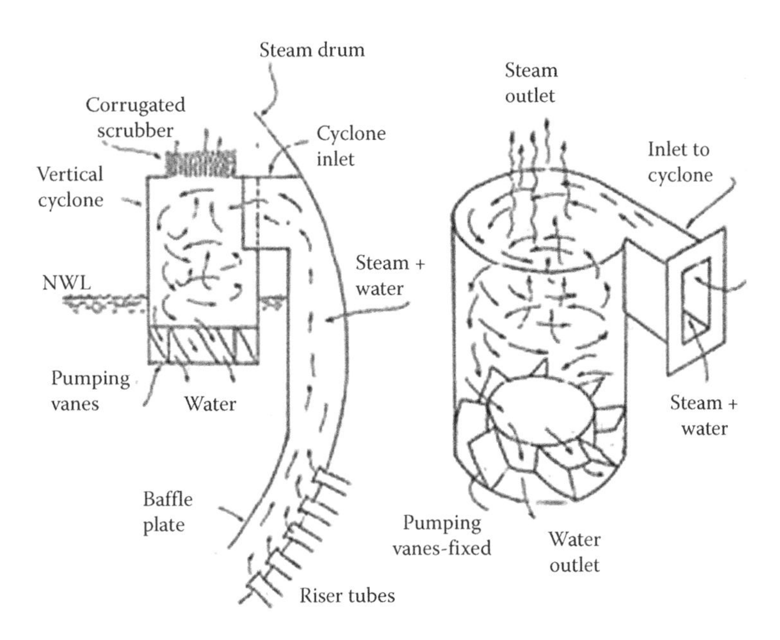

FIGURE 4.21
Steam separating equipment inside the steam drum.

The most important requirement in steam purification is to ensure that steam is uniformly distributed to all the separators. Likewise, steam should also be withdrawn from the dry box in a uniform manner.

An isometric view of cross-section of a steam drum with all internals is shown in Figure 4.23. Note that the drum is made of two different thicknesses, with the lower half being thicker, as it has lower ligament η.

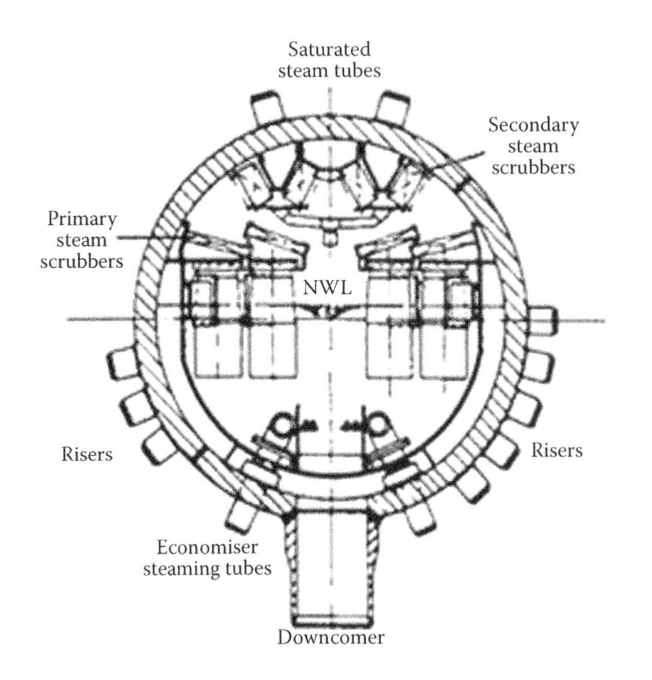

FIGURE 4.22
Steam separation with four rows of cyclones in a high-capacity boiler with large steam drum.

4.38 Drum Pressure (*see* Boiler Pressures)

4.39 Drum Slings

A steam drum is supported only at its two ends either by

- Slings from the top suspension girders, which is the most common or
- Bottom saddles resting on the supporting beams at drum floor

Drum slings can be in the form of either

- U rods or solid/bar slings
- Plate or laminated slings

U rods or bar/solid slings, shown in Figure 4.24, are used for total weights of up to 200 te, which is adequate for most industrial and smaller utility boilers. Solid rods are made of high-tensile bar material in 50–150 mm diameter depending on the load. Usually two or three lengths are welded to form the required single length. The welding followed by stress relieving and radiography, are all carried out with utmost stringency.

For higher loads, laminated or plate slings fabricated from high-tensile plates are used. Typical weight that a pair of plate slings can support can be as high as 1000 te.

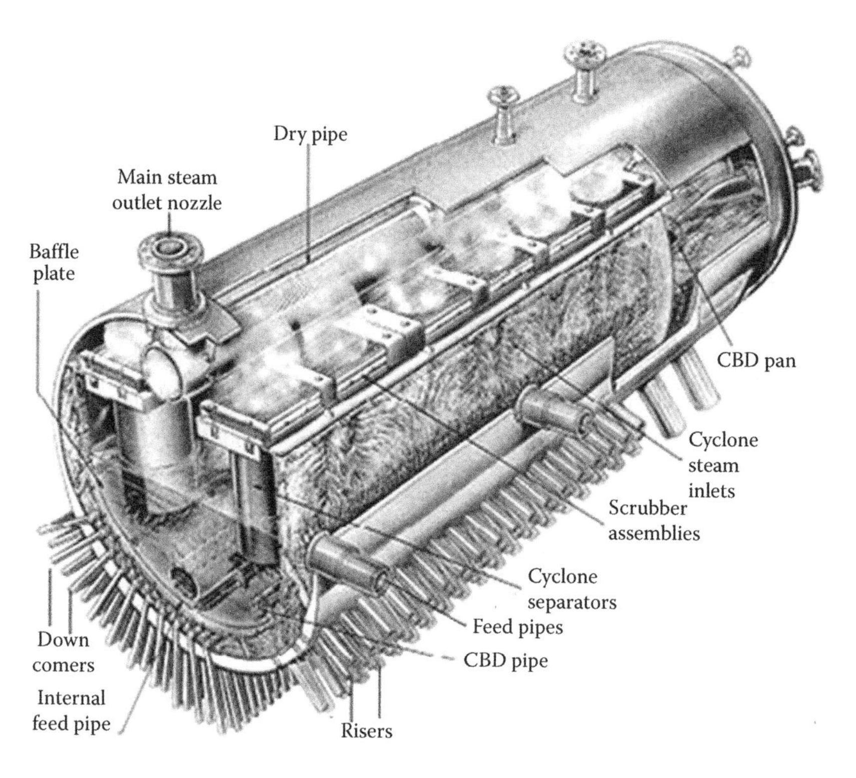

FIGURE 4.23
An isometric view of all the drum internals and external connections in a steam drum.

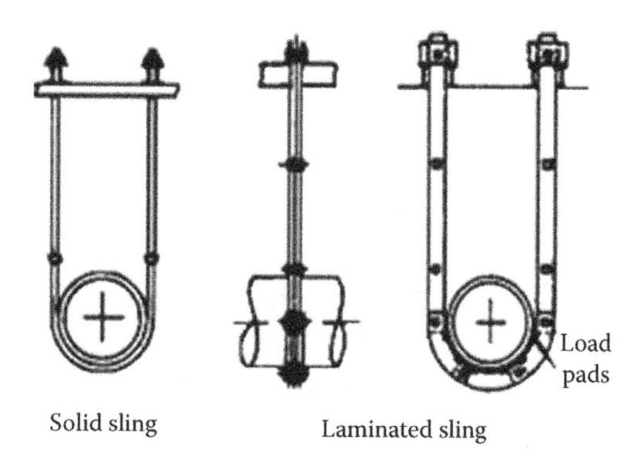

FIGURE 4.24
Bar and laminated drum slings.

Since the plate slings are made in several pieces, they fit the drum profile more closely than rod slings, *see* Figure 4.25 which shows both types.

4.40 Drum Supporting Method

In top-supported boilers, steam drum is usually hung from the top suspension girders by means of drum slings. Sometimes it is also supported on two beams by drum saddles welded on either end of the shell or dish ends. The water drum is not supported.

In bottom supported boilers, the drums are supported by large bore down-comers from the same level as the bottom headers. Alternatively, when the BB is long and the bottom drum is almost at the base level, as in the case of package or stoker-fired boilers or horizontal Hrsgs, the support is given to the bottom drum. The top drum is not supported but allowed to expand upwards.

Steam or water drum, whichever is supported, is further anchored usually at the midpoint so that it can expand axially on either side along with the furnace. The anchoring prevents a drum from moving in a direction perpendicular to its axis. This way the expansion of all PP is guided and controlled with respect to anchor point. The anchoring of the drum can also be done at either end.

4.41 Drum Water Treatment (*see* Water)

4.42 Dry Air (*see* Combustion Calculations in Combustion)

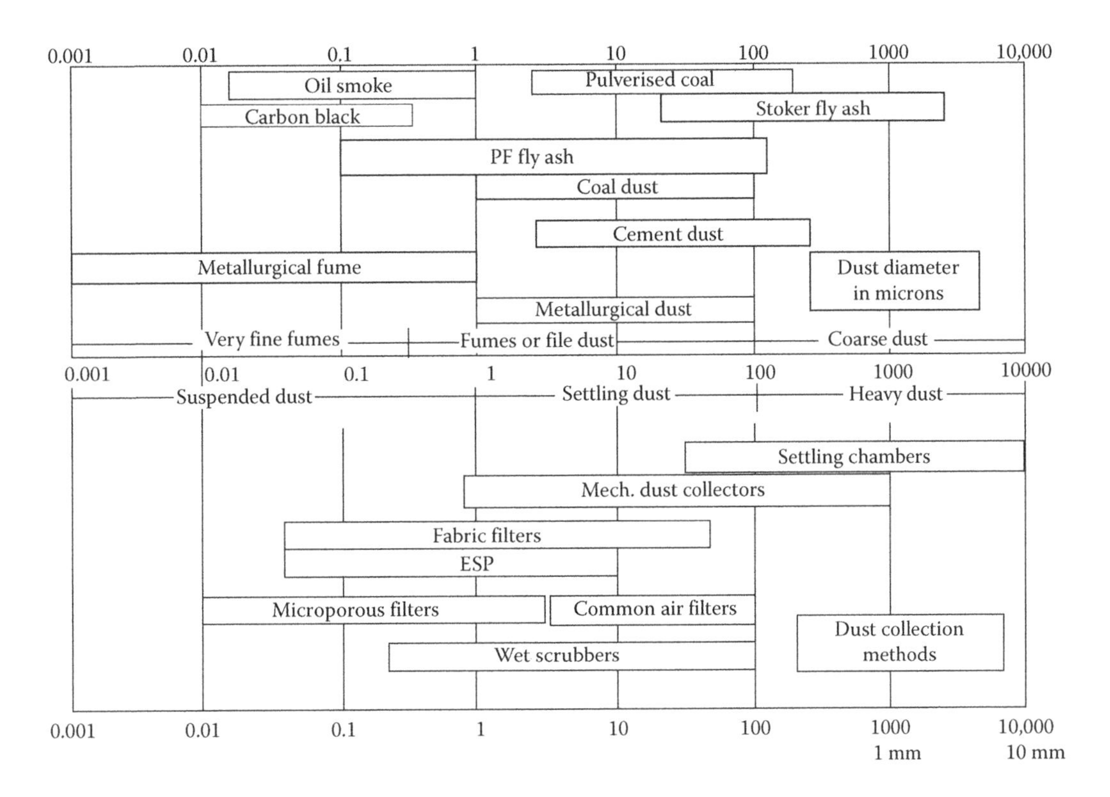

FIGURE 4.25
Dust gradation and dust collection.

4.43 Dry Bulb Temperature T_{db} (*see* Psychrometric Chart)

4.44 Dry Wall Flow or Film Boiling (*see* Flow Types in Vertical and Horizontal Tubes)

4.45 D-type Boilers (*see* Package Boilers in Oil and Gas Fired Boilers)

4.46 Ductility (*see* Metal Properties)

4.47 Dulong Formula (*see* Coal Calorific Value *under* Coal)

4.48 Dust Collection Equipment

Dust in flue gas is mainly some fine ash and unburnt C elutriated from the combustion chamber by flue gases. Suspension firing in stokers, fluidised-bed firing and PF firing all contribute to high levels of dust burden in flue gas.

NG and many other manufactured gases from refinery are clean and dust-free but gases such as BFG and COG are dusty as they are derived from solid fuels. Heavy FOs produce fine C in flue gases, called soot.

All these dust emissions are required to be captured before the flue gases are emitted into atmosphere.

Dust collection equipment is also required to meet the specified **opacity** limit which is a measure of clearness of stack. A beam of light is passed through the flue gases and checked as to how much is either absorbed or scattered. This limit is usually ~20%. See Opacity of stack.

Dust, smoke, fumes and mist are all based on decreasing particle size. The collection principle and equipment has to be different for different grades of dust (as shown in Figure 4.25).

Different combustion systems generate different types and amounts of ash in flue gas. Stoker firing produces 10–40% of fly ash with plenty of grit. PF firing produces between 70% and 90% of fine fly ash from 7 to 12 μm. FBC generates 90–95% of fly ash along with sizeable $CaSO_4$ if the desulphurisation is carried out.

Dust collection equipment can be categorised as

1. Mechanical collectors for low and medium collection η
   a. Inertial separators
   b. Cyclone dust collectors
2. Filters for very high collection η
   a. Electro filters/electrostatic precipitators
   b. Bag/cloth/fabric filters
3. Wet scrubbers for medium to high-collection η and special applications

These are described in detail under respective headings. A summary of the salient features of all the equipment is given in Table 4.2.

Also the collection ηs of the various dust collecting mechanisms is given in Figure 4.26 for a range of dust sizing.

4.48.1 Mechanical Dust Collectors

Mechanical collectors for low- and medium-collection η are of two types:

a. Inertial separators
b. Cyclone dust collectors

4.48.1.a Inertial Separators

They use inertia and gravity to separate dust particles from gas. Baffle plates, inside ash hoppers, in the gas turns and channel type separators in CFBC boilers of U-beam design (see Figure C-6) are typical examples. With 30–70 mm wg of pressure loss 50–70% overall collection η is realisable with most small particles of <10 μm escaping. They are mostly used as pre-collectors. Appropriate material selection for the baffle plates is important to save them from erosion when ash is aggressive.

TABLE 4.2
Dust Collection Methods and Their Areas of Application

Equipment	Type	Particle Size (μm)	Pressure Loss (mmH_2O)	Collection η (%)	Equipment Cost	Operating Cost	Space Needed
Gravity dust collector	Baffle chamber	>50	10~15	40~60	Small	Small	Very small
Inertial dust collector		>10	30~70	50~70	Small	Small	Small
Centrifugal collector	Large cyclone	>10	50~150	<85	Medium	Medium	Small
	Multi-clone	>5		<95	Medium	Medium	Small
Scrubbing collector	Venturi scrubber	>0.5	300~900	80~99	Medium	High	Medium
Filter dust separator	Bag filter	>0.2	100~200	>99	Medium or higher	Medium or higher	Large
Electrostatic precipitator		<2.0	10~20	>99.9	High	Small to medium	Much larger

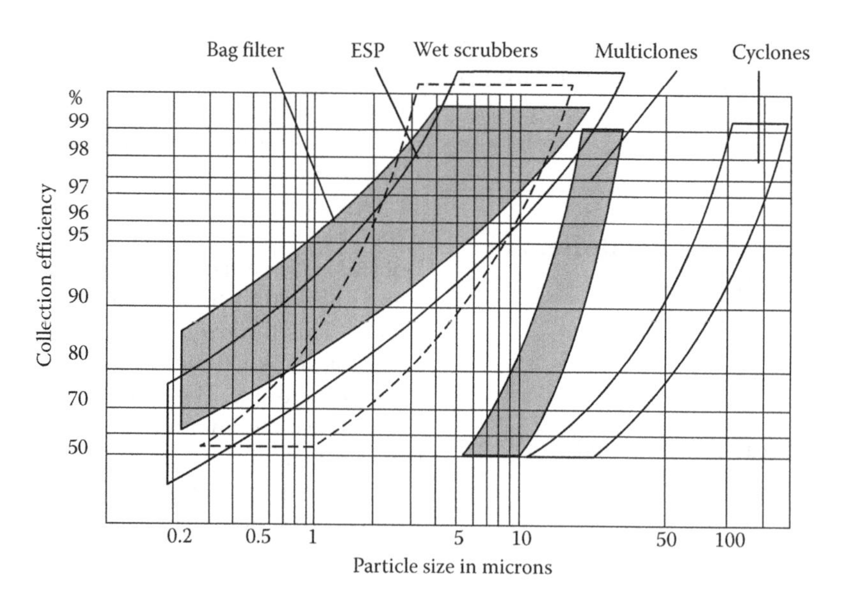

FIGURE 4.26
Collection efficiencies of various types of dust collection equipment.

4.48.1.b Cyclone Separators

Separation of dust by centrifugal action is the principle behind the cyclone operation.

Dusty gas is accelerated to a velocity of ~25 m/s and given a near circular motion along the periphery of the cyclone when the dust progressively separates and falls down along the outer walls. The swirling jet is given a gentle downward spin on sloping walls and finally led outwards by the centrally located outlet pipe.

With 50–150 mm wg of pressure drop a collection η of 85–95% is achieved with almost all sub-10 μm dust not getting captured.

Cyclones were the only dust catchers in many boilers some 4–5 decades ago when the emission rules were a lot more benign. Their main use nowadays is for pre-collection of ash mainly in FBC boilers. For some time they were used as pre-collectors before ESPs in PF boilers ostensibly to reduce the dust burden to make ESPs more compact. This hope was belied as the size of ESPs was governed more by the time taken by the fine particles and not so much by dust loading.

Cyclones can be of large or small diameter. Large cyclones can vary from 1 to 10 m diameter or more and are used in 1, 2, 4 and 6 numbers. Typical large diameter cyclone is shown in Figure 3.4.

Small diameter cyclones or multi-clones are employed in large numbers in a unit for higher η and more compact arrangement (Figure 4.27). Owing to smaller diameter of 200–500 mm, the centrifugal action is better and hence the collection η is higher. The attendant pressure losses are also more. Cyclones are made of white CI or Ni-hard with hardness >400 BHN to withstand wear because of high level of inlet dust or abrasive ash.

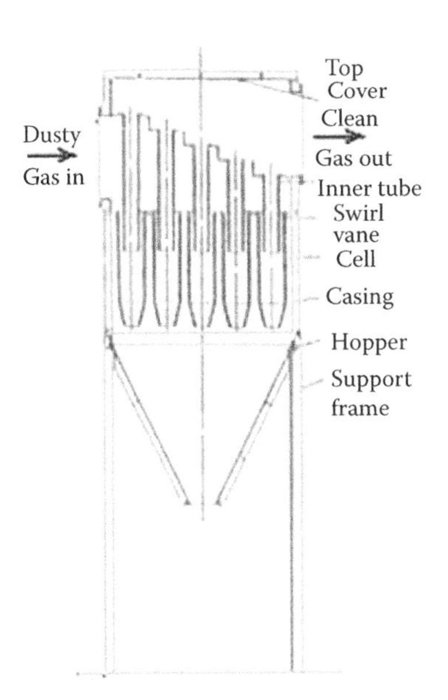

FIGURE 4.27
Multi-clone type centrifugal dust collector.

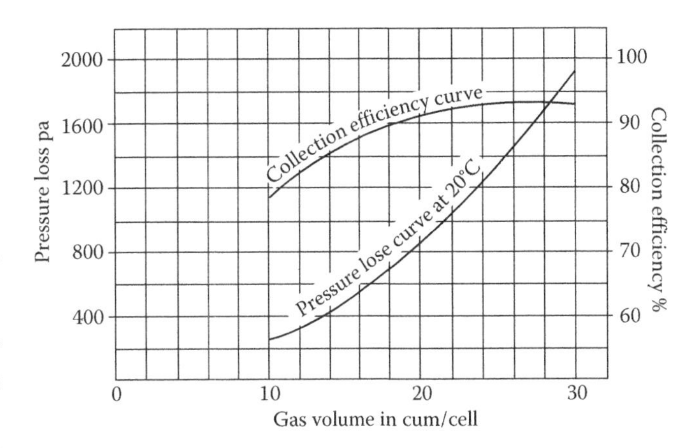

FIGURE 4.28
Performance of a multi-cyclone dust collector.

Both types of cyclones offer the advantages of constructional simplicity, compactness, low cost and negligible maintenance. Running costs are medium to high as the pressure drops are 50–125 mm wg. The main drawback is that the collection η for very fine particles of <5 μm is poor resulting in unacceptable stack opacity. Also with high inlet dust loading, the multi-clones tend to choke. Typical performance of multiclones is depicted in Figure 4.28.

Multi-clones are more suited for coal applications. Large diameter cyclones can take heavier dust loads and withstand very high temperature of order of 900°C with adequate refractory lining as in the CFBC boilers. They are also better suited for bio fuels which have soft ash needing lower velocities and larger turns to prevent dust re-entrainment.

4.48.2 Filters

4.48.2.a Electro Filter or Electrostatic Precipitator

Electrostatic precipitator (ESP) or electro filter is a high η dust collector which is commonly used behind both industrial and utility boilers.

ESP is a large air tight duct in which emitting electrodes and collecting plates are hung from the top and impressed with very high voltage DC power (as shown in Figure 4.29). This high voltage charges the dust particles negatively in slow moving gas stream and migrates them to the collecting electrodes that are periodically rapped by hammers. Rapping helps to dislodge the dust which gets collected in the bottom ash hoppers. This, in brief, is the way an ESP works.

ESP is large electro mechanical equipment. The mechanical items in an ESP are

- Housing with inlet and outlet funnels and perforated inlet and outlet screen
- Ash hoppers of adequate storage capacity with electrical heating pads

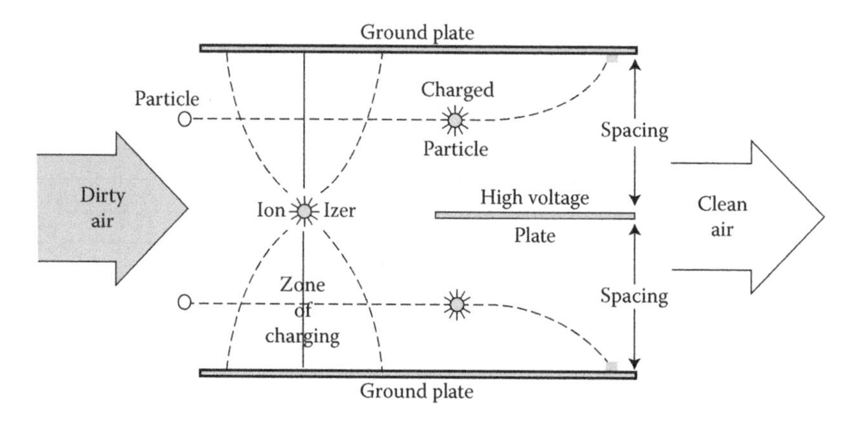

FIGURE 4.29
Operating principle of an ESP.

- Supporting structure for the housing with suitable expansion arrangement.
- Electrode assembly with proper supporting arrangement and insulations.
- Rapping system—either mechanical or magnetic.
- Insulation and lagging.
- Galleries and ladders for proper access.

The electrical system consists of transformer rectifier (TR) sets placed on the roof with cabling running from the junction box to TR sets. Control system consists of electronic controls to vary the parameters for optimum collection. A typical large ESP for boiler application is shown in Figure 4.30. This has rigid discharge electrodes with magnetic impulse gravity impact (rapping in ESP) (MIGI) rapping from the top for both collecting and discharge electrodes.

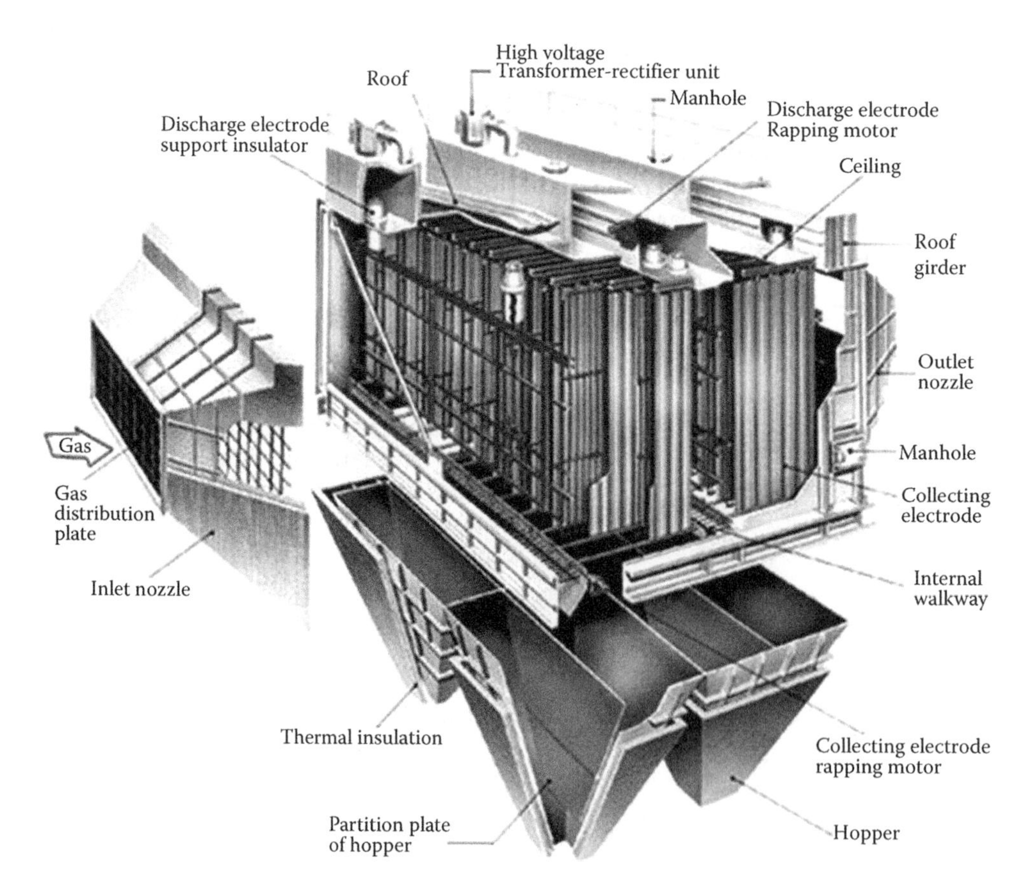

FIGURE 4.30
Typical large ESP and its parts.

ESPs are characterised by

a. Type of emitting electrodes—weighted wire, rigid or rigid frame:

   Each is progressively made more robust. For abrasive and heavier dust loading, rigid frame designs are to be accepted as the most suitable, at least in the larger utility sizes. Weighted wire is essentially a spring with a weight at the bottom. Rigid electrode is any specially shaped electrode, including spiked rods, made stronger than the spring. Rigid frame is the encasement of the electrodes in structural frames for even better strength. Weighted wire design is normally limited to lengths of 9 m. The rigid frames can be as tall as 15 Mtrs. Figure 4.31 shows different types of electrodes.

b. Type of rapping—mechanical or MIGI

   Mechanical rapping consists of striking the discharge electrodes, collecting electrodes and screen plates periodically by swinging hammers which are rotated by rapping shafts located at various places along the gas flow. Rapping can be at the top, middle or bottom of the plates.

   MIGI rapping is the hitting of collecting electrodes from the top by vertical plungers which are controlled by solenoid coils. The former is European while the latter is American practice, each with its own set of features.

c. Spacing of collecting electrodes across the gas flow, narrow or widely spaced

   Closely spaced electrodes (200–250 mm) is now increased to 400 mm or above in large ESPs as the voltage in TR sets is steadily increased from <50 to >110 kV peak. Wide spacing helps full manual access for inspection and cleaning of the plates. ESPs for modern power plants, both small and large, are adopting wide spacing of electrodes.

d. Number of fields

   With emission levels getting progressively lower and lower ESPs in power plants are rarely built with <3 fields. As many as seven fields are needed for very low dust emission of ~30 mg/Nm^3 or even less with high ash coals.

e. Number of streams—single or multiple

   For small- and medium-sized boilers it is normal to have single stream of ESP. When mechanical limitations are reached double or even multiple streams are employed for larger gas flows.

Advantages and Limitations of ESP

The overwhelming popularity of ESPs is primarily due to

- Low-pressure loss (<25 mm WG)
- Low-power consumption
- Minimum maintenance

The collection efficiencies can easily exceed 99.9% and the range of dust collected is very wide from submicron particles of 0.05 μm to as large as 10 μm and more.

High initial cost and large space requirement are the main drawbacks.

Collection η of particles <2 μm is also poor.

Collection η of ESP

Collection η of an ESP is the ratio of dust collected to the inlet dust.

$$\eta = \frac{(\text{Inlet dust} - \text{Outlet dust})}{\text{Inlet dust}} \quad \text{or} \quad \frac{1 - \text{Outlet dust}}{\text{Inlet dust}} \tag{4.13}$$

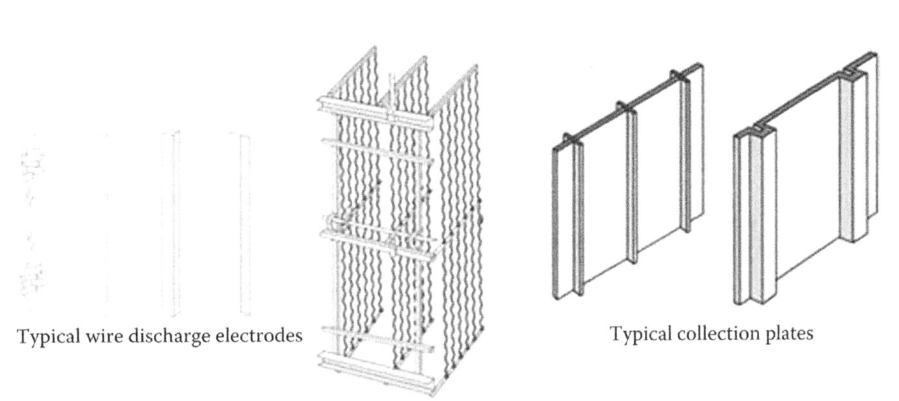

FIGURE 4.31
Discharge and collecting electrodes of ESPs.

Collection η is thus *a*% of inlet load. In other words, an ESP is a *constant percentage collector* unlike a bag filter which is a *constant emission collector.*

Collection η improves with

a. *Specific collection area (SCA),* that is, area for unit gas volume in $m^2/Am^3.s$
b. *Migration or drift velocity,* that is, velocity of charged particle as it moves towards collecting electrode in m/s.

Migration velocity is governed by ash resistivity. Lower the ash resistivity, higher is the migration velocity and better is the collection η. It also means a smaller collection area and a compact ESP.

Ash resistivity varies from typically 10^8 to 10^{14} Ω cm. Ash from low-S coals has an ash resistivity >2×10^{12} Ω cm in the temperature range 120–220°C which is very difficult to collect and calls for large ESP for maintaining good collection η. As ash resistivity reduces with temperature, *ash conditioning* is sometimes done for improving performance.

Deutsche–Anderson equation is the classical equation used to determine the collection η of the precipitator under ideal conditions and is expressed as

$$\eta = 1 - e^{w(A/Q)} \quad (4.14)$$

where

η = collection η of the precipitator
e = base of natural logarithm = 2.718
w = migration velocity, cm/s (ft/s)
A = the effective collecting plate area of the precipitator, m^2 (ft^2)
Q = gas flow through the precipitator, m^3/s (ft^3/s)

This equation has been used extensively for many years to calculate theoretical η. Whereas the equation is scientifically valid a number of operating parameters can cause the results to be in error by a factor of 2 or more as it neglects three significant process variables. It assumes that

- The dust re-entrainment to be zero during the rapping process.
- The particle size and, consequently, the migration velocity are uniform which is not true as larger particles generally have higher migration velocity rates than smaller particles.
- The gas flow rate is uniform everywhere across the precipitator and that particle sneakage (particles escape capture) through the hopper section does not occur. Particle sneakage can occur when the flue gas flows down through the hopper section instead of through the ESP chambers, thus preventing particles from being subjected to the electric field.

Therefore, this equation is used only for making preliminary estimates of precipitator η. Realistic projections of η are obtained by using modified Deutsch–Anderson equation.

ESP versus Bag Filter

For a comparison with bag filter please see the description under the heading of bag filter.

4.48.2.b Bag Filter/Cloth Filter/Fabric Filter

Bag filter is a large airtight duct in which collecting bags are so arranged that the dusty gases are compelled to pass through them and leave behind the dust on the bags. The bags are periodically de-dusted when the pressure drop across the bags reaches the upper limit because of the chokge. The real filter is the thin layer of dust that is settled on the bags and not the bag itself. This layer which is closely adherent to the bag is, therefore, intentionally not dusted off the bag surface.

Bags for boiler flue gases are made of fibre glass and synthetic felt like proprietary materials, such as Ryton, Nomex, Daytex, Teflon and so on depending on the ash to be collected. Bags are cylindrical in shape, stitched lengthwise in cloth and closed from one side.

Depending on the type of cleaning of the bags are the filters classified into three types, namely

1. Shaker filters
2. Reverse gas filters
3. Pulse jet filters

In 1 and 2 above, the bags are open at the bottom from where the dusty gases enter and leave behind the dust on the inside of the bags. Mechanical shaking is employed in shaker filters which imposes severe stresses on the bag and reduce their life. Shaker filters are, therefore, not popular.

In reverse gas filters, the section needing cleaning is isolated and the flow of cleaned gases is reversed through it to shrivel the bags and remove the dust.

In pulse jet filters, the filtration action is opposite. The bags are open at the top from where the cleaned gases exit after depositing the dust on the outside of the bags. Periodically the bags are blown with HP compressed air stream to expand the bags and remove the dust. Bags in Pulse jet filters, therefore, are required to be stronger.

Figure 4.32 describes the action of reverse and pulse jet filters.

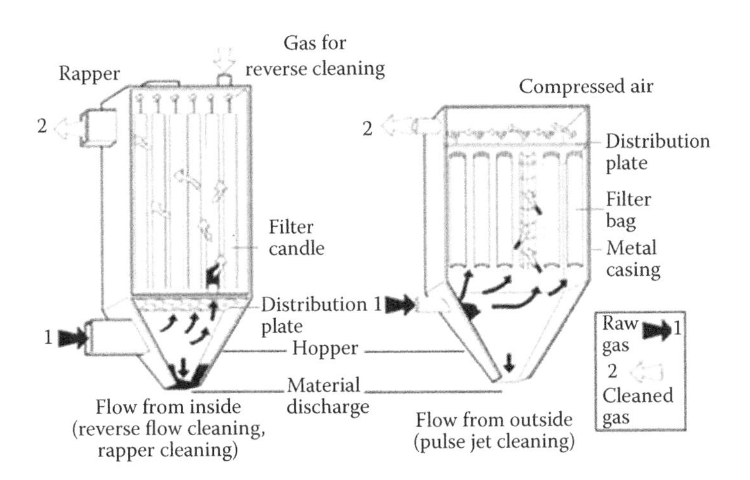

FIGURE 4.32
Reverse and pulse jet fabric filters.

Reverse gas filters are also called as bag houses while pulse jets are called as bag/fabric or cloth filters.

Advantages and Limitations

Bag filters enjoy the advantages of

- Constant high dust collection η (>99.9%)
- Relatively lower installed cost
- Smaller footprint
- Superior collection as very fine particles of <2 μm can also be captured (not possible with ESP)

The main disadvantages are

- Bag replacement is needed every 2–3 years depending on the abrasiveness of ash
- High gas pressure drops of up to 150 mm wg calling for high auxiliary power
- Not particularly suitable for biofuels with very light fly ash which can deposit the burning char and destroy the bags

Bag Filters versus ESP

Comparison with ESP is inevitable. In general ESP

- Needs higher initial cost
- Occupies larger area
- Consumes much lower auxiliary power
- Is practically is maintenance free
- Is capable of withstanding much higher temperature ~400°C against bag filters at 160–180°C (230°C with fibre glass bags).

Although ESPs have been a standard with boiler installations bag filters are gaining acceptance where extremely low-emission levels of the order of 10 mg/Nm^3 are to be met provided the bags can withstand the abrasion of ash. Also, bag filters are a better choice if the flue gas is known to be having very fine dust having size <2 μm.

Bag Filter Performance

The primary parameters deciding the performance are

- Air to cloth (A/C) ratio
- Gas pressure drop
- Fabric life

A/C ratio captures as to *how much air is passing per unit cloth surface* and is the most important sizing factor. It is expressed in Am^3/m^2/h. It is also called the *filtration velocity* expressed in m/min or fpm. Lower the ratio or velocity higher is the filter size. Typical A/C ratio or filtration velocity for

Reverse gas filters is 0.45–0.7 m/min (1.5–2.3 fpm)
Pulse jet filters is 0.90–1.20 m/min (3–4 fpm)

These are for coal firing and on net basis (net area = total cloth area–area under cleaning)

Reverse Gas Filters

In *reverse gas filters*, as shown in Figure 4.33, the bags are typically of 200–300 mm diameter and 7–11 m long. The cleaned gases are made to flow back through a section of bags needing to be cleaned. The cleaning operation is gentle allowing longer bag life and use of woven material.

Pulse Jet Filters

In pulse jet filters, depicted in Figure 4.34, the cleaning forces are heavier. Fresh air (free of oil, dust and M) at high pressures of 5–8 barg is blown through

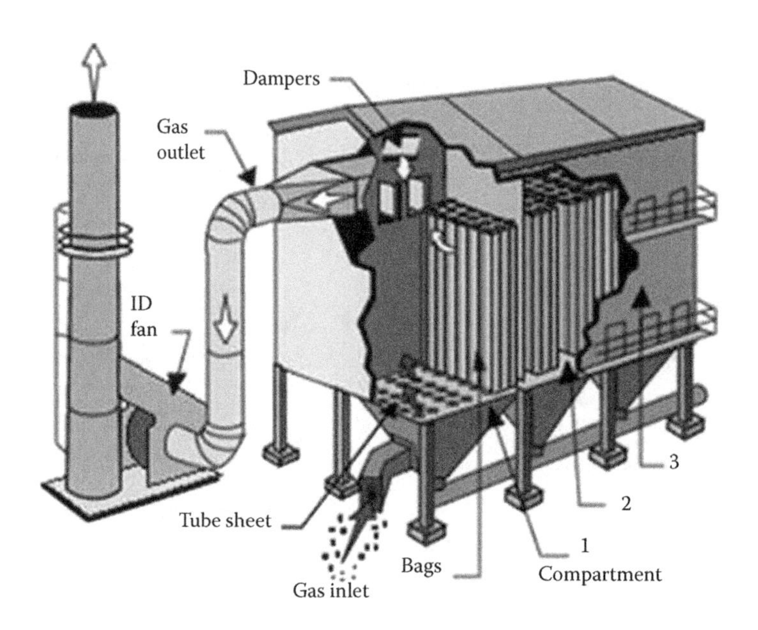

FIGURE 4.33
Reverse gas filter or bag house.

the bags demanding an expensive felted material and stronger support cages for bags. The bags are smaller 130–150 mm in dia and shorter, 5–9 m long. A/C ratio is also significantly higher making the pulse jet filers very compact.

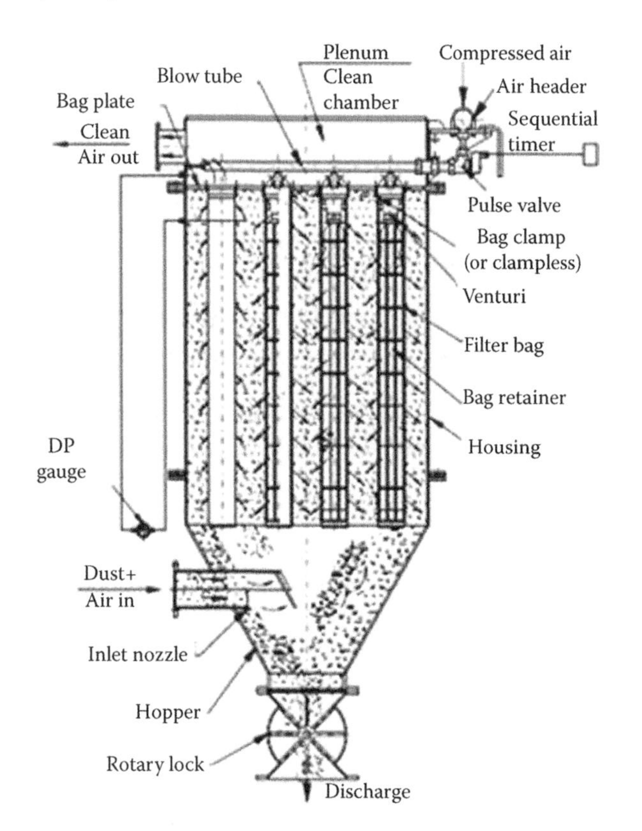

FIGURE 4.34
Pulse jet filter or bag/fabric or cloth filters.

Reverse jet filters have been popular worldwide despite larger area requirement and higher initial cost because the bag life is longer and replacement cost is lower. The pressure drops are also lower. Over the years, improvements have been made for pulse jet filters to narrow this gap.

4.48.3 Wet Scrubbers

Conventional dust collection equipments directly abstract dust form flue gases and precipitate in dry form. But scrubbers employ an intermediate medium such as water spray which impounds the dust particles in the flue gas and separates as wet slurry, which is then suitably treated before disposal. Wet scrubber is thus a two-stage separation phenomenon.

The advantage of wet scrubber is its extreme compactness even after the slurry treatment arrangement is taken into account. The serious drawbacks are

- High power consumption
- High treatment cost of the slurry
- Corrosion protection needed for all equipments downstream of the scrubber

If S is present in the fuel, the sulphurous gases coming in contact with water makes the slurry acidic.

Wet scrubbers have been popular with pulping industry behind small- and medium-size BLR boilers in the former times. As the flue gases coming in contact with BL pick up its malodour, wet scrubbers have been discontinued for long. They are frequently used for the

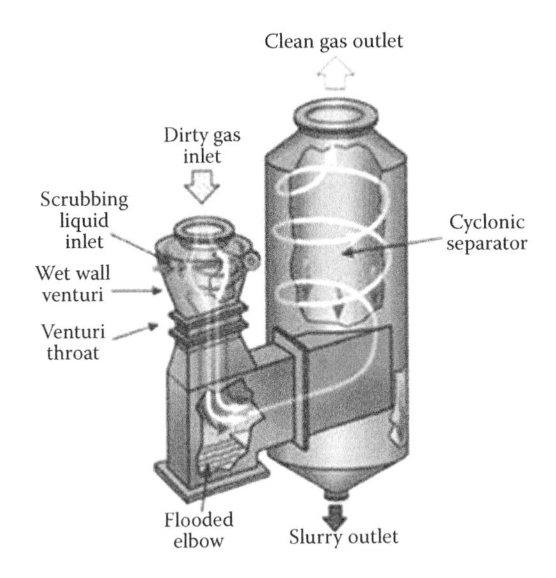

FIGURE 4.35
Wet scrubber.

boilers firing biofuels, particularly when space is at premium. With no S in most biofuels, corrosion issues are almost absent.

A wet scrubber is a combination of venturi scrubber and a cyclone separator, as shown in Figure 4.35. Flue gases entering the scrubber are accelerated to velocities of 30–60 m/s in the throat of the venturi where water is sprayed through several nozzles. The fine droplets of water coming in intimate contact with dust settle out of the gas stream. The dust remaining with flue gas then rotates several times in cyclone separator where it gets collected on the separator wall because of the centrifugal force. Many times, there is also a wall flow to wash the dust. The cleaned gases exit from the top while the slurry is collected at the bottom for further treatment. Collection ηs are comparable of ESP and bag filter at >99.9% but the cost of power for meeting the gas Δp across the scrubber and for pumping water are much higher than in bag filters. For control of the collection η, an adjustable damper is usually provided at the venturi throat.

4.49 Dynamic Classifier (*see* Vertical Mills in Pulverisers)

5

E

5.1 Economiser (*see also* Backend Equipment in Heating Surfaces *and* Pressure Parts)

Eco is a gas–water HX to heat the incoming fw with flue gases before fw reaches the steam drum. In case of SC boilers, water from Eco is led to the water walls as there is no drum. It is so named as it 'economises' on the use of fuel by cooling the hot flue gases to acceptable economical levels. It is the last water-cooled HS in a boiler. Being a low-temperature HX, Eco needs a lot of HS, which, in turn, demands more space, supporting structure and soot blowing, making bulky.

Hot flue gases are ~400°C at the exit of Evap in bidrum boilers or ~700°C at the exit of SH/RH banks in single-drum boilers. When Eco is followed by AH, as is the case with most power boilers, cooling of gases in the Eco is modest and depends on the gas temperature needed at AH inlet to suit the heating of air appropriately.

When there is only Eco and no AH, as in many process boilers, the optimum cooling of flue gases varies with the incoming fwt and S in fuel and is limited by the low temperature or **dew point corrosion** of steel. It is usually between 130°C and 160°C or even higher for coal and oil, depending on the fuel S. It can be as low as 110°C for NG firing in industrial boilers and even lower at 75°C in Hrsgs. See Figure 1.14 showing the permissible metal temperature versus S in fuel in the topic of ash corrosion.

Ecos are susceptible to both inside (water side) and outside (gas or ash side) corrosion.

Even traces of dissolved O_2 in fw can cause **pitting corrosion** on the inside of the tubes, which is prevented by deaeration followed by O_2 scavenging of the fw.

External corrosion is caused by the condensation of H_2SO_3 acid of the flue gases on tubes. This is prevented by maintaining fwt appropriately high, depending on the S in fuel.

Eco banks have to be guarded against ash fouling by maintaining adequate lateral spacing depending on the fuel. Ash in flue gas at temperatures <600°C is almost entirely friable in nature and does not foul Ecos as in SHs. **Mass or lane soot blowing** is adequate to keep the tubes clear of the dust. Clear lateral tube spacing is rarely less than 19 mm (¾″) even for NG and is increased to 38 (1½″) or even 50.8 mm (2″) for high-ash-coals. The tube banks along the gas flow are rarely more than 1.8 m deep from soot bowing consideration. Ecos in NG-fired boilers do not require SBs and naturally the tube banks can be deeper as dictated by accessibility considerations.

Ecos also have to be protected against gas-side erosion by limiting the flue **gas velocities** in tube banks appropriately. At the same time, high-gas-velocities are essential, as the heat transfer at these low gas temperatures is almost entirely by convection. Moreover, it is highly necessary to keep the surfaces compact to be economical. Satisfactory experience with the fuel is the best guide for sizing. Typical gas mass velocities range from 5.4–6.8 kg/m^2 s (4000–5000 lbs/ft^2 h).

Eco tubes are invariably made of CS as it is always sub-cooled water, with its excellent cooling abilities, which flows through the tubes. The metal temperatures can rarely exceed 350°C. In SC boilers where there is a combination of very high p & t, the tubes are made of higher metallurgy like SA 213 T2 or T11.

The most common tube diameters are 38.1, 44.5 and 50.8 mm OD (1½, 1¾ and 2″). 31.8 and 38.1 mm OD (1¼ and 1½″) is popular in Hrsgs and SC boilers. Large-diameter tubes like 63.5 and 76.2 mm OD (2½ and 3″) are used in vertical tube Ecos of Hrsgs and other WH boilers like BLR and MSW boilers.

Eco classification Ecos are classified in several ways:

a. By flow arrangement
b. Steaming and non-steaming Eco
c. Bare tube and finned tube Ecos
   1. Helical or spiral finning
   2. Plate or rectangular finning
   3. Longitudinal finning
   4. Stud finning or pin/gun studding
   5. Steel tube Ecos with CI gills
d. Inline and staggered arrangement of tubes
e. Horizontal and vertical Ecos
f. CI Ecos

a. *By flow arrangement:* By way of water and gas flow arrangement, they can be classified as *counter, parallel,* and *cross-flow* Ecos. Most of them are in counter-flow arrangement to minimise HS. Also, fw must always flow upwards to prevent vapour formation leading to flow disturbance. In exceptional cases where fw flow has to be downward, special care is taken, like fitting of ferrules at tube inlets. Moreover, Δp of water must be quite high, at not less than 2 bar, across the Eco banks.

b. *Steaming and non-steaming Eco:* Most Ecos are non-steaming, where fw exit temperature from Eco is at least 30°C lower than the saturation temperature of drum water. This difference, known as the **approach temperature**, prevents water hammering.

   Steaming Eco is the one where the rise in fwt is >67% of the difference between the fw inlet and (drum) saturation temperature.

$$t_2 - t_1 > 0.67(t_{sat} - t_1) \tag{5.1}$$

   where t_1 and t_2 are the inlet and outlet water temperatures in Eco and t_{sat} is the saturation temperature of drum water.

   Actual steaming in the Eco is normally restricted to 5% by weight and higher levels are also occasionally employed when fw quality is very good. Special care is needed to make the steam-laden water gently enter the steam drum without steam hammering.

c. *Bare tube and finned tube Ecos: See also Extended surfaces:* Most Ecos are bare tube because of the simplicity of construction and ease of support. See Figure 16.5 which shows several coils of bare tube Eco stacked in shop floor ready for despatch.

   On account of a wide difference in the heat transfer rates between water and gas films, finning of tubes and thus extending the HS make an eminent sense in Ecos to make them compact. Despite additional costs of finning, they can be economical on the whole. There are mainly five types of finning:

   1. *Helical* or spiral *finning*, where high-frequency welding of thin fins is done continuously on tube surface on close spacing. Fins are generally 19–22 mm tall and 0.75–3.0 mm thick on a spacing of 75–315 fins/m. This type of finning is popular with gas-fired boilers and Hrsgs. More details are available under circular fins in **Extended Surfaces**. Figure 5.1 shows a bundle of spirally/helically finned tubes in the shop floor.
   2. *Plate or rectangular finning,* where typically 3 mm fins are resistance welded to the tubes on a spacing of 13 or 25.4 mm (½–1″) (as shown in Figure 5.2). This construction is for dusty fuels like coal. Some more details are given in rectangular plate fins under Extended Surfaces.

      A typical modular steel tube Eco with rectangular welded plate fins is shown in Figure 5.3. Such shop-fabricated assemblies are quite popular with process boilers of small and medium sizes.
   3. *Longitudinal finning* was popular for horizontal Ecos of coal- and **oil-fired boilers** where ~6 mm strips were welded along the tube axis on either side. The tubes are usually arranged in staggered pattern. They have not been particularly successful in both coal- and oil-fired boilers. See Longitudinal Fins under Extended Surfaces and Figure 5.15. However, such finning has been very successful in **vertical Eco** tubes with gas flow along the tubes in longitudinal manner,

FIGURE 5.1
Continuous helical or spiral finned tubes.

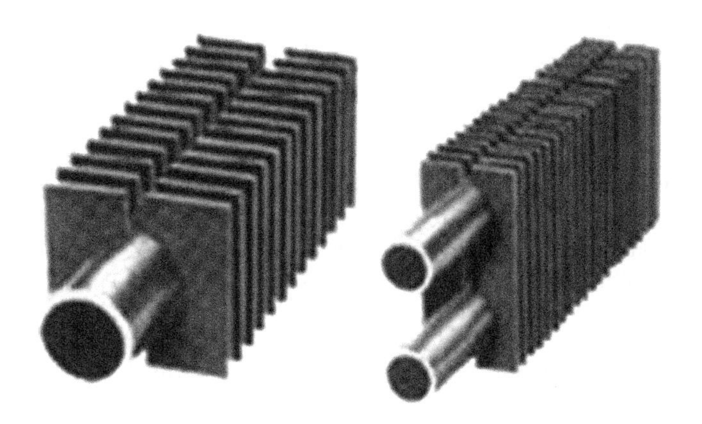

FIGURE 5.2
Rectangular welded plate fins on steel tubes.

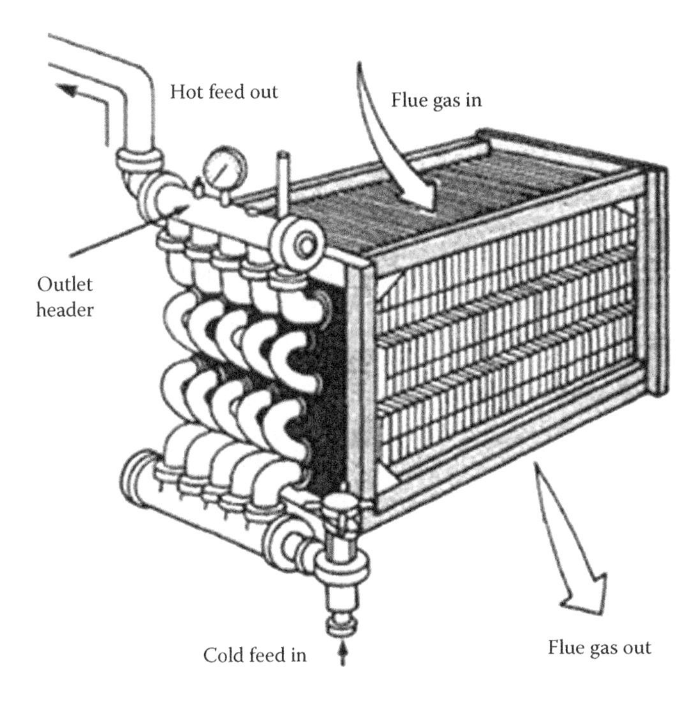

FIGURE 5.3
Modular steel tube economiser with rectangular plate fins.

FIGURE 5.4
Pin studded tubes.

as the tubes are stiffened by fins and overall HS is reduced. See Figure 18.4 and description in BLR boiler construction under the topic of Recovery Boilers.

4. *Stud finning* or *pin/gun studding* of Ecos is quite popular with clean fuels like NG that do not foul the surfaces. The closely spaced pins provide a good amount of extended surface making a tube bank quite compact but the gas Δp is usually higher. With fouling fuels like coal, it is rarely used in Ecos. Pin studding is largely replaced by spiral finning these days. Figure 5.4 shows a close-up view of pin-studded tube bundle.
5. *Steel tube Ecos with CI gills* In applications where heavy external corrosion and/or erosion is expected, CS tubes are shrunk fitted with spigotted CI gilled sections. The interlocking spigot design of the CI sleeves ensures perfect thermal and mechanical integrity with the inside CS tube in such a way that it is totally isolated from the corrosive gases. See Figure 5.5.

This construction is heavy and expensive, but highly reliable for very erosive and corrosive applications. Great care in fitting of gills is needed as the entire success is dependent on the integrity of the fit between tube and gill. Metal-to-metal contact is essential. No air entrapment is permitted between tube and gill, lest the heat transfer should be impaired and certain amount of HS rendered ineffective. Usually, the tubes are ground to fine finish in centreless grinders and the CI gills are heated and shrunk fitted to ensure that the joint is sound as desired.

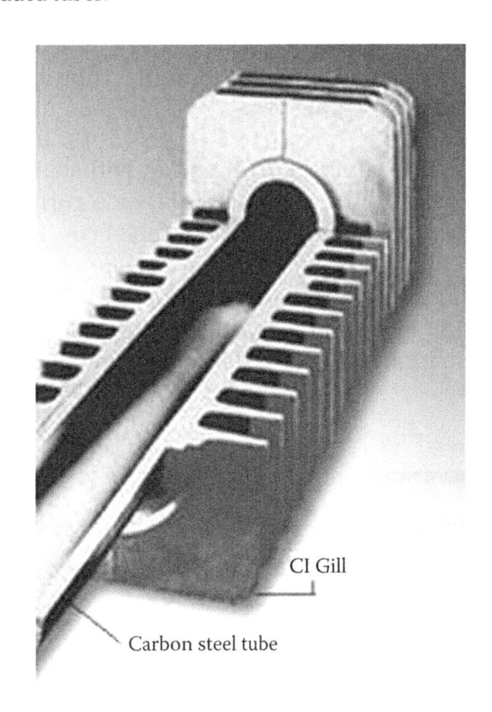

FIGURE 5.5
CI gilled tube element.

The gills can be square or round.

As the CI gills are not exposed to water pressure, there is no limitation to use this design even for boilers of the highest pressure unlike the pure CI gilled Ecos.

d. *Inline and staggered arrangement of tubes:* Figure 8.26 in Heating Surfaces shows the difference in the two methods of arranging the tubes. A staggered arrangement always yields a more compact HX and is usually employed for clean fuels for fear of erosion. Inline arrangement is most common for abrasive fuels like coal.

e. *Horizontal and vertical Ecos:* Horizontal fully drainable Ecos are the most common for both industrial and utility boilers which are shown in Figure 16.5. However, fuels which produce sticky ash deposits, even at low temperatures, are best served by vertical Ecos, with gas flow along the tube axis to minimise ash accumulation on the tubes. BLR boilers and MSW boilers are some examples for employing **vertical tube Ecos**. See Figure 17.5 and description.

f. *CI Ecos:* For small LP boilers operating at <17.5 barg it is possible to install gilled tube Ecos made entirely of CI. Both the tube and fin are made integral.

Besides being compact, CI Ecos can withstand waterside corrosion on the inside and gas-side corrosion on the outside much better than steel tube Ecos. However, there is a limitation in terms of pressure withholding capacity due to the inherent porosity of CI. Also, the CI construction tends to be heavier. Small process boilers with not so high operator skills can benefit by CI Ecos.

5.2 Effective Projected Radiant Surface (*see* Furnace in Heating Surfaces)

5.3 Efficiency (η) of Boiler

Boiler η is a matter of utmost importance to all. Naturally, there should be standard methods of stating η and norms to establish the claims. While there are minor variations in practices adopted in various countries and for different types of boilers, the principles remain nearly the same. These are described for conventional boilers in the following paragraphs. The popular testing codes are ASME performance test code (PTC) 4, British standards (BS) 2885 and German standards (DIN) 1942.

For boilers burning waste fuels, the guarantee considerations are slightly different. In most cases, the boilers are mainly expected to have the capability to burn all the waste fuel generated safely and reliably with η accorded secondary consideration. In cases like bagasse-fired boilers where the waste fuel can be regulated and excess fuel can be stored, boilers are run conventionally and η can be measured on somewhat similar lines as normal boilers.

For **Whrbs** and **Hrsgs**, however, the considerations are even more different and significant changes are to be made to the normal procedures outlined below. There is no combustion in most cases and measurement of heat input is not possible. A boiler, in many instances, is treated as part of the total process. If the overall steaming and other process results are met, the boiler is deemed to have fulfilled its performance. Sound mechanical integrity and modulation in step with main plant process dynamics is perceived more important than meeting certain η parameters in isolation. It is essential to have a deeper and clearer understanding on normal and guaranteed performances between the supplier and owner here than in conventional boilers, as the codes cannot cover all cases.

This topic of η of boilers is further explained under the following sub-topics:

1. Definitions of η
2. Acceptance and routine testing
3. Boiler η calculation
4. Break up of losses in η calculation
5. Heat loss calculation
6. η Testing of boiler by simplified method
7. η Testing by regular method
8. η Testing of Hrsgs

5.3.1 Definitions of η

A boiler is basically an *energy converter* transforming the chemical energy of fuel into heat energy of steam. Boiler η captures the effectiveness of this energy conversion.

$$\text{Boiler } \eta = \frac{\text{Heat output in steam}}{\text{Heat input}} \tag{5.2}$$

$$= \frac{\text{Heat input} - \text{losses}}{\text{Heat input}} \tag{5.3}$$

$$= 100 - \frac{\text{Heat loss}}{\text{Heat input}}\% \tag{5.4}$$

When reheating is involved, in the numerator of the above equation, heat given to RH should be added to the heat output, that is, RH steam × (heat of hot RH – heat of cold RH) in Equation 5.2.

5.3.1.1 Heat Credits

Heat input into the boiler envelop is the hv of the fuel, either **Gcv** or **Ncv**. Heat credits are additional heat inputs, which are sometimes considered, when auxiliary power consumed in fans and mills is sizable and a part of energy is returned to the system in the form of heat. In **CFBC** boilers, for instance, the FD and PA fan heads are high enough to heat the incoming air by 20–30°C, which merits consideration for heat credit. Sensible heat of fuel above the datum temperature is also a heat credit.

5.3.1.2 Various Shades of Efficiencies

There are many ways of expressing the η of a boiler:

- *Gross η on Gcv* is a popular way of denoting boiler η, where heat input is derived from $W_f \times$ Gcv of fuel, W_f being the fuel fired (not fuel burnt which is less by the unburnt fuel amount).
- *Gross η on Ncv* is derived when heat input is reckoned on Ncv basis, that is, $W_f \times$ Ncv.
- *Net η* is the η on net basis, that is, when auxiliary power is deducted from the heat output.
- Net η can also be expressed on Gcv or Ncv basis.
- η on Ncv basis.

For arriving at the gross boiler η on Ncv basis, multiplying the η by Ncv/Gcv fraction is a rough and ready method, which is nearly correct and good enough for normal purposes. If a more accurate figure or loss break up is desired

- Losses 1a, 1c, 2, 3, 4 have to be adjusted by Ncv/Gcv ratio.
- In 1b alone, adjustment has to be done for heat carried away by water vapour as given in Equation 5.10.

5.3.2 Acceptance and Routine Testing

In industry and power plants, boilers are tested off and on to verify their performance, which is known as routine testing. Such testing is less rigorous, less time consuming and inexpensive as the purpose is at best only to identify and implement certain performance improvement measures. It is an internal affair of the plant with no statutory third-party involvement.

Acceptance testing is more accurate, elaborate and expensive as there is a third-party involved. Commercial settlements are based on the results of acceptance tests. Naturally, they have to be more rigorous and often an independent testing team is employed to avoid possible disagreements. Acceptance test is carried out when the boiler is purchased. Also, when some major repairs or improvements are undertaken, and the effects have to be quantified acceptance tests are done.

5.3.3 Boiler η Calculation

Commercially, this is perhaps the most important of all boiler calculations, since boilers are bought or sold on this consideration. Therefore, all persons connected with boilers ought to be familiar with this basic calculation.

Boiler η can be established by either the *direct or the indirect* method.

In *direct method* (as per BS) or *input–output method* (as per ASME)

$$\text{Gross}\,\eta = \frac{\text{Net output \%}}{\text{Total input}} \tag{5.5}$$

This method needs accurate measurement of steam and fuel flows. While it gives a fairly good idea of the boiler η, as in routine testing or for small boilers with oil or gas firing, it is not good enough for acceptance testing or for solid fuel-fired boilers since the margin of error is rarely less than 3%. The main reasons for not using this method are that

a. Flow measurements are low in accuracy
b. Errors in measurement directly reflect as errors in η in roughly the same proportion

In *indirect* (as per BS) or *heat loss* (as per ASME) method

$$\text{Gross}\,\eta = 100 - \frac{\text{Heat loss \%}}{\text{Heat input}} \tag{5.6}$$

Here, instead of the output, the heat losses are estimated per unit fuel. This method is universally adopted as the inaccuracy can be <0.25% as

- No flow measurements are involved but measurement of mainly temperatures pressures and flue gas analysis are carried out
- Measuring errors affect the final η only marginally

Hrsg testing however is usually by direct method by the procedure laid out in PTC 4.4 as explained under the topic of Heat Recovery Steam Generators.

5.3.4 Break up of Losses in η Calculation

% Gross η of boiler = 100 − Σ losses 1–5% where (5.7)

1. Stack loss consists of
   a. Dry gas loss (L_{dg})
   b. M loss (L_m)
   c. Humidity loss (L_h)
2. Unburnt loss (L_{ub})
3. Radiation loss (L_r)
4. Sensible heat loss
5. Unaccountable loss (L_u)

1. *Stack Losses*: This accounts for 70–80% of the total losses in a boiler. These losses are minimised by cooling the flue gases to as low a temperature as possible without encountering low-temperature corrosion in the back end equipment.
   a. Dry gas loss forms 60–70% of stack losses. Reducing the final gas outlet temperature and excess air are keys to minimising these losses.
   b. M loss: This varies from 8% to 20% of stack losses for most fuels. Higher M and higher H_2O in fuel increase the M loss.
   c. Humidity loss is usually negligible being <0.1%.
2. *Unburnt loss*: Incompleteness of fuel burning is what is measured. Excess air employed is the most important consideration here. While it is nearly 0 for oil and gas, when burning coal it is 1–2% for PF and CFBC, 4–6% for BFBC and 4–10% for stoker.

   Unburnt losses and stack losses have to be balanced, as shown in Figure 5.6. Both of them are dependent on excess air. Increased excess air undoubtedly reduces unburnt loss but the stack losses go up. Optimum excess air places the operation in the zone of maximum η.
3. *Radiation loss*: This is the loss from the boiler setting due to radiation and natural convection. Usually, this is <1% for small boilers and <0.3% for large utility boilers.
4. *Sensible heat loss*: This loss is due to the heat carried away by bottom and fly ash. This is applicable only in firing of solid fuels.
5. *Unaccountable or unmeasurable loss*: These are small losses which are difficult to measure. These losses include
   - Effects of sulphation and calcination reactions in FBC boilers
   - Unstated instrument tolerances and errors
   - Heat carried away by the atomising steam
   - Any other immeasurable losses

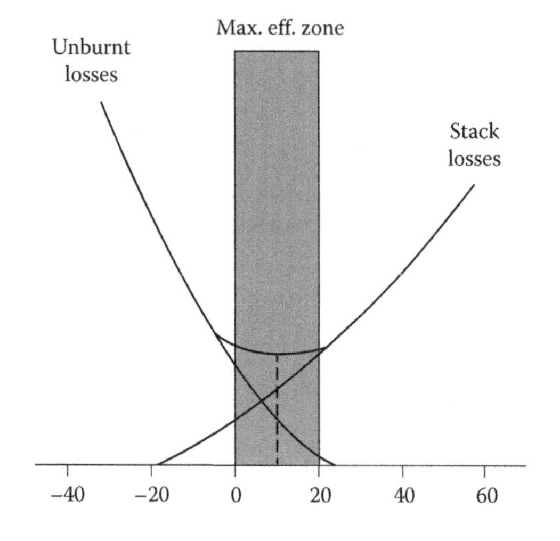

FIGURE 5.6
Variation of boiler unburnt and stack losses with change in excess air.

Losses due to sulphation and calcination are calculable while others are difficult. It is normal for buyer and seller to agree on this matter in advance.

5.3.5 Heat Loss Calculation

5.3.5.1 Stack Losses Are a Measure of How Efficiently

a. The flue gases are cooled without running the risk of low-temperature corrosion
b. The flue gas quantities are kept to a minimum by optimising excess air for the fuel and preventing tramp air leakages

Stack losses, as mentioned earlier, typically form about 70–80% of total losses.

Of the above, the M losses vary from 8% to 20% for most fuels depending upon the fuel M, fuel H_2, excess air and exit gas temperature. The dry gas losses form the rest as the humidity loss is rather small at <0.1%.

$$\text{a. Dry gas loss } (L_{dg})\% = \frac{W_g(T_g - T_a) \times C_p \times 100}{\text{Gcv/Ncv}} \tag{5.8}$$

$$\text{b. M loss } (L_m)\% = \frac{(9H_2 + m) \times (H_s - h_a) \times 100}{\text{Gcv}} \tag{5.9}$$

$$\text{or on Ncv basis } \% = \frac{(9H_2 + m) \times (H_s - h_a - 578) \times 100}{\text{Ncv}} \tag{5.10}$$

$$\text{c. Humidity loss } (L_h)\% = \frac{W_h \times (H_s - H_a) \times 100}{\text{Gcv/Ncv}} \tag{5.11}$$

where

W_g = weight of gas leaving the system in kg/kg of fuel burnt (not fired)
W_h = weight of M in air in kg/kg of fuel burnt (not fired)
T_g = temperature of flue gas leaving the system in °C
T_a = temperature of air entering the system in °C
= ambient in most cases less air heated from external source
C_p = mean specific heat of gas between T_g and T_a in °C
W_a = weight of air in steam AH kg/kg
H_2 = % hydrogen in fuel by weight
m = % M in fuel by weight
H_s = enthalpy of steam at T_g and 0.07 bar (1 psia) in kcal/kg = 583 kcal/kg
h_a = enthalpy of water at T_a in kcal/kg

5.3.5.2 Unburnt Loss

Unburnt Loss (L_{ub}) is a measure of how much fuel is left out of burning process in the firing equipment for the excess air chosen. μ of heat release of firing equipment is measured by the amount of C burn up, which is

$$C_b = (100 - \%C \text{ in the residue}) \div \% \text{ total C at inlet} \quad (5.12)$$

C burn up varies a great deal with different types of fuels, firing equipment, excess air and ash content. C_b is

- Nearly 100% for gas and oil
- 98–99% for CFBC and PF, 90–95% for BFBC and 80–90% for stoker in case of coal

TABLE 5.1
Unburnt Losses with Various Fuels in Different Firing Equipments

Method of Firing	C Burn Up	Unburnt Loss as % of Gcv
Gas	100%	0%
Oil	~100%	~0%
PF and CFBC coal	>98%	<2%
BFBC coal	85–90%	4–6%
Stoker coal	80–90%	4–10%
Stoker—bagasse	~98%	~2%

The finer the fuel sizing and the greater the turbulence, the lower is the excess air requirement and the higher the C burn up. C burn up is also influenced by VM in fuel.

Unburnt loss is generally about 1–2% for PF and CFBC, 4–6% for BFBC, 4–10% for stoker on coal and 2% for bagasse. See Table 5.1. The correct way to estimate this loss is to calculate from the proprietary data of the firing equipment manufacturer and tally with the previous field results to see if it needed any correction.

5.3.5.3 Radiation Losses Are Small

They are <1% and become smaller as the boiler size and the watercooling of furnace increase. ABMA graph, shown in Figure 5.7, is generally followed despite its drawbacks as it gives an approximate value very quickly.

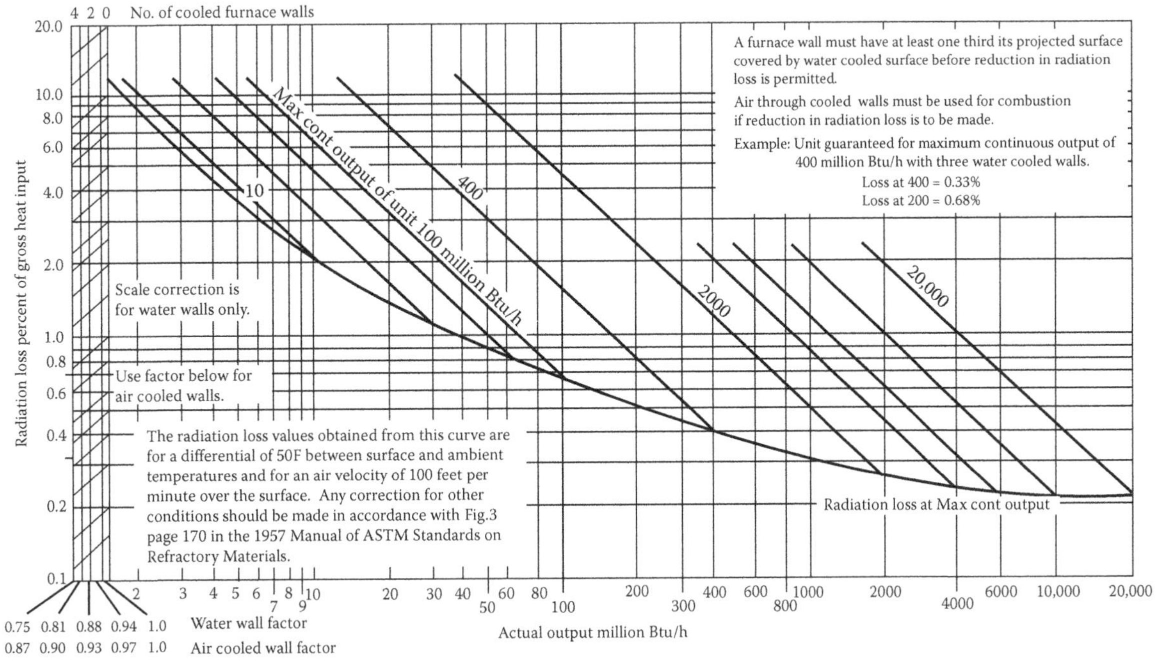

FIGURE 5.7
ABMA chart of radiation losses (PTC 4.1).

This graph was evolved decades ago on the following principles:

- This is a comprehensive loss on account of both radiation and convection conductance for the entire boiler, that is, the main boiler setting, ducting and milling plant.
- The surface to ambient temperature difference is taken as 50 °F (~27°C) and the air velocity 100 fpm (~0.5 m/s).
- A furnace wall has at least one-third of its projected surface covered by water-cooled walls before reduction in radiation loss is permitted.

There are several limitations in this graph and yet it is very popular because of its convenience and the low value of radiation loss. For example, as per this graph, radiation loss from a compact oil-fired boiler is same as a large PF-fired boiler of the same evaporation, since it is based on the boiler output and not on the surface area. Likewise, FBC boilers were not in the picture when this graph was evolved. Radiation losses from big high-temperature cyclones cannot be accounted for here. Suitable additional loss amount may have to be considered.

In the present code PTC 4, this graph is withdrawn and the code recommends the radiation loss estimation by measuring the surface temperature and area individually for various parts.

5.3.5.4 Sensible Heat Loss of Ash

Sensible Heat Loss of Ash for solid fuels can be measured by knowing the split of bottom and fly ash and their respective temperatures. Representative collections are required to be made prior to testing to ascertain the fly ash collection in various hoppers, so that an agreement can be reached in advance on the ash split. Gas temperatures at various hoppers are recorded during the test to arrive at sensible heat loss.

In FBC, the sensible heat losses are high, as the bed material also gets discharged along with ash. The losses are even higher when limestone is added for desulphurisation

5.3.5.5 Unaccountable Losses

Unaccountable Losses are those very losses that cannot be exactly quantified and are small enough to be put together and assigned a reasonable value. They comprise usually of

- Unstated instrument tolerances and errors
- Any other unmeasurable losses

TABLE 5.2

Normally Accepted Figures for Lu, MM and Tolerances

Fuel	Firing	Lu%	MM%	Tolerance
Gas	Burner	0.25	0.25	+0/−5.0% of losses
Oil	Burner	0.25	0.25	
Coal	PF	0.25	0.5	
	Grate	0.5	1.0	+0/−6.5% of losses
Agro-fuels	Grate	1.0	1.0	

When the effects of sulphation and calcination reactions in case of FBC boilers are clubbed in unaccountable losses, the values go up quite substantially.

Typical example for an unmeasurable loss is the heat carried away in oil firing by the atomising steam.

The unaccountable losses considered usually for various fuels are as given in Table 5.2.

Manufacturers' margin (MM) and tolerance are reckoned only at the time of boiler sale. When the calculated η figures are to be converted into guarantees by boiler makers, MM is considered, as a protection. These are somewhat subjective.

- Manufacturers' margin is a safety figure taken into account for any inaccuracies of design, manufacturing and erection that can take place inadvertently. It is also to account for any unexpected variation in performance during testing.
- Tolerances are for any errors and omissions as well as instrument tolerances normal to the performance trials. These are usually subject to mutual agreement between buyer and seller. If tolerance is not allowed, the calculated figures are reduced suitably and minimum η is offered.

It is best to keep tolerances out of reckoning and evaluate boilers on minimum η basis as practices vary in different countries.

5.3.6 η Testing of Boiler by Simplified Method

The simplest method of η testing is the input–output or direct method but it is fraught with errors as explained earlier and is rarely used.

Methods of η testing for acceptance have to be rigorous, as they have commercial implications, particularly at the time of handing over of the plant by the manufacturer. When the plants are evaluated closely at the time of contracting, when η figures are committed to their second decimal of accuracy, the η tests at the time of handing over have also got to be equally rigorous.

The ASME PTC 4.1(1964 and reaffirmed in 1991), which has been withdrawn since 1998, is still popular with industrial and smaller utility boilers as it lays down an *abbreviated test procedure* more suitable for

- Routine testing of all boilers
- Acceptance testing of industrial boilers

In the abbreviated test:

- Only the major losses are considered.
- Chemical heat of the fuel (without heat credits) is taken as the input.
- Humidity loss, being small, is ignored.

For utility boilers, *Long Test*, considering all losses and heat credits, is recommended but mostly the abbreviated procedure is adopted for its practical approach and simplicity. Besides lesser the time taken, and the cost incurred for boiler testing is also lower.

Broadly, the main measurements in the heat loss method for calculating η are

- The temperatures (fw in, SH out, RH in, RH out, exit gas and ambient air)
- Pressures (SH out, drum, SH in, RH out, fw)
- Flue gas analysis ahead of ID fan preferably by Orsat analyser
- cvs of ash and fuel

Steam flows of SH and RH are also measured to prove the guaranteed flows, but they do not figure in calculating η. They are measured to ensure that the flow conditions are met.

The salient points in the performance testing of boilers are as follows. The test codes explain all the requirements in detail and must be referred to for procedures, precautions and calculations.

- The boiler should be handed over to the manufacturer a few weeks before the test as required for setting the boiler parameters correctly and tuning the performance and controls. Larger the boiler, longer is the tuning time. For industrial boilers, usually a week or two should be sufficient.
- Setting up test instrumentation, particularly the TCs, gas sampling lines and ash collecting probes is time consuming. This can take typically 6 weeks in a large utility boiler. By using more gauge instrumentation instead of test instrumentation, this time can be cut down.
- After maintaining equilibrium as specified, performance testing should be quite short. It should be >4 h as per code and <6 h by practice, as running the boiler with constant load is not easily practicable in either a power or a process plant. Also, there should be no soot blowing and blowdown.
- Peak and minimum output demonstration tests are kept outside this main test period.
- Where required the testing on alternate fuels or testing for part loads are scheduled for different days.
- The duration of the test, the procedure, the scope, the fuels, the tolerances and so on should be agreed in advance as laid out in the code.
- All the instruments are required to be of test quality and are to be calibrated before test. It is normally necessary to perform trials for a couple of days to get systems, the instruments and the man power properly tuned to the testing.
- Fuel and ash samples are taken every hour. Other test readings are taken usually at 15 min intervals.

It is quite a big task to isolate a boiler for testing purposes for a week or two, depending on the extent of testing, as it entails a lot of expense and effort. Many times, therefore, the formal testing is not undertaken, when the operating staff is convinced that the boiler has clearly outperformed.

Calculation of η from the Performance Test Results:

The principle behind the calculation of η is as follows:

- Losses 1(a and b) and 2 are calculated from test results.
- Loss 3 is read from the ABMA chart.
- Loss 4 is agreed before the test.

For calculating the unburnt loss (Loss 2) in the case of PF-fired and FBC boilers, the ash distribution between fly ash and bottom/bed ash is often agreed upon, based on the boiler operation thus far, to make things simpler. This is because the loss is <2% and the sensitivity of error in ash distribution is very small in comparison to the effort required to collect the data correctly.

The code gives the following formulae for arriving at the losses in 1a, 1b and 2 using

- Flue gas analysis (CO_2, CO, O_2, N_2% by volume in dry flue)
- Fuel analysis (C, H, S% in fuel)
- C in flue gas (C_g) and ash in% by wt
- Temperature of exit gas and inlet air
- Gcv or Ncv in kcal/kg

The following formulae are used for calculating the losses 1a, 1b and 2 on Gcv basis:

$$1a - \text{Dry gas loss } (L_{dg})\% = W \times 24\,(T_g - T_a) \div \text{Gcv} \quad (5.13)$$

where W is the weight of dry flue gas/kg of fuel

$$W = \frac{11CO_2 + 7(CO+N_2) + 8O_2}{3(CO_2 + CO)}\left(C_g + \frac{S}{183}\right) \quad (5.14)$$

C_g is the wt of carbon in flue gas/kg of fuel

$$1b - M \text{ loss } (L_m)\% = [(100 - T_a) + 539 + 0.5(T_g - 100)] \times (9H_2 + m) \div Gcv \quad (5.15)$$

$$2 - \text{Unburnt loss } (L_{ub})\% = \%\text{ash in fuel} \times C_a \times 80.78 \div Gcv \quad (5.16)$$

where C_a is % combustibles in fly and bottom/bed ash.

PTC 4.1 provides for a certain tolerance when it specifies measurement errors for all types of instruments used in testing and the corresponding error is calculated as η. This works out to about ±0.3%.

Table 5.3 lists the accepted measurement error of each variable and its effect on the test result. The maximum effects are in the analyses of coal by **Bomb calorimeter** and gas by **Orsat analyser**, respectively.

TABLE 5.3

Measurements and Tolerances as per Heat Loss Method

S. No.	Measurement	Measurement Error (%)	Error in Calculation of Boiler η%
1	Gcv coal	0.5	0.03
	Gcv oil/gas	0.35	0.02
2	Orsat analysis	3.0	0.3
3	Exit gas temperature	0.5	0.02
4	Inlet air temperature	0.5	0
5	UA of coal—C, H_2	1.0	0.1
6	H_2O in fuel	1.0	0

Note:

1. All % figures are ± for both error and η.
2. Air and gas temperature measurements are with calibrated devices.

TABLE 5.4

Salient Points of Difference between PTC 4.1-1964 (1991) and PTC 4-1998

S. No.	Parameter	PTC 4.1	PTC 4	Remarks
1	Gross versus fuel η	$100 - \frac{\text{Heat losses }\%}{\text{Fuel input + Heat credits}}$	$100 - \frac{\text{Heat losses - Heat credits}}{\text{Fuel input}}$	Fuel η replaces gross η of PTC 4.1. Fuel flow is directly calculable
2	Test uncertainty	Not considered	Uncertainty values listed in 1.3-1 and level to be agreed	This is mainly driven by the economical instrumentation and data sampling procedures chosen. Section 4 of PTC 4 provides background and guidance in these matters
2a	Radiation and unaccountable losses	ABMA radiation loss curve to be used	Losses are accurately evaluated ABMA curve not used	
3	Reference temperature	Usually ambient. Heat credits can be neglected	25°C to be adopted and heat credits and losses are calculated	With a fixed reference temperature, all results are on the same basis and are directly comparable
4	Output	Energy absorbed by working fluid	Energy absorbed by working fluid that is not recovered within the specific gravity (SG) envelope. For example, steam to SCAPH	
5	Objectives		Better covered	Test objectives or performance parameters
6	Corrections		Better defined	Corrections to design conditions
7	Flue gas	Based on measured O_2 and CO_2	Based on measured O_2 both wet and dry	For η calculations
8	Calculation of losses/credits	Specific heat	Enthalpy	In PTC 4, the enthalpy of dry air, dry flue gas, M vapour and fuels are defined by equations (curve fits) as opposed to difficult-to-read figures
9	Fuel analysis	Mass basis	Conversion from volumetric to mass is also provided	

TABLE 5.5
Differences in Measurements of Losses and Credits between PTC 4.1 and PTC 4

Loss	PTC 4 Item	PTC 4.1	PTC 4.1 Abbreviated Test	H/C_p Potential Difference	Notes
1a	Dry gas loss	Yes	Yes	Yes	—
1b	M loss from H_2 and H_2O	Yes	Yes	No	No change
1c	M in air loss	Yes	No	No	No change
	H_2O vapour in fuel loss	No	No	—	New item
2	Unburnt carbon loss	Yes	Yes	—	No change
3	Surface radiation and convection loss	Yes	Yes	—	As noted above, the calculation procedure is different, and PTC 4 generally yields greater loss for coal-fired units
4	Sensible heat of residue loss	Yes	No	Yes	Users may agree on residue split if not measured.
	Hot air quality control equipment loss	No	No	—	New item. Added to distinguish the loss due to the hot AQC equipment separate from the steam generator
	Entering dry air and M in dry air credit	Yes	No	Yes	—
	Sensible heat in fuel credit	Yes	No	Yes	—
	Auxiliary equipment power credit	Yes	No	—	No change
	Losses and credits for sorbent reactions	No	No	—	Have been added and are different from prior industry values published by the ABMA

5.3.7 η Testing by Regular Method

PTC 4.1 was superseded and withdrawn in 1998 and replaced by a new code PTC 4, which overcomes many deficiencies and makes testing more accurate and elaborate. Table 5.4 brings out the essential differences between the two codes and is reproduced from my book *Boilers for Power and Process* (CRC Press, 2009).

Assuming that tests are performed with the same type of instruments to the same levels of accuracy, the results by PTC 4.1 and 4 would vary marginally because

- It is the fuel η as per PTC 4 and gross η as per PTC 4.1 with slight differences in the way the heat credits are treated.
- PTC 4 defines the reference temperature, makes heat credit accounting compulsory and adopts certain minor improvements in the calculation methods.

PTC 4 is undoubtedly a more comprehensive test code that is based on modern measurement, data reduction and calculation with uncertainty analysis included and is, therefore, superior to PTC 4.1.

There are certain differences in the estimation of losses and credits between the two codes, as shown in Table 5.5. This table is also taken from my previous book.

5.3.8 η Testing of Hrsgs (see Heat Recovery Steam Generators)

5.4 Efficiency, Ligament

A tube sheet is a steel sheet with holes drilled in certain regular pattern for the attachment of tubes. It is considered as flat plate subject to bending and shear.

- Ligament is the area of metal between the holes in a tube sheet.
- Ligament η is a factor to obtain the lowered strength of the plate due to the drilling of holes.

Ligament η is needed to calculate the thicknesses of pressure parts like drums, headers and tube sheets.

The following symbols are used in the formulae for calculating ligament η:

p = longitudinal pitch of adjacent openings (mm)
p' = diagonal pitch of adjacent openings (mm)
p_1/p_2 = alternate pitch between openings (mm)
d = diameter of openings (mm)
n = number of openings in length p_1
η = ligament η

The three types of ligaments are

1. *Longitudinal*: located between the front holes and lengthwise holes along the drum
2. *Circumferential*: located between the holes and encircle the drum

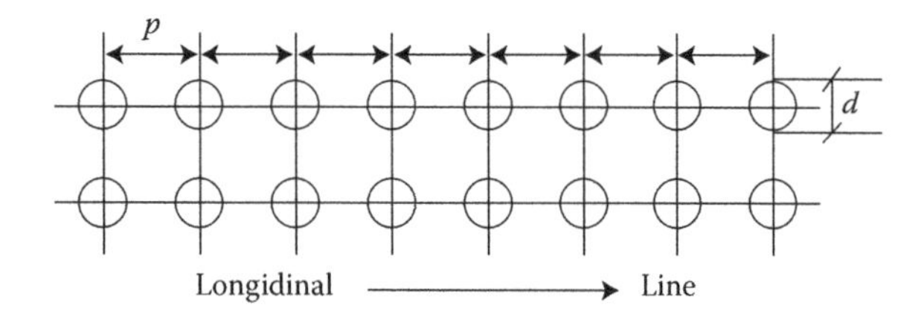

FIGURE 5.8
Uniform spacing of tube holes.

3. *Diagonal*: a special case because they are located between the holes and are offset at an angle to each other

The rules of ligaments are applicable to groups of openings in cylindrical pressure parts that form a definite pattern. These rules also apply to openings not spaced to exceed two diameters centre to centre.

Longitudinal ligament η is given for uniform spacing/pitching p, as shown by Figure 5.8 by the equation

$$\eta = \frac{p - d}{p} \tag{5.17}$$

In case of alternate spacing/pitching p_1 and p_2, as shown in Figure 5.9, average pitching is considered in the same η formula where

$$p = \frac{p_1 + p_2}{2} \tag{5.18}$$

Diagonal η comes into play with staggered pitching, as shown in Figure 5.10.

To obtain the ligament η diagram in Fig. PG-52-1. may be used from Section I of ASME code or Fig. UG-53.5, Section VIII-1.

For calculating the circumferential η, tube pitching perpendicular to the longitudinal line along the circumference is to be considered.

5.5 Elasticity (of Fluids) (*see* Fluid Properties *under* Fluid Characteristics)

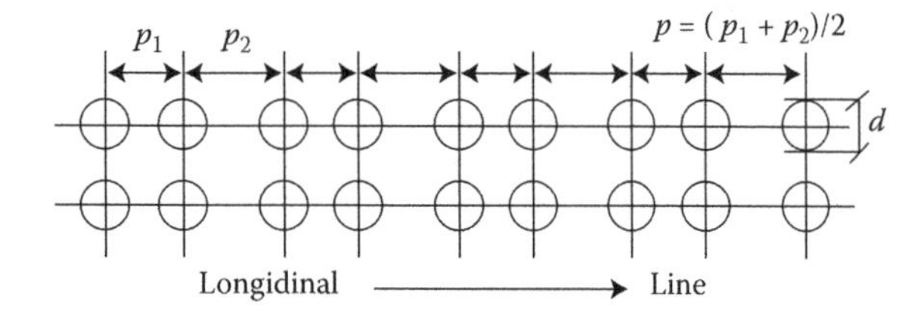

FIGURE 5.9
Alternate spacing of tube holes.

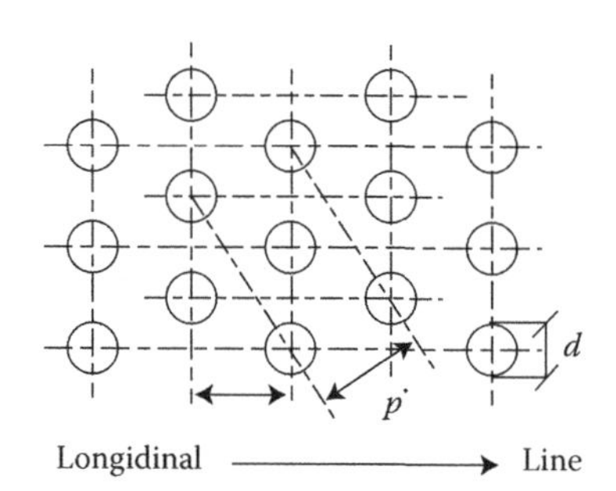

FIGURE 5.10
Staggered pitching of tube holes.

5.6 Electro-Filters (*see* Electrostatic Precipitators *under* Dust Collection Equipments)

5.7 Emissivity (*see* Radiation Heat in Heat Transfer)

5.8 Energy (*E*) (*see* Thermodynamic Properties)

5.9 Electromatic Safety Valve (*see* Safety Valves *under* Valves)

5.10 Electronic Water Level Indicators (*see* Remote Water Level Indicators)

5.11 Electrostatic Precipitators (*see* Dust Collection Equipment)

5.12 Emissions and Control (*see also* Acid Rain)

This is a very large and complex topic. In addition to acid rain, all words in bold here may be referred to under respective headings for more detailed description of the matter.

Emissions from fossil fuel-fired boilers can be categorised as the following, which have been increasingly curtailed all over the world for the past 40 years, by the ever-tightening environmental protection laws:

- **PM**
- **NO_x, SO_x, CO**
- **VOC**
- Hazardous air pollution (HAP) namely, HCl and HF, Hg, trace metals and trace organic compounds.

PM is collected in large boilers by either ESP or bag filters. Even as low a limit as 10 mg/NM^3 is achievable.

NO_x reduction is achieved at the combustion stage by

a. **LNB**
b. **SOFA**
c. **Reburning**
d. **FGR**

Additional reduction can be achieved in the post-combustion stage by **SCR** or **SCNR**. All these techniques are elaborated under the topic of nitrogen oxides (NO_x) in flue gas. In case of FBC boilers, the problem of NO_x is not an issue as the low combustion temperatures do not produce NO_x beyond limits in most cases.

FGD is the way to capture SO_x emissions as explained under the topic of **Desulphurisation**. Gas scrubbing in wet or dry form is the principle of FGD. In case of FBC boilers, almost all the desired desulphurisation is achieved during the combustion process itself with inbed addition of limestone.

TABLE 5.6

Typical Emission Norms in Advanced Countries

Pollutant		Unit	Limit	Method
Acidic gases	HCl	mg/nm^3	10	On-line
	HF	mg/nm^3	1	On-line
	SO_2	mg/nm^3	50	On-line
	NO_x	mg/nm^3	200	On-line
Dust		mg/nm^3	10	On-line
Dioxins and furans		ng/nm^3	0.1	Twice a year

Control of CO and VOC is done by optimising the combustion parameters and adopting prudent combustion techniques that provide the 3Ts (time, temperature and turbulence) in adequate measure.

HAP is controlled by the action of FGD and dust filters together.

Emission norms in different parts of the world are different. Table 5.6 shows typical limits observed in the developed countries.

5.13 Endurance Limit or Fatigue Strength (*see* Metal Properties)

5.14 Energy (*E*) (*see* Thermodynamic Properties)

5.15 Engler, Degrees for Fuel Oils (*see* Fuel Properties in Liquid Fuels)

5.16 Enthalpy (*H*) (*see* Thermodynamic Properties)

5.17 Entropy (*s* or ϕ) (*see* Thermodynamic Properties)

5.18 Enthalpy–Entropy Diagram (*H–s* Diagram) (*see* Entropy in Thermodynamic Properties)

5.19 Equivalent Diameter (De) (*see* Diameter, Equivalent)

5.20 Erosion (*see* Wear)

5.21 Electrically Resistance Welded Tubes (*see* Boiler Tubes)

5.22 Evaporation of Boiler (*see* Boiler Rating *and* Boiler Steam Low)

5.23 Evaporation of Water (*see* Properties of Steam and Water)

5.24 Evaporator Bank (*see* Heating Surfaces)

5.25 Evaporator (Pulp) (*see* Kraft Pulping Process)

5.26 Excess Air (*see* Combustion)

5.27 Expanded Drums (*see* Steam Drum in Pressure Parts)

5.28 Expansion Joints (*see* Draught Plant)

5.29 Extended Surfaces

This topic of extended surfaces is covered under the following sub–topics:

1. Fins or gills
2. Circular fins
3. Rectangular plate fins
4. Longitudinal fins
5. Fin η

5.29.1 Fins or Gills

In the design of HX or tube banks, whenever there is a large difference in the heat transfer coefficients on either side of the tube, it may be economical to employ finned tubes over plain tubes. The **convective heat transfer** from a surface can be substantially improved if extensions can be added to increase the surface area. There are many ways in which the surface of the tube can be extended with protrusions called *fins or gills*. Fins can be on the inside or outside of the tubes, wherever the heat transfer rate is lower. In the tube banks of boilers, they are always on the outside of tubes as gas side heat transfer is lower. Figure 5.11 shows various types of fins in use.

The most important prerequisite for using fin tubes is the cleanliness of the fin surface. A dirty fluid defeats

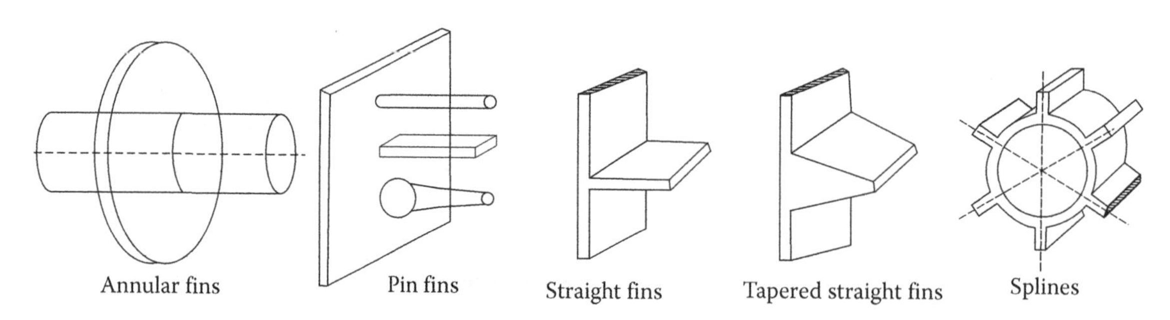

FIGURE 5.11
Various types of fins to create extended surfaces.

the very purpose of providing an extended surface as it would build dust deposits on the fins and make them ineffective. A dusty fluid, on the other hand, may also erode the projecting fins if the dust is erosive.

Another very important parameter which can sabotage this whole concept is the welding integrity between the tube and fin. It is vital that the fitting of the fin on tube is flawless and the welding is very good without any defect or entrapment so that the conduction of heat from fin to tube is smooth. Fin has to be fully **welded** to the tube and the weld should be generous so that the heat transfer is without interruption and the η of fin is optimum.

The main advantages of fin tube exchangers are compactness and reduced Δp.

Boilers with clean flue gases, such as with those fired with light oil and NG, are ideally suited for using finned tubes. In conventional boilers, Ecos are more or less the only surfaces where the differential in heat transfer rates across the tubes are large enough to adopt finned tube construction. In coal-fired boilers also, finned tube Ecos are employed profitably, when ash in coal is low and is known to be clearly non-abrasive. Most Whrbs handle dusty and abrasive gases and are no candidates to use finned tubes. On the other hand, modern Hrsgs, extracting large amounts of heat from very clean medium hot exhaust gases, cannot ever be built with η, economy and compactness without employing finned tubes to the maximum.

In boiler practice, fins on tubes are almost always transverse or annular. Longitudinal fins for Ecos were popular some time ago, but it has been discontinued as the long-term record has not been favourable. Pitching of fins, fin thickness and fin height are the factors that need careful selection. Satisfactory operating experience is the best guide.

5.29.2 Circular Fins

Solid high-frequency continuous helically finned tubes are the most popular among all types in conventional boilers. For Eco application, the tubes and the fins are of low CS execution. Tubes are usually 31.8, 38.1, 44.5 and 50.8 mm od. Staggered arrangement is suitable only for clean gases like NG and to an extent distillate oil no. 2. Fins are 19–22 mm high with spacing and thickness increasing as the gases get dustier. Table 5.7 summarises the application range. It is important that the tubes are kept free of fouling by restricting the bank depths to ~1.5 m and employing suitable soot blowers.

In Hrsgs, both serrated and solid fins are used, which are shown in Figure 5.12. Serrated fins have much greater HS than solid fins but are more prone to fouling. They are employed only when the gases are known to be very clean and non-fouling, which is the case with Hrsgs behind GTs using NG or conventional boilers firing NG.

FIGURE 5.12
Serrated and solid circular fins.

TABLE 5.7
Spacing of Solid Fins on Tubes and the Range of Application

Fuel		NG	Light Distillate	Heavy FO	Coal
Fins per	Inch	8	4	3	2
	Meter	315	157	117	78
Fin thickness (mm)		0.75–1.5	0.75–3.0	1.25–3.0	1.25–3.0
Staggered?		Yes	No	No	No

5.29.3 Rectangular Plate Fins

Rectangular welded plate fins have been employed for a long time now even for solid fuel firing. These fins are resistance welded with force, as shown in Figure 5.13. Unlike in spiral fin arrangement, here clear gas lanes are formed and therefore better cleaning is claimed.

Fins are of low CS, resistance welded and are square or rectangular in shape. Fins are typically 3 mm thick and spaced at 13 and 25.4 mm (½ and 1″) apart depending on the fouling tendencies. See Figure 5.14.

Gas velocities are limited to 15 m/s for clean gases. They are used in non-corrosive environment with gas, oil and coal firing even with moderately high dust loads.

As the HS that can be packed in the same volume is about nine times more, the resulting Eco becomes very compact and the draught losses reduce considerably. The requirement of SBs also reduces. Here, rake-type SBs are required for keeping the surfaces clean. In coal applications, there is always a possibility of fly ash plugging unless the SBs are used at proper intervals. For high coals with abrasive ash, plain tubes are considered a safer bet.

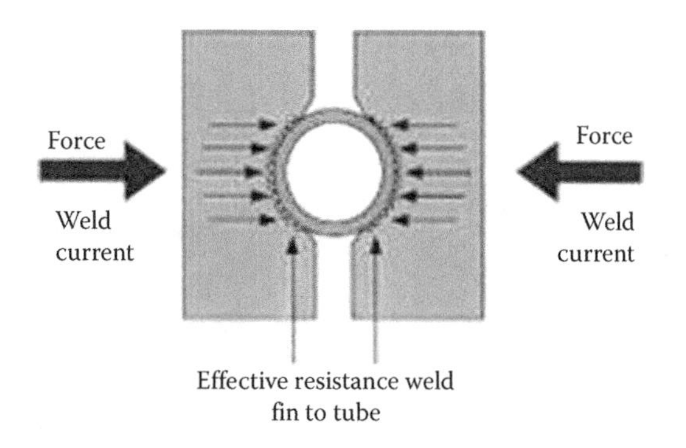

FIGURE 5.13
Method of fixing the rectangular plate fins on tube.

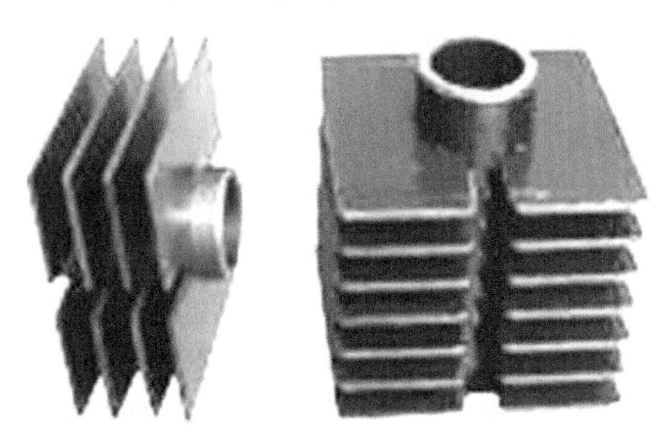

FIGURE 5.14
Tube with rectangular welded plate fins.

5.29.4 Longitudinal Fins

Ecos with longitudinal fins, as shown in Figure 5.15, were built in the past for PF boilers. Even though the tubes were in staggered arrangement, the location of thick and tall fins (6 × 50 mm typically) was expected to deflect gases and provide erosion protection to tubes. Ecos became compact in this arrangement. But many installations experienced erosion with coal firing and also weld failures in the long run, particularly with oil-fired boilers. Such arrangement has more or less lost favour now.

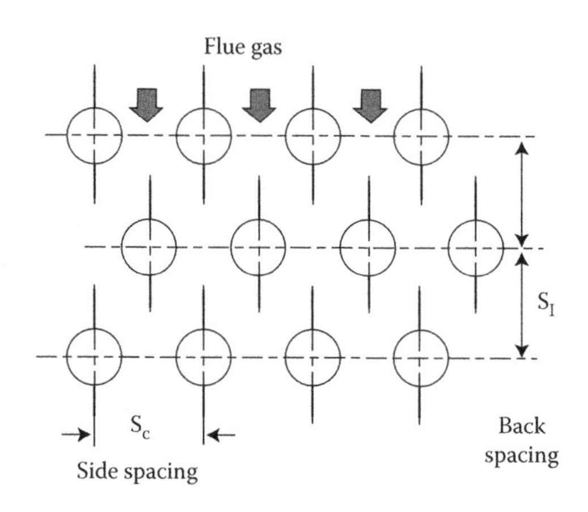

FIGURE 5.15
Longitudinal strip fins on tube.

5.29.5 Fin Efficiency

The HS added by the fin is not as efficient for heat transfer as bare tube surface owing to resistance to conduction through the fin. The effective heat transfer area of a fin tube is the sum of

$$\text{HS of the plain tube} + \text{fin area on both sides} \times \text{fin } \eta \quad (5.19)$$

Fin η indicates how much of heat abstracted by fins theoretically is actually transferred to the tube. Fin efficiency is found from mathematically derived relations, in which the film heat transfer coefficient is assumed to be constant over the entire fin and temperature gradients across the thickness of the fin have been neglected.

5.30 Explosion in Boiler (*see* Operation and Maintenance Topics)

6

F

6.1 Fabric Filter (*see* Dust Collection Equipment)

6.2 Fanning's Friction Factor (*see* Moody's Factor)

6.3 Fans

In a boiler air is drawn from the atmosphere, pushed through the ducts to the combustion chamber where it reacts with fuel to turn into flue gas which is then extracted through the flues and dust catchers before exhausting into the atmosphere.

This entire scheme is called the *draught system* and all items put together, *draught plant.*

This topic is divided into the following sub-topics:

1. Furnace draught
2. Boiler fans
3. Fans versus compressors
4. Fans and boiler performance
5. Fan construction
6. Fan control
7. Fan laws
8. Fan operation (series and parallel)
9. Fan power
10. Fan static head and η
11. Fan types

6.3.1 Furnace Draught (*see also* Draught)

When combustion air, sucked from the atmosphere, is pushed all the way to the stack and the furnace experiences positive pressure, it is termed *FD operation,* which is the case with most oil- and gas-fired boilers. Hrsgs also operate under FD but combustion air is substituted by exhaust gas from GT. Many process Whrbs also operate with FD conditions.

When combustion air is pushed to the furnace and after combustion the flue gases are pulled all the way to the stack, furnace pressure is neutral and it is called *balanced draught operation.* Most solid fuel firing is with balanced draught.

If there is only pulling of air and gas all the way, with furnace in negative pressure, it is *induced/natural draught operation.* In large modern boilers, ID firing is rare these days. Small boilers with tall stacks in the olden days used to operate under ID. Locomotive boilers were ID boilers.

6.3.2 Boiler Fans

- *FD fans* are for sucking the fresh air and pushing it up to the furnace or stack.
- *ID fans* are for sucking the flue gases from the furnace and discharging to the atmosphere through the stack.
- *Primary air (PA) fans* are for providing part of combustion air at a higher pressure than FD air, like in the case of air to the mills in PF boilers.
- *Secondary air (SA) fans* are for delivering part of combustion air at higher pressure in the combustion chamber for burning of VM and agitation of flue gases as in grate firing or FBC.
- *Gas recirculation (GR) fans* are for extracting low temperature flue gases, usually from Eco inlet or exhaust, to feed them back to the combustion chamber for either NO_x or SH temperature control.
- *Start-up fan* is needed in industrial boilers, operating in island mode, for 'black start duty' when nil or inadequate power is available for running the fans. Small fans are provided exclusively for start-up purpose which are driven by STs or diesel engines.

6.3.3 Fans versus Compressors

Fans are large-volume LP air/gas compressing machines. They are volumetric devices that move air or gas from one place to another, overcoming the resistance placed

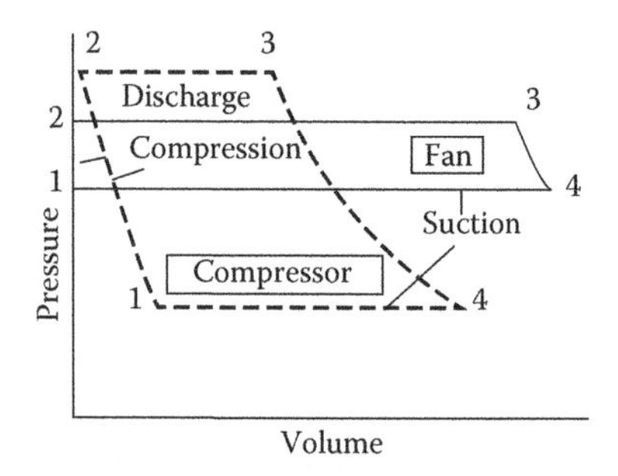

FIGURE 6.1
Comparison of fan and compressor characteristics.

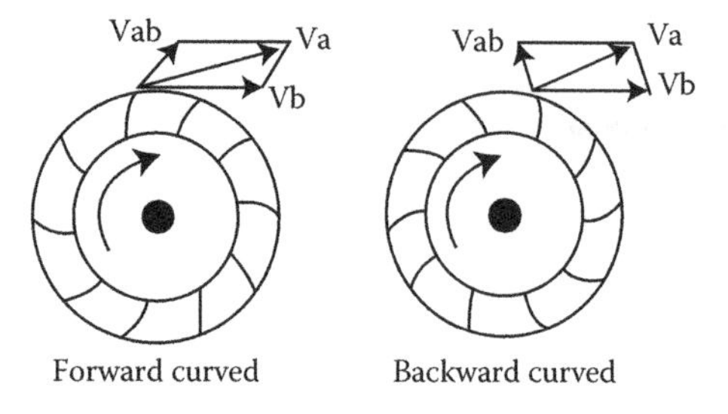

FIGURE 6.2
Forward and backward curved blading and their effects on discharge velocities.

in between. They generate a certain volume in cum and pressure head in mm wg. Though similar in function, fans are diametrically opposite to the compressors which produce HP and low volumes (as shown in Figure 6.1). Fans generate a head of ~2000 mm wg (0.2 barg) maximum while blowers can go up to 2 barg while compressors begin at ~2.5 barg.

6.3.4 Fans and Boiler Performance (see *also* System Resistance of Boiler)

Fans form the most important auxiliary of any boiler as the boiler performance is critically dependent on the fit and performance of the fans. Combustion η, auxiliary power consumption and boiler dynamics are all directly connected with the fan performance. Noise from the fans is also an indirect performance measure. It is imperative, therefore, that the process design parameters of the fan along with mechanical specifications are very carefully developed by the boiler maker in close association with the fan maker. Fan test codes such as PTC 11 come to the aid of both in choosing the correct class of fans and their performance guarantees. Closer the fan manufacturing tolerances, tighter are the performance guarantees.

6.3.5 Fan Construction

The two dominant types of construction for boiler fans are

1. Centrifugal
2. Axial

depending on the flow of air with respect to the fan axis.

Centrifugal fan construction is defined by the blade shape in the rotor. There are essentially three basic shapes of blade, namely, *forward curved*, *radial* and *backward curved* blades, with several variations in between.

Forward blading imparts a greater absolute velocity to the fluid while backward blading does the reverse as shown in Figure 6.2. As a result, the forward-bladed fans operate at a lower tip speed while the backward-bladed fans operate at a higher tip speed. Most importantly

- Forward blading has a tendency to accumulate dust on the blades and hence is more suitable for clean application.
- Backward blades have self-cleaning property and are suitable for dusty applications like flue gases from coal.

Various bladings and their efficiencies are shown in Figure 6.3:

- Backward blading also enjoys the advantage of non-overloading power characteristics and higher η of ~75–80%. It is best suited for all fans in a boiler.
- The η can be further improved to ~90% by aerofoil type of blading.
- The radial-bladed fans have the advantage of not accumulating any dust and thus being suitable for gases having high dust loads. The efficiencies are lower around 72–75%.
- Open blading radial fans with even lower efficiencies of ~65% are best suited for most abrasive duties, such as the re-injection of fly ash.

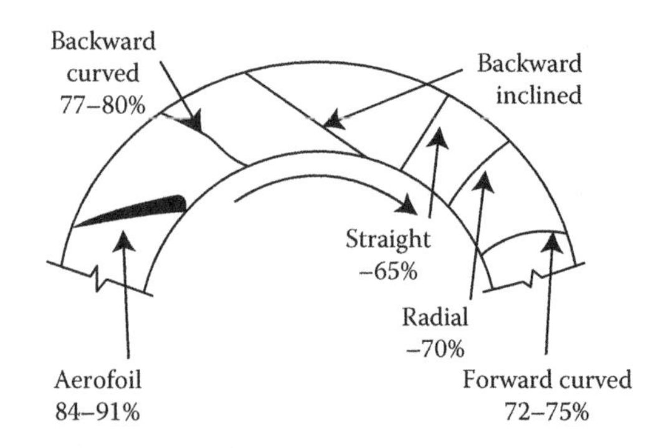

FIGURE 6.3
Forward, backward and radial blading with variations in between.

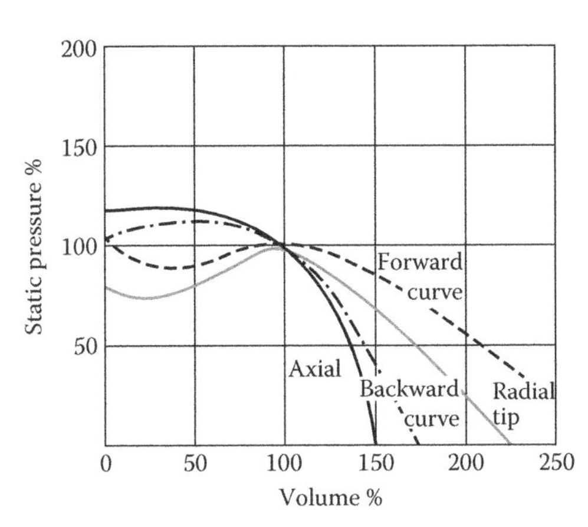

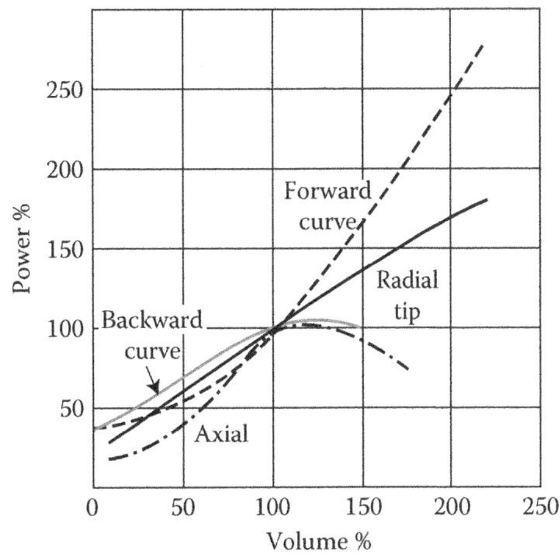

FIGURE 6.4
H–Q and P–Q characteristics of various types of fans.

The characteristics of fans with various bladings are shown in Figure 6.4.

Axial fans are of either single or two stages. They are highly efficient over a wide operating range with self-limiting power characteristics. But they are also highly prone to erosion and are more expensive. Their use is usually limited to the FD fans in large utility boilers. They are also used for ID fans if flue gas is not abrasive.

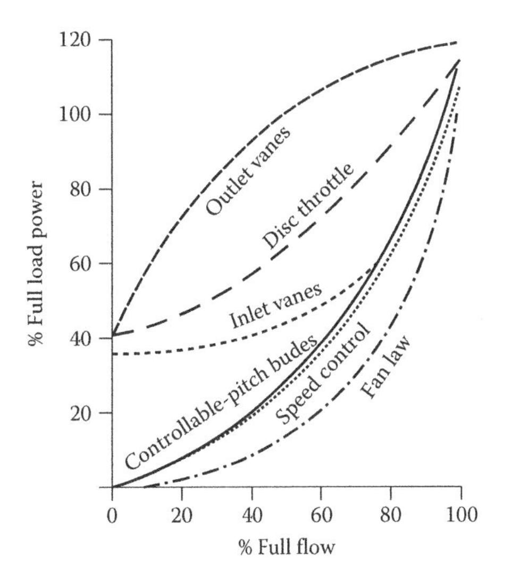

FIGURE 6.5
Various methods of fan control.

6.3.6 Fan Control

Fan control is directly related to the control of the boiler. It is essential, therefore, that the control is smooth, efficient (consuming less power) and spread over a wide range.

Axial flow fans are always controlled by variation blade angle. In centrifugal fans, there are two ways of control, namely *damper control* and *speed variation*. Fan control by speed variation is undoubtedly more efficient but is also more expensive. Figure 6.5 compares the damper and speed controls.

Outlet damper control is the least efficient as the fluid is throttled after raising it to full pressure. The control is simple but coarse and is only for small fans. Many times this is used along with the inlet vane control (of fans) (IVC) for enhancing the range and fineness of control.

IVC and differential damper control (of fans) (DDC): IVC, also called the *vortex damper*, is the most popular type of fan control because of the simplicity of construction and relatively highly efficient control. The aerodynamically shaped inlet vanes, as depicted in Figure 6.6, are located at the eye of the fan impeller and provide a spin to the fluid in the direction of wheel rotation, imparting the same effect as reduction of speed. Greater the spin, lower is the outlet pressure

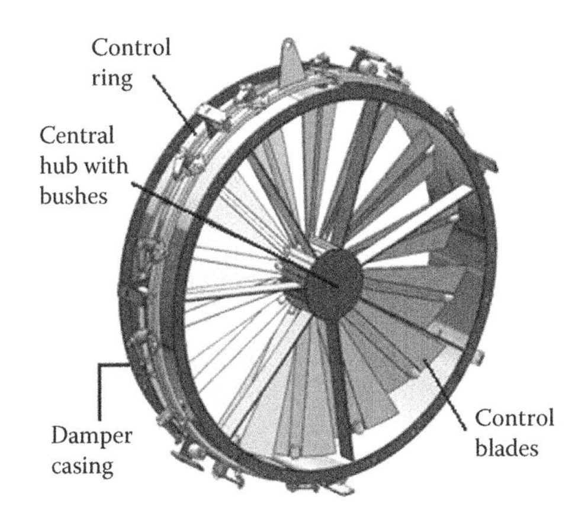

FIGURE 6.6
Inlet vane control (IVC) assembly.

and discharge volume. Inlet vane control enjoys the advantages of

- Being compact and economical
- Having stepless variation
- Having low hysteresis
- Being greatly amenable to automatic control

FD, PA and SA fans are routinely controlled by IVC in industrial boilers.

However, dusty gases are unsuitable for IVC as the vane shapes are eroded after some time.

Differential dampers at the entry to the inlet box of the fan (not at the eye of the impeller) are used with dusty gases. Differential damper is a multi-louver damper with aerofoil-type blades for good and efficient flow control. While DDC is rugged and can withstand the erosive gases, it is not as efficient as IVC. The spinning of gas is not aerodynamically as good as with IVC that is located at the impeller eye. ID fans are usually controlled by DDC in most industrial boilers.

Control by speed variation is justified in large installations or with boilers operating over wide load ranges when the savings in power can neutralise the high expenses involved in adopting speed variation. Several methods are employed such as

- Variable speed hydraulic couplings or fluid drives
- Variable frequency drive (VFD)
- Slip ring motors
- DC motors
- Drive turbines

VFDs have been gaining greater acceptance for both small and large boilers because of their simplicity, minimal O&M costs and progressive reduction in cost.

Drive turbines are attractive for sizes >50 kW in process plants where steam is available. ST drives then become doubly attractive as they consume only raw steam (and not power which is converted from steam) and save big power because of control of fans by speed.

Variable-speed hydraulic couplings are also popular but are steadily yielding place to VFDs.

6.3.7 Fan Laws

Fan laws can be stated as follows:

a. For the same fan at constant ρ

Discharge volume is proportional to speed, that is, $Q \propto N$ (6.1)

Static head is proportional to speed², that is, $P \propto N^2$ (6.2)

Horsepower is proportional to speed³, that is, $HP \propto N^3$ (6.3)

b. For geometrically similar fans of different sizes

Discharge volume is proportional to diameter³, that is, $Q \propto D^3$ (6.4)

Static head is proportional to diameter², that is, $P \propto D^2$ (6.5)

Horsepower is proportional to diameter⁵, that is, $HP \propto D^5$ (6.6)

6.3.8 Fan Operation (Series and Parallel)

Operation of a pair of fans can be either in

- Series when the pressure needs to be built up or
- Parallel when the volume needs to be built up

as shown in Figure 6.7.

PA fan taking suction from FD duct is an example of series operation. A pair of FD, ID, PA and SA fans

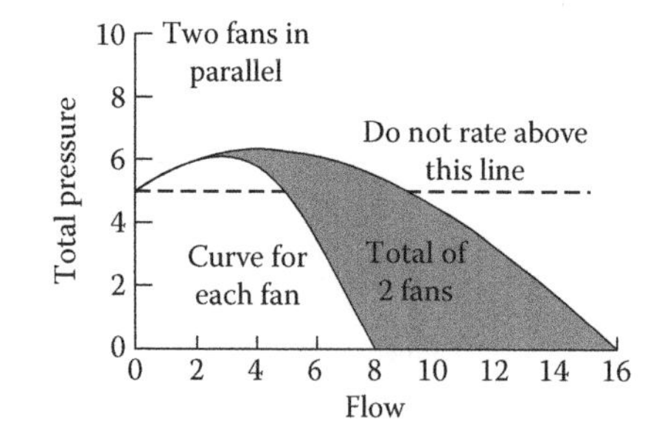

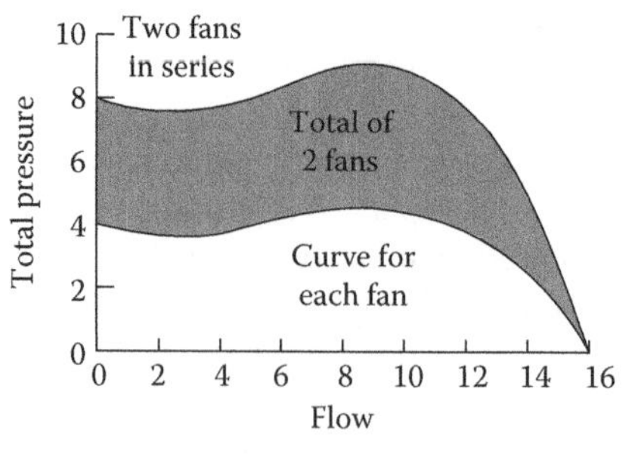

FIGURE 6.7
Parallel and series operation of a pair of centrifugal fans.

sharing half the load each is an example of parallel operation. Dissimilar fans or even axial and centrifugal fans can be in series operation. But for parallel operation, the fans have to be identical lest the fans should work in unstable mode and start hunting.

6.3.9 Fan Power

Fan power consumption can be expressed as

$$HP = \frac{kVH}{\eta c} \tag{6.7}$$

where k = compressibility factor
V = inlet volume in cum/h or cfm
H = pressure rise across the fan in mm or in wg
c = constant 2743 (SI) or 6354 (fps)
η = fan sstatic η%

$$kW = \frac{kVH}{\eta c} \tag{6.8}$$

where V is in cum/s, H in kPa and $c = 1$.

Compressibility factor is the ratio of the total pressures developed by a fan with an incompressible and compressible fluid with all other conditions remaining the same (Table 6.1).

6.3.10 Fan Static Head and η

Static head developed by fans is the useful pressure head in the discharge duct which can be measured by manometer. When the velocity head of the stream is also added, then it is the total head developed by the fan. The total head can be measured by a pitot tube.

It is the static head which is of practical relevance.

The ηs are also accordingly total/mechanical η and static η. They are both measured as % of energy input to the fan. Static η takes into account only the static pressure of the fan while the total/mechanical η considers the total head developed.

It is the static head which is available to the system, similarly, it is the static η which is used by the manufacturers to specify the fan performance.

$$\text{Static } \eta = (\text{static head in wg} \times \text{cfm}) / (6354 \times \text{shaft HP}) \times 100\% \tag{6.9}$$

$$= (\text{static head in mm wg} \times \text{m}^3/\text{h}) / (2743 \times \text{shaft HP}) \times 100\% \tag{6.10}$$

TABLE 6.1
Compressibility Factors for Various Pressure Ratios of Fans

Compressibility factor	1.0	1.05	1.1	1.15
Pressure ratio	1.0	0.98	0.97	0.95

$$\text{Mechanical } \eta = (\text{total head in mm wg} \times \text{cfm}) / (6354 \times \text{shaft HP}) \times 100\% \tag{6.11}$$

$$= (\text{total head in mm wg} \times \text{m}^3/\text{h}) / (2743 \times \text{shaft HP}) \times 100\% \tag{6.12}$$

Shaft HP is the power at the input of the shaft which is the motor/prime mover input power, less the transmission losses. Bearing and other losses are part of the fan losses.

6.3.11 Fan Types

Fans are of mainly two types depending on the flow of fluid with respect to the fan axis:

- In *centrifugal fans*, the fluid flow is *perpendicular* to the shaft. Fluid is forced to rotate in fan housing and develop pressure. Fluid accelerates from the root of the blade to the tip and collects in scroll housing. See Figure 6.8.
- In *propeller fans*, the flow is axial (Figure 6.9) but the pressure generated is too small for use in boilers. There is practically no casing to guide the air. They are popular in ventilation. See a short description on propeller fans under axial fans.

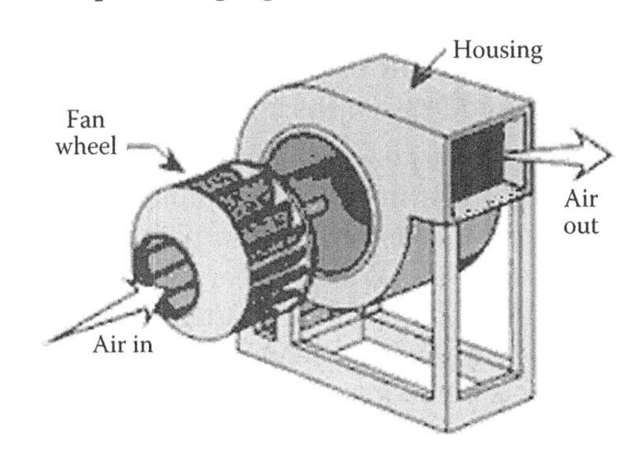

FIGURE 6.8
Centrifugal fan.

FIGURE 6.9
Propeller fan.

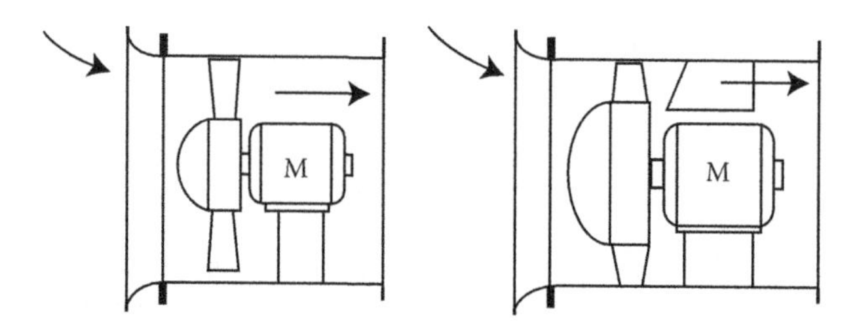

FIGURE 6.10
Tube and vane axial fans.

- In *axial fans* also, the flow of fluids is *parallel* to the fan axis. Axial fans are, in a way, propeller fans but with proper housing to provide definite control.
- *Vane axials* have guides either in front or rear of fan blades to direct the flow and enhance η while *tube axials* have none. See Figure 6.10.

Based on constructional features, there can be fans of several types.

6.3.11.1 By Support

Centrifugal fans can be supported in two ways: *overhung* and *between bearings/simply supported.*

Overhung design is good for small fans for clean air application. The impeller is mounted on the fan shaft which is supported by a pair of bearings on the drive side. With no bearings on the non-drive end, the fan shaft is shorter, which makes the fan compact and economical. This design cannot be used for dusty gases for possible shaft imbalance and vibrations in due course when dust deposits take place. See Figure 6.11.

Impeller on motor shaft is another type of overhung design employed for extremely small fans such as mill seal air or scanner fans and so on. Only cold air application is possible for fear of damage of the motor bearings. See Figure 6.12.

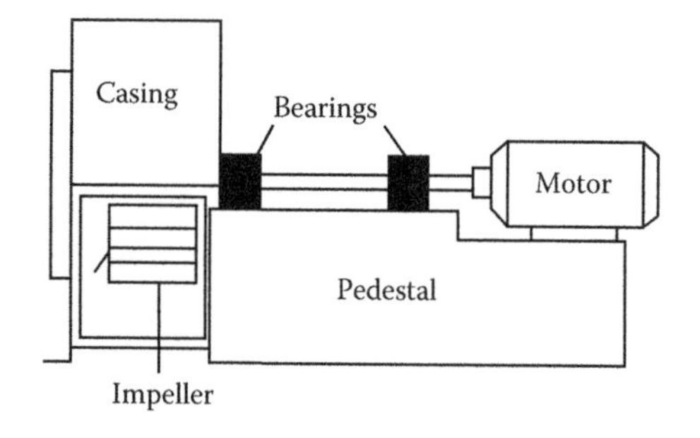

FIGURE 6.11
Overhung design of centrifugal fan.

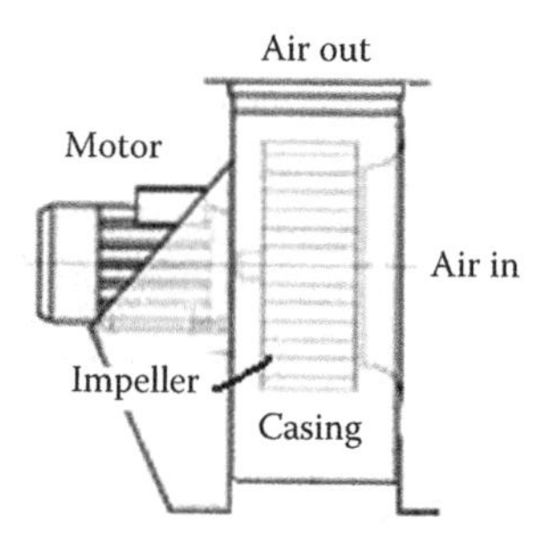

FIGURE 6.12
Impeller-on-motor shaft design of centrifugal fan.

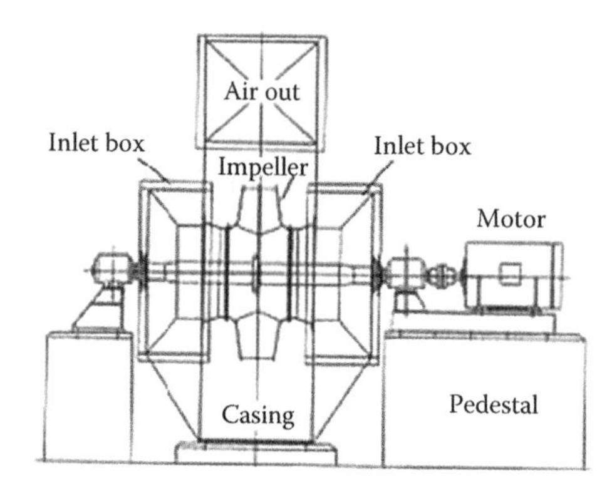

FIGURE 6.13
Double inlet simply supported centrifugal fans.

Simply supported design employs bearings on either side of the fan shaft, making the arrangement more robust and suitable for dusty gases and large volumes. See Figure 6.13.

6.3.11.2 By Number of Inlets

Centrifugal fans can have either single or double inlets.

Single inlet fans are economical up to ~50 cum/s (~100,000 cfm). Single entry creates an axial thrust along the direction of flow and calls for big thrust bearing. See Figure 6.14. It needs a single inlet ducting and control damper and is thus cheaper.

Double inlet fans with double width impellers are suitable for larger capacities. No thrust bearings are required as the axial forces equalise. These fans are more stable and can withstand vibrations and dusty applications better. See Figure 6.13. This type of fan needs double inlet ducting and two inlet control dampers.

6.3.11.3 By Type of Drive

These fans can have direct, belt or dual drives.

Direct drive is the simplest and most popular. It calls for little maintenance but the fan may have to operate at off peak η since the drive rpm may

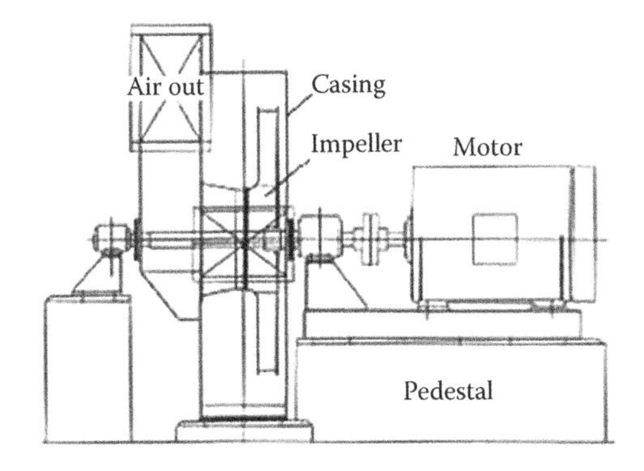

FIGURE 6.14
Single inlet overhung design centrifugal fan.

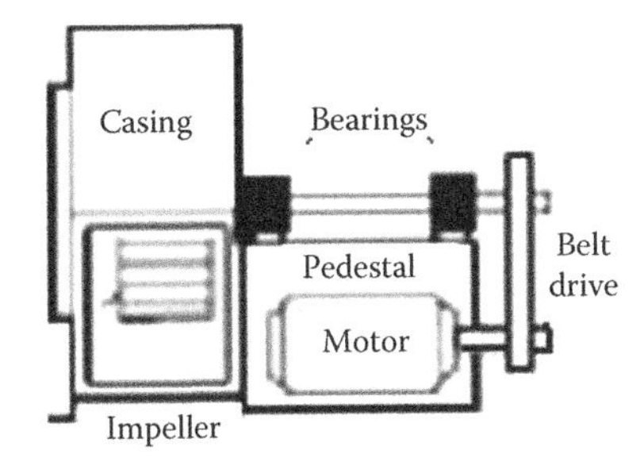

FIGURE 6.15
Belt-driven centrifugal fan.

be at variance to the optimum rpm. Direct drive is the only solution for large fans. Figures 6.13 and 6.14 both illustrate such drives.

Belt drive is often employed for smaller fans up to 150 kW. Fan speed is adjusted to operate at the optimum η. The arrangement occupies less floor space but the belt slackening and wear is to be considered. See Figure 6.15.

Dual drive is popular in certain process industries, like refineries and petrochemicals, with motor and ST on either side of the fan. Motor is usually used for start-up and ST for regular driving. This arrangement also enhances the availability. But the fan is more expensive with shaft extending on both sides.

6.4 Fatigue Strength or Endurance Limit (*see* Metal Properties)

6.5 Feeders (*see* Fuel Feeders)

6.6 Feed Water (*see* Water)

6.7 Feed Water Temperature (*see* Boiler Steam Temperatures)

6.8 Ferritic Steels (*see* Metals)

6.9 Ferritic Austenitic Steels (*see* Metals)

6.10 Fibre Glass (*see* Insulating Materials)

6.11 Field-Erected Boilers (*see also* Field-Erected Boilers in Oil- and Gas-Fired Boiler)

Field-erected boilers are the opposite of package boilers. Large sizes of these boilers do not permit full or substantial shop assembly. Many sub-assemblies of the boiler are shop assembled to reduce the field erection time and effort and also improve the overall finish of the boiler. Approximately 40% of total work needs to be done at site.

Most of the boilers firing solid fuels are field-erected as are the large oil- and gas-fired boilers.

6.12 Filling of Boiler (*see* Operation and Maintenance Topics)

6.13 Film Boiling (*see* Properties of Steam and Water)

6.14 Fins (*see* Extended Surfaces)

6.15 Finned Tubes (for Hrsgs) (*see* Heat Recovery Steam Generators)

6.16 Fire Bricks (*see* Refractories)

6.17 Fire Clay Refractories (*see* Refractories)

6.18 Fire Tube Boiler (*see* Smoke Tube Boilers)

6.19 Fittings (for Boilers) (*see* Auxiliaries)

6.20 Fixed Carbon (*see* Coal)

6.21 Flame Monitors/Scanners

The purpose of flame scanners is to monitor the health of the flame as long as the burner is in firing mode and help take the safety step of cutting off fuel in case of flame failure. It is a vital part of BMS. Figure 6.16 depicts the action in a simplified way.

Flame scanners were introduced only in the 1950s and 1960s with the advent of reliable photovoltaic cells.

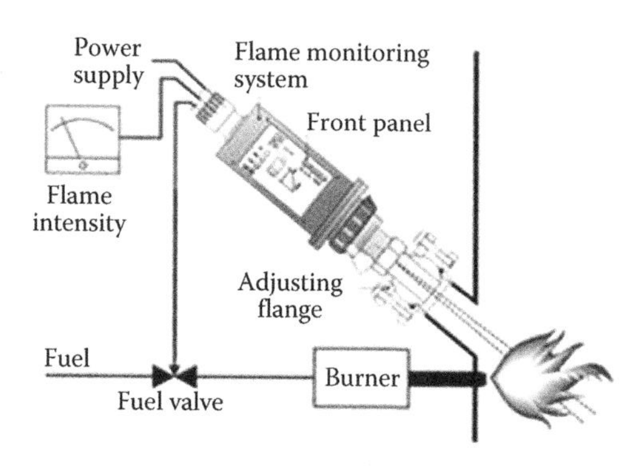

FIGURE 6.16
Action of a flame scanner.

Flame interruption followed by burner restarting, mainly due to temporary suspension in fuel flow, is always highly hazardous, often leading to explosions. With the continuous monitoring of flames with modern scanners, burner firing has become far safer.

All flames produce electromagnetic waves. Which span the entire spectrum of UV, visible and IR ranges. Oil and gas flames produce radiation in the UV range while coal and other solid fuels, including refractory, produce in IR range. Early scanners were able to detect UV radiation only and hence were suitable mainly for oil and gas flames. Scanners that followed later were able to detect IR radiation also.

While monitoring a single flame is quite easy and reliable in all multiple burner arrangements interference from the adjacent flame or refractory cannot be avoided leading to erroneous results. The modern scanners can detect both UV and IR radiations and in addition, detect the radiation at the root of the flames. All flames produce typical flickers as they emanate from the burners. Detection of the flicker as well as the main body of flame is a more definite method of recognising any flame despite the other disturbances. With such advanced scanners, it is possible to scan the multiple flames with multiple fuels and also deal with complex situation like burners positioned above grates.

Selection of an appropriate scanner is most important in any type of burner firing. A wide variety of scanners are available from simple UV scanner to the complex UV + IR + flame flicker types to suit all combinations of fuels, firing modes, burners and so on.

The importance of proper installation of scanner cannot be underestimated, and a suitable location has to be chosen (as shown in Figure 6.17).

Whenever gas electric igniters are used, a separate flame scanner for the igniter in addition to sighting the main flame, is also needed. Such an igniter and the scanner are often packaged together.

A flame scanner assembly consists of a viewing head, flexible cable, mounting flange and output terminal (as shown in Figure 6.18).

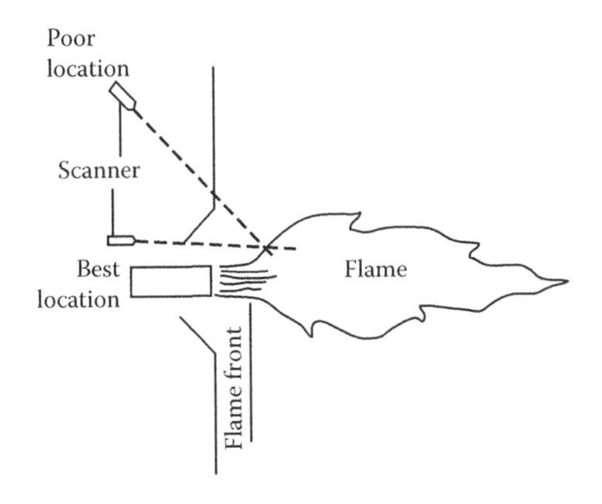

FIGURE 6.17
Good location for a flame scanner.

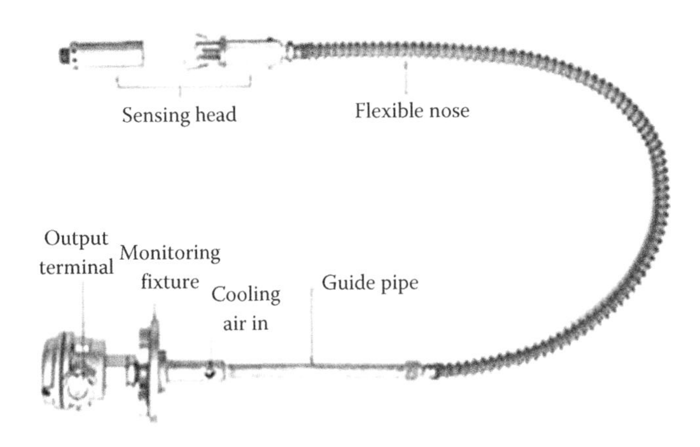

FIGURE 6.18
Components of a flame scanner system.

6.22 Flame Supervisor Safety System (*see* Operation and Maintenance Topics)

6.23 Flash Point (*see* Liquid Fuels)

6.24 Flow Measurement

Quantification of the flowing fluids is flow measurement.

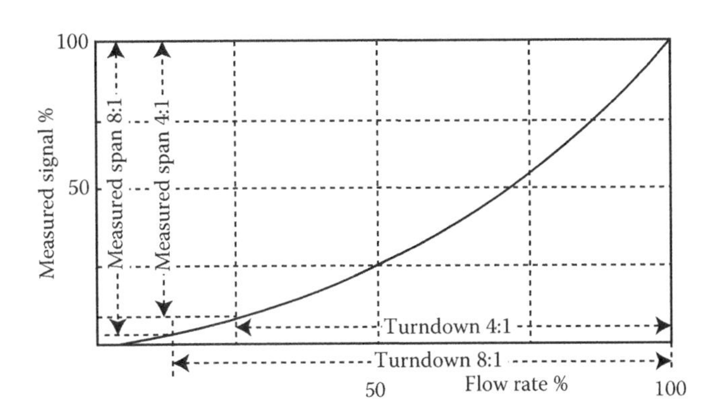

FIGURE 6.19
Flow versus pressure drop relationship in differential pressure flow meters.

There are many types of flow measuring devices based on several different principles, of which the following Δp flow meters can be considered as the basic devices, most commonly employed in boiler practice. In the order of increasing permanent pressure losses, these are

- Pitot-static tubes (see in Pitot-Static Tubes also)
- Orifice plates (see in Orifice Plates also)
- Flow nozzles (see in Flow Nozzles also)
- Venturi tubes (see in Venturi Tubes also)

In Δp meters, the flow is calculated by *measuring pressure drop over an obstruction* inserted in the flow. As per Bernoulli's equation, the Δp is a function of the square of flow velocity. This is shown in Figure 6.19. Note that the loss across measuring device reduces sharply at lower loads as per square law. Naturally, for higher turndown, in order to provide measurable Δp at the lowest load, the pressure loss at full load has to be quite high. In other words, higher the permanent loss in the instrument better is the accuracy and range of measurement and naturally more expensive is the system.

In boiler houses usually:

- Pitot tubes are used for measuring air and gas flows
- Orifice plates for oil, flow nozzles for steam
- Venturi tubes for air and coal–air mixtures

6.25 Flow Types in Vertical and Horizontal Tubes

Flow of steam and water is of great importance in boilers. In single phase flows, as encountesred in SH, Ecos and in steam and water lines, flow pattern is simple and there is no difference between vertical and horizontal disposition of tubes and pipes. Two-phase flow, experienced in boiler circulating system, is more complicated and there are differences in the flow patterns between the two.

6.25.1 Flow in Vertical Tubes

When rising water is steadily heated in a vertical tube, there is a progressively increasing steam formation (as shown in Figure 6.20b through f). Water entering the heated tube undergoes these overlapping stages smoothly one after the other.

The following are the various stable flow regimes. Stage a is pure water and stage g and h are pure steam. The stages in between, namely b to f, are in two-phase flow and the mixture stays at the saturation temperature. At stage a, water temperature is lower and at stage g and h, the temperature of steam begins to rise:

a. Water is in *sub-cooled* state with zero steam content. The mass fraction of vapour is zero.
b. On application of heat, water reaches saturation temperature and small bubbles begin to form inside the tube surface. The mass fraction of vapour is ~0.5%.
c. With more absorption of heat, the bubbles become larger and coalesce to form *scum*. The mass fraction of vapour in scum flow is ~3%.

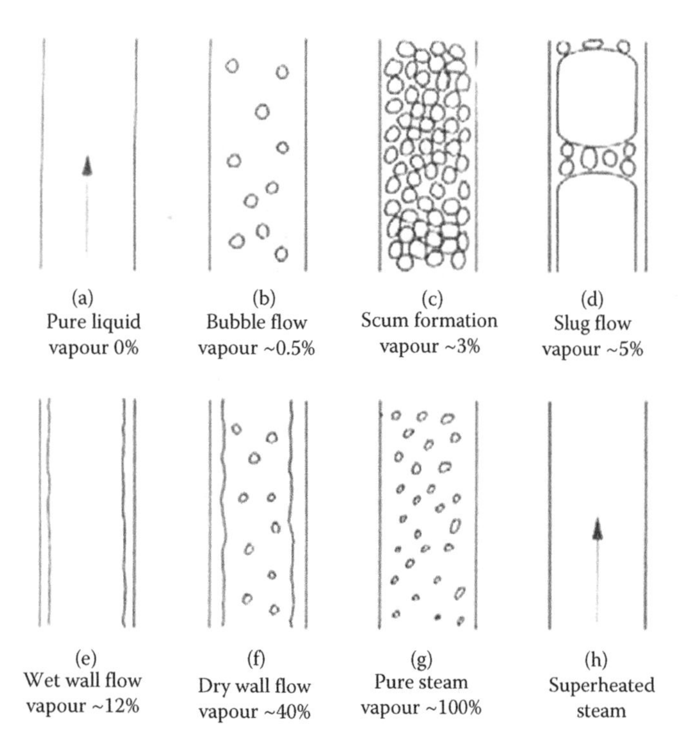

FIGURE 6.20
Various stages of two-phase flow in heated vertical tubes individually depicted.

d. With more heat and increasing velocity of the two-phase mixture, there is an increasing coalescence of bubbles to form large *slug* plugs. The spaces between the slugs are filled with scum. The mass fraction of vapour in slug flow is ~5%.
e. There is now so much vapour on further heating, that the liquid is forced to the wall which forms a thin film there. The flow is turbulent. This type of flow is called *wet wall flow or ring flow*. The mass fraction of vapour is ~12%.

   This is same as *nucleate boiling* with the water film absorbing the heat to keep the tube wall cool at almost the same temperature as the water film.
f. In the *dry wall flow*, the water film has disappeared from the tube wall with the addition of some more heat. All the remaining water is now in the form of fine droplets suspended in steam vapour (mist). The mass fraction of vapour now is >40%. The tube wall is cooled by steam film and the wall temperature is now higher than earlier.

   This is the same as *film boiling*. If the heat flux is high, the tube wall will get sharply overheated leading to tube failure on account of DNB.
g. All water has now become steam. The mass fraction of vapour is 100%.
h. Further heat addition goes to increase the steam temperature and make it (superheated) for the first time.

Figure 6.21 shows all the flows arranged sequentially to show how actually steam forms in a vertical tube when heated from outside nearly replicating the boiler tube in action.

6.25.2 Flow in Horizontal Tubes

In a steadily heated horizontal tube, the flow patterns are different and gravity comes into play. Also, when the Δρ is high between two fluids, segregation takes place. Different horizontal flow regimes are depicted in Figure 6.22.

In a boiler tube because of the ρ difference between steam and water, all the steam bubbles migrate to the top of the tube and slide along the tube wall. Bubble flow and annular flow are not of great relevance:

a. At higher velocities of >1 m/s, the steam bubbles join together and move along the tubes with water in an elongated bubble flow pattern.
b. At low velocities of <0.5 m/s, the flow is even more asymmetrical and unstable. The water and steam flow separately in tubes in stratified flow pattern.

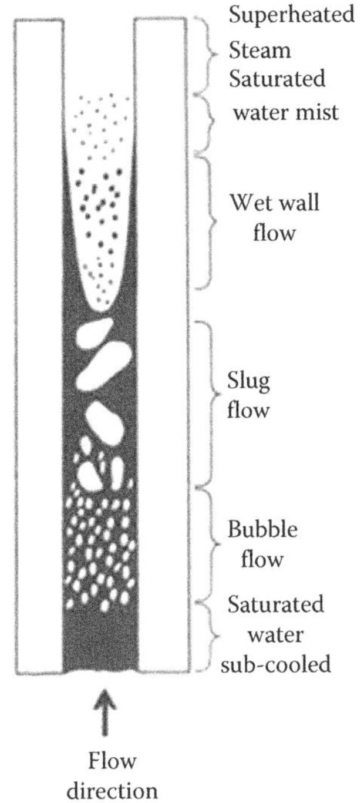

FIGURE 6.21
Two-phase flow in a heated vertical tube with all stages sequentially arranged.

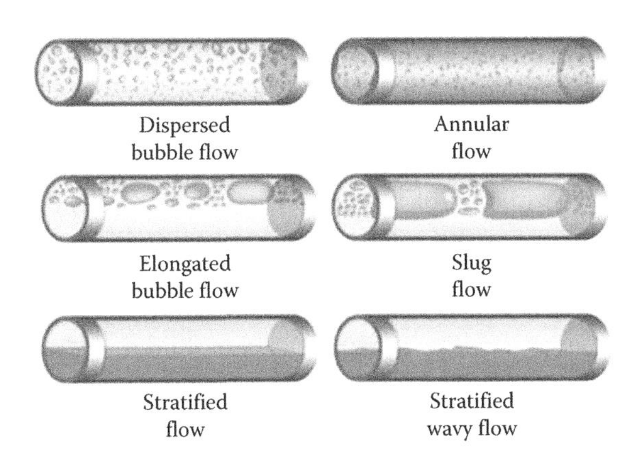

FIGURE 6.22
Liquid flow patterns in heated horizontal tubes.

Choked flow condition (*see also* Cavitation and Critical Flow and Pressure (for Nozzles))

It is a state when the flow rate of fluids cannot be increased even if the downstream pressure is decreased. The phenomenon is *different in liquids and gases.*

In liquid applications, it is caused by vapour bubbles formed by cavitation or flashing, thereby choking the flow passage. It is due to the venturi effect. Cavitation is quite noisy and can be sufficiently violent to physically damage valves, pipes and associated equipment.

In the case of gases, choke flow is caused when the flow velocity reaches the maximum, namely, sonic proportions and further reduction in downstream pressure cannot increase the gas flow. The mass flow rate can only be increased by increasing the upstream pressure, or by decreasing the upstream temperature.

6.26 Flow Nozzle (*see* Steam Flow Measurement in Instruments)

6.27 Flue Tube Boilers (*also* Fire/Smoke Tube/Shell Type Boilers) (*see* Smoke Tube Boilers)

6.28 Flue Gas (*see* Properties of Air and Flue Gas)

6.29 Flues and Ducts (*see* Draught Plant)

6.30 Flue Tube Boilers (*see* Smoke Tube Boilers)

6.31 Flue Gas Desulphurisation (*see* Desulphurisation)

6.32 Flue Gas Recirculation or Recycling (*see* Gas Recirculation)

6.33 Fluid Characteristics

This topic is divided into the following sub-topics:

1. Fluid
2. Types of fluid
3. Fluid properties
4. Viscosity
   a. Absolute/dynamic viscosity
   b. Kinematic viscosity

6.33.1 Fluid

Fluid is a substance that lends itself to continuous deformation (flow) under the impressed shear stress. Both liquids and gases are fluids. The main difference between a liquid and a gas is that the volume of a liquid remains definite because it takes the shape of the surface on or in which it comes into contact, whereas a gas occupies the complete space available in the container in which it is kept.

6.33.2 Types of Fluid

An ideal fluid is one that is incompressible and has no viscosity or surface tension.

Strictly, ideal fluids do not exist, but running fluids like *water and air with low viscosity can be treated as ideal.*

Real fluid is one that is opposite of ideal fluid.

Newtonian fluid is one in which the *shear stress is proportional to the shear strain* (velocity gradient). The stress at each point is linearly proportional to the strain. It is a linear relationship and the line passes through the origin. The constant of proportionality is *viscosity*. The concept is analogous to Hooke's law for solids. *All gases are Newtonian, as are most common liquids such as water, HCs and oils.*

Non-Newtonian fluid is the opposite.

Compressible fluid is one which reduces in volume under pressure. Its ρ changes with pressure.

Incompressible fluid is one which experiences no reduction in volume or ρ with change in pressure.

Ordinarily, gases are compressible and liquids are incompressible. The pressure referred here is not extreme pressure.

Figure 6.23 depicts the stress–strain relationships in various types of fluids.

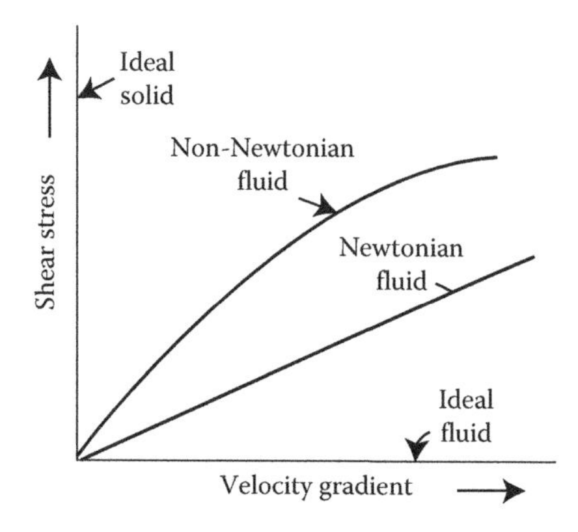

FIGURE 6.23
Relationship between stress and strain for various types of fluids.

6.33.3 Fluid Properties

The following is a list of important properties of fluids encountered in boilers:

- *Mass density* is the *mass of the fluid per unit volume* expressed in kg/cum or lbs/cft.
- *Specific weight* is the *weight per unit volume* of the fluid which is mass × gravitational force expressed in N/cum or lbs/cft.
- *Specific volume* is the *volume occupied by unit mass of fluid in* cum/kg.
- *Relative density* or SG is the *ratio of mass densities of fluid and water at STP conditions*, that is, 4°C and 760 mm Hg.
- *Viscosity* is a *measure of resistance to flow* of fluids. The viscous force is due to the intermolecular forces acting in the fluid. The flow or rate of deformation of fluids under shear stress is different for different fluids due to the difference in viscosity. Fluids with high viscosity deform slowly.
- *Compressibility* is the property of fluid to *undergo a decrease in volume when external pressure is applied.*
- *Elasticity* is the ability of fluid to *regain its original volume after the external pressure is released.*
- *Vapour pressure or equilibrium vapour pressure*: Molecules of vapour escape from the surface of liquid to fill the closed space of the container above until the pressure equalises and equilibrium is established. This is called the vapour pressure of the liquid. This is the equilibrium pressure experienced by vapour from a liquid or solid at a specified temperature.

The equilibrium vapour pressure is an indication of a liquid's evaporation rate. It relates to the tendency of particles to escape from the liquid (or a solid). A substance with a high vapour pressure at normal temperatures is often referred to as volatile.

- *Surface tension* is a phenomenon in which the *surface of a liquid acts like a thin elastic sheet*, where the liquid is in contact with gas or air.
- *Capillary action or capillarity* is the *ability of liquid to flow against gravity in narrow spaces* like thin tubes, in porous materials like paper or some non-porous materials such as liquefied C fibre. This is due to the attraction of intermolecular forces between the liquid and solid surrounding surfaces.

6.33.4 Viscosity (*see also* Kinematic Viscosity of Oils *under* Liquid Fuels)

Viscosity is the *resistance of fluid to flow*. Fluid flow is a continued and permanent distortion because of the impressed shear stress.

This resistance to flow is because of the internal friction which is on account of:

a. Molecular cohesion in liquids or
b. Molecular transfer from one layer to the other in gases

In liquids, molecular cohesion, and in gases, molecular transfer, predominates.

With rise in temperature, therefore, the viscosity lowers in liquids and rises in gases.

Viscosity is expressed in two forms:

a. Absolute or dynamic viscosity
b. Kinematic viscosity

a. *Absolute/Dynamic Viscosity* (μ)

Absolute/dynamic viscosity is the *force that exists when unit area of fluid is moving at unit velocity at a unit distance from a similar area placed parallel to it*. It is the *ratio of shearing stress to the rate of deformation*.

The unit is poise (Ns/10 m^2), named after the French physician Jean Louis Marie Poiseuille, but for convenience, centipoises (1/100 poise = Ns/1000 m^2) is used.

The absolute/dynamic viscosity of water at 20°C is ~1 centipoise (cP).

b. *Kinematic Viscosity* (ν)

Kinematic viscosity is a *measure of a fluid's resistance to flow and shear under the forces of gravity*.

Kinematic viscosity of a fluid is its *dynamic viscosity divided by* ρ, that is, $\nu = \mu/\rho$. The unit is stoke, named after the Irish scientist George Gabriel Stokes, expressed as $m^2/10^4\ s^2$, but for convenience, centistokes (1/100 stoke = $m^2/10^6\ s^2$) is widely used.

Kinematic viscosity of water at 20°C is ~1 centistoke (cS).

6.34 Fluid Flow

This topic is covered under the following sub-topics:

1. Types of fluid flow
2. Study of fluids
3. Vena contracta
4. Fluid flow patterns in HXs
5. Bernoulli's equation
6. Velocity head
7. Darcy–Weisbach formula for head loss
8. Moody's friction factor
9. Piping or ducting losses

6.34.1 Types of Fluid Flow

Fluid flow is the science of fluids in motion.

Steady and unsteady flows:

- Steady-state flow is when the fluid properties (temperature, pressure and velocity) at any single point in the system *do not change over time*. The system mass flow rate is always constant and hence there is no accumulation of mass within any component in the system.
- Unsteady flow is the opposite of steady flow.

Uniform and non-uniform flow:

- The flow velocity is of the *same magnitude and direction at every point* in uniform fluid flow. And the opposite is non-uniform flow.
- This means that fluid flowing near a solid boundary will be non-uniform—as the fluid at the boundary must take the speed of the boundary, which is zero. However, if the size and shape of the cross section of the stream of fluid is constant, the flow is considered uniform.

Compressible and incompressible flow:

- *Compressible flow* is a fluid flow in which ρ *does not remain constant*, such as the flow of gases through nozzles.
- *Incompressible flow* is a fluid flow in which ρ remains constant like in gases flowing in fans. Liquids are incompressible.

Laminar, turbulent and transition flows (*see also* Reynold's Number):

- Laminar or streamline flow, occurs when a *fluid flows in parallel layers*, with no disruption between the layers. It is the opposite of turbulent flow. In non-scientific terms, laminar flow is 'smooth', while turbulent flow is 'rough'. Transitional flow occurs when laminar flow changes to turbulent.

6.34.2 Study of Fluids

Fluid mechanics is the science that deals with the *behaviour of fluids* either at rest or in motion.

Fluid statics is the study of fluids that are at absolute or relative rest.

Fluid dynamics is the study of fluids in motion.

Fluid kinematics is the study of fluids *under translation, rotation and deformation* without considering the causing force and energy

Computational fluid dynamics (CFD) is a branch of fluid mechanics where numerical methods and algorithms are used for analysis of problems involving fluid flows with the aid of high-speed computers.

6.34.3 Vena Contracta

Vena contracta is the point in a fluid stream where the *diameter of the stream is the least*, such as in the case of a stream issuing out of a nozzle (orifice). The maximum contraction takes place at a section slightly on the downstream side of the orifice, where the *jet is more or less horizontal* (as shown in Figure 6.24).

Coefficient of contraction C_c = area at vena contracta/area of orifice with a typical value of 0.64.

6.34.4 Fluid Flow Patterns in HXs (*see also* Log Mean Temperature Difference [LMTD or Δtm])

In heat exchangers, there are three types of fluid flow arrangements:

Parallel flow—where the fluids enter and exit in the same direction

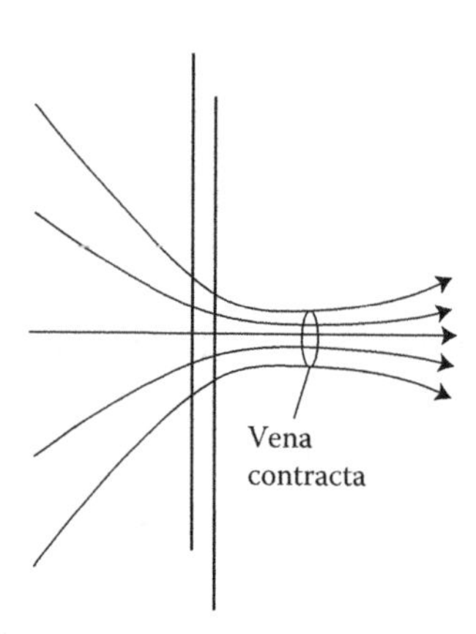

FIGURE 6.24
Vena contracta.

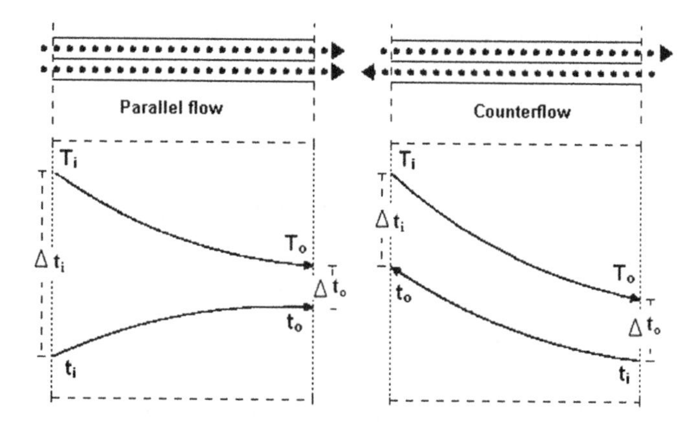

FIGURE 6.25
Fluid flows in parallel and counter flow in heat exchangers.

Counter flow—where the fluids enter and exit in the opposite direction

Cross flow—where the fluids enter and exit perpendicular to each other

Parallel and counter flow patterns are illustrated in Figure 6.25. In cross flow, the two fluids flow at right angles to each other like in a single bank of TAH.

6.34.5 Bernoulli's Equation

Bernoulli's equation is essentially an *energy conservation law*.

It is the most basic and universally applied law in fluid flow.

Bernoulli's equation states that for an ideal fluid the sum of pressure, velocity/kinetic and datum energy in a flowing fluid is constant and mutually convertible.

Stated differently, for an ideal fluid (incompressible and non-viscous) and steady flow (with uniform velocity across the cross section), the total energy comprising pressure, velocity/kinetic and datum energy is constant at any point in the fluid.

$$p_1 + \frac{V_1^2}{2g} + Z_1 = p_2 + \frac{V_2^2}{2g} + Z_2 \qquad (6.13)$$

where p is the pressure head in m (pressure in kg/m^2 × sp. vol. in m^3/kg)

V is the velocity in m/s

Z is the elevation in m

6.34.6 Velocity Head

Velocity or kinetic energy of a fluid is the dynamic energy because of its motion and expressed by its velocity head. $V^2/2g$. Units for velocity head are in/ft/m of fluid. It is usually converted into head of water and expressed in mm/inches wg.

6.34.7 Darcy–Weisbach Formula for Head Loss (*see also* Moody's Diagram of Friction Factor)

This formula is for calculating the head loss flow of fully developed, incompressible, steady flow of fluids through a pipe.

The flow of liquid through a pipe is resisted by

a. *Viscous* shear stresses within the liquid
b. *Turbulence* that occurs along the internal walls of the pipe, created by the *roughness* of the pipe material

This resistance is usually known as *pipe friction* and is measured in ft or m head of the fluid. The term head loss is also used to express the resistance to flow.

Many factors affect the head losses in pipes, namely the

- Viscosity of the fluid being handled
- Bore of the pipes
- Roughness of the internal surface of the pipes
- Changes in elevations within the system
- Length of travel of the fluid
- Resistance through various valves

Darcy–Weisbach equation: $h_f = f\,(L/d) \times (v^2/2g)$ (6.14)

where h_f = head loss in m/ft
f = a dimensionless friction factor
L = length of pipe work in m or ft
d = inner diameter of pipe work in m/ft
v = velocity of fluid in m/s or fps
g = acceleration due to gravity in m/s² or ft/s²

Friction factor f is taken from Moody's diagram which is 4 times the Fanning's friction factor.

6.34.8 Moody's Friction Factor

Friction factor f or λ in Darcy's formula of pipe friction can be read from Moody's diagram which is given in Figure 6.26.

From the diagram it is evident that the friction factor f is given by $64/R_e$ in the laminar flow ($R_e < 2000$) which is highly indeterminate in the transition zone governed by the relative roughness in the turbulent zone.

Fanning's friction factor is also widely used which is 1/4th of Moody's factor. In the laminar zone, f is given by $16/R_e$ in the case of Fanning.

6.34.9 Piping or Ducting Losses

In any fluid flow in pipes and ducts, there are mainly six types of flow losses which are given in Table 6.2.

6.35 Fluid Velocities (*see also* Draught Plant)

The two sets of working fluids in a boiler are water and steam, and air and gas. Velocities of water and steam determine the sizes of integral and external piping and air and gas determine the sizes of flues and ducts.

Table 6.3 provides a range of water and steam velocities. The lower end of the range is chosen for keeping the pressure losses low. The higher end is for short runs of piping, extreme cases and when there is a limitation of available economical pipe size.

Velocities for air and gas are tabulated under flues and ducts under draught plant.

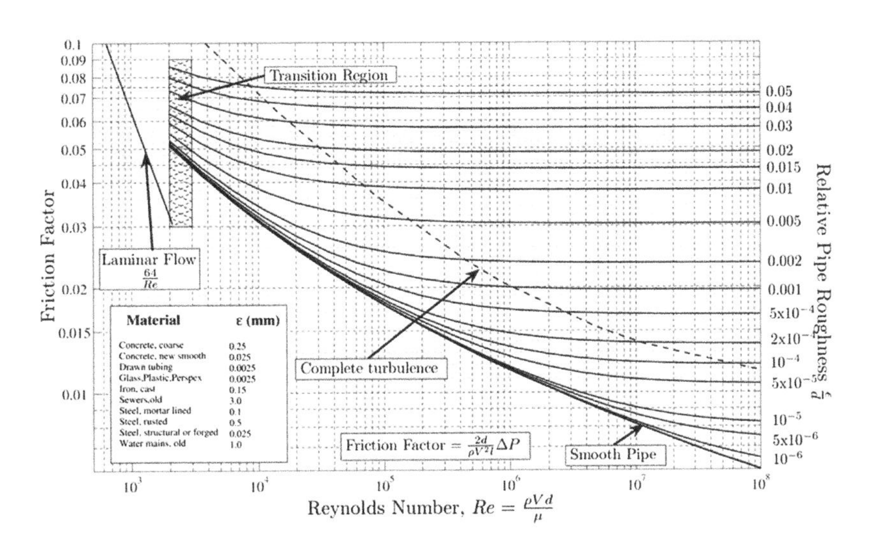

FIGURE 6.26
Moody's diagram for Reynold's number versus relative roughness and friction factor.

TABLE 6.2

Calculation of Piping or Ducting Losses

S. No	Loss Due To	Picture	Formula
1	Loss of head due to friction (major loss)	L, V	$\frac{fl}{d} \cdot \frac{V^2}{2g}$
2	Loss due to pipe fittings like bends, elbows, valves and so on	—	$K\frac{V^2}{2g}$ K is found experimentally
3	Loss due to sudden enlargement	V_1, V_2	$\frac{(V_1 - V_2)^2}{2g}$
4	Loss due to sudden contraction	V_2	$\left(\frac{1}{C_c} - 1\right)^2 \cdot \frac{V^2}{2g}$ Generally taken as $0.5 \cdot \frac{V^2}{2g}$
5	Loss due to entry to a pipe	V	$0.5 \cdot \frac{V^2}{2g}$
6	Loss at exit of the pipe	V	$\frac{V^2}{2g}$

6.36 Fluidisation

State of fluidisation is the one that lies between *static* and *entrained/transport* states.

Static bed: When air is forced through an enclosed bed of solids spread on a bottom plate it flows through the voids in the bed without causing disturbance to the bed. As the quantity of air is increased, the velocity and the Δp through bed increase somewhat linearly until the bed starts to expand and get lifted. This is static bed.

Fluidisation: At some point, called the *minimum fluidisation point*, the bed begins to be lifted off the bottom plate. For a bed of coal particles of <10 mm, the fluidisation point is at ~1 m/s of bed velocity. This point is called the *onset of fluidisation*.

The bed velocity is the velocity of air or gas leaving the bed given by fluid volume/bed area.

Bubbling bed: As flow is increased further, bubbles of air are formed and the bed height increases. The particles in the bed begin to swirl within the expanded bed. The bed is flat and vigorously churning. This is called a *bubbling bed* as the process resembles the bubbling of heated water.

Circulating bed: With further increase of air flow, the bed expands to fill the chamber above and form a cloud of particles. The cloud is dense at the bottom and lean at the top with some particles actually leaving the chamber while the majority flows down to the bottom along the walls of the chamber. Elutriated particles can be returned to the bed to set up a circulation of bed particles. This happens at bed velocities >~4 m/s when the mean particle diameter is kept low at 0.5–1 mm and it is termed *full circulation bed or fast fluidisation circulating bed*.

Expanded or turbulent bed: This is a stage in fluidisation that lies between the bubbling and circulating beds.

Entrainment: At >~7 m/s, the phenomenon enters into a transport regime when the particles simply leave the chamber without returning to the bed along the walls, marking the end of fluidisation process. This is termed as beginning of *entrainment or transport flow*.

TABLE 6.3
Range of Fluid Velocities Employed in Pipes

Nominal Pipe Size in mm		Average Velocity in m/s <50 mm	50–150 mm	>200 mm
Saturated Steam				
Saturated steam at sub-atmospheric pressure		—	10–15	15–20
Saturated steam at 0–1 kg/cm² (g)	~0–2 bar	15–20	17–30	20–30
Saturated steam at 1.1–7 kg/cm²	~2–8 bar	15–20	20–33	25–43
Saturated steam over 7 kg/cm² (g)	>8 bar	15–25	20–35	30–50
SH Steam				
SH steam at 0–7 kg/cm² (g)	~0–8 bar	20–30	25–40	30–50
SH steam at 7.1–35 kg/cm² (g)	~8–37 bar	20–33	28–43	35–55
SH steam at 35.1–70 kg/cm² (g)	~37–72.5 bar	22–33	30–50	40–61
SH steam over 70 kg/cm² (g)	>72.5 bar	22–35	35–61	50–76
Water at Pump Suction				
Pump suction, condensate		—	0.4–0.6	0.6–0.7
Pump suction, boiler feed		—	0.6–0.9	0.6–0.9
Pump suction, general service		0.6–0.9	0.7–1.3	0.9–1.5
Water at Pump Discharge				
Pump discharge, condensate		0.9–1.2	1.2–2.1	1.5–2.2
Pump discharge, boiler feed		1.0–1.2	1.5–2.1	1.8–2.4
Pump discharge, general service		0.9–1.0	1.5–2.4	1.5–2.4
Viscous Liquids				
Pump suction, viscous liquid		0.3	0.3–0.4	0.4–0.5
Pump discharge, viscous liquid		1.0	1.0	1.2–1.4
Air				
Compressed air		7–10	10–15	18

This entire process is depicted in Figure 6.27. In practice, at each succeeding stage

- The particle size gets finer
- Bed voidage increases
- The combustion chamber becomes slimmer and taller

Combustion can take place in all these regimes. Grate firing is combustion in static bed while PF firing is combustion in entrained or transport flow. The combustion that takes place in the fluidised bed regime is the FBC:

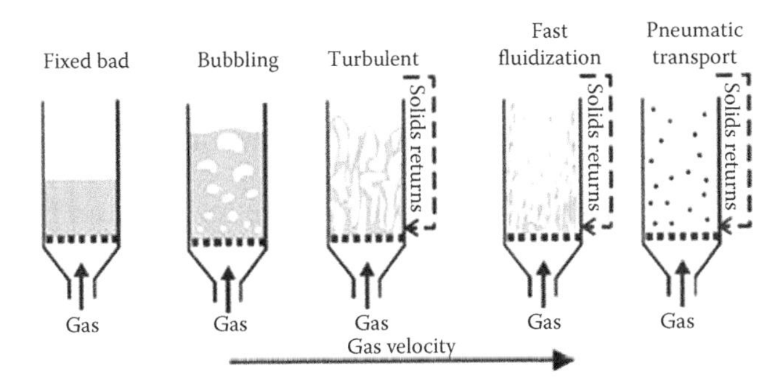

FIGURE 6.27
Fluidisation regimes between the static bed and full entrainment.

Bubbling FBC in the velocity range of ~1–3 m/s
Circulating FBC in the velocity range of ~4–6 m/s
Expanded/turbulent bed FBC in the velocity range of ~2–4 m/s

It is worth noting that

- In the entire velocity range of ~1–6 m/s, the behaviour of the bed is very similar to that of a fluid and hence the name fluidised bed.
- FBC is characterised by nearly constant Δp in bed as opposed to the case of static bed and entrained flow where the Δp increases and decreases with load, respectively.

Figure 6.28 depicts the Δp in bed with increasing fluidisation velocity. In the bubbling and circulating parts of the FBC, the Δp in the bed is nearly flat. The reason is that once the fluidisation sets in, the bed keeps expanding (voidage increases) and the Δp remains practically constant.

In the static bed, Δp increases, almost linearly, with velocity. On the contrary, in the transport regime, the voidage further rises leading to drop in Δp across the bed.

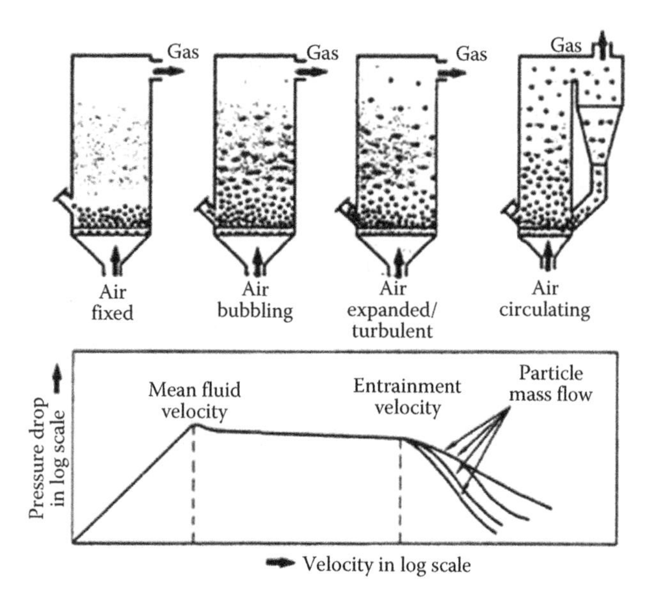

FIGURE 6.28
Bed pressure drop versus air/gas velocity at various stages of fluidisation.

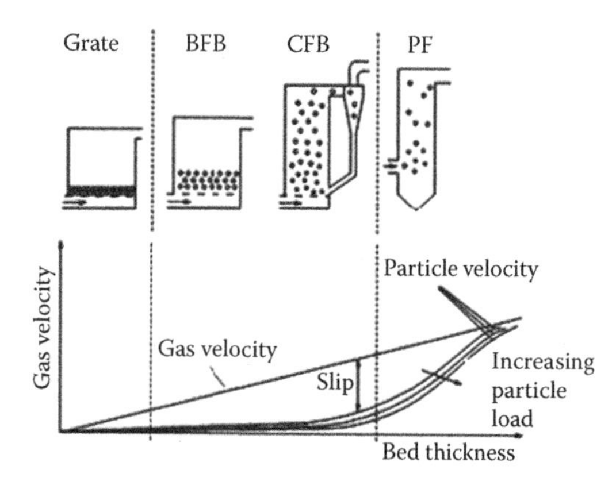

FIGURE 6.29
Increasing slip velocity in fluidised bed regime.

The difference between the velocities of fluid and bed material is known as *slip velocity*. With increase in bed velocities, the slip velocity also progressively increases in the fluidisation regime (as shown in Figure 6.29). However, in transport regime, as the particles start attaining the speed of fluid, the slip velocity again reduces.

Material properties of fluid and bed material, namely, the viscosity, ρ and particle size decide the velocities at which the various regimes set in.

6.37 Fluidised Bed Combustion

After grate firing and PF firing, FBC is the next major breakthrough in firing of solid fuels. As the name indicates, FBC is the combustion of fuel when it is in fluidised mode. The popularity of FBC has been growing since its commercialisation in the early 1980s. It has, since then, marginalised grate firing for coal and lignite on the one side and cannibalised PF firing for boilers up to 150 MWe and more on the other. The most salient benefits of FBC are

1. *Environmental friendliness*: The popularity of FBC boilers is mainly due to the fact that FBC is a *cool* and *clean* process.
   - *Cool* because the combustion takes place in the temperature range of ~800–900°C (producing minimum amount of **NO_x**).
   - *Clean* because **desulphurisation** is effected within the combustion chamber during the combustion process itself.

   In most cases, therefore, no additional denitrification (deNO_x) and **FGD** (deSO_x) plants are required. The environmental compliance of the FBC boilers is quite unmatched.
2. Fuel *flexibility*: In the same manner, the fuel flexibility of FBC boilers is also quite extraordinary. This is primarily because the fuels enter a vigorously churning mass of hot (800–900°C) inert material and get to reside and re-circulate there for a reasonable amount of time, enabling almost all types of solid fuels to burn out.

   The ability of FBC to burn difficult and diverse fuels is outstanding as it can burn:
   - Hard-to-burn fuels, such as **anthracite**, **pet coke** and low-volatile coal
   - High-M coals like **lignites** and **peat**
   - Various types of **sludge**
   - High-**burden or ballast** (M + A) coals and so on
3. *Multi-fuel* firing: In addition to wide fuel variation, FBC also exhibits very good multi-fuel firing ability. The same combustor can burn very dissimilar fuels quite easily. Many unburnable fuels like pulp mill sludge, anthracite culm, pet coke and so on can be burnt efficiently.
4. *High* combustion η: C burn up η is quite high in FBC system. In BFBC, it is ~92–96% and in CFBC it is ~96–99+% depending upon the fuel.

The shortcomings of FBC boilers are few and can be stated as

- Slightly higher power consumption
- Inability to seamlessly turn down to low loads in BFBC
- Slightly inferior dynamics compared to PF firing

TABLE 6.4
Capabilities of Grate, BFB and CFB Modes of Combustion

Firing Equipment	Capacity (tph)		Lcv of Fuel (kcal/kg)		Full Load Bed Velocity (m/s)	Combustion η on Coal (%)	Typical Boiler η on Coal (%)
	Max	Min	Max	Min			
Grates	200	10	4000	1500	<1	80–90	74–78
BFBC	250	10	4700	500	2–3	92–96	80–84
CFBC	1800+	70	9400	1200	4–6	98–99+	85–88

There are three types of FBC boilers in operation based on the full load bed velocities employed, namely

- BFBC boilers with ~2–3 m/s
- CFBC boilers with ~4–6 m/s
- Expanded bed FBC boiler with ~3.5–4.5 m/s

The bed velocities would be lower at lower loads.

Table 6.4 captures the capabilities of different types of combustion. It can be observed that while CFBC has scalability, BFBC can burn much inferior fuels. With a lot of positive features for CFBC, it is expected that the range of applications will further increase expanding the appeal of CFBC.

Figure 6.30 diagrammatically shows the combustors of the two technologies of BFBC and CFBC. The expanded bed comes in between the two, with cyclone like CFB but with a (deeper) bed in the combustor like BFB.

The combustion fundamentals in all the three types of FBC are similar. Cool and clean combustion produces low emissions, which in most cases requires no further gas cleaning treatment. Bed temperature, excess air and Ca/S ratio are the three most important combustion parameters that need to be monitored. Figure 6.31 shows the effect of these three factors on the NO_x, SO_x and CO levels in the exhaust gases. A bed temperature of ~850°C is a good balance to optimise SO_2 and CO. Likewise, an excess air level between 15% and 20% yields low CO and SO_2. Ca/S ratio of 1.8–2.5 is a good compromise in CFBC boilers. In BFBC, it has to be higher as the residence times are lower. But NO_x levels have a tendency to go up.

The three types of FBC modes are compared and contrasted in Table 6.5.

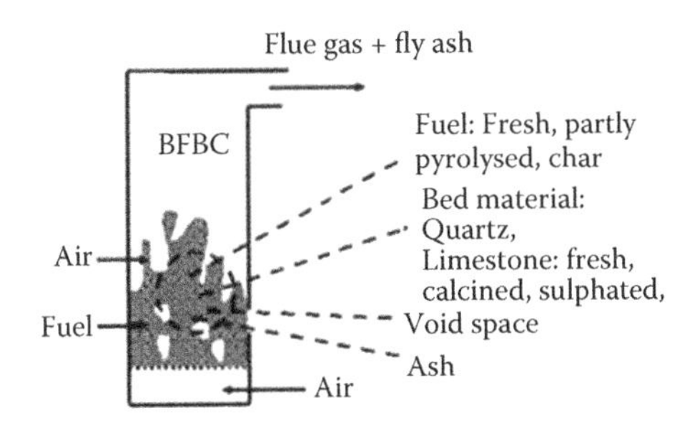

FIGURE 6.30
Modes of combustion in BFBC and CFBC.

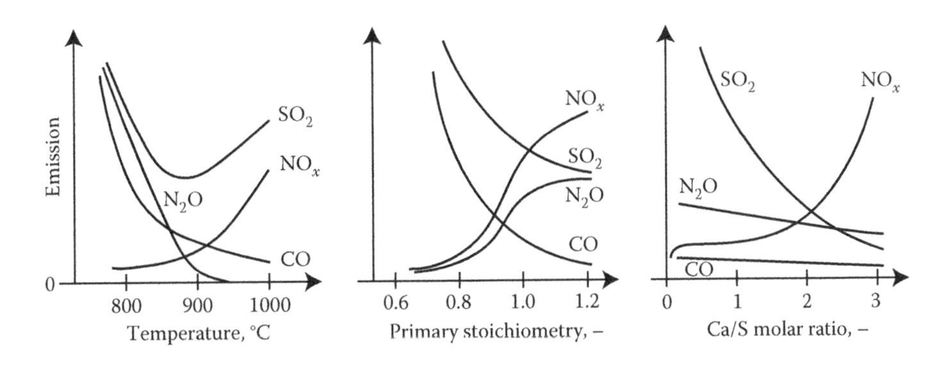

FIGURE 6.31
Effect of FBC parameters on emissions.

TABLE 6.5

Comparison of Bubbling, Expanded and Circulating Fluidised Bed Boilers

S. No.	Parameter	BFBC	Expanded Bed	CFBC	Comments
1	Boiler	Short and stout	In between	Tall and slender	
2	Range (tph) with coal	<150	100–800	100–2000	
3	η	Good	Better by ~4–6% points	Better by ~4–6% points	
4	C burn-up for coals	92–96%	98–99%	98–99%	Higher efficiency due to longer residence time
5	η on GCV (coal)	Approx. 84%	Approx. 88%	Approx. 88%	Higher unburnt losses in BFB
6	Part load η	Poor	Nearly same, up to 50% load	Nearly same, up to 50% load	
7	Bed velocity at full load	Low 1.8–3.0 m/s (~6–10 fps)	Medium 3.5–4.5 m/s(~11–15 fps)	High 4.5–7.0 m/s (~15–25 fps)	
8	Average bed particle size	0.5–1.5 mm	~0.75 mm	0.1–0.3 mm	
9	PA:SA ratio for coal	90:10	60:40	60:40	No staged combustion in BFB
10	Residence time (s)	2.5–3.0	4.0–5.5	4.0–5.5	Max. BFB shorter
11	HRR MW/m^2 $kcal/m^2$ h Btu/ft^2 h	~1.5 $\sim 1.3 \times 10^6$ ~475,000	~5 MW/m^2 $\sim 4.3 \times 10^6$ $\sim 1.6 \times 10^6$	~5–7 MW/m^2 $\sim 4.3–6 \times 10^6$ $\sim 1.6–2.2 \times 10^6$	
12	Bed temperature	Variable	Constant	Constant	
13	Bed control	In bed tubes or open bottom for bio-fuels	Ash and at times gas recirculation	Ash and at times gas recirculation	
14	Turn down	1:3 Stepped	1:3 Stepless	1:3 Stepless	Stepless makes boiler control very simple
15	Boiler response	Sluggish	Very responsive	Very responsive	
16	Fuel flexibility	Good	Better	Better	
17	Fuel sizing for coal	6.0–8.0 mm Inbed, <20 mm overbed	8.0–10.0 mm	6.0–8.0 mm	
18	Fines in coal <1.0 mm	<20%	<40%	<40%	Tolerance for fines low in BFBC
19	Tube erosion	Bed tubes	—	Furnace tubes	
20	Refractory erosion	No	No	Mostly overcome	
21	Auxiliary power	Base ~12–14 kw/MWth/h	Higher by ~15–20%	Higher by ~60–80%	BFBC aux. power is the least
22	Emission, NO_x	<200 ppm	<100 ppm	<100 ppm	No staged combustion in BFB
23	Emission, CO	<250 ppm	<100 ppm	<100 ppm	
24	Emission, SO_x	<300 ppm	<100 ppm	<100 ppm	For high S coal of ~3% S
25	Limestone needed	High	Low	Low	High due to very low ash recirculation
26	Desulphurisation	~85%	~95%	~95%	
27	Bed ash	Many points Messy system	Less points Simple system	Less points Simple system	

6.37.1 Pressurised Fluidised Bed Combustion

Pressurised fluidised bed combustion (PFBC) is the FBC carried out under pressurised conditions. Combustion air is compressed and admitted into the PFBC combustor along with fuel where the combustion takes place similar to AFBC. The resulting gases are dedusted in cyclones and filters which are also under pressurised conditions. All the advantages of **FBC** are available in PFBC also, such as

- Clean combustion (low-temperature combustion and in-furnace **desulphurisation**)
- Fuel flexibility
- Fuel variability
- The ability to burn high ash/high M coals and inferior fuels

In addition, there are two other attractions which make PFBC worth pursuing despite the additional expense which are not possible in AFBC, namely

1. The extreme compactness of the plant is a great advantage, as the compressed air and gas volumes are much lower. This is a relatively

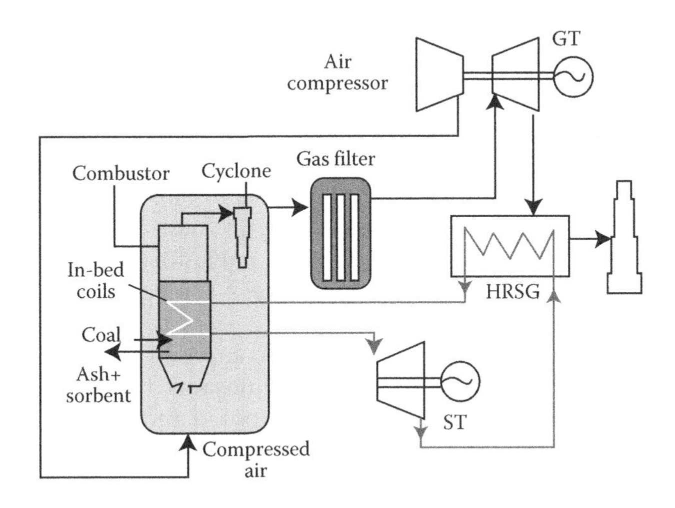

FIGURE 6.32
Simplified schematic of PFBC in combined cycle mode.

smaller benefit compared to the possibility of η advancement explained below.

2. PFBC is not equipped with backend equipment like conventional systems but made amenable to CC system incorporating both GT and ST with attendant increase in overall η. A typical schematic arrangement of a PFBC-based CC plant is shown in Figure 6.32.

The additional cost of PFBC is recoverable only because of the possibility of **CC** operation on solid fuels with its enhanced cycle η.

As with AFBC, both bubbling and circulating bed configurations are possible. Currently, all the commercial-scale operating units use bubbling beds and hence the acronym PFBC normally refers to pressurised bubbling bed units. Pressurised circulating fluidised bed combustion (PCFBC) demonstration unit was considered but no GT was available for the CC configuration.

In PFBC, the combustor and hot gas cyclones are all enclosed in a PV. Both coal and sorbent are fed and ash removed across the pressure boundary. For hard coal applications, both coal and sorbent can be crushed together and fed as paste with ~25% water. The combustion temperature of 800–900°C generates lower NO_x than PF firing but N_2O is higher. SO_2 emissions can be reduced by the injection of a sorbent like limestone or dolomite. Its subsequent removal with ash is quite the same as in AFBC.

Units operate typically at 1–1.5 MPa of gas pressure. The pressurised coal combustion system produces steam in conventional heat transfer tubing in boiler to generate power from ST. The combustor also produces hot gas which is supplied to GT. Gas cleaning is a vital aspect of the system as is the ability of the GT to cope with some residual solids. The need to pressurise the coal, sorbent and combustion air and thereafter to depressurise flue gases and remove ash introduces some significant operating complications. The combustion air is pressurised in the compressor section of the GT. The proportion of power from the ST:GTs is ~80%:20%.

PFBC and power generation by the CC route involves some unique control considerations. The combustor and GT have to be properly matched through the whole operating range. GTs are special, as they operate at lower gas temperatures of ~900°C as the maximum gas temperature from the PFBC is limited by ash fusion characteristics. No ash softening should take place and no alkali metals should vapourise (to avoid their re-condensation later in the system). Hence, a GT with HP ratio with compression inter-cooling is used to offset the effects of the relatively low GT gas temperature inlet.

Higher process pressure results in several advantages, in addition to CC operation and higher combustion rate:

- HRR per unit bed area is much greater in PFBC.
- The increased pressure and corresponding higher air/gas ρ allow much lower fluidising velocities (~1 m/s) which reduce the risk of erosion for immersed bed coils.
- At elevated pressures the heat released within the combustors increases and bed depths of 3–4 m are required to accommodate the HX area necessary for the control of bed temperature. At reduced load, bed material is extracted, so that part of the heat exchange surface is exposed.
- The combined effect of lower velocity and deeper beds results in greatly increased in-bed residence time which reduces emissions of SO_x and improves combustion η. In fact, the deeper beds allow 50% of the total residence time to be in the bed where it is more effective compared to 10–15% in the shallow atmospheric bubbling beds.
- Since air mass flow is proportional to the ρ the high air/gas ρ results in much lower bed plan area. For the same mass flow, a bubbling bed PFBC at 12 bar with a superficial velocity of 1 m/s would require 28% of the bed plan area of a bubbling AFBC.

Considerable effort has gone into the development of PFBC during the 1990s. PFBC is used on a commercial scale in Sweden and Japan with coals of good quality. The demonstration plant of 130 MWe (>224 MWth co-generation) was built in Vartan near Stockholm, Sweden in 1991 and has been meeting all the stringent environmental conditions. Another demonstration plant of 80 MWe capacity has been operating in Escatron, Spain using black lignite with 36% ash. The third demonstration plant of 70 MWe at TIDD station, Ohio, USA was shut down in 1994 after an 8-year demonstration period in which a large amount of useful data and experience were obtained. A 70 MWe demo plant has been operated at Wakamatsu from 1993 to 1996. The United Kingdom has gathered a large amount of data on an 80 MWe PFBC plant in Grimethrope during its operation from 1980 to 1992. Two larger units have started up in Japan at Karita and Osaki. These are of 360 and 250 MWe capacity, respectively. The Karita unit uses SC steam. The size of the PFBC is tied with the capacity of the GT.

Second-generation PFBC units give an η of ~45% in CC mode offering ~5% η enhancement besides lower emissions.

6.38 Fluidised Bed Ash Coolers (*see* Circulating Fluidised Bed Combustion)

6.39 Fluidised Bed Heat Exchanger (*see* Circulating Fluidised Bed Combustion)

6.40 Fly Ash (*see* Ash: Bottom *and* Fly)

6.41 Foaming and Priming (*see* Operation and Maintenance Topics)

6.42 Forced Circulation (*see* Circulation)

6.43 Forgings (*see* Materials of Construction)

6.44 Fossil Fuels

Fossil **fuels** are so called as they are *derived from fossils*, which were formed millions of years ago, during the time of the dinosaurs. They are fossilised organic remains that, over millions of years, have been converted to *coal, oil and gas* (as shown in Figure 6.33). Since their formation takes so long, these sources are also called *non-renewable*.

These fuels are made up of decomposed plant and animal matter. When plants, dinosaurs and other ancient creatures died, they decomposed and were buried, layer upon layer under the ground. Their decomposed remains gradually changed over the years. It took millions of years to form these layers into a hard, black rock-like substance called **coal**, a thick liquid called oil or **petroleum**, and **NG**—the three major forms of fossil fuels. Fossil fuels are high in C.

Fossil fuels are usually found below ground. Coal is either mined or dug out while oil and NG are pumped

FIGURE 6.33
Formation of fossil fuels over the geological times.

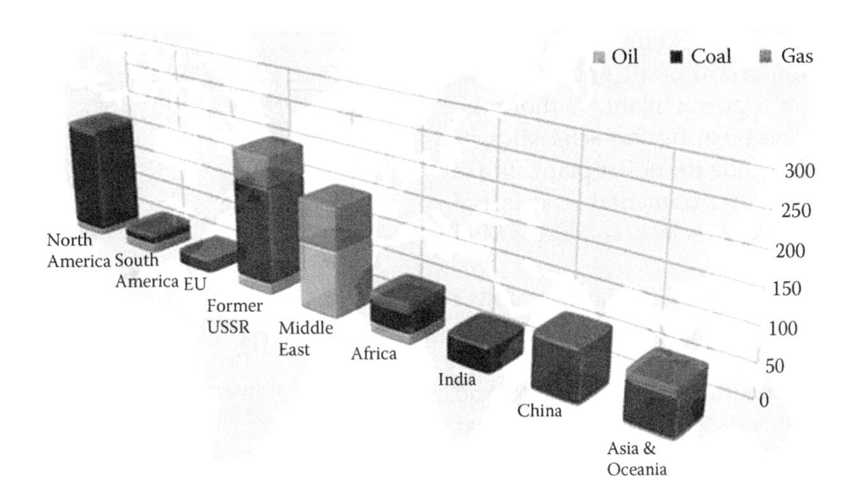

FIGURE 6.34
Reserves of fossil fuels around the world.

out. Coal is widely distributed and is easier to locate than oil and gas.

Fossil fuels take millions of years to make, but burn and disappear in seconds. It is best, therefore, not to waste fossil fuels as they are not renewable. Every year, millions of tonnes of coal are consumed as energy. This has led to global warming (**GHG** effect) and the depletion of resources. At present, the worldwide burning of fossil fuels releases billions of tonnes of CO_2 (measured as C) into the atmosphere every year.

Burning any fossil fuel means pollution of some sort. Mining and exploration for fossil fuels can cause disturbance to the surrounding ecosystem. The burning of fossil fuels emits oxides of S and N into the atmosphere.

Spread of fossil fuels across the globe is not uniform. Coal is the most widely spread fuel with recoverable deposits in over 70 countries. Ironically, petroleum is concentrated in those nations where the governance is not particularly strong or stable. Global fuel reserves across the leading production centres is well brought out in Figure 6.34 which is taken from the World Coal Institute of UK. European and South American continents have low energy reserves while the United States, former USSR and the Middle East have the best resources. The Middle East, interestingly, has only petroleum reserves and no solid fuels.

6.45 Fouling (*see* Slagging and Fouling)

6.46 Fourier's Law of Heat Conduction (*see* Conduction Heat in Heat Transfer)

6.47 Frame Machines in Gas Turbines (*see* GT Types in Gas Turbines)

6.48 Freeboard (*see* Fluidised Bed Combustion, Bubbling *and* Circulating)

6.49 Free Burning Coals (*see* Caking Coals in Coal)

6.50 Free Convection (*see* Convection in Heat Transfer)

6.51 Fuels (*see also* Fossil Fuels *and* Coals of the World in Appendix 3)

A conventional fuel (not nuclear fuel) can be considered as a *mixture or compound of one or more* **combustibles** which, when heated to its ignition temperature, is capable of releasing heat energy in a combustion reaction with air.

The entire power plant is built around the fuel and it is certainly the most important of all parameters as it governs every aspect of a power plant. A thorough understanding of the fuel, its burning characteristics, its transportation inside and outside the boiler plant, its ash generation and disposal, the environmental aspects and so on is necessary for the design, selection and O&M of the boiler and power plant.

Coal has been historically the main fuel for power plants in most parts of the world except where oil and gas are available in abundance in regions like the Middle East, Venezuela and so on. For some time in the 1960s, oil had become abundant but the oil crisis of the early 1970s has changed all that permanently, making oil an expensive fuel. There was a temporary dip in oil prices towards the end of the 1990s only to rise again from the early 2000s. Coal has reasserted its position as the most competitive and dependable fuel from the beginning of the twenty-first century even though its prices rose sharply between 2003 and 2008. The balance had been disturbed in the last couple of decades of the twentieth century with **NG** displacing coal as power plant fuel, even in areas where gas was not popular before, mainly on account of

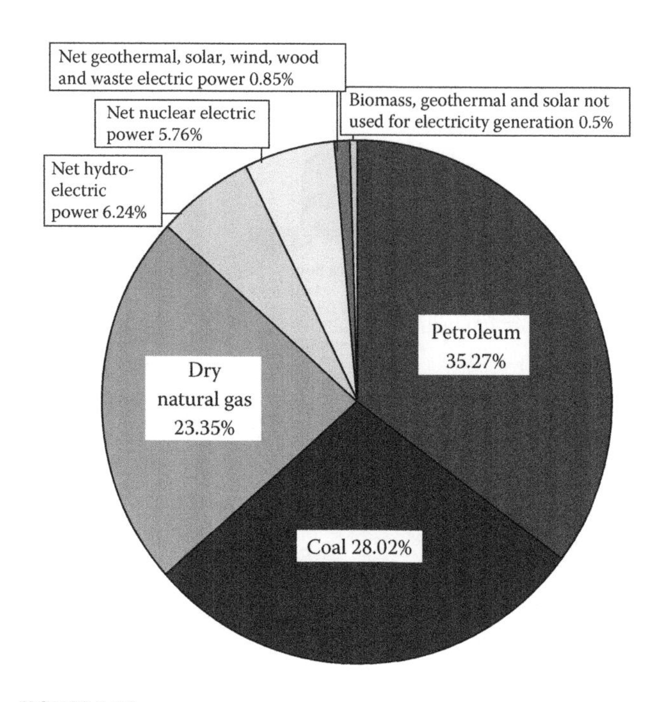

FIGURE 6.35
World fuel consumption patterns. (From Environmental Impact Assessment Scoping Report of BP 2006.)

- Greater gas production at more competitive pricing.
- Easier gas availability because of more gas pipes and shipping of liquefied NG (**LNG**) coupled with
- Bigger **CCPPs** operating at very high overall η
- Built-in very short periods due to faster deliveries of **GTs**

A lot of power generation has come up on NG. Now coal is finding favour again with NG prices firming up. Just as the fortunes of coal started improving, ironically the worldwide awareness about **GHG** is again making the picture a little hazy for the future of coal.

However, for industry the fuel scene is a little different. Both power and steam are required to be produced competitively, but to a smaller scale as compared to the utilities. Prime fuels are important only when substantial captive or cogenerated power is to be generated. At other times, there are various waste or manufactured fuels which are required to be used by the industry. Fuel cost is not always as overwhelmingly important to the industry as it is to utility, as power and steam are only a couple of the several inputs to the end product.

World fuel consumption is a mixed bag, as shown in Figure 6.35. FO, coal and NG are the three dominant energy sources. While petroleum dominates in transport sector, for power generation, however, it is coal which is the most important fuel producing over 70% electricity worldwide.

Fuels are classified in several ways as described in a separate topic below.

The estimates of the fuel reserves and how long they would last have often gone wrong, particularly in the developing countries, where the data is still being updated because of the excavation activities having started relatively more recently. However, there is a general consensus that coal would last for about 120 years at the present rate of consumption.

Fuel production, transportation and pricing get inextricably mixed up with geo politics and display extreme volatility. This is particularly characteristic of crude oil and to some extent NG. Coal availability and prices have been fairly predictable until recently. With big coal-based power generation by China and India under way, coal shortages have been experienced with attendant sharp rise and volatility in prices.

The only certainty seems to be the ever-rising prices of fuels. This makes it economically viable now to mine difficult fuels in remote areas. Deep water drilling and extraction from **oil shale** are examples of mining fuels which were considered to be too expensive some years ago. Newer fuels which were reckoned to be unviable some time back also find markets now. **CBM** is an example.

It has also not been possible to predict the behaviour of the fuel market. The large utilities worldwide therefore have found it necessary to diversify the fuels and their sources. Also, to gain further economy, multi-fuel firing is being demanded of the boilers lately wherever feasible.

While the fuel decides the type and shape of the boiler, the fuel ash governs the size and disposition of HSs. The effects of ash are most severe in coal-fired boilers as it is coal that has the highest ash in all fuels. It is not enough if there is good coal for burning: it should also have benign ash that will not slag or foul or cause erosion or corrosion. A deep understanding of ash is as important as that of any fuel to design and operate a boiler.

6.52 Fuel Classification (*see also* Coals of the World in Appendix 3)

Fuels can be classified in the following manner based on their origin:

- *Fossil fuels*: These are various fuels extracted from earth, such as **coal**, **FO** and **NG**. Formed over millennia by the forces of earth, they are fuels which are *non-renewable.*
- *Bio- or agro-fuels*: **Agro-**, bio- or vegetable fuels are essentially organic wastes generated by various crops, forest and purpose-grown trees. They are *renewable* and are eco-friendly to burn.
- *Waste fuels*: These are usually by-products such as some bio-fuels (like **bagasse**), process gases and spent liquors. They can also be **municipal waste**.
- *Manufactured or synthetic fuels*: These are produced in dedicated plants. Examples are fuels like **producer gas, town gas, LPG and so on**.
- Fuels can also be classified by their state.
- *Solid fuels* such as **coal**, **lignite**, **peat**, **wood**, **bagasse** and so on.
- *Liquid fuels* such as **crude oils**, **FOs**, **BL** and so on.
- *Gaseous fuels* such as **NG**, **producer gas**, **CO gas** and so on.

Prime fuels are fuels that are used most. Fossil fuels continue to be prime fuels despite a large increase of the other fuels in absolute quantity.

6.53 Fuel Distributors or Spreaders

Fuel distributors are needed to spread the solid fuels on to the grate or the bed (of FBC boiler). Fuel stored in the bunker is extracted and fed to the fuel distributors

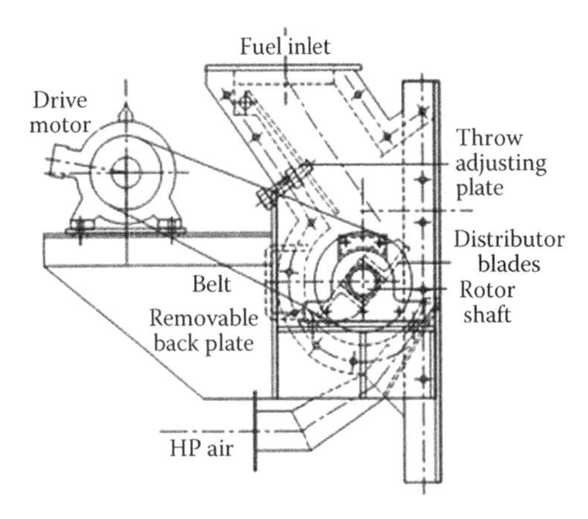

FIGURE 6.36
Typical rotary mechanical spreader/distributor.

by the fuel feeders as per the load demand of the boiler. The distributors spread the fuel evenly in the furnace; variation of the fuel quantity is done by the fuel feeders.

Fuel distributors are for feeding nutty fuels like coal or fluffy fuels like bagasse, rice husk, RDF and similar bio-fuels which burn partly in suspension in spreader firing or BFBC. They are

a. Mechanical rotary distributors
b. Pneumatic distributors

a. *Rotary mechanical distributors/spreaders:* A typical spreader is shown in Figure 6.36. They
   - Operate usually between 600 and 900 rpm throw the coal over a trajectory of ~3–7 m.
   - Can be either of over-throw or under-throw design with the former being more popular. Under-throw design which provides a more flat trajectory is supposed to be better for burning coals high in M and fines. They are popular with **FBC** boilers.
   - Can have either two sets or four sets of blades. Each blade is nearly sinusoidal in shape to spread coal wide in the desired area. The blades are attached to the rotor and have 3–5 plates bolted to them.
   - Are exposed to direct heat of the furnace and, therefore, are susceptible to distortion when the boiler is banked or the distributor is stopped. It is therefore, normal to have a water jacket for the shafts. Also, the distributors are kept running all the while.
   - Are usually available in two sizes of 500 and 750 mm nominal width which can handle 4 or 6 tph of coal. The centre distances are usually ~1 m.

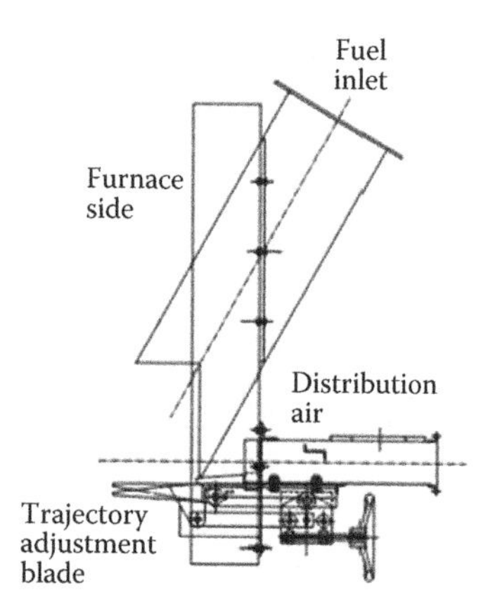

FIGURE 6.37
Pneumatic fuel distributor for agro-fuels.

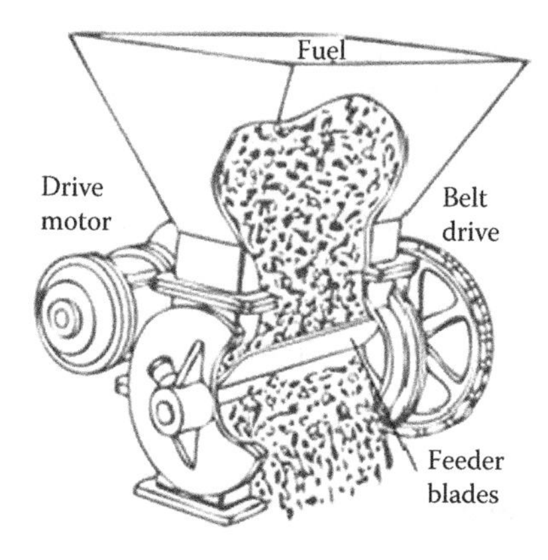

FIGURE 6.38
Rotary coal feeder.

- Have adjustments to vary the trajectory by controlling the angle of entry to the rotor.
- Are provided with air tuyers and air nozzles to blow the coal fines into the furnace and prevent them from cascading on to the grate.

b. Pneumatic distributors:

Shown in Figure 6.37 they are chutes with a flat plate and air nozzle fitted at the bottom at the entry to the furnace. The trajectory plate has an adjustment to control the length of throw. HP air is admitted above the plate to carry the fuel into the furnace. A rotating damper varies the air quantity in a sinusoidal way so that the fuel is spread fairly uniformly over the entire length of the grate.

6.54 Fuel Feeders

Fuel feeders are *transport links between fuel storage and fuel preparation or firing equipments*. They are devices, located at the bottom of fuel bunkers, for regulating the discharge of fuel as per the load demand of the boiler. As they are in the fuel circuit of the boiler, the feeders have to be mechanically very robust and there cannot be any outage. Also, they have to be highly responsive and vary the output precisely. Speed variation in response to the fuel signal is the most usual way of feeder control. There are several types of fuel feeders which are indicated below in the ascending order of output:

1. Rotary feeders
2. Screw feeders
3. Drag link/drag chain feeders
4. Belt feeders
5. Plate feeders

All the above feeders transport the fuel horizontally from the bunker except for rotary feeder which discharges vertically.

6.54.1 Rotary Feeder

Rotary feeders are for smaller capacities. Nutty fuels like **coal** and low-M **lignites** are suited for these feeders for capacities up to ~10 tph. The tightly fitted vane-type rotor rotating slowly within the casing provides smooth discharge and prevents back flow of flue gases into the bunkers during periods of puffing. See Figure 6.38.

The main drawback is that with high surface M in coals during rainy days, the vanes tend to pack up. Some amount of steam heating of the casing and shaft is done to substantially overcome the problem. This design of feeder on the whole can be termed as unsuitable in areas of high precipitation.

Rotary feeders are used in stoker, BFBC and small PF boilers.

Rotary feeders are also popular for discharging bio-fuels like **bagasse** from the bunkers. Being very voluminous it is usual to provide generous sizing for fuels like bagasse. A large drum fitted with serrated plates to facilitate the discharge of stringy material rotates loosely in the casing (as shown in Figure 6.39). The feeders are usually 600 and 750 mm in width and discharging 8–11 tph of bagasse. In large boilers, they can be as wide as 2000 mm with a drum size of 1000 mm (as shown in Figure 6.39).

6.54.2 Screw Feeder

Screw feeders are also popular for conveying and feeding of **coals** and **bio-fuels** in limited capacities of ~10 tph. It is normal to provide expanding screws to prevent jamming in case of bio-fuels. Hard facing of the flights is also recommended for longer life.

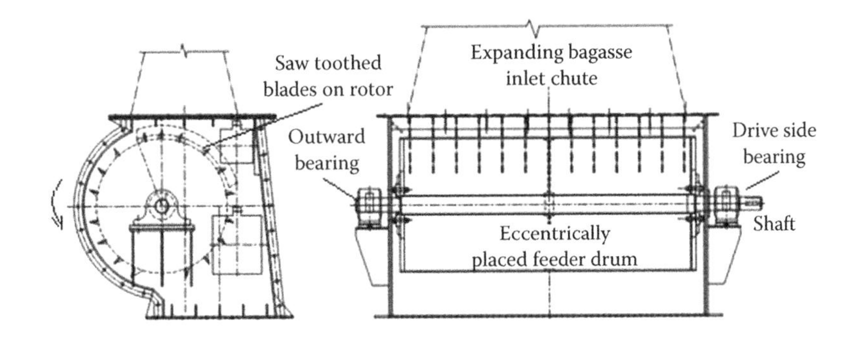

FIGURE 6.39
Rotary bagasse feeder.

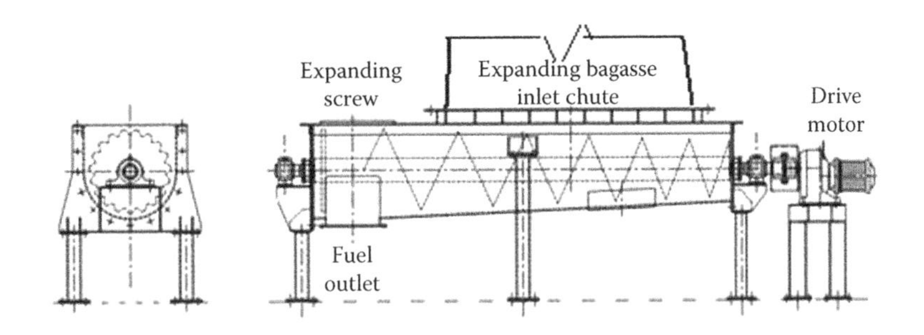

FIGURE 6.40
Screw feeder for bagasse.

A typical screw feeder for bagasse and bio-fuel is shown in Figure 6.40.

6.54.3 Drag Link/Volumetric Feeder

Drag link/chain feeder contains a pair of forged chains riding on a pair of sprockets on the drive and driven shafts. The chains are interconnected by links. In smaller sizes, the construction can be simplified by having a single chain and sprocket assembly.

There is a top and a bottom plate on which the chain assembly moves. Fuel, as it descends from bunker, is moved by the chain from the top plate to the bottom plate and then to the discharge chute (as shown in Figure 6.41).

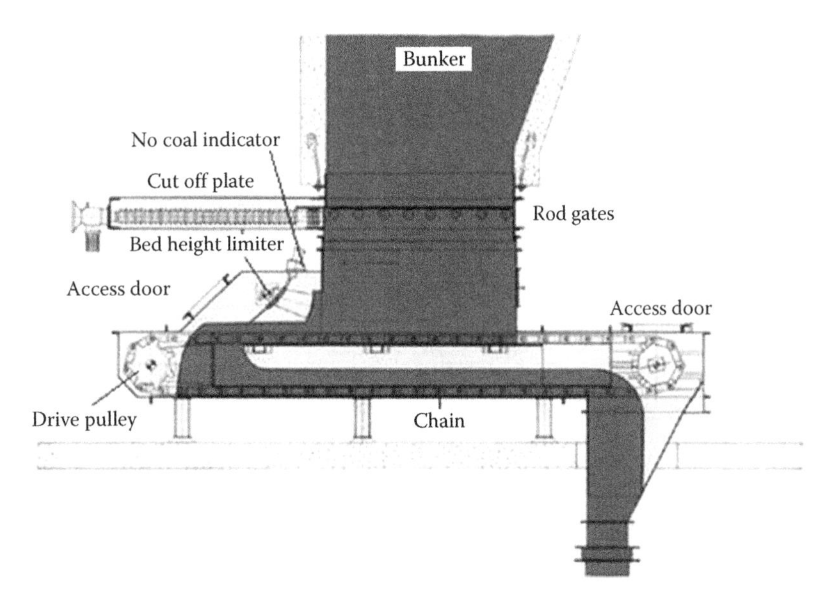

FIGURE 6.41
Drag chain volumetric feeder. (From Maschinenfabrik Besta & Meyer GmbH & Co. KG, Germany, with permission.)

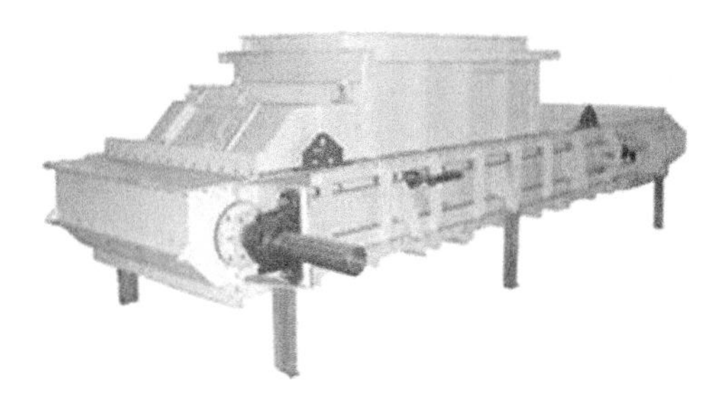

FIGURE 6.42
Drag chain feeder in assembled condition. (From Maschinenfabrik Besta & Meyer GmbH & Co. KG, Germany, with permission.)

At the bunker outlet, a motorised cut-off plate is provided for isolation and a rod-type gate for choosing the appropriate bunker opening. Coal bed height is held fixed and hence speed variation regulates the output. A tightening arrangement ensures that the chain stays flat on the plates. An emergency fuel exit is usually provided at the bottom of the feeder to evacuate the fuel fast in case of fire. A no-coal indicator is also fitted. Figure 6.42 shows the feeder in the assembled condition. Figure 6.42 also provides a comparison with more complicated construction of the plate feeders described in lignite firing section.

For **PF** and **CFBC** boilers, this type of feeder is very popular for feeding coal and lignite into the mills.

- They come in typical standard widths of 600, 800, 1000 and 1200 mm handling approximately 30–100 tph of coal (~40–130 m^3/h).
- These feeders are usually volumetric. But they can also be made gravimetric.
- They are also suitable for lignite for relatively smaller units as the fuel ρ is low and volume high. Plate feeders are more appropriate for lignites.
- These feeders can withdraw from either single or double discharge compartment design of bunkers.
- Bunker heights can be ~20 m and gross discharge length a maximum of 5 m.
- Pressurised construction to ~1500 mm wg (~2.15 psig) is possible.

6.54.4 Belt/Gravimetric Feeder

Belt feeder employs a troughed endless belt with corrugated edges running between a set of drive and driven shafts. Coal from bunker is carried to the discharge chute. The gravimetric feeder weighs material on a length of belt between two fixed rollers using a weighing roller as depicted in Figure 6.43. Belt speed is determined by a tachometer attached to the motor shaft. A microprocessor multiplies the speed and weight signal to arrive at the feeder output. The microprocessor matches the feeder output to the demanded output by adjusting the feeder motor speed. The action of the microprocessor is shown in Figure 6.44. A belt tensioning device helps to keep the belt taut all the time. A scraper belt at the bottom collects and drops the spillage into the discharge chute.

Gravimetric feeders provide good accuracy and reliability. They are usually made in 2.1 and 3.6 m (7 and 11 ft) in length in standard widths of 750, 900, 1050 and 1200 mm (~30, 36, 42 and 48 inch) for boiler applications. Maximum discharge is limited usually to 120 tph on coal.

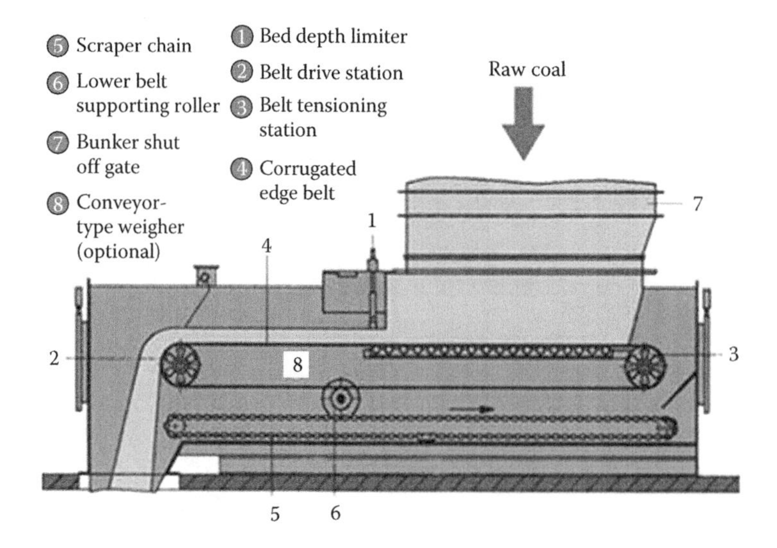

FIGURE 6.43
Gravimetric belt feeder. (From Hitachi Power Europe GmbH, Germany, with permission.)

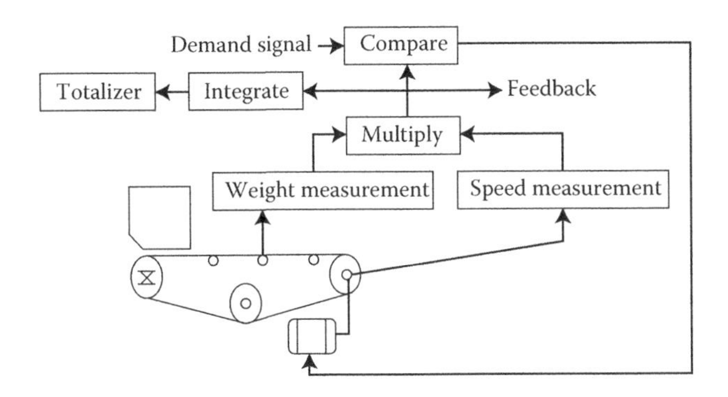

FIGURE 6.44
Microprocessor-based control of a gravimetric coal feeder.

Gravimetric and Volumetric Feeders

Gravimetric and volumetric feeders are both used in large industrial and utility boilers with the former more popular in the United States and the latter in European markets. With growing pressure to reduce emissions, the use of gravimetric feeders is likely to increase. Owing to the heterogeneity of coal, gravimetric feed systems are advantageous over volumetric when it comes to achieving optimised combustion in PF-fired boilers with lower emissions. There is no effective way to measure kJ/Btu flow and this value must be inferred from either volume or weight flow. The volume flow measurement will be in error by the variation in both ρ and cv, whereas the weight flow signal will be in error only to the extent of the variation in hv. The gravimetric feeder is vital if an accurate flow rate is to be maintained.

6.54.5 Plate Feeder (*see* Lignite)

6.55 Fuel Firing Modes

A boiler may be capable of firing a single or multiple fuels—alone or simultaneously, continuously or intermittently, all the time of the year or during season, for base duty or peaking and so on. Depending on the duty, fuels are classified in the following manner:

a. *Base, main or prime fuel* is the fuel on which the boiler is designed to generate steam all or most of the time. The prime guarantees are based on this fuel. In a PF boiler, for instance, design coal is the prime fuel even though there is a provision to fire other fuels like heavy and light FOs.
b. *Alternate fuel* is the fuel which generates steam in the absence of main fuel. The firing equipment and the boiler have to be sized appropriate to the amount of duty that is required to be met by the alternate fuel. Usually all guarantees are required on alternate fuels also.

   In dual firing, like typical oil- and gas-fired boiler, if FO is the main fuel, gas is the alternate fuel and vice versa.

   Another example is the fuel used during the off season in a boiler using a seasonal fuel. FO or coals/lignites are the typical alternate fuels in bagasse-fired boilers needed to maintain generation during off season.
c. Auxiliary *fuel* is a fuel that needs to be burnt in the boiler as and when it is available. Usually, the amount of this fuel available is only a fraction of main fuel. Normally, it is required that all provisions for its burning in the boiler to the extent of its availability are to be made with scarcely any additional guarantee.

   A typical example of an auxiliary fuel is the pith in pulp mills or coke breeze in coke plant and so on.
d. *Supplementary fuel* is the fuel which is fired to meet any specific duty in a boiler while normal generation is obtained by the main fuel. Typically, a light FO is used as supplementary fuel in Hrsgs to obtain short time peak evaporation.
e. *Start-up fuel*: In most boilers, start-up fuel is needed to start the fuel and take it up to a certain range before it is possible to introduce the main fuel. This is because the main fuels are usually heavier, needing a hot furnace setting for proper ignition. Also, many times there may not be adequate controllability to admit very small amounts of fuel as needed at the start.

   In stoker and BFBC-fired boilers, start-up fuel is any light FO, NG or charcoal.

   In PF and CFBC boilers, it is any light FO or gas.

6.56 Fuel Oils (*see* Liquid Fuels)

6.57 Fuel Ratio (*see* Coal Classification *under* Coal)

6.58 Furnace (*see* Pressure Parts *and* Heating Surfaces)

6.59 Furnace Air Staging (*see* Separated Overfire Air)

6.60 Furnace Safeguard Supervisory System (*see* Operation and Maintenance Topics)

6.61 Furnace Draught (*see* Fans)

6.62 Fusibility of Fuel Ash (*see* Ash Fusibility)

7

G

7.1 Gas Bypassing

There are mainly four situations in which the word 'gas bypassing' is used in boilers:

1. Despite adequate provisioning of HS, anticipated amount of flue gas cooling does not take place in the boiler, leading to reduced **thermal η** and loss of heat. Essentially, gases here are bypassing the HS. This happens when part of the flue gases go around the tube banks because of excessive end gaps or dislocation of gas deflection baffles. This is called gas bypassing the tube banks.

   This matter can be easily overcome in most cases by suitably blocking the end gaps (between the tube banks and the casing), sealing all the gaps in gas deflection baffles and preventing gases skirting around the HSs.
2. In **two-pass utility boilers** with twin/**divided passes** in the rear pass, a gas bypass duct is provided between the passes to divert a part of the flue gases. The purpose is to control the RHOT and avoid spraying water into RH system as it leads to a fall in cycle η. This arrangement of gas bypassing is shown in Figure 7.1.

   The drawback attributed to this arrangement is the likely jamming of bypass dampers as they span the entire width of the boiler. This can be substantially overcome by having dampers external to the boiler setting.
3. **AH** gas bypassing, along with air bypassing, during start-up and low-load running of boilers is a fairly standard practice adopted to keep the low-temperature section of AH above the dew point to save from **low-temperature corrosion**.
4. Gas bypassing through the **bypass stack** is practised in Hrsgs during initial start-up to save the internals from sudden rush of high-temperature exhaust gas from GT. It can also provide the flexibility of running the GT independent of Hrsg. See Bypass Stack in Draught Plant (Section 4.34.d).

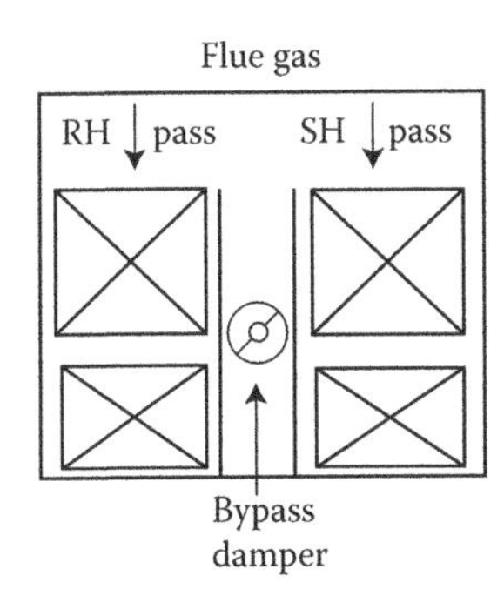

FIGURE 7.1
Gas bypassing in the rear pass of two-pass RH boiler.

7.2 Gas Cleaning (*see also* 1. Dust Collection Equipment, 2. Desulphurisation, 3. Nitrogen Oxides in Flue Gas, 4. Reburning, 5. Selective Catalytic Reduction, 6. Selective Non-Catalytic Reduction, 7. Separated Overfire Air)

Flue gas contains

a. **PM**-like fine ash and unburnt C and
b. noxious gases like **CO, SO_x** and **NO_x**

which are to be removed by means of gas cleaning to acceptable limits prescribed by the local pollution authorities before emitting the gases into the atmosphere.

Gas cleaning is an all inclusive term which refers to the act of removing dust as well as noxious gases.

- Dust removal is relatively simple and carried out before the stack by employing suitable **dust collection equipment**.
- Removal of SO_x involves **desulphurisation** either in furnace or before the stack.
- Removal of NO_x is quite involved and can be done in two stages.
  a. NO_x formation can be minimised by adopting various techniques at combustion stage, such as employing **LNB, SOFA, reburning** and so on.
  b. Post-combustion NO_x can be removed by certain **SCR** and **SNCR** reactions.

NG is almost the only prime fuel that requires no flue gas cleaning, as the fuel has neither dust nor S, and the NO_x produced is very low.

7.3 Gas Electric Igniters (*see* Igniters)

7.4 Gaseous Fuels

Gaseous fuels are combustible gases which release their energy on burning.

Gaseous fuels can be fossil, manufactured or waste fuels.

NG is the prime fossil gaseous fuel which is perhaps the most ideal fuel. Its burning properties are very good with fast, easy and clean burning creating no soot as it has no ash. Naturally, the boilers burning NG are the most compact. Its contribution to **GHG** effect is negligible. It has to be transported by pipelines or else by liquefying and transport by ships in the form of **LNG**. **CBM** is another natural fossil gas which has been exploited lately.

Manufactured gases available are **producer gas, town gas** and so on.

There are several waste gases like **coke oven, blast furnace, refinery, CO gas**, and so on.

By and large, gases are relatively easy to handle and burn. All high cv gases need extremely careful handling to prevent explosions.

Wobbe index

Interchangeability of fuel gases is measured by the Wobbe index. It compares the energy output of different gases during combustion. The Wobbe index is essential for analysing the impact of a fuel changeover and is a common specification for appliances that use gas and also that transport gas.

Wobbe index = Gcv/√SG of gas and is commonly expressed in Btu/scft or MJ/S m^3 (1000 Btu/scft = 37.3 MJ/S m^3).

There are three well-accepted families of fuel gases based on the Wobbe index:

- Family 1 covers manufactured gases like town gas and syngas.
- Family 2 covers **NG** (with high and low ranges).
- Family 3 covers **LPG**.

Combustion equipment is typically designed to burn an FG within a particular family: H_2-rich town gas, NG or LPG.

Typical ranges of the Wobbe index for the three families are ~22–30, 39–55 and 73–87 MJ/S m^3.

Various types of fuel gases for burning in boilers or GTs, as listed, are briefly described below:

1. Methane (CH_4)
2. NG
3. BFG
4. COG
5. Carbon monoxide (CO) gas
6. Producer gas
7. Refinery gas
8. CBM
9. Shale gas

7.4.1 Methane (CH_4)

CH_4 is a gaseous organic compound which is colourless and odourless that occurs in NG (called fire damp in coal mines) and from bacterial decomposition of vegetation in the absence of O_2. It is the simplest member of the paraffin HCs and burns readily forming CO_2 and H_2O if supplied with enough O_2 for completion of combustion or CO if the O_2 is insufficient. Mixtures of 5–14% CH_4 in air are explosive and have caused many mine disasters.

The chief source of CH_4 is NG, but it can also be produced from coal. CH_4 is used widely as a fuel in homes, commercial establishments and factories. As a safety measure, it is mixed with trace amounts of an odorant to allow its detection. It is also a raw material for many industrial materials, including fertilizers, explosives, chloroform, carbon tetrachloride and carbon black and is the principal source of methanol.

7.4.2 Natural Gas

By way of chemical composition NG is a mixture of HC gases that occur usually along with crude oil deposits. It is a non-toxic, colourless, highly flammable gas, containing principally CH_4, varying from 70% to 95% by volume and C_2H_6 varying from 5% to 15%, both of which are gaseous under atmospheric conditions. There are also other HCs like C_3H_8, C_4H_{10}, C_5H_{12} and C_6H_{14}. In addition, NG contains other gases like CO2, H_2, H_2S, N_2 as well as noble gases like He and Ar. Because NG and formation water occur together in the reservoir, gas recovered from a well contains water vapour, which is partially condensed during transmission to the processing plant.

The physical properties of NG include colour, odour and flammability. The principal ingredient is CH_4, which is colourless, odourless and highly flammable. However, some of the associated gases in NG, especially H_2S, have a distinct and penetrating odour, and a few ppm are sufficient to impart a decided odour. NG is much lighter than air. If there is a leakage in the pipeline, NG escapes to higher levels harmlessly.

NG also has a very narrow combustion range. It will only ignite or burn when the concentration is between

4% and 15% of NG in air. Concentrations outside this range will not burn.

NG is more ubiquitous than oil. NG is formed underground by decomposition of organic materials in land plants and aquatic animals. NG is found along with crude oil in most cases but it can also occur separately in sand, sand-stone and limestone deposits.

Its high cv and flammability together with nearly no ash makes it almost an ideal fuel for domestic, industrial and power plant use. On combustion, it produces lower CO_2 and NO_x, making it an environmentally more agreeable fuel.

In power plants, its use is mainly in **industrial boilers** and **GTs**, more in the latter. Its benefits are

- No storage facilities at plant site are required as NG is supplied to the plant by pipelines.
- **Dust collection** and ash disposal issues are absent.
- Boiler **controls** are extremely simple, responsive and seamless.
- Boilers are very compact as the HSs can be arranged very closely with no fear of fouling. NG boilers are therefore cheaper.

On extraction of NG, it is usually treated for the removal of unwanted gases before piping to the customers.

NG varies in its properties depending on the source.

- Its ρ varies from about 0.7 to 1.1 kg/cum. The higher the CH_4, the lower is the ρ.
- The cv also varies widely from

  ~750 to 950 kcal/cum or 85 to 105 Btu/cft or 870 to 1104 kW/cum

  ~7000 to 10,000 kcal/kg or 12,600 to 18,000 Btu/lb or 29.3 to 41.9 MJ/kg

  These are at ISO conditions of 15°C and 760 mm Hg.
- The gas temperature supplied at the plant gate usually varies from 5°C to 25°C even though it comes out of the well from 0°C to 75°C.
- Even though it is colourless and odourless at the well head, it is usual to inject traces of mercaptan (an S-bearing compound) to provide odour for leak detection.

NG occurs in several ways:

- *Associated gas*: When NG is trapped above crude oil in the underground crude oil wells, it is called associated gas. Naturally such gas is high in heavy HCs.
- *Wet gas*: NG, found in the deep condensate wells ~3000 m, that exists along with liquid HCs is known as wet gas. Wet gas is rich in heavy HCs such as C_5H_{12} and C_6H_{14}.
- *Dry gas*: There are underground wells where NG exists alone without any crude oil. Gas from such well is called dry gas and it is substantially free of heavy HCs.
- *Sour gas*: When H_2S is in excess of ~4 ppm by volume, NG has a typically foul odour and is known as sour gas.
- *Sweet gas*: When NG is treated for the removal of mercaptan and H_2S, the resulting gas is sweet gas. Presence of H_2S induces toxicity to the gas. Also, it is injurious to certain metals. Sweetening of gas is done to meet some special requirements.

7.4.3 Blast Furnace Gas (*see also* Whrbs in Steel Plant *under* Waste Heat Recovery Boilers)

In conventional large integrated steel plants or pig iron plants, iron ore is reduced to raw or pig iron in the blast furnaces. Blast furnace is a tall furnace in which the iron ore (Fe_3O_4), limestone ($CaCO_3$) and metallurgical coke (C) are charged at the top while hot HP air at about 600–700°C is forced through tuyers at the bottom. The air and charge come into intimate contact within the furnace as they travel in opposite directions. Coke reacts with O_2 in air to form CO, generating some heat. CO then reduces Fe_3O_4 at this high temperature to form Fe and CO_2. $CaCO_3$ reacts with the products to remove the impurities and in this process more CO_2 is formed.

BFG emanating from the top of the blast furnace contains largely CO and CO_2 along with N_2 and H_2 with traces of CH_4. H_2 is derived from high-temperature dissociation of H_2O while N_2 is from the air blast. BFG at the exit of blast furnace is

- Hot, at temperatures of 150–425°C
- Dusty, laden with iron oxide particles at ~115 g/dry sm^3 or 50 grains/dscf
- At HP of 500–750 mm wg

The gas pressure continuously varies because of the turbulence inside the blast furnace.

BFG has large N_2% which makes it heavy, low in cv and slow to ignite and burn. It is a highly lethal gas as the % of CO is very high.

The gas properties are given in Table 7.1. Other notable characteristics of BFG are as follows:

- It burns with a faintly bluish and almost non-luminous flame.
- Support gas flame is needed for ignition and to ensure completeness of combustion.

TABLE 7.1

Properties of Blast Furnace Gas

Properties	Units	Values	Comments
Composition		Range	Volume dry
N_2	%	55–65	Very high
CO	%	25–35	Very high
CO_2	%	5–15	
H_2	%	1–3	
CH_4	%	<5	
Density	kg/nm³	1.15–1.3	Heavy gas
Gcv on volume	MJ/sm³	~2.9 3–5.45	Very low
	kcal/sm³	~700–1300	
	Btu/scf	~80–140	
Gcv on weight	MJ/kg	~2.30–4.60	
	kcal/kg	~550–1100	
	Btu/lb	~1000–2000	
Ignition	°C	~800	Very high

- Extreme care in transport and burning is needed to prevent gas leakages which are disastrous.

The high level of iron oxide dust makes the BFG very fouling in nature. The gas lines, burners, furnace and the convection banks are all subject to fouling. BFG is usually cleaned in a dry dust catcher to reduce the suspended solids to ~7–35 g/dscm (3–15 grains/dscf) before sending to the boiler. Sometimes it is further cleaned in dry or wet separator to reduce the dust load to ~0.2–1.4 g/dscm (0.1 to 0.6 grains/dscf). Wet cleaning lowers its temperature and increases the humidity of the gas.

7.4.4 Coke Oven Gas (see *also* Whrbs in Steel Plant *under* Waste Heat Recovery Boilers)

Coke is produced by destructive distillation of suitable grades of coking coal in coke oven batteries by pyrolysis (heating out of contact with air). The COG so generated is ~15% of the weight of coal char. Raw coal is placed in the ovens and externally heated until all volatiles are driven out. By-products like ammonium sulphate ($(NH_4)_2SO_4$), phenol or carbolic acid (C_6H_5OH), naphthalene ($C_{10}H_8$), light oils and tars are recovered before COG is sent for combustion.

COG mainly consists of H_2, CH_4, C_2H_4 and CO with small amounts of CO_2, N_2, O_2 and heavy HCs. The composition of the gas varies considerably depending on the coal used in ovens and length of time taken for coking. High-temperature carbonisation gives a higher yield of low Gcv gas and low-temperature carbonisation the opposite. Table 7.2 summarises the COG properties.

COG is almost always washed and cooled before sending to the gas holders to strip impurities like tar, dust, benzol, H_2S and so on.

TABLE 7.2

Properties of Coke Oven Gas

Properties	Units	Values	Comments
Composition		Range	Volume dry
H_2	%	25–60	
CH_4	%	25–50	
C_2H_4	%	3–13	
CO	%	5–12	
ρ	kg/nm³	0.45–0.65	Light gas
Gcv on volume	MJ/sm³	~17.8–30.1	
	kcal/sm³	~4250–7400	
	Btu/scft	~470–830	
Gcv on weight	MJ/kg	~38.1–49.0	
	kcal/kg	~9100–11700	
	Btu/lb	~16400–21000	

A substantial portion of the gas produced is used for heating of the coke ovens and the remainder in boilers for steam and power generation.

Owing to the high amount of H_2, COG burns freely. The impurities in the gas tend to deposit themselves in the pipelines and burners. It is normal to provide large gas ports in the burners for easy access for cleaning.

7.4.5 Carbon Monoxide Gas (see *also* Carbon Monoxide Boilers *under* Whrbs in Oil Refineries)

TCC (thermofour catalytic cracker) or **FCC** (fluid catalytic cracker) are the two catalytic cracking processes of petroleum in the refineries. The catalysts used in the crackers are required to be regenerated by scrubbing them clean to remove the coke deposited on them in the cracking process.

This regeneration of the catalyst is performed by burning away the C in the regeneration tower with the help of large quantities of HP air at about 2 atg. Though large, this air has to be kept to a minimum, in fact at sub-stoichiometric levels, to contain the resultant temperature and prevent the destruction of catalysts. It is also to minimise compressor power. The combustion, thus, produces a large amount of CO and hence this waste gas is termed as CO gas.

Leaving the regenerator at as high a temperature as 600–700°C the CO gas naturally contains a large amount of sensible heat of ~50%. Gcv is <350 kcal/m³ and the CO content is 4–10% by volume. Catalytic dust carried over is in traces and it is <40 microns in size. The ignition temperature of CO gas is 610–660°C and stable combustion can take place if the adiabatic flame temperature is ~980°C. An auxiliary fuel with 10% burner capacity generally suffices this need.

CO to CO_2 oxidation takes place by burning the gas in CO boiler and generating steam and power. This is consumed in the refinery itself. Besides producing process

TABLE 7.3

Properties of CO Gas

Properties	Units	Range %	Typical
Constituent		By Volume	
CO	%	4–10	8.7
CO_2	%	6–12	7.2
O_2	%	0.1–0.6	0.6
N_2	%	65–80	66.6
H_2O	%	8–16	16.9
SO_2	%	0.1–0.3	—
HC	%	Traces	—
Total	%		100.0
Gas temperature	°C	600–700	640
Pressure	mm wg	800–1500	
Catalyst			
Dust	g/nm³	0.2–1.5	
Size	microns	<40	
Ignition temperature	°C	610–660	High
Gcv	MJ/nm³	<1.5	Very low
	kcal/m³	<350	~50% Sensible heat
	Btu/scft	<40	

steam and power, the CO boilers incidentally attenuate the noise and discharge CO_2 instead of toxic CO. Typical analysis of CO gas is shown in Table 7.3.

7.4.6 Producer Gas

Producer gas is an FG consisting mainly of flammable gases CO and H_2 with small amounts of CH_4 and inert gases CO_2 and N_2.

It is prepared in a furnace or Gen in which air and steam are forced upward through a burning fuel of coal or coke which results in partial combustion. The C of the fuel is oxidised by the O_2 of the air from below to form CO. The N_2 of the air, being inert, passes through the fire without change. Steam adds H_2 also.

Producer gas has a low cv because it has a large inert content of N_2 and CO_2.

Producer gas composition from biomass is typically

CO : 18–22%
H_2 : 18–22%
CO_2: 11–13%
CH_4: 2–4%
N_2 : 39–51%

It is widely used in industry because it can be made with cheap fuel. When producer gas contains H_2, it is also a source material for the manufacture of synthetic NH_3.

7.4.7 Refinery Gas (*see also* Crude Oil *or* Petroleum *under* Liquid Fuels)

Refinery gas (RG) is a gas produced in petroleum refineries in the process of distillation, cracking and reforming with gas being withdrawn from any point in the system.

RG principally consists of paraffins with low boiling point, namely CH_4, C_2H_6, C_3H_8, C_4H_{10}, C_5H_{12}, and small amounts of olefins, namely C_2H_4 and C_4H_8. Minor quantities CO, CO_2 and H_2 may be present along with O_2 and N_2. SO_2 may also be there if it has not been removed from the original crude.

It is a light gas with ρ varying from 0.45 to 0.65 kg/cum.

It has high Gcv of ~17.8–31.0 MJ/NM³ or 4250–7400 kcal/NM³ on volumetric basis or ~38.0–49.0 MJ/kg or 9100–11700 kcal/kg on weight basis

Its main use is in the generation of steam and power in refineries. It is also piped to the industries nearby if in excess. RG being a high-cv gas like NG, is burnt very similar to NG in the same type of boiler.

7.4.8 Coal Bed Methane, Coal Mine Methane, Coal Bed Gas or Coal Seam Gas

NG, primarily CH_4, which occurs naturally in the fractures and matrix of coal beds, is called CBM. The gas is generated *in situ* during coalification and is adsorbed on the internal surface area of coal. It is also known as CMM when it is released during mining operations. It is retained by coal beds in the sub-surface in near liquid state.

Its presence was long known as it was a serious safety hazard in underground coal mines. It is a 'sweet gas' as there is no H_2S associated. Unlike in NG, CBM is free of heavier fractions like C_3H_8 and C_4H_{10}.

CH_4 from unmined coal seams is recovered through drainage systems constructed by drilling a series of vertical or horizontal wells directly into the seam (as shown in Figure 7.2). Water must first be drawn from the coal seam in order to reduce pressure and release the methane from its adsorbed state on the surface of the coal and the surrounding rock strata. Once dewatering has taken place

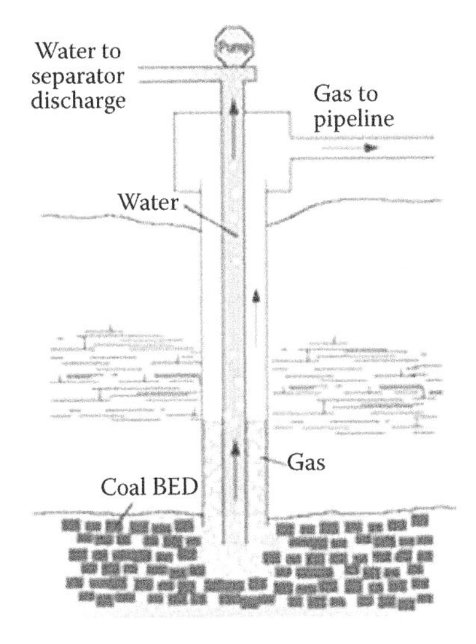

FIGURE 7.2
Extraction of coal bed/coal mine methane (CMM).

and the pressure has been reduced, the released methane can escape more easily to the surface via the wells.

The choice of vertical or horizontal wells is dependent on the geology of the coal seam. In the case of seams at shallow depths, vertical wells have been traditionally used. These vertical systems often use layers of fracture wells, which drain the methane from fractures in the coal seam produced as a result of the increased pressure created during the dewatering process. At these shallow depths, the combination of high permeability and LP make the vertical systems ideal as extra methane flow enhancement is not required and the structure of the vertical and fracture wells remains stable.

The largest CBM resource bases lie in the former Soviet Union, Canada, China, Australia and the United States. However, much of the world's CBM recovery potential remains untapped. CBM production now accounts for approximately 8% of the total NG production in the United States, which is the largest producer of CBM at present. CBM is one of the promising energy sources of the future.

7.4.9 Shale Gas

Shale gas refers to NG that is trapped within shale formations which are fine-grained sedimentary rocks that can be rich sources of petroleum and NG.

Gas is stored in shale in three different ways:

1. Adsorbed gas is gas attached to organic matter or to clays.
2. Free gas is gas held within the tiny spaces in the rock (pores, porosity or micro-porosity) or in spaces created by the rock cracking (fractures or micro-fractures).
3. Solution gas is gas held within other liquids, such as bitumen and oil.

Shale gas is distributed quite widely across the globe. But its exploitation has been limited. The United States is the leading country that produces large amounts of shale gas. This is one of the promising energy sources of the future.

7.5 Gas-Fired Boilers (*see* Oil- and Gas-Fired Boilers)

7.6 Gas Laning

Gas laning is the creation of lanes by missing out tubes here and there from an otherwise systematic tube formation in a bank as depicted in Figure 7.3. This has to be avoided as it promotes an increased gas flow through these lanes (called 'laning') which leads to (a) **erosion** and (b) overheating of the faces of tubes on either side of the lane. Rectification of this defect is rather difficult.

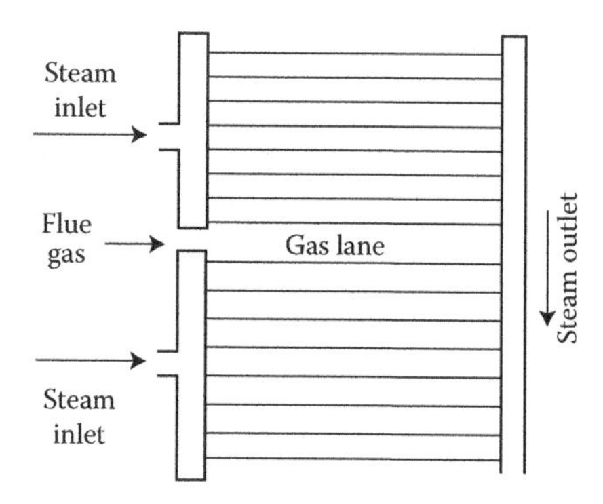

FIGURE 7.3
Example of gas laning.

7.7 Gas Recirculation or Recycling and Gas Tempering

Gas tempering and recirculation are powerful tools to alter the heat absorption pattern in a boiler without interfering with combustion. Flue gas from AH or Eco exits at 250–400°C and can be reintroduced into the furnace (as shown in Figure 7.4). There will be an increase in gas quantity and lowering of gas temperatures without any change in total heat of gases which alters the heat absorption pattern. Flue gas can be introduced into the furnace either

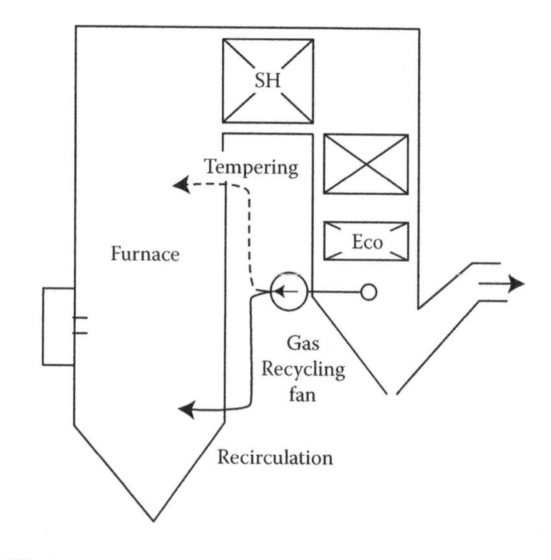

FIGURE 7.4
Scheme of gas recirculation and gas tempering in a boiler.

- Below the centre line of heat input (CHI) as *GR* or
- Close to the furnace exit as in *gas tempering*

In GR or FGR, the cold flue gas mixes with the gas from combustion resulting in a drop of furnace temperature. As furnace heat transfer is entirely by radiation, this reduced temperature leads to a fall in furnace absorption. Therefore, there is practically *no change in* **FEGT**.

- With higher gas quantity and nearly the same FEGT, all the surfaces downstream experience greater heat absorption because of higher convection. The change is progressively greater towards lower temperatures. Heat from the furnace is thus moved to the convection surfaces. In other words, the *heat from the front is moved to the rear of boiler.*
- Reduction in furnace temperature has a profound impact on reducing **NO_x** levels in flue gas without reducing boiler η. NO_x values can drop to <20 ppm corrected to 3% O_2 when burning NG where uncontrolled NO_x readings are generally in the area of 80–120 ppm.
- Increase in flue gas quantities has another interesting and profound advantage when firing dissimilar fuels producing widely varying gas quantities. If the same full steam temperature is desired both from oil and coal firing, where the gas quantities from oil are some 15% lower, instead of increasing the HS FGR can be adopted. Increased gas mass flow would restore the effectiveness of HS as in coal firing rendering the boiler compact and cost effective.
- Gas recycling is greater at low-loads as the SH/RH temperatures are lower. As the furnace absorption is lowered, GR is also used for controlling the furnace temperatures both for regulating metal temperatures and **slagging**.

In *gas tempering*, the cold flue gas is admitted into the upper furnace before the exit plane after the combustion zone. This results in only a small change in furnace heat absorption but a *good reduction in FEGT* as the combustion gas gets diluted with the tempering gas.

- Increased gas flow and reduced gas temperatures contribute to a reduction in heat absorption in the secondary superheater (SSH), nearly no change in the RH and some increase in primary superheater (PSH) and good increase in Eco. Heat is redistributed SH downwards without much change to the furnace.
- Furnace and RH heat absorption remain almost the same regardless of the amount of gas tempering. Its use is more in reducing the secondary SH temperature and also to reduce the fouling in SH area.

In general, in both types of gas recycling, greater the amount of gas recycled, greater is the redistribution of heat. About 5% of flue gas by weight is considered as minimum for preventing back flow. Recycling of gas helps in extending the **control range** of SH and RH. In place of increased HS and attemperation, FGR may be adopted beneficially.

Fans for recycling of gas experience a very heavy duty dealing with high-temperature gas very dusty in case of coal firing in producing reasonable HPs. This is an important consideration in evaluating the gas recycling option.

7.8 Gas Side Corrosion (*see* Ash Corrosion)

7.9 Gas Turbines (*see* Turbines, Gas)

7.10 Gas Turbine Cycle (*see* Thermodynamic Cycles)

7.11 Gas Velocities

Inside a conventional boiler, flue gas flows through the furnace and various tube banks of **SH, RH, Eco** and **AH** before discharging into the atmosphere though the stack. All the while it is giving its heat away to steam and water while it is continuously getting cooled. Heat is transmitted by both **radiation** and **convection**, with the latter playing a greater role as the gas cools down.

Convection heat transfer is mainly governed by gas velocity$^{0.66}$ and it is essential to adopt as high a gas velocity as possible, consistent with **erosion** and vibration levels. Fly ash erosion matters mainly for solid fuels

TABLE 7.4

Norms for Lateral Tube Spacing and Maximum Gas Velocities in Tube Banks for Various Fuels

	Pendant and Horizontal SHs				Economisers		
	Max Gas Velocity		Clear Spacing Across Gas Flow		Max Gas Velocity		Clear Spacing
Fuel	m/s	fps	Front (mm)	Rear (mm)	m/s	fps	Across Gas Flow (mm)
Gas	30	100	50	50	36	120	19
Oil distillate	30	100	50	50	27	90	25
Oil residual	18	60	100–150	50	18	60	25–32
PF low/soft ash	15–18	50–60	200	75–150	15–18	50–60	38
PF abrasive ash	12–14	40–45	250–400	100–150	12–15	40–50	50
SS coal	18	60	100–150	50	18	60	32
Bagasse/biomass	15–18	50–60	175	63.5	15–18	50–60	32
Coal in BFBC	10	35	150	50	12–15	40–50	32–38
Coal in CFBC	10	35	200	50	12	40	38–50

and takes place at much lower velocities. Tube side erosion is proportional

a. Directly to the dust loading of gas
b. To nearly its gas velocity 3.5 times

For oil- and gas-fired boilers it is the usual vibration and noise that limit the flue gas velocities. They are naturally much lower in solid fuel-fired boilers. They are further lower if ash in fuel is more and ash is abrasive.

The ultimate size of a boiler is substantially governed by the gas velocities adopted. Higher the gas velocities, more compact and competitive is the boiler but with greater fan power consumption. And lower gas velocities make the boiler more expensive but less prone to erosion along with lower auxiliary power.

In addition to the gas velocities tube spacing is also important to prevent ash bridging.

Table 7.4 indicates the permissible gas velocities for various types of boilers along with minimum spacing across the gas flow. In practice, the actual velocities and spacing adopted are decided by previous experience with similar fuel and firing.

7.12 Gauge Glasses (*see* Water Level Indicators)

7.13 Glass Wool (*see* Insulating Materials)

7.14 Grain Size of Steel (*see* Metallurgy)

7.15 Graphitisation in Steel Pipe

Graphitisation is the conversion of C in metal to graphite at high temperatures. **CS** pipes used in the steam service, particularly in the area of welds, are known to experience graphitisation at temperatures >425°C (~800°F). The codes, therefore, advise use of SA 53 and 106 steel pipes for carrying steam to <425°C. This phenomenon is not experienced for tubes used inside the boiler where they are swept by flue gases and CS tubes are freely used for higher temperatures.

7.16 Grashof's Number (*see* Dimensionless Numbers in Heat Transfer)

7.17 Grates

This topic is elaborated under the following sub-topics:

1. General
   1.1. Grate
   1.2. Grate types
2. Types of burning
   2.1. MB

   2.2. Spreader burning
   2.3. Semi-pile burning
3. Types of fuel feeding
   3.1. Underfeed grates
   3.2. Overfeed grates
4. Ash discharge
5. Grate combustion

7.17.1 General

7.17.1.1 Grate

Grate is either a fixed or moving surface at the bottom of a furnace on which the specified solid fuel burns. Grates are flat or inclined. Fuel is usually fed from the top of the grate even though it can be from bottom in some small units. Always PA is provided from the bottom and SA from the top. There are several types of grates. Almost all the solid fuels can be burnt on one or the other type of grate.

A grate acts as (1) carrier of fuel and ash and (2) distributor of air.

7.17.1.1.1 Grate and Stoker

The terms 'grate' and 'stoker' are used interchangeably in the industry. Strictly, a grate together with the fuel distribution arrangement is a stoker. Fuel distributor in SS firing alone is also often called a stoker.

7.17.1.1.2 Working of Grate

Solid fuel is delivered on top of the grate either by gravity or by flight, as in suspension firing. A part of the air is admitted from the bottom at low velocities (<1 m/s through the fuel bed), constituting 60–80% of the total air, PA in most cases is sub-stoichiometric. PA permeates the entire bed and provides combustion air. The ash layer formed on the fuel is effectively scrubbed away to expose fresh surface by the gentle agitation that PA creates as it goes through the bed.

Balance combustion air as SA is provided from the top at HP, usually from both front and rear walls, to penetrate and agitate the flames for an effective completeness of combustion. It also provides intense turbulence in the furnace chamber to prevent stratification of gases. The gas flow and the temperature at the furnace exit are made more uniform this way.

Grates are the earliest form of **solid fuel** firing devices going back in history to well over 150 years. Of the numerous designs of the grates, only a few are surviving today. The mechanical construction has been steadily worked upon for several decades to reach near perfection and saturation. Grate clips or grate bars, which are the main building blocks of the grates, are more or less fully developed. Usually grate bars have a typical construction on the underside, with deep ribs, so that they can withstand rough use and more importantly, get cooled by the undergrate PA. That is how the grate bars stay fairly cool at ~400°C in coal firing even while there is a raging fire on the top of grate surface. Mechanical robustness of a grate is of utmost importance as any problem will call for a shutdown of the boiler, which is very expensive.

7.17.1.2 Grate Types

Grates are divided primarily into (1) stationary and (2) mechanical grates.

7.17.1.2.1 Stationary Grates

Stationary grates with hand firing, such as those used in the former locomotive boilers for coal, are practically extinct and are of historical interest today.

Stationary grates with semi-pile firing such as PHG and IWCG are quite popular for wood and other bio-fuels.

7.17.1.2.2 Mechanical Grates

For the present-day requirements of burning the poor-quality prime fuels efficiently, fuel combinations and waste/bio-fuels, mechanical grates are highly suitable.

Different grate designs suit different fuels and applications.

Mechanical grates can be classified as

- Dumping grates (DG)
- Chain/travelling grates (CG/TG)
- Oscillating/pulsating/step grates (PG)
- Vibrating grates (VG)

Each of these grates is described separately.

Niche area of Grate Operation:

Grates were the only firing equipments in early years for firing solid fuels. TGs with SSs ruled supreme in the 1940s and 1950s generating steam for both power and process plants. With the development of other technologies like **PF** in the 1920s, **FBC** in the 1970s, along with the deteriorating fuel situation and growing boiler sizes, stoker firing has been totally sidelined and relegated to only special applications today. However, they are indispensable in burning a variety of **agro-fuels** and solid wastes.

7.17.2 Types of Burning

Fuel burning on grates can be classified as

1. MB when fuel burns entirely on the grate
2. Spreader burning when fuel burns partly in suspension and partly on the grate
3. Semi-pile burning when fuel burns on the grate in the form of a thin ribbon

7.17.2.1 Mass Burning

MB usually has gravity feeding. CGs/TGs and RGs are used in MB. DGs are unsuitable. Figure 7.5 shows a typical chain or travelling grate arrangement with gravity feeding with CAD. In MB

- Front and rear arches are usually needed to deflect hot gases and heat to help both in ignition and in combustion.
- Undergrate air compartments are for distribution of PA as needed for combustion along the length of the grate. Undergrate pressure required for MB is higher at ~55 mm wg as against ~40 mm wg for spreader firing of coal due to higher bed thickness.
- HP SA at ~600–750 mm wg at suitable angle is needed for agitation of bed.
- ~85% of combustion is on the grate with coals and ~15% above in furnace.
- Carryover of ash from furnace varies from ~15% to 30%, with lower levels for coals and higher for bio-fuels.
- Performance is highly dependent on coal being reasonably good, that is, decent levels of
    - fuel sizing and fines (100% <32 mm and 35% <3 mm or 15% <1 mm)
    - A content (<25% as fired)
    - M content (<18% as fired)
    - VM (>25% on daf basis)

7.17.2.2 Spreader Burning

Spreader burning has fuel distributors for throwing the fuel evenly on CGs/TGs or DGs. Figure 7.6 shows a typical SS with TG. Spreader firing is a truly two-in-one combustion as it takes partly in furnace and partly on grate, releasing nearly twice the heat for the same grate area. The resulting boiler is more compact.

- As combustion takes place in furnace in suspension to the extent of 40–60% for bio-fuels and 35–50% for coals, the fuel bed is thin on the grate (50–80 mm) and combustion more vigorous.
- Load response is much faster than in MB.
- A seamless 3:1 load turn down is normally possible.
- Furnace arches are usually not required. But HP SA of 600–750 mm wg ranging from 20% to 30% of total air is provided for assisting suspension firing and providing thorough agitation.
- For bio-fuels which are very light and prone to high carry–over, a part of SA is diverted as TA to provide an air curtain in the upper furnace.
- In CAD, ash is usually carried towards the front side. In coal firing, the spreader throws the larger pieces of coal towards the rear side and the smaller ones at the front. Forward travel gives the larger pieces more time for combustion.
- In suspension firing, the fly ash carried over is quite high at 20–40%.

Figure 7.7 shows the combustion on DG.

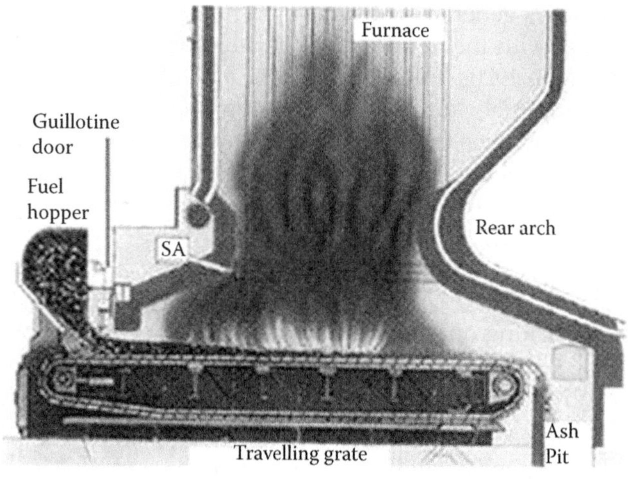

Furnace
Guillotine door
Fuel hopper
SA
Rear arch
Ash Pit
Travelling grate

FIGURE 7.5
Mass burning in gravity-fed chain/travelling grate.

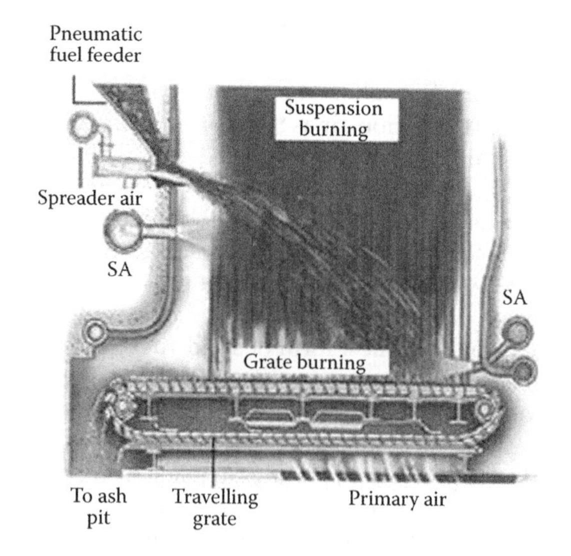

FIGURE 7.6
Spreader firing of bio-fuel on travelling grate.

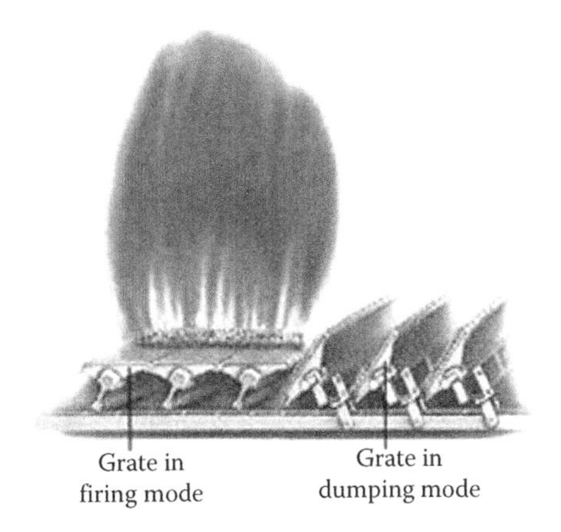

FIGURE 7.7
Combustion on dumping grate.

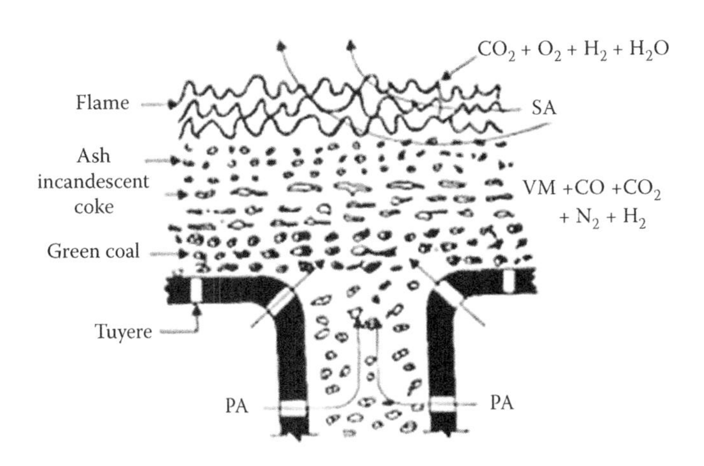

FIGURE 7.9
Undergrate feeding of coal.

7.17.2.3 Semi-Pile Burning

Semi-pile burning has the fuel delivered by gravity on a step grate, as shown in Figure 7.8. Combustion pattern of semi-pile is very similar to that of MB on a TG. However, the combustion at the bottom of the step grate is very close to classical pile burning. The pile is accumulated on a DG which is periodically dumped. IWCGs are employed in modern boilers in place of step grates which are simple, inexpensive and less efficient. The refractory roof is also replaced by HP SA ~600–750 mm wg directed at the pile and a rear arch to radiate the heat on to the pile which burns out the char efficiently.

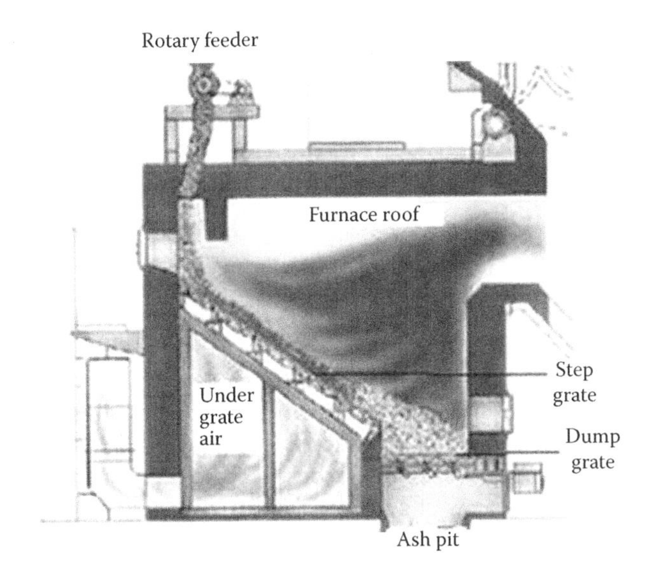

FIGURE 7.8
Semi-pile burning in inclined grate followed by mass burning on a dump grate.

7.17.3 Types of Fuel Feeding

Based on the type of feeding, stokers can be classified as

- Underfeed grates
- Overfeed grates

7.17.3.1 Underfeed Grates

Are fed with both air and coal from the bottom of the grate, as shown in Figure 7.9. Combustion proceeds from bottom up and the ash rises to the top and finally discharges from the sides. The boiler capacities are very small. They have practically disappeared from the industrial market, surviving only in small pockets here and there.

7.17.3.2 Overfeed Grates

Are fed with the fuel from the top and air from bottom of the grate, as shown in Figure 7.10. Combustion is from the top layer downwards. Ash is discharged from either the bottom or the end. CGs/TGs, DGs and PGs are all examples of overfeed grates.

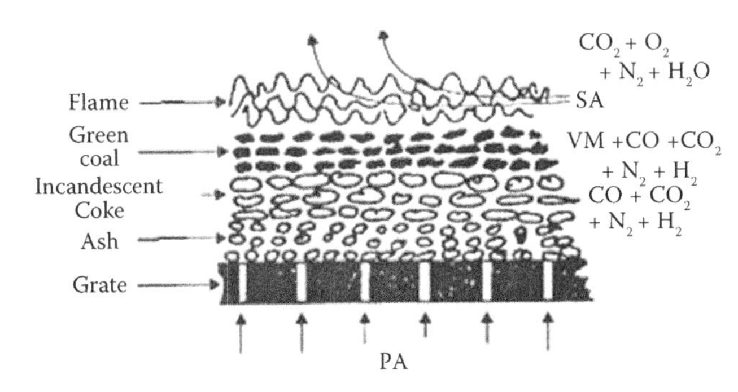

FIGURE 7.10
Overgrate feeding of coal.

7.17.4 Ash Discharge

Ash is discharged from a grate either continuously (CAD) or intermittently (IAD).

CAD is superior as it renders stability to the boiler operation as well as the operation of the process or power generation plant. Overall boiler η is higher.

IAD is slightly less efficient, as during de-ashing

a. Loss of green fuel is inevitable.
b. Over-firing the operating part of the grate to maintain evaporation causes more unburnt loss.

Although combustion η claimed may be the same for both, the actual on-load η in an IAD grate is always lower by ~2% points.

TGs and RGs are of CAD while DGs and PHGs are of IAD.

7.17.5 Grate Combustion

7.17.5.1 Gravity Feed

Combustion of coal on a grate fed by gravity proceeds in a very classical fashion in the following stages:

- Drying and devolatilisation on receiving the furnace radiation
- Ignition due to heat from burning gases of furnace and radiation from arches
- Combustion of char or FC on grate

The combustion process on a grate is illustrated in Figure 7.11.

After M and VM are distilled away in the first 500 mm from coal, the fuel bed, now consisting of only coke, slowly expands as the FC/char begins to burn. Simultaneously, an ash layer starts building up on top of the bed until all the FC turns into ash. There is a 600–750 mm of burn-out section at the end of the grate where enough time is given for the last traces of unburnt C trapped in the ash layer to burn. Typical time taken in furnace for the combustion of bituminous coals on a TG with gravity feed is ~20 min. The coal layer varies from 150 to 200 mm—higher-ash coals need higher bed thickness. The ash layer at the discharge varies from ~100 to 150 mm in thickness.

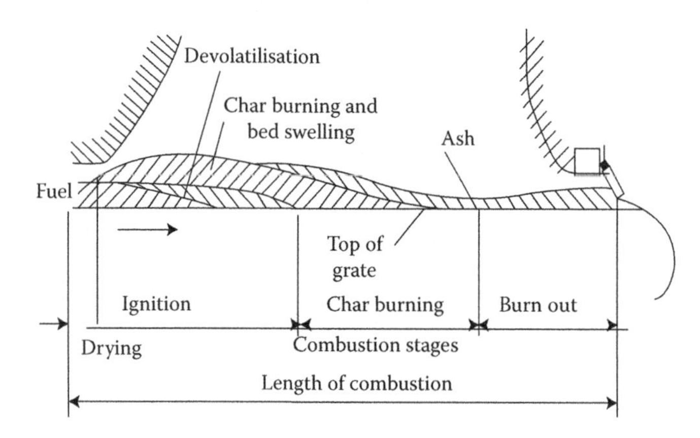

FIGURE 7.11
Burning process of coal on grate.

7.17.5.2 Spreader Feeding

In suspension firing resulting from spreader feeding, this classical combustion mode is slightly altered as drying and devolatilisation take place in the furnace during the flight of the coal particle and only char burns out on top of the grate.

7.17.5.3 Moistening of Coal

In coal firing in all grates, moistening of coal is needed to the extent of ~10% of surface M to make the

a. Bed porous during combustion by having streams of steam
b. Fuel fines stick to the lumps for complete combustion

7.17.5.4 Uniform Fuel Bed

It is absolutely necessary to have a uniform bed of fuel without segregation into lumps and fines so that airflow is uniform all over, which alone can result in good combustion. But segregation of coal is a normal occurrence during its travel from yard to bunker. **Anti-segregation** or *parabolic chutes* and **traversing chutes** are the usual methods adopted, which are explained later. Alternatively, a bunker bottom is split into several parts to feed each feeder. Agro-fuels do not segregate as much as coals, being bulky and fluffy.

7.17.5.5 Pros and Cons of Grate Firing

Pros and cons of grate firing are as listed below. The physical size of grate poses a limit on maximum evaporation. It is to be remembered, however, that grates are best suited for small and medium range of boilers for certain fuels. PF cannot be made in smaller sizes. While FBC overlaps with grates it may not be the best fit for certain fuels.

The advantages of grate firing are

- Simplicity of construction
- Low auxiliary power consumption
- Easy operation

- Good fuel flexibility
- No fear of explosion

The disadvantages are that

- Grates are maintenance prone, particularly the moving grates, thus lowering the boiler availability
- Combustion η is lower compared to FBC and PF firing
- Grate size limits boiler evaporation
- Sensitivity of performance to fuel sizing is high
- Grates have moderate fuel flexibility and fuel variation tolerance
- Grate firing may not be able to meet tight environmental requirements

7.18 Grate Types

The major types of grates are covered under the following sub-topics:

1. Grate, Pinhole (PHG)
2. Grate, Dumping (DG)
3. a. Grate, Chain (CG)
   b. Grate, Travelling (TG)
4. Grate, Reciprocating (RG
5. Grate, vibrating (VG)

7.18.1 Grate, Pinhole

PHGs fall in the category of stationary grates with thin pile or thin bed burning. They can be either flat PHGs at a nominal inclination of 5–7° or IWCGs at slopes of ~35° to the horizontal (shown in Figure 7.8) depending on whether the M in fuel is ~50% or even ~65%. PHGs, being stationary, are not suited for MB. They are also unsuitable for coal firing, but are good for speader burning of wood and bio-fuels like bagasse, which can be distributed on the grate quite easily with the help of spreaders. Large boilers of even 300 tph have been built for wood firing.

High-temperature **Cr steel grate castings** are bolted on to the furnace floor to create a flat stationary grate surface. The castings on their underside are machined and ground to fit the contour of the tubes perfectly, for a good metal-to-metal contact. That way the castings remain relatively cool at <150°C even as combustion takes place above. At regular intervals, castings with steam nozzles are inserted in the grate. This is to

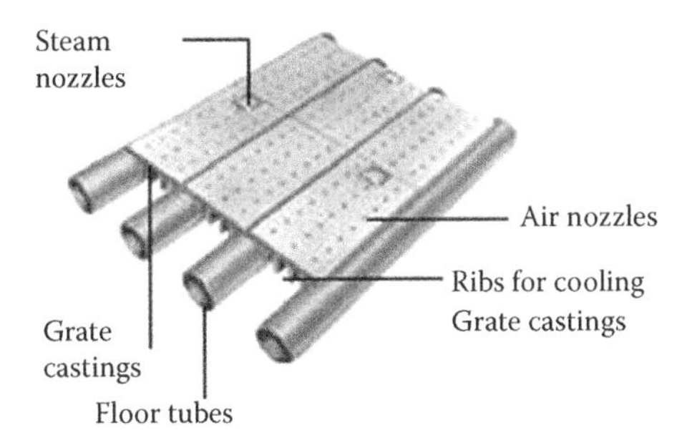

FIGURE 7.12
Casting of a pinhole grate.

help periodically blow ash towards the discharge end with the help of steam. Figure 7.12 shows typical PHG castings bolted to the floor tubes. The ribs provided underneath the castings help to dissipate the heat to the incoming PA. Steam jet arrangement can also be seen on two castings.

The combustion ratings of TGs and PHGs are very similar. TGs are more versatile while PHGs are better suited for low-ash-bearing bio-fuels like wood, bagasse and so on. Oil and gas can be the auxiliary fuels in PHGs but ash-bearing solid fuels like coal, even with medium level of ash, have to be ruled out as ash disposal is a problem. TG, on the other hand, can fire coal as well and is thus much more flexible. But maintenance of PHGs is low as the grate is stationary with no moving parts.

For really high M fuels exceeding 55% of M, IWCG is required with a long arch to deflect the hot gases and provide radiation. Fuels with M as high as 65% can be burnt without auxiliary oil or gas support.

7.18.2 Grate, Dumping

Ash is dumped periodically and hence the name DG. In construction and grate ratings, it bears a great similarity to TG with SS with burning rates at about 70–80% of TG as the grate has no movement for CAD. For the same reason, this grate is not suitable for MB but only for spreader burning. As the return strand is altogether avoided, DGs are much lighter and cheaper.

Figure 7.13 shows a typical one section of a small DG in closed and open positions.

DG for firing coals with low ash of <15% on dry basis is suitable for small boilers. They are more popular for

FIGURE 7.13
Small manual dumping grate assembly in shop floor in closed and dumping positions.

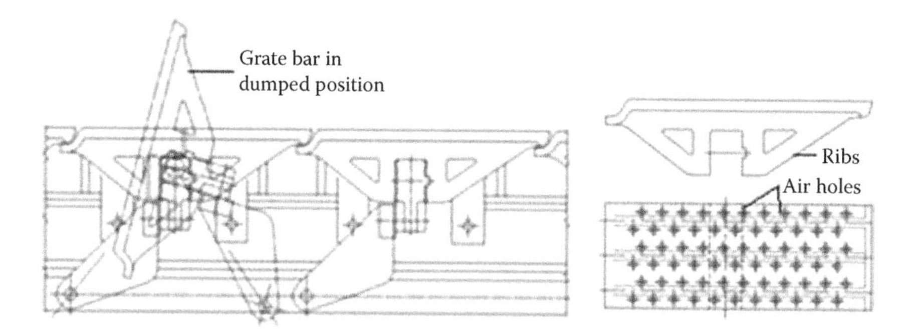

FIGURE 7.14
Castings of a dumping grate with operating lever.

bio-fuels such as bagasse where boilers up to ~150 tph are built. For higher capacities, TGs with SS firing is the right answer. As DGs are made in several sections, when one section is under ash discharge, the fuel to the other sections is fed more so that the evaporation is maintained.

Figure 7.14 shows a pair of grate bars in closed and open positions along with the details of a typical grate bar. Note that combustion air enters the fuel bed always from the bottom of the grate similar to the CGs/TGs.

A typical small SS boiler with DG for process steam using bio-fuel is shown in Figure 7.15.

The salient features of DGs are

- They are always made in sections of 1.5–2 m in width with independent fuel distributor and dumping arrangement. Grate with minimum two sections is built so that there is steam generation even when one section is being dumped.

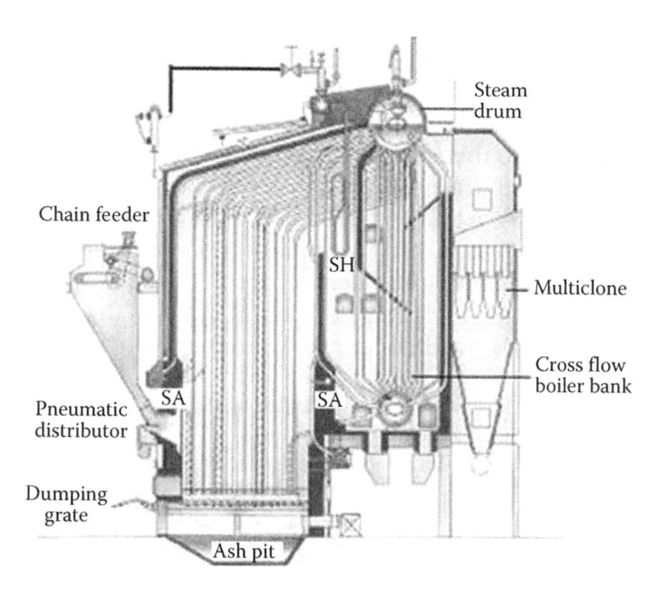

FIGURE 7.15
SS dumping grate boiler for bio-fuel.

- The minimum grate length is ~2.4 m and the usual increments are at ~225 mm depending on the length of great castings.

Likewise, the minimum grate width is approximately 1.5 m.

The maximum net length is ~5 m and gross length is ~5.5 m to suit the manual raking tools to tend the grate.

- HRRs are ~70% of the releasing rates of SSs for coal. But for bio-fuels, it can be nearly 85–90% mainly because of less ash.
- The maximum HRR for bio-fuels are
  - With cold air—2.5 MW/m^2 or ~2.17 × 10^6 kcal/m^2/h or 800,000 Btu/ft^2 h
  - With hot air—2.85 MW/m^2 or ~2.44 × 10^6 kcal/m^2/h or 900,000 Btu/ft^2 h
- Typical turn down is 3:1.
- The hot air temperature limit is ~ 200°C in CS execution and can be higher with AS.

To sum up

- The DGs are still the favourites of the industry to burn fuels like **bagasse** and low-ash coal where steam pressure fluctuations and loss of green fuel during the grate dumping periods is acceptable.
- A combination of suspension (with about 40–60% heat release in furnace) and grate firing makes the DGs release much higher heat than gravity-fed CGs.
- DGs are very simple and inexpensive. No return strand, as in TGs, makes them very compact, light, cheap and easy to maintain.
- The dumping effects get smoothened out as the number of grates increase in a boiler.

- Reasonably high HRR, of around 70–80% of TGs, are achieved with bagasse and similar uniformly sized bio-fuels.
- They are unsuitable for firing coal of medium to high-ash or fuels with some clinkering tendencies, as there is no CAD from the grates.

Very popular with sugar and biomass plants, bagasse-fired boilers as large as 150 tph have been built with multiple DGs.

7.18.3.a Grate, Chain

CGs and TGs are similar in ratings, arrangement and operation. Both of them are endless steel belts with CAD and are used in both mass and suspension burning. Naturally, the words CG and TG are used interchangeably ignoring the differences in their constructional features.

CGs are made of **forged steel** links connected together by pins and other fasteners. The grate links experience the tensile forces. Naturally, the CG shown in Figure 7.16 cannot be made as large as TG.

TGs, on the other hand, employ chains and T/bulb bars between them to form a basic frame which experiences tensile forces. The grate bars are merely fitted over the T or bulb bars and are free to swing. They are made of **heat-resistant castings** and experience no force. Naturally, the sizes to which the TGs can be built are much larger than the CGs. Also life expectancy of TGs is higher as the grate bars are not subject to tensile forces as in the links in CGs.

When it comes to combustion, both of them behave in the same way. The grate ratings are very similar. CGs are lighter and cheaper as the number of components is less.

Typical combustion rates in MB mode are 160–185 kg/m^2 h (33–38 lb/ft^2 h) for arch and archless settings of furnace. Ash discharge rates are limited to 350 kg/m^2 h (230 lb/ft^2 h)

FIGURE 7.16
Typical chain grate.

7.18.3.b Grate, Travelling

TG is an endless steel belt carried by the front and rear shafts. It may be driven by either of the shafts. Combustion takes place on the top surface with air fed from the bottom. It can be used both for mass and spreader burning with nearly no change in construction. In MB with gravity feeding, the grate moves from front to the rear while in spreader burning with overthrow feeding, it moves from rear to the front.

There are two types of travelling grates based on mechanical construction, both of which are popular:

a. Catenary type
b. Chain type

1. In *catenary*-type TG, the shaft centres are held constant and the extra length of chain is kept slinging at the bottom to form a natural catenary. The self-weight of the bottom strand keeps the top part taut. No additional chain tensioning is needed. The shafts are solid and limited to about 150 mm diameter. Pairs of chains, spaced at ~1.2 m, form grate sections with T-bars or bulb bars running between them and carrying grate bars. While the T bars are subject to flexure, the grate bars experience no tensile forces.

   There are two types of grate bars—plate type and bar type.

   - The plate-type bars are high-temperature castings of varying widths up to ~300 mm in length sliding over the T-bars.
   - The bar-type castings are actually a combination of the T-bar and the grate bar and are directly bolted to the carrier chains.

   Venturi-shaped air nozzles are drilled in both types of castings to provide high air pressure drop to provide uniform distribution of air across the entire grate. This way the partitioning and control of air are avoided.

2. In *chain-type* travelling grates, the shaft distance is adjustable by means of a chain tensioning arrangement. The bottom strand of the chain is supported on rails. The upper and lower strands of the chain are held parallel ~1000 mm apart to allow for access and provision of air dampers and seals. The shafts are usually made of hollow pipes and the sprockets are large. The grate bars are normally link types, as shown in Figure 7.17. Air holes can be through the castings or at their sides. In this design, air can be controlled in the individual compartments unlike in the catenary design.

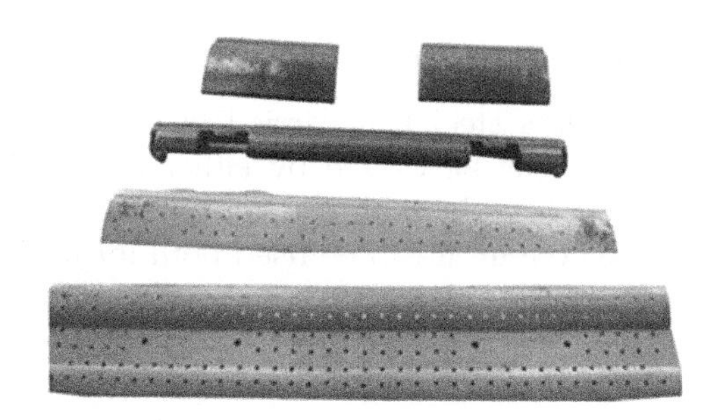

FIGURE 7.17
Grate bars/castings of travelling grates.

Mechanically sound construction is of paramount importance for obtaining optimum performance, reliable service, high availability and long life from all grates. Any outage calls for boiler shut down which is very expensive. Robust construction of each part with appropriate metallurgy is the key for a trouble-free life.

Grate bars or grate links are the main building blocks of any grate. The geometry and tolerances are of utmost importance as they have to perfectly mesh with each other with no uncontrolled air leakage. At the same time, they should not foul with each other when they turn over the front and rear sprockets. Deep ribs under the grate bars make them mechanically strong and permit them to lose the heat to the undergrate air and thus keep them cool. Appropriately designed grate bars always stay at <450°C even with hot air. They are usually Cr–Ni castings with ~1% alloying.

Chains and links are the other crucial set of components in a grate. See Figure 7.18. Usually high tensility chains of ~500 MPa (32 t/in^2) are employed to cater to jamming conditions. Also, the chains should not elongate during the operation, as differential elongation between the chains can lead to the twisting of T-bars and other problems.

Air nozzles or *air passages* should constitute between 6% and 10% of the total grate surface to provide adequate pressure drop for uniform distibution of combustion air. More air passage is required for coals and less for bio-fuels as there is more SA in bio-fuels. Usually air pressure loss of ~40 mm WG takes place across the grates.

Cold or hot air can be supplied with hot air as high as even 230°C. For temperatures >200°C, alloy grate bars and T-bars are usually required. For coal firing, the normal upper limit is 160°C to prevent caking or coking.

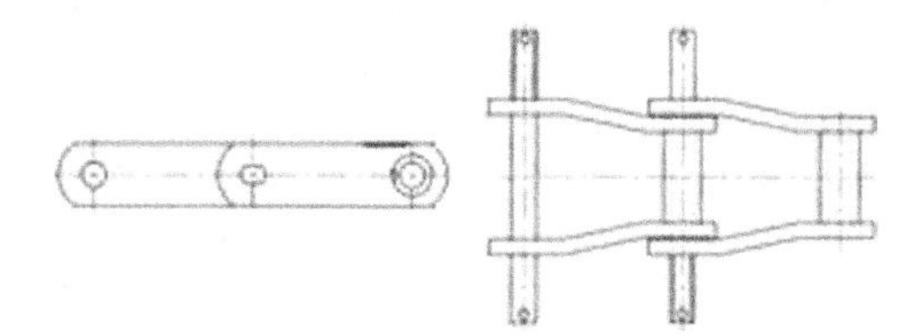

FIGURE 7.18
Typical chain and link assembly of travelling grate chain.

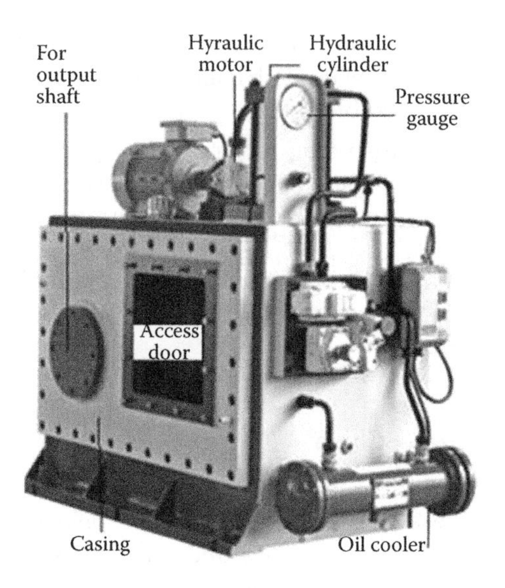

FIGURE 7.19
Hydraulic grate drive.

Grate area is limited mainly by mechanical considerations. Catenary grates are generally limited to ~5 m in width and chain-type grates to ~6 m. Twin grates, with drives on either side, are used for greater width. The length of the grate rarely exceeds 8 m shaft centre distance. Grates, as small as 2 m in width, have been built.

Grate drives can be either mechanical or hydraulic. Both of them are nearly the same in terms of initial and running cost, power consumption and even reliability.

Figure 7.19 shows a typical hydraulic grate drive. Reciprocating motion of the piston of the hydraulic cylinder is converted into rotary motion of the drive shaft by means of a pawl and ratchet mechanism housed inside the casing. The motion imparted to the shaft is not continuous as in mechanical drive but inching type, whenever the piston moves the ratchet.

Figure 7.20 depicts a mechanical grate drive where a motor rotates the grate shaft through a planetary gear box and flexible couplings.

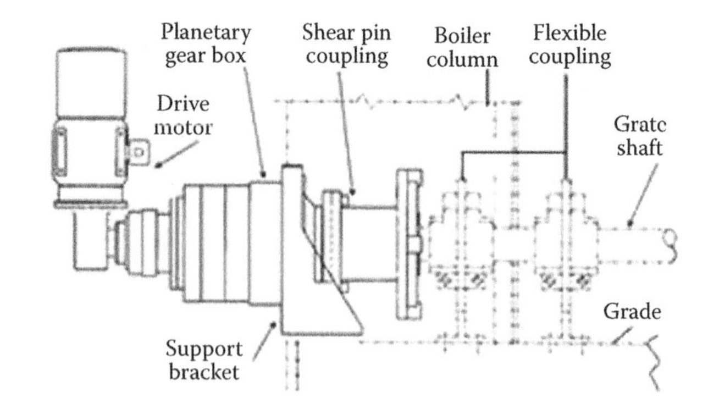

FIGURE 7.20
Mechanical grate drive.

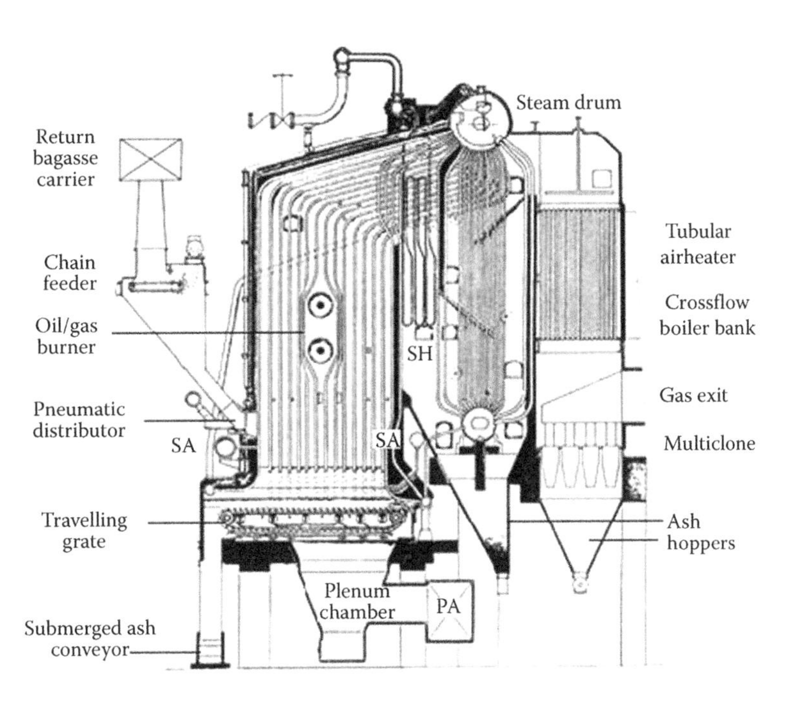

FIGURE 7.21
Biomass with supplementary oil/gas-fired SS travelling grate boiler.

Figure 7.21 shows a typical SS-fired TG boiler equipped with drag chain feeder and pneumatic spreader for firing a bio-fuel like bagasse.

SS which combines the pneumatic or mechanical type of fuel spreading on CG or TG is the most versatile of all the stokers with

- The highest HRR
- CAD
- Very good load response because of suspension firing
- Full ACC
- Exceptional fuel flexibility and
- Very high multi-fuel firing capability

However, SSs have certain limitations mainly, that they cannot

- Burn high-ash coals (>35% ash) with good η as unburnt losses run high.
- Effectively burn lignites of high M (~55% and above).
- Deal with high S in coal.
- Have low NO_x coal combustion to meet the emerging norms.

They were the backbone of the industry because of their simplicity and versatility until the FBCs unseated them with their

- Superior low-temperature combustion technique (with the resultant **low NO_x**)
- In-furnace **desulphurisation** (making high-S coals burn easily)
- Greater versatility with high-ash and inferior coals than SS
- Very simple O&M due to practically no moving parts

As the environmental regulations are getting tighter all over the world, it has progressively pushed SS firing to the burning of various bio-fuels that have no S and which burn at low temperatures because of their high M. Such fuels being fluffy do not lend themselves to burning in FBC satisfactorily and easily. Bagasse and bio-fuel burning is still an area where SSs still rule supreme.

7.18.4 Grate: Reciprocating, Pulsating, Step or Pusher (*see also* MSW Boilers Mass Burning)

RGs are also called *step/pulsating* or *pusher* grates. Usually the grates are inclined and the grate bar arrangement resembles steps and hence the name step grates. The forward motion of fuel is by the pushing movement of the grate bars unlike the carrying action in TGs. Also, the travel of air is substantially parallel to the grate bar and not so much perpendicular. The pushing action helps to penetrate and agitate fuel mass and expose fresh surface for combustion. These grates, thus, are suitable for bulky

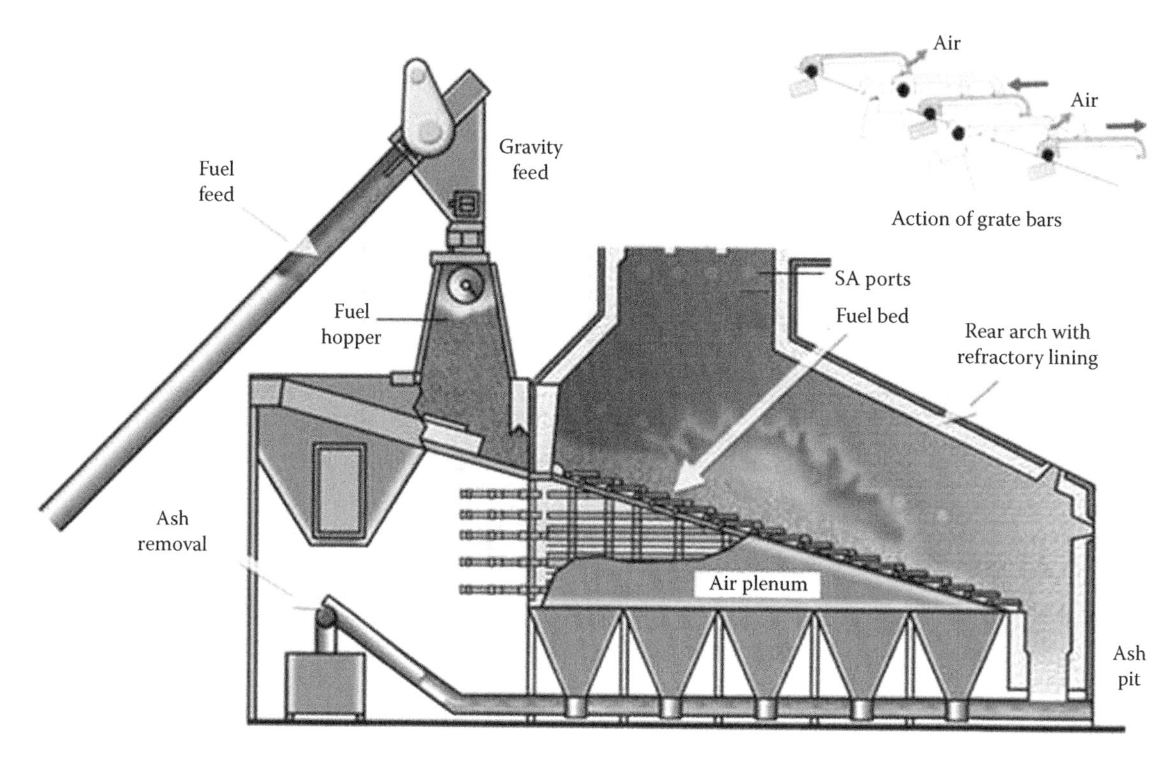

FIGURE 7.22
Typical small reciprocating grate for burning bio-fuels.

fuels like bagasse, various bio-fuels and MSW. A typical small reciprocating grate is shown in Figure 7.22 with an inset depicting the action of reciprocating grate bars.

The grate bars are arranged at an inclination ranging from 0 to 30° to horizontal depending on the fuel quality. For normal bio-fuels, the grate is set at a single inclination and is gravity fed. As the grate is directly below the inlet hopper, the fuel gently slides on to the first section of the grate which is fixed and then pushed forward by the reciprocating action of the grate bars. As the fuel descends further on the grate, the combustion goes through the three stages of drying, ignition and burning of char sequentially along the length. Finally, the ash rolls over to the ash pit at the end of travel. For better combustion, the grate bars are grouped and moved at different speeds. The adjacent grate bars move in opposite directions to give a scissor-like action. Alternate grate bars can also be made stationary to suit the fuel.

The tumbling of fuel belt over the grate bars is helpful in exposing new surface for accelerated combustion. At the same time, this continuous agitation of bed does not help burning of coals, as the semi-molten spongy ash is pushed around to accumulate coal and fine particles and turn into clinker. Clinker formation is further accelerated in these grates because of continuous tumbling. The reciprocating action of the grate bars may not be able to push and discharge the big clinker into the ash pit unlike the TG which carries it positively. The clinker, when it grows large enough to straddle two grate bars, merely rocks because of the reciprocating action but does not roll down. This prohibits the grate from firing coals with medium and high ash.

Deep front and rear arches are needed for deflecting flue gases on the green fuel as well as radiating the heat to the fuel bed. The throat formed above the grate helps to retain the gases close to the bed briefly and improve combustion.

Grate bars in RGs are heavier than those in TGs and DGs. They can be thin or wide depending upon the application and location. But they are always long because of the scissor-like action between them. Substantial air flow is from the front side of the bars, unlike in the other grate bars where air is always from the bottom. Grate bars for bio-fuels are usually long and thin, typically 300 × 25 mm. Figure 7.23 shows different types of grate bars for reciprocating grates.

Fuel sizing is not so important in gravity-fed RGs as the entry is by a free fall. Most are gravity fed but RGs can also be fed by SSs. Then, fuel sizing matters.

For very high M, uneven and low-quality fuels like MSW, RG is the only solution. Usually, there are three grates working in unison. The grates are made at three different decreasing slopes, for drying, ignition and burning of char, respectively. All of them have different drives to vary the movement of fuel. This way, the fuel is treated optimally in various combustion zones. The grate bars are large and heavy to withstand the weight of fuel and the erosion forces of the trash matter in the fuel.

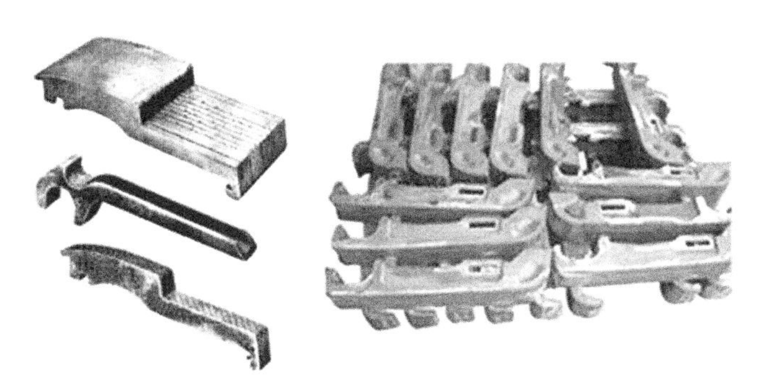

FIGURE 7.23
Long and thin grate bars in reciprocating grates for bio-fuels.

Grates for MSW tend to be very large and expensive:

- Large because the volume of poor quality fuel to be burnt is huge and
- Expensive because the metallurgy of the grate bars has to be of high AS or ss to withstand the corrosive forces of the fuel

In some cases where the overheating of grate bars is expected, in addition to alloying the grate bars, water cooling is employed. Such grates are highly expensive and made by very few. The underside of a typical heavy-duty grate bar for burning MSW is shown in Figure 7.24.

To summarise

- PGs or RGs are most ideally suited for high- to very-high-M bulky fuels like MSW or very light and fluffy bio-fuels unsuitable for spreading.
- With such fuels, the reciprocating action of the grate bars helps in uncovering the fresh fuel and creating air passages, thus improving the combustion η. Also, such action helps to plough the thick bed of low Gcv fluffy fuels like no other grate.
- Single or multiple grates in series can be used depending on the fuel M. For MSW where fuel M can exceed 70%, a steeper grate is used in the first stage for fuel drying, followed by combustion grate at a reduced inclination.
- CAD and single strand of grate with no return section are positive features of the RGs.
- The periodic agitation of the fuel bed on the grate, on the other hand, can contribute to clinker formation with coal.

A typical large MSW mass-fired boiler for high p & t equipped with step grate is shown in Figure 7.25. In this case, it is a single grate and not a combination of three.

FIGURE 7.24
Reverse side of the large-sized grate bar for reciprocating grate for MSW.

7.18.5 Grate, Vibrating

VG is a PHG whose under frame can be vibrated back and forth with the aid of an external drive to help ash and fuel move forward. The leaf springs connecting the base frame and the under frame of the grate provide this flexibility.

This grate is suited for solid fuels with a little ash like many **bio-fuels, bagasse, peat** and **RDF**. These grates can be either water or air cooled. Likewise they can be horizontal or slightly inclined (typically at 6° to horizontal) depending upon the boiler layout. Figure 7.26 shows horizontal water-cooled vibrating grate (VG) at the bottom of a furnace.

The main advantage of VG is in the position of the drive which offers no restriction to the grate width. This permits the construction of larger units. Boilers as large as 300 tph have been built with VGs while ~200 tph is the limit with TGs. Water-cooled VGs also claim 10–15% higher HRR compared to TGs. Simplicity of construction and less number of components make VGs very reliable pieces of equipment capable of trouble-free operation.

Coal firing cannot be done directly on the VG but can be performed with PF burners in a co-firing mode.

Anti-Segregating Chute

In grate firing, it is essential that solid fuels like coal or lignite are spread on the grate as uniformly as possible across the width (and also along the length in case of spreader firing), so that the burning is uniform and not patchy. Otherwise, the fines burn off quickly while lumps end up as unburnt fuel making the combustion highly irregular and inefficient.

The remedy is to receive fuel of uniform consistency in the inlet hopper from the bunkers. Since the fuel is always in segregated condition in the bunkers, direct pipe connections merely manage to transter the segregation from bunkers to hoppers. The proper method is to employ anti-segregation chutes between the two.

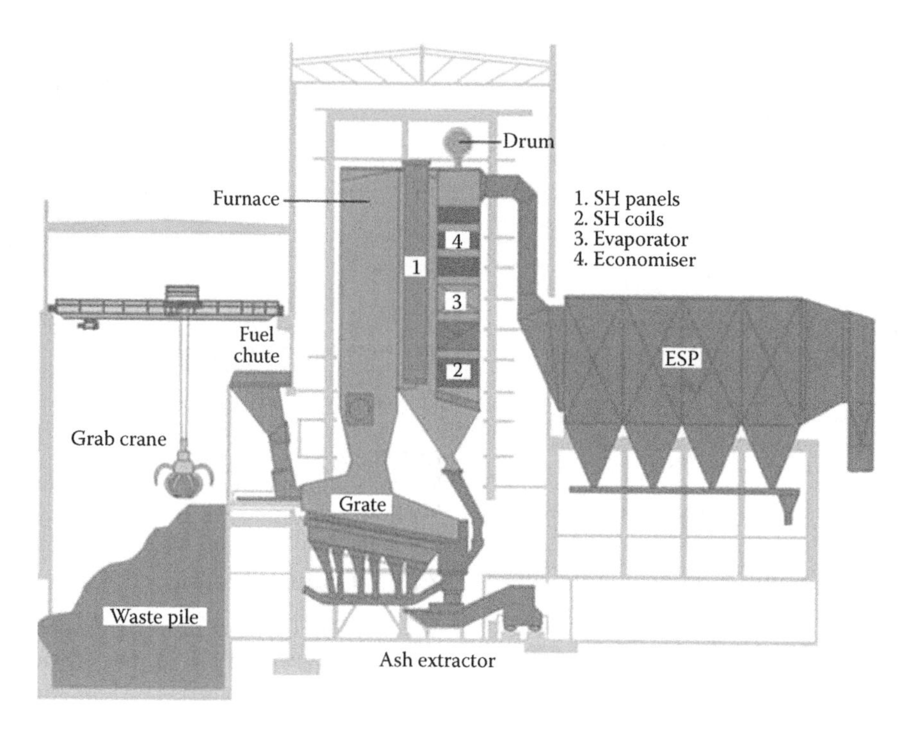

FIGURE 7.25
HP boiler for burning MSW on reciprocating grate.

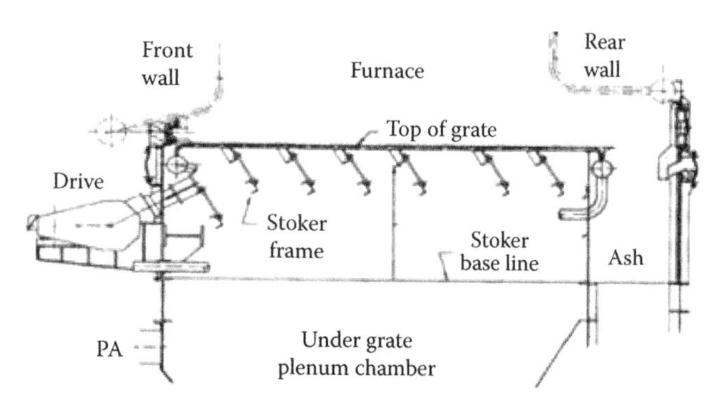

FIGURE 7.26
Flat water-cooled vibrating grate.

There are two types of these chutes:

- Traversing chutes which move from one side of the grate to the other while feeding the fuel hopper.
- Fixed parabolic chutes of 3–5 m width between the bunker and hopper which have a special parabolic shape at the bottom. As coal or lignite pieces roll down the parabolic shape of the chutes, they spread out and become more uniform.

Anti-segregating chutes are not needed for agro-fuels which are too bulky to cause problems of segregation typical to coals.

7.19 Green Liquor (*see* Kraft Pulping Process *and* Recovery Boiler)

7.20 Green House Gases

Any gas in the atmosphere that absorbs and emits radiation within the thermal infrared range is called GHG. GHGs, in decreasing order, are

Water vapour (H_2O)
Carbon dioxide (CO_2)
Methane (CH_4)
Nitrous oxide (N_2O)
Ozone (O_3)
Hydrofluorocarbons (HFCs)
Perfluorocarbons (PFCs)
Sulphur hexafluoride (SF_6)

These GHGs are essential to keep our planet warm and comfortable. Without them, our earth would be too cold and uninhabitable. At the same time, increase in GHGs warms up the planet with disastrous consequences.

The GHG effect is attributable to the increasing consumption of fossil fuels, particularly FO and coal that increases the emissions of CO_2. CO_2 restricts the sun's heat most from escaping from the atmosphere, thus giving rise to global warming. Since the time of the Industrial Revolution, the levels of CO_2 in the atmosphere are estimated to have gone up from ~280 to ~390 ppm with a corresponding increase in the temperature by ~3°C. This is a serious matter and concerns all.

Global warming potential (GWP) is a measure of how much a given mass of GHG is estimated to contribute to global warming. GWP is calculated relative to CO_2 and it is always accompanied with time interval.

7.21 Grindability Index (*see* Hardgrove Index)

7.22 Grinding (*see* Pulverising)

7.23 Gross Calorific Value (*see* Thermodynamic Properties)

7.24 Gross Steam Generation (*see* Steam Generation)

7.25 Gross Thermal η (*see* Definitions of η *under* Efficiency of Boiler)

7.26 Guillotine Dampers (*see* Dampers in Draught Plant)

8

H

8.1 Hardenability (of Steel) (*see* Metallurgy)

8.2 Hardness of Steel (*see* Metallurgy)

8.3 Hardness of Water (*see* Water)

8.4 Hardness Measurements of Metals (*see* Testing of Materials)

8.5 Hardgrove Index

HGI is an empirical measure of the ease of grinding coal, a property that is vital for milling. It is an index showing the relative hardness of that particular coal compared to the standard soft coal with an index of 100. The parameters that add to the difficulty in grinding are

- Hardness
- Fibrous nature of coal
- Presence of sticky and plastic materials

ASTM D409 Standard Test Method for Grindability of Coal by the Hardgrove-Machine Method gives the procedure.

The underlying principle is that the work done in pulverising is proportional to the new surface generated.

Hardgrove machine, named after the inventor R.N. Hardgrove, is shown in Figure 8.1. It is a miniature pulveriser where the amount of grinding energy can be measured. It consists of a stationary grinding bowl with a horizontal track in which run eight steel balls each of 25 mm diameter. The balls are driven by an upper grinding ring which is rotated at 19–21 rpm. The upper grinding ring is connected to a spindle which is driven by a small ac motor of 0.25 hp through reduction gears. Weights are added to the driving spindle so that the total vertical force of the balls due to weights, shaft, top-ring and gear is equal to 29 ± 0.225 kg. The machine is fitted with an automatic counter for stopping the machine after 60 ± 0.25 revolutions.

The test method involves grinding a sample of 50 g of air-dried and sized coal of ~0.6–1.2 mm in the test mill for 60 revolutions and comparing the –75 microns fraction generated with the standard curves.

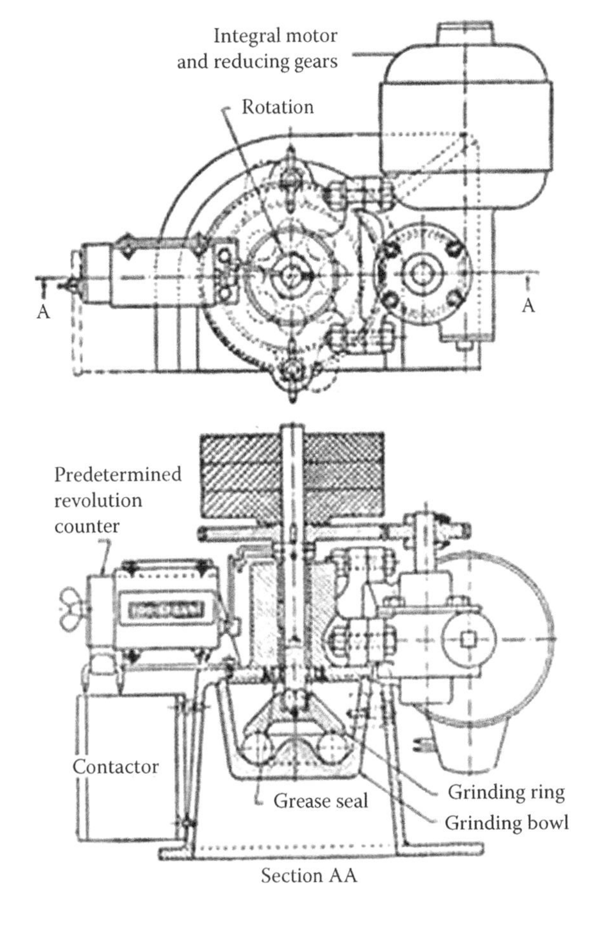

FIGURE 8.1
Hardgrove machine.

A general relationship exists between grindability of coal and its rank, as shown by the degree of change in the natural series from brown coals and lignites to anthracite. Coals that are easiest to grind (having the highest HGI) are those of ~14–30% VM content on dry MMF basis. Coals of either lower or higher VM content are more difficult to grind.

The two important characteristics of HGI are that it is

- Empirical. It has relatively low reproducibility and repeatability which can lead to ambiguity in evaluating mill performance and coal properties.
- Non-linear. Coals with low values of HGI are more difficult and high values are much easier to grind.

Pulveriser base capacities are founded on HGI of 50 or 55. The variation of mill capacity and power with HGI is quite significant, as shown in Figure 8.2. With increasing HGI, the mill output rises and power consumption reduces. The graph also shows how the fineness affects the mill output. For the same mill output, the product fineness increases.

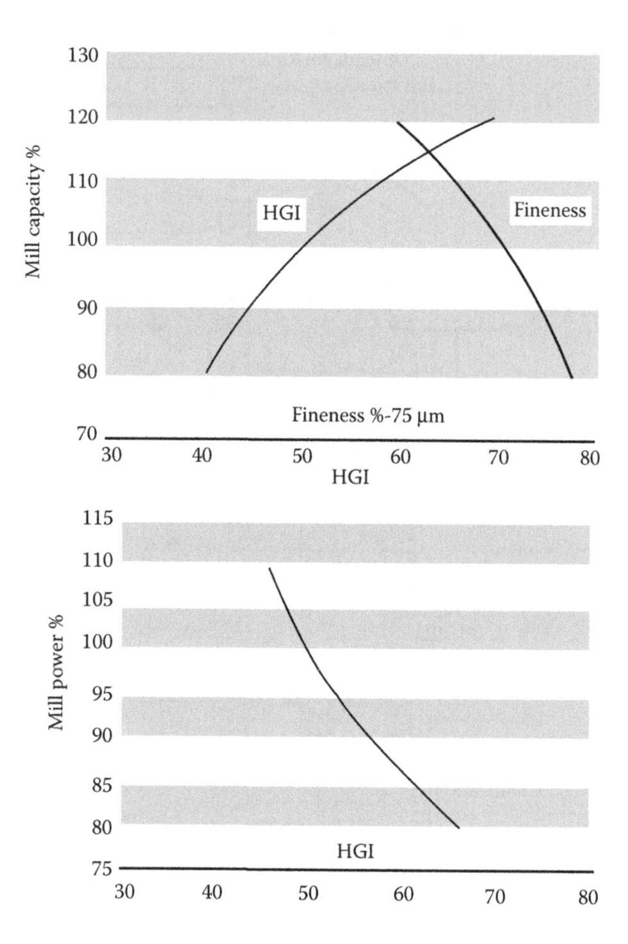

FIGURE 8.2
Variation of mill capacity and power with HGI.

Essentially, there is an envelop of not so well-defined contours of coal properties, within which HGI is reliable. There are known instances of coal pulveriser capacity being inadequate when using some coals with rather unusual properties, a fact that has come to be discovered as the range of internationally traded coals increased.

While HGI portrays the relative difficulty in pulverising, it does not provide direct information on the likely **wear** rates of mill grinding components.

8.6 Headers (*see* Pressure Parts)

8.7 Heat or Heat Energy (*see* Thermodynamic Properties)

8.8 Heat Balance (*see also* Definition of Efficiency in Efficiency of Boiler)

Heat balance for any balanced system states that

$$\text{heat input} = \text{heat output} + \text{heat losses} \quad (8.1)$$

In a conventional boiler, for unit fuel

- Heat input is by the cv of fuel, its sensible heat above the datum and heat credits if any.
- Heat output is the net heat added to:
  a. fw to turn it into main steam
  b. RH steam to heat from cold to hot RH steam
  c. Any other heating like external SH and so on
- Losses are stack, unburnt, sensible, radiation and unaccountable loss and also heat loss in blowdown, which are fully elaborated under the topic of Efficiency of Boiler.

Heat balance is the first step for estimation of η and fuel fired and so on.

Shown in Figure 8.3 is a heat and flow balance arrangement of an RH boiler depicted in the form of a Sankey diagram.

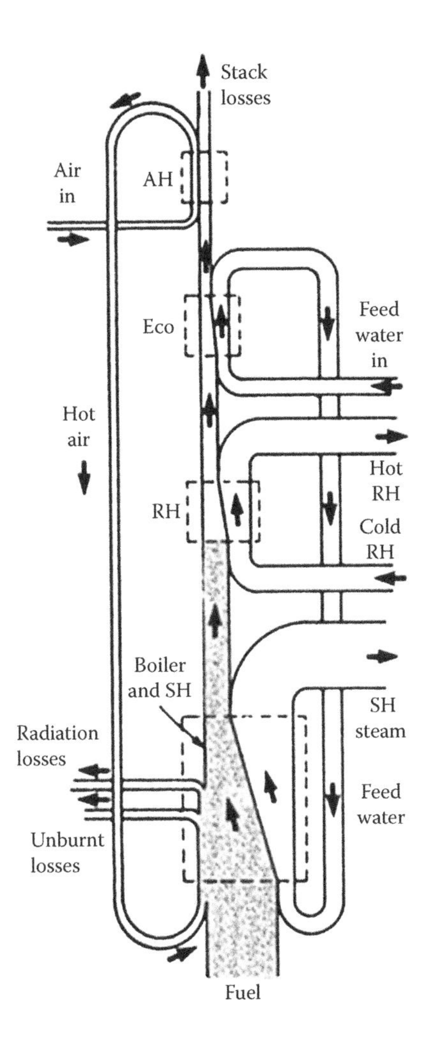

FIGURE 8.3
Sankey diagram for heat balance in an RH boiler.

8.9 Heat Credits (*see* Heat Loss Calculation in Efficiency of Boiler)

8.10 Heat Distribution to HSs (at Various Pressure Levels) (*see* Industrial Boilers)

8.11 Heat Duty of Boiler (*see* Boiler Heat Duty)

8.12 Heat Exchangers in Boiler (*see also* Fluid Flow Patterns in HXs *and* Log Mean Temperature Difference [LMTD or Δtm])

HX is an equipment built for transferring heat from one medium to another in an optimally efficient manner. After the combustion chamber or **furnace**, a boiler is nothing but a series of HXs transferring the heat of flue gases primarily to steam, water and combustion air. These HXs are termed as **SH, RH, BB, Eco** and **AH** in boiler parlance. HX takes place with both fluids flowing in **parallel, counter** or **cross flow** patterns to suit the situation.

8.13 Heat Flow Equation (*see* Thermodynamic Properties)

8.14 Heat Flux (*see* Thermodynamic Properties)

8.15 Heat Input into Boiler (H_{in}) (*see also* Heat Release Rate)

Heat input into the boiler is gross heat given to the boiler, which is the total fuel × Gcv of fuel + heat credits in MJ, kcal or Btu. Heat credits are any heat inputs other than from the fuels. Fuel flow is either per hour in British and mks units or per seconds in SI units. Accordingly, it is in Btu/h, kcal/h or kcal/s.

$$\text{Heat input} = H_{in} = W_f \text{ (weight of fuel fired)} \times \text{Gcv of fuel} + \text{heat credits} \quad (8.2)$$

Heat input is different from heat released and available (HR and A) which is based on fuel net calorific value (NCV), including the heat addition in AH and adjusted for the unburnt C and furnace radiation losses (assumed to be half of the total). Ambient temperature is taken as the datum to calculate this heat.

$$\text{Total (HR and A)} = H_o = \text{fuel fired} \times \left[(\text{NCV} + H_a) - \left\{ \frac{R}{2} + U_{bl} \right\} \right] \quad (8.3)$$

where
R is the radiation loss
H_a is the heat added to air
U_{bl} is the unburnt loss

HR and A provides a realistic picture of the heat input which is genuinely available for converting to steam.

HR and A is the first step in calculating all the furnace parameters like residence time, heat flux, FEGT and so on.

Heat input is different from Heat release, which is the heat released in the furnace by the combustion of fuel. See HRR below.

8.16 Heat Loss Account (*see* Break up of Losses in Efficiency of Boiler)

8.17 Heat of Combustion (*see* Combustion)

8.18 Heat of Evaporation (*see* Properties of Steam and Water)

8.19 Heat Output of Boiler (*see* Boiler Heat Duty)

8.20 Heat Pipe AH (*see* Airheater)

8.21 Heat Rate (*see* Thermodynamic Properties)

8.22 Heat Release Rate

In boiler parlance HRR applies to furnace and firing equipment.

By heat release is meant the total heat produced by combustion of fuel (W_f) in furnace in MJ, kcal or Btu per unit time.

$$\text{Total heat release } H_{in} = W_f \times \text{Gcv} \tag{8.4}$$

HRR in Furnace (*see also* furnace sizing in Heating Surfaces)

HR and A is the starting point for calculating the HRRs of furnace.

HRR in furnace can be either plan area or volume based, both of which are needed to define a furnace capability. For different fuels and different firing systems, there are established norms to be observed for both plan and volumetric HRR which result in efficient combustion with no fouling and slagging.

HRRs are the most important parameters for sizing a furnace.

Cleaner the fuel, higher is the permissible HRR and smaller is the furnace.

Volumetric HRR is another way of expressing the furnace residence time.

HRR in Firing Equipment

For the firing equipment heat released is the gross heat without any adjustments.

Capacity of a burner is normally expressed in HRR in MW_{th}, kcal/h or Btu/h, that is

$$\text{HRR/burner} = \frac{H_{in}}{\text{number of burners}} = \frac{W_f \times \text{Gev}}{\text{number of burners}} \tag{8.5}$$

Capability of grates is expressed in HRR per unit area such as MW/m^2, kcal/m^2 h, Btu/ft^2 h

$$\text{HRR of grate} = \frac{H_{in}}{\text{grate area}} = \frac{W_f \times \text{Gcv}}{\text{grate area}} \tag{8.6}$$

8.23 Heat Released and Available (*see* Heat Input)

8.24 Heat Treatment of Steel (*see* Metallurgy)

8.25 Heat Loss Method in Boiler Testing (*see* Efficiency of Boiler)

8.26 Heat Resistant Castings (*see* Castings in Materials of Construction)

8.27 Heating Surfaces (*see also* Pressure Parts)

HS in boiler is the surface through which heat transfer takes place between flue gases and water/steam. HSs in a boiler are mainly the PPs and AH.

This topic of HSs is covered under the following sub-topics:

1. Designating HSs
2. HSs in boiler
3. Effectiveness of boiler HSs

1. Furnace
   1a. Furnace cleaning
   1b. FEGT
   1c. Furnace pressure
   1d. Furnace tubes
   1e. Furnace size or volume
   1f. Furnace wall construction
2. Superheater (SH)
   2a. SH classification
3. Reheater (RH)
4. Boiler/convection/evaporation surface
   4a. Boiler bank, convection bank or Evap bank (BB)
   BB construction
   4b. Wing wall Evap and division wall
   4c. Bed coils
5. Eco
6. Airheater (AH)
7. Backend equipment

8.27.1 Designating HSs

Boiler HSs are designated in several ways:

- *Commercial or gross HS* is the total amount of surface provided in the boiler and is an indication of the amount of PP steel expressed in m^2/ft^2. Its utility is more for comparison. Surfaces outside the gas passes are also included.
- *Actual or net HS* is the surface, out of the above, which is actually swept by the gases and participates in heat transfer.
- *Effective or adjusted HS* is the surface, out of the above, which is actually used in heat transfer equations. Effective HS = net HS × effectiveness factor which is proprietary in nature, basically taking into account factors like cleanliness, actual surface swept by gases, emissivity, effect of any tube attachments and so on.

8.27.2 HSs in Boiler

The main HSs in a conventional boiler are

1. Furnace
2. SH
3. RH
4. BB
5. Eco
6. AH

The following are worth noting:

- In BFBC boiler, there is additionally the bed HS.
- In CFBC boiler, there may be FBHE instead.
- In Hot cyclone CFBC boilers cyclones may be cooled by SH or evaporator tube panels.
- Hrsgs AH is absent. Usually, a preheater is present to heat the condensate.
- In BLR boilers also, AH is not present.

8.27.3 Effectiveness of Boiler HSs

How effective is the unit area of HS depends entirely on its location in the boiler. This varies considerably, as shown in Figure 8.4, where heat absorbed by various HSs and their corresponding location is given. HS in furnace is the most efficient and AH is the least efficient but in terms of cost, it is the reverse. In actual fact, however, the surfaces are practically not interchangeable; only some minor adjustment may be possible, in the backend, namely, Eco and AH. Backend banks are for low-end heat extraction and are usually placed in the rear pass in conventional boilers. Lower the desired final exit gas temperature, larger is the backend HS. Careful balancing is required to be done between the

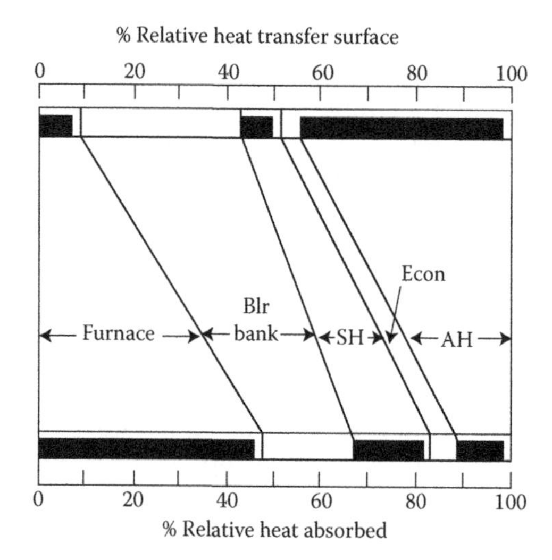

FIGURE 8.4
Reducing effectiveness of heating surfaces with flue gas cooling.

additional cost and fuel savings, to decide the optimum final exit gas temperature.

8.27.3.1 Furnace

Boiler furnace is an enclosure surrounded by fully or partly water cooled furnace walls in which the

a. Fuels are given adequate time for completion of combustion.
b. Gases are cooled adequately for inserting further HSs beyond *furnace exit plane.*

Usually, furnace accommodates firing equipment within its enclosure. Firing equipment is a **grate**, set of **burners** or **bed of combustor**. In many process **Whrbs**, there may not be any combustion and yet a furnace is provided for cooling of gases and removal of dust. Only if the waste gases are clean and cool the furnace is dispensed with and the gases enter the HX tube bundles for dissipation of heat. **Hrsg** behind GT is an example of a boiler without furnace.

In larger boilers, additional HSs in the form of

a. **Division walls**
b. **Platens**
c. **Wing walls**

may have to be introduced in furnace for achieving the desired cooling of flue gas. They can be both saturated or superheated HS.

Except in cases like **CFBC** or **BLR** boilers, where there is a large flow of ash/smelt on the furnace walls, in all other cases, heat to furnace walls is almost entirely by radiation. Flue gas cooling in furnace is a function of the HS provided in the furnace walls, as radiant heat transfer recognises only the flat surface (excluding the curvature of the tubes). The HS of a furnace is always the projected surface.

- *Projected radiant surface* (*PRS*) can be defined as the *furnace HS that participates in cooling of gases.* In addition to the PHS of the enclosure walls, projected surface of the exit plan is also added.
- *Effective PRS (EPRS)* is the *adjusted PRS used for actual calculation* purposes. EPRS takes into account the type of walls (**membrane, tangent, spaced**, etc.), type of fuel, extent of refractory of walls, possible slagging and so on.

Modern furnaces have minimum refractory and maximum water cooling to increase the overall availability of boilers. The aim is to minimise erosion and furnace slagging caused by the presence of refractory. However, refractory cannot entirely be eliminated in any furnace. Certain minimum amount for lining of manholes, hand holes, peep holes, burner throats and so on is inevitable. Further, a proper refractory lining is required in the following cases:

- With staged combustion, as in CFBC boilers, the lower portion of furnace has to be protected against corrosion because of the prevailing reducing conditions. This is achieved by having a layer of castable refractory over closely spaced pin studs, as shown in Figure 8.5. Fin spacing is required to be quite close, as the pins have to transfer heat away from the furnace, besides keeping the refractory layer cool.
- In BLR boilers, furnace tubes are to be protected from the erosive and corrosive forces due to the constant flow of hot molten smelt. Also, the lower part of furnace is under reducing conditions due to staged air flow. This protection is achieved by covering the lower furnace tubes

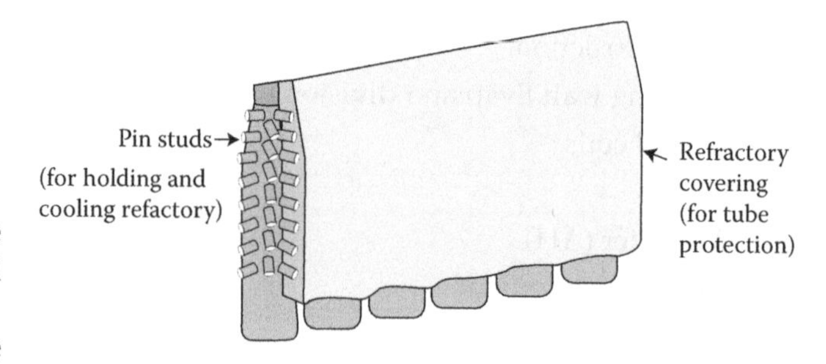

FIGURE 8.5
Refractory layer over pin studs in a furnace wall.

with refractory (usually plastic chrome ore (PCO)) over densely placed pin studs similar to the construction shown in Figure 8.5.

- In package boilers, the floor tubes which are at small inclination have to be protected against overheating due to high heat flux from above, by suitable tile flooring. Steam bubbles formed in the floor tubes, due to intense radiation, tend to coalesce and form a steam blanket inside the tubes leading to their overheating. The floor tubes are to be prevented from seeing the heat by placing refractory tiles on them. See Figures 15.11 and 15.12.
- Furnace walls around SS and BFBC boilers have to be protected, many times, against erosion from the impinging coal particles by **pin studding** and covering with **alumina refractory**.

This aspect is elaborated further under the topic of Furnace Wall Construction.

8.27.3.1.a Furnace Cleaning (see Operation and Maintenance Topics)

8.27.3.1.b Furnace Exit Gas Temperature or Furnace Outlet Temperature

Furnace exit gas temperature/furnace outlet temperature (FEGT/FOT) is the average gas temperature at the exit of furnace plane.

FEGT is a very important temperature in the boiler. It is the starting point for sizing of all further heat transfer surfaces. Higher FEGT is the result of smaller furnace, which, in turn, yields smaller HSs downstream. The reverse happens with lower FEGT. At the same time the risk of **slagging** and **fouling**, **high-temperature corrosion** and overheating of SH/RH tubes escalate with higher FEGT. Correct FEGT is, thus, a balance between compactness of boiler HXs and tube overheating.

The optimum FEGT is usually ~50–100°C below the **IADT** of the fuel under reducing conditions. As a boiler has to deal with several shades of a fuel over its lifetime, experience from similar working installations are usually considered where available, particularly for large coal-fired boilers.

Neither accurate estimation of nor measurement FEGT is not possible because of the ever-changing conditions in boiler furnaces, which has by no means a simple geometry. The size and the number of flames, emissivity, flame temperatures and cleanliness of furnace walls are all constantly changing.

There are proprietary graphs and calculation methods with each boiler maker to arrive at the FEGT. Broadly, it is estimated by subtracting the heat absorbed in furnace from the **HR and A** (H_o). Heat absorbed is a function

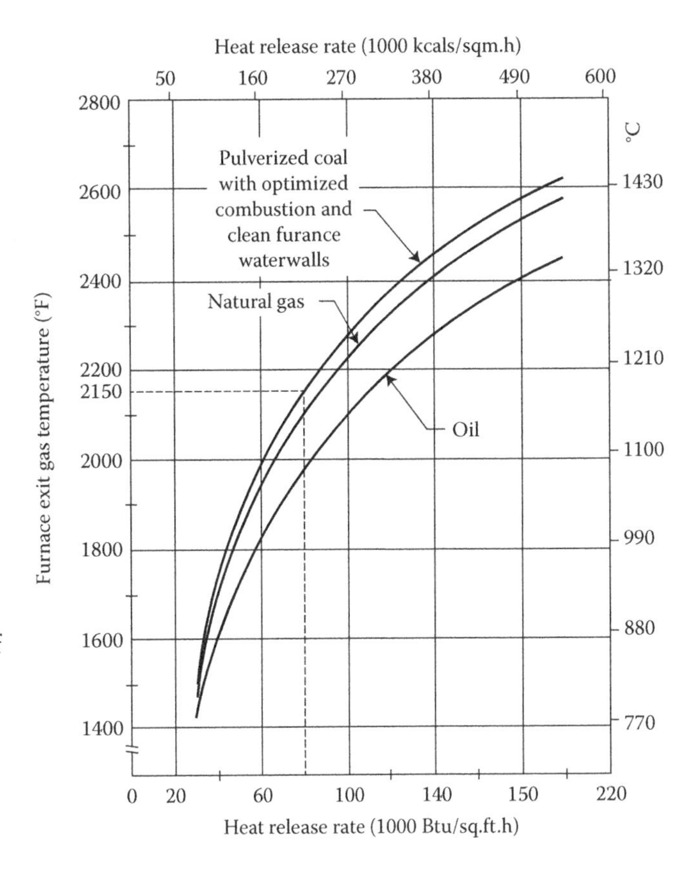

FIGURE 8.6
FEGT in a PF fired boiler with multiple fuels.

of the heat flux. The FEGT so obtained is only accurate to the extent of ±50°C. A wide variation in FEGT from end to end as well as top to bottom of the exit plane is inevitable.

A typical FEGT graph of a PF-fired boiler is given in Figure 8.6 where FEGT with oil and gas firing is also indicated. Lower FEGT with FO is due to the high **emissivity** of oil flame resulting in high **radiant heat transfer** in furnace.

8.27.3.1.c Furnace Pressure

A furnace can be subjected to slight negative or heavy positive pressure depending on whether the boiler is in ***balance draft*** or ***pressurised*** mode of operation. See Draught. In both cases, the furnace is subject to mild fluctuations. Occasionally, the disturbance can even cause furnace puffing.

Furnaces are usually designed to withstand fluctuations of ±200 mm (8″) wg. With large boilers, having sizeable Eco and AH and gas cleaning systems, ID fan suction pressure may be quite significant. In a flameout situation or sudden trip of FD fan, the abrupt termination of fuel and/or air supply can cause a sudden escalation in negative pressure which is called ***implosion***. Maloperation of dampers can also cause a similar phenomenon. The results of an implosion are quite serious as it would lead to an inward collapse of ducts and furnace. See Explosions and Implosions under the topic of Operation and Maintenance for more details. In such situations, the furnace setting is usually designed for fluctuations of ±500 mm wg.

Buckstay and Tie Bar (see also Tie Bars/Tie Channels*)*

The furnace is always strengthened by means of buckstays. Buckstays are binding members on all four sides of a furnace spaced every 2–4 m to provide necessary rigidity to furnace to keep it in shape at all times. They resist both outward and inward bending of furnace tubes due to explosion and implosion forces, respectively. See Figure 8.7.

Buckstays are attached to the furnace wall by means of **tie bars**, which in turn are connected to the wall tubes by tie clips. The arrangement of connecting the adjacent buckstays by means of corner connections provides for free outward expansion of furnace.

Buckstays can be either *floating* or *fixed*. The floating buckstays are fixed to the furnace walls and they move up and down along with expansion and contraction of the furnace. Fixed buckstays are rigidly connected to the structure and allow the furnace tubes to move up and down. Fixed buckstays are normally employed in smaller boilers.

8.27.3.1.d Furnace Tubes

The furnace experiences the highest gas temperature in the whole boiler. Yet the material of construction for most furnaces in sub-c range of boilers is only CS. In HP boilers, sometimes C–Mo (~0.3–0.5%) tubes like SA213 T1 or DIN 17175 15Mo3 are used, but it is to reduce the tube thicknesses.

The high heat transfer rate of boiling water is the main reason why CS tubes can be used in furnaces despite the highest gas temperatures. Figure 8.8 shows the progressive temperature drop from furnace to water inside the tubes. The various resistances to the flow of heat are shown in Figure 8.8. Conductance by water film is significantly more than gas film. Also, the conductance of metal is even higher.

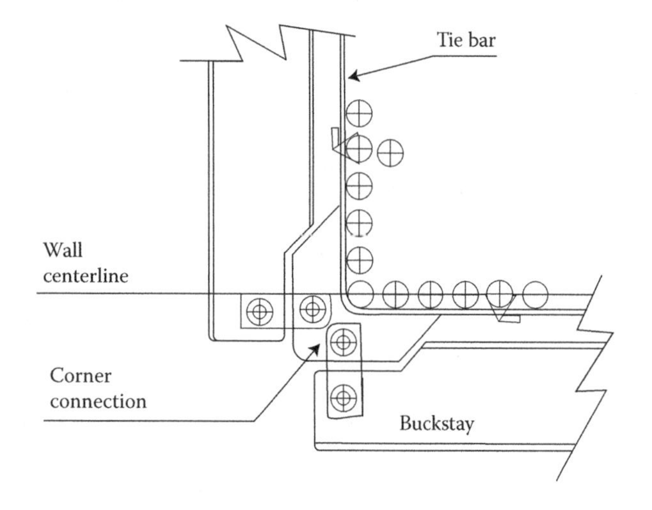

FIGURE 8.7
Floating buckstay assembly.

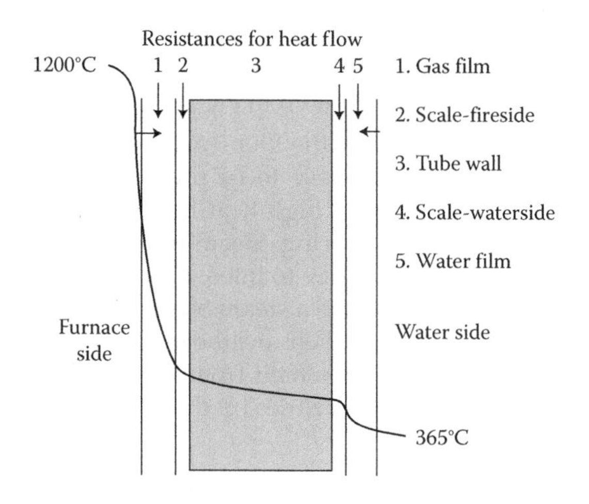

FIGURE 8.8
Resistance to flow of heat form flue gas to water in tube.

In SC boilers, where water directly turns into steam in furnace, Cr–Mo (typical SA 213 T11) instead of CS tubes has to be employed, particularly in the upper regions, as the steam content will be higher with correspondingly lowered cooling effect.

Tube diameters vary from 38.1 to 76.0 mm OD (11/2″ to 3″) with tube thicknesses ranging from 3.25 mm (10 swg) to 6.0 mm and higher. Most boiler codes do not permit <3.25 mm, from strength and durability considerations. Membrane wall furnace tubes are rarely <4.0 mm minimum thickness as the tubes are to be welded with fins.

Package boilers can use smaller-diameter tubes of 38.1 and 50.8 mm (11/2″ to 2″) od as the furnaces are short. Tall boilers generally tend to adopt 63.5 and 76.2 mm (21/2″ to 3″) od tubes. But this is flexible. With assisted and forced circulation, tube sizes are generally lower by a step.

The fin thickness in membrane wall is usually 6 mm. In large PF boiler furnaces, fins of 8 mm may be used due to high heat flux.

From a heat transfer point of view, tube diameter is of no concern. Larger tube is better only from circulation and strength considerations. Also less number of tubes are required. But tube thickness is more. Smaller tube is more flexible and thinner but more tubes are needed. On the whole, small tubes tend to yield less overall weight.

In areas of high heat flux particularly in large HP boilers, such as around the burner zone, ribbed/rifled tubes may have to be used to prevent DNB.

8.27.3.1.e Furnace Size or Volume (see also Heat Release Rates)

Furnace volume is the size of furnace in cum or cft. It is measured from the top of the grate, fluidised bed, ash pit or the floor to the exit plane where the next surface namely, the SH or screen starts.

Residence Time

Furnace volume provides the *residence time* for the fuel particle to burn out. Residence time is the *notional time taken by the fuel from its point of entry into furnace till its point of exit of the furnace.* The point of entry may be top of the grate, top of air nozzles in fluid bed or the mid-point of the burner pattern. The point of exit is the mid-point of the furnace exit plane.

Residence time is calculated by dividing the effective furnace volume (from the point of heat input to the centre of exit plane) by the gas volume, assuming the gases to be uniformly at FEGT. *Residence time should exceed the time required for the fuel particle to burn out,* which depends mainly on

- The size of the fuel particle
- Turbulence in firing equipment and furnace
- The temperature inside the furnace
- The amount of char in fuel

Solid fuels naturally require more residence time than liquid fuels , which in turn need more than gaseous fuels. Table 8.1 gives the range of residence times in seconds for various fuels in different combustion equipment.

Furnace Sizing

Furnace should be large enough to

- Complete combustion
- Generate adequate steam
- Provide proper furnace wall heat absorption with no overheating at any place
- Prevent slag formation

For PF furnaces, the slagging consideration of fuel ash demands a larger furnace, which may result in relatively low furnacewall absorption rates. But heat flux and absorption around the burner zone will be naturally very high.

TABLE 8.1

Furnace Residence Times in Seconds for Various Fuels

	Coals	Agro-Fuels	FOs	NG
Stoker	2.2–3.0	2.5–3.5	–	–
Burner; PF/oil, gas	1–2	–	<1	<1
BFBC	2.5–3.0	–	–	–
CFBC	4–5.5	–	–	–

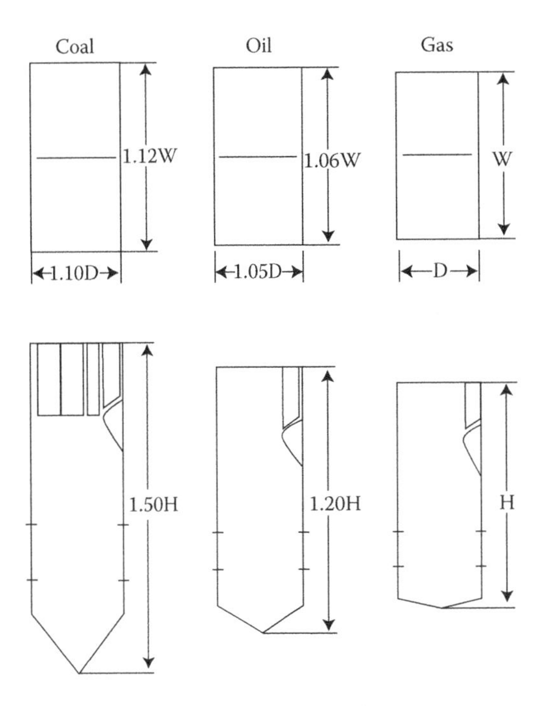

FIGURE 8.9
Comparative furnace sizes for coal, oil and gas.

In FO firing combustion is rapid, radiation is intense and localised heat absorption high within the active burning zone. Furnace can therefore be much smaller. It is usually made a little larger than what is required for minimum residence time in order to maintain the desired FEGT and metal temperatures. This is particularly true for package boilers.

With NG firing, the furnace can be even smaller. Unlike in oil firing, hot spots are not present and the heat absorption is more uniform mainly due to lower emissivity of flames.

Typical comparative sizes of furnace are given in Figure 8.9 for the three prime fuels.

8.27.3.1.f Furnace Wall Construction

There are many types of wall constructions in furnaces of modern boilers having evolved over several decades, starting from simple uncooled brick walls in the bottom supported boilers in the small olden boilers. These constructions are

- Membrane wall
- Tangent tube
- Flat stud and refractory backed
- Spaced tube
- Tube and tile
- Pin studded and refractory covered

In the construction of most furnaces, a combination of wall constructions is needed. While it is the membrane

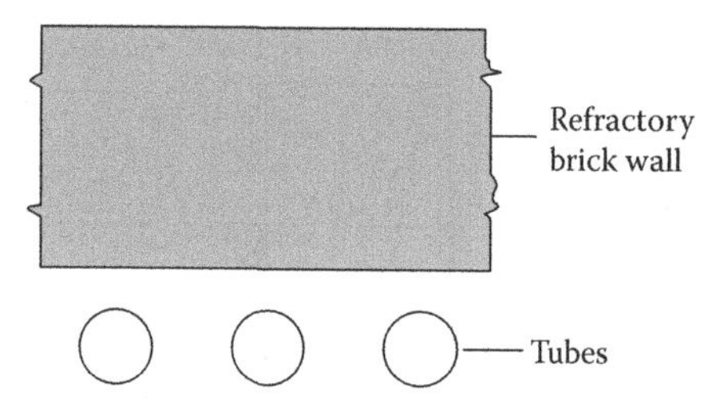

FIGURE 8.10
Early efforts at water cooling of brick-walled furnace.

panel which predominates in most parts in modern boilers, the other constructions also have their place.

The first effort at water cooling of furnace wall was simply to place furnace tubes in front of the bricks to reduce the incidence of heat to protract the life of bricks. As understanding of circulation and WT improved, the water cooling of walls increased. By the 1950s, full water cooling with tangent walls had become common in large boilers with big improvement in the availability of furnaces. Figure 8.10 shows such an early effort at water cooling of the furnace wall where a thick brick wall has a few tubes placed separately in its front. Brick walls in the early boilers had to carry the load of the boiler besides withstanding the heat.

a. *Membrane or mono wall* (also described under PPs) is a set of tubes welded to each other with membrane bars/strips between them. The bars vary from 6 to 60 mm in width depending upon the heat flux. 6 mm thick strips are usually adequate for most duties, demanding 8 mm thickness only for high heat flux applications, like PF boiler furnaces. Tubes vary in size from 38.1 to 76.2 mm (11/2″ OD to 3″ OD).

   Figure 8.11 shows the cross section of a membrane wall along with insulation and casing. See also Figure 16.3 for a cross-sectional view of the membrane panel.

   Exceptions apart, furnaces of modern times are almost entirely made of membrane wall construction because of the following advantages:

   - There are practically no air or gas leakages, contributing to slightly improved boiler η.
   - Insulation and lagging on the external walls is simpler and also easier to apply.

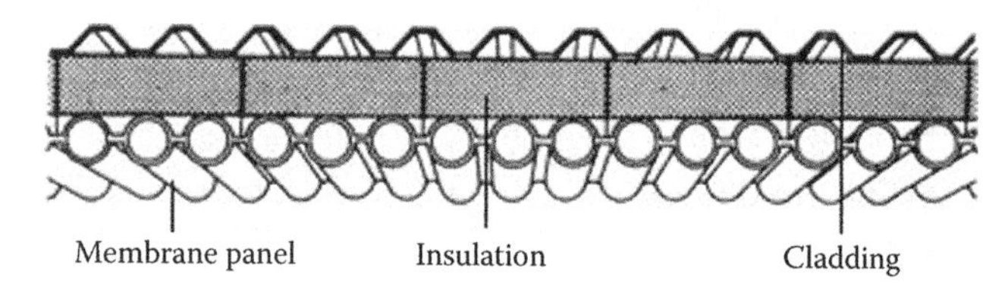

FIGURE 8.11
Membrane or mono wall membrance.

   - Erection is much easier and faster as the walls are substantially built in manufacturing shops.
   - Tube replacement in a running boiler is much easier and faster.
   - Fit and finish of shop made panels is far superior.
   - Capacity to withstand furnace puffs is much better.
   - There are fewer tubes for the same furnace area due to the membrane bars between tubes.
   - The tubes are usually smaller in diameter and therefore thinner.

   Large and efficient boilers are practically inconceivable without membrane wall furnaces.

b. Tangent walls (also described under PPs) are constructed with individual tubes lifted into their position during erection. Less than 1 mm gap is provided between tubes to account for tube tolerance on its od. On the outside, many times a veneer casing of about 3 mm thickness is welded for the alignment of tubes. Insulation and lagging is built on veneer casings. This type of construction, which was popular in the 1950s and the 1960s before the advent of membrane wall manufacturing facilities, was the first step in creating fully water-cooled furnaces. In terms of efficacy of water cooling, both the membrane wall and tangent well constructions rank the same. But in practice, the tangent tube construction suffered from many disadvantages namely

   - The entire tangent wall was to be painstakingly built tube by tube during the erection stage, making the construction very long and laborious.
   - Minor air and gas leakages were inevitable, contributing to minor loss of η.
   - Insulation was relatively more difficult to apply.
   - Repair of the furnace tube was quite complicated and time consuming. The affected tube has to be jacked out of position for cutting and welding which is quite difficult.
   - Capacity to withstand furnace puffs is not as good as membrane panels.
   - For the same area, the number of tubes in tangent construction is more.
   - With only tubes to fill all the space, the tube diameters are larger with proportionately thicker walls. The overall PP cost for walls is higher.

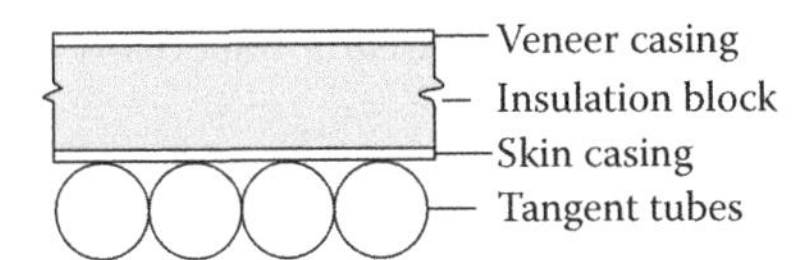

FIGURE 8.12
Tangent tube wall.

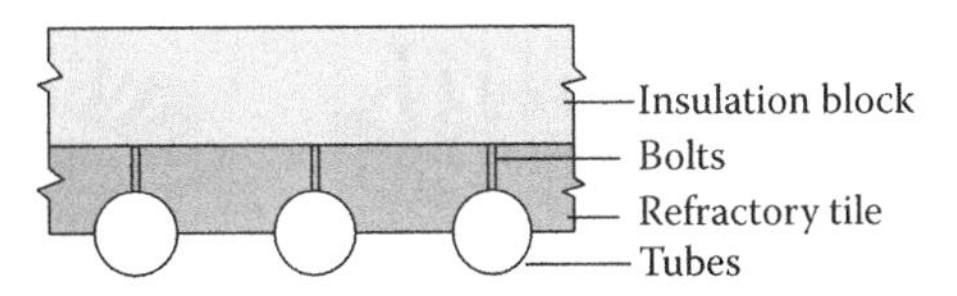

FIGURE 8.14
Tube and tile wall construction.

See Figure 8.12 for details of tangent tube wall construction. To protect the skin casing from corrosion, the void between the casing and tangent tube is filled with castable refractory in the case of oil firing. See also Figure 16.4 for an isometric view of furnace corner with tangent wall construction showing the tie bar and buckstay.

c. *Flat studded and refractory backed tube* construction shown in Figure 8.13 is employed whenever alternate tubes in membrane or tangent walls are needed to be removed, mainly for admitting tubes from SH, Eco and other banks. The main furnace is rarely built with this construction as the flat studs may get overheated and destroy the integrity of the wall. Typically roof, rear wall casing and walls in the second pass are some of the areas where this construction is used.

d. *Spaced tube* construction was the early effort to introduce water cooling in furnace walls to reduce refractory maintenance and increase the firing rate. Several types of spaced tube constructions were developed of which *tube and tile construction* was the most enduring, as thin tiles could be bolted to the tubes on the outside. The load of the tile and insulation could be easily taken by the furnace tube.

   The spaced tube construction represented a significant improvement over the uncooled refractory wall even though problems like air and gas leakages, differential expansion between refractory and tube, significant maintenance, tendency for furnace fouling and so on still continued. This construction is still favoured, particularly in many Whrbs, where re-radiation of heat from furnace is required for combustion stability. For very high M fuels also, it is needed for the same reason. The other incidental advantage realised in many Whrbs, which usually employ LP or MP steaming conditions, is the opportunity to use thinner tubes like 3.25 mm (10 swg) which is not admissible in membrane walls, which requires ~4 mm from welding considerations. See Figures 8.14 and 8.15 for details of typical tube and tile and spaced tube constructions of walls.

FIGURE 8.15
Spaced tube construction of walls.

e. Pin *studded and refractory covered walls* are needed when sub-stoichiometric conditions prevail in the furnace, The refractory, held in place over tubes by the closely spaced pin studs, shields the tubes from reducing conditions and also takes the brunt of erosive forces. Such construction is adopted in CFBC, BFBC and BLR boilers in the lower part of furnace. Pin studs help to cool the refractory layer and also transfer heat to tubes. Figure 8.5 depicts the isometric view of such construction.

Figure 8.16 shows a typical tube with pin studding on the furnace side and further covered with castable refractory for their protection. Flat studs are also seen in Figure 8.16. Figure 8.17 shows the cross section of the individual tube.

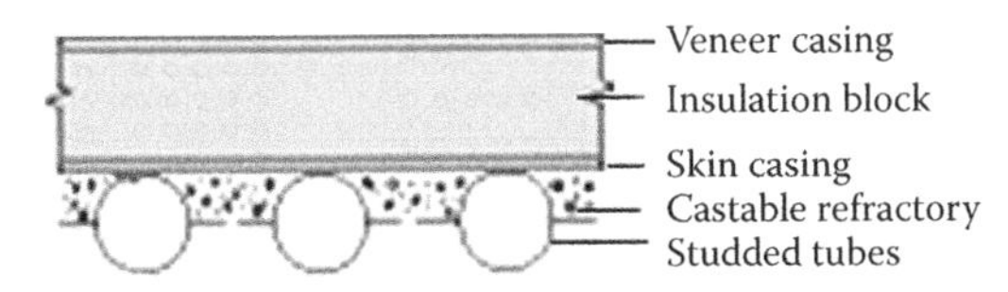

FIGURE 8.13
Flat studded wall.

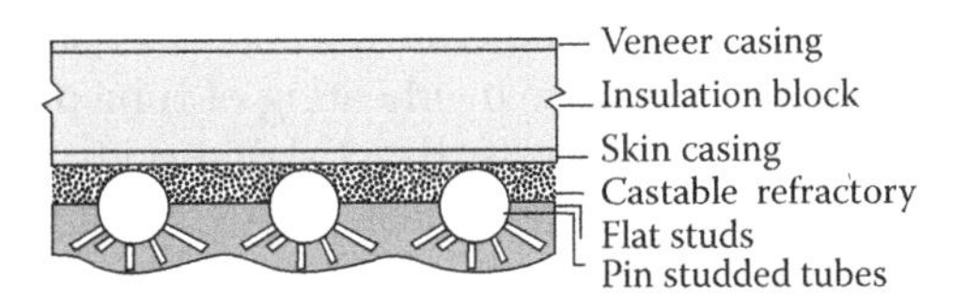

FIGURE 8.16
Pin studded tube wall.

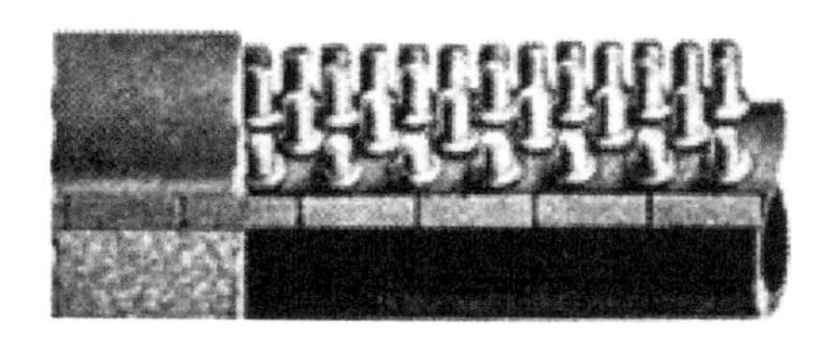

FIGURE 8.17
Cross section of pin studded tube covered with refractory.

8.27.3.2 Superheater

SH is the *HS which imparts superheat to the saturated steam*, which is separated and purified in the steam drum of sub-c boiler. In SC boilers however fw turns into steam progressively in the furnace. Since high steam temperatures are to be attained in SH, these banks are placed in conventional two-pass boilers

- Either at the top of the furnace as widely spaced radiant surfaces in the form of **wing wall** or **platen** SH or
- As closely spaced convection **pendant** or horizontal surfaces downstream of the furnace exit plane extending often into the next pass

In **Hrsgs**, several **Whrbs** and many field-erected single drum type oil/gas-fired boilers, there is no radiant but only convection SH.

SHs in **FBC** boilers are located in unconventional areas.

- In BFBC boilers, they are placed as the **bed coil SHs** as final SH. See Bed Coil in BFBC boilers under the topic of Bubbling Fluidised Bed Combustion for more details.
- In CFBC boilers, part of SHs can be placed as **FBHEs** in the return ash circuit of cyclones or for cooling the cyclones with steam. See Circulating Fluidised Bed Combustion for more details.

Higher the temperature zone that SH occupies, lower is the HS required but needs superior metallurgy usually and vice versa. Compactness and higher-quality metals are to be suitably balanced to keep the overall costs low.

SH sizing and disposition involve difficult compromises and need intelligent selection from among divergent parameters. Under-surfacing reduces the cycle η (due to low steam temperature) and over-surfacing reduces the SH life (due to overheating of tube material) and increases wasteful costs. Exact sizing with all fuels under all loads is not an easy task, particularly for high steam temperatures.

Broadly, there can be three considerations in SH design:

1. Disposition of surfaces
2. Selection of tube materials based on the metal temperature
3. Δp of steam and gas

1. Disposition of SH surfaces

The three most important parameters for HS disposition are the gas side

- **Ash erosion**
- **Ash corrosion**
- **Slagging and fouling**

Gas side erosion determines the spacing of the tubes across the boiler width perpendicular to the gas flow. Erosion is mainly a function of the

- Amount of ash in fuel
- Its abrasiveness
- Flue gas velocity[3.5]

For a given FG, side erosion is controlled by limiting the gas velocities, as given in Table 7.2 in the topic of Gas Velocities. Also, the side and back spacing of pitch across and along the gas flow is important from the point of ash buildup.

For the same limit of gas velocity, the tube pitching can become narrower as the gases cool down which is shown in Figure 8.18. This is because as ash gets cooled it progressively becomes solidified and loses its ability to build-up.

High temperature corrosion of SHs is a serious matter where high gas and tube temperatures coincide. Low-melting alkali metals (Na and K) and V, besides S and Cl_2 are serious offenders. Placing SH inside the furnace, locating SBs suitably and choosing appropriate metallurgy for tubes is all governed by the corrosive constituents of the fuel.

Typical spacing in mm
600 300 150 100
Platens Sec. SH Pry. SH Boiler bank

FIGURE 8.18
Narrowing lateral tube spacing with cooling flue gases.

Slagging and fouling of SHs is the *progressive buildup of semi-molten ash deposits*. Molten ash deposits in furnace (slagging) and crystalline deposits in SH (fouling) are primarily caused by low-melting ash in fuel. Furnace sizing has to be large enough to cool the flue gases to at least <50°C of **IADT** under reducing conditions. Thereafter, the spacing of all SH panels should be wider than normal to prevent ash build-up. This will mean lower heat transfer and higher HS and naturally higher cost.

2. Selection of tube materials
Steam in SH and RH actually works like its coolant. Higher the surrounding flue gas temperature, higher should be the steam mass velocity to keep the tube metals as close to the steam temperature as possible. Table 8.2 provides guidelines for steam mass flow velocities.

Based on the estimated tube metal temperature, after taking into account the transfer of heat due to radiation and convection, the material for the tube is selected based on the guidelines in Table 8.3. Boiler codes provide exact permissible limits.

3. Pressure drops (Δp) of steam and gas
Δp in SH is parasitic in nature and adds to pumping pressure of BFP. While it must be restricted to the minimum, care should be taken to see that there is no flow stagnation or flow reversal in SH coils at low loads causing a possible tube overheating. Usually, between the two consecutive headers, Δp of 1.5–2.0 bar has to be maintained. The approximate Δp of steam in the total SH is 8–10% of the SOP.

TABLE 8.2

Norms for Mass Flow Velocities of Steam in SH and RH Tubes in Convection Zone

Steam Temperature °C		Range of Mass Velocity in Convective Banks	
	°F	kg/m² s	lbs/ft² h
370–425	700–800	330–460	250,000–35,0000
425–480	800–900	530–660	400,000–500,000
480–540	900–1000	660–800	500,000–600,000
540–590	1000–1100	950 and above	700,000 and above

TABLE 8.3

Tube Materials to Suit Various Wall Temperatures

S.No.	Material	Specification	Limiting Mean Tube Wall Temperature (°C)
1	CS	SA178/192 210A1	455–480
2	Mo steel	SA213 T1	510
3	Cr–Mo steel	SA213 T11	565–595
4	Cr–Mo steel	SA213 T22	575 595
5	High Cr steel	SA213 T91	595
6	SS	SA213 TP-304H	~650

Likewise, there is a need to minimise the draught loss across the SH to reduce the auxiliary fan power. Adopting inline tubes and reducing gas velocities are the means usually employed.

Gas Δp is usually not a concern in SH as steam Δp as gas drop is normally quite small.

8.27.3.2.a SH Classification

There are several types of SHs and they can be classified by way of

1. Design
   A. Parallel, counter and mixed flow
   B. Radiant, convective, combination
   C. Single and multi-pass
2. Construction
   A. Vertical, horizontal and combination
   B. Pendant and platen
   C. Drainable and non-drainable
   D. In-line and staggered
   E. Plain tube and finned tube
   F. Single and multi-tubes per loop

To describe an SH adequately, both design and construction description need to be indicated. Figure 8.19 plots the water and steam path in a large two-pass boiler which also helps to understand various types of SH arrangements.

(1-A) *Parallel, counter and mixed flow*
Depending upon the direction of flow of steam and gas in SH, a bank can be in parallel, counter or mixed flow arrangements based on whether the flows are in

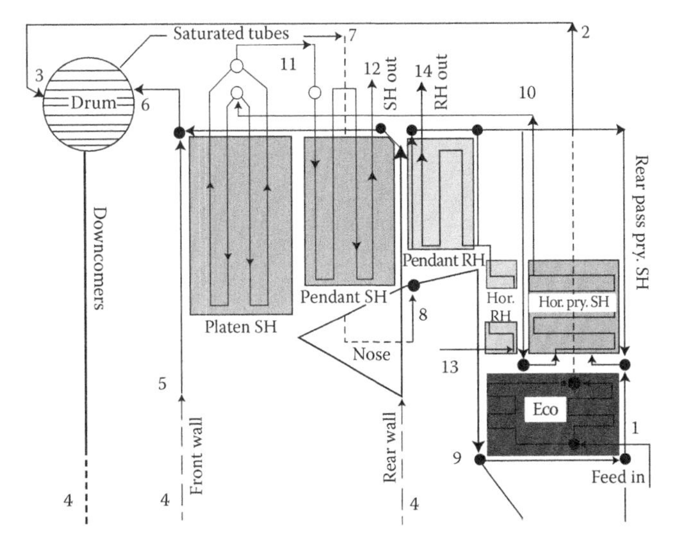

FIGURE 8.19
SH and RH nomenclature and location in a two-pass utility boiler.

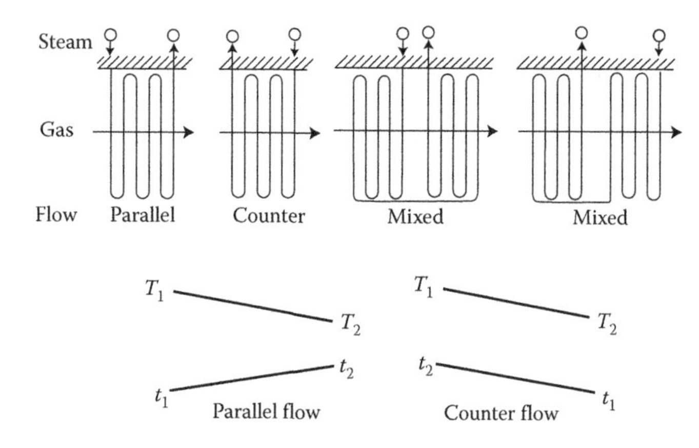

FIGURE 8.20
SH and RH classification based on flow arrangements.

the same, opposite or mixed directions (as shown in Figure 8.20).

Figure 8.21 shows a typical mixed flow arrangement of a convective SH.

Counter flow arrangement is the most effective and yields the most compact tube bank, as the **LMTD** is the highest. Tube metal temperature will also be the highest at the outlet of the SH, as the highest gas and steam temperatures coincide.

If the metal temperatures have to be contained, a parallel flow arrangement is better, as the coolest gas meets the hottest steam at the steam exit, but the HS would be more.

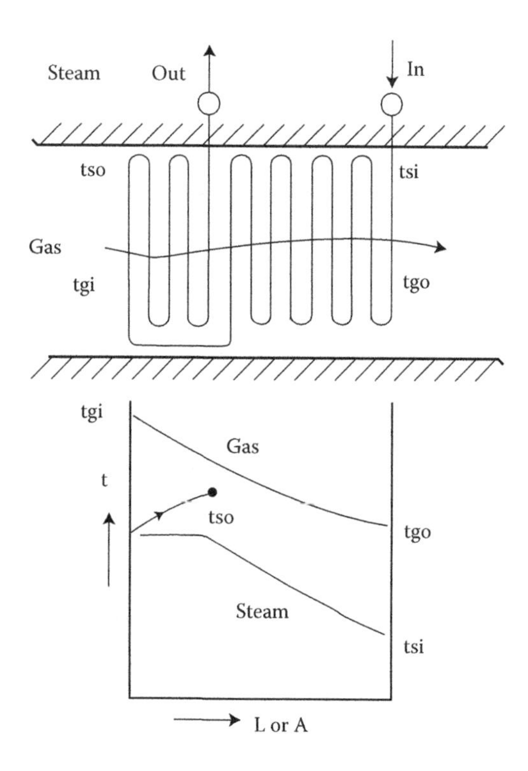

FIGURE 8.21
Mixed flow arrangement in SH and RH.

In a mixed flow arrangement, the benefits of both accrue. In most SHs, mixed flow arrangement is followed.

In most high temperature SHs with final steam >500°C, it is normal to place the finishing SH in parallel flow or mixed flow. Platen SHs are also almost always in parallel flow. RH is in counter flow in the first part and mixed flow in the second. The other SH banks are mostly on counter flow.

(1-B) *Radiant, convection SHs and their combination*
When heat is received by SH tube bank substantially from furnace radiation, it is radiant SH and when it is from the hot gases by convection, it is convection SH.

Radiant SHs are widely spaced and placed in furnace, either hung from the roof as *vertical platens* or supported from the front panel as in *wing wall SHs*. They can be *horizontal platens* also, supported from both walls or suspended from furnace roof.

Convective SHs are placed downstream of the furnace exit plane on closer pitch, which progressively narrows as the gases cool down.

Heat transfer in radiant SH is proportional to $T_1^4 - T_2^4$ where T_1 and T_2 are the absolute temperatures of the heat source and SH tube, respectively. At lower boiler loads, both the temperatures do not reduce proportionally to the steam flow. Furnace temperature reduces less than the steam temperature. Therefore, for unit amount of steam, heat absorbed at part loads with lower gas flows, is in fact higher, resulting in an increased steam temperature.

On the contrary, at lower boiler loads, the steam temperatures reduce in convective SHs somewhat sympathetically, as the heat absorbed is proportional to the gas mass flow$^{0.66}$.

The two SHs, thus, display a diametrically opposite behaviour with load, as shown in Figure 8.22. Together, they tend to give a fairly flat characteristic, which is most desirable as it reduces the attemperation and the need for superior metallurgy.

Large SHs of modern boilers are unthinkable without a combination of radiant and convective SHs. Figure 8.23 shows the arrangement of convective and radiant SHs in an SS-fired boiler.

(1-C) *Single-pass and multi-pass SHs*
Certain minimum steam mass flow or mass velocity is essential, as given in Table 8.2 to promptly remove the heat incident on the tubes and keep them cool. In most boilers, this is realised in a tube bank spread across the width of the entire boiler by choosing appropriate tube spacing and diameter. This is called *single steam pass* arrangement.

However, with low Gcv and slow burning fuels, such as bio-fuels, boiler width is much larger for relatively modest amount of steam generated. To fulfil the steam mass flow requirements, the steam may have to be admitted in 1/2 or 1/3rd the number of tubes. Accordingly,

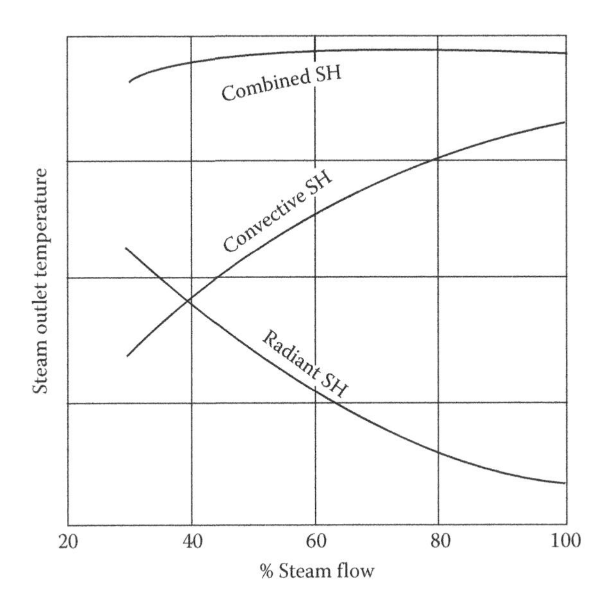

FIGURE 8.22
Steam temperature variation with load in convective and radiant SHs.

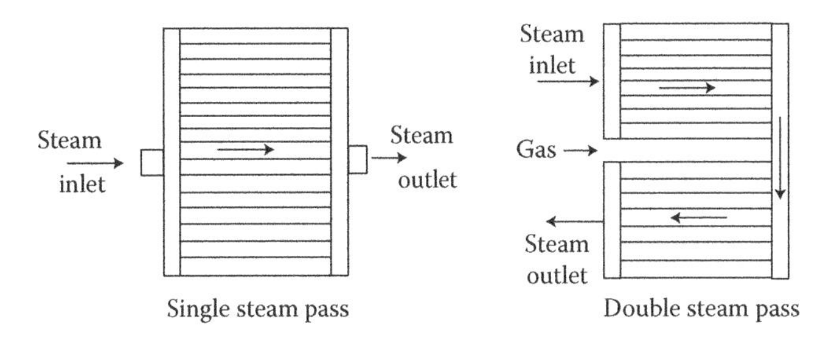

FIGURE 8.24
Single- and two-pass steam flow arrangements.

the entire SH tube bank has to be divided into two or three steam passes, with steam going back and forth. This is called *multi-pass steam flow* arrangement.

In Figure 8.24, single and double steam pass arrangements have been portrayed. In a two-pass arrangement, steam inlet and outlet are at the same elevation in the boiler but at opposite ends. It is possible to have a single header common for both inlet and outlet with a diaphragm in the middle to separate the flows.

(2-A) *Vertical, horizontal and mixed arrangement*
Depending on the boiler geometry, the SH/RH tubes can be arranged vertically or horizontally or in combination. Vertical SH tubes are easy to suspend from the roof and the tubes are free to expand downwards. The disadvantage is that the steam condenses in tubes on cooling of boiler and while restarting significantly slows down the start-up. This is because the condensed water has to gently go out of the tubes without causing vapour lock needing controlled heating and consequently slower start-up.

Horizontal SHs/RHs are fully drainable and therefore amenable to quick starting. Supporting horizontal tubes is relatively more difficult as they have to be stacked up one over the other or they need vertical tubes for support. Typical horizontal coil and the supporting arrangement is shown in Figure 8.25. Here, the weight of the coil is transferred to the side wall tubes by means of lugs. Several other support designs are also possible.

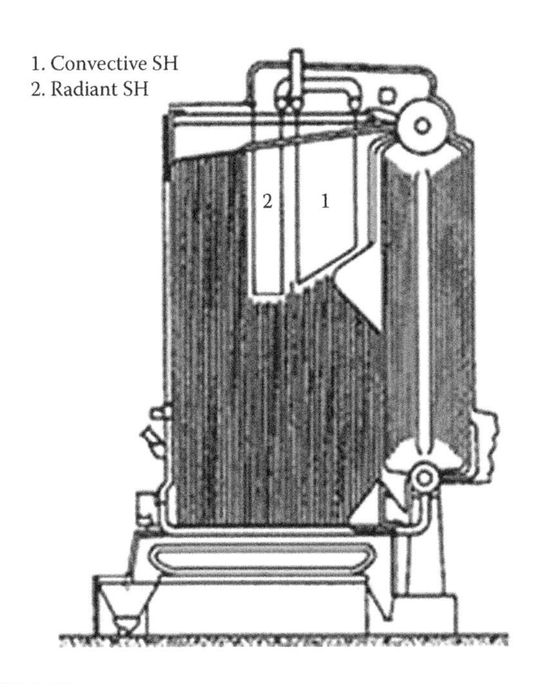

FIGURE 8.23
Radiant and convective SH coils in a stoker-fired boiler.

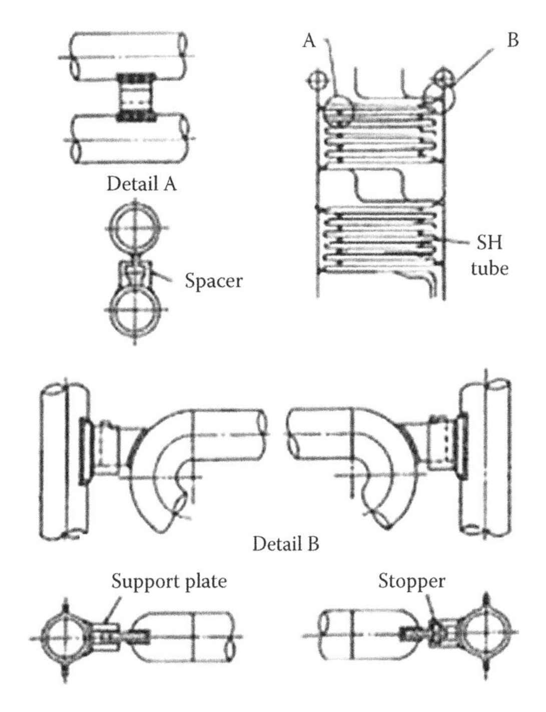

FIGURE 8.25
Horizontal SH and RH coils and supports.

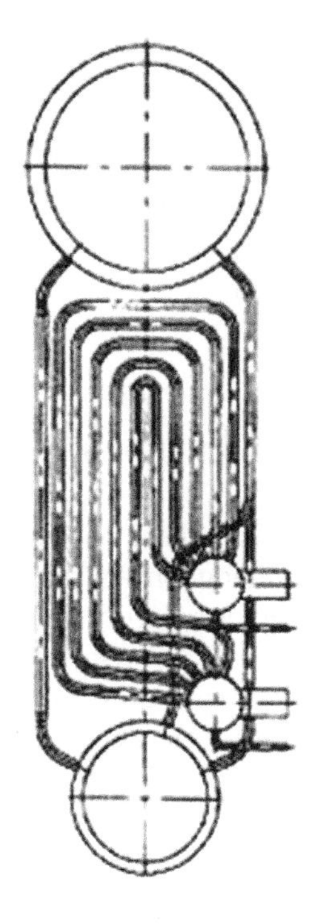

FIGURE 8.26
Vertical inverted loop drainable SH in a package boiler.

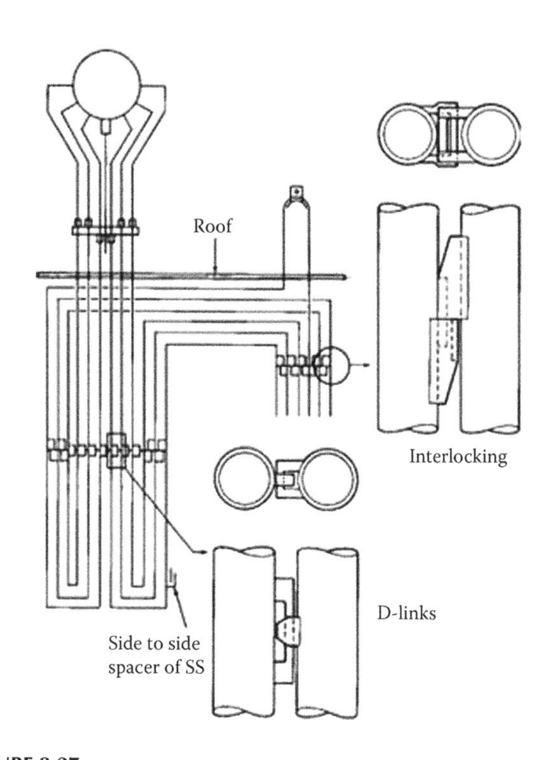

FIGURE 8.27
Typical vertical pendant SH in a large utility boiler. (From The Babcock and Wilcox Company, USA. With permission.)

Most of the times, with gas flow going up and then down in most boilers vertical and horizontal SHs are combined. Sometimes, the boiler geometry does not permit such arrangement. Tower-type boilers, for example, can have only horizontal banks. In some package boilers where gas flow is horizontal, such combination is hardly possible; it is mostly vertical SHs (as shown in Figure 8.26).

(2-B) *Pendant and Platen SHs*
Pendant SH as the name suggests is a loop-type SH. In vertical arrangement, this looks like a pendant and hence the name. However, horizontal pendants are also possible, as shown in Figure 8.25. Unlike in the platen SH, where the adjacent tubes touch each other to form a plate to facilitate easy flow of molten ash, in the pendant SH the adjacent tubes have suitable gap. This is because a pendant SH is not placed in furnace but in a convection-pass, where it receives heat by convection and intertube radiation. Certain minimum gap, typically a clear 20 mm between tube to tube on the outside, should exist along the gas flow for the development of a boundary layer. Spacing of the pendants across the gas flow is dependant on the permissible gas velocities and the tendency of ash to bridge. See Gas Velocities.

A typical vertical pendant SH of a large utility boiler is shown in Figure 8.27. Spacers for alignment of tubes can be seen here. The purpose of the spacers is to keep the adjacent tubes of the pendant in line along the gas flow. There is no load being transferred in most cases through the tube spacers as the pendant is hung from the top. For keeping the coils equidistant from each other perpendicular to the gas flow, lateral spacers are employed. As cooling is a problem in lateral spacers, usually a thin wrap around steam tube is used for this task.

Platen and wing wall SHs, placed in the upper part of the furnace at the end of combustion zone, receive heat entirely by radiation. Heat flux, being placed in furnace, is quite high. The spacing has to be quite wide apart, usually not <400 mm in coal- and lignite-fired boilers, for the platens to 'see' the flame. In large utility boilers, it is not uncommon to have platens spaced even 2 m apart. Very high steam mass velocities, typically around 1350 kg/m^2 s (1×10^6 lbs/ft^2 h) are employed to cool the tubes and contain the metal temperatures.

Platens are prone to accumulate slagging deposits and they require heavy steam blows from SBs to dislodge the deposits and keep them clean. Figure 18.15 shows the action of retractable SB on a vertical platen. Platen SHs can be placed vertically as in two-pass boilers or horizontally as in tower-type boilers. See Figure 16.9 which shows both types of platens. Typical wing wall SH is similar to the wing wall Evap described later.

(2-C) Drainable and non-drainable SHs
Drainable SH tube banks are always preferred from the point of quick start-up and corrosionfree layup. Horizontal arrangement of SH lends itself to draining as seen in Figure 8.25. Tower-type boilers, with all surfaces

fully drainable due to horizontal disposition, have faster dynamics.

Vertical tubes have to be inverted to make them drainable, as shown in Figure 8.26. They are normally employed in small package and industrial boilers. Relatively small SH can be obtained from them and for large SH non-drainable vertical SHs are inevitable.

(2-D) In-line and staggered arrangement of tubes

Staggered arrangement improves the heat transfer and reduces the HS requirement. With clean fuels like NG and light oil, staggered arrangement is workable. But for most fuels, in-line arrangement is needed to prevent ash buildup, even though it demands more HS.

Figure 8.28 shows both the tube arrangements.

(2-E) Plain tube and finned tube (see also Extended Surfaces)

SHs are normally made with plain tubes and SBs provided for removal of ash deposits. **Extended surface** tubes are only employed for clean fuels where SBs are not required. Finned tube SHs are most common in Hrsgs. They are also found in package boilers firing clean fuels.

(2-F) Single-tube and multi-tube per loop

To attain the recommended steam mass velocity from tube metal cooling considerations, tubes may be arranged in single or multiple numbers in a loop (as shown in Figure 8.29). Multiple tubes per loop are often needed in highly rated boilers as steam generation over unit width of boiler is quite high. Particularly at high temperatures, where the pendants and platens are required to be spaced apart from fouling requirements, the necessary area to steam can be provided only with multiple tubes per loop. A good example is the pendant SH depicted in Figure 8.25 where there are six tubes per loop. A single-tube SH would have excessive pressure drop due to high mass flow, which is eased only by multi-tube arrangement.

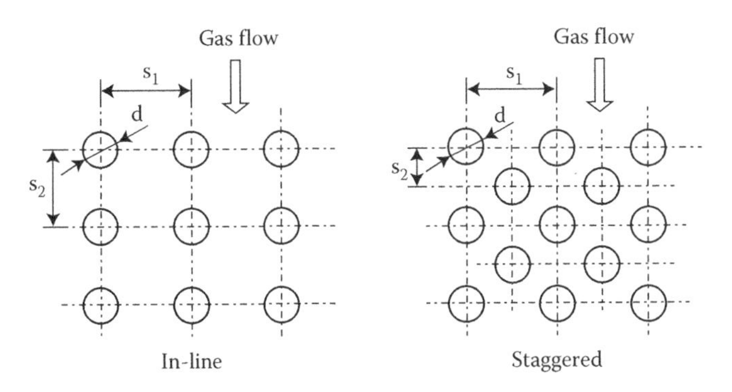

FIGURE 8.28
In-line and staggered arrangement of tubes in a tube bank/bundle.

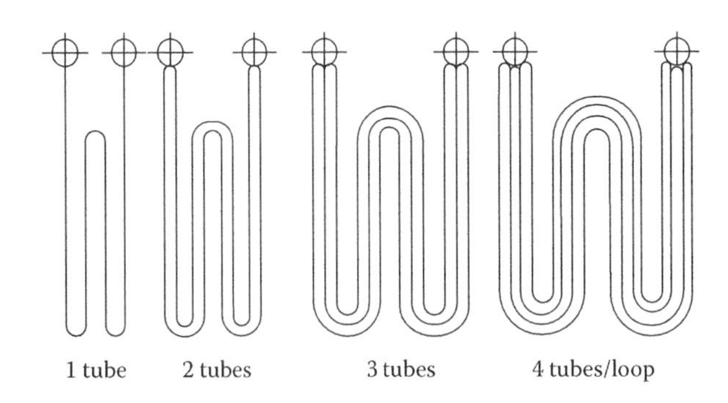

FIGURE 8.29
Single and multiple tubes per loop.

8.27.3.3 Reheater

The design and construction principles of RH are not much different from those of SH. RH heat duty is naturally much smaller than SH duty.

- HS is much smaller in comparison, as RH receives steam from ST with reasonable SH.
- Steam pressure drop in RH has to be restricted to a minimum to maintain high cycle efficiency. For that
  - Tubes are generally of larger diameter, ~50.8 mm OD or more.
  - RH is almost always accommodated in a single-tube bank, with some occasional exceptions both from pressure drop considerations.
- As high temperature has to be attained at the outlet, RH bank is placed at the furnace outlet in high gas temperature zone.
- RH is always made fully or substantially drainable.
- As steam pressure is lower, tube thicknesses are also considerably lower.
- Tube materials are always of high metallurgy because of the highest temperatures to be attained.
- In many instances, RHOT is kept ~20–30°C higher than SOT to take advantage of lower RH pressure to gain high cycle η.

RH is rarely employed at ST sizes <100 MWe and SOP <100 bar. This is because the additional cost of ST and complete RH arrangement (consisting of the cold and hot RH pipes between boiler and ST, RH assembly inside the boiler, valves and mountings, attemperator assembly and gas control arrangement) add up to a tidy capital investment. Higher cycle η has to balance this higher cost and it happens only for sizes >~100 MWe. Technically, there is no limitation for having RH even at lower sizes.

In larger Hrsgs like frame 9FA or equivalent RH is possible to be built, as the TG size is ~120 MWe. The exhaust gas temperature of GT is also >600°C permitting both SH and RH to attain temperatures as high as 570°C.

RH pressures are usually ~1/3 to 1/4 of the SOP. In conventional boilers, the RH flow is ~2% lower than main steam flow due to the leakage in the HP casing of ST. In the case of Hrsgs, however, RH flow is in fact greater than main steam flow, as IP steam is generated at the same pressure as RH and added to RH flow for further heating.

8.27.3.4 Boiler/Convection/Evaporation Surface

Evaporation surfaces in boilers are distributed mainly among the following:

1. **Furnace**
2. **BB**
3. **Division wall**
4. **Wing wall**
5. Bed coils in **BFBC** boilers
6. Coils of **FBHE** in certain CFBC boilers
7. **Cyclone** walls of **CFBC** boilers where they are water cooled

BB and furnace are the two main evaporating surfaces in a conventional boiler. For balancing of surface and volume requirements division walls are often required to be inserted in furnace. In single-drum boilers, at times, wing wall evaporators are provided for the same effect. Almost all BFBC boilers firing high Gcv solids employ inbed Evap coils for bed temperature control. In hot cyclone CFBC boilers where FBHE are given for steam temperature control, some Evap coils are also placed in the FBHE bed. Also, in many CFB boilers, hot cyclones are cooled by water taken from Evap circuit.

Boiling water, at saturation temperature corresponding to the drum pressure, circulates in all these surfaces. Heat transfer in furnace, division walls and wing walls is almost entirely by radiation while in BB it is nearly all by convection. Even though the water circulating in both is at the same temperature, the tube metal temperatures are some 10–20°C higher in furnace as they are exposed to direct combustion with higher heat flux. In the bed coils, heat transfer is substantially by convection from bed solids. In cyclone walls, it is by the conduction of the gas facing refractory layer.

8.27.3.4.a Boiler Bank, Convection Bank or Evaporator Bank

In the older boilers, working at lower ps & ts with relatively good solid fuels, it was not uncommon to have two, three or even four boiler banks in a single boiler. See a typical five-drum four-bank boiler in Figure 8.30.

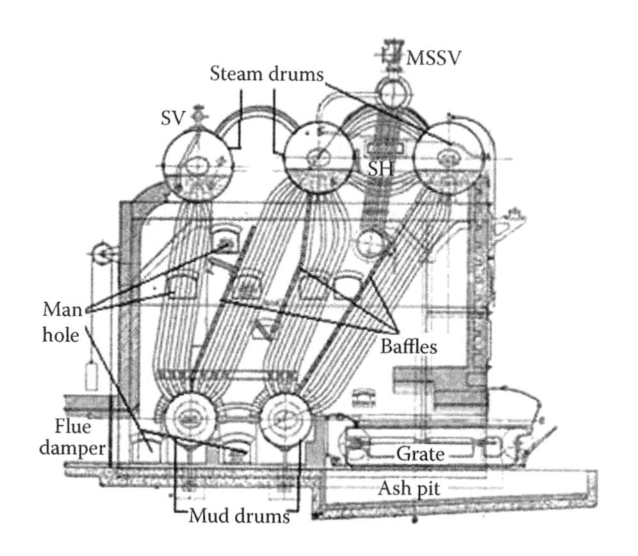

FIGURE 8.30
Multi-drum multi-bank bottom supported LP/MP boiler of early twentieth century.

With the increase in boiler p and t, improvements in water quality and the need to meet more stringent NO_x requirements, the furnace and Eco have grown in size at the expense of BBs.

Most modern boilers have a single BB between top-steam drum and bottom/water/mud drum. Exception to this can be small **'O'** and **'A' type boilers** with two BBs. A BB serves as a big reservoir of water and energy, making the boiler operation a bit sluggish compared to single-drum boilers but also quite stable. This turns out to be quite benign to operational deficiencies because of large water-holding capacity.

BB is the most expensive HS after furnace if we ignore SH and RH.

Due to large HS in BBs, the gas temperature drop is sizable as seen in Figure 8.31.

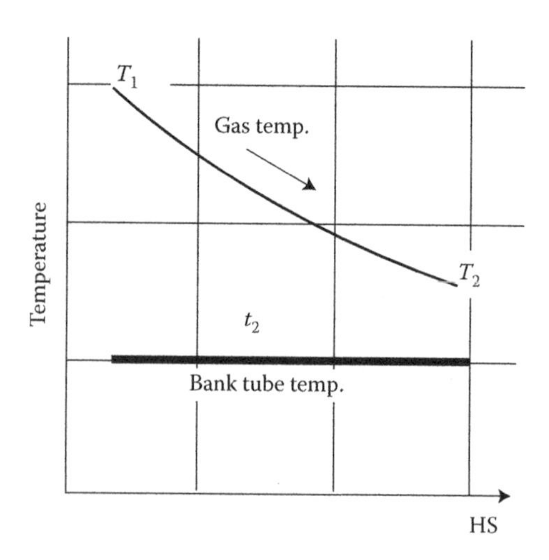

FIGURE 8.31
Gas temperature drop over a boiler bank.

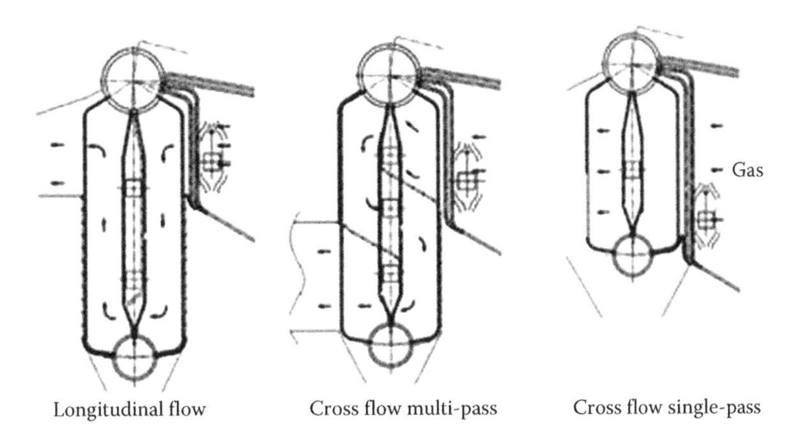

FIGURE 8.32
Different boiler bank configurations in a two-pass boiler.

Gas exit temperature (T_2) for BBs should always be <450°C to avoid

- Possible flow reversal in the downcomer set of tubes leading to circulation and tube overheating problems
- Refractory lining of gas duct at exit

For coal-fired boilers $T_2 - t_2$ is ~100–120°C for single-pass and ~100°C for multi-pass BBs.

BB construction

BBs can be arranged with gases flowing

a. Perpendicular to the drum axis, as in most two or three pass boilers, with gas flowing vertically up and down
b. Along the drum axis, as in many package boilers, with gases flowing horizontally

In the case of (a), the gases may flow

- Normal to the tubes in cross flow pattern or
- Along the tubes in longitudinal flow pattern

Single pass cross flow is adopted if the dust loading in gases is high or the dust is abrasive. But multipass cross flow is only for clean gases.

Due to high gas velocities in the longitudinal arrangement, the gas should be relatively clean. In both the arrangements, single or multiple passes are possible.

In the case of (b), the flow gases are always in cross flow pattern. It is essential that the gases are very clean, as in NG or light FO firing, lest the dust from gases should separate and build-up in the bank.

Both these arrangements are shown in Figure 8.32.

Generally, the colder 1/3 of tube section is made into downcomer part by suitable baffling inside the steam drum. If this is found insufficient unheated downcomers are arranged external to the BB.

BBs can also be arranged in near horizontal disposition like in vertical Hrsgs or in several Whrbs boilers where the gas flow is in the vertical direction. The tubes are inclined at ~5–7° to the horizontal to facilitate a smooth flow of circulating water. They can also be made horizontal with saving of space but a suitable circulating pump has to be incorporated in the circuit to ensure flow of circulating water under all loads.

8.27.3.4.b Wing Wall Evaporator and Division Wall

A typical wing wall Evap is shown in Figure 8.33. Both wing wall Evap and division wall are radiant surfaces exposed to direct flames and placed in very high gas

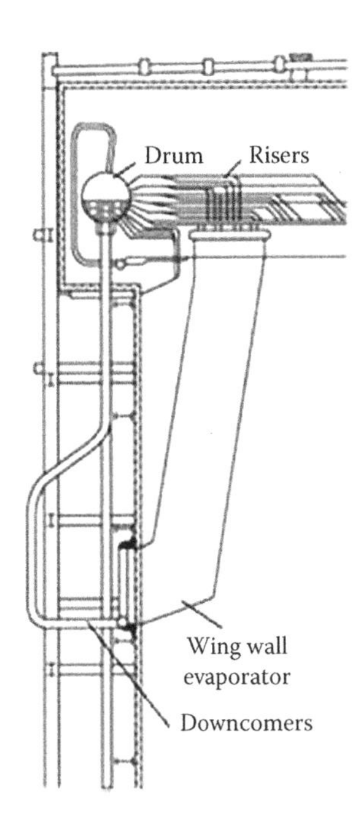

FIGURE 8.33
Typical wing wall evaporator.

temperatures. They receive almost their entire heat from radiation which is in sharp contrast to BBs that get their heat almost entirely from convection.

See under the topic of PPs for more description on division wall.

8.27.3.4.c Bed Coils

See Bed Coil in BFBC boilers under the topic of Bubbling Fluidised Bed Combustion.

8.27.3.5 Economiser (see *in Section E and* see also Backend Equipment below)

8.27.3.6 Airheater (see *in Section A and* see also Backend Equipment below)

Backend equipment

Eco and AH, either together or alone, are referred to as backend equipment, since they are the last of the heat traps extracting heat from flue gases, after they have been cooled in BB and SH/RH coils, respectively. In Hrsgs, the last heat trap after Eco is in the form of *condensate pre-heater* instead of an AH. Condensate at ~30–40°C coming directly from condenser without any LP fw heating is used for cooling the flue gases.

Choice of Eco or AH depends on the merits of each case.

- In process steam boilers where the inlet fwt is quite low, between ~85°C and 120°C, it is usual to have only Eco as necessary gas cooling can be achieved as hot combustion air may not be needed in most cases.
- But in power plants, where regenerative feed water heater (FWH) raises the inlet temperatures from 140°C to 310°C or more, it is imperative to have both Eco and AH to attain desired exit flue gas temperature.
- In Hrsgs, only Ecos are employed as there is no combustion or combustion air.
- When fuels are rich in M, such as the bio-fuels, AHs are invariably incorporated to raise the temperature of the combustion air, which improves greatly the combustion speed and quality.
- In PF boilers, AH is absolutely essential as the mills need hot air for drying and transporting the PF.
- In BLR boilers, AHs are not used as they get fouled with dust in no time.

Action of Eco and AH is quite different even though both extract heat from flue gases. See Figure 8.3.

- While Eco transfers the heat of flue gases directly to fw, AH transfers it to the combustion air which, in turn, gives it to flue gases after participating in combustion reactions in an indirect and long route.
- With hot air from AH, the furnace runs hotter, FEGT a little higher and NO_x generation is a little more. The duct layout with AH is also a bit more elaborate.

Temperature limits of operation of Eco and AH are as follows:

- At the cold end, the low temperature, cold or dew point gas side corrosion is the main consideration in sizing of both AH and Eco. See Figure 1.14 in ash corrosion for the limits.
- At the hot end, it is preferred that Eco water outlet temperature should be ~30°C less than the saturation temperature of the drum water to avoid water hammering. If the steaming Eco is employed, the steam should be generally limited to ~5%.
- Hot air from the AH is limited by permissible tolerance levels of the downstream equipment such as grate, mill or the burner.

Improvement in thermal η is quite appreciable in both.

- Every 22°C (40°F) of drop in final exit gas temperature increases the boiler η by ~1% point leading to sympathetic reduction in fuel consumption (as shown in Figure 8.34).
- While it pays to maximise the backend to raise the η, it should be noted that the addition of HS is also large, as it is low-grade heat that is being captured. See 'effectiveness of HSs' at the beginning of the topic and Figure 8.4.
- The fan power also rises disproportionately as the HSs required are large for marginal cooling of gas.

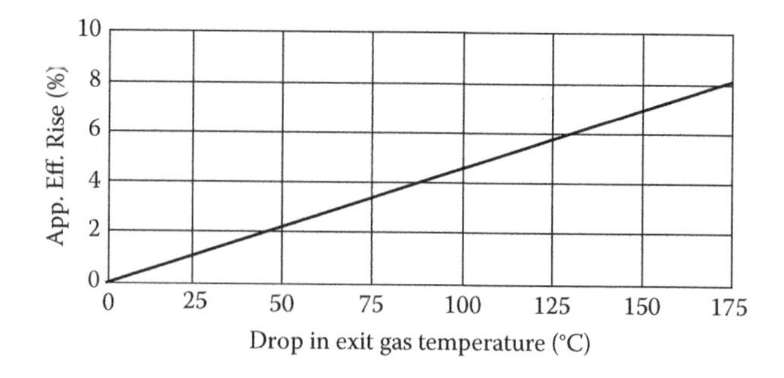

FIGURE 8.34
Boiler efficiency rise versus drop in gas exit temperature.

A careful balancing is therefore required, taking into account the prevailing fuel and power costs, in arriving at the optimum sizing of the backend equipment. In societies with high fuel and power costs (they go hand-in-hand), it is normal to aim for optimum cooling of gases by investing in large backend.

In constructional aspects, the two surfaces are quite different.

- While the Eco is a PP, AH is a non-pressure part (NPP) item.
- Heat transfer in Eco is 3–4 times more than AH. It is therefore compact.
- AH is bulkier and when the elaborate air and gas ducts are also considered, AH may turn out to be more expensive in many cases. Besides the fan power is also generally higher.

Sequencing of Eco and AH is a matter of convenience and economy. The classical arrangement is to place the Eco first, followed by AH, as in most PF-fired boilers where the fw inlet temperature is quite high. When such is not the case the Eco and or AH are both split into two and placed in the most cost-effective sequence.

8.28 Heat Recovery Steam Generators

Hrsgs are the *only WH* boilers *in the utility boiler* space. All other **Whrbs** belong to process industry. Both are compared in a separate topic titled Whrb versus Hrsg.

Hrsgs are *relatively new*. They started as small and simple WH recovery boilers behind the small **GTs** of the 1950s. As the GTs grew in size, the oil- and gas-fired package boilers were adapted with suitable alterations as the first step. Further explosive growth in GT sizes saw Hrsgs also sympathetically evolving into distinct designs of their own, bearing scarcely any resemblance to the earlier models or conventional boilers. It is perhaps because of the smaller companies that developed Hrsgs without the overbearing influence the large utility boiler makers that these boilers are strikingly simple and starkly different from the designs of conventional boilers.

GTs exhaust their spent combustion flue gases from 450°C to 630°C (780°F to 1165°F) with gas flows going as high as 2700 tph (~740 kg/s or 1630 lbs/s). The exhaust gases contain ample O_2 at 15–16% by weight as 220–300% excess air is used in firing fuels in GTs. Hrsgs are required to cool these gases and generate steam for power and process.

Figure 8.35 shows a typical **CCPP** using GT and Hrsg in an isometric view in (a) and flow schematic in (b).

The topic of Hrsg is covered under the following sub-topics:

Hrsgs versus conventional boilers
Design issues of Hrsgs
Hrsg scope
Fired and unfired Hrsgs
Finned tubes for Hrsgs
Thermal and mechanical design aspects of Hrsg
Horizontal and vertical Hrsgs
Efficiency testing of Hrsgs

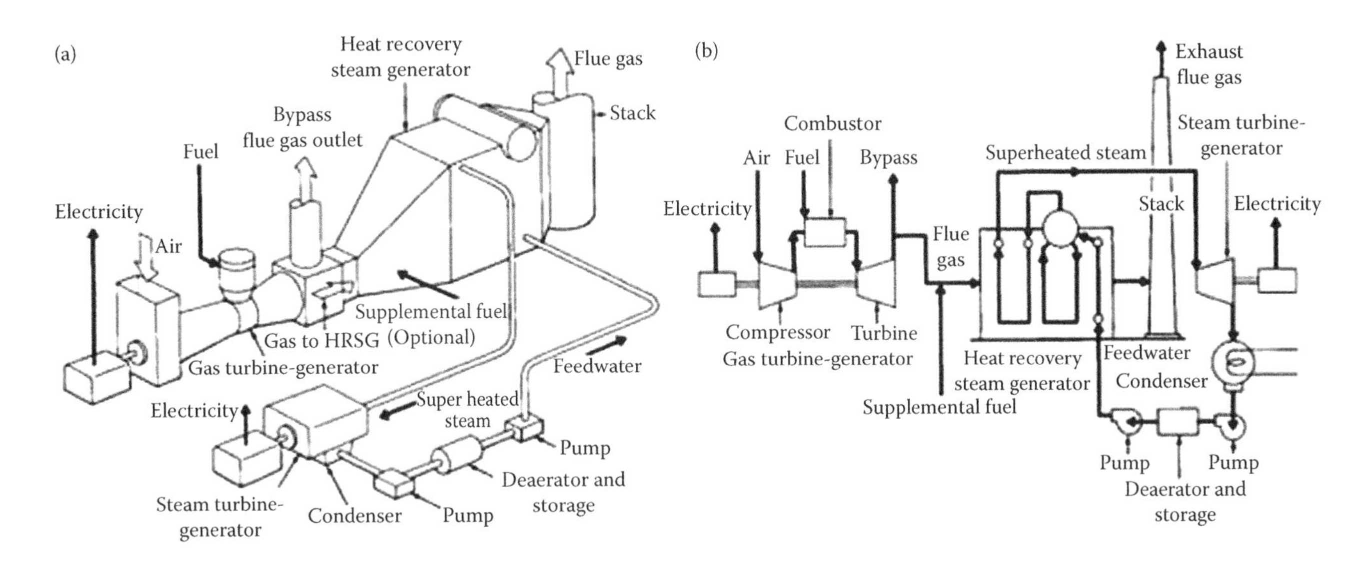

FIGURE 8.35
Combined cycle power generation using GT, Hrsg and ST—(a) isometric and (b) schematic presentations. (From The Babcock and Wilcox Company, USA. With permission.)

8.28.1 Hrsgs versus Conventional Boilers

Hrsgs are quite different from the conventional boilers as explained below:

- Hrsgs are required to extract heat from the clean hot exhaust gases. As such, there is no combustion unless **supplementary combustion** is introduced. In most cases therefore, there is no furnace.
- At times, supplementary or auxiliary firing is done respectively to either increase the steam output or maintain the steam output even when GT is down. Only in such cases, space has to be provided for accomodating duct burners and their flames.
- There are normally no fans required in Hrsgs as the GT exhaust gases are at sufficient HP and there is no firing.
- FD fan is provided only when auxiliary firing has to be done to maintain the steam output even when GT is down. This is called *fresh-air-firing* and the FD fan, a *fresh air fan*.
- Hrsgs also do not have any AHs as there is no firing and hence no combustion air.
- The exhaust gases being highly clean, almost all tubes in Hrsg are of closely spaced **finned** construction to maximise the HS. Plain tubes are used only if there is a fear of exceeding tube metal temperature with finning.
- The tubes are always placed in staggered or triangular construction to make the HXs compact.
- Tube pitching is standardised by each manufacturer and the entire database is built on that tube spacing, which happens to be the prime dimension.
- Exhaust gases are mostly non-corrosive or only at times mildly corrosive. It is possible to cool the gases all the way to 75–80°C and maximise the heat recovery.
- In CC, the LP fw heating is not done in separate heaters like in a steam cycle. Condensate is heated in Hrsgs in preheater section.
- Steam is generated at a single pressure only in very small Hrsgs for process and Cogen requirement. In most cases, there are two pressure levels. In case of pure power generation there are three levels. When the STs are >~100 MWe, RH is also employed.
- RH steam flows are usually higher than main steam flows as the steam generated at IP level in Hrsg is added to the RH steam from turbine. This is quite unlike the conventional boilers where the RH steam is some 5% lower than SH steam.
- **Attemperation** in a conventional sense is not applicable to the Hrsgs. Hrsg abstracts WH from GT and no overheating of steam takes place.
- Likewise, there is no specified **load turndown** in Hrsgs like in conventional boilers.
- There is no **peak duty** either. GTs are already designed to their optimum firing temperature and any peak firing will result in shortening of life of hot parts of GT.

8.28.2 Design Issues of Hrsgs

As the GTs are sold in standard frame sizes, with practically no customisation, it should be expected that the Hrsgs behind them can also be standardised. Unfortunately, it is not the case because

- The steaming conditions vary depending on the CC design parameters.
- Fuels and excess air used in GTs vary quite a bit, resulting in wide fluctuations in the GT exhaust flows.
- GT being a **volumetric machine**, its output and exhaust are highly dependent on the ambient conditions.

With variations in the gas flow and pressure levels Hrsgs are required to be engineered afresh every time.

As Hrsgs are required to maximise heat extraction from the exhaust gases within the available exhaust gas pressure the sizing has to be the most optimum with no latitude for error for gas pressure drop. If there is a shortfall of steam temperature, it cannot be made up as there will be no space and gas pressure to accommodate the additional HS. This is because it is ultra-low-grade heat at Hrsg cold end and the addition of surface will be very large unlike in conventional boilers. Available gas pressure drop also poses serious limitation.

Surfacing of Hrsg is thus an exact science with *all design loaded upfront* with no scope for rectification at a later stage.

8.28.3 Hrsg Scope

The Hrsg scope begins at the outlet flange of the **bypass stack** or the exhaust duct of GT and extends all the way to the stack. **Stack** is always an integral part of Hrsg. It is in fact supported on Hrsg in the vertical design.

On the PP side, the Hrsg consists of drums, various tube banks and interconnecting piping similar to conventional boilers. Figure 8.38 can be considered as a representative scope sketch.

8.28.4 Fired and Unfired Hrsgs

Hrsgs are mostly unfired. In the CC mode, the cycle η is optimum when there is no additional firing. This is because when firing is done in Hrsg, the fuel is used only in the steam cycle and not in the GT cycle. In other words, instead of deriving 45–55% of lower heating value (Lhv) from the fuel, only 30–35% is realised, making the operation less efficient.

Firing in Hrsg is resorted to in two specific conditions:

1. *Supplementary firing* for generating a little more steam from the unit and correspondingly a little more power from the ST. From cycle η considerations this is done intermittently to meet peak loads for short periods. In CC mode, it is not efficient to resort to supplementary firing as explained above. Also, firing extra fuel is expensive in an Hrsg both because of hardware and process needs. Firing of additional fuel is done in duct burners in the inlet duct. The related points are as follows:
   - For supplementary firing, additional fuel is needed but no additional air is required as the GT exhaust gases have adequate O_2 at 15–16% and ample excess gas.
   - There is an additional gas pressure drop due to duct burner and the upstream flow straighteners for which the GT exhaust pressure available should be adequate.
   - The fuel used can be different from the main fuel of GT but flue gases from the supplementary fuel should be clean and compatible with the closely spaced finning on the tubes.
   - As the supplementary firing raises the flue gas temperatures to ~800°C, the downstream duct has to be internally insulated and the metallurgy of the first rows of tubes has to be improved.
   - This high-temperature regime also increases the gas pressure drop.
   - RH and SH temperatures tend to be higher which improves efficiency of the ST fractionally.
2. *Auxiliary firing* is normally done in Hrsgs in Cogen or process mode to maintain the process steam even during the GT outage. Usually, the capacity of auxiliary firing is 100% of Hrsg steam output. During the GT outage, there is no exhaust gas flow to provide O_2 for combustion and hence fresh air fans are required along with **duct burners**. Precautions like insulation of duct downstream of the duct burner and improving the metallurgy of tubes is also required here. If the fresh air or FD fan is difficult to install, an ID fan can also be installed depending on the complexities of the layout.

8.28.5 Finned Tubes for Hrsgs

Hrsgs are made almost entirely with finned tubes except for a few rows at the beginning in the high temperature units. These are made of plain tubes to avoid tube overheating.

Finned tubes of Hrsgs are made of *high-frequency resistance welded fins* which have a continuous metal-to-metal bonding between the fins and the tubes. Such strong welding is necessary not only to transfer all heat from the fin to the tubes but also to keep the Δt between fins and tube to a minimum. In Hrsgs, this is of vital importance as the cooling and heating of the Hrsgs is extremely rapid during start-ups and shut downs. When the GT is started, the gas temperature reaches 60–70% of the full temperature between 5 and 10 minutes. Unless the bonding is strong, there are possibilities of the fins separating from the tubes.

8.28.5.a *Fin Types* (see also *Extended Surfaces*)

There are two types of circular fins employed in Hrsgs. They are also described in circular fins under the topic of Extended Surfaces.

1. *Serrated fins:*

   Serrated fins shown in Figure 8.36a provide more HS per unit length and are also cheaper. Owing to the serrations, they are suited only for

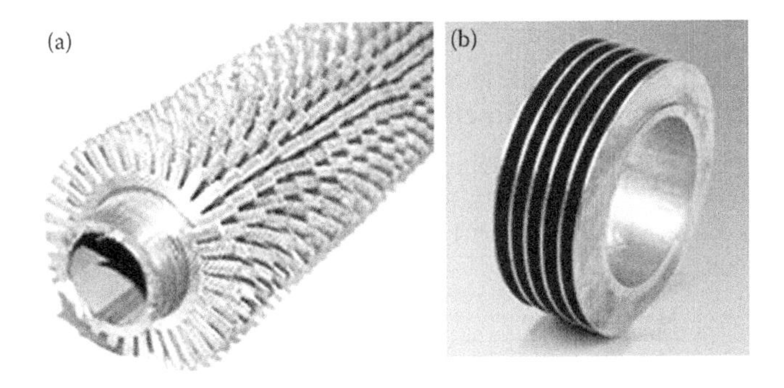

FIGURE 8.36
Tubes with (a) serrated and (b) solid fins.

clean fuels like gases as they do not pack up the gaps between the serrations.

2. *Solid fins*

   Shown in Figure 8.36b these are for slightly fouling fuels. They are also employed in high-temperature sections of SH and RH as the fin cooling is better.

8.28.5.b Fin Thickness

Fin thicknesses vary from 0.8 to 1.5 mm with a maximum pitching of ~310 fins/m with fins of 0.8 mm width. More fin height makes the tip temperature higher. It is customary to limit the fins to 15 mm height. Fin thickness is varied as per fuel—thick fins for heavy fuels and thin fins for gases.

8.28.5.c Tube and Fin Materials

Selection of tube materials is primarily governed by the mean metal temperature of the tubes like in conventional boilers (refer Table 8.3). As there is no incidence of radiant heat, tube metal temperatures are ~10°C higher than the water temperature in Eco and Evap, and ~35°C higher than the SH and RH temperatures.

Limiting temperature of fins is given in Table 8.4. Any combination of tube and fin is possible as welding them is not a problem.

8.28.6 Thermal and Mechanical Design Aspects of Hrsg

The differences with conventional boilers and certain design aspects of Hrsgs have already been elaborated. The following are in addition to those points:

- Due to the restricted discharge pressure of gas at the discharge flange of GT, the gas flow in Hrsg is always unidirectional without any turns, as in conventional units. All the convection surfaces are arranged in counter flow to gas flow to optimise on the HS. Only the tubes in the first bank may be arranged in parallel flow in high-temperature Hrsgs for lowering tube metal temperatures.
- *Pinch point* plays a very important role in sizing an Hrsg. **Pinch point** is a point below which cooling of gas with Evap surfaces becomes less economical due to disproportionately large requirement of HS. Modern designs believe in having a pinch point between ~8°C and 14°C (15–25°F) above the saturation temperature.
- A similar gap of ~8–14°C (15–25°F) is required to be maintained between the water and gas temperatures at the outlet of the Eco which is known as the ***approach** temperature*. This is to prevent steam in the Eco during normal running. However, steaming is unavoidable during **cycling** duty which is becoming more popular.
- *SH approach temperature* is the temperature difference between outlet of the SH bank and incoming flue gas. This is recommended to be maintained between ~22°C and 33°C (40–60°F) for optimising HS of SH.

TABLE 8.4

Fin Materials and Their Limiting Temperatures

Material of Fin	Limiting Fin Temperature °C	Limiting Fin Temperature °F
CS	480	900
11% Cr steel	650	1200
18–8 Cr–Ni steel	870	1600
25–12 Cr–Ni steel	1040	1900

The three concepts are well captured in Figure 8.37.

- Hrsg is a combination of one, two or three boilers in a single envelop operating at one, two or three levels of p and t. Excepting in small units, two pressure levels (2P) is normal.

  With every additional pressure level, there is a gain of ~4% points in cycle η.

- Three pressures levels (3P) are common in power cycles of 100 MW and above (Fr 6FA or equivalent).

  The three pressure levels are designated as HP, IP and LP for high, intermediate and low pressures.

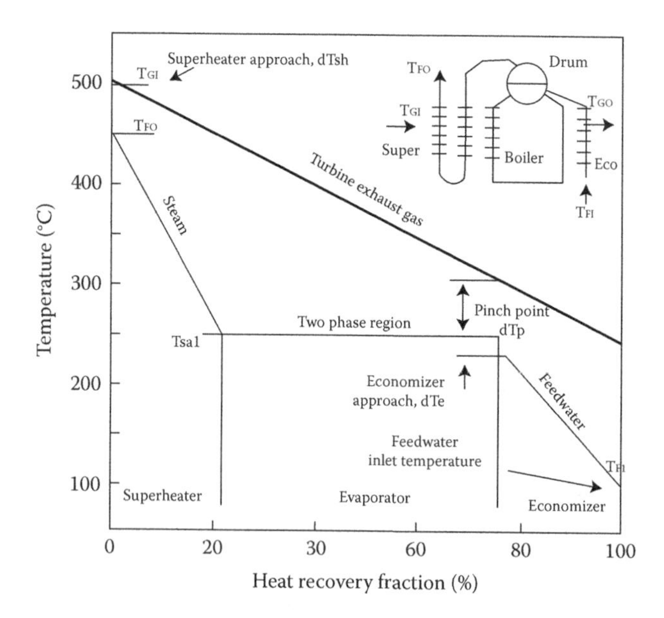

FIGURE 8.37
Pinch point and approach temperatures in Hrsg design.

- HP steam is for power generation. HP level has been steadily on the rise in line with the exhaust gas temperature of GT. Presently, 170 bar and 570°C is quite common in Hrsgs behind large GTs like 9FA.
- IP steam is also for power generation used for injection steam into the GT. Usually, IP levels are between 25 and 35 bar with temperatures up to 450°C.

  In RH boilers, IP steam is added to the cold RH steam from ST.
- LP steam is for condensate heating or deaeration between 3 and 6 bar.

- Gas exhaust temperatures for Hrsgs are much lower than that of conventional boilers as the incoming fw is directly the condensate at 30–40°C as against a minimum of 105°C of deaerated and heated fw in conventional boilers. In NG firing, the gas exit temperature can be as low as 75–80°C with big preheater banks as the last heat trap. Such low-exit gas temperature is attractive only when fuel costs are high.
- SH and RH banks employ the same tube materials as conventional boilers. Use of ss is almost prohibited both due to its excessive elongation (~30% higher than CS) and the fear of stress corrosion cracking. Fins of all metallurgies are used except ss. Fin material should be high in Cr and Ni for withstanding high gas temperatures.
- No **attemperation** needs to done in Hrsg. However, attemperators are provided at the exit of SH and RH to protect tubes from escalation of metal temperatures during plant cycling conditions.
- Bank depths are limited to ~3 m from consideration of adequate access.
- With clean flue gases Hrsgs do not need SBs. However, when FOs are used SBs may be required. Soot formation on tubes can lead to sharp degradation in performance.
- In conventional boilers, water-cooled boiler walls are employed in hot zone and steel casings in cooler zones. In Hrsgs only steel casing is used from beginning to the end.
  - Casing insulation is done either on the
  - Inside (cold casing) when gases are hot at >480°C or
  - Outside (hot casing) when gases are at a temperature of <480°C.
  - A combination of cold and hot casing is also used on many occasions.
  - Cold casing is more expensive as the insulation needs to be protected against gas flows by AS liners.
  - **Glass wool** and **slag wool** are useful only up to 500°C but up to 650°C sometimes in blanketed form.
  - In fire ducts downstream of duct burners, when the temperatures reach as high as 950°C with supplementary/auxiliary firing **ceramic fibre** blankets are used for insulation.
- ***Weather dampers*** are peculiar to Hrsgs. See Weather Dampers under the topic of Draught Plant. As the Hrsg stacks are rather large and short weather dampers are required for
  - Preventing ingress of water from rain or snow into the boiler internals in case of vertical Hrsgs
  - Conserving heat when the unit is required to be stopped off and on in the case of cycling duties.
- Stack of Hrsg is a little different from the stack in conventional boilers.
  - Stack in Hrsg runs usually much cooler than in conventional boilers
  - It is an integral part of a vertical Hrsg as it is always mounted on its top. It is independent and self-standing in case of horizontal design.
  - As the gas volumes are much larger, it is normal to employ high-gas-velocities of 20–25 m/s, thanks to the clean gases, to reduce the stack diameter. Even then the stack is quite stout.
  - Stack height is merely required to exceed the neighbouring structures and not really meant for dispersion of pollutants, as there are none in the clean Hrsg gases.
  - Hrsg stack is required to be insulated from outside to avoid condensation as M can cause corrosion.

8.28.7 Horizontal and Vertical Hrsgs

The modern Hrsg is either horizontal or vertical in design as designated by the gas flow. See Figure 8.38. As there are no turns in the gas flow, like in conventional boilers, all the tubes are either vertical or horizontal, respectively.

The Hrsgs trace their origin to the packaged oil- and gas-fired boilers. The horizontal designs for Hrsgs came from the United States and vertical ones from European boiler practices, each with its own set of advantages and limitations. To the final user, both Hrsgs would cost

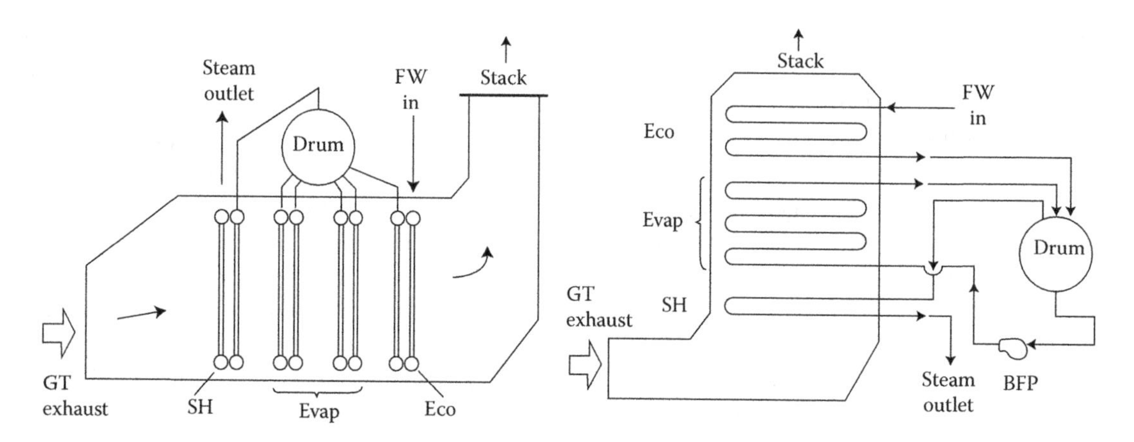

FIGURE 8.38
Concept of horizontal and vertical Hrsgs.

approximately the same with some differences in the matter of layouts and dynamics. The horizontal design with natural circulation and a little more of water-holding capacity tends to be more operator friendly while the vertical designs with more flexibility in tube arrangements are better suited for cyclic operations. As the industry is maturing and traditional boundaries of business are transgressed, the horizontal designs are now made with better tube flexibility and vertical designs are adopting natural circulation, blurring the differences substantially.

8.28.7.a Horizontal Hrsgs

Figure 8.39 shows a schematic arrangement of a 3P+RH type of large horizontal Hrsg with supplementary firing and Figure 8.40 shows a typical cross section of a

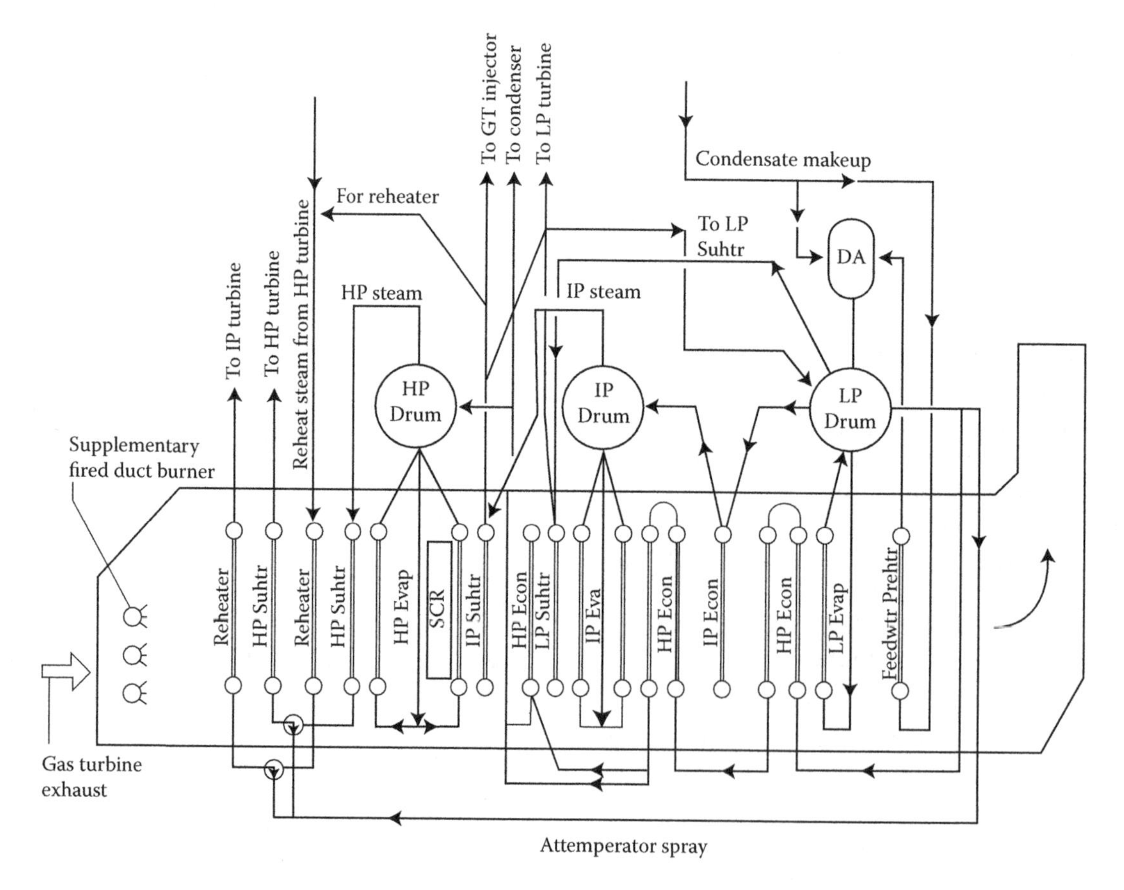

FIGURE 8.39
Schematic of a large horizontal 3P+RH type Hrsg with supplementary firing.

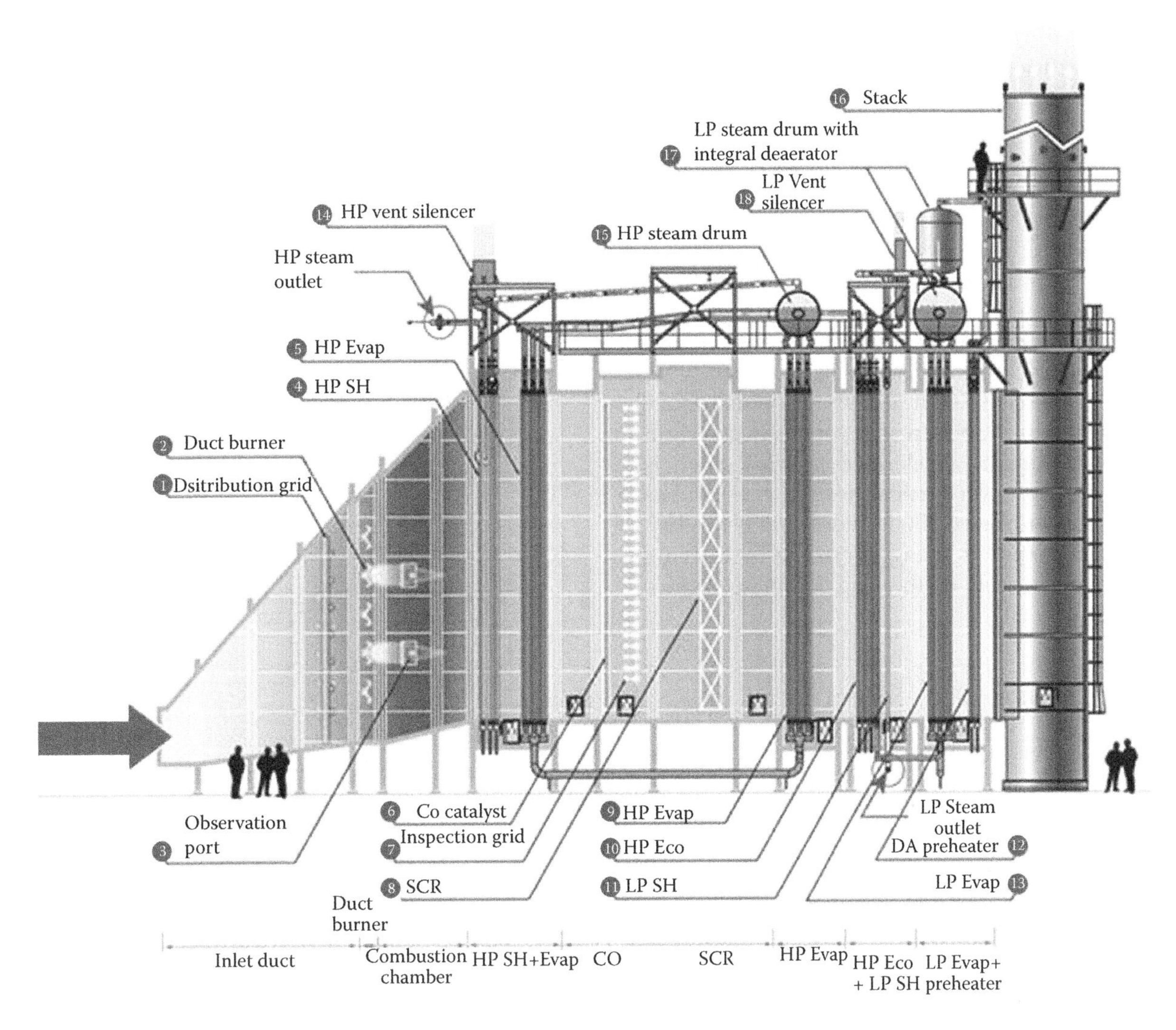

FIGURE 8.40
Cross section of a typical horizontal 2P Hrsg with supplementary firing and catalyst chamber.

2P Hrsg with supplementary firing and also catalyst elements.

- Horizontal designs are **natural circulation**, bottom supported with an independent stack supported from the ground.
- The **circulation ratio** is generally in the range of 6–8 in the HP circuit and higher in IP and LP circuits. There is no need for circulating pumps, thus saving some auxiliary power.
- Being a natural circulation boiler, the tubes are larger in diameter, normally 38.1, 44.5, 50.8 and 63.5 mm od (1½″, 1¾″, 2″ and 2¼″). Naturally, the marginally lower heat transfer coefficients due to larger tubes calls for slightly larger HS and thicker tubes. All this contributes to higher overall weight of tubes.
- But the drums tend to be smaller than in vertical design. This is due to the absence of swell effect of water in drums, as several independent vertical SH tubes enter the boiler drum. This is unlike in a vertical design where the legs of SH are all joined to form coils. Steam from all legs gets accumulated and causes swell when it enters the drum under pressure fluctuations.
- Horizontal Hrsgs tend to occupy slightly more floor space as the tube banks are placed one behind the other. The footprint increases further if the catalyst section for **NO_x** control is also required to be included.
- As the boiler is of low-set, in many cases, draining and blow off may be a problem unless assistance of a pump is provided.

- Serious limitation of horizontal Hrsg design is the stiffness of the tube elements which are fixed in top and bottom headers. During the start-up conditions, the flue gases from GT rush into Hrsg at high temperatures and high volumes reaching full load conditions in a matter of minutes. The tubes are subject to rapid heating. Unless there is enough flexibility, the tubes tend to rupture at the ends within a short time of operation, because they are rigidly welded to top and bottom headers. The reverse happens when GT shuts down. Cyclic operation of CCPP with daily start and shut downs induce severe stresses and stress reversals in the rigid tubes leading to accelerated failures. Such is not the case with vertical design as the tubes form serpentine coils with adequate flexibility. This deficiency is being overcome by altering the construction of the end joints to induce flexibity.

8.28.7.b Vertical Hrsgs

Figure 8.41 shows a schematic diagram of a classical 3P+RH type Hrsg with **assisted circulation** and Figure 8.42 shows the cross section of a typical simple 2P Hrsg for Cogen application.

- Vertical Hrsgs are top-supported boilers with a stack mounted on the Hrsg. They can be treated as tower-type boilers with the furnace removed. In fact many detailed constructional features are common between the two.
- The circulation ratios are quite low at 1.9–3.0 requiring assistance in circulating water, at least in the HP circuit in many cases. Designs are available today which avoid using of pumps up to design pressures of ~130 bar. However, induced circulation is needed during start-ups for which the help of **BFP** is taken.
- Vertical Hrsgs occupy slightly smaller floor space but are taller.
- The structure is heavy as the modules as well as the stack have to be supported. The effect of wind load is more, as a unit becomes taller. The platforms are also more.
- The tube diameters are lower at 31.8, 38.1 and 44.5 mm od (1¼″, 1½″, 1¾″). The heat transfer is better and HS is slightly lower. The tube thicknesses are also lower. Together, they contribute to lower weight of the tube banks.
- Small and thinner tubes in serpentine construction provide a lot of flexibility. Quick start-ups

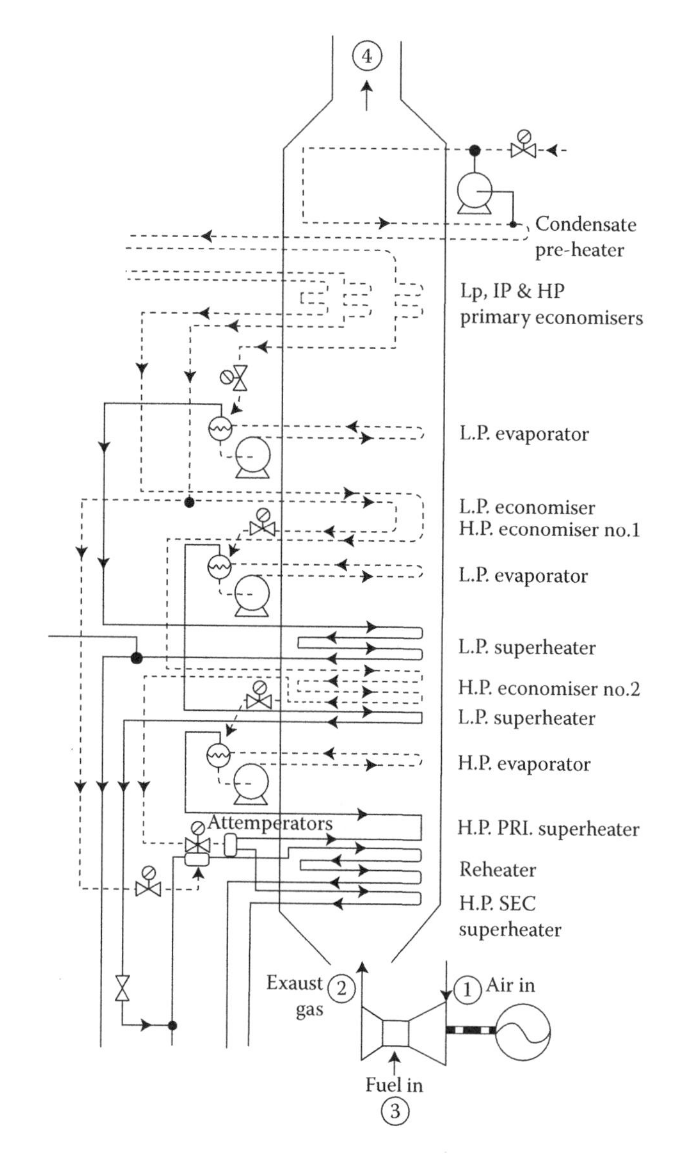

FIGURE 8.41
Schematic of a vertical 3P+RH Hrsg in CCPP application.

and shut downs and cyclic duty pose little problem in this design.

- However, the drums are larger and hence thicker, as they have to cater to the **swell effect**.
- Blow down and draining are both easier as the bottom modules rarely start at <6 m above ground.
- All modules must be shop assembled to reduce the construction time. Even then it takes a little longer to erect a vertical Hrsg as compared to the horizontal.

Table 8.5 provides a comparison of the two designs of Hrsgs.

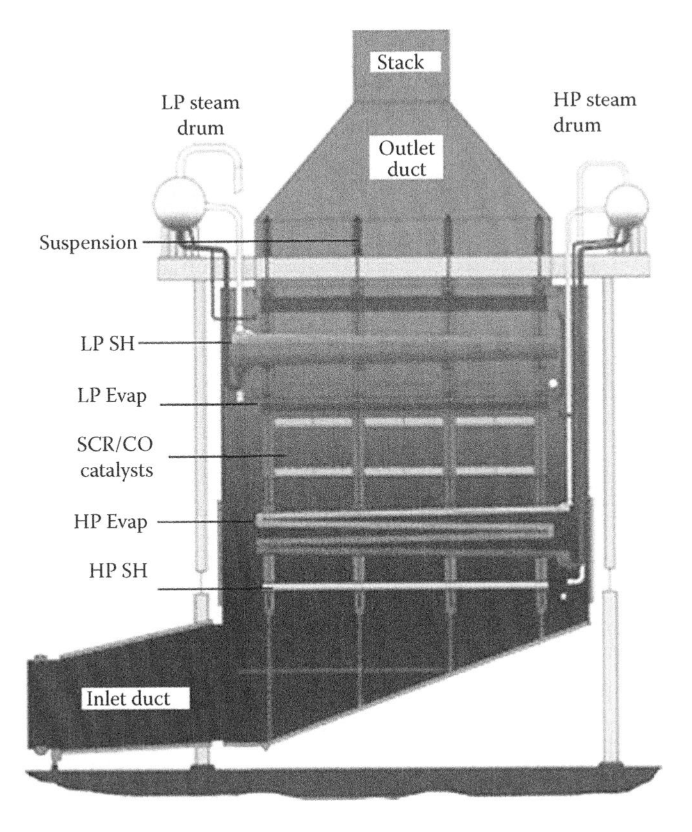

FIGURE 8.42
2P vertical Hrsg for Cogen application.

TABLE 8.5
Comparison of Horizontal and Vertical Hrsgs

Hrsg Type	Horizontal	Vertical
HS	Slightly higher	
Drums		Slightly larger
Structure	Lighter	Heavier
Plot plan	Larger	Smaller
Erection time		Slightly longer
Supplementary firing		
- Inlet duct	Easy	Easy
- Between the banks	Easy	More difficult and expensive
Cycling duty	Difficult	Easy
η	Same	Same
Total installed cost	Equal	Equal

8.28.8 Efficiency Testing of Hrsgs

There are major differences in the way the Hrsgs and conventional boilers or even Whrbs are tested. The most important differences on which Hrsg test code PTC 4.4 is developed are that

a. Hrsg is treated as part of a CCPP instead of a standalone steam Gen.
b. No separate η guarantees are given for Hrsgs and guarantees for total CCPP alone are applicable.

TABLE 8.6
Measuring Tolerances in the Direct Measurement of Various Process Parameters

	Variable	Variation
(a)	Water flow to econ.	±2%
(b)	Econ. recirculation flow	±3%
(c)	DeSH water flow	±4%
(d)	Blowdown flow	±4%
(e)	Fuel flow to gas turbine	±2%
(f)	Supplemental fuel flow	±2%
(g)	GT power output	±2%
(h)	GT to boiler	±10°F
(i)	Stack gas temperature	±10°F
(j)	Water temp. to econ.	±10°F
(k)	Steam temp. at deSH exit	±10°F
(l)	Ambient temperature	±5°F
(m)	Barometric pressure	±1%
(n)	Steam pressure	±2%
(o)	Flow	
	Air	±2%
	Hrsg gas	±2%
	GT exhaust	±2%

c. The performance test is a direct test and not an indirect test as in conventional boilers.

It follows that the adequacy of performance is basically to be demonstrated to the overall CCPP supplier and not to the client as independent equipment. It is sufficient to demonstrate that the Hrsg

a. Flue gas pressure loss is within the design limits.
b. Steaming conditions of HP, IP and LP circuits are as per design conditions at the specified loads with the GT exhaust gas being supplied at the specified quantities and temperatures.

The test as per PTC 4.4 is done with the following in mind:

- fw inlet and steam outlet pressures and temperatures are noted.
- Hrsg gas inlet and exhaust pressures and temperatures are noted.
- Direct test method is adopted and performance test is conducted for not less than 2 h.
- The plant is made to run at nearly the same load as the test load, for some time before and after the test, to attain a steady state condition.
- No blow down is done during test.
- Drum levels are maintained the same at the beginning and end of the test.
- Frequency of test readings can be at more than 10 min for all primary readings except for flow

meter readings, which have to be recorded every 5 min.

- Code specifies the measuring tolerances which are given in Table 8.6.

8.29 Heat Released and Available (*see* Heat Input into Boiler)

8.30 Heat Transfer

Heat transfer is the flow of thermal energy from higher to lower level.

There are three classical modes of heat transfer: (1) conduction, (2) convection and (3) radiation.

Individually, each of these heat transfer modes is simple but in most applications they occur simultaneously, with the added complexity of irregular shapes, dynamic conditions and so on. This makes estimation of heat transfer a very difficult subject.

8.30.1 Conduction Heat

Conduction heat is the *flow of heat by physical contact* between neighbouring molecules from a *higher to lower temperature* region to equivalise the temperatures. It takes place *mostly in solids* and also to some extent in liquids and gases, without any bulk motion of the bodies,

a. From one part of a body to the other
b. From one body to another

Energy flows without significant molecular movement due to the combination of lattice vibrations and free flow of electrons. Temperature gradient is essential for conduction and the resulting heat flow is defined in Fourier's law of heat conduction.

In boiler, examples of conduction are

- Heat flow from flue gases to water or steam across metal wall of a tube
- Loss of furnace heat to the surroundings through insulation
- Heat loss from hot pipes through insulation and so on

8.30.1.1 Fourier's Law of Heat Conduction

Fourier's law is an *empirical law* based on observation.

It states that the rate of flow of heat flux by conduction through a homogenous solid in any direction is directly proportional to the temperature gradient present in that direction.

In other words, heat flow is directly proportional to the

a. Area of the section set at right angles to the direction of flow
b. Temperature difference along the path of heat flow

$$\frac{q}{A} \alpha \frac{t_1 - t_2}{l} \quad (8.7)$$

$$\frac{q}{A} = \frac{k(t_1 - t_2)}{l} \quad (8.8)$$

where

q = rate of heat flow in watts or kcal/h or Btu/h

A = surface area perpendicular to the heat flow in m^2 or ft^2

l = thickness of body in m/cm or ft/in

$t_1 - t_2$ = temperature gradient in °C or °F

k = thermal conductivity in watts/m/°C, kcal/m²/°C/cm, Btu/ft² h/°F/in or Btu/ft² h/°F/ft

8.30.2 Convection Heat

Convection is the *molecular movement within fluids.* Convection heat is the flow of heat from the surface to the surrounding fluid by conduction and mixing of fluid molecules. *Convection does not take place in solids* as it requires motion of particles. It happens when heat flows

- From one point to another within the fluid due to its own motion
- From one fluid to another when fluids are mixed
- Between a solid and fluid due to a relative motion between them

There is *always mass transfer involved in convection* along with heat due to the movement of particles.

Some of the examples of convection in boilers are

- Heat flow from flue gases to tube banks
- Heat loss from the lagging of wall or pipe

Convection can take place either in (1) single phase or in (2) two phases.

8.30.2.1 Single-Phase Convection Can Be Either Forced Convection or Free/Natural Convection

a. *Forced convection* results when the fluid moves over the surface due to an external force. The velocities and resultant heat transfer rates are both higher.

All heat transmission in tube banks of boilers is by combination of inter-tube radiation and forced convection. There are several modes in which the convection heat transfer takes place from gas to tube and thereafter from tube to fluid inside, depending on how both the fluids are flowing with respect to the tube bank and the geometrical details of the bank. Equations for convection heat transfer vary considerably depending on the tubes being placed normal or parallel to the gas flow. Likewise, equations for heat transfer inside the tubes vary depending on the flow being laminar or turbulent.

Free/natural convection takes place with only natural force at work, such as gravity, expansion, buoyancy and so on. The velocities and heat transfer rates are both low.

Free convection is a self-sustained flow driven by the presence of a temperature gradient (as opposed to forced convection where external means are used to provide the flow). As a result of the temperature difference, the ρ of the field is also not uniform. Buoyancy will induce a flow current due to the gravitational field and the variation in the ρ of the field.

Air and gas are important natural convection media in boilers. Natural/free convection conductance is given by

$$\mathrm{Ufc} = C(\Delta t)^{0.33} \quad (8.9)$$

where C is a coefficient depending upon the shape and position of heat transfer surface and Δt is the temperature difference between the surface and air/gas. C varies from 0.18 to 0.22 in fps units.

8.30.2.2 Two-Phase Convection Can Take Place as in the Case of Boiling or Condensation

a. *Boiling*: When liquid absorbs heat from a surface which is at a higher temperature than its saturation temperature, change of phase from liquid to vapour takes place. This is called boiling.
b. *Condensation:* When the vapour gives out heat in contact with a surface which is lower than its saturation temperature, there is a change of phase as the vapour turns into liquid. This is called condensation.

Convection can be sub-divided as free/natural convection or forced convection.

8.30.3 Radiation Heat

Radiation heat is the flow of heat entirely by *electromagnetic waves* from a hot to cold surface without any medium in between. Every surface emits electromagnetic waves in all directions.

A classical example of radiation in boilers is flow of heat from flue gas to circulating water in boiler furnace.

(1) *Absorptivity*, (2) *transmissivity* and (3) *reflectivity* are the three fractions of the heat incident on a surface and being absorbed, transmitted or reflected, respectively. The sum of the three fractions is always unity.

A *black body* absorbs the entire radiation incident on it with no part reflected or transmitted. No actual bodies are completely black.

Emissivity is the ratio of heat absorbed by a body to be emitted by a black body at the same temperature.

A number of commercial surfaces have emissivity of 0.8–0.95 and behave like black bodies.

Similarly, a *white body* fully reflects and *transparent body* fully transmits the entire incident radiant energy while an *opaque body* does not transmit any.

Under equilibrium with its surroundings, the emissivity of a surface equals its absorptivity which is known as *Kirchoff's law of radiation.*

Emitting power is the radiant flux from a surface at a particular temperature.

Radiation emitted by a *black body* is proportional to the fourth power of its absolute temperature known as *Stefan–Boltzmann law.*

8.30.3.1 Kirchoff's Law of Thermal Radiation

Kirchoff's law in thermodynamics states that, at equilibrium, the emission from a heated object is equal to its absorption, that is, *emissivity of a body or surface equals its absorptivity.*

As a corollary, the *emissivity cannot exceed* 1 and it is not possible to thermally radiate more energy than a black body.

It also means that a poor reflector is a good emitter and a good reflector is a poor emitter.

8.30.3.2 Stefan–Boltzmann's Law of Radiation

This law states that the radiation emitted by a black body depends on its surface area and the fourth power of its absolute temperature.

$$q \propto T^4 = \sigma T^4 \quad (8.10)$$

For a grey body with a certain emissivity, it is

$$q = \sigma € S T^4 \quad (8.11)$$

where q = rate of heat flow in W/s.

σ = Stefan–Boltzmann constant = 5.6703 x 10^{-8} W/m^2 K^4

S = Surface area of the body in sqm

T = Absolute temperature T in °K
ϵ = Emissivity

When two black bodies at temperatures T_1 and T_2 (where $T_1 > T_2$) exchange heat, the heat flow from body 1 to body 2 is

$$q_{12} = F_{12}\sigma S_1(T_1^4 - T_2^4) \quad (8.12)$$

8.30.3.3 *Radiation: Luminous and Non-Luminous*

Luminous radiation in a boiler is the heat transfer from flames to the surroundings which typically takes place in a furnace. This is governed substantially by Stefan–Boltzmann's law. In actual practice, proprietary empirical data is used for determining the heat transfer in boiler furnaces.

Non-luminous gas radiation is the heat transfer, usually at more than 400°C, from gases like CO, CO_2, SO_2 and N_2 in the inter-tube gaps in SH, RH and BBs. These gases are selective radiators. They emit and absorb only in certain wavelength bands which are outside the visible range.

Cavity Radiation

Cavity is the open space between two successive tube banks in boilers. Cavities are needed to separate banks when they become too deep to clean by SBs or access properly for maintenance. They are also needed to provide access doors and install SBs. Cavities are rarely <500 mm.

Heat transfer in a cavity is entirely by non-luminous radiation to all the surrounding surfaces.

8.30.4 Combined Heat Transfer

When heat is transferred by more than one mode of classical heat transfer, it is called combined heat transfer and an overall heat transfer coefficient is evolved to estimate the heat movement. Heat transfer Q is then expressed as

$$Q = UA\Delta T \quad \text{in kcal/s} \quad (8.13)$$

where U is the overall heat transfer coefficient kcal/m² K s
A is the area of surface in m²
ΔT is the temperature differential in K

A typical example is the combined heat transfer from flue gas to steam in an SH/RH through the tube metal. Heat is transferred by convection through the boundary layers developed on either side of tube wall while it is transferred by conduction across the tube metal.

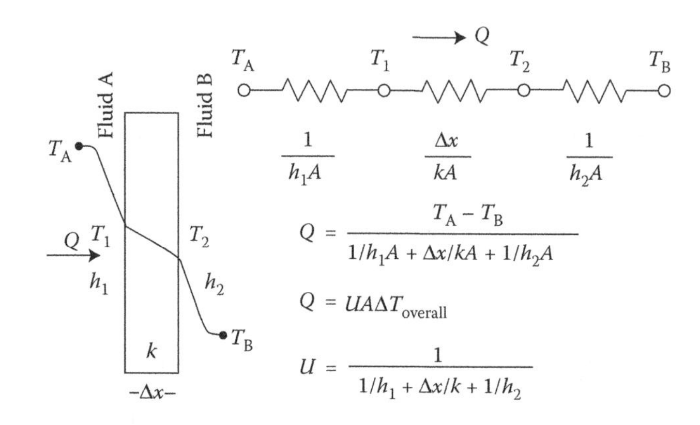

FIGURE 8.43
Combined heat transfer.

There are two convective heat transfer coefficients (h_1 and h_2) for the two fluid films inside (steam) and outside (gas) the tubes. Thermal conductivity (k) and thickness (Δx) of the tube wall must also be accounted for. The relationship of the overall heat transfer coefficient to the individual conduction and convection terms is shown in Figure 8.43.

When heat is transferred by hot gases to tubes carrying steam or water, it is the combined heat transfer of radiation and convection.

8.30.5 Dimensionless Numbers

Dimensionless number is one that represents a property of a physical system but which is not measured on a scale of physical units such as mass, distance and time. It is a ratio of various physical properties (such as ρ or heat capacity) and conditions (such as flow rate or weight) of such nature that a resulting number has no defining units.

Dimensionless numbers are also known as non-dimensional parameters.

There are numerous dimensionless numbers of which the following six numbers can be considered as the most important and useful in the area of heat transfer and fluid flow.

1. Reynold's number (Re)
2. Prandtl number (Pr)
3. Nusselt number (Nu)
4. Stanton number (St)
5. Grashoff number (Gr)
6. Rayleigh number (Ra)

8.30.5.1 *Reynold's Number (Re)* (see also *Moody's Friction Factor*)

Re is a dimensionless number representing the *ratio of inertial forces to viscous forces* in fluid flow.

It indicates the type of fluid flow whether viscous/ laminar or turbulent.

The viscous and inertial forces in flow are roughly proportional to velocity and velocity^2, respectively.

It follows that the loss of pressure head is roughly proportional to v in laminar flow and v^2 in turbulent flow, respectively.

Osborne Reynolds (1842–1912) demonstrated that a flow exists in three regimes—*laminar, transition* and *turbulent*, as shown in Figure 8.44—depending on the predominance of either viscous or inertial forces. For this, he made use of a simple experiment of passing a mixture of water and dye of the same SG through a horizontal glass tube.

$$\text{Re} = \frac{\text{Inertial force}}{\text{Viscous force}} \tag{8.14}$$

$$= \frac{\rho v^2}{\mu v/D} \tag{8.15}$$

$$= \frac{\rho v D}{\mu} \tag{8.16}$$

$$= \frac{GD}{\mu} \tag{8.17}$$

where

G = Mass velocity in kg/m^2 h
D = Pipe diameter in m
μ = Absolute fluid viscosity in kg/m h
ρ = Density in kg/cum
v = Velocity in m/s

At low fluid velocities (Re < 2000), it is the viscous forces that are dominant, making it a **viscous** or **laminar flow**. At higher velocities (Re > 4000), it is the inertial forces that are dominant, making it a turbulent flow. In between, it is the transition that's captured in Moody's diagram (Figure 6.26) which shows the relationship between Reynold's number and relative roughness as well as friction factor.

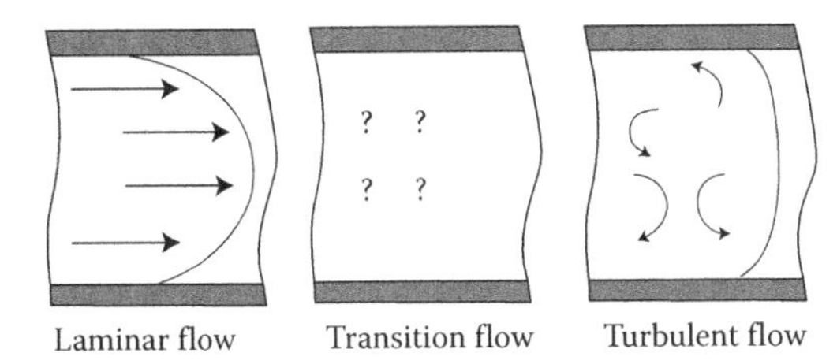

FIGURE 8.44
Types of fluid flow—laminar, transition and turbulent.

8.30.5.2 Prandtl Number (Pr)

Named after the German physicist Ludwig Prandtl (1875–1953), Pr is a dimensionless number used in the study of diffusion in flowing systems, essentially *to gauge the ease of heat flow through a fluid, particularly in forced and free convection.*

It is the ratio of the momentum diffusivity or kinematic viscosity to the molecular or thermal diffusivity.

It represents the temperature gradient similarity.

Symbolised Pr, it is also known as Schmidt number 1 (N_{Sc}).

$$\text{Pr} = \frac{\text{Kinematic viscosity}}{\text{Thermal diffusivity}} \tag{8.18}$$

$$= \frac{\mu/\rho}{k/\rho C_P} \tag{8.19}$$

$$= C_P \frac{\mu}{k} \tag{8.20}$$

8.30.5.3 Nusselt Number (Nu)

The dimensionless Nusselt number Nr, named after Wilhelm Nusselt (1882–1957), is the ratio of convective heat transfer to conductive heat transfer normal to the boundary layer.

$$Nu = \frac{UcD}{K} \tag{8.21}$$

where

Uc = Convection conductance
D = Pipe diameter in m
K = Thermal conductivity in w/m/°C

Nusselt number close to unity, means that convection and conduction are of similar magnitude, is characteristic of 'slug flow' or laminar flow.

A larger Nusselt number corresponds to more active convection, with turbulent flow typically in the 100–1000 range.

Nusselt number can be expressed as the product of the Stanton, Reynolds and Prandtl numbers:

$$\text{Nu} = \text{St Re Pr} \tag{8.22}$$

$$\frac{UcD}{K} = \left\{\frac{Uc}{\rho Vcp}\right\}\left\{\frac{\rho VD}{\mu}\right\}\left\{\frac{C_P\mu}{K}\right\} \tag{8.23}$$

8.30.5.4 *Stanton Number (St)*

Stanton is the ratio of heat absorbed to that available for absorption in a fluid

$$St = \frac{Uc}{CpG} \tag{8.24}$$

where
- Uc = Convection conductance
- Cp = Specific heat at constant pressure
- G = Mass velocity

8.30.5.5 *Grashof Number (Gr)*

Named after the German engineer Franz Grashof (1826–1893) this dimensionless number is *ratio of buoyancy to viscous forces acting on fluid*, usually in cases of *natural convection*.

It is the ratio of product of inertial and buoyancy force to the square of viscous forces.

The role of Gr in free convection is not unlike that of Re in forced convection.

Higher the Gr, greater is the free convection coefficient.

Nusselt number is a function of the Grashof number and the Prandtl number alone.

$$Nu = f(Gr, Pr) \tag{8.25}$$

8.30.5.6 *Rayleigh Number*

In many instances, it is better to combine the Grashof number and the Prandtl number to define a new parameter, the Rayleigh number

$$Ra = Gr \times Pr \tag{8.26}$$

The most important use of the Rayleigh number is to *characterise the laminar to turbulence transition of a free convection boundary layer flow*. For example, when $Ra > 10^9$, the vertical free convection boundary layer flow over a flat plate becomes turbulent.

When the Rayleigh number is below the critical value for that fluid, heat transfer is primarily in the form of conduction; when it exceeds the critical value, heat transfer is primarily in the form of convection.

8.31 Heating/Calorific Value (hv/cv) of Fuels (*see* Thermodynamic Properties)

8.32 High Temperature Corrosion (*see* Ash Corrosion)

8.33 Hog Fuel (*see* Agro-Fuels)

8.34 Hooke's Law (*see* Metal Properties)

8.35 Hoppers

The purpose of boiler hoppers is to collect ash at the bottom of tube banks. They are also provided to facilitate gas turns in flues as the dust separates and collects at the bottom.

Most hoppers operate at <450°C and are made of CS. At higher temperatures, such as below the BB, they are lined with either calcium silicate blocks or refractory tiles depending on the erosive nature of flue gases. Alloys or ss hoppers are somewhat unusual.

The valley angles of hoppers should exceed the angle of repose of ash at that condition to prevent bridging of outlet funnels due to build up of ash on the sides. Also, the ash outlet ports should be generous to prevent choking, particularly during shut down conditions as ash is highly hygroscopic in nature.

8.36 Horizontal Mills (*see* Pulverisers *and* Beater Mills in Lignite and Firing)

8.37 Horse Shoe Furnace (*see* Pile Burning)

8.38 Hot Gas Generators

Hot gas generators (HGG), either one or two in number, are employed in CFBC boiler design for providing hot gas during start-up of the boiler with oil as start-up fuel. See the description in Section 2.38.10 and Figure 3.12 of HGG in heating the bed for start-up under the topic of Circulating Fluidised Bed Combustion.

The purpose of HGG is to deliver high-temperature gases by mixing a lot of air to dilute the hot gases produced by the oil burners. Several variations in design are possible to meet this end.

In conventional design of HGG, it is a horizontal empty vessel lined with refractory having oil burner on one end and hot gas discharge to the windbox on the other. The vessel is of double shell construction with annular gap provided for passage of air, which helps in shell cooling as well as conservation of heat. Part of the primary air is diverted into the circular burner of the HGG to produce a short and bushy flame. Typically, excess air levels can be as high as 70%. The remaining air is channelised around the shell in the annular air passage before mixing with hot gases after their combustion.

HGG is used only during the start-up of the boiler to burn oil and generate hot gas at ~900°C (~1650°F) which heats the bed positively from the bottom until it attains the **ignition temperature** of the main fuel. This varies from 400°C to 600°C (750°F to 1110°F) depending upon the FC of the coal. As the bed is heated from the bottom with very hot gas at 900°C, bed heating is rapid which contributes to the shortening of start-up time of the boiler.

8.39 Humidity (*see* Psychrometric Chart)

8.40 Hybrid Boilers (*see* Smoke Tube Boilers)

8.41 Hydraulic Diameter (*see* Equivalent Diameter)

8.42 Hydraulic or Hydrostatic Testing

HT is performed on a pressure holding part, pressure holding component or completed boiler at the end of manufacturing, fabrication, erection and repair to simulate actual p and t holding condition and check for the leakages of both parent material and the welds. HT is, thus, spread over several stages to assure the integrity of the material and fabrication and timely detection of any fault. HT is considered as a very reliable and effective way of inspection not subject to interpretation.

Any boiler PP experiences both p and t conditions in its actual working while HT can only test for pressure. To create higher levels of stress that would have been induced because of the temperature effect, the HT is performed at 1.5 times the **design pressure**.

Tubes, pipes, castings and forgings are the main PPs bought from the steel mills. All of them are hydraulically tested before purchase. As HT is a very cumbersome and time-consuming procedure to be individually performed for several hundred pieces, it is normal these days to perform online eddy current testing instead for mainly tubulars, which is accepted by most codes.

Fabricated boiler components such as the **drums, headers, pipes, panels, coils, pendants, platens** and others are all subject to HT before despatch to site at 1.5 times the design pressure when the final test at site is done at 1.25 times. It may be possible to waive testing of some components in the shops but in such cases the site testing of completed unit is required to be done at 1.5 times the design pressure of the boiler. Boiler shops have a separate HT enclosure with pump and draining facilities to carry out HT on various completed PP assemblies.

The salient features of the HT performed at site after PP erection are

- Unlike shop HTs which are routine, the HT at site is a major milestone, as it marks the successful completion of all PP erection and permits further work, like refractory and insulation erection.
- Water for HT should be clean, clear and preferably softened for all surfaces that will be drained immediately after the test. Non-drainable banks like SH and RH are to be filled with demineralised (DM) water or condensate, treated with 10 ppm of NH_3 for pH control and 500 ppm of NH_4 for O_2 control.
- Water temperature should be >20°C to keep the tube metal temperatures above the dew point of the surrounding air to prevent condensation of M that can obscure the small leakage from detection. Also, it should be low enough for comfortably touching the PP. Too high a water temperature promotes flash steam in a leak, making the task of leak detection rather difficult.
- Venting at all levels should be carried out to fill all parts with water as any entrapment of air will prevent raising full pressure.

- Another important precaution to be taken is to fit the hydrostatic plugs in the SVs to prevent the seats from damage and gag them (at ~70–80% lift pressure) to avoid lifting of valves as the pressure rises above the set valve pressures.
- While boiler filling is first done by **fill pump** or **BFP**, pressure raising is done with the help of HT pump which is normally a small reciprocating pump. Inspection is carried out continuously as the pressure is raised.
- Water is drained to a suitable level and repairs carried out in case of detection of any leak. On reaching the full boiler test pressure it is held for a sufficiently long time to carry out a thorough inspection and to detect any minute leaks.
- As the official HT is required to be witnessed by an external inspector, either from the state department or the inspection agency, an internal HT precedes the final HT.

8.43 Hydrogen Embrittlement (*see* Water)

9

I

9.1 Ignifluid Boilers

Ignifluid technology developed and patented in the 1950s by the former Babcock Atlantique of France is the forerunner of the modern **expanded-bed CFBC**. It was truly the first commercialised FBC boiler system gaining popularity across many continents.

This technology was developed at a time when the pollution norms were not so stringent and large *SS*-fired boilers faced capacity limitations on coal at ~150 tph. The idea was to counter industrial *PF*-fired boilers with an FBC stoker for captive and cogen market.

The combustion temperatures would be anywhere from 900°C to 1200°C depending on the fusion point of ash in coal, with only staged combustion. **Fluidising velocity** was ~4 m/s. This is the range of temperatures achieved in the *high-temperature fluidisation*, as there were neither the bed tubes nor ash re-circulation to lower it to the classical 850°C. The molten ash at this temperature would mildly soften and come together to form clinker, which tended to settle out of bed because of its higher ρ. The inclined **CG**-like combustor not only helped in differential distribution of air along the length of the bed, but also discharged the clinker into the ash pit. The combustor would be roughly one-fifth of the width of an equivalent **TG** making the boiler quite narrow and tall with no inert bed material but only hot semi-molten coke at the bottom of the combustor. The bed was highly reactive and responsive. The load changes were extremely rapid. With no appreciably high carryover of fly ash than in the SS-fired boiler, the PP construction and layout was very simple. The boilers resembled the TG-fired boilers in most ways.

When the oil crisis of the 1970s struck markets, this innovative technology was fully mature and ready to move up in size to meet the small utility boiler range. Boilers as large as 380 tph/100 MWe were built. The boilers were made to fire very wide range of fuels, starting from **anthracite** to **lignite**.

As the pollution norms began tightening, the limitations started showing up glaringly.

- Even for low ash bearing coals, $\mathbf{NO_x}$ levels could not be lowered to <300 mg/Nm^3 as against 200 mg/Nm^3 that the competing FBC technologies could demonstrate.
- Likewise, the sorbent blown through the SA nozzles on the side walls could desulphurise only to the extent of 60% against 85–95% by the new FBC boilers.
- It gained popularity with low ash coals but with high ash coals, the unburnt losses were as high as 6%, compared with <4% for BFB and <1% for CFB.

By the mid-1990s, this technology lost out to the new FBCs mainly because of the reasons of poorer environmetal compliance and lower efficiency. The parent company was also in no shape to invest and modernise.

A typical ignifluid boiler installation is shown in Figure 9.1.

9.2 Igniters

An igniter is an essential component of every **burner**. The combustion process of the burner is set into motion by the igniter.

Every igniter has two functions namely, to provide for the main flame with

- The ignition energy for light up
- Heating support either intermittently or continuously the for stabilisation

NFPA 85E (Standards for Prevention of Furnace Explosion in PF-fired multiple burner boiler furnace) classifies the igniters as continuous, intermittent and interrupted types as laid out in Table 9.1

Igniters can be stationary or retractable. Stationary igniters are simple and maintenance-free but need adequate supply of scavanging air to keep them clean and cool. Retractable igniters are therefore preferred, particularly for large burners. Portable igniters are used for small duties in package boilers.

Figure 9.2 captures the action of a typical igniter which is a gas–electric igniter here.

Broadly, there are four types of igniters for burners for boiler application.

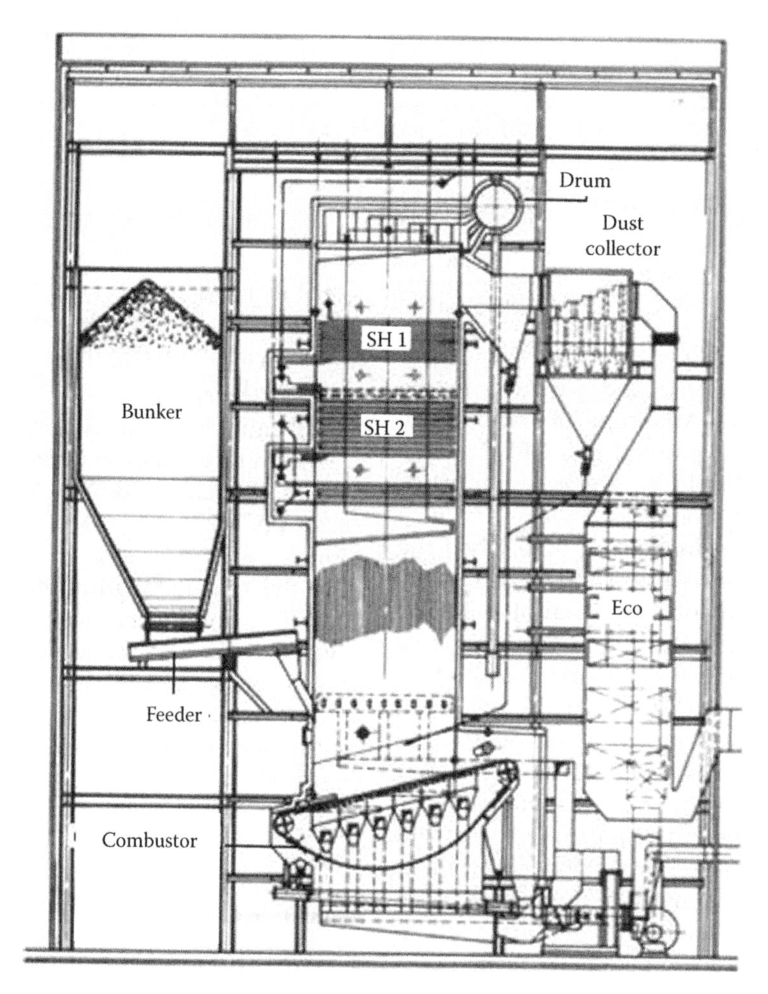

FIGURE 9.1
Typical Ignifluid boiler installation.

9.2.1 Gas–Electric Igniters

Gas–electric (GE) igniters are by far the most common of all four types. Here, an electric spark ignites a gas flame, usually NG or LPG, which in turn ignites the main flame. The gas flame is usually large, ~500 mm in diameter and 1500 mm long which assures a positive light up of the main flame. Also, continuous or intermittent supports can be given to the main flame when required. The igniter needs gas supply at pressures of 75–175 mm (3″–7″) wg. The capacity of the igniter is ~1% or more of the burner duty. A separate flame monitor or flame rod is built into the system to view the igniter flame independently.

If any high cv FG is available in the plant, GE igniters are the first choice. They are clean, reliable, durable and almost maintenance-free.

Figure 9.3 shows a typical GE igniter.

9.2.2 High-Energy Arc Igniters

When there is no gas in the plant, high-energy arc (HEA) igniters are used to ignite the main flame directly

TABLE 9.1
Igniter Classification as per NFPA 85E

Class	1	2	3	3 special
Igniter Type	Continuous	Intermittent	Interrupted	Direct Electric
Load	>10%	4% to 10%	<4%	
Duty	Ignite and support under any burner light off or operating condition	Ignite under light off and support under low load/adverse condition	Ignite and turn off after timed trial. Continuous supporting prohibited	High-energy igniter capable of directly capable of igniting main burner

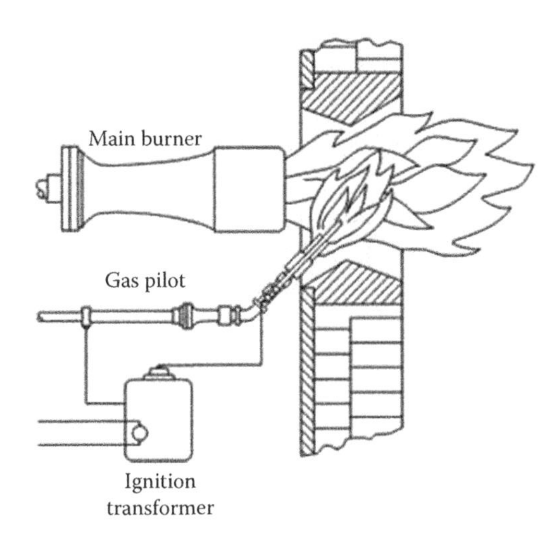

FIGURE 9.2
Action of igniter.

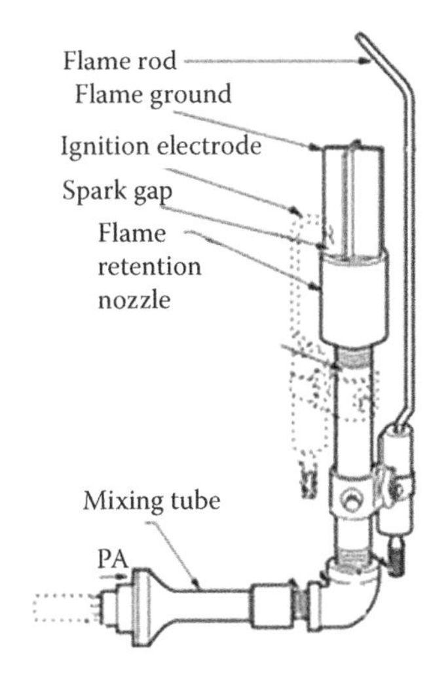

FIGURE 9.3
GE igniter.

by means of an electric arc. Even low-grade FOs can be lighted up safely. The only requirement is that the electrode should be free of any C and other deposits to ensure a strong spark is needed for a safe light up. The output power level can be as high as 2–5 kW of peak power per spark which is adequate to ignite even heavy residual oil. The ignition energy is released in five or six rapid pulses per second. HEA igniter system consists of an exciter, flexible cable, a spark rod, a tip and a retractor.

Figure 9.4 shows a typical HEA igniter.

9.2.3 Hot Carbon Rod Igniters

This type of igniter is also for direct ignition, even for heavy fuel oil (HFO). There is no sparking. It works on the principle of heating a C rod to a very high temperature of 2500°C. The electrode is typically a 300-mm long and 6–8 mm diameter special C rod with a Cu–Ni coated outer surface. When high current at low voltage is passed through the rod, within seconds, it reaches a high temperature when the outer surface burns off and exposes ~25 mm of white hot C which ignites any fuel. There is a loss of C and the electrode life is typically ~250 starts.

9.2.4 Spark Electric Igniters

These are for small burners. The power output is <300 W. The initial and running costs are both low. These igniters are suited for lighting up gases and light oils.

Table 9.2 gives a comparison of all the four types of igniters.

9.3 Ignition Temperature (*see* Combustion)

TABLE 9.2
Comparison of Igniters

Type of Igniter	Spark Electric	Gas–Electric	High-Energy Arc	Hot Carbon
Voltage, V	5000–10,000	5000–10,000 +oil/gas	1500–2000	10–20
Power, W	300	300+	2000–5000	3000
Burner capacity	Small	Medium and large	Medium and large	Large
Initial cost	Low	Average	Average	Average
Running cost	Low	Average	Average	High

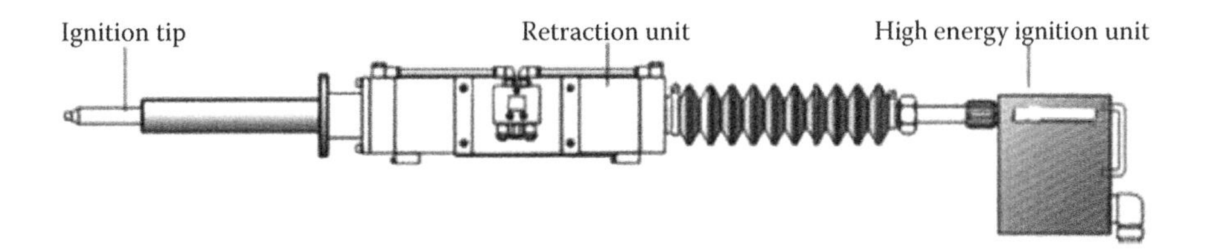

FIGURE 9.4
HEA igniter.

9.4 Implosion (*see* Operation and Maintenance Topics)

9.5 Incineration

Incineration is the combustion of solid waste fuels. The organic constituents burn to produce heat and flue gas. The inorganics burn to produce heat and ash. When the heat is recovered from an incineration process it becomes a waste-to-energy (WTE) plant.

9.6 Inconel

Inconel steels are a group of ~20 specialty alloys of Special Metals Corporation that use higher amounts of Ni and Cr% than in ss along with other elements in small quantities. Inconel alloys are in a group of metals known as the Ni-based super alloys. They are quite expensive and therefore usually reserved for applications where ss is inadequate.

One outstanding characteristic of high-Ni alloys is their good resistance to a wide variety of corrosive forces. With few exceptions, Inconel alloys do significantly better than martensitic, ferritic or austenitic ss in corrosive environments.

The other area is their ability to withstand very high temperatures and yet retain their strength as they are extremely resistant to oxidation.

Inconel is increasingly used in the **MSW** boilers in WTE plants for **cladding** of furnace and other tubes to withstand high-temperature corrosion. They are also used in high-temperature sections of SH and RH tubes in **SC** boilers.

The chemical composition of some of the common Inconel steels is shown in Table 9.3.

9.7 Indirect Firing (in Pulverised Fuel Boilers) (*see* Pulverised Fuel Firing)

TABLE 9.3

Composition of Some Commonly Used Inconel Steels

Inconel	Ni	Cr	Fe	Mo	Nb	Co	Mn	Cu	Al	Ti	Si	C	S	P	Bo
600	72.0	14.0–17.0	6.0–10.0				1.0	0.5			0.5	0.15	0.015		
617	44.2–56.0	20.0–24.0	3.0	8.0–10.0		10.0–15.0	0.5	0.5	0.8–1.5	0.6	0.5	0.15	0.015	0.015	0.006
625	58.0	20.0–23.0	5.0	8.0–10.0	3.15–4.15	1.0	0.5		0.4	0.4	0.5	0.1	0.015	0.015	
718	50.0–55.0	17.0–21.0	Balance	2.8–3.3	4.75–5.5	1.0	0.35	0.2–0.8	0.65–1.15	0.3	0.35	0.08	0.015	0.015	0.006

9.8 Industrial Boilers (*see also* 1. Boiler, 2. Boiler Classification and 3. Utility Boilers)

Industrial boilers, in the first place, have steaming conditions that are more modest than **utility boilers**, with evaporation <400 tph, SOP <150 bar, SOT <565°C and with no RH. They supply process steam and/or power (either captive or cogen power) to the industry and are not connected to the power grid.

They are further characterised by the following features. They

- Come in a wide variety of configurations to meet a range of steaming conditions and fuels.
- Burn a variety of fuels—various prime, waste and manufactured fuels found in an array of industries they serve, as well as bio-fuels.
- Also produce steam for district heating in cold countries.
- Are often required to have multi-fuel firing flexibility to use the available fuels, optimise the use of the cheapest fuel or use the fuel of the season in case of bio-fuels.
- Are expected to integrate with the dynamics of the main plant seamlessly when used for generating captive power or in cogen mode.

Fuel situation is getting progressively complex with waste fuels from industry increasing, as process industries are becoming larger, more complex and more varied. With increasing attention that biomass firing is receiving the bio-fuels turned over to boilers are also increasing. These developments are increasing the variety in the industrial boiler field.

Industrial boilers are quite different from utility boilers as explained in detail under the heading of utility boilers.

9.8.1 Classification of Industrial Boilers

By way of boiler construction industrial boilers can be divided into two main groups:

a. Bi-drum boilers
b. Single-drum or radiant boilers

By the type of fuel and firing equipment

1. Oil/gas-fired boilers
2. Grate-fired boilers
3. FBC boilers
   a. BFBC
   b. CFBC
   c. Expanded bed
4. PF boilers
5. Boilers with pile burning
6. Hrsg behind GTs
7. Whrb in process industries

9.8.1.1 Bi-Drum Boilers

Modern day industrial and captive/cogen boilers are made in either single- or bi-drum design. In the former times when the fuels were better, steaming conditions were modest and pollution norms were simpler, the boilers relied on smaller furnaces and larger BBs. Multiple steam and water drums with several convection banks formed the bulk of HS in those boilers, as shown in Figure 8.30. As these underlying conditions have significantly changed, the furnaces became larger and BBs have shrunk. Boilers are now made with a single BB or none at all, as in many European designs.

Bi-drum boilers are the most popular of all boiler designs and are highly standardised. A variety of designs are evolved to suit many fuels. All package boilers are bi-drum type. They form the back bone of industrial boilers both for both process and power.

- Bi-drum boilers are suitable for drum pressures up to ~120–140 bar depending on the ability to expand the bank tubes in the thick drums.
- With a good water-holding capacity, the bi-drum boilers are more forgiving for certain levels of operational deficiencies.
- Boiler dynamics are slightly sluggish in comparison with the single-drum boilers having no BB.
- For the same steaming conditions the bi-drum boilers tend to be a little heavier.

9.8.1.2 Single-Drum or Radiant Boilers

These boilers, generally used for drum pressures >60 bar, are not only popular for larger HP industrial and cogen applications but also completely dominate the utility market.

As the boiler pressures rise the proportion of latent heat in the total heat of steam decline. Figure 9.5 shows how precipitously the latent heat drops (to nearly 25%!) as the boiler conditions change from 10 bar saturated to 185 bar 540/540°C. Correspondingly, the need for Evap surface sympathetically reduces.

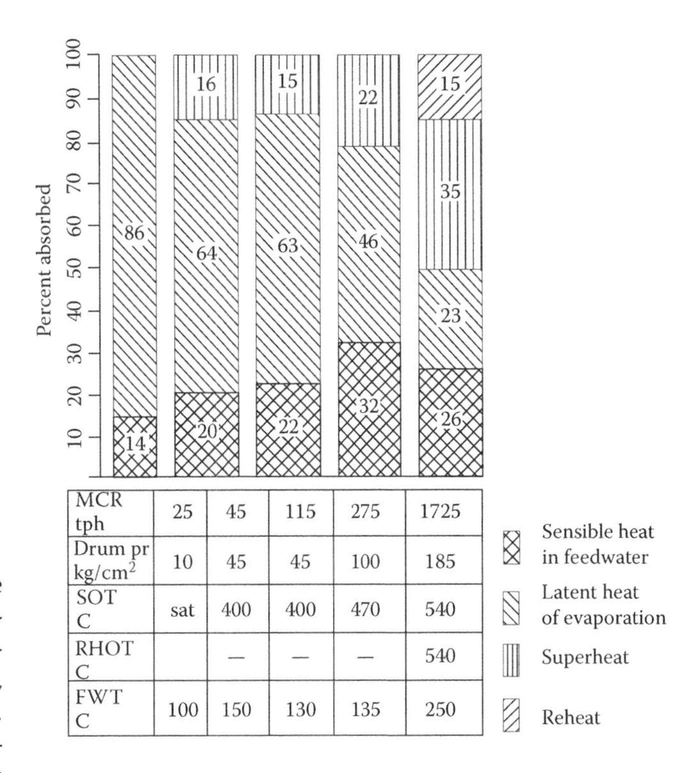

FIGURE 9.5
Reduction of latent heat in steam with rising pressures.

In Table 9.4 the same fact is captured in terms of HS. From 45% at 14 barg saturated the HS of BB reduces to 10% at 125 barg and 540°C or even 0% beyond that. With rising pressure it is a natural progression from bi-drum to single-drum boilers.

Radiant boilers: Single-drum boilers are also called radiant boilers because the heat transfer is substantially by radiant heat to SH. HP bi-drum boilers are also often referred to as radiant bi-drums for the same reason.

Single-drum boilers have lesser water holding and hence lower thermal inertia making the design amenable to faster start ups and shutdowns. The overall boiler response is quicker. Approximately 60–120 bar can be considered as an overlapping range for single- and

TABLE 9.4
Increasing SH and reducing Evaporator Surfaces with Rising Steam Pressures

Boiler Type	Bi-drum	Bi-drum	Bi-drum	Single Drum
Drum pr, kg/cm^2 g	14	42	125	125
SOT, °C	sat	400	540	540
FWT, °C	115	115	175	175
Furnace HS%	55	58	64	64
BB HS%	45	30	10	0
SH HS%	0	12	26	26
Eco HS%	0	0	0	10

bi-drum boilers. In that range bi-drums generally work out a little heavier as there is an additional drum and convection bank. Single-drum boilers may demand a slightly better instrumentation and water quality. There is no comfort of large water holding to accommodate some level of operational errors.

9.8.2 Industrial Boilers Can Also Be Classified on the Type of Fuel and Firing Equipment

9.8.2.1 Oil/Gas-Fired Boilers: (see also *Oil and Gas-Fired Boilers*)

Oil- and gas-fired boilers are burner fired. The boilers are compact as the fuels burn very quickly and there is no significant ash to contend with. Both package- and field-erected boilers are possible. These boilers are amenable to extensive standardisation because of the substantial uniformity of the fuel characteristics.

9.8.2.2 Grate-Fired Boilers (see also *Grates and Grate Types*)

Grate firing is the most common way to burn a variety of solid fuels to generate small to medium amounts of steam required for the industry to meet their process and power needs. A grate can be broadly described as *a set of shaped perforated plates located at the bottom of a furnace on which the fuels are fed from above (or even below sometimes) and air is admitted from below to enable combustion take place partly on the grate and partly above in the furnace.* The grates can be stationary, moving or reciprocating.

Workhorses of yesteryears, the fortunes of grates have been severely affected by the rise of BFBC boilers and the tightening of pollution norms, on both SO_x and NO_x.

Grates can be classified as (1) stationary and (2) mechanical grates.

1. *Stationary grates with semi-pile firing* such as pin hole (**PHG**) and inclined water-cooled grates (**IWCG**) are quite popular for wood and other bio-fuels.
2. *Mechanical grates* are highly suitable for burning efficiently the poor-quality prime fuels, fuel combinations and waste/bio-fuels.

   The mechanical grates can be classified as
3. **DG**
4. **CG/TG**
5. **PG/RG**

In all these firing modes, the bi-drum boilers are the industry standard for pressures up to ~60 bar and sometimes even up to 120–140 bar and the single-drum boilers at higher pressures. However single-drum boilers are also popular in certain markets like Europe in lower pressures.

Bottom support for boilers is possible and economical up to ~40 to 60 tph.

9.8.2.3 Fluidised-Bed Combustion Boilers (see also *Fluidised-Bed Combustion*)

BFBC, **CFBC**, Circofluid or **cold CFBC** system and **IR CFBC** boilers

There are three types of atmospheric FBC boilers—(a) BFBC, (b) CFBC and (c) expanded or turbulent fluidised bed—which are characterised by low-temperature combustion and in-furnace desulphurisation where required, meeting the air pollution norms fairly effortlessly. FBC is a genuine breakthrough in combustion technology in several decades. Commercialised in the late 1970s in response to the tightening air pollution norms and escalating oil prices, the FBC boilers have grown out of their teething problems and firmly established themselves by mid-1990s.

a. BFBC operate in the lower bed velocity regimes of fluidisation. The full-load air velocities employed in the bed are between 2 and 3 m/s at MCR to expand the bed to ~1200 mm of height. The minimum velocity required for fluidisation is ~0.8 m/s which is met even at the reduced boiler loads. With abrasive fuels the full load bed velocities are reduced to even 1.5–2 m/s to minimise the erosion of bed coils. Evap and sometimes SH coils are placed inside the expanded bed to abstract heat and contain the bed temperature when firing coals. The bed is composed of inert material or ash from coal and contains only ~3% of fresh fuel. Limestone is added to the bed only if desulphurisation is needed.

   The HRR in bed is comparable with the release in SS and hence the footprints of the two match closely. Likewise, with residence time requirement of ~2.5 s, the height of BFBC boiler is only slightly lower than the height of equivalent SS-fired boiler.

   For coal firing BFBCs are made to sizes as large as 150 tph, making it a perfect replacement of SS and suitable for both process and small power. Before the development of CFBC much larger boilers have been built for coal firing in 1980s at a few locations. With CFBC capable of addressing the higher sizes, these days the BFBCs for coal are accepted up to ~150 tph if the pollution norms are met. Large BFBCs of even ~400 tph are built in Scandinavia and other pulp-producing countries where the pulping process produces large amounts of sludge having high M. Heat of the bed helps in the combustion of such low Gcv fuel having high M. Such boilers have no bed coils as the M is adequate to keep the bed cool. See BFBC boilers with **open bottom** under the heading of BFBC boilers.

b. CFBCs or full circulation boilers, however, employ ash for cooling the bed together with staged combustion to contain the bed temperature and produce lower emissions. Fresh fuel and limestone, if needed, are added to the bed which is composed of inert material or coal ash. The residence times are higher between 4 and 5 s and velocities of fluidising air are also more at 4–7 m/s making the boilers very tall and slender. At these velocities there is no clear bed but only a cloud of particles filling the entire free board. CFBC boilers are taller than PF boilers and slightly stouter.

   Compared with the BFBC the C burn up and desulphurisation efficiencies are higher, load dynamics are superior and reliability is better (as there are no worn out or punctured bed tubes to periodically replace). The economical range of CFBC is upwards of 100 tph and going as high as ~2000 tph, making it straddle captive/cogen and utility boiler markets.

   CFBC is seen as a potential-utility boiler, rivalling PF, both in terms of efficiency and capacity. At present large sizes of utility boilers are made on the basis of CFBC principle, but for fuels generally not easily burnable in PF firing with ease, such as **pet coke**, coals with high S, **anthracite** and so on or for fuel combinations. The fractionally higher power consumption and lower thermal efficiency of CFBC at the present stage of development places it unfavourably with respect to PF for conventional fuels.

c. Expanded or turbulent fluidised beds employ full-load gas velocities of 3.5–4.5 m/s which are between the BFBC and CFBC range and have a clear expanded bed of inert material. Like in the CFBC the bed temperature control is by the combination of ash re-circulation and staged combustion. Developed a few years after both BFBC and CFBC were established, the expanded beds tried to combine the simplicity of BFBC and superior efficiency and dynamics of CFBC with good success. They are now beginning to break away from their industrial boiler mould and enter utility range.

9.8.2.4 Pulverised Fuel Boilers (see also *Pulverised Fuel Firing and Pulverised Fuel Boilers*)

PF firing is the most dominant of all the firing techniques and PF boilers constitute perhaps the single largest type. PF firing is employed to burn a variety of coals. CFBC has begun to slowly challenge and share the market with PF for the last couple of decades.

In large utility range PF is the undisputed choice for normal coals. With over eight decades of experience PF boilers have burnt almost every type of coal across the world—a record in which CFBC has to do much catching up. However in industrial range CFBC is fast displacing PF as industry values fuel flexibility and multi-fuel firing ability more than a fractionally higher efficiency. It is a matter of time before PF firing exits industrial market altogether.

PF-fired boilers have the highest combustion efficiency at 99+% which is fractionally higher than CFBC. The power consumption is marginally lower in spite of the additional milling plant in PF boilers. But in respect of fuel flexibility, multi-fuel firing capability and low NO_x generation PF is much inferior to CFBC while in-furnace S capture is not possible. A conventional PF-fired boiler can be cheaper than CFBC for coal, but when low NO_x burners are required to be installed along with reduced heat load in plan area and $deSO_x$ units are also to be added, the CFBC works out more economical in many cases.

PF boilers for industry, working at slightly lower p & t, come in highly standardised single- and bi-drum configurations.

Shown in Figure 9.6 is a typical industrial top supported non-RH PF boiler with front wall circular burner firing with conventional layout of all vertical mills located in front of the boiler.

9.8.2.5 Boilers with Pile Burning (see *Pile Burning*)

Pile burning is one the early methods of firing, mainly developed for burning bio-fuel wastes such as **wood bark**, saw dust, wood shavings, **bagasse** and so on. The method derives its name from the way the fuel is formed into a pile for burning. Pile burning can be in thick or thin piles, the latter being more modern and efficient. **Horseshoe furnaces** and **PHGs** are representatives of the two systems, respectively.

Thick pile burning is explained under the heading of Pile burning while the thin pile burning is covered in GFG under PHGs. See also **IWCGs** which are used in wood firing when M is high in the range 50–65%.

9.8.2.6 Heat Recovery Steam Generators Behind Gas Turbines (see *Heat Recovery Steam Generators*)

Hrsg is the only WH boiler in the utility sector. It is relatively new but has already grown into a big business because of the rise of gas based power generation. The

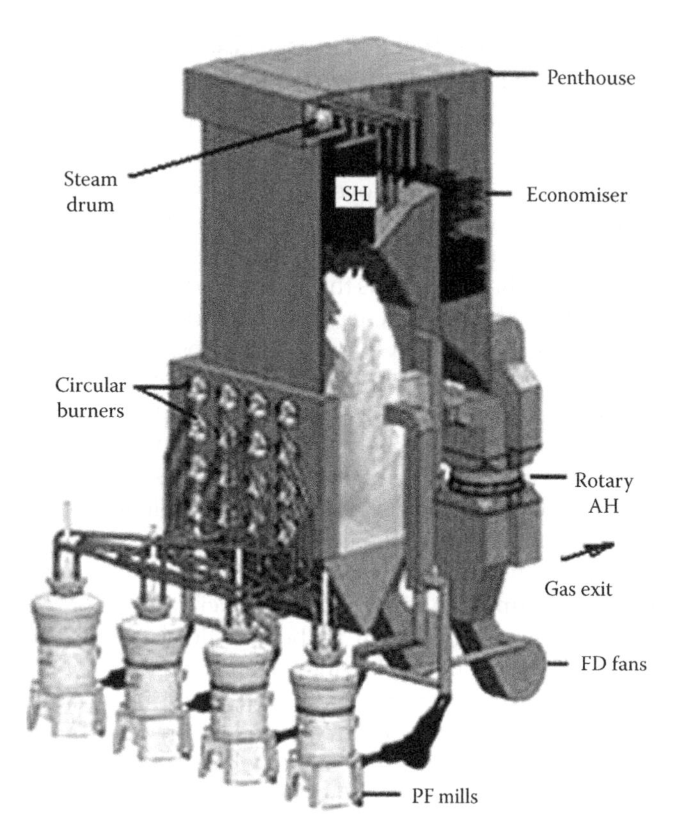

FIGURE 9.6
Single-drum industrial PF boiler with 4 mills and rotary AH.

sizes have also dramatically scaled up in a short time keeping pace with the growth of GTs. Hrsgs serve both industrial and utility power plants. Being a WH recovery unit even the biggest Hrsg is only of the size of a large industrial boiler. Its construction is entirely different from the conventional boilers being based on finned tubes and a succession of tube banks.

9.8.2.7 Waste Heat Recovery Boilers in Process Industries (see Waste Heat Recovery Boilers)

Process industry abounds in **WH**. Recovery of heat depends on the grade of heat and quantum, whether high, medium or low. Heat recovery is usually uneconomical from small quantities of low-grade heat. Also it depends on the mode of utilisation and the distance involved. In cement industry where no process steam is usually needed heat recovery has to be only for power generation making the scheme unviable in most cases. With the ever rising costs of fuels and power WH recovery is becoming more and more viable with passage of time.

Whrbs come in small to large sizes depending on the size of process plant, grade of heat and the amount and also the cleanliness of the flue gases, straddling the fire tube, combo and water tube designs. The applications are numerous. Only prominent Whrbs of mostly water tube design are covered under the topic of Whrb.

9.9 Inertia in Boilers

Resistance of boiler to load change is called inertia. It is a function of total weight consisting of boiler PPs and water holding. Higher this weight larger the inertia and slower is the response. For better dynamics single-drum boilers and smaller tubes (consequently thinner) are therefore preferred. Load response of SC boilers is superior to drum boilers because of the absence of the heaviest part such as drum and less tube weight accompanied by the reduction in water weight.

9.10 Initial Ash Deformation Temperature (*see* Ash Fusibility)

9.11 Input–Output Method (in Boiler Testing) (*see* Efficiency Testing)

9.12 Instruments for Measurement

There are several streams of solids and fluids in a boiler and most of them get measured at some place or the other in their passage through the boiler. Steam and water, air and gas, fuels and ash and chemicals and blowdown are the important streams and are common to almost all boilers. It is normal to measure and record their pressure, temperature, flow and analysis as appropriate. In case of steam, its purity and sometimes quality is also measured.

Test and commercial instrumentation are the two types. Test instruments are portable, many times delicate, need skilled operators to use and interpret and hence unsuitable for continuous normal use. Commercial instruments, however, are a little less accurate but rugged and reliable. Many times they can also be used for test purposes after re-calibration, as their accuracy levels are being steadily improved.

There are numerous instruments in use and their number is steadily increasing, but there are certain important and primary measurements which are typical to boiler practice. Only those are described here.

1. Air flow measurement
   a. Orifice meter/plate
   b. Venturi meters
   c. Aerofoils

2. Steam flow measurement
   Flow nozzle:
3. Flue gas analysis by Orsat apparatus
4. Flue gas O_2 analysers—Zirconia type
5. Measurement of fuel cv by bomb calorimeter
6. Viscosity measurement of oils
   a. Saybolt seconds
   b. Redwood seconds
   c. Engler, degrees for FOs
7. Velocity measurement
   a. Pitot tube
   b. Pitot-static or Prandtl tube
8. Pressure measurement
   a. Manometers
   b. Mechanical pressure gauges
   c. Piezometer
9. Steam purity measurement
10. Steam quality measurement
11. Temperature measurement
   a. Thermocouples (TC)
   b. Thermometers
   c. Pyrometers
12. Opacity measurement

9.12.1 Air Flow Measurement

Measuring of airflows in boiler is essential for controlling combustion. The principle behind airflow measurement is to measure the Δp across a calibrated restriction inserted in the air duct. There are three types of devices which are commercially used in boilers, namely

a. Orifice plates
b. Venturi meter
c. Aerofoils
d. Peizometer

Pitot tube is also used for air measurement. But being delicate, it is used more as a test and calibration than for permanent measuring device. Tubular AH can also be used for flow measurement as it can provide reliable and meaningful Δp.

All the devices aforementioned are sized to produce a Δp of ~75 mm wg (3″) at full airflow. There are well-established standards which give the constructional details of the equipment and the accuracy levels achievable. They are calibrated for various airflows and calibration charts are prepared in such a way that a Δp directly indicates the flow through the duct.

TABLE 9.5

Comparison of Flow Measuring Devices

Measuring Instrument	Orifice	Venturi	Aerofoil
Measuring difference, mm wg	75	75	75
Permanent loss, mm wg	40	22	18
Relative first cost	1	~1.5	~2
Comments	Used often in tempering air duct	Requires longer duct length than aerofoil	Multiple aerofoils reduce length

Orifice plate has the lowest cost and the highest permanent loss. Aerofoil is nearly twice as expensive with half the permanent loss. Venturi meter is mid way. Table 9.5 gives a more detailed comparison.

The type of flow measurement to be installed is usually guided by the duct layouts and space availability. Approximately 10 upstream and 5 downstream equivalent diameters (d_e) of straight length are required for all devices for flow straightening. Larger boilers generally use venturies and aerofoils while smaller boilers, where loss of power is not so significant, employ orifice plates. By adopting multiple aerofoils the straight length requirements can be cut down which is the reason for its greater popularity despite higher cost. Figure 9.7 gives an indication of pressure losses/recoveries in various devices.

Flow measurement entails power loss. In larger boilers the permanent pressure drop adds to the recurring fan power. It is essential that this loss is kept to a minimum even if it means employing a more expensive device.

9.12.1.a Orifice Meter/Plate

Orifice meter consists of accurately made flat plate, usually of ss with a circular hole having sharp edge drilled in it, with a pressure tap each on upstream and downstream. There are, in general, three methods of placing the taps. The discharge coefficient of the meter depends on the position of taps. The tap location is called

1. Flange location—Taps located ~25 mm upstream and downstream from face of orifice.
2. *Vena Contracta* location—Taps located 1 pipe dia (bore) upstream and 0.3–0.8 pipe dia downstream from face of orifice.
3. Pipe location—Taps located 2.5 times nominal pipe dia upstream and 8 times nominal pipe dia downstream from face of orifice.

A typical orifice plate arrangement within a pipe is shown in Figure 9.8. The tap location is at the vena contracta.

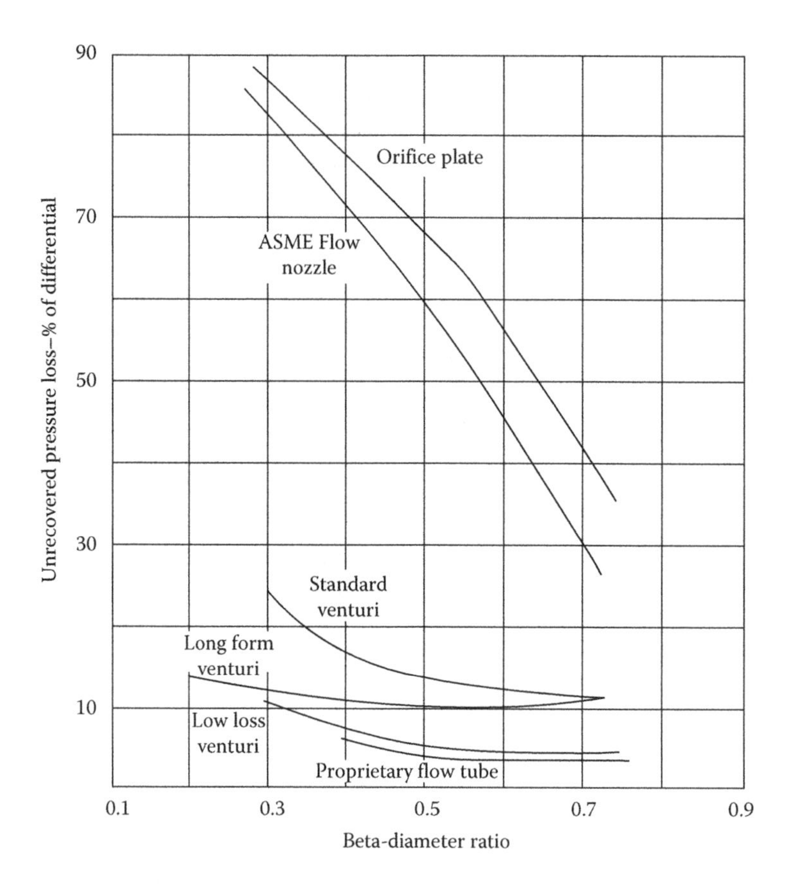

FIGURE 9.7
Permanent pressure loss in various flow measuring devices.

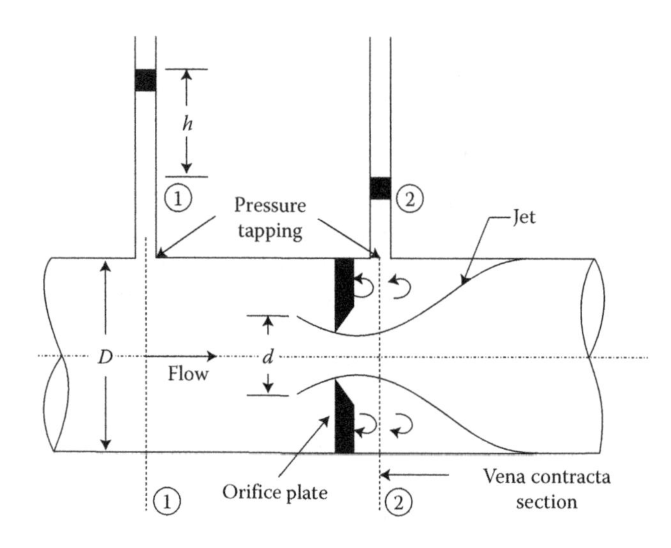

FIGURE 9.8
Arrangement of orifice plate in a pipe for flow measurement.

The discharge coefficient—c_d—varies considerably with changes in area ratio $(d/D)^2$ and the *Re*. c_d of 0.60 may be taken as standard, but it varies noticeably at low values of the *Re*.

The salient aspects of the orifice meters are that

- The pressure recovery is limited and the permanent pressure loss depends primarily on the area ratio. For an area ratio of 0.5, the head loss is ~70–75% of the orifice differential.
- They are suitable both for clean and dirty liquids and also some slurry services.
- The rangeability is 4–1.
- The pressure loss is *medium*.
- Typical accuracy is 2–4% of full scale.
- They require straight length of minimum *10* upstream diameters.
- The viscosity effect is *high*.
- The relative cost is *low*.

9.12.1.b Venturi Meters

A venturi tube is essentially a long converging–diverging nozzle carefully proportioned and accurately made with smooth and concentric contours to reduce friction losses to a minimum.

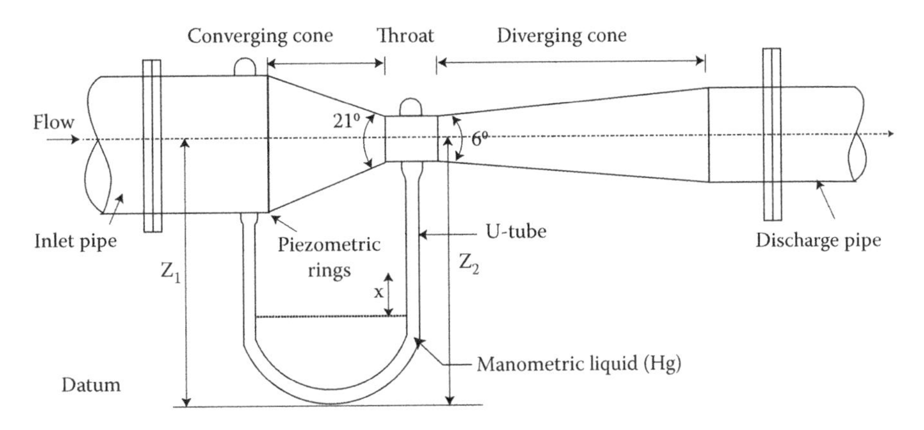

FIGURE 9.9
Venturi meter for flow measurement.

The fluid is accelerated through a converging cone of angle 15–21° and decelerated through a diverging cone of 5–7°. The pressure difference between the upstream side of the cone and the throat is measured which provides a signal for the rate of flow. See Figure 9.9 which shows a venturi in a round pipe. A venturi can also be in a rectangular duct with two sides maintained parallel.

The divergent cone is where the fluid slows down and most of the kinetic energy reverts to pressure energy. There is no vena contracta. The flow area is minimum and velocity maximum in the throat.

A discharge coefficient $c_d = 0.975$ can be indicated as standard, but the value varies significantly at low values of the *Re*.

The pressure recovery is much higher for the venturi meter than for the orifice plate which makes it ideally suitable where small pressure heads are available.

The other salient features of venturi meters are that the

- Venturi tube is suitable for clean, dirty and viscous liquids and some slurry services
- Rangeability is *4–1*
- Pressure loss is *low*
- Typical accuracy is 1% of full range
- Required upstream straight pipe length is *5–20* diameters
- Viscosity effect is *high*
- Relative cost is *medium*

9.12.1.c Aerofoils

Aerofoils are very popular for measuring the air flows in square and rectangular ducts, mainly because of the reduced straight length requirements, despite being more expensive than venturies.

When an aerofoil is introduced in a rectangular duct a half-venturi is formed between the straight edge and aerofoil with a converging–diverging flow pattern. The Δp between the throat and inlet is indicative of flow. In most parameters it is nearly the same as venturi. Aerofoils are very sensitive to the flow disturbances and need turning vanes upstream for straightening the flow. A straight duct length of at least two times the length of aerofoil is needed (as shown in Figure 9.10). Multiple

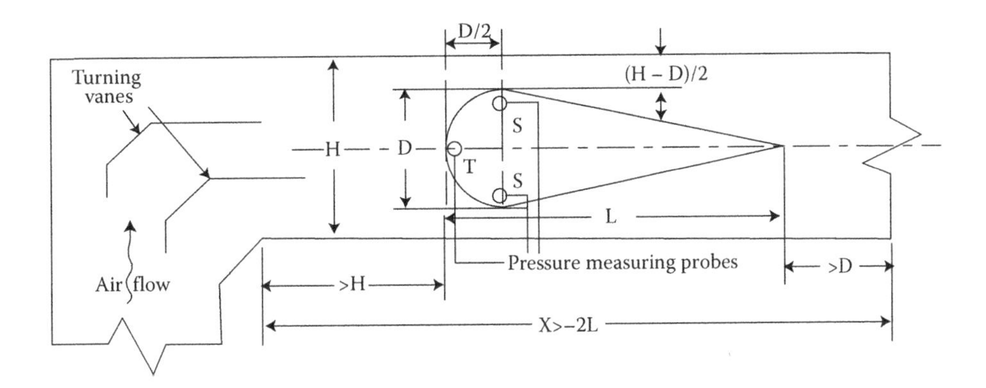

FIGURE 9.10
Aerofoil for flow measurement in ducts.

aerofoils, usually 3, are used in case this duct length cannot be met. The duct has to be suitably partitioned by plates with pipe stiffeners at both ends.

Air and gas flow measurement in boilers is done by orifice, venturi or aerofoil depending on the suitability. A comparison of the three measuring devices is presented in Table 9.5.

9.12.1.d Piezometer (see Pressure Measurement)

9.12.2 Steam Flow Measurement

Orifice meters and flow nozzles are the two common methods of steam flow measurement with the former being more popular in small boilers despite higher pressure losses.

Flow Nozzle

Like orifice meter a flow nozzle also works on the principle of Δp measurement. It is relatively simple, cheap, accurate and available for many applications in many materials.

Flow nozzle consists of a well proportioned and machined convergent nozzle, with no divergent nozzle like in a venturi. The energy recovery downstream of the converging nozzle is not as high and the dissipation of energy is high. The coefficient of discharge cd is influenced by only the converging nozzle and not the flow downstream. It is comparable with venturi.

The other salient features are

- It is suitable for both clean and dirty liquids
- The rangeability is 4–1
- The relative pressure loss is *medium*
- Typical accuracy is 1–2% of full range
- Required upstream pipe length is minimum 10 diameters
- The viscosity effect *high*
- The relative cost is *medium*

A typical flow nozzle is shown in Figure 9.11

9.12.3 Flue Gas Analysis by Orsat Apparatus

Orsat apparatus, as test instrument, is very handy and accurate (up to ± 0.5%) analyser for flue gas analysis by volume on dry basis. For acceptance tests of boiler, Orsat analyser is often insisted upon. CO_2, CO and O_2 are measured and N_2% is obtained by difference.

The apparatus consists of a water jacketed burette or eudiometer (A) graduated up to 100 cc and connected to an aspirator bottle (B) by a flexible rubber pipe (C). Flasks 1, 2 and 3 contain solutions of

- Potassium hydroxide (KOH)
- Freshly prepared Pyrogallic acid (trihydroxybenzene)
- Cuprous chloride ($CuCl_2$)

for absorbing CO_2, O_2 and CO (also ethylene and acetylene) gases, respectively. The duplicate flasks behind the main flasks are packed with small gas tubes to enhance the absorption rate. There is the main isolating cock D for gas retention in the system along with E, F and G, the three isolating cocks for the flasks.

The procedure consists of

- Aspirating 100 cc of flue gas with help of aspirator bottle
- Absorbing the gas in the three absorbents in a definite sequence
- Noting the absorption by difference in eudiometer

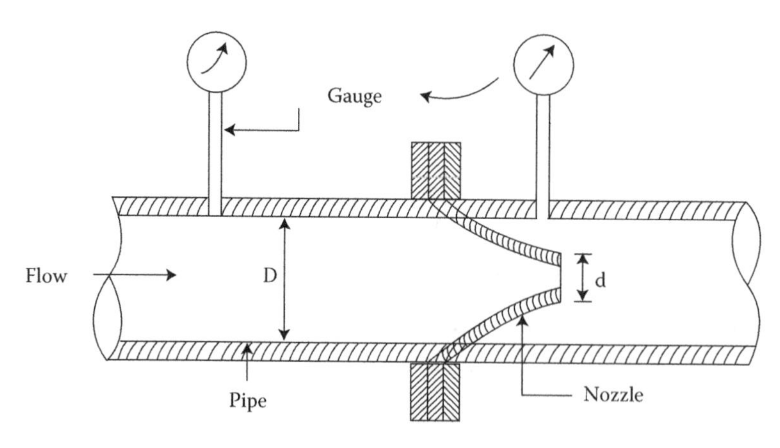

FIGURE 9.11
Flow nozzle for steam flow measurement.

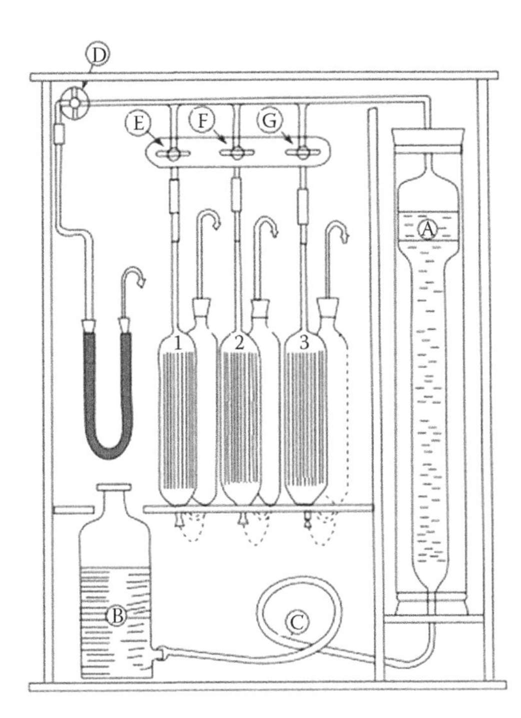

FIGURE 9.12
Orsat analyser for measurement of gas composition.

Great care and strict adherence to the procedure along with correct preparation of absorbents to the specified concentration are required to obtain error-free results. Often a suction pump is used to obtain gas sample as aspirator alone is not adequate to suck from gas ducts.

Figure 9.12 shows the picture of a typical Orsat analyser.

9.12.4 Flue Gas O_2 Analysers—Zirconia Type

O_2 measurement in flue gas is very important for monitoring boiler efficiency. There are several types of instruments to measure O_2 in flue gas, namely

- **Electrochemical**
- **Paramagnetic**
- **Polarographic**
- **Zirconium oxide/zirconia (ZrO_2)**

of which the last one is the most appropriate for the dust-laden high-temperature boiler flue gases.

ZrO_2 sensors use a solid-state electrolyte fabricated from stabilised ZrO_2, which is plated on opposing sides with platinum that serves as the sensor electrodes. For the sensor to operate properly, it must be heated to ~650°C when the Zr lattice becomes porous, allowing the movement of O_2 ions from a higher concentration to a lower one, based on the partial pressure of O_2. To create this partial pressure differential, one electrode is usually exposed to air (20.9% O_2) whereas the other electrode to the sample gas. The movement of O_2 ions across the ZrO_2 produces a voltage between the two electrodes, the magnitude of which is based on the partial pressure differential of O_2 created by the reference gas and sample gas. The pros and cons of the sensor are that

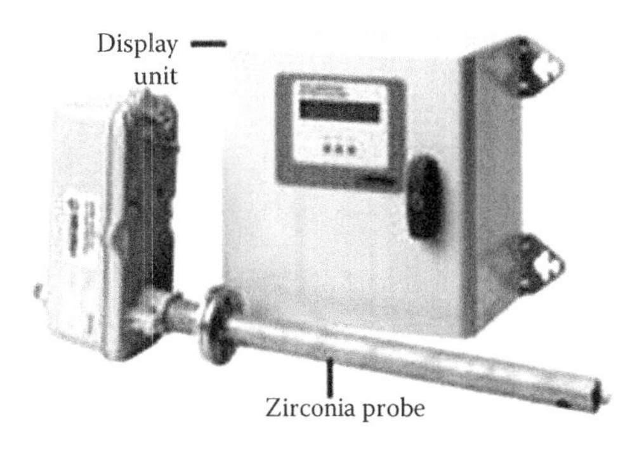

FIGURE 9.13
Oxygen analyser of zirconia type.

- It exhibits excellent response time characteristics.
- The same sensor can be used to measure 100% O_2, as well as ppb concentrations.
- Probe life is shortened by on/off operation in the high-temperature operation.
- It is unsuitable for trace measurements when reducing gases (HCs of any species, H_2 and CO) are present in the sample gas as they react with the O_2, consuming it before measurement itself thus producing a lower than actual reading. The magnitude of the error is proportional to the concentration of reducing gas.

ZrO_2 analyser is the 'defacto standard' for *in situ* combustion control applications. Figure 9.13 shows the zirconia probe and analyser/indicator. The measurement range varies from 0.01% to 25% O_2 with accuracy of ± 2% or better.

9.12.5 Measurement of cv of Fuel by Bomb Calorimeter

Heat changes in chemical reactions (combustion) are measured with calorimeters. Depending on the need different types of calorimeters are used. Reactions taking place involving gases at constant volume are carried out in a closed container with rigid walls that can withstand HP such as a bomb calorimeter. Heavy steel is used in making the body of the bomb calorimeter which is coated inside with gold or platinum to avoid oxidation and is fitted with a tight screw cap. There are two electrodes R_1 and R_2, which are connected to each other through a platinum wire S, which remains dipped in a Pt cup just below it (see Figure 9.14).

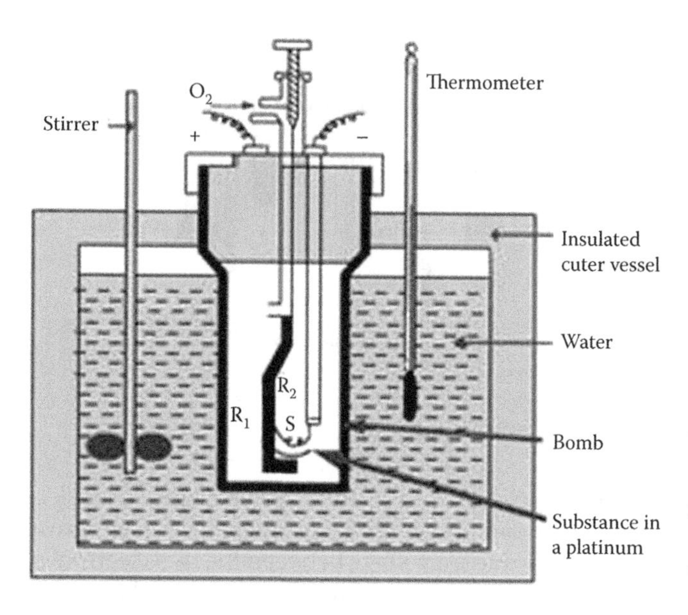

FIGURE 9.14
Bomb calorimeter for cv measurement.

Few grams of the sample are taken in the platinum cup. The vessel is then filled with excess of O_2 at a pressure of ~15–19 m of Hg and sealed. It is now dipped in an insulated water bath provided with a mechanical stirrer and a thermometer, sensitive enough to read up to 0.01°C (Beckmann's thermometer).

The initial temperature of water is noted and the reaction (combustion) is initiated by passing electric current through the Pt wire. The heat evolved during the chemical reactions raises the temperature of water which is recorded from the thermometer. By knowing the heat capacity of the calorimeter and also the rise in temperature, the heat of reaction or combustion at constant volume can be calculated.

Bomb calorimeter can be used to measure cv of the following groups of materials

- Building materials
- Solid fuels such as coal, coke and so on
- Liquid fuels such as gasoline, kerosene, FO and GT fuels
- HC fuels
- Food, supplements and crops
- Waste and refuse
- Combustible materials and so on

9.12.6 Viscosity Measurement of Oils

Viscosity measurement of FO and lubricating oils is important in boiler houses. Kinematic viscosity is measured indirectly by counting the time taken by oil to flow out of a specified viscometer cup at the required temperature, which can then be converted to cS using appropriate formulae.

9.12.6.a Saybolt Seconds

Saybolt seconds, popular in American practice, is the time taken in seconds for the oil at a particular temperature to flow out of 60 cc cup of the viscometer through a specified orifice. It is commonly measured at 100°F, 150°F and 210°F (37.8°C, 65.6°C and 98.9°C, respectively). The test method is prescribed in ASTM D88 (see Figure 9.15).

For relatively lighter oils, SSU test is employed. SSU is normally applicable up to 600 s.

For more viscous oils, Saybolt Second Furol (SSF) is used in which the orifice is a little larger.

SSU readings are roughly 10 times the SSF readings for the same oil.

To convert SSU to stokes, the following equations are used:

$$\text{Stokes} = 0.0026\ \text{SSU} - 1.95/\text{SSU} \text{ for } 32 < \text{SSU} < 100 \quad (9.1)$$

$$= 0.0020\ \text{SSU} - 1.35/\text{SSU} \text{ for SSU} > 100 \quad (9.2)$$

To convert SSF to stokes, the following equations are used:

$$25 < \text{SSF} < 40 \text{ s, stokes} = 0.0224(\text{SSF}) - 1.84/(\text{SSF}) \quad (9.3)$$

$$\text{SSF} > 40 \text{ s, stokes} = 0.0216(\text{SSF}) - 0.60/(\text{SSF}) \quad (9.4)$$

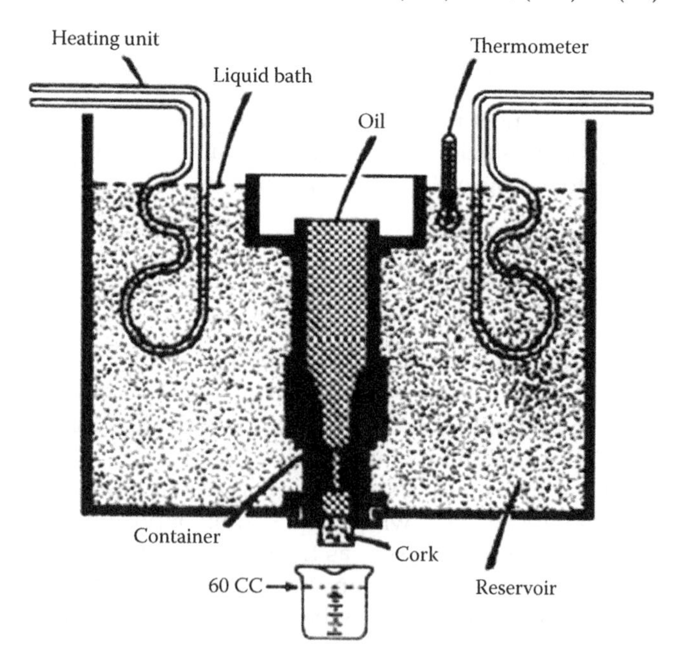

FIGURE 9.15
Saybolt viscometer for viscosity measurement of oils.

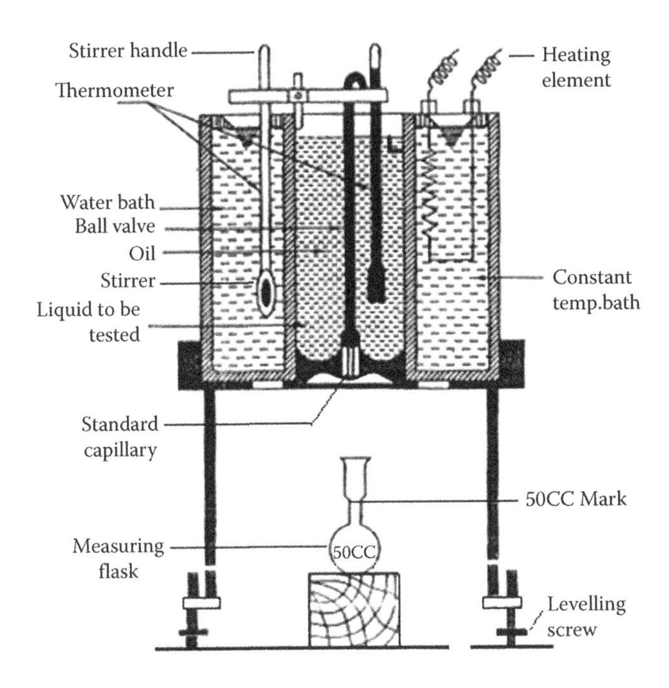

FIGURE 9.16
Redwood viscometer for viscosity measurement of oils.

9.12.6.b Redwood Seconds

Redwood seconds is a unit of kinematic viscosity, named after the English chemist Sir B Redwood and commonly used in Britain and other countries with British influence.

It is the time taken in seconds for 50 cc of the sample liquid to flow through a Redwood viscometer, depicted in Figure 9.16.

Viscometer no. 1 or no. 2 is used depending on whether the time of flow of oil at the desired temperature is > or <2000 s.

The difference between the two viscometers is the diameter of the orifice which is, 1.62 mm ø × 10 mm long for Redwood no. 1 and 3.5 mm ø × 5 mm long for no. 2. Thus, Redwood no.1 readings are 10 times Redwood no. 2.

The viscosity in centistokes is given roughly by the formula

$$0.260t - (0.0188/t) \tag{9.5}$$

where t is the flow time in seconds.

9.12.6.c Engler, Degrees for Fuel Oil

Degree Engler is still occasionally used in the UK. It is a measure of kinematic viscosity based on comparing a flow of the liquid under test to the flow of water.

Viscosity in Engler degrees is the ratio of the time of flow of 200 cc of the fluid under test to the time of flow of 200 cc of water at the same temperature in a

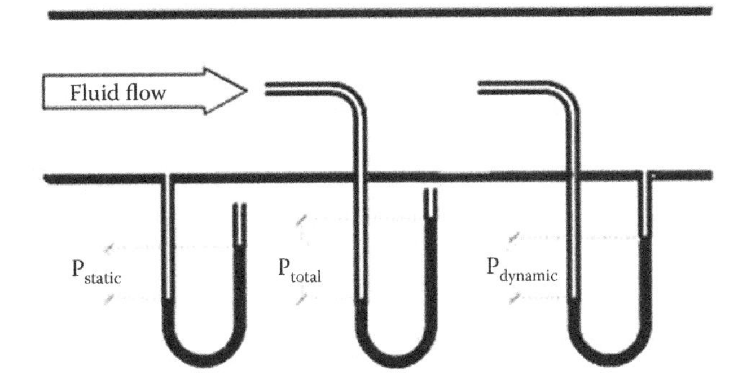

FIGURE 9.17
Static, dynamic and total pressures of fluid flowing in a pipe.

standardised Engler viscosity meter. The base temperature is usually 20°C, though sometimes 50°C or 100°C is also used.

9.12.7 Velocity Measurement

9.12.7.a Pitot Tube

Pitot tube is a simple instrument of pressure measurement which is used for obtaining the fluid velocity by measuring the pressure difference between total and static pressure.

It is a simple long capillary tube bent at right angles which measures the total pressure when inserted along the flow direction with its opening facing the flow.

The pressure so measured is the *total* or *stagnation* pressure, which is the sum of *static* and *dynamic* pressures (as shown in Figure 9.17). From Bernoulli's equation

$$p_{\text{total}} = p_{\text{static}} + \rho \underline{v}^2 \tag{9.6}$$

$$V = \sqrt{\frac{2(P_{\text{total}} - P_{\text{static}})}{\rho}} \tag{9.7}$$

$$= \sqrt{\frac{2p_{\text{dynamic}}}{\rho}} \tag{9.8}$$

where pressure p is in Pa, density ρ in kg/cum and velocity v is in m/s.

Static pressure can be easily found by providing a tapping on the pipe or duct.

Flow can be calculated from velocity by multiplying with internal area of the pipe.

9.12.7.b Pitot-Static or Prandtl Tube

Pitot-static or Prandtl tube, shown in Figure 9.18 as a and b, is essentially a pair of capillary tubes concentrically placed one inside the other with holes made in a way

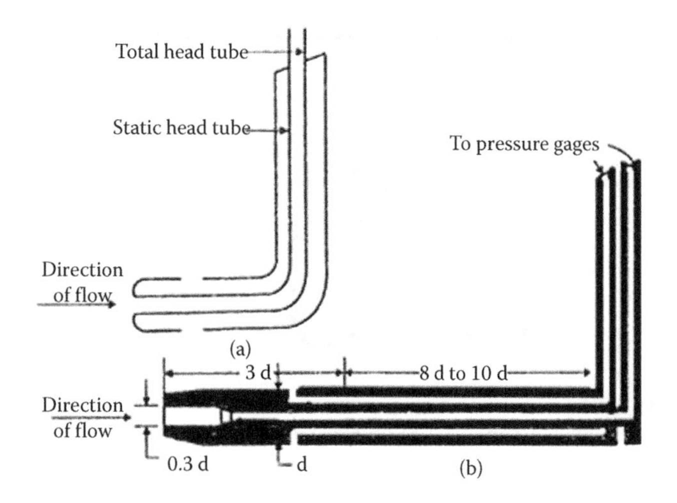

FIGURE 9.18
(a) Pitot and (b) Prandtl tubes.

that the dynamic pressure (difference between the static and total pressures) can be measured with a manometer. This tube needs to be placed along the fluid stream with holes in the total head tube facing the flow.

9.12.8 Pressure Measurement

Pressure is measured by, as shown by Table 9.6, by

a. Manometers by balancing against a column of liquid or
b. Mechanical gauges by balancing against a mechanical element like dead weight, spring or diaphragm and so on

9.12.8.a Manometers

Manometers are bulky and are usually used for low pressures such as air and gas pressures in flues, ducts, hoppers and so on. They are simple, direct and accurate.

TABLE 9.6
Overview of Pressure Measuring Instruments in Boilers

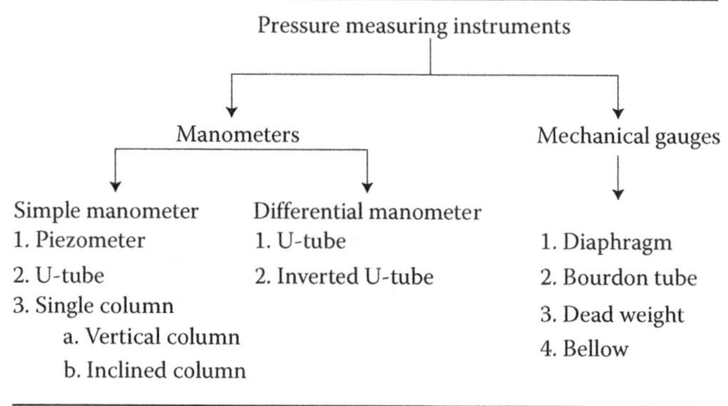

The result is given in mm/in of liquid head. The principle of manometer is explained in Figure 9.19.

9.12.8.b Mechanical Pressure Gauges

Mechanical gauges are sturdy, compact and transportable. They are used for measuring high pressures obtained in steam, water, oil and gas. Out of the four types of mechanical gauges the Bourdon gauge is the most popular which is illustrated in Figure 9.20. The working principle of Bourdon gauge is quite simple. Bourdon tube is filled with a liquid which is in contact with the fluid in pipe or vessel across a diaphragm. Increase in fluid pressure compresses the Bourdon fluid and increases the length of the Bourdon tube which is translated into a direct reading.

9.12.8.c Piezometer

Piezometer is a simple form of manometer with one end open to atmosphere and the other end connected to the point of pressure measurement. It is used in boiler practice to measure air flows in small fans in industrial

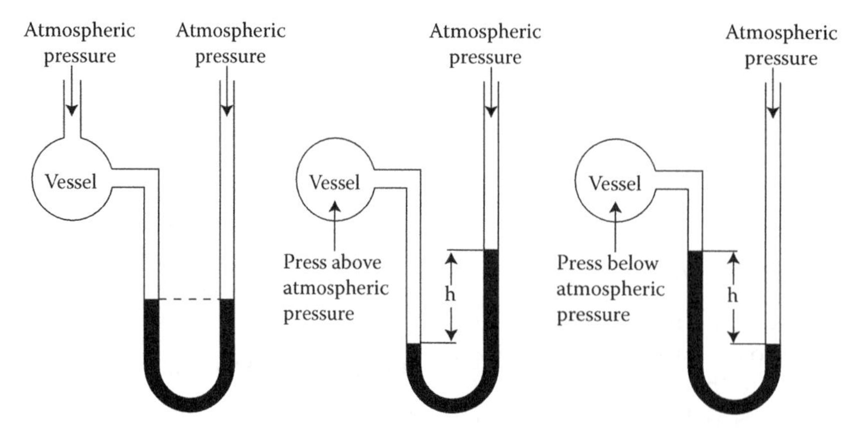

FIGURE 9.19
Pressure and vacuum measurements in manometers.

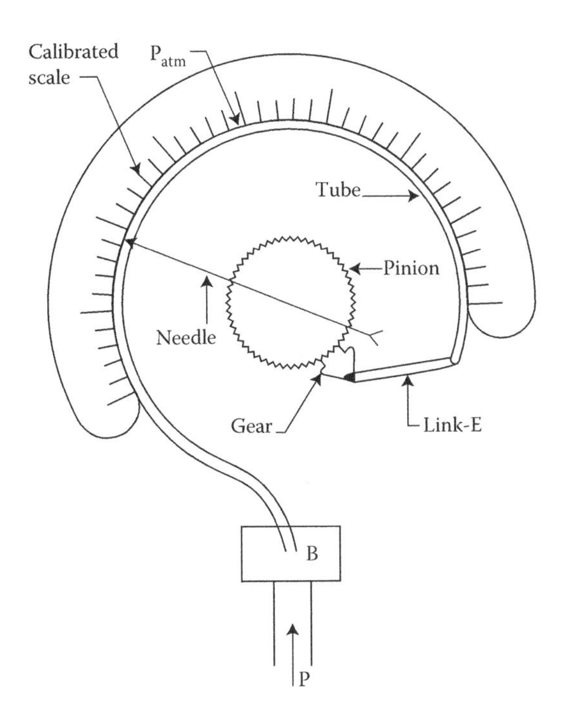

FIGURE 9.20
Bourdon-type pressure gauge.

boilers. A piezometric ring is made of small tube, such as 6 mm NB and placed at the inlet throat of the fan. The ring has a number of small holes of typically 1–2 mm to sense and average the (low) pressure. The small depression in air pressure at the fan inlet is measure and thereafter converted into flow. The main advantage is that the permanent air loss in flow measurement is negligible. However, the turndown is usually limited to ~1:3 because of the limitation in Δp.

9.12.9 Steam Purity Measurement

Modern STs with high η and long times between overhauls have extremely low tolerance for solids in steam. Accurate measurement of steam purity is, therefore, highly desired. Steam purity is expressed as solids in ppb.

Solid or gaseous impurities can be present in steam.

- Solids are usually dissolved in water droplets and are present as dust. Most soluble chemical constituents of fw are converted to Na salts by water conditioning. Hence most solids present in steam are Na salts, with minor amounts of Ca, Mg, Fe and Cu.
- Gaseous constituents commonly found in steam at <140 bar/~2000 psig are NH_3, CO_2, N_2, SiO_2 and amines. Only SiO_2 is actually problematic with its deposits in SH and ST while the others merely interfere with the measurement of steam purity.

In commercial practice, there are three types of on-load steam purity measurements each with its own distinct advantages, namely

- *Conductivity method* for steam purity >0.5 ppm or 50 ppb
- Sodium flame photometer for >0.001 ppm or 1 ppb
- Na ion analyser for as low as 0.1 ppb

9.12.9.a *Specific Conductance or Conductivity Method (see also Conductivity [for Total Dissolved Solids in Measurement Steam and Water])*

Impurities induce conductivity to steam and water, which are otherwise non-conductors of electricity.

1 ppm solids corresponds to ~0.55 μS/cm. Purity levels of <1 ppm can be measured by this simple conductivity meter. However, extreme care is necessary to avoid ingress of dissolved gases, namely CO_2 and NH_3 which introduce large errors in readings. Degassing devices are therefore normally incorporated.

This is one of the most commonly used and inexpensive methods where the specific conductance, measured in μS or μmho, is proportional to the concentration of ions in the sample. When bw is carried over in steam its dissolved solids contaminate and increase the conductivity of steam sample. Measurement of this increase provides a rapid and reasonably accurate method of steam purity.

The limitation of specific conductance is that gases such as CO_2 and NH_3 ionise in water solution and interfere with measurement by increasing conductivity. This interference is more in a high-purity steam sample. It is therefore used for samples containing purity of 0.5–1 ppm when the specific conductance is in the range 1.0–2.0 μS. To obtain a proper measurement of dissolved solids the influence of each gas must be determined and conductivity readings must be corrected. Equipment is available to degas a sample before measurement of conductance.

9.12.9.b *Na Tracer Method or Sodium Flame Photometer*

Solids carried over from bw to steam are substantially Na salts. From the traces of Na measured in the photometer the amount of carryover can be inferred knowing the solids of bw. Accuracy level of 6 ppb solids is claimed by suppliers for portable units. Each 0.1 ppm of sodium corresponds to ~0.3 ppm of solids.

Na tracer technique is more expensive but a more accurate method. This method is based on the fact that the ratio of total solids to Na in the steam as well as bw is the same for all pressures <165 bar or 2400 psig. Therefore, when the ratio of total solids to Na for bw is known, the total solids in the steam can be accurately assessed by measuring its Na content. Because Na constitutes approximately one-third of the total solids in most boiler waters and can be accurately measured at extremely low concentrations, this method of steam purity testing has been very useful in a large number of plants.

9.12.9.c Na Ion Analyser

An ion electrode similar to a pH electrode is used to measure the Na content of the steam sample.

A regulated amount of an agent, such as NH_4OH, is added to a regulated amount of condensed steam sample to raise pH and eliminate the possibility of H ion interference. A reservoir stores the conditioned sample and feeds it at a constant flow rate to the tip of the Na ion electrode and then to a reference electrode. The measured electrode signal is compared with the reference electrode potential and translated into Na ion concentration, which is displayed on a meter and supplied to a recording device.

The instruments operate in a concentration range of 0.1 ppb to 100,000 ppm of Na ion with a sensitivity of 0.1 ppb. Re-calibrated on a weekly basis, these instruments are valuable for continuous, long-term testing and monitoring.

Even though they measure total contamination they do not show rapid changes in Na concentration because of a lag in electrode response and the dilution effect of the reservoir. Peaks exceeding guarantee limits or a known carryover range may not show up affecting interpretation of test results.

The exact ratio of Na to dissolved solids in the bw and consequently in the steam can be determined for each plant but is ~1:3 for most plants (i.e. for each 0.1 ppm of Na in the steam there is ~0.3 ppm of dissolved solids).

Experience has shown that solids levels as low as 0.003 ppm can be measured with either shipped bottle samples or in-plant testing.

9.12.9.d Steam Sampling

All these measurements are affected by the quality of steam sampling. Steam sampling pipes have to be as small and short as possible and made of ss to avoid induction of possible corrosion products. Sampling nozzles are recommended by the ASTM Standard D 1066 and ASME PTC 19.11. The nozzles have ports spaced in such a way that they sample equal cross-sectional areas of the steam line. Iso-kinetic flow is established when steam velocity entering the sampling nozzle is equal to the velocity of the steam in the header. This condition helps to ensure representative sampling for more reliable test results.

9.12.10 Steam Quality Measurement (*see also* Steam Quality or Dryness of Steam and Throttling)

Throttling calorimeter, depicted in Figure 9.21, is a simple and accurate device to determine the *quality of steam*, that is, % water in total weight of steam. It is different from purity of steam which indicates the amount of solids carried over by steam when separated from water.

Throttling calorimeter *measures the amount of M* in steam. Steam is drawn from a vertical main through a sampling nipple, passed around the first thermometer cup, then through a 3 mm orifice held in a disc between two flanges, and lastly around the second thermometer cup and to the atmosphere. Thermometers are inserted in the wells, which are filled with Hg or heavy cylinder oil. The instrument and all pipes and fittings leading to it are thoroughly insulated to minimise radiation losses. The exhaust pipe is kept short to prevent back pressure below the disk. Steam velocity changes are kept to a minimum.

When steam passes gently through the orifice from a higher to a lower pressure with no appreciable variation in velocity no external work is done. And, if there is no loss from radiation, heat in the steam on either side of orifice should be exactly the same. Knowing the steam pressure in the main and the reading of the first thermometer the heat in steam is computed. From steam tables the temperature reading of the second thermometer is estimated. Shortfall in the second thermometer reading is indicative of the M in steam, which can be calculated.

The type of calorimeter illustrated here cheap, simple and accurate for pressures <45 bar having M in steam of ~5%. Accuracy levels obtained can be from 0.25% to 0.5%. For higher pressures and greater accuracy, steam purity in ppm is usually measured in preference to steam quality.

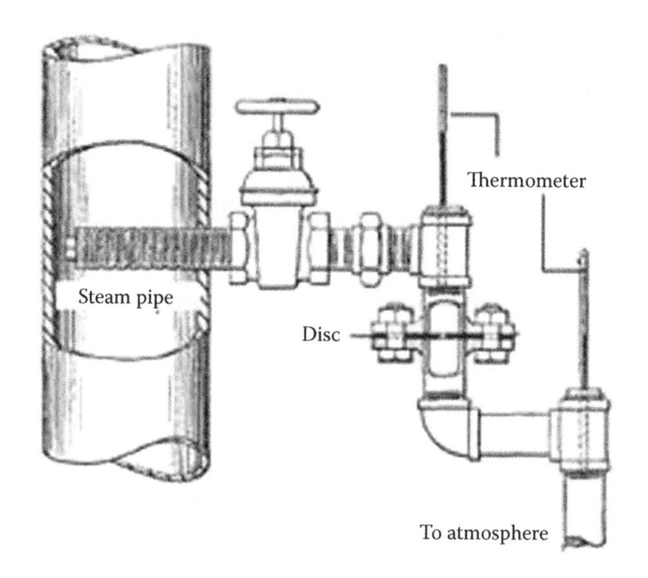

FIGURE 9.21
Throttling calorimeter for measurement of steam dryness.

TABLE 9.7
Metal Combinations and Temperature Measuring Ranges in TCs

		Temperature Range			
		Continuous Range		Maximum	
Thermocouple	ISA Notation	°C	°F	°C	°F
Copper–constantan	T	−185–340	−300–650	590	1100
Iron–constantan	J	−18–760	0–1400	980	1800
Chromel–constantan	E	−185–870	−300–1600	980	1800
Chromel–Alumel	K	−18–1260	0–2300	1370	2500
90% Platinum–10% rhodium to platinum	S	−18–1425	0–2600	1455	2650

9.12.11 Temperature Measurement

9.12.11.a Thermocouples

The most widely used industrial thermometer is the TC.

It was noted by Thomas Seebeck in 1822 that

- A voltage difference (emf) appears when a wire is heated at one end.
- No emf appears if both ends are at the same temperature regardless of the level of temperature.
- There is no flow of current if the circuit is made with wire of the same material.

A TC consists of two electrical wires of dissimilar metals joined together at both ends to form a circuit. It produces a small unique voltage when one end is heated and the other is held constant (at a known temperature). This voltage is measured and interpreted by a TC thermometer.

TCs can be made from a variety of metals. They cover a temperature range—185–1425°C (–300–2600°F). Commonly used metal combinations are listed in Table 9.7 along with their limiting temperatures. Selection of TC is based on the ability to withstand oxidation attack at the maximum service temperature. All TC materials tend to deteriorate at their maximum limits.

- Constantan is an alloy of Ni and Cu in the ratio of 45% and 55%, respectively
- Chromel is an alloy of Ni and Cr in the ratio of 90% and 10%, respectively
- Alumel is an alloy of Ni and Al, Si and Mn in the ratio of 95% and 5%

The popularity of TCs is because of their accuracy, durability, versatility, speed of response and cost effectiveness.

TCs are available in different combinations of metals or calibrations. The four most common calibrations are J, K, T and E. Each calibration has a different temperature range and environment, although the maximum temperature varies with the diameter of the wire used in the TC. **ASTM** E230 provides all the specifications for most of the common industrial grades.

TC wires are housed inside a metallic (ss or Inconel) tube called TC probe whose wall is referred to as the sheath. Inconel sheath supports higher temperature ranges but ss is often preferred because of its broad chemical compatibility. The tip of the TC probe is available in three different styles—grounded, ungrounded and exposed (as shown in Figure 9.22). With a grounded tip the TC is in contact with the sheath wall. A grounded junction provides a fast response time but it is most susceptible to electrical ground loops. In ungrounded junctions, the TC is separated from the sheath wall by a layer of insulation. The tip of the TC protrudes outside the sheath wall with an exposed junction. Exposed junction TC is best suited for air measurement.

For measuring the temperature of a solid surface the entire measurement area of the sensor must be in contact with the surface which is difficult when working with a rigid sensor and a rigid surface. Since TCs are made of pliable metals, the junction can be formed flat and thin to provide maximum contact with a rigid solid surface. These TCs are an excellent choice for surface measurement.

Chordal TC

Chordal TCs are for measuring the tube metal temperature of particularly furnace tubes where pad-type TCs cannot be protected from the erosive gases.

Furnace tube metal temperatures are close to the inside water temperatures rarely exceeding 40°C under normal conditions even for SC boilers. As the deposits develop inside the tubes metal temperatures tend to increase sharply (as shown in Figure 9.23).

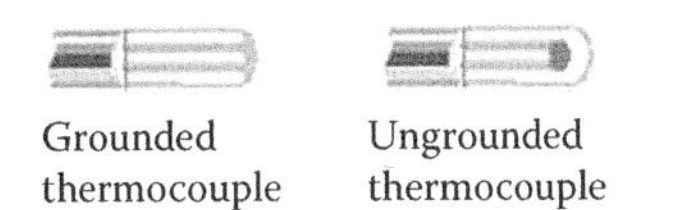

FIGURE 9.22
Tips of TCs.

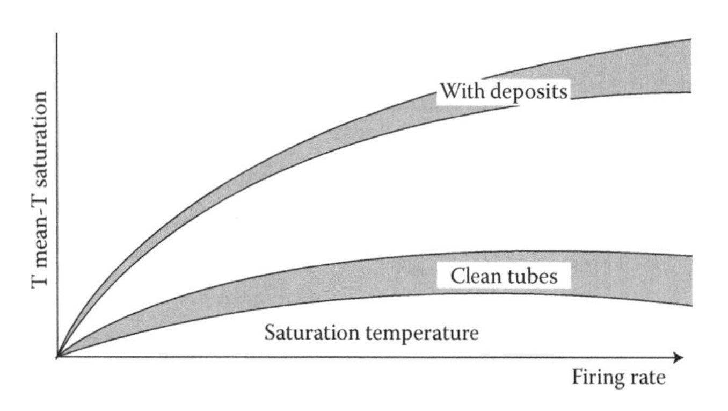

FIGURE 9.23
Rise in furnace tube metal temperatures with the formation of inside deposit.

Chordal TCs are fixed on representative furnace wall tubes and monitored for temperature increases on usually weekly basis at about the same boiler load. Any abnormal temperature rise is indicative of build up of internal deposit. Appropriate remedial measures are planned with no fear of boiler break down because of tube failures.

High Velocity and Multiple-Shield High-Velocity TCs (HVT and MHVT)

It is highly desirable to measure accurately gas temperatures at furnace exit and in SH region, particularly in large boilers.

- It is important to know how the HSs are behaving in reality so that corrective steps can be initiated where necessary.
- In two-pass boilers with non-drainable SHs close monitoring of FEGT is needed to prevent overheating of SH tubes because of excessive gas temperatures, particularly during start up.
- Also the overall status of the furnace cleanliness can be inferred by the record of gas temperatures over a period of time.

However, at these levels bare TC or optical/radiation pyrometers do not provide reliable readings. As the sensing element is at a lower temperature than the flue gas, the radiant heat of gas is not adequately captured, which results in the temperature indicated being lower than actual.

The way to overcome this is to suck the flue gases at the same velocity (as in the inside of the chamber) over the hot junction of a TC which is shielded by a pipe. The pipe is made retractable and the assembly fixed on the side wall. It can be inserted to any length inside the chamber to map the temperature profile.

The HVT probe with its water cooling and air aspirating arrangements appears like a retractable SB (as shown in Figure 9.24).

Figure 9.25 shows the essential difference between HVT and MHVT. In the MHVT at the inlet of the probe

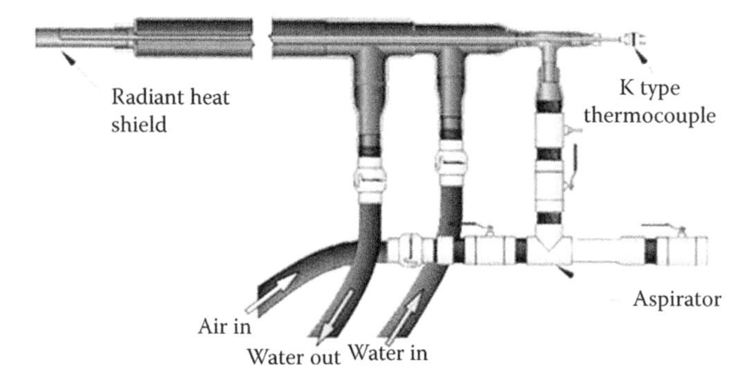

FIGURE 9.24
Shielded high-velocity TC for measuring furnace and SH gas temperatures.

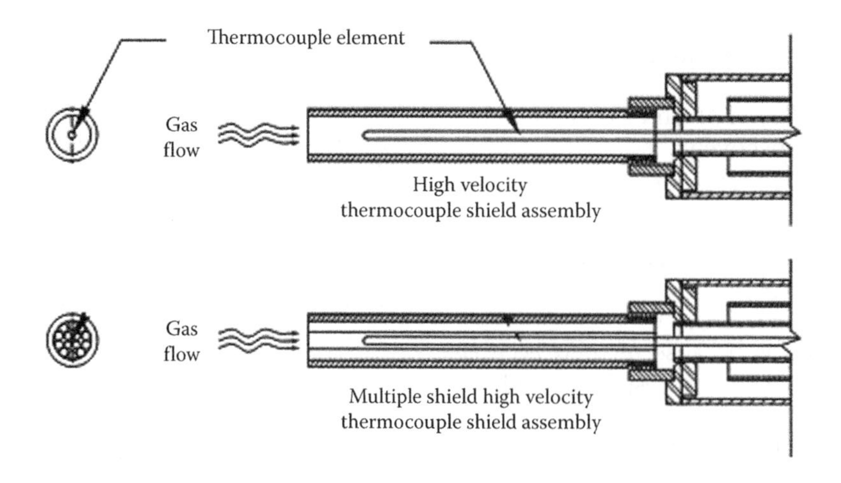

FIGURE 9.25
Difference between HVT and multiple shield HVT TCs.

several streams of gas flow are created and more shields are added to improve the gas flow pattern so that the temperature measured is even more close to the actual. MHVT temperatures are higher than HVT and can be practically the same as the calculated. The difference is perceptible at temperatures >870°C (1600°F) where the effect of radiant fraction of the heat is considerable (see Figure 19.1). In coal and other dirty applications MHVT probes are not feasible because of small gas openings which get clogged.

9.12.11.b Thermometers

A thermometer is an instrument to measure and indicate the temperature of a specific application or condition. It is commonly known as Hg-in-glass thermometer and works on the elementary principle of expansion of Hg on application of heat. Clinical thermometer is the most commonly used thermometer. By far it provides the best accuracy in the range −40°C to + 400°C (−40–750°F).

An industrial thermometer is built very ruggedly to withstand some amount of rough use. Yet it is delicate in comparison. Figure 9.26 shows a typical unit. Hg pollution is slowly restricting its use.

It is only a local instrument installed at the point of measurement with limited visibility. In a boiler plant it is used for measuring ambient conditions and some low-temperature points in water. It is useful for calibration and as test instrument.

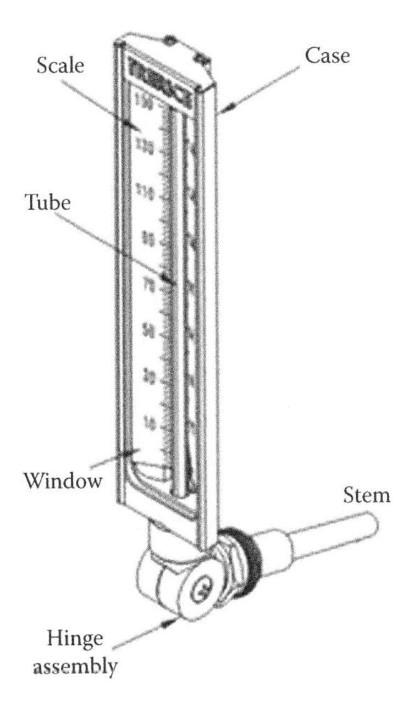

FIGURE 9.26
Mercury in glass thermometer of industrial class.

9.12.11.c Pyrometers

Pyrometers are non-contacting temperature-measuring instruments for measuring high temperatures beyond the range of thermometers. They are principally of two types, namely optical and radiation. For measuring gas temperatures both are not of use.

Optical Pyrometer

Optical pyrometer is for measuring relatively high temperatures as in heating furnaces, steel mills and iron foundries and so on and operates by measuring radiation from the body whose temperature is to be measured. It measures the temperature of glowing bodies by comparing them visually with an incandescent filament of known temperature, whose temperature can be adjusted. Optical pyrometers work on the basic principle of using the human eye to match the brightness of the hot object to the brightness of a calibrated lamp filament inside the instrument (as shown in Figure 9.27). The optical system contains filters that restrict the wavelength sensitivity of the devices to a narrow wavelength band ~0.65–0.66 μm (the red region of the visible spectrum). Optical pyrometer can achieve accuracy of ±0.5% of the temperature being observed.

Radiation Pyrometer

In radiation pyrometer also there is a lens system at work. Part of the thermal radiation emitted by a hot object is intercepted by a lens and focused onto a thermopile. The resultant heating of the thermopile causes it to generate an electrical signal (proportional to the thermal radiation) which can be displayed on a recorder. This principle is depicted in Figure 9.28. A thermopile

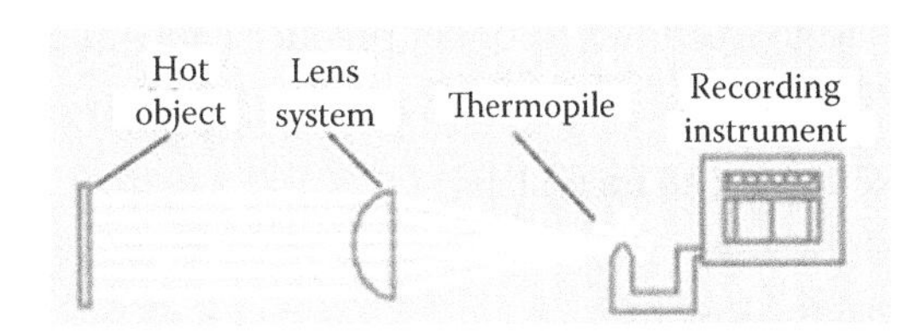

FIGURE 9.27
Operating principle of an optical pyrometer.

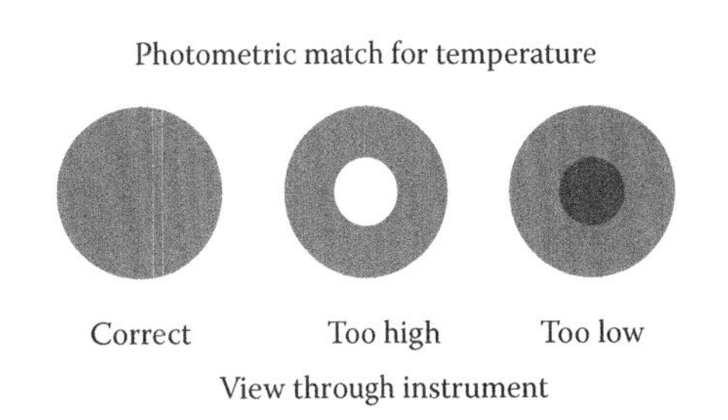

FIGURE 9.28
Temperature measurement by photometric matching in a radiation pyrometer.

is an electronic device, composed of TCs connected usually in series that convert thermalenergy into electricalenergy. Thermopile does not measure the absolute temperature but generates an output voltage proportional to the temperature difference or gradient. This is suitably calibrated to directly indicate the temperature.

9.12.12 Opacity Measurement (*see* Opacity of Stack)

9.13 Insulating Materials

Some of the important aspects of the insulating materials are as below.

- Unlike the refractory materials, which are substantially mineral based and therefore capable of withstanding the highest temperatures and erosion, the insulation materials are non-mineral (except for ceramic fibres), can stand up to a temperature of no higher than ~1650°C and do not have any abrasion resistance.
- The insulating materials, however, possess much lower heat conductivity, several times lower than the refractory materials.
- The insulating materials are fluffy and hold lot of air which is the main reason for the reduced heat flow.
- k for insulating materials rises steeply with temperature unlike the refractories.

Insulation materials in boiler practice come in mainly four forms:

- Reformed shapes/slabs
- Mattresses
- Plastic cement
- Loose fill

The insulating materials used in boilers are

1. Calcium silicate in block forms
2. Mineral/slag wool
3. Glass wool or fibre glass
4. Ceramic fibre
5. High-temperature plastic

9.13.1 Calcium Silicate

Also popularly called Cal-Sil, Ca_2SiO_4 is also known as calcium ortho-silicate and sometimes formulated 2CaO.

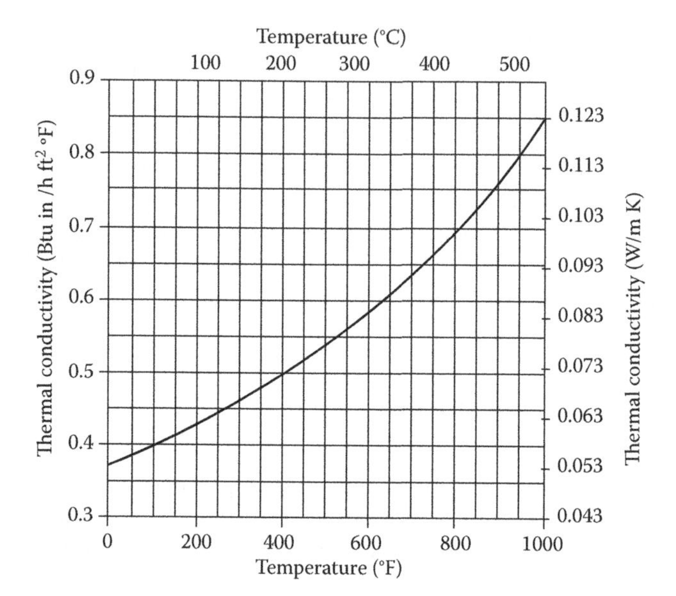

FIGURE 9.29
Thermal conductivity of calcium silicate.

SiO_2. It is a white powder with a low bulk ρ and high physical water absorption derived from limestone and diatomaceousearth. It is a safe substitute to asbestos for medium temperature insulation and competes against rockwool as well as proprietary insulation materials such as pearlite mixture and vermiculite bonded with sodiumsilicate.

Ca_2SiO_4 is characterised by light weight, low k, high temperature and chemical resistance. The relation between temperature and k is indicated in Figure 9.29.

This is very popular that block insulation up to 650°C which comes in preformed shapes or blocks. It is particularly suitable for piping and valves. The blocks are ideal for insulating hoppers and ducts from inside for metal protection where there is no abrasive ash in flue gases.

9.13.2 Mineral or Slag or Rock Wool

This is made from blowing the slag produced in steel mills into fine fibre by steam or air jets. Mineral wool has practically displaced the glass wool, which contains glass fibres in place of slag fibres, from human health considerations. Besides, it is also slightly cheaper.

Mineral wool is M and corrosion resistant, fire retardant and chemically inert with good insulating properties.

- In *blanket* form the wool is held in place by wire mesh on one side or both sides depending on the application and is used up to a temperature of 650°C. For vertical walls of furnace resin bonding of the mattress is usually preferred as the wool is held in place without settling at the

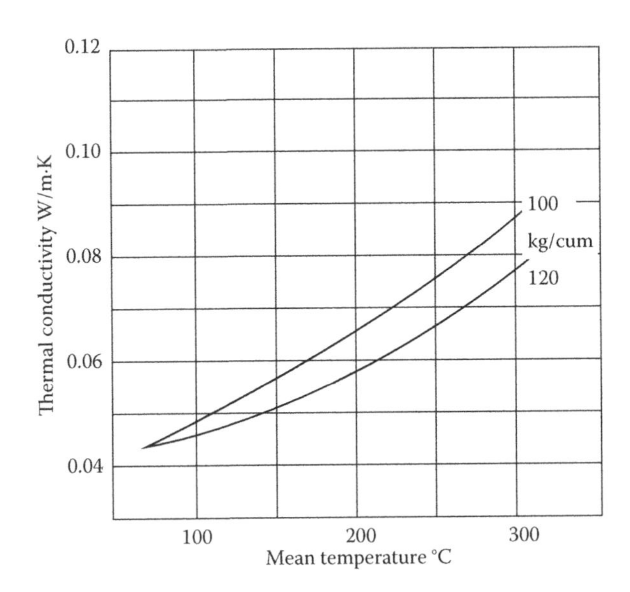

FIGURE 9.30
Thermal conductivity of resin-bonded mineral wool blanket.

bottom because of the slight vibrations in the walls. Typically densities of 100–140 kg/m^3 are employed for normal applications.

Typical thermal conductivity versus temperature graph for a resin bonded mattress wth steel mesh on either side is shown in Figure 9.30.

- When clay is used as binder and the wool is moulded at high temperature and pressure, mineral wool blocks can be formed which can be used up to 1050°C. These blocks are used for insulating the boiler and membrane walls.

9.13.3 Glass Wool or Fibre Glass

This is an insulating material arranged in texture like wool made from fibre glass. Glass wool is produced in rolls or in slabs, with different thermal and mechanical properties. Air captured and held in the glass fires is what provides the insulating property. It also results in a low ρ which can be varied through compression and binder content.

Fibre glass and mineral wool are nearly the same in terms of k and other properties. Glass wool is much lighter. But glass wool suffers from the disadvantage of being labour and environment unfriendly because of the sharp needles of glass. Besides causing injuries its fine dust is also polluting. It is also slightly more expensive. It is practically now overtaken by mineral wool in most markets.

9.13.4 Ceramic Fibre Linings

These are produced by melting the alumina-silica (Al_2O_3–SiO_2) china clay and blowing air to form fibres

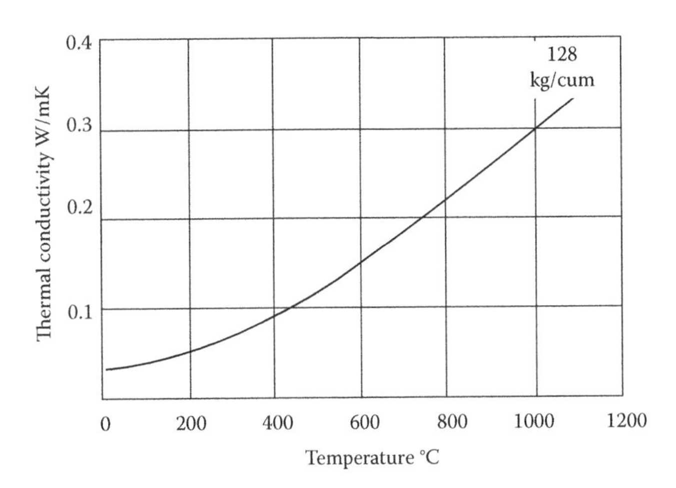

FIGURE 9.31
Thermal conductivity of ceramic wool blanket.

of 50–100 mm long and 3 μm in diameter. They can be offered in block or mattress form capable of withstanding very high temperatures. The blocks are a substitute for insulating bricks and the mattresses can be used for higher temperatures. Their acid resistance is good but has lower alkali resistance. Ceramic fibre is useful when temperatures as high as 1650°C are encountered. This material is reserved only for some special applications due to high cost.

Typical k versus temperature graph for ceramic wool blanket is given in Figure 9.31 and composition of 60–70% SiO_2 and 25–40% CaO + MgO with $Al2O_3$ < 0.3%. Densities varying from 50 to 160 kg/m^3

9.13.5 High-Temperature Plastic Cement

This insulating plastic cement, made from mineral wool and clay, is suitable up to ~1050°C for use on irregular shapes and for filing gaps in block insulation.

Table 9.8 puts together important properties of some insulating materials in general used frequently in boilers. Manufacturer's catalogues and data sheets give more accurate information on each of these products which is more authentic.

9.14 Insulating Castables (*see* Refractories)

9.15 Integral Piping (*see* Pressure Parts)

TABLE 9.8

Properties of Insulating Materials

Insulating Material	ρ, kg/m³	Limiting temperature, °C	Thermal Conductivity, k						Specific Heat, Cp	
			W/mk	@°C	W/mk	@°C	W/mk	@°C	kJ/kg·k	@°C
Asbestos, loose	575	800	0.205	100	0.216	300	0.230	500	0.82	20–100
Glass wool	100	~500	0.059	100	0.080	200	0.095	300	0.796	0–100
Mineral wool	200	850	0.053	100	0.065	200	0.089	300	0.837	0–100

9.16 Intermittent Blowdown (*see* Blowdown in Water)

9.17 Iron–Carbon Diagram (*see* Metallurgy)

9.18 Internal Re-Circulation Circulating Fluidised-Bed System (*see also* Circulating Fluidised-Bed Combustion)

This is an expanded-/turbulent-bed CFB technology developed by B&W of the USA based on principle of Studvik of Sweden. Here the cyclones are replaced by the combination of inertial separators (patented as U-beams) at the discharge end of first pass and mechanical dust collector (MDC) in the second pass after Ecos. The positioning of the U-beams is such that the collected dust rolls down along the rear wall creating an 'internal re-circulation' dispensing with the entire loop seal arrangement. As a conventional cyclone with ~150–160 mm wg Δp is replaced by U-beams + MDC which consume only ~100 mm wg, the power consumption is significantly reduced. This is in addition to the lower air power consumption because of lower fluidising velocities. The advantages of this design are as follows:

- This is the most compact and most conventional of all CFBC boilers
- With no cyclones, this is the only technology which can retrofit an existing PF boiler
- The power consumption is the least among all CFBs
- The refractory content is also the least
- Cold start-up times are ~4–5 h as against 10–12 h in a classical CFB
- The load response is very good because of the least amount of refractory

The U-beams, typically ~8 m long, are required to withstand very high gas temperatures of ~900°C and experience heavy dust loads. They are no doubt made of high quality ss materials such as SS 309H, SS 310H and RA 253 MA. In spite of this, if the gas temperatures in combustor should escalate because of bed temperature fluctuations, the U-beams can suffer distortion leading to permanent damage to the alignment. This will increase the problems further. Maintaining bed temperatures strictly under control is most essential in this design. Although the refractory is less, some special varieties are needed for withstanding the heavy ash flows adequately. Refractory erosion is a problem that needs to be overcome particularly in high ash coals. The main features of this technology are given in Table 9.9.

Figure 9.32 shows a typical single-drum IR-CFB boiler of 680 tph, 19.7 MPa, 560°C (1,500,000 lbs/h, 2850 psig, 1040°F) high p & t unit designed for coal.

The salient constructional features of this design are summarised as follows:

- The design is like a conventional **bi-** or **single-drum** boiler fitted with CFB combustor at the bottom of the furnace and an inertial U-beam separator hung at the top in the same place as a vertical **platen** SH.
- There are well-defined zones in the bed when fluidised, namely, dense primary zone and

TABLE 9.9

Salient Design Parameters of IR-CFB Boilers

Parameter	Value	Value
Bed velocity	3.7–4.3 m/s	12–14 fps
Bed temperature	840–950°C	1545–1740°F
PA as% of total air	55–70%	55–70%
Excess air	15–25%	15–25%
Gas velocity in furnace	5.5–6.7 m/s	18–22 fps
Gas velocity in U-beams	~8 m/s	25 fps
Δpin U-beams in wg	<25 mm	<1 in
Collection η in U-beams	97–98%	97–98%
Δp in MDC in wg	<78 mm	<3″
Collection η in MDC	~90%	~90%
Second-pass enclosure	Membrane	Membrane

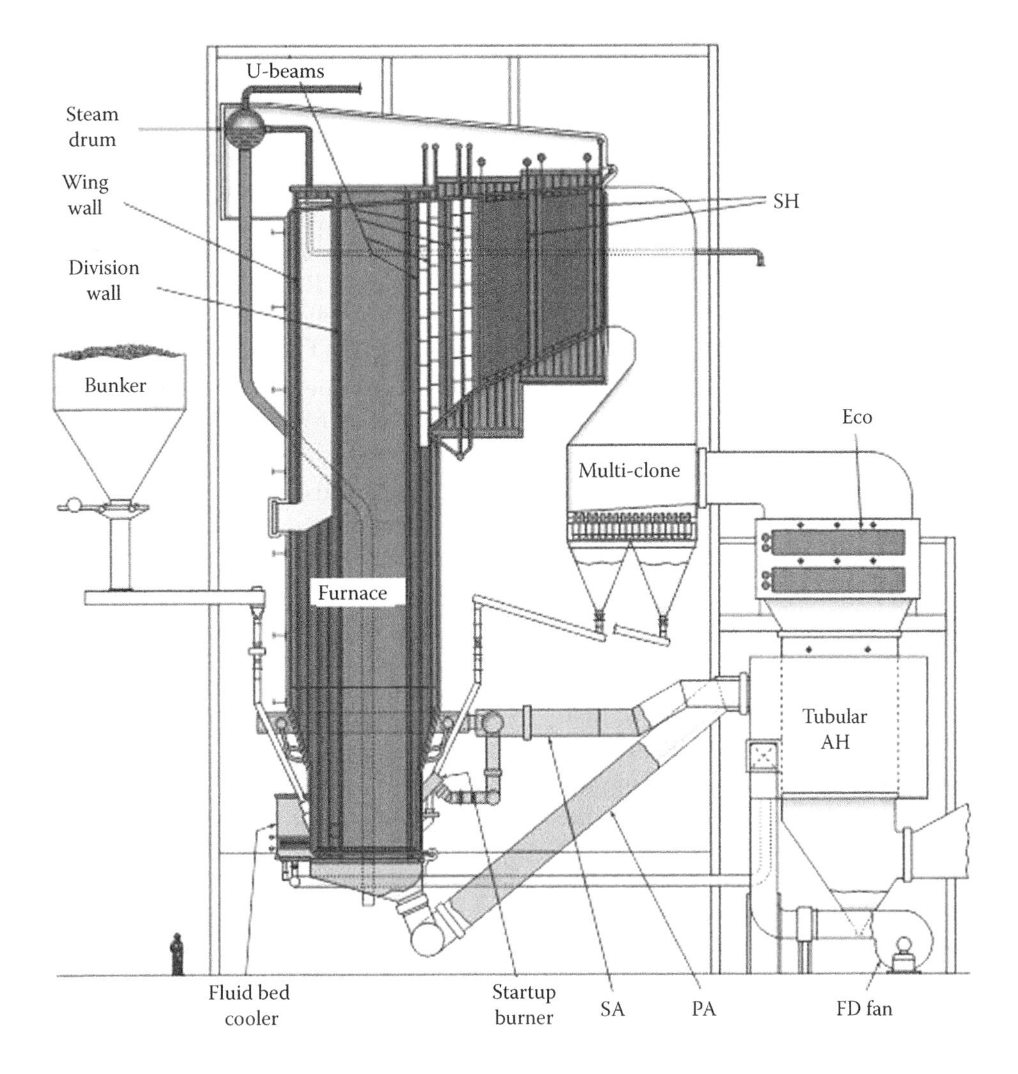

FIGURE 9.32
Typical single-drum HP non-reheat IR-CFB boiler for CPP. (From The Babcock and Wilcox Company, with permission.)

disengaging zone. OFA is provided at the end of disengaging zone.

- Boiler works on balanced draught operation with the furnace at slightly negative pressure above the expanded bed.
- To prevent erosion and corrosion against the impinging bed material and the prevailing reducing conditions in the lower combustor, respectively, the furnace tubes are lined with refractory up to the TA nozzles.
- PA nozzles, of ss 304 material, are small and closely spaced (typically 100 mm or 4″ square pitching with ~100 nozzles/m^2) and are of similar construction as nozzles of BFBC.
- Overbed burners or lances heat the bed.
- Fuel feeding is by gravity. Feeding chutes are placed at an inclination of 60–65° to horizontal. This makes the system simple unlike in many hot cyclone designs where the furnace is under positive pressure needing pressurised fuel feeding.
- Bed ash discharge is at the bottom of the furnace with 200 (8″) NB ash pipes. A range of 2–4 ash pipes are employed for most industrial boilers.
- To cool the gases to the desired level in the furnace division or wing wall HSs are provided.
- A two-stage dust collection is employed. U-beams perform primary collection at ~97.5% efficiency. To enhance the collection efficiency an **MDC** with efficiency of >90% is located after BB or Eco in a cooler zone at an elevation suitable for pneumatic conveying of dust. The collectors are made of hard castings with hardness of 550 **BHN** to withstand erosion of dusty gases spinning at high speeds in excess of 20 m/s.

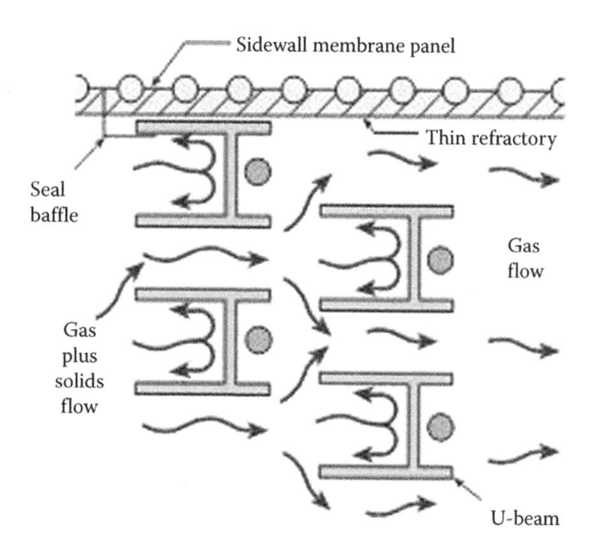

FIGURE 9.33
Cross-section of U-beams. (From The Babcock and Wilcox Company, with permission.)

Alternatively, ESP ash from the first field where particles are ~70 μm is re-circulated.

- Return leg and L-valve, used for returning the ash collected at the bottom of U-beams in the earlier designs, is now superseded. Two sets of U-beams are located at furnace exit where ~75% of the solids are collected and roll down over the rear wall. Four sets of beams are placed after a gap where 90% of the incoming solids are collected in a solid storage hopper with discharge on to the rear wall.
- U-beams, typically ~8 m long, are made of ss materials such as TP 309H, TP 310H or RA 253 MA depending on the prevailing conditions to withstand high temperature and force from heavily dust-laden gases at velocities of 6–8 m/s. RA 253 MA is typically used for pet coke application. Cross section or U-beam is shown in Figure 9.33.
- SHs are pendant type while Ecos are horizontal. First row of Eco tubes are protected against erosion by suitable tube shields.
- The enclosure of the second pass is made of membrane wall like in hot cyclone design as gases are at temperature exceeding 450°C.
- This IR of ash has dispensed with the return leg altogether, with corresponding reduction in refractory and air blower for L-valve. Ash re-circulation rate is high. Bed temperature is controlled by regulating the return of cold ash from MDC.

9.19 Iron Castings (*see* Castings in Materials of Construction)

10

J

10.1 Joule Cycle (*see* Brayton Cycle)

10.2 J-Valve (*see* Loop Seal in Fluidised-Bed Combustion, Circulating)

11

K

11.1 Killed Steel (*see* Deoxidation of Steels)

11.2 Kinematic Viscosity (υ) (*see* Fluid Characteristics)

11.3 Kirchoff's Law of Thermal Radiation (*see* Radiation Heat *under* Heat Transfer)

11.4 Kraft Pulping Process (*see also* Recovery Boiler)

Paper is made from wood pulp which is extracted from cellulosic materials like wood, bamboo or other similar material. Approximately 85% of paper made is by the *Kraft* or *Sulphate* process which is schematically presented in Figure 11.1. Kraft is the Swedish and German word for 'strong'. Sulphate process was so called as it produces long fibres which are strong, helping to make tough paper. This process was patented in 1884 and has not been economically viable without a BL recovery boiler. This is an alkaline process which is different from the acidic *Sulphite* process of making pulp.

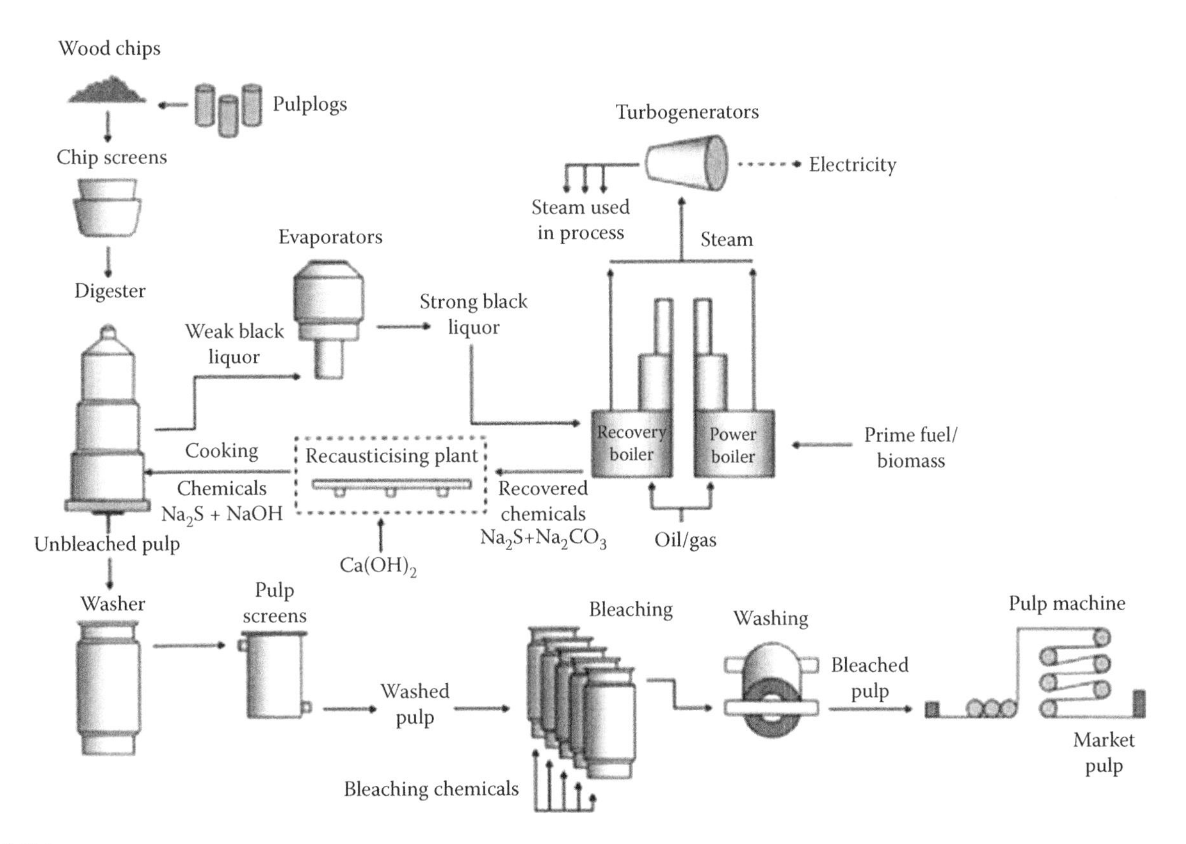

FIGURE 11.1
Kraft pulping process.

Wood logs are debarked, made into small chips in chipper house, screened and fed into digesters (batch or continuous) for cooking. The cooking operation is performed by steam and chemicals ($NaOH + Na_2S$) under medium p & t. The aqueous solution of chemicals, called the *cooking* or *white liquor* reacts to dissolve *lignin*, the binding material in wood, to help cellulose to separate out into pulp. It also dissolves other organic matter of the wood and together with lignin it constitutes nearly half of wood by weight. Pulp is called the *brown stock* at this stage because of its colour. It is processed further by screening, bleaching and washing before turning it into paper grade white pulp.

The spent cooking liquor, containing lignin and some woody matter, is now termed as *weak BL* with 15–18% dissolved solids, which is too lean for combustion. Therefore, it is taken to multiple effect evaporators (MEE) for concentrating to 60 to even as high as 80% solids, depending on the liquor viscosity. The *strong BL* is turned into *heavy BL* after adding Na_2SO_4 for making up the losses in digester. The liquor is then heated in steam heaters for easy pumpability before it is sent to BLR boiler at 70–75% concentration. If BL is concentrated to ~55–60% in the MEE, further concentration is done in *cyclone or cascade-type direct contact evaporators* using the heat of flue gases. This practice is going out of fashion as the flue gases pick up malodour of the BL and spoil the environment.

In the BLR boiler, the strong BL is first dried in the furnace and subjected to reducing conditions so that Na_2SO_4 is reduced to Na_2S. The reducing condition in furnace is obtained by staging of combustion air. The organic matter of BL, dissolved from wood in digester, is burnt to produce heat. Although this heat is recovered in the form of steam, the inorganic Na compounds are turned into *smelt*. The smelt, containing mainly Na_2CO_3 and Na_2S, is recovered from the hearth by tapping in red hot condition and mixing with causticising liquor in dissolving tank to make it into *green liquor.*

Green liquor is turned into white cooking liquor by *causticising* treatment with slaked lime where $Ca(OH)_2$ converts Na_2CO_3 to NaOH. The $CaCO_3$ sludge formed from causticising plant is heated in lime kiln to expel CO_2. The resulting CaO is rehydrated to $Ca(OH)_2$ for further use again.

12

L

12.1 Lagging

Lagging or cladding is the outer-most covering of a boiler to protect the insulation from damage. In former times it used to be cloth covering held tightly by calcium silicate or bitumen sealing layer. The purpose was to protect the inner insulation from seepage of M and to give an aesthetic look to the boiler.

These days lagging is in the form of thin metal sheeting, either galvanised iron (GI) or Al, both being weather resistant. Besides a much better appearance of the metal sheeting, it also gives an added protection to the insulation materials as metal can withstand human load when people have to necessarily walk on the horizontal surfaces some times for access.

GI is stronger but lacks the lustre of Al. Usually for furnace plain GI sheeting is used in the thickness of 1 mm (19 **swg**) and Al in 1.2 mm (18 swg) plain or 0.9 mm corrugated. For pipes 22 or 24 swg (0.7 or 0.56 mm) are used for ease of bending. Both CI and Al work out about same in price. Al is gaining a greater acceptance progressively because of its better appearance.

12.2 Laminar Flow (*see* Fluid Flow Types)

12.3 La Mont Boiler (*see* Once-Through Boiler)

12.4 Lane Blowing (*see* Sootblowers)

12.5 Laws of Steam and Water (*see* Thermodynamic Properties)

12.6 Laws of Thermodynamics (*see* Thermodynamic Properties)

12.7 Latent Heat (*see* Properties of Steam and Water)

12.8 Life Extension

Boilers are designed for a nominal life of 30 years. High-temperature components such as SHs, RHs and AS headers, however, are designed for 10 years governed by creep-strength. These are usually replaced for another fresh spell of 10 years.

Towards the end of the nominal design life of ~30 years, the boiler plant availability steadily reduces and the reliability becomes uncertain due to unscheduled and unforeseen stoppages, mainly arising out of PP failures. With sharp escalation in the costs of new equipment, in most cases, it makes little economic sense to put up a new replacement plant discarding the existing one. However, if there has been no major technological change or discontinuity in fuel, it makes an eminent case to undertake life extension of the boiler.

Life extension of a boiler involves at the end of about 30 years

- Undertaking an appropriate residual/remnant life assessment (RLA) study
- Follow up with the necessary repair and replacement work as revealed in RLA

Reduction in life in a component is reflected usually in its accelerated thinning. Many a time it may be reaching a point of failure while looking apparently alright, thus escaping attention. This loss of thickness is caused in general by (a) oxidation, (b) wear and (c) corrosion. Parts exposed to high p & t, high-velocity dusty gases, acidic water and so on are high on the list of fast deterioration. High-temperature PPs placed in the gas-passes and milling/firing equipments are most prone to accelerated damage.

Systematic O&M practices reveal which parts need frequent replacement and repair. But there is no holistic study of the rate of deterioration of parts or of the residual life of those components which have not failed. Repairs are done as and when needed.

On the contrary, RLA takes an overall view by studying all wear-prone parts and predicts the residual life of components likely to fail irrespective of their current health. RLA is an advanced study which draws on the advancements in the measuring techniques and metallurgical know how. Thicknesses and surface conditions of various affected PPs are measured to understand the severity of duty and corresponding wear rate to predict the remaining life. The real technology lies in the RLA, conducted with the help of some special instruments and assessments derived on the basis of a large body of in-house knowledge.

What follows the RLA as repair work is not just the repairing task but also the replacement of most wear prone components, even those looking apparently acceptable. The purpose is to enable the boiler to provide a relatively uninterrupted operation for the next stage of its life as if it were a new unit. Appropriate guarantees for performance and working are provided for the expected fresh lease of life.

12.9 Ligament Efficiency (*see* Efficiency, Ligament)

12.10 Lignite and Firing

This topic is expanded under the following sub-topics:

1. Lignite properties
2. Brown coals
3. Plate belt feeders
4. Beater mills
5. Lignite and brown coal firing
6. Lignite-fired boilers

12.10.1 Lignite Properties

The word *Lignite* traces its origin to the Latin word *Lignum*, meaning wood.

Lignite or brown coal, the carbonaceous fuel intermediate between **coal** and **peat**, is brown or yellowish in colour, woody in texture and soft.

Lignites are high in VM and M as they are formed from plants rich in resin.

Lignites, on daf basis, have Gcv of 10–20 MJ/kg (~2350–4700 kcal/kg or 4230–8460 Btu/lb), with M ranging from 30% to 75% and ash from 2% to 20%. Lignites with high M of 50% and higher found in Europe are called *brown coals* because of the appearance. Despite high M, lignites give an appearance of being dry. They are used almost exclusively as fuel for steam–electric power generation. Figure 12.1 gives the properties of lignites from some of the major fields across many countries.

Lignites are lighter, that is, the bulk ρ is less and naturally the energy ρ is also low. This makes long distance transportation uneconomical. Usually 50 km from mine-mouth is the maximum distance considered as a limit for transportation. Hence, they are not traded in fuel markets of the world.

They are prone to **spontaneous combustion** because of high O_2 and VM making large stock piles and bunkers dangerous. Prolonged storage is also undesirable, not only from fire hazard consideration but also from fuel degradation point; VM tends to distil off the fuel causing a drop in the cv.

Two types of **mills** and **feeders** are employed in lignite preparation in power plants.

- Low M lignites with ~30% M can be ground in conventional **vertical pulverisers** using PA at high temperatures for drying. They can also be transferred from bunkers to mills by conventional **drag feeders** and **belt feeders** as the volumes are relatively low because of their higher Gcv.
- With higher M, as in brown coals, the amount of fuel to be handled increases disproportionately because of lower Gcv. Simultaneously, the amount of transport air and its temperature rises sharply. **Beater mills** become necessary with flue gases drawn from the furnace for drying the fuel during pulverisation. Higher the M in lignite, easier it is to grind as it tends to break down into powder by itself after the M is driven off. HGI is rarely <100 and has no relevance for lignites. **Plate feeders** are needed for in-plant movement from bunker to mill.

CG, SS, BFBC and CFBC are all well suited to burn lignites besides PF firing.

High VM and lower **ignition temperature** (typically ~400°C) makes lignite a very easy fuel to ignite and burn after the M is removed. Salient points to be observed while firing lignites are as follows:

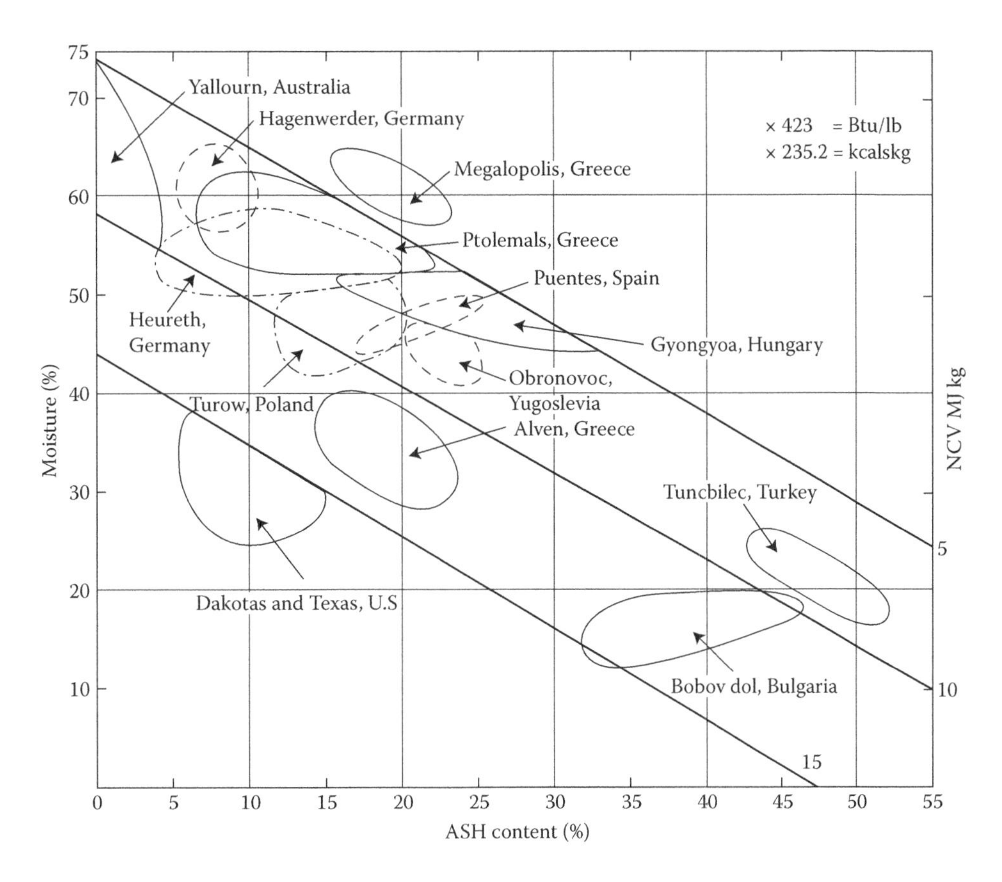

FIGURE 12.1
A, M and Lcv of lignites from major fields.

- O_2 content in lignite is high and the combustion air requirement is, therefore, lower.
- High M in fuel produces more flue gas requiring larger furnaces and ID fans.
- As the **ash-softening temperature** is lower, the spacing of the SH/RH panels has to be wider.
- All this makes the boiler significantly bigger than coal firing units; higher the fuel M larger the boiler.

The main use of lignite is in power generation. Germany, Eastern Europe, Greece, Turkey, Russia, Australia, South Africa, the USA, China, South Korea and India are some of leading producers of lignite. Lignite is widely distributed across the globe. It is next only to coal in terms of production and power generation.

12.10.2 Brown Coals

Brown coals are high M lignites with M ranging from 50% to 75% and having slight brownish or yellow colour. They are generally found close to the surface and are readily amenable to strip or open-cast mining. Brown coals are predominant in Germany, East Europe, Greece, Turkey, Croatia, Russia and Australia and large power stations are built on this fuel. Mining of brown coal is steadily falling in Germany and the neighbouring countries while its production is rising in almost all the remaining regions.

Typical analysis of German brown coal is given in Table 12.1.

12.10.3 Plate Belt Feeders

Very large boilers have been built with poor fuels such as brown coal and shredded peat. The feeding equipment

TABLE 12.1
Typical Analysis of German Brown Coals (Lignites)

						Daf Gcv		
% by wt	VM%	C%	H%	O%	S%	MJ/kg	kcal/kg	Btu/lb
Brown coal	45–65	60–75	5.8–6.0	17–34	0.5–3	<28.47	6700	12,040

has to be different from those for coal for these soft and friable fuels. Moreover, being of low cv with low ρ, the amount and volume of fuel to be fired is much larger for the same steaming. As the number of mills is normally limited to eight from layout considerations, a very large volume of fuel has to be transported by each feeder from bunker to mill.

All this rules out the possibility of using conventional coal feeders, both drag link and belt feeders in most cases, purely from volume considerations. Drag link type of feeder is simply unsuitable for transporting lignite and peat which are not nutty and strong to be pushed by the feeder flights. Feeders would face jamming in no time as the fuel crumbles and packs under the flights.

Plate belt feeder is the most appropriate equipment. It is a steel belt feeder with the belt made up of several strips of plates which can swing. The fuel, instead of being dragged by steel chain, is carried on the belt. These feeders are big and very long. They demand high skills of engineering and fabrication. Naturally, they are the most expensive of all the fuel feeders.

Figure 12.2 shows a typical belt feeder suitable for double-outlet bunkers. Similar to the case of drag link feeders the output control is by speed variation of the drive. The fuel lands on the top plate of the feeder which has, in many cases, a roller arrangement underneath to facilitate the smooth forward movement of fuel. Even though the fuel is not dragged on the bottom plate but carried on the belt, the young fuels have tendencies to break down and generate considerable fines. Hence, an additional cleaning chain is provided to drag all the escaping fines from the bottom plate of the casing into the discharge hopper. Suitable scrapers are provided to keep the plates clean. A twin screw arrangement at the discharge end directs the fuel to the inlet of the mill. The entire arrangement needs to be robust to provide a trouble-free service.

A plate feeder clearing a twin bunker is a very common arrangement with lignite and peat. The feeder is no doubt longer because of the twin inlets but it facilitates provision of steep angles and large outlets in fuel bunker, both of which are needed for the free flow of fuel without hang ups. This way it leads to a very elegant and economical bunker arrangement. Plate belt feeders for lignite and peat are much heavier and more complicated than drag chain feeders for coal as can be seen from the isometric view in Figure 12.3.

12.10.4 Beater Mills

Beater wheel mills or beater mills for short are the mills employed for pulverising young coals such as the high M lignites, brown coals and **peat**.

Although **coals**, **anthracites** and low M **lignites** are hard and nutty, the young fuels are wet and made of smaller pieces. It is in fact the M which holds together these fuel particles in the same way as wetness in mud forms it into lumps. Removal of M in itself is good enough to loosen the lump into pieces and a gentle impact thereafter to disintegrate into powder. It is not necessary to reduce it to the same level of fineness as for coal because the residual M in any case will explode on entering the furnace to reduce fuel powder to even finer material.

A beater wheel mill is primarily a combination of a beater and fan. Hot gases from furnace, suitably mixed with hot air from AH exit, are sucked by the fan action. Flue gas is preferred to hot air because

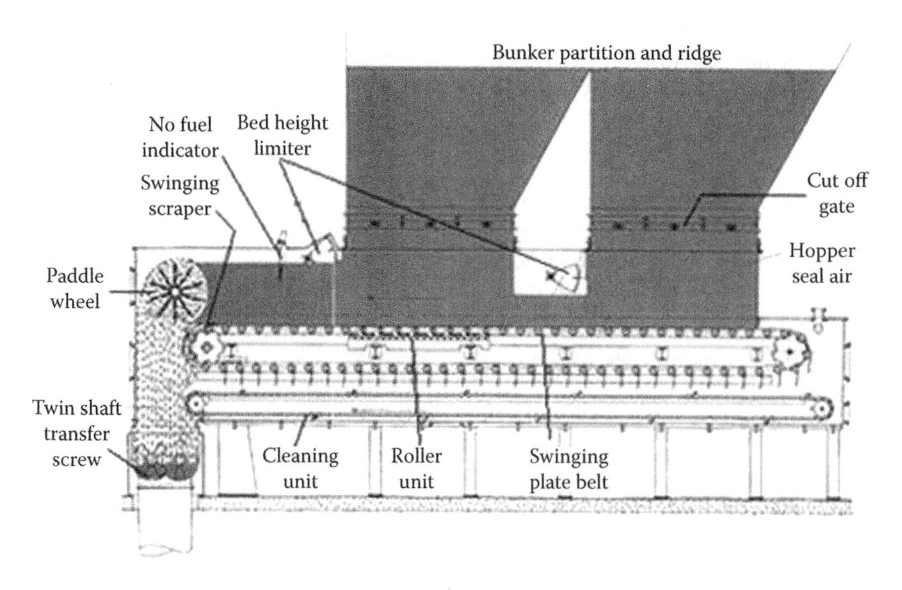

FIGURE 12.2
Plate belt feeder with twin fuel inlets. (From Maschinenfabrik Besta & Meyer GmbH & Co. KG, Germany, with permission.)

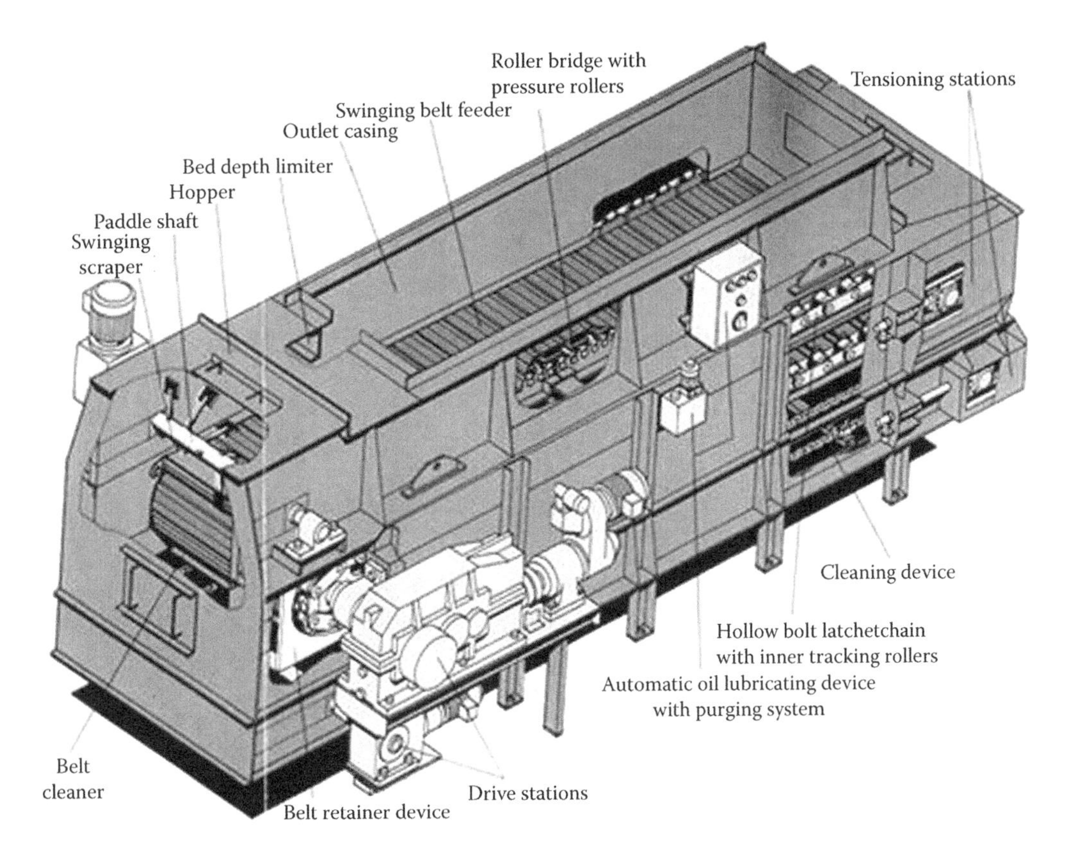

FIGURE 12.3
Isometric view of a plate belt feeder. (From Maschinenfabrik Besta & Meyer GmbH & Co. KG, Germany, with permission.)

a. The temperature required can be as high as 1000°C for high M in fuel.
b. Air can cause spontaneous ignition of these fuels with high inherent O_2 content.

Lignite feeders discharge fuel into the suction duct through which hot gases are sucked. Lignite gets heated and loses the M quickly. The hot dry fuel then enters the fan at temperatures of 120–200°C where it is given a gentle blow by the fan wheel to reduce to fine powder. The gases carry the powder directly into the furnace in many cases. Sometimes when the fuel M is highly pulverised lignite and gas enter a classifier where the larger particles are returned to the suction of the fan and the fine particles go to the burners. Usually, the fuel fineness is in the range 50–60% <90 μm. This arrangement of direct firing is shown in Figure 12.4 where there are three rows of burners one above the other.

Figure 12.5 shows the mill and discharge piping arrangement where a classifier is placed at the mill discharge to recycle large lignite particles. This is typical of high M lignites or brown coals.

There are two types of beater wheel mills to suit

a. Medium M fuels (up to 40%)
b. High M fuels (up to 70%)

A typical medium M mill is shown in Figure 12.6, where the fan wheel is preceded by beater wheels with swing hammers. Medium M fuels are usually harder and more abrasive requiring heavier beating. They also need lower gas temperature for drying. For the same capacity, these mills are smaller than mills for high M. The dried fuel is pulverised in the beater section and discharged by the fan pressure to the burners. Usually there is no classifier as the beater section of the mill is adequate to ensure the necessary fineness.

For high M fuels, the beater mills required are much bigger in size because of large quantities of high-temperature flue gases needed for fuel drying. Discharge volumes are even larger as high amounts of fuel M are distilled into it. There is no separate section of impact hammers. The beater wheel itself acts as both an impactor and fan. Normally classifiers are required at the discharge. It is usually an overhung design so that the inlet section can be wheeled away to provide full access to the beater. Figure 12.7 shows a typical beater wheel mill suitable for high M fuels. These mills are also known as *wet fan mills*.

In both designs, the ducts and beater mills are suitably lined to protect against **erosion**. The beater wheel also is given heavy lining for blades to improve life.

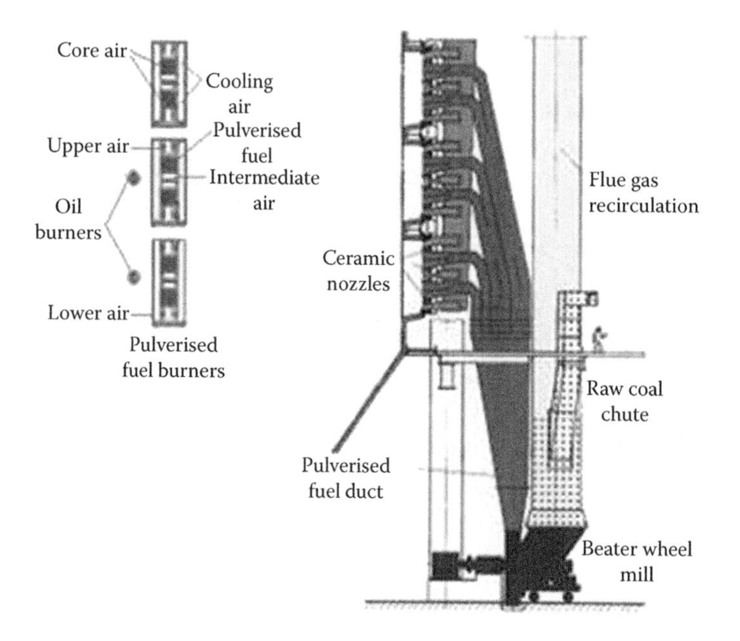

FIGURE 12.4
Direct firing of medium-moisture lignites.

Broad features of the beater mills are as follows:

- Mill speeds vary from 400 to 800 rpm
- The capacities range from 10 to 140 tph of fuel for high and 20–200 tph for low M mills
- Usually, mill outlet temperatures are ~120°C
- The wheel diameters vary from 2000 to 4000 mm
- Motor sizes are from 100 to 1600 kW for high and 150–2000 kW for low M mills

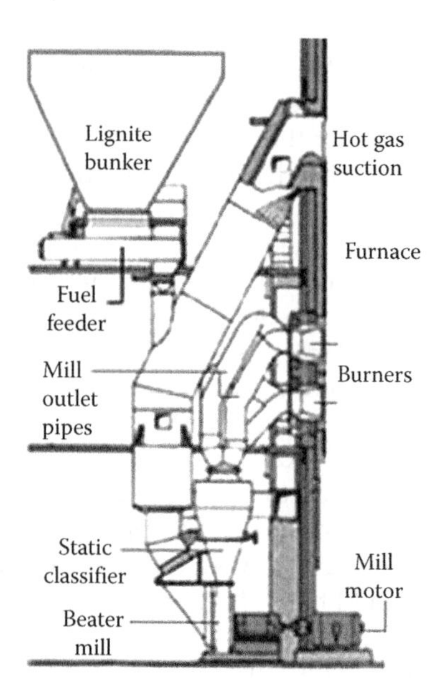

FIGURE 12.5
Direct firing of high-moisture brown coals.

- Specific power consumption varies from 4 to 16 kWh/te

12.10.5 Lignite and Brown Coal Firing

For lignites with <40% M, the mills and burners used are similar to those of coal firing with appropriate modifications. Both wall and tangential firing can be adopted.

For higher M lignites with >40%, the milling and firing equipments are entirely different. Beater mills are employed where flue gas from furnace is used for fuel drying. Substantially dried lignite along with hot moist flue gases is introduced into the furnace through burners. A typical brown coal burner is illustrated in Figure 12.8. Either four or eight sets of burners are employed which discharge the fuel tangentially towards an imaginary circle. The burners are not tiltable. The mills are placed at the four corners and on the four sides for four- or eight-burner arrangement, respectively. A typical lignite-firing arrangement is shown in Figure 12.9.

12.10.6 Lignite-Fired Boilers

Very large lignite-fired boilers were built in the last couple of decades. Approximately 1100 MWe is the largest boiler and built according to Benson technology employing SC conditions. The following are the salient features of brown coal firing.

- Once the M is driven out the combustion is rapid.
- Large amount of vapour in the air–fuel mixture depresses the combustion temperatures and

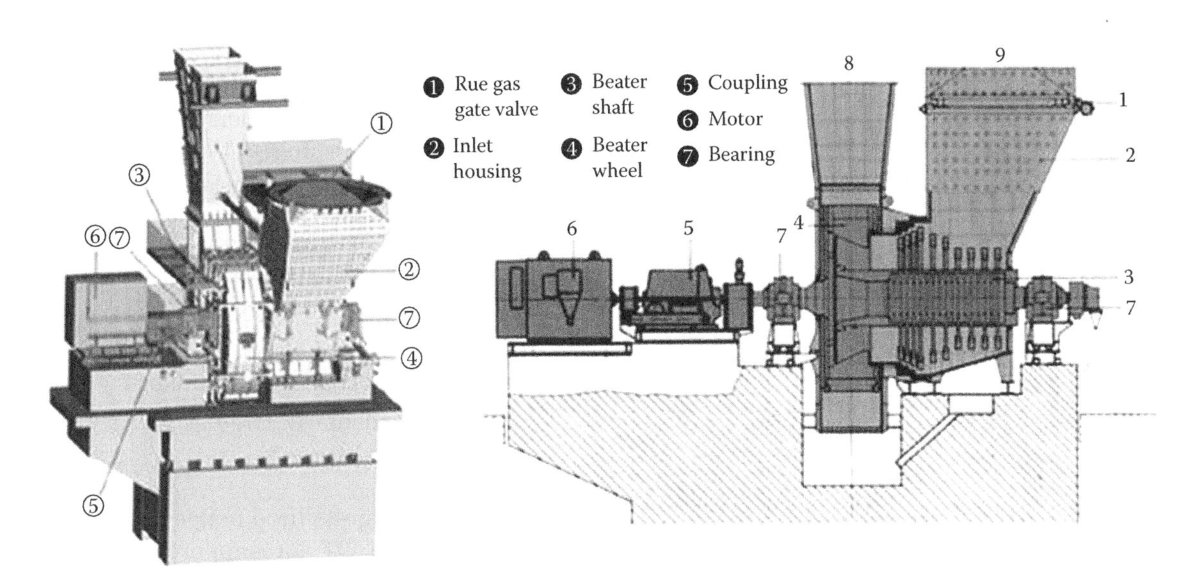

FIGURE 12.6
Beater mill for medium moisture in fuel—isometric and cross-sectional views. (From Hitachi Power Europe GmbH, Germany, with permission.)

correspondingly NO_x generation. With modern burners NO_x levels of <200 mg/nm^3 are achieved dispensing with costly air staging and deNO_x plants.

- However, because of the low flue gas temperatures inherent in lignite firing, the boilers need a lot more of HS and are consequently much larger compared to equivalent coal-fired boilers.
- The milling plant is also a lot different as the fuel is very soft, needing no grinding but only gentle tapping, to make it break down. Beater mills of appropriate type are employed.
- Flue gas from furnace, tempered by hot air from AH, is drawn into the beater mills for fuel drying. Fuel pipes carry the dried and sized lignite along with M-laden flue gases and admit them uniformly on all four sides in one to three levels into the furnace through large burners.
- This makes furnace construction quite at variance from PC boilers.

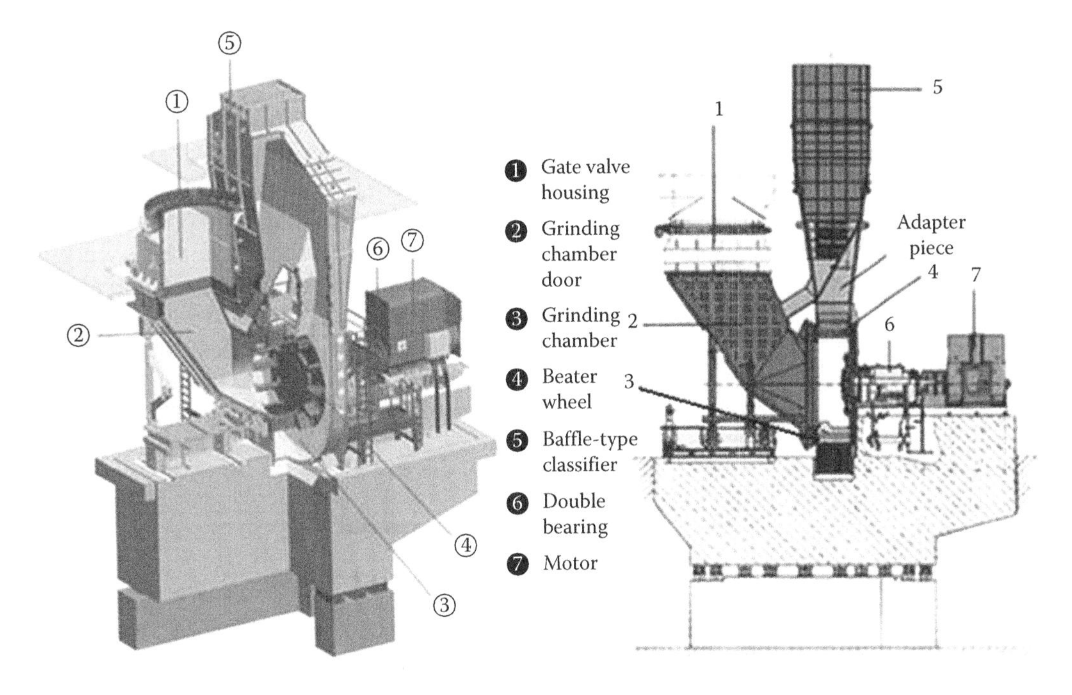

FIGURE 12.7
Beater mill for high moisture in fuel or wet fan mill—isometric and cross-sectional views. (From Hitachi Power Europe GmbH, Germany, with permission.)

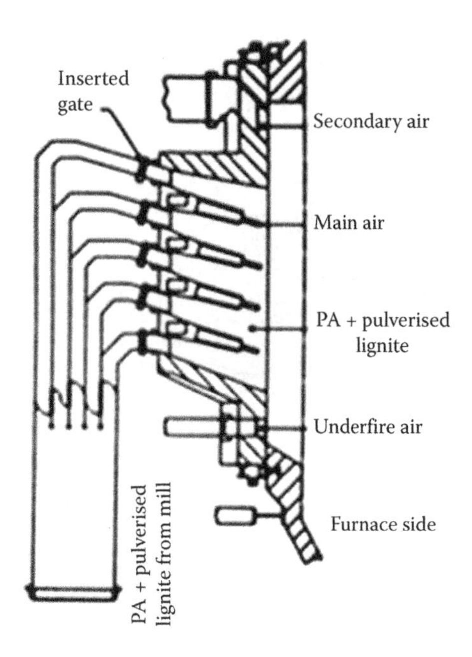

FIGURE 12.8
Burner for brown coal firing.

- The fuel feeding equipment is much bigger in size. Usually plate-type feeders are needed as drag chain feeders are found inadequate in larger boilers.
- Large gas volumes flowing over large convective HSs are highly erosive to the tubes, particularly when ash content in lignite is high. Tower-type boilers are preferred to two-pass construction as the gas flows are normal to tubes with no gas turns that can separate the ash and cause tube erosion.

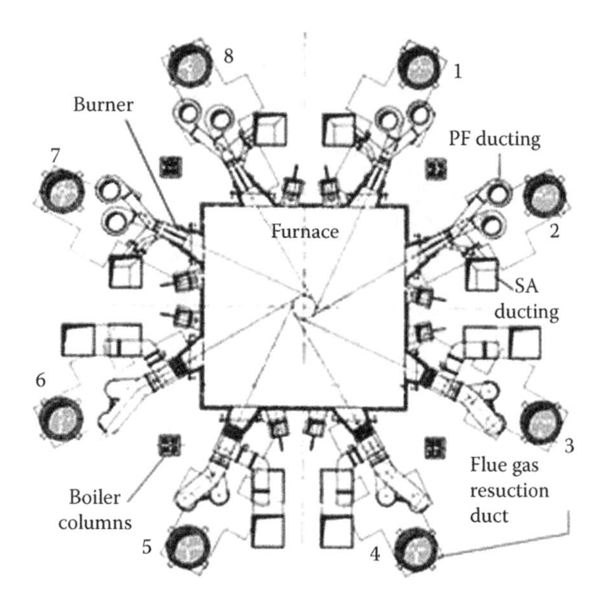

FIGURE 12.9
Eight beater mills around the furnace of a brown coal-fired boiler.

- Owing to high chances of spontaneous combustion ~10–16 h of storage in bunkers which is common with PF boilers is simply not safe. Usually bunkers hold ~4–6 h of lignite. Fuel required is more in volume as the fuel is lighter and low in cv. The feeding system sizes are therefore larger and have to run for longer time as bunkers are small.
- SBs are needed to keep the HSs clean even though the slagging and fouling issues are not as severe as in coal-fired PF boilers.
- AHs are particularly large as the flue gas volumes are high.

A typical large lignite-fired boiler is shown in Figure 12.10. It is a large OT top-supported fully drainable tower-type boiler with three levels of lignite burners and eight beater mills for high M brown coal. All HSs are in the first pass with only rotary AH in the second pass.

12.11 Limestone

Limestone is composed of calcite ($CaCO_3$). It is a calcareous sedimentary rock formed at the bottom of lakes and seas with the accumulation of shells, bones and other calcium-rich materials. Over the millennia, several layers are built up adding weight and turning it into solid stone because of heat and pressure. Rock which contains more than 95% of $CaCO_3$ is known as high-calcium limestone.

Physically, limestone is quite impervious, hard, compact, fine to very fine grained calcareous rock of sedimentary nature. The physical properties and chemical analysis are given in Tables 12.2 and 12.3, respectively. Being a natural stone the properties vary quite significantly and the following are only indicative.

Acidic products should not be used on limestone as calcareous stones readily dissolve in acid.

Limestone is of interest to a boiler engineer as it is added to coal for in-bed **desulphurisation** in FBC boilers. It is also used in **FGD plants** for removal of sulphur oxides from flue gases.

12.12 Liquid Fuels

Liquid fuel is a term given to various types of liquid materials that provide energy on combustion. Liquid fuel is also used in referring to a type of rocket propellant.

Most liquid fuels used in boilers are petroleum based such as the crude oil or various types of FOs. Petroleum

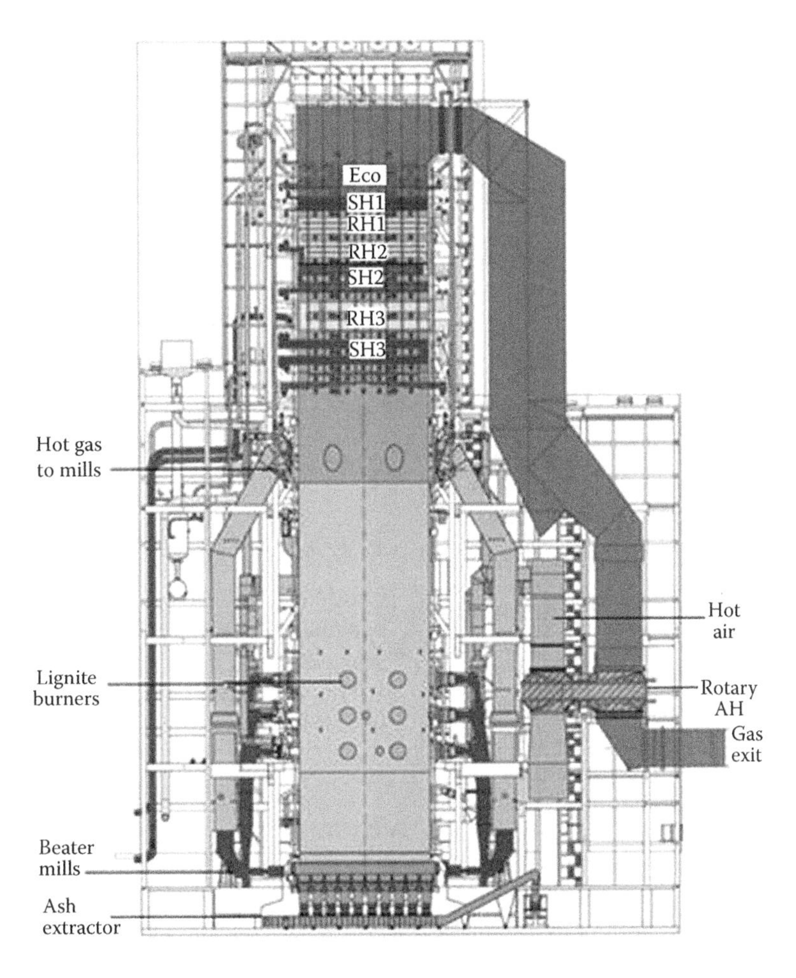

FIGURE 12.10
A large tower type SC brown coal-fired boiler with eight mills and three levels of firing. (From Hitachi Power Europe GmbH, Germany, with permission.)

TABLE 12.2

Physical Properties of Limestone

Hardness	3–4 on Moh's Scale
ρ	2.5–2.7 kg/cm^3
Compound strength	60–170 N/mm^2
Water absorption	<1%
Porosity	Quite low
Weather impact	Resistant

TABLE 12.3

Chemical Analysis of Limestone

CaO	38–42%
SiO_2	15–18%
Al_2O_3	3–5%
MgO	0.5 to 3%
$FeO + Fe_2O_3$	1–1.5%
Alkalies	1–1.5%

products in general are all composed of two major elements namely, C and H. The combination of these two elements is called an **HC**.

Even processed fuels such as **LPG** or **LNG**, which are important fuels, are petroleum based. Crude petroleum is burnt for steam generation only infrequently as the lighter fractions depress the flash point and present a fire hazard. Once the lighter fractions are removed the same oil becomes safe for handling and combustion.

Pulp mills have a significant liquid fuel called **BL**, which is actually an intermediate process chemical that has to be necessarily burnt for recovering chemicals. There are also several **sludge** materials, which are process wastes, which are burnt in the boilers mostly as supplementary fuels.

Petroleum-based liquid fuels are most convenient to handle and burn. They are the easiest of all fuels to transport, store and combust. Fuel oils have the highest energy ρ among all fuels, that is, maximum energy is packed in the least volume.

With the increasing popularity of CCPPs FO-based steam generation has gone out of fashion on η consideration and large FO-based utility boilers have not been built for over a couple of decades now, except in refineries for consuming the heavy bottoms.

The HCs mostly found in boiler FOs fall into four main classes, namely

- Paraffinic
- Aromatic
- Naphthenic
- Olefinic

Paraffinic HCs (CNH) are lower in SG than aromatic HCs of the same boiling point, while naphthenic and olefinic compounds are intermediate in ρ. Their (2*N* + 2) resistance to chemical change or oxidation is very good. These HCs are clean burning, and thus are desirable in distillates such as gas oil or diesel oil.

Aromatic HCs (CNH) possess a much higher SG than the other three classes. Aromatics are very stable under heat and are chemically active to a moderate degree. The (2*N* – 6) aromatic compounds contain a higher proportion of C than the other HC types. Owing to this characteristic, they have a tendency to smoke, which somewhat limits their use in diesel engines.

Naphthenic HCs (CNH-ring type) are extremely stable, cyclo-ring compounds and in many cases have more stability than the paraffins. These HCs are more commonly found in heavy 2*N* marine FOs rather than distillate oils.

Olefinic HCs (CNH-straight chain) are more chemically active than the other three classes of HCs. Olefins are subject to oxidation or polymerisation, forming gums. Olefins are not present in large amounts in straight-run distillates, but are found in large quantities in cracked FOs.

Table 12.4 gives a list of relevant ASTM standards for FOs.

This topic is divided mainly into two sub-topics, namely

a. Fuel properties
b. Major liquid fuels

TABLE 12.4

List of ASTM Standards on Fuel Oils

D56	Test method for FlA point by tag closed-cup tester
D93	Tests methods for point by Pensky–Martens closed tester
D189	Test method for Conradson carbon residue of petroleum products
D396	Spec for fuel oils
D445	Test method for kinematics viscosity of transparent and opaque liquids (and the calculation of dynamic viscosity)
D524	Test method for Ramsbottom carbon residue of petroleum products
D4530	Test method for determination of carbon residue (micro method)

12.12.1 Fuel Properties

Properties of importance to boiler engineers are listed below.

1. API gravity
2. Carbon residue
3. Flash point
4. Kinematic viscosity of oils
5. Pour point
6. Sulphur in oil
7. Ash
8. Sediment

12.12.1.1 API Gravity (see also *Specific Gravity*)

Crude oil ρ is one of the most important characteristics, because FOs are purchased by volume as barrels, gallons or cum.

The oil industry conventionally uses API gravity which is jointly developed by API and NIST.

The formula for API gravity is

$$°API = \frac{141.5}{\text{Specific gravity @15/15°C}} - 131.5 \quad (12.1)$$

$$\text{Alternatively, SG@15/15°C} = \frac{141.5}{131.5 + °API} \quad (12.2)$$

From this, it follows that

- Higher the API gravity, lighter is the fuel.
- @ 10°API, oil has an SG of 1.
- Most of the oils vary between 10°C and 70°C API (SG of 1–0.7).
    - Light FOs have >31.1°API (<870 kg/m^3).
    - Medium FOs have >22.3° and <31.1°API (870–920 kg/m^3).
    - –HFOs have <22.3°API (>920 kg/m^3).
    - Extra HFOs have <10.0°API (>1000 kg/m^3). This is typically bitumen.

12.12.1.2 Carbon Residue

The C residue of a fuel is the *tendency to form C deposits under high temperature conditions in an inert atmosphere.* It is the stuff that remains after thermal degradation of an organic compound.

It may be expressed as

- Ramsbottom carbon residue (RCR) determined by ASTM D524

- Conradson carbon residue (CCR) determined by ASTM D189 or
- Micro carbon residue (MCR) determined by ASTM D4530

Numerically, the CCR value is the same as that of MCR.

The C residue value is considered to give an approximate indication of combustibility and deposit forming tendency of a fuel. As the CCR increases, typically the asphaltene content of an HFO also increases. The combination of higher CCR and asphaltene increases the centrifuge sludge and burden in fine filter requiring more frequent centrifuge desludging and cleaning/replacement of filter element. Higher CCR also lowers the Gcv and Ncv (on weight basis) of an HFO.

12.12.1.3 Flash Point

Flash point of a substance is *the lowest temperature at which enough fluid can evaporate to form a combustible concentration of gas*. The vapours so formed ignite when exposed to a tiny flame. Flash point is determined to minimise fire risk during normal storage and handling. It is no indication of burning property.

Flash point is different from *fire point*, which is the lowest temperature at which the vapours will burn continuously when a tiny flame is brought near it. Fire point is higher than flash point by 5–40°C

For emergency fuels external to the machinery space, the flash point must be >43°C.

The normal maximum storage temperature of a fuel is 10°C below the flash point, unless special arrangements are otherwise made.

Flash point is an indication of *how easily a substance may ignite*. Materials with higher flash points are less flammable or hazardous than chemicals with lower flash points.

Flash point is determined for HFOs by Pensky–Marten flash point testing apparatus which essentially consists of

- An oil cup of ~50 mm diameter and 55 mm depth with level marking rings inside. The lid of the cup contains four standard openings with one each for introducing the thermometer, stirrer, air and oil flame.
- There is a shutter at the top of the lid. By moving the shutter the aperture in the lid opens to admit flame on top of the oil inside.
- The oil cup is supported over an air bath, which is heated by a burner.
- There is a pilot burner arrangement to light the flame.

Figure 12.11 shows the apparatus.

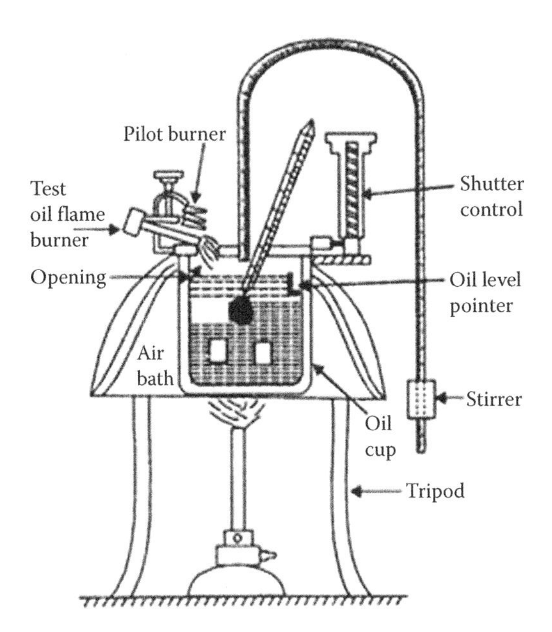

FIGURE 12.11
Pensky–Marten apparatus for flash point determination for FOs.

12.12.1.4 *Kinematic Viscosity of Oils* (see also *Viscosity in Fluid Properties*)

Kinematic viscosity of FOs is measured by viscometers and expressed in seconds or degrees as described in Viscosity measurement of oils under Instruments for measurement. A ready viscosity conversion chart is given in Table 12.5.

Figure 12.12 gives the variation of viscosity with temperature for all FOs.

12.12.1.5 Pour Point

Pour point is *the lowest temperature at which a liquid starts to flow under test conditions*. But it is not necessarily the temperature *at which it stops flowing*.

Storage tanks maintain FO temperature at ~3°C higher than the pour point to assure flowability.

Pour point also represents the pumpability.

12.12.1.6 Sulphur in Oils

In varying forms and concentrations S occurs in all crude oils. When the crude is distilled S derivatives tend to concentrate in the heavier fractions, leaving the lighter fractions with relatively low S contents.

S in FO is responsible for **low-temperature corrosion** of **back-end equipment**. As the S content in FOs raises above 3% the problem of condensation of corrosive vapours on the cold surfaces of Eco and AH becomes increasingly difficult to control.

S also reduces the cv of oil by ~70 kcal/kg (126 Btu/lb) for each% by weight.

TABLE 12.5
Fuel Oil Viscosity Conversion Table

Kinematic Viscosity cSt @ 50°C	Approximate Viscosity Equivalents for Fuel Oils				
	Redwood 1 Sec @100°F	Engler⁰ @ 50°C	SSU @ 100°F	SSF @ 100°F	cSt @ 82.2°C
30	200	4.1	230	26	11
40	280	5.3	320	35	13.5
60	440	7.9	510	53	18
80	610	10.5	700	72	22.5
100	780	13.2	900	93	26.5
120	950	15.8	1120	116	30
150	1250	19.8	1400	147	35
180	1500	23.7	1750	175	40
240	2200	31.6	2370	257	49
280	2500	36.9	2850	300	55
320	2900	42.1	3300	345	61
380	3600	50.0	4100	420	69
420	4100	55.0	465	480	74
460	4600	61.0	5200	540	79

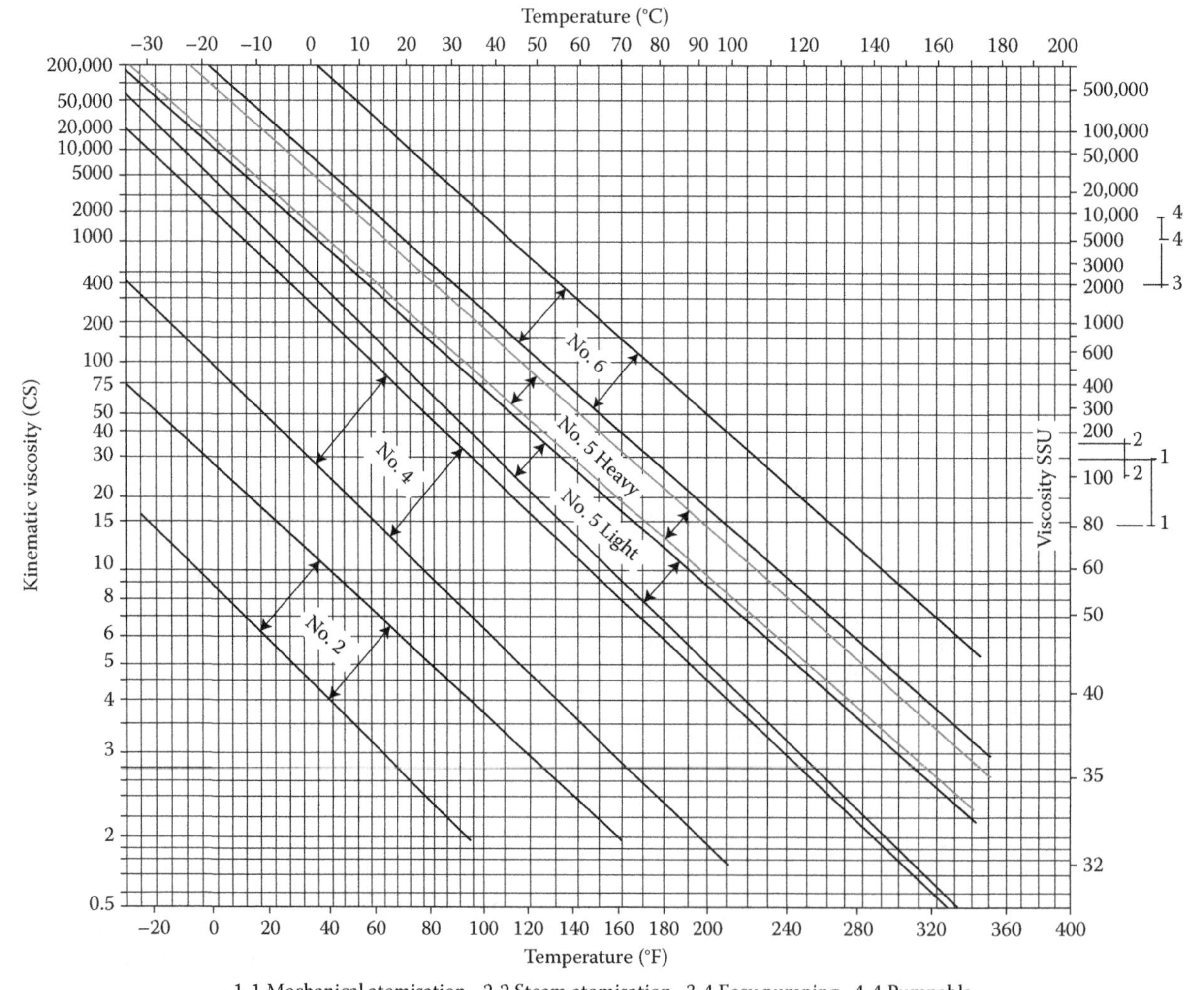

FIGURE 12.12
FO viscosity variation with temperature.

12.12.1.7 *Ash* (see also *Oil Ash*)

Ash in HFO includes

- The inorganic metallic content, in dispersed and dissolved form, such as V, Ni, Fe, Na, K or Ca
- Other non-combustibles
- Solid contamination such as sand, rust and catalyst particles

High-temperature corrosion of SH and RH is caused by the metallic ash content.

12.12.1.8 *Sediment*

This is another contaminant, mainly *consisting of rust, scale, weld slag, dirt and other debris* that gets introduced in storage or in pipeline transport, most of which can be *removed by settling, straining or filtration or centrifuging.*

As the SG and the viscosity of FOs increases, the sediment level also increases. Sediment originates primarily from transport and storage contamination. This sediment is mainly a result of a slowing of the natural settlement process. Increase of both water and sediment of HFO decrease the usable energy in the fuel. Both of them can be settled out to a large degree during heated storage. However, as the SG of HFO approaches that of water the settling process becomes less and less effective. The effectiveness of purification/clarification treatment is also reduced. Sediment removal is essential to reduce high ash or particulate contamination of a fuel since both can contribute to increased deposits, corrosion or abrasive wear.

12.12.2 Major Liquid Fuels

Major liquid fuels described here are listed below.

1. Crude oil or petroleum
   a. *Atmospheric or straight-run distillation*
   b. *Vacuum distillation*
   c. *Thermal cracking*
   d. *Catalytic cracking*
2. Bitumen/asphalt and coal tar
3. Fuel oils
4. Refuse fuels from oil refinery
5. LNG
6. LPG
7. Naphtha
8. Oil/tar/bituminous sands
9. Oil shale
10. Orimulsion
11. Sludge

12.12.2.1 *Crude Oil or Petroleum*

Crude oil or petroleum is a naturally occurring inflammable organic liquid beneath the earth surface, either under land or sea. It is a complex mixture of a variety of HCs along with traces of nitrogenous and sulphurous compounds.

It is formed for over 600 million years from incompletely decayed plant and animal remains buried under thick layers of rock. Under high p & t of the geological conditions, petroleum is formed underground and includes both crude oil and NG, as they are both mixtures of HCs.

Crude oil varies from oil field to oil field in colour and composition, from a pale yellow low-viscosity liquid to heavy black consistencies.

Among the leading producers of petroleum are Saudi Arabia, Russia, the USA, Iran, China, Norway, Mexico, Venezuela, Iraq, the UK, the UAE, Nigeria, Sudan, Indonesia and Kuwait. The largest reserves are in the Middle East.

The main use of crude oils is to produce FOs in the process of distillation and cracking. Crude oil by itself is seldom used in boilers or GTs as the lighter fractions depress the flash point and present a fire hazard. With the removal of lighter fractions, oil becomes safe both for handling and combustion.

FO-based power generation by the ST cycle route has gone out of fashion with greater availability of cheaper NG and the increasing efficiencies and sizes of GTs and Hrsgs.

There are four methods of refining crude oil to produce a variety of petroleum-based oils. These are elaborated below.

a. Atmospheric or Straight-Run Distillation

This is the oldest and most common refining process and consists of boiling crude oil at atmospheric pressure in a fractionating tower up to temperatures <371°C (700°F). At higher temperature, HCs begin to crack and form materials which are not desired at this point in the process. As the various constituents of crude oil vapourise at different rates, the lighter, more volatile gases rise high in the tower before condensing and being collected. The heavier, less volatile gases condense and are collected lower in the fractionating tower. This is shown in Figure 12.13.

b. Vacuum Distillation

Here the pressure is maintained at partial vacuum in the distillation/condensation tower so that the residual fuel from an atmospheric process yields additional heavy distillates and further concentrates the impurities and C in its residual oil or vacuum bottoms having very high viscosity. Vacuum distillers produce residual oils that are feed stocks for other refinery processes.

Vacuum distillation bottoms are further refined by the use of a secondary process such as viscosity breaking. Vacuum bottoms are heated to a higher p & t for cracking

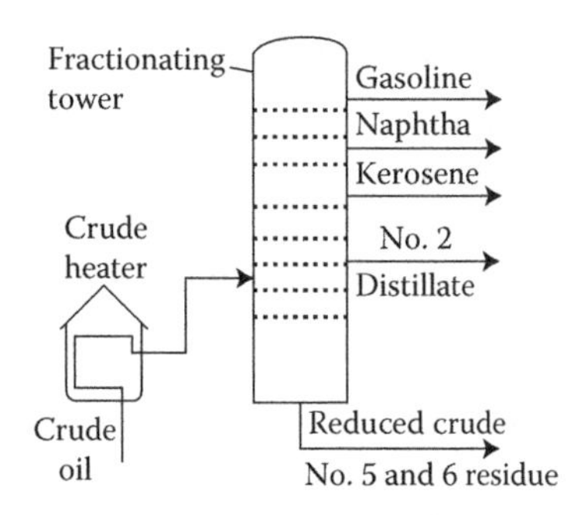

FIGURE 12.13
Atmospheric or straight distillation.

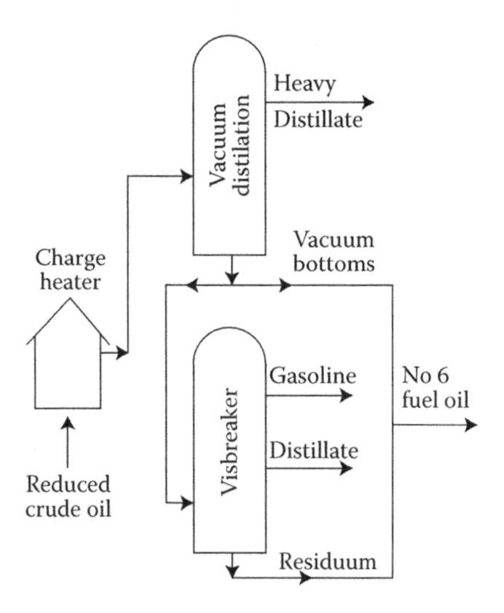

FIGURE 12.14
Vacuum distillation.

(not as high as in the thermal cracking) to break down to a residuum which is considerably lower in viscosity. Whereas this is utilised as HFO with little or no blending with a lighter distillate, it has increased SG, and less desirable characteristics, such as high Conradson C and high asphaltenes. It is usually less stable and less compatible with other residuals than the original feed stock. This process is illustrated in Figure 12.14.

c. Thermal Cracking

Straight-run residual feed stock is heated to high p & t in the reaction chamber where heavy, large, long chain oil molecules are cracked or broken, producing both short-chain and additional long-chain molecules. These cracked products are then vapourised in the flash chamber and flow to the fractionating tower where they are condensed at different levels to the various products (as shown in Figure 12.15). Thermal cracking increases the yield of high-quality distillate fuels and reduces the yield of residual fuel. Like vis-breaking, this process also yields residual FOs with a high SG and high S, V, Conradson C/asphaltenes content together with

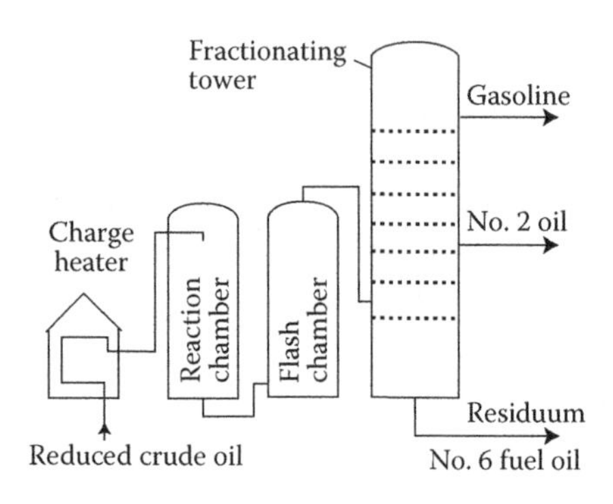

FIGURE 12.15
Thermal cracking.

poorer stability and compatibility. This results finally into cracked residual fuel which is more difficult to burn, has a lower cv (on weight basis), higher sludge and impurities content.

d. Catalytic Cracking

Catalytic cracking, including FCC, employs chemical catalysts to increase the yield of the thermal cracking process. The presence of powdered catalyst in the feed stock allows the breakdown of long-chain molecules into lighter short-chain HC molecules at lower p & t. The fine powdered catalyst in the charge stock can carry over into the resultant-cracked residuum. This powder termed as 'catalyst fines' or 'cat cracker catalyst', is very hard and extremely abrasive such as fine sand, and can rapidly increase the wear of injection pumps, fuel valves, injectors, piston rings, piston ring grooves, liners, stuffing box seals and turbocharger blading. Whenever a straight-run residual oil has been processed through FCC unit to produce cracked residuum, there has been a very high probability that some of these abrasive catalyst particles, comprised of Al_2O_3 and/or SiO_2, will be carried over. This process is illustrated in Figure 12.16.

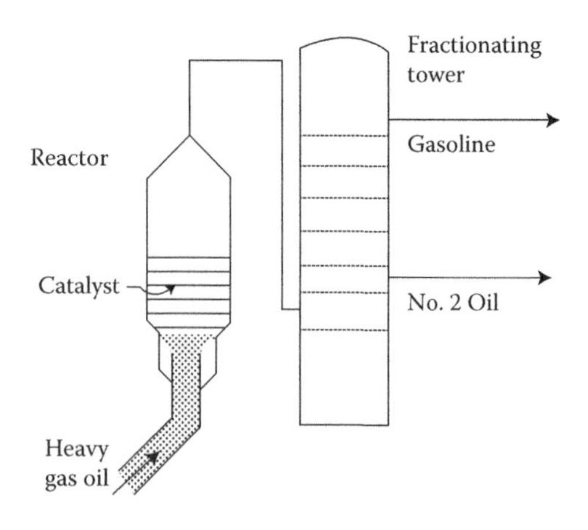

FIGURE 12.16
Catalytic cracking.

12.12.2.2 Bitumen/Asphalt and Coal Tar

Bitumen or asphalt is a mixture of highly viscous organic liquids. Bitumen is black, sticky and soluble in CS_2 (carbon disulfide).

- *Crude bitumen* is a naturally occurring form of petroleum. It is so thick and heavy at room temperature that it has to be heated or diluted before it can flow.
- *Refined bitumen* is the residual fraction or bottoms obtained by fractional distillation of petroleum. It is the heaviest fraction and the one with the highest boiling point, boiling at ~525°C (977°F).

Although bitumen and asphalt mean the same, coal tar (also called tar for short) is a similar-looking product but with entirely different origin and constituents. Coal tar is the principal liquid product when coal is carbonised to make coke or gasified to make coal gas. Coal is the heated in the absence of air at temperatures ranging from ~900° to 1200°C (1650–2200°F).

Being inflammable, coal tar is the one that is many times used in boilers as fuel, mostly for disposal, when it cannot be sold more profitably. At room temperatures coal tar is very viscous. It needs considerable heating to make it suitable for pumping and atomisation. Thereafter, it burns just like FO in the same equipment. Typical analysis is as given Table 12.6.

At one time coal tar was the major source of organic chemicals, most of which are now derived from petroleum and NG.

The main use of bitumen and coal tar is for surfacing the roads and runways. It is also a good roofing material.

12.12.2.3 Fuel Oils

FO is a liquefied petroleum product that can be burnt to generate heat and power. It is the fraction obtained from crude oil distillation—either a distillate or residue.

FOs used in power plants, ships and other industrial applications are mainly termed furnace oils which are the residues from distillation, suitably blended with other petroleum fractions to produce the desired viscosity and flash point.

12.12.2.3.1 Low Sulphur Heavy Stock

LSHS is a residual fuel which can be used in place of FO. The main differences are that LSHS has

- Higher pour point at ~66°C
- Higher cv (1–2% higher than HFO)
- Lower viscosity at handling temperatures
- Lower S content of <1%

Higher pour point requires more heating of LSHS in tanks and pipe tracing arrangements to keep it hot at all stages and is maintained at ~75°C. Special care is also taken so that no 'boil over' of the product takes place in the storage tank.

The main advantage lies in its low S content with reduced risk of corrosion both at high and low temperature and emission of lesser quantity of S oxides. Higher cv is generally beneficial as the consumption is lower than that of FO.

12.12.3.2 Classification of Fuel Oils

ASTM D396 classifies FO into six grades, numbering 1, 2, 4, 5(light), 5(heavy) and 6 according to the boiling points ranging from 175°C to 600°C with increase in viscosity and lowering of prices.

- Number 1 and 2 are middle distillate fuels for domestic and small industrial burners. They are variously referred to as distillate FOs, diesel FOs, light FOs, gasoil or just distillates. Number 1 is similar to kerosene and is the fraction that boils off right after gasoline.
- Number 4 is heavy distillate fuel for commercial and industrial burners. It is usually a blend of distillate and residual FOs, such as number 2 and 6. It may be just a heavy distillate.
- Number 5 light and heavy and number 6 are residual FOs (RFO) or furnace FOs (FFO) used in industrial burners. Numbers 5 and 6 are what remain of the crude oil after gasoline and the distillate FOs are extracted through distillation.

  Light and HFOs are produced by blending of distillates and RFO.

TABLE 12.6

Properties of Coal Tar

	C	H	S	O	N	H_2O	Viscosity	Flash Point	SG	Gcv Dry
w/w%	89.9	6.0	1.2	1.8	0.4	0.7	900 SSU @1220F	69°C 156°F	1.07	39 MJ/kg 9305 kcal/kg 16750 Btu/lb

TABLE 12.7

Physical Properties of FOs

Grade of Fuel Popular Designation	No. 1 Light	No. 2 Med	No. 4	No. 5 (Light)	No. 5 (Heavy)	No. 6 (Heavy) Bunker C
Use	Domestic	Domestic				Industry
Flash point (min) °C	38	38	55	55	55	65
Pour point (max) °C	0	−7	−7	—	—	—
Water and sediment vol%	trace	<0.1	<0.5	<1.0	<1.0	<2.0
C residue on 10% bottoms %	<0.15	<0.35	—	—	—	—
A% by wt (max)	—	—	0.1	0.1	0.1	—
Distillation temps. 10% pt °C 90% pt °C	<215 <288	282–338	—	—	—	—
Viscosity range SSU@38°C SSF@50°C	—		45–125	150–300	350–750 (23–40)	900–9000 45–300
API gravity (min)	35	30	—	—	—	—
SG@15/15°C (min)	0.8017	0.876				
Sulphur (max) %0.5	0.5	0.7	No limit	No limit	No limit	No limit

Preheating is usually needed for handling and proper atomisation.

Number 6 is the most common FO for use in boilers and is also termed as Bunker C oil.

Tables 12.7 and 12.8 provide the properties of FOs and their analyses generally as classified in ASTM 396.

In Table 12.7 while SGs of grades 1 and 2 are given for grades 4, 5 and 6 they are not specified as they vary with the source of crude and extent of cracking and distillation.

12.12.2.4 Refuse Fuels from Oil Refinery

Refinery operations generate a number of by-products like tar, petroleum coke and so on.

The liquid products are called *refinery sludge* with high SG and varying amount of solids, mostly consisting of oil coke. The composition and source of crude oil influences the properties of sludge.

12.12.2.4.1 Acid Sludge

Acid sludge is the most troublesome with its frequently varying and widely changing properties. SG varies from 4 tol 4 API. Viscosity is indeterminate. It contains a large amount of dilute H_2SO_4 acid, sometimes as much as 40%. Gcv varies from 18.4 to 40.6 MJ/kg, ~4400–9700 kcal/kg or ~8000–17,450 Btu/lb.

Alkaline sludge is more benign. It is a bit more consistent in its composition and less troublesome to burn.

12.12.2.5 Liquefied Natural Gas

LNG is a relatively new fuel steadily gaining popularity since the mid-1960s.

TABLE 12.8

Chemical Analysis and Properties of FOs

UA by wt%	No. 1	No. 2	No. 4	No. 5	No. 6
S	0.01–0.5	0.05–1.0	0.2–2.0	0.5–3.0	0.7–3.5
H	13.3–14.1	11.8–13.9	10.6–13.0	10.5–12.0	9.5–12.0
C	85.9–6.7	86.1–88.2	86.5–89.2	86.5–89.2	86.5–90.2
N	0–0.1	0–0.1	—		—
O	—	—	—	—	—
A	—	—	0–0.1	0–0.1	0.01–0.5
Other properties					
API gravity	40–44	28–40	15–30	14–22	7–22
SG@15/15°C	0.825–0.806	0.887–0.825	0.966–0.876	0.9725–0.922	1.0217–0.922
Pour point °C	−18 to −45	−18 to −40	−23 to +10	−23 to +27	−9 to +30
GCV MJ/kg	45.76–46.18	44.59–45.93	42.52–45.12	44.19–44.23	40.49–44.17
kcal/kg	10,930–11,030	10,650–10,970	10,155–10,777	10,555–10565	9670–10,550
Btu/lb	19,675–19,855	19,170–19,745	18,280–19,400	19,000–19,017	17,405–18,990

NG is cooled to –162°C to condense it into liquid and make LNG, when it occupies <1/600 times the original volume. The reduction in volume makes it convenient and cost efficient to transport over long distances either through pipelines, by specially designed cryogenic sea vessels called LNG carriers or cryogenic road tankers.

LNG is odourless, colourless, non-corrosive and non-toxic.

Its hazards include flammability, freezing and asphyxia.

NG is the most ideal fuel but for its bulk and difficulty in transportation. These deficiencies are remedied by LNG. The energy ρ of LNG is ~60% of diesel oil. It has a ρ of 400–500 kg/cum, depending on its temperature, pressure and composition which is nearly half of water.

A tonne of LNG = ~1500 cum of NG of cv 8900 kcal/cum or ~2.5 tonnes of coal of 5300 kcal/kg.

NG is composed primarily of CH_4 (typically, at least 90%), but may also contain C_2H_6, C_3H_8 and heavier HCs. Small quantities of N, O, CO_2, S compounds, and water may also be found in 'pipeline' NG. The liquefaction process removes the O_2, CO_2, S compounds and water. The process can also be designed to purify the LNG to almost 100% CH_4.

At the consuming end LNG is re-gasified and converted into its original state and delivered to the plant in pipelines just like NG.

LNG is a relatively expensive fuel because it needs

- Liquefaction and loading facilities at the production end
- Transportation by pipelines or ships
- Re-gasification and distribution facilities at the consuming end

LNG is a well-traded fuel in the world markets with about 20 exporting and importing nations. Algeria, Libya, Qatar, South East Asia and Alaska are the major exporters exporting to Japan, South Korea, Taiwan, China, India and so on.

The main use of LNG in power plants is in firing the GTs.

Figure 12.17 depicts the entire LNG value chain.

12.12.2.6 Liquefied Petroleum Gas

Liquefied petroleum gas is also called LPG, *LP gas* or *Auto gas*.

It is mainly used for domestic heating, appliances and automobiles. In boiler plants, it is a start-up fuel.

LPG is refined from FO or NG. It is basically HC with a substantial amount of C_3H_8 and some C_4H_{10} as main constituents. Hence, the characteristics of C_3H_8 are usually taken as a close approximation to those of LPG.

LPG is produced by a blend of butane and C_3H_8 readily liquefied under moderate pressure.

Unlike the NG, LPG vapour is heavier than air and it settles down in low-lying places. For ease of detection, a mercaptan odourant is intentionally added.

12.12.2.7 Naphtha

Naphtha is a colourless, highly volatile, flammable liquid mixture of HCs distilled from petroleum (petroleum naphtha) and occasionally coal tar (coal tar naphtha) or NG. Figure 12.13 shows naphtha production in the atmospheric crude distillation unit (CDU).

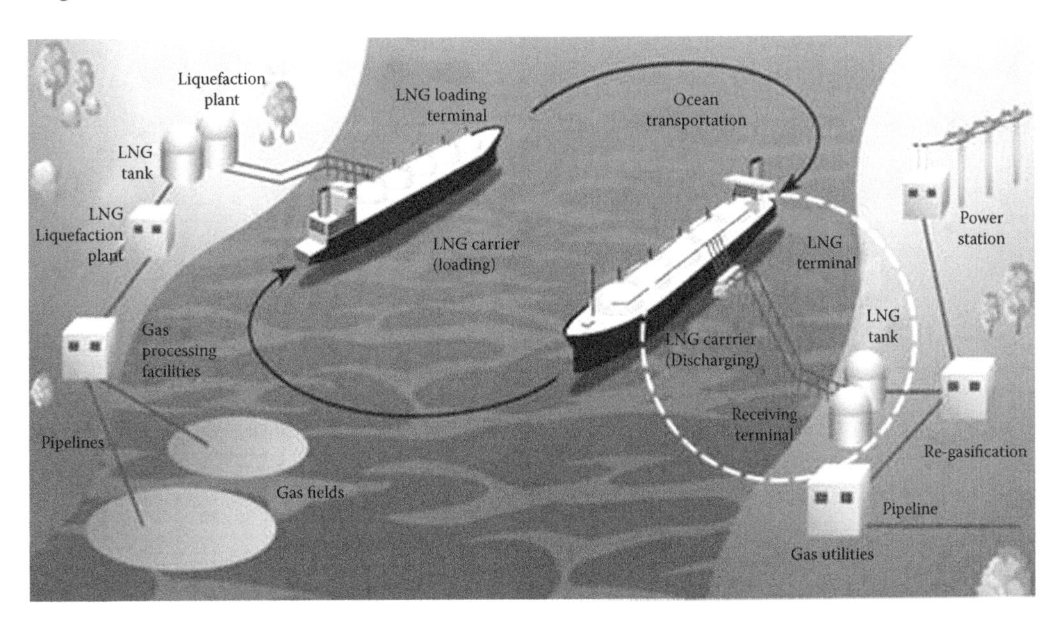

FIGURE 12.17
Overview of LNG cycle.

It is a general term for spirits having low boiling points ranging from as low as 30°C to as high as 260°C.

The main commercial use is as solvents, thinners and carriers.

Technically, petroleum and kerosene can also be considered as naphtha.

Naphtha is used as feedstock in fertiliser and petrochemical industries. It is also used to a limited extent as fuel in boilers and GTs.

Naphtha has a GCV of >44 MJ/kg, ~10,500 kcal/kg or 18,900 Btu/Lb.

Its flash point is usually <23°C.

S is <0.15% and fouling metals such as V, Na and Pb are almost absent.

12.12.2.8 Oil/Tar/Bituminous Sands

Oil sands or tar sands are naturally occurring bituminous sands, a type of unconventional petroleum deposit, discovered in many countries but are found in extremely large quantities in Canada and Venezuela. The sands contain mixtures of sand, clay, water and bitumen. Bitumen is a dense and extremely viscous form of petroleum which is sticky and black in colour. In Canada it is often described as petroleum that exists in the semi-solid or solid phase in natural deposits. In Venezuela, because of its warm climate, the bitumen is more fluid and less rocky and oil sands are termed extra heavy petroleum.

Oil sands have begun to be considered as part of oil reserves of the world only recently after the oil prices have gone up significantly and newer technologies have been commercialised for their exploitation. As of now extraction of liquid fuels remains a highly energy intensive and GHG generating process. But as the deposits are large it holds a prospect for future.

12.12.2.9 Oil Shale

Oil shale generally refers to any sedimentary rock that contains solid bituminous materials (called kerogen). When heated by pyrolysis the rocks release petroleum-like oils in adequate amounts that can burn without any additional processing. Hence oil shale is known as 'the rock that burns' (see Figure 18.7).

Many millions of years ago oil shale was formed by the deposition of silt and organic debris on lake beds and sea bottoms. Heat, pressure and time transformed the materials into oil shale in the same as oil formation. However, the heat and pressure were less.

Oil shale can be mined and processed to generate oil. But oil extraction from oil shale is complex and more expensive. As oil substances in oil shale are solid and cannot be pumped directly out of the ground, the oil shale must be heated after mining to a high temperature (*retorting*) to separate the liquid for collection. Alternatively, in *in situ retorting* the oil shale is heated while it is still underground and then pumped to the surface (as shown in Figure 12.18). This process is still in the early stages of commercialisation.

As the price of petroleum has risen oil shale has gained attention as an energy resource. However, oil shale mining and processing raises a number of environmental concerns, such as the use of land and water, waste disposal and wastewater management, GHG emissions and air pollution.

Oil shale has been used as fuel and as a source of FO in small quantities for long. Few countries currently

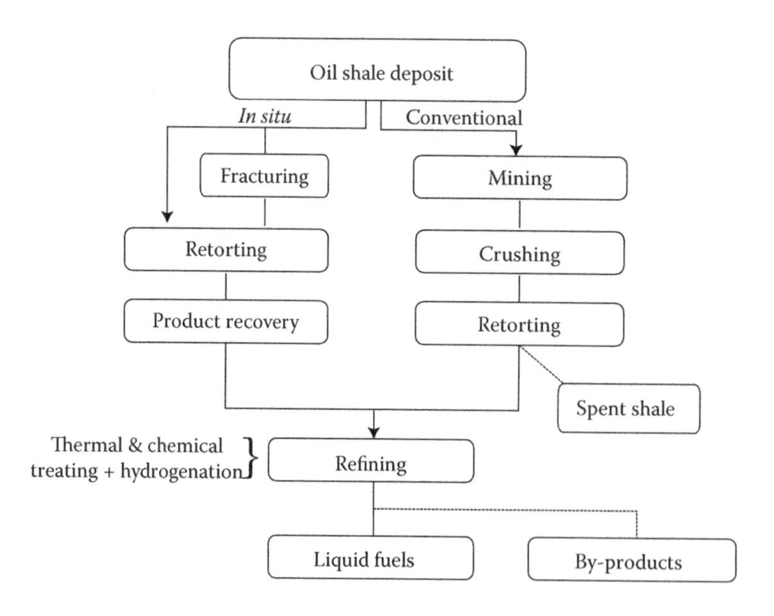

FIGURE 12.18
Chain of events from oil shale mining to conversion to liquid fuels.

produce FO from oil shale on a significantly commercial level. Many countries do not have sufficient oil shale resources, but in those countries that do have significant oil shale resources, the oil shale industry has not developed because the cost production of FO from oil shale has been much higher than from crude oils. This has inhibited the development of better technologies that may reduce its cost.

Industry can burn oil shale directly as a low-grade fuel for power generation and heating purposes and can use it as a raw material in chemical and construction-materials processing.

Oil shale is found in many parts of the world but by far the largest deposits are in the USA in the Green River Formation, which covers portions of Colorado, Utah and Wyoming. Estonia and China have well-established oil shale industries while Brazil, Germany, Israel and Russia also utilise oil shale.

12.12.2.10 Orimulsion

Orimulsion® is a registered trademark for a bitumen-based fuel that was developed for industrial use by British Petroleum (BP) and Petroleos de Venezuela SA (PDVSA).

The bitumen occurs naturally and is obtained from the world's largest deposit in the Orinoco Belt in Venezuela where the reserves of bitumen are estimated at >50% of the world's expected oil reserves.

Extreme high **viscosity** of raw bitumen makes it unsuitable for direct use in boilers. Orimulsion made by mixing the bitumen with ~30% fresh water and a small amount of alcohol-based surfactant behaves like FO. With large reserves of bitumen Orimulsion as a fuel for power generation, has numerous advantages as it

- Can be priced competitive to the internationally traded coal
- Is relatively easy and safe to produce, transport, handle and store
- Is easy to ignite and has good combustion characteristics
- Can be used in boilers designed to run on coal or HFO, with suitable modification

Orimulsion is currently used as a commercial boiler fuel in power plants worldwide (e.g. Canada, Denmark, Japan, Italy, Lithuania, the UK and China).

Emissions with Orimulsion firing are controllable with existing commercial air pollution-control technologies. For **NO_x control** use of combustion control technologies, such as **LNB** and **re-burn**, and post-combustion control technology, such as **SCR** are employed.

TABLE 12.9

Properties of Orimulsion

General Properties	
M in fuel by wt	29.5%
Gcv, MJ/kg	29.77
kcal/kg	7110
Btu/lb	12,700
Density, kg/m^3 @ 15.5°C	1010.3
Flash point, °C	>65
Median droplet size, μ	11
UA on Dry Basis (%wt)	
C	60.1
H	10.1
O	26.4
N	0.35
S	2.85
A	0.2
Ash Composition	
V (ppm)	310
Ni (ppm	75
Na (ppm)	19
Mg (ppm)	370

Properties of Orimulsion are broadly as given in Table 12.9 which are close to those of HFO. Briefly, Orimulsion has high water and S, with low ash content which has heavy metals in it. Owing to high V in ash boiler HSs are susceptible to **high-** and **low-temperature corrosion**. The manufacturing technique of Orimulsion has been improved over the years to make it close to FO.

The use of Orimulsion is on decline of late. The main reason is the political volatility. Besides, if the temperature drops to <30°C there are chances of de-emulsification leading to solidification of fuel in pipe line which cannot be reversed.

12.12.2.11 Sludge (see also *Bubbling Fluidised-Bed Combustion Boilers with Open Bottom* under *Bubbling Fluidised-Bed Combustion Boilers*)

Sludge is some *specific waste in semi-solid form.*

It is a generic term for describing

- Solids separated from suspension in a liquid or
- A residual semi-solid material left from industrial, WT or waste WT processes

Sludge also denotes the sediment of accumulated minerals in a steam boiler which is periodically blown off.

In municipal and industrial WT it is a semi-liquid mass of accumulated settled solids deposited from the treatment process. Most WT plant sludge is inert and can be used as components in some manufactured products and as a base for fertiliser. It is also called 'residual solids' (see Figure 19.7).

Industrial sludge is any sludge that is not domestic wastewater sludge. This includes wastewater sludge from manufacturing or processing of raw materials, intermediate products, final products or other activities that include pollutants from non-domestic wastewater sources.

Pulp mills produce reasonably large amounts of sludge which are also called 'bio solids'. Pulp mill sludge is a complex and changeable mixture of dozens or even hundreds of compounds, just like mill wastewater. Some are well known such as heavy metals, dioxin and other organochlorines. Some, created by the bacteria in the treatment ponds, are probably unknown to science.

Sludge has been historically used as land fill material which is getting fast discontinued due to increased pollution awareness and general rise in cost of fuels. It is being more profitably used by the industry by resorting to cofiring in their own power boilers. Sludge can be mixed with the main fuel with minor changes to firing equipment and boiler. The advantages are

- Savings of main fuel
- Reduction in NO_x levels if the main fuel is coal
- Savings in disposal costs

Cofiring is the most easily done in stoker-fired boilers. Cofiring in PF boilers requires major modifications to burners or as in many cases a complete change of burners. Fuel preparation is also more.

It is the BFBC boilers which can tackle sludge most efficiently and effectively with least amount of support fuel. In BFBC boilers sludge firing is not cofiring but main firing. Large BFBC boilers are built with no in-bed tubes but with open bottom construction to deal with high M low cv fuel like sludge (see Figure 2.25).

Typical analysis of sludge is given in Table 12.10.

The burden in fuel is very high, varying from 55% to 70% and Gcv on daf basis quite low from ~12.77 to 15.1 MJ/kg, ~3050–3600 kcal/kg or ~5500–6500 Btu/lb making it a very difficult fuel to burn. S and N are both low.

TABLE 12.10

Properties of Typical Sludge from Pulp Mill

Constituent	C	H	N	O	S	A	M	GCV
w/w%	14.70	1.80	0.35	17.50	0.05	18.10	47.50	4.885 MJ/kg 1222 kcal/kg 2100 Btu/lb

12.13 Liquid Penetrating Test (*see* Testing of Materials)

12.14 Liquefied Natural Gas (*see* Liquid Fuels)

12.15 Liquefied Petroleum Gas (*see* Liquid Fuels)

12.16 Log Mean Temperature Difference (*see also* Fluid Flow Patterns in Fluid)

In HX applications, the inlet and outlet temperatures are commonly specified based on the fluid in the tubes. The temperature change that takes place across the HX from the entrance to the exit is not linear. A precise temperature change between two fluids across the heat exchanger is best represented by the log mean temperature difference (LMTD or Δt_m).

LMTD is the thermal driving potential for heat transfer in flow systems and HX.

In any heat exchanger heat transfer from hot to cold is given by

$$Q = U \times A \times \Delta t_m \quad (12.3)$$

where Q is the heat transferred in watts, kcal/h or Btu/h.

A is the area of the surface over which heat transfer takes place in m² or ft².

Δt_m is the mean temperature difference in °K, °C or °F.

If the temperature difference at the two ends of the heat exchanger were Δt_1 and Δt_2

$$\Delta t_m = \text{LMTD} = \frac{\Delta t_1 - \Delta t_2}{In(\Delta t_1/\Delta t_2)} \quad (12.4)$$

LMTD in parallel and counterflow arrangements of heat exchangers, such as the **SH**, **RH**, **Eco** and **AH** banks, is given below. The flow arrangement is shown in Figure 12.19 (see also Figure 8.20). Tubular AHs are however usually placed in cross flow arrangement

$$\text{LMTD in parallel flow} = (T_1 - t_1) - (T_2 - t_2) \quad (12.5)$$

$$\log_e(T_1 - t_1)/(T_2 - t_2)$$

$$\text{LMTD in counterflow} = (T_1 - t_2) - (T_2 - t_1) \quad (12.6)$$

$$\log_e(T_1 - t_2)/(T_2 - t_1)$$

LMTD in cross flow is the same as in counterflow but multiplied by a correction factor Fc which is read from Figure 12.20.

12.17 Louvre Dampers (*see* Dampers)

12.18 Low Cement Castables (*see* Refractories)

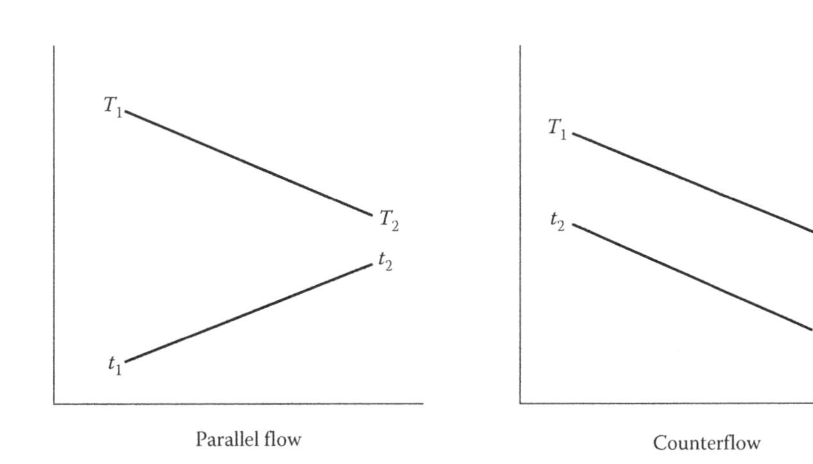

FIGURE 12.19
Parallel and counterflow arrangements of fluids in heat exchangers.

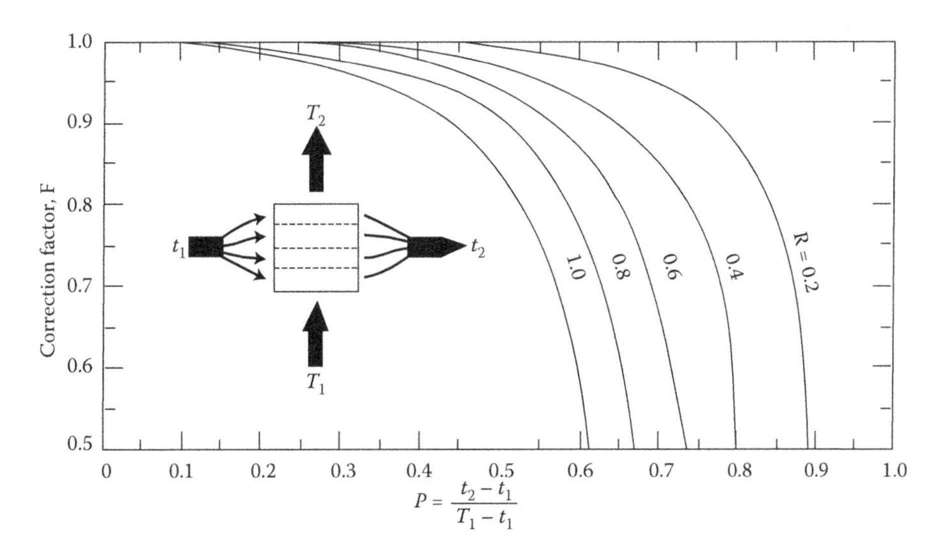

FIGURE 12.20
Correction factor Fc for LMTD of cross flow.

12.19 Low NO_x Burners (*see also* NO_x in Flue Gas)

LNBs are, in a way, modified axial flow burners with

- Reduced excess air
- More split in air streams
- FGR if needed

All the above factors work concurrently to reduce both **fuel NO_x** and **thermal NO_x** in furnace. LNBs are the most cost effective and the easiest of all NO_x reduction measures.

- Lowered excess air reduces the tendency to form NO_x because of scarcity of O_2.
- Further, as the air stream is split the combustion takes place in stages and this protracted combustion does not permit flame temperatures to reach their peak.

Thus, the flame is starved of both temperature and free air, suppressing the NO_x forming tendency. These measures in themselves are adequate to secure over 50% NO_x reduction.

LNB is a mature technology that has been improved over time. This technology is available for all fuels and all types of firings, such as **wall** or **tangential burners**.

A typical wall mounted LNB for **PF** firing is shown in Figure 12.21.

LNBs are designed to control fuel and air mixing in each burner to create longer and more number of split flames. High-flame temperature is thus reduced which results in less NO_x formation. The improved flame structure also reduces the amount of O_2 available in the hottest part of the flame, thus improving burner η. All the three stages, namely, combustion, reduction and burnout are achieved within a conventional LNB.

- In the initial stages, combustion occurs in a fuel rich, O_2-deficient zone where the NO_x are formed.
- A reducing atmosphere follows where HCs are formed which react with the already formed NO_x.
- In the third stage internal air staging completes the combustion but may result in additional NO_x formation. This however can be minimised by completing the combustion in an air lean environment.

LNBs can be combined with other primary measures such as **SOFA**, **re-burn** or **FGR**. Plant experience shows that the combination of LNBs with other primary measures is capable of achieving NO_x removal in excess of 70%.

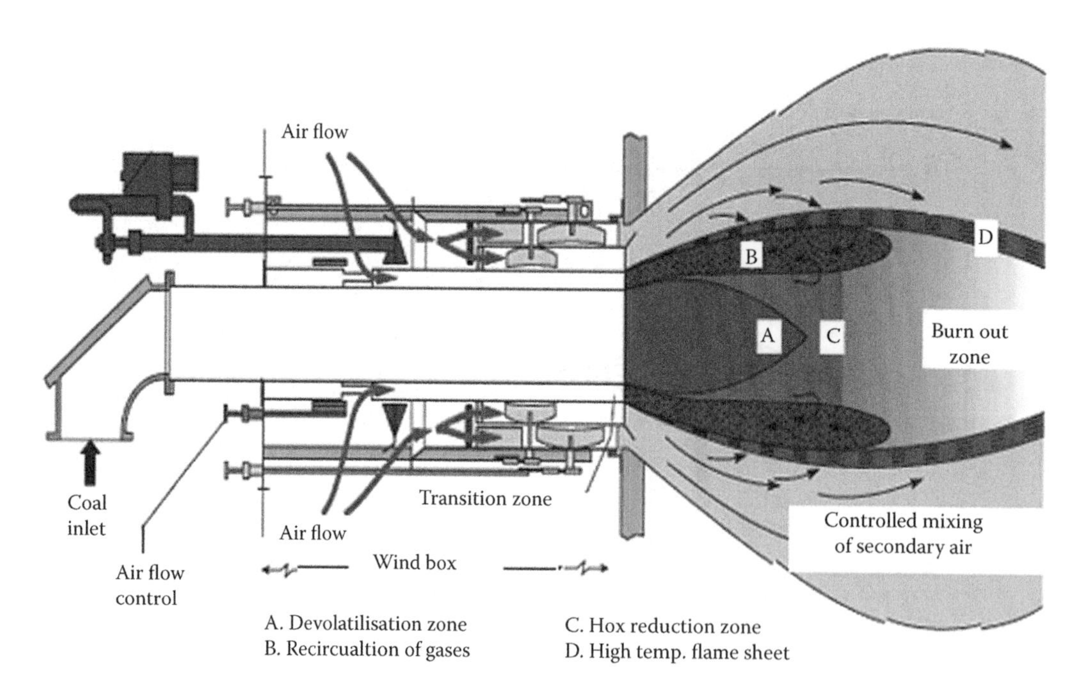

FIGURE 12.21
Wall-mounted low NO_x burner (LNB).

12.20 Low Sulphur Heavy Stock (*see* Fuel Oils *under* Liquid Fuels)

12.21 Low Temperature Corrosion (*see* Ash Corrosion)

12.22 Luminous Radiation (*see* Radiation in Heat Transfer)

12.23 L-Valve (*see* Loop Seal in Circulating Fluidised-Bed Combustion)

13

M

13.1 Magnetic Particle Testing (*see* Non-Destructive Testing in Testing of Materials)

13.2 Malleability (*see* Metal Properties)

13.3 Manometer (*see* Pressure Measurement *under* Instruments for Measurement)

13.4 Manufacturing Topics

13.4.1 Tube Expansion

Tube expansion is a very important and a long-drawn activity, at least in **industrial boilers**, which have usually BB and TAH. In **utility boilers**, many times there may be no tube expansion as there is no BB and the TAH is replaced by RAH.

Tube expansion is a cold working process. It is essential that the **hardness** of tube be as low as possible for the ease and success of expansion. BB tubes, being thicker than AH tubes, are **annealed** at both ends.

The tubes are expanded at both their ends in the case of **BBs** and **TAHs**. In LP boilers, tubes are also sometimes expanded in the **headers**. Tube holes are made fractionally larger than the tube od such that, when tube is expanded by force to fill the tube hole, it is the friction that holds the two together. Experience has shown that ~20 mm is the minimum straight distance needed for holding this type of joint. Accordingly, the minimum thickness of the flat shell plate is maintained at 20 mm and when the plate is curved it is made suitably thicker. Tube expansion is followed by bell mouthing to arrest the possible axial movement of tubes. In case of boiler drums, additionally, one or two grooves are made in shell plate so that the tube material flows into the grooves during expansion and locks itself in place. This is required to prevent the loosening of tubes during transportation.

The tubes must protrude inside the drum by 8–10 mm for satisfactory bell mouthing. They should also be spaced apart in such a way that there is a minimum distance of ~12 mm between the bell mouthed ends as needed by expanders. Usually, the tubes are given a seal run for pressures above 80–100 bar. The maximum pressure for holding the tubes by tube expansion is usually ~100–120 bar, at which pressure, the drums would have already reached a thickness of ~150 mm. Reliably expanding tubes over such long straight length is not so easy.

Figure 13.1 shows the cross section of a tube expander while Figure 13.2 shows the whole assembly. Both of them are without their drives. Figure 13.1 gives an idea of how the tube is expanded and flared.

13.4.2 Tube Swaging

Tube swaging is a forming process using dies to either increase or decrease the size of the tube ends. It is

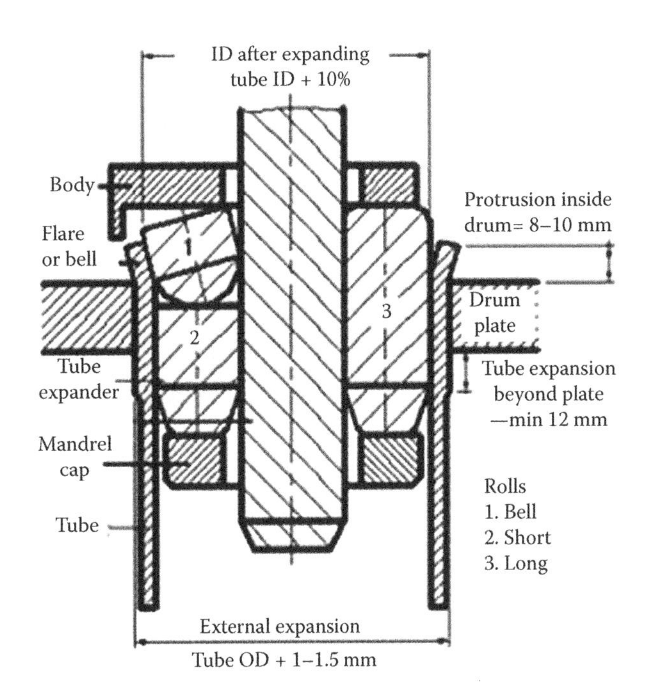

FIGURE 13.1
Tube expansion.

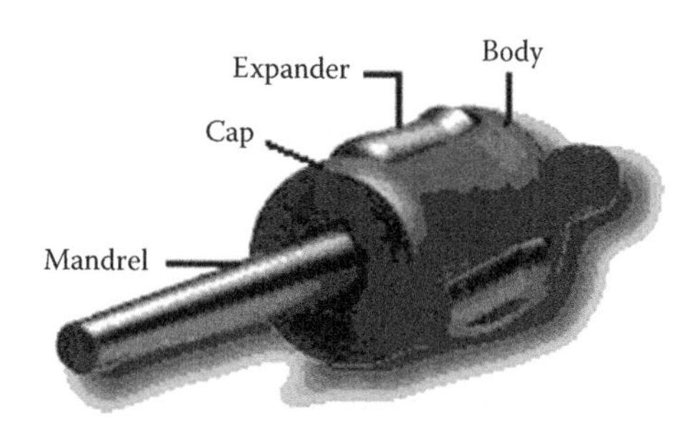

FIGURE 13.2
Tube expander.

usually a cold working process and therefore causes work hardening of the tube. It can also be done at high temperature by hot working process.

BB and furnace tubes are many times swaged to reduce the tube od at the ends so that the **ligament η** of the drum or header increases, thereby reducing their thickness. Bidrum boilers of 100 bar would not be possible without the BB tubes swaged at their ends. A typical swaged end of a tube is shown in Figure 13.3.

Swaging can be done by

- Squeezing the tube through a confined die similar to the wire drawing process or
- Rotary swaging by rolls where two or four split roll dies are rotated over the tube while closing them progressively until the required diameter is achieved

Swaging is limited to normally 25% (63.5–50.8 mm) reduction in diameter. If more swaging is required, an intermediate **stress relieving** is required for reducing the hardness of the tube and avoiding surface cracking. Swaging is followed by annealing of the swaged ends to remove the stresses developed because of cold working and induce softness and ductility.

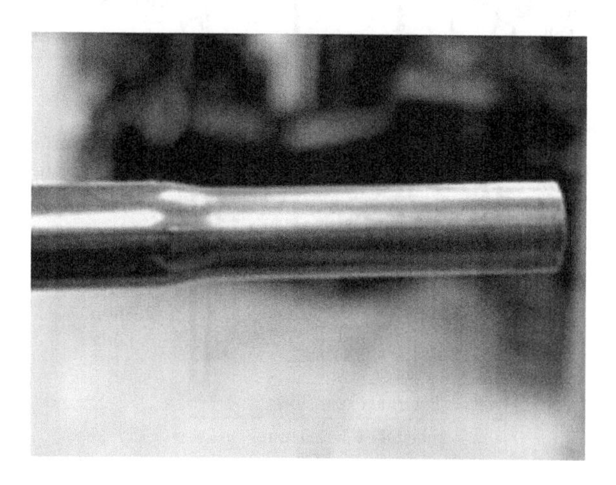

FIGURE 13.3
Swaged end of a tube.

13.5 Mass Blowing (*see* in Rotary Sootblowers *under* Soot Blowers)

13.6 Mass Burning (*see* Types of Firing in Grates)

13.7 Mass Velocity (*see* Thermodynamic Properties)

13.8 Materials of Construction

The core of the boiler is essentially the PPs. Boiler materials are, therefore, high-temperature PP steels which are

1. Plates
2. Tubes
3. Pipes
4. Castings
5. Forgings

13.8.1 Boiler Quality Plates

Large cylindrical pressure-holding components like drums, RH headers, blowdown tanks and so on are required to be made from BQ plates.

BQ plate is primarily a very high-quality, fully deoxidised, mild steel (MS) plate of medium to high thickness, suitably alloyed where necessary to withstand high p & t. The essential requirements are

- Very good weldability
- Assured and verified high-temperature properties
- High tensile strength even at high temperatures

BQ plates are available in three classes of metallurgy

1. CS (low, medium and high C)
2. Low and medium AS
3. High AS—not in ASME steels

13.8.1.1 CS Plates

Low CS has good weldability while high CS has good tensile strength.

The American practice prefers

- CS for both LP and HP applications and
- Thick CS to thinner AS drums

The European method is to avoid C > 0.25% and adopt AS plates for HP and have thinner drums.

High CS plates of SA 515, 516 and 299 are extensively used for generally thickness up to ~40, <100 and >100 mm for the three steels, respectively.

13.8.1.2 Low- and Medium-AS Plates

SA 302 is a very popular low-alloy American steel employed for high ps & ts. For SH and RH headers of large diameter, when they are required to be rolled out of plates, medium- and high-AS plates are regularly employed.

13.8.1.3 High-AS Plates

High ASs are not listed in ASME standards but are routinely used in Europe to reduce the drum and header thicknesses. A thin drum is helpful in cycling duty of boilers.

Alloy plates are preferred in European practice. Table 13.1 gives a selection of BQ plates available in American Specs.

The plates are usually procured in normalised condition for rolling of shells, except in case of pressing dish ends when annealed plates are procured. BQ plates are to be **fully killed** with no inclusions:

13.8.2 Boiler Tubes

Heat transfer in a boiler is almost exclusively by the tubes. Accordingly, every boiler employs tubes of different sizes, thicknesses and metallurgy.

13.8.2.1 Tube versus Pipe

To state it crudely, a tube or a pipe is a rod with a hole. In boiler industry, a tube is characterised by the following features:

- Tubes are specified by their od and thickness (indicated by **wire gauge** SWG or BWG).
- The maximum od of tubes rolled is 127 mm (5″).
- Tolerances on thickness are much closer than in pipes.
- The C content and hardness are lower as tubes are to be bent to close radii.

Even though tubes and pipes look alike, there are serious differences, not permitting interchangeability:

- Tubes are for conveying **fluids** and simultaneously transfer heat while pipes are only for conveying.

TABLE 13.1

Boiler Quality Plates to ASME Specification

	BQ Plates to ASME Specification (SA)						
Plate Specification	Nominal Composition	Max. C %	UTS ksig	YS ksig	UTS MPa	YS MPa	Max. °C
1. CS							
285 Gr C	C	0.22	55	29	380	200	482
515 Gr 70	C	0.35	70	38	480	265	538
516 Gr 70	C	0.35	70	38	480	265	538
299	C	0.30	75	40	515	275	538
2. Low ASs							
302 Gr B	1¼%Mn– ½Mo	0.25	80	50	550	350	538
3. Medium ASs							
387 11 cl 11	1¼%Cr– ½Mo	0.17	75	45	515	315	566
387 22 cl 2	2¼%Cr–1%Mo	0.17	75	45	515	315	577

Note: 1. Gr: Grade, cl: Class. 2. Brief details of popular American drum plates are provided here for a quick reference. For accurate and current details, the latest Specs must be referred to. 3. Tensile strength reduces as the plate thickness increases. Tensility given here refers to higher thicknesses. Tensile and yield strength values given are the minimum values. Figures in MPa are rounded off.

- Tubes are limited to 127 mm od while pipes come in much larger size of 1220 mm od and above.
- Tubes are always specified by od while pipes are designated by nominal bore (NB/nb) up to 305 mm (12 in) and by od beyond that.

In a boiler, **HSs** are made of tube while **integral** and **external piping, large-bore downcomers** and **headers** are made of pipe materials.

13.8.2.2 Tube Classification

Based on the method of manufacture, all boiler tubes fall into two categories:

1. ERW tubes made by folding and welding steel strips (skelp)
2. Seamless tubes produced by piercing a red hot billet with a plunger

The advantages of ERW tubes are

- The inside diameter (id) and od are perfectly concentric, making it ideal for expanded tubes.
- The inside surface is very smooth, thereby reducing the pressure losses.
- Lower cost.

1. ERW tubes:
As ERW tubes are made by longitudinally bending and welding of skelp, tube thickness is limited to ~6 mm from welding consideration skelp. The acceptability of ERW CS tube is very high but AS is low. Often, it is not selected for **BB** tubes which require flaring during tube expansion for fear of opening up of the weld seam. This hesitation may not be justified today after the modern mills have established that the welding process has been thoroughly upgraded and made very reliable.

2. Seamless tubes:
Seamless tubes are made in two qualities, namely hot finished seamless (HFS) and CDS.

HFS tubes, when reheated to appropriate temperature and re-rolled to closer finish, CDS tubes are produced. CDS tubes have closer tolerances, comparable to ERW, but are more costly than HFS. Only when lower Δps are required like in SH and RH tube banks in large boilers, CDS tubes are employed. Otherwise, the first preference is always HFS.

The advantages of the seamless tubes are that thicknesses >6 mm are available and there is no fear of opening of weld seam. However, the concentricity of inside and outside diameters is not as good as in ERW. Generally, seamless tubes are marginally more expensive.

ERW or Seamless?

This is largely a matter of practice in each market. The European market has an overwhelming preference for seamless tubes while in the other markets, both are accepted on merit. Generally, ERW tubes find acceptance for LP and MP boilers <70 mm. In many markets, ERW tubes are preferred wherever tubes have to be expanded because of better concentricity. Membrane panels made with ERW tubes are also gaining wider acceptance as the ERW tube-making practices have been perfected over the years.

13.8.2.3 Tube Materials

The metallurgy of tubes is temperature dependent. Accordingly, tubes are made in the following execution:

1. CS
2. Low AS
3. High AS
4. ss

Table 13.2 gives the details of all these classes of tubes along with the limiting temperature. While making the tube selection, it is normal practice to

- Prefer low CS tubes as their weldability is superior
- Avoid using ss tubes to the extent possible, as they are not only expensive but suffer from higher elongation and stress corrosion and they need great care in welding.

With every upgrade of metallurgy, tubes become more expensive by 20–40%. Unless there is a compensating thickness reduction upgradation to higher metallurgy is normally avoided, unless temperature so warrants.

13.8.2.4 Special Tubes

a. Bimetal/clad/composite tubes

   When aggressive wear, corrosion or very high metal temperatures are to be resisted, bimetal tubes are used usually with CS on the inside and ss (304/316) on the outside. While this combination is the most common and economical, low AS with high Cr cladding is also employed to counter severe corrosion. Typical applications are the furnace and SH tubes in BLR boilers and WTE plants burning RDF or MSW.

TABLE 13.2
Boiler Tubes ASME Specification

	ASME Tube Specifications (SA)						
Material	**Nominal Composition**	**Temperature Limit (°C)**	**UTS kpsi**	**YS kpsi**	**UTS MPa**	**YS MPa**	**Use**
1. CS							
Low tensile							
178 A (ERW)	C(<0.18)	510	47	26	324	179	1,2,3
192	C(<0.18)	510	47	26	324	179	1,2,3
Medium tensile							
210 A-1	C(<0.27)	510	60	37	414	255	1,2,3
High tensile							
210 C	C(<0.35)	510	70	40		276	2,3
178 C (ERW)	C(<0.35)	510	60	37	414	255	2,3
178 D (ERW)		510	70	40		276	1,2,3
2. Low AS							
209 T1	C–½Mo	524	55	30	379	207	1,2,3
209 T1A	C–½Mo	524	60	32	414		2,3
213 T2	½Cr–½Mo	552	60	30	414	207	1,3
213 T11	1¼%Cr– ½Mo	566	60	30	414	207	3
213 T22	2¼% Cr1%Mo	602	60	30	414	207	3
3. High AS							
213 T 9 1	9Cr–1Mo–V	649	85	60	586	414	3
4. Stainless Steel							
213 TP-304 H	18Cr–8Ni	760	75	30	517	207	3
213 TP-321 H	18Cr–10Ni–Ti	760	75	30	517	207	3
213 TP-347 H	18Cr–10Ni–Cb	760	75	30	517	207	3
213 TP-316H	16Cr–12Ni–2Mo	760	75	30	517	207	3
213 TP-310 H	25Cr–20Ni	816	75	30	517	207	3

Note: 1. Safe maximum outside tube wall metal temperatures given here are on the basis of oxidation resistance. Design codes decide the temperature limits. For CS tubes usage beyond 454°C, when permitted by code, special inspection is required for 100% η. 2. Numbers in 'Use' column refer to: (1) Furnace walls exposed to high heat; (2) Other enclosures not exposed to high heat; (3) SH, RH, Eco. 3. Steels with suffix H are modified to suit high-temperature duties of SH and RH by increasing the C by 0.02%; 4. Thickness as per SA standards is usually the minimum thickness with no negative tolerance.

An example of composite tube wall is shown in Figure 13.4.

b. Omega tubes (see Omega Tubes)
c. *Ribbed tubes* (see Boiling or Evaporation in Properties of Steam and Water)

13.8.2.5 Chromising

Chromising is high-temperature (~900°C) pack or gaseous diffusion of Cr into the surface of a component to enhance high-temperature corrosion and oxidation resistance.

CS and several ASs like T11, T22 and T91 can be chromised. Depending on the base metal Cr-rich layer of ~400–800 microns can be created on the tube having ~30–40% Cr. At the surface, a thin layer of 15–25 microns of Cr_3C_2 (chromium carbide) is formed.

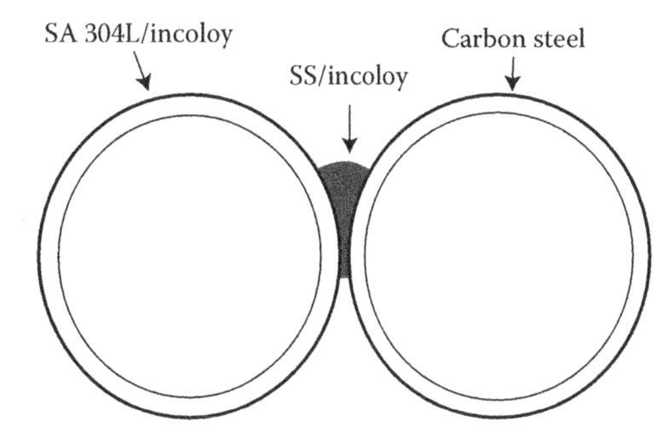

FIGURE 13.4
Composite tube wall.

Highly corrosive environment of BLR and MSW firing boilers may require chromised tubes.

13.8.3 Pipes (see *also* Schedule of Pipes)

Similar to tube, pipe is also a rod with hole but tubes and pipes are quite different as explained in the topic on Tubes. Pipes are used where conveyance of fluids is required with no **heat transfer**.

In boilers, pipes are used for

1. Downcomers and headers which are placed inside or outside the gas stream
2. Integral piping, namely, drain, vent, blowdown, SB, attemperator and interconnecting piping
3. fw, main steam and RH steam pipes

Pipes of only seamless class of execution are employed in boilers because of the combination of p & t. Pressure pipes of HC industry, such as line pipes and submerged arc (welding) (SAW) pipes are not used. Table 13.3 lists the popularly used BQ pipes of American Specs.

13.8.4 Castings

There are a variety of castings in any boiler:

- They can be for ordinary or pressure duty for normal and high temperatures.
- They are not preferred for pressure-holding duty as the structure is comparatively porous and weak.
- But intricate shapes and thick walls are easy to obtain in castings.
- They are also cheaper and easier to produce.

Castings are of two types, namely, *iron castings* and *steel castings*.

- Ash doors, man/peep hole doors and various grate and mill parts are made of iron castings. The majority of iron castings are Cr castings capable of withstanding temperatures up to 600/700°C. Several castings are required ranging from a few grams to a few kilograms for air nozzles, grate links, grate castings, SB hangers, sleeves of gilled tube Ecos and so on.
- Steel castings are relatively stronger. Valve bodies are made of steel castings. Other than for valve bodies or similar shapes, steel castings are not employed for pressure duty in boilers. Large steel castings are employed in pulverisers to withstand the grinding forces. SH and RH lugs are usually made of high-temperature steel castings that can hold out to vibration forces.

There are also heat-resistant steel castings required to withstand temperatures >600/700°C.

TABLE 13.3

Pressure Pipes to ASME Specification

	ASME Specification SA							
	Nominal Specification	Temperature Limit (°C)	UTS kpsi	YS kpsi	UTS MPa	YS MPa	Use
1. CS							
Medium tensile							
106 Gr B	C(<0.30) ©	510	60	35	414	241	1,2,3
High tensile							
106 Gr C	C(<0.35) ©	510	70	40	482	276	1,2,3
2. Low AS							
335 P1	C–½Mo	468	55	30	379	207	1,2,3
335 P12	C–½Mo	566	60	30	414	207	2,3
335 P11	1¼%Cr ½Mo	566	60	30	414	207	1,2,3
335 P 22	2¼% Cr 1%Mo	593	60	30	414	207	2,3
3. High AS							
335 P 9 1	9Cr–1Mo–V	649	85	60	586	414	2,3

Note: 1. ©—Limited to 427°C (800°F) for pipes outside of boiler setting for fear of graphitisation by steam. 2. Thickness as per SA standards is usually the minimum thickness with no negative tolerance. 3. Uses: (1) Furnace wall headers and pipes exposed to high heat; (2) Eco headers and pipes not exposed to high heat; (3) SH and RH headers and pipes.

Ferritic steels are suitable up to 900°C with the chief alloying element being Cr.

Ferritic austenitic steels are used for improved heat resistance and toughness on account of addition of Ni along with Cr.

Austenitic steels have Ni in excess of 9% with much superior toughness and shock resistance.

SH and RH lugs and burner impellers/swirlers experience temperatures as high as 1000–1500°C which require much higher levels of Cr and Ni. They have typically a composition of

- 25% Cr–12% Ni to A351 CH20
- 25% Cr–20% Ni to A351 CK20

In oil-fired boilers which use oils with high V and S, the parts are to be made with even higher Cr and Ni to resist ash corrosion in addition to temperature. The castings employed typically contain

- 50% Cr–50% Ni
- 60% Cr–40% Ni

Table 13.4 gives the properties of castings as per American Specs.

13.8.5 Forgings

Forgings are better suited for pressure duty because of their dense structure. Several PPs are made out of forgings in a boiler. The following is a typical list of such forgings:

- Nozzles for drums, headers and pipes
- End closers for headers
- Pipe fitting such as elbows, bends, tees, reducers, expanders and so on
- Pipe and header attachments such as weldolets, pads, thermowelds and so on
- Special forgings such as tube joints

Table 13.5 gives a list of pressure forgings to ASME Specs.

13.9 Maximum Continuous Rating (MCR) (*see* Boiler Ratings)

13.10 Maximum Operating Pressure (*see* Allowable Pressure in Boiler Pressures)

13.11 Mean Metal Temperature (*see* Temperature in Pressure Part Design)

13.12 Mean Temperature Difference (*see* Log Mean Temperature Difference [LMTD or Δtm])

TABLE 13.4

Boiler Quality Pressure Castings to ASME Specifications

	Nominal Specification	Temperature Limit (°C)	UTS kpsi	YS kpsi	UTS MPa	YS MPa
1. CS						
Medium CS						
SA 216-WCA	C(<0.25)	510	60	30	414	207
High CS						
SA 216-WCB	C(<0.30)	510	70	36	482	248
2. Low AS						
SA 217 WC6	1¼%Cr ½Mo	593	70	40	482	276
SA 217 WC9	2¼% Cr 1%Mo	602	70	40	482	276
3. ss						
351 CH 20	25Cr-12Ni	816	70	30	482	207

Note: UTS and YS indicated are minimum values.

TABLE 13.5

Boiler Quality Pressure Forgings to ASME Specifications

	Nominal Specification	Temperature Limit (°C)	UTS kpsi	YS kpsi	UTS MPa	YS MPa
1. CS						
SA 105	C(<0.35)	510	70	36	482	248
2. Low AS						
SA 182 F12	C–½Mo	468	70	40	482	276
SA 336 F12	C–½Mo	566	70	40	482	276
SA 182 F11	1¼%Cr ½Mo	566	70	40	482	276
SA 182 F22	2¼% Cr 1%Mo	602	75	45	517	310
3. High AS						
SA 182 F 91	9Cr–1Mo–V	649	85	60	586	414
4. ss						
182 F304 H	18Cr–8Ni	760	75	30	517	207
182 F 316 H	16Cr–12Ni–2Mo		75	30	517	207

Note: UTS and YS indicated are minimum values.

13.13 Membrane Panel/Wall (*see* Pressure Parts *and* Furnace in Heating Surfaces)

13.14 Metals

Basic metals employed in boiler construction covered here are

1. Cast iron
2. Steel

13.14.1 Cast Iron

Cast iron (CI) is an alloy of Fe–C–Si with C ranging from 2% to 4.3% and Si from 0.5% to 3.0%. Lesser amounts of Mn are usually present along with traces of impurities like P and S.

It is a hard, brittle and non-malleable alloy.

CI is made from remelting of pig iron with scrap and alloying elements in cupola, air furnace or electric furnace. It is cast into moulds for a variety of products.

Depending on the application, C and Si contents may be reduced to the desired levels and other elements may be added to the melt, before the final form is produced by casting.

CI is brittle except in malleable irons.

CI has low melting point, good fluidity, castability, excellent machinability, resistance to deformation and wear resistance. It has wide applications and is used in pipes, machines, automotive parts and so on. CI can be cast into complicated shapes quite easily.

Compared to steel, CI has

- Inferior strength, malleability, toughness and ductility but
- Better fluidity and lower cost

Use of CI in boilers has progressively declined over time. It is practically no longer used in boiler PPs. But it is used extensively for stoker, pulveriser and beater mill parts, manhole and peep hole castings and so on. Sleeves of gilled tube Ecos are made of CI.

CI Classification

There are four basic types of CIs, of which the first two are the most important:

- Grey CI
- White CI and chilled CI
- Malleable iron
- Ductile/nodular CI

13.14.1.1 *Grey CI*

It is characterised by graphite microstructure which is grey in appearance. Grey CI has

- Less tensile strength and shock resistance than steel

- Higher compressive strength comparable to low CS
- Good machinability
- Good resistance to wear

Grey CI is the most widely used of all cast metals. C is in free state in grey CI, in notches and discontinuities in iron. The strength of CI increases as the size of graphite crystal reduces and the amount of Fe_3C (cementite) increases. Good machinability of grey CI is because of free graphite crystals which act as lubricant for cutting tool.

Grey CI is available in a wide range of tensile strengths varying from ~1500 to 6500 kg/cm^2. Alloying elements like Cr, Ni, V, Mo and Cu are used.

13.14.1.2 White and Chilled CI

Owing to faster cooling and lower Si content, C precipitates as cementite (Fe_3C) rather than graphite. Fe_3C is relatively in large particles usually in a eutectic mixture with martensite. This is called white CI.

Many parts in coal mills, cement and clinker grinders are made of chilled CI.

White CI is too brittle for structural use but finds application where hardness, abrasion resistance and cost are important.

Impellers, shell liners, balls in tube mills and grinding media in pulverisers are in white CI.

Chilled CI differs from white CI only in the method of manufacture, when it is cast against metal chills. Rapid cooling of areas adjoining the chills promotes formation of cementite. This results in high resistance to wear and abrasion.

White CI can also be made by using high % of Cr, which is a strong carbide forming element. High-Cr white CI alloys can be sand cast to large dimensions. Spider rings and bull rings in vertical coal mills are examples of such large-sized castings.

White CI is characterised by the prevalence of carbide, promoting

- High compression strength
- Hardness
- Wear resistance

13.14.1.3 Malleable Iron

Malleable iron is essentially white CI heat treated at ~900°C so that graphite separates much more slowly into spheroidal particles rather than flat. Iron becomes malleable because in this spheroidal condition 'C' no longer forms planes of weakness.

In general, the properties of malleable iron are more like that of MS.

Malleable iron has got good ductility and excellent machinability. It has high resistance to atmospheric corrosion.

13.14.1.4 Ductile/Nodular CI

It is also called *nodular CI.* Ductile iron contains small amounts of Mg to slow down the growth of graphite precipitate allowing C to precipitate as spheroidal particles. The properties are similar to malleable iron but parts can be cast with larger sections. Also, it is suitable for making pressure castings.

13.14.2 Steel

Pure iron with 99.9% purity is useful only for electrical and chemical industries as a sintering material. When C is added, the resulting alloy is steel. C varies from 0.2% to 2.1% by weight depending on the grade.

In addition to C, for mechanical applications, Fe has to be dosed with other elements to impart properties like hardness, toughness, wear and corrosion resistance and so on. Further, when other alloying elements such as Mn, Cr, V, W and so on are added, myriad varieties of steels result to meet infinite needs.

Steel is produced from pig iron by the oxidation of its impurities with air, O_2 or iron oxide, as the impurities have greater affinity for O_2 than Fe. There are five processes of converting pig iron to steel ingots, namely

1. Basic open hearth
2. Basic oxygen
3. Acid open hearth
4. Acid Bessemer
5. Electric arc

The manufacturing process has no particular influence on the properties of steel.

Unlike CI, steel can be made in a variety of forms, such as tubulars, flat products like sheets and plates, long products like bars, structurals and strips, castings, forgings and so on.

Steel Classification

Based on the mechanical properties, steel can be classified broadly on the following criteria:

1. CSs based on the % of C
2. ASs based on the % of alloying elements
3. Killed, semi-killed and rimmed steels based on the level of deoxidation as explained in Metallurgy
4. Fine grain and coarse grain steels based on grain size elaborated in Metallurgy
5. Based on manufacturing method

6. Deep, shallow and non-hardening steels based on hardenability given in metallurgy

13.14.2.1 Carbon Steels

Steel is an alloy of Fe and C. CS is where the alloying element is only C when only minor elements like Mn, Si and C may be present.

- C has a great influence on the properties of steel.
- It increases hardness, tensile strength, hardenability and fatigue resistance.
- It decreases ductility, malleability, toughness, machinability, formability and weldability.

Based on C percentage, the CS can be loosely divided into three types:

a. Low CS or MS with C ranging from 0.008% to 0.3%
    - MS is soft, ductile, malleable and tough
    - Very well machinable and weldable
    - Non-hardenable, but surface hardness can be increased through carburising

    MS is the most extensively used of all CS, nearly 80%. The majority of boiler PPs and structural steels are of low-C variety because of their excellent weldability and adequate strength.
b. In medium-C steels, C varies from 0.3% to 0.6%. These are called machinery steels.
    - Their weldability is low
    - They are shallow hardening type
    - They need to be hot worked as cooled working is difficult

    In boilers, machined items such as fasteners and shafts alone are made of medium-C steels.
c. High-C steels have C ranging up to 0.62%. They are also called tool steels.
    - They are hard, brittle and wear resistant
    - They are difficult to machine and weld
    - They are hardened by heat treatment to high hardness levels with good depth of hardening

    In boilers, their use is limited to components in various auxiliaries such fans, mills, feeders, shafts and so on.

13.14.2.2 Alloy Steels

When alloying elements are added to steel ranging from 1% to 50% by weight, the resultant is AS.

Low ASs have alloying elements <8% and *high ASs* have >8%.

Alloying is done to CS to enhance the properties of strength, hardness, wear resistance, hardenability, toughness and so on. Further heat treatment is usually required to achieve the desired properties. The usual alloying elements are Bo, Cr, Mo, Mn, Ni, Si and V.

Low-alloy low-C (0.1–0.3% C) steels have good weldability, although some of the alloying elements may tend to reduce weldability and formability. *High-strength low-alloy* (HSLA) steels have high strength but reduced weldability.

ss are among the most commonly used high ASs. A minimum of 10% Cr content is required to resist corrosion.

13.14.2.3 Stainless Steels

ss are high ASs with Cr content ranging from 10% to 30% and Fe >50%. C is usually present in amounts varying from <0.03% to >1.0% in certain martensitic grades. These have superior corrosion resistance because of the formation of a characteristic invisible and adherent Cr-rich oxide surface film.

There are three groups of ss based on their grain structure, namely, ferritic, martensitic and austenitic. There is also another group known as precipitation hardening steels which are a combination of austenitic and martensitic steels.

1. *Ferritic ss*: These ss contain no Ni and only Cr up to 30%. They are also called *stainless irons* because of their low C.
    - These are ferromagnetic, non-heat treatable.
    - They have good corrosion and heat resistance. They have good resistance to stress corrosion cracking.
    - They have good ductility and formability, but high-temperature mechanical properties are relatively inferior to austenitic ss.
    - Toughness is limited at low temperatures and in heavy sections.

    They are relatively inexpensive. The most widely used ferritic ss is grade 430, offering general-purpose corrosion resistance, often in decorative applications.
2. *Martensitic ss*: These steels also contain only Cr except for two grades (414 and 431) where ~2% Ni is also present. They possess a martensitic crystalline structure in the hardened condition. Cr is <18%, while C may exceed 1.0%. The Cr and C contents are adjusted to ensure a martensitic structure after hardening. Excess carbides may be present to enhance wear resistance or as in the case of knife blades, to maintain cutting

edges. High-temperature springs, ball valves and seats are some of the applications in boilers.

- They are also magnetic. Unlike ferritic ss, they are heat treatable by quenching and tempering.
- They are not as corrosion resistant as the other two grades.
- They have the highest hardness level among all grades of ss.

Grade 410 with 12% Cr is the most widely used martensitic ss with exceptional strength. It is a low-cost, heat-treatable grade suitable for non-severe corrosion applications.

3. *Austenitic ss*: These steels contain Ni and at times Mn, in addition to Cr. Cr–Ni steels are the most widely used ss and are known as 18–8 (Cr–Ni) steels. They have austenitic, face-centred cubic (fcc) crystal structure. Austenite is formed through the generous use of austenitising elements such as Ni, Mn, and N. Cr varies from 16% to 26%; Ni is commonly <35%.

- These steels are non-magnetic and non-heat treatable.
- They are usually annealed and cold worked. They cannot be hardened except by cold working.
- They have excellent corrosion and heat resistance with good mechanical properties over a wide range of temperatures.
- They have excellent toughness at high and low temperatures as low as –270°C.
- Their resistance to sulphurous gases is not high if Cr carbide is present in the grain boundaries.

The other interesting points are

- The most common 18–8 steel accounting for more than half the total ss produced world over is grade 304.
- The Cr–Ni ratio can be modified to improve properties. Mo is added between 2% and 3% to improve corrosion resistance in 316 steel.
- 25%–20 Cr–Ni ss-like grade 310 can be used up to ~1100°C without excessive scaling.
- When post-weld annealing is impractical after welding, two grades like 304L and 316L are employed which have a very low C content of <.03%.

Precipitation Hardness Steels:
These are 18–8 Ni–Cr alloys containing one or more precipitation hardening elements such as Al, Ti, Cu, Nb and Mo. The precipitation hardening is achieved by a relatively simple ageing treatment of the fabricated part.

These steels have both high strength and corrosion resistance. High strength is, unfortunately, achieved at the expense of toughness. The corrosion resistance of precipitation hardening ss is comparable to that of the standard AISI 304 and AISI 316 austenitic alloys. The ageing treatments are designed to optimise strength, corrosion resistance and toughness. To improve toughness, the amount of C is kept low.

Chemical composition of the most common ss is presented in Table 13.6.

13.15 Metal Properties

TABLE 13.6

Composition of Various Stainless Steels

	Nominal Composition (%)					
ASTM No.	**C max**	**Mn max**	**Si max**	**Cr**	**Ni**	**Other**
Ferritic Steels						
430	0.12	1.0	1.0	16.0–18.0		
Austenitic Steels						
304	0.08	2.0	1.0	18.0–20.0	8.0–10.5	
310	0.25	2.0	1.5	24.0–26.0	19.0–22.0	
316	0.08	2.0	1.0	16.0–18.0	10.0–14.0	2.0–3.0 Mo
321	0.08	2.0	1.0	17.0–19.0	9.0–12.0	5 × C min Ti
347	0.08	2.0	1.0	17.0–19.0	9.0–12.0	10 × C min Cb–Ta
Martensitic Steels						
410	0.15	1.0	1.0	11.5–13.5		

13.15.1 Steel Properties

The most important properties of steel which are of great relevance to the boiler are as follows:

1. Tensile and yield strength
2. Ductility and malleability
3. Toughness and brittleness
4. Hardness and hardenability
5. Creep and stress rupture
6. Fatigue strength

Each of these is explained under their respective headings below.

*13.15.1.1 Tensile and Yield Strengths (*see also *Tensile Test in Testing)*

Tensile strength or *ultimate tensile strength* (UTS) is the ability of material to withstand the outward axial loading or pull.

Compressive strength is the opposite and is the ability to withstand the inward axial load or push.

The material is under *tension* when it is pulled and under *compression* when it is pushed.

Load per unit area is termed as *stress* (in N/mm^2 or psi). The material starts to elongate from the time the load is applied. This elongation is known as *strain* (mm/mm or in/in).

Stress–strain relationship of materials is very vital because it provides information on the behaviour of the metal under loading conditions and provides data on yield strength, tensile strength and ductility. This is shown in Figure 13.5.

Elongation is linear and proportional to load up to the *elastic point*; the material returns to its original state when the tensile load is removed without any permanent deformation.

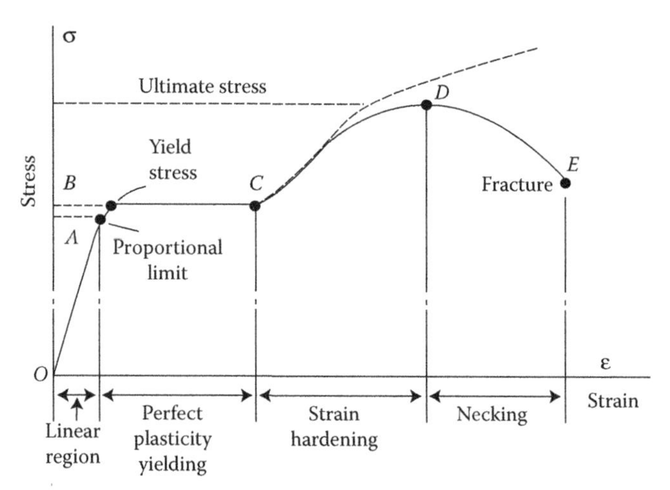

FIGURE 13.5
Stress–strain relationship for steel materials.

Hooke's law of elasticity captures this aspect and states that, in the elastic range of materials

- Elongation is proportional to the tensile load applied or strain is proportional to the stress.
- Stress = the strain × modulus of elasticity.

Modulus of elasticity, tensile modulus or Young's modulus

$$E = \frac{\text{stress}}{\text{strain}} \quad \text{in N/mm}^2 \text{ or psi,} \qquad (13.1)$$

which is one of the basic properties of any material.

When any material is loaded beyond the elastic limit, it enters the *plastic range,* where permanent deformation sets in and the material does not return to the original shape. This plastic elongation is accompanied by uniform reduction in thickness. The material elongates with no further addition to the load. This yielding of material is called *yield point.* Since this point is difficult to establish reliably, specified percentages of plastic elongation are adopted, such as 0.1, 0.2 and 0.5% of the original length.

Yield strength (YS) is the load required to produce the specified plastic elongation expressed in N/mm^2 or psi. YS or the yield point is also defined as the amount of stress that a material can undergo before moving from elastic deformation into plastic deformation. Yield stress is the corresponding load divided by the original area of the specimen to produce the specified elongation.

With further increase of load, the specimen elongates excessively with continuous thinning. At some point, there is a neck formation in the middle of the specimen for the ductile materials before the eventual breaking.

UTS is the breaking load divided by the original area of the specimen.

In brittle materials however, the necking does not take place resulting in an abrupt rupture of the specimen.

Typical values for A36 structural steel are

$UTS = 400 \times 10^6 \ N/mm^2$

$YS = 250 \times 10^6 \ N/mm^2$

$E = 200{,}000 \times 10^6 \ N/mm^2$

13.15.1.2 Ductility and Malleability

Ductility is the ability of a material to plastically undergo elongation without fracture. It is percentage elongation of the specimen up to the point of rupture or the % of area reduction with respect to the original length or area.

As the ductility increases, the strength of material decreases. Strong metals tend to be hard and brittle. Malleability is the ability of material to undergo flattening under compression, by hammering or rolling, without rupturing. While both ductility and malleability refer to the ability of plastic deformation, *ductility is under tension and malleability is under compression.* It is not necessary that ductile material should be malleable.

13.15.1.3 Toughness and Brittleness

Toughness is the property of a material that enables it to absorb and distribute within itself relatively large amounts of energy (both stresses and strains) of repeated impacts and/or shocks, and *undergo considerable deformation before fracturing or failing.* It is the *resistance of a material to fracture when stressed* which is measured by the amount of energy per unit volume that it can absorb before rupturing in J/m^3 or in lbf/in^3.

High toughness requires both high strength and high ductility. Strength is how much force the material can support, while toughness is how much energy a material can absorb, before rupture. A material is both strong and tough if it ruptures under high forces, exhibiting high strains; but a brittle material is only strong with limited strain values and not tough.

To measure toughness, a pendulum is swung from a particular height against a sample at the bottom and its swing is measured as in the Charpy and Izod tests as explained in impact tests.

Toughness is the opposite of brittleness. Brittleness is the tendency of a material to fracture upon the application of a relatively small amount of force, impact, or shock without undergoing any plastic elongation or strain. It is that characteristic of a material that is manifested by sudden or abrupt failure without appreciable prior ductile or plastic deformation. This fracture absorbs relatively little energy, even in materials of high strength.

13.15.1.4 Creep

Creep is a property of material to *undergo deformation when subject to high level of stresses that are below the yield point, at sustained elevated temperature.*

The rate of deformation or creep strain is a function of the material properties, time, temperature and stress. The deformation may become so large that a component no longer is capable of performing its function. Creep strain is time-dependent deformation. Usually creep strain becomes noticeable at ~30% of the melting temperature of the metals. Creep is important only where there is a combination of high temperature and high stress such as in SH and RH tubes.

Creep has three stages of elongation before the eventual failure. Under stress, there is an immediate minute elongation in the elastic range of any material. The creep comes into picture with high stress and temperature acting over a long period. In initial stage called *primary creep* or stage I, the elongation is relatively high. Thereafter, it slows down to near constant, attributed to the balance between work hardening and thermal softening, called the *secondary or steady state creep* or stage II. In the *tertiary creep* or stage III, the strain rate increases exponentially or the strain accelerates, because of necking phenomenon which leads eventually to creep failure. This is depicted in Figure 13.6.

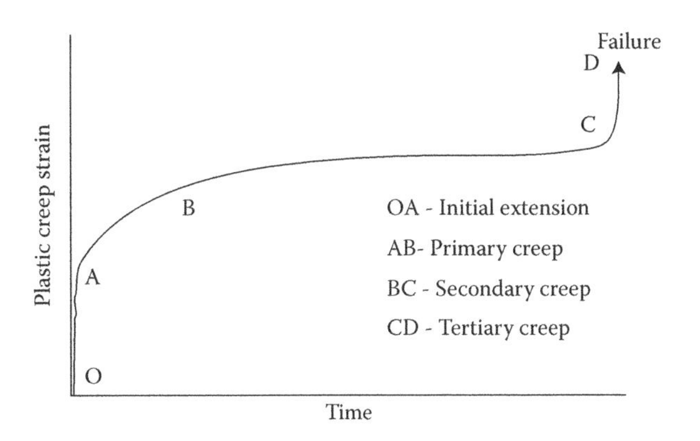

FIGURE 13.6
Stages of time-dependent creep strain.

Creep is measured as 0.1% and 0.01% of elongation in 1000 h which is expressed as 1% elongation in 10,000 and 100,000 h, respectively, for certain material at a particular temperature. Creep test is carried out as per ASTM E139–70 by applying constant load to the tensile test specimen maintained at a constant test temperature. In creep testing, the main goal is to determine the minimum creep rate in stage II or secondary creep

There are a number of factors that affect creep:

- Mo, W and, to an extent, Cr and V help to improve creep resistance.
- C also improves creep resistance up to 550°C.
- Austenitic steels and special alloys have superior creep resistance as compared to low- and high-alloy ferritic steels.
- Coarse-grained steels fare better than fine grained steels.
- Si, Al, Cr and Ni improve creep resistance by enhancing the surface stability which is impaired by the surface reactions.

13.15.1.5 Stress Rupture

Stress rupture is the sudden, complete failure of a specimen held at a specific temperature under a definite constant load for a given period of time.

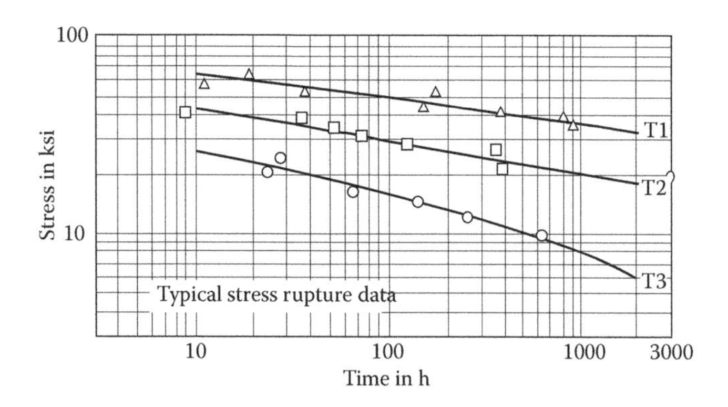

FIGURE 13.7
Stress rupture charts for a steel for various temperatures.

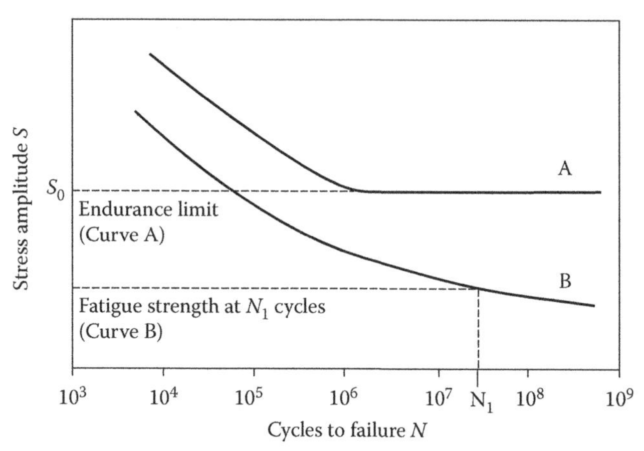

FIGURE 13.8
S–N curve (stress–cycles of reversals) for a steel at a specified temperature.

Stress rupture testing is similar to creep testing except that

- The stresses used are higher than in a creep test.
- Stress rupture testing is always done until failure of the material.

Stress rupture tests are used to determine the time to cause failure. Test data is plotted on a log–log scale as in the chart in Figure 13.7. A straight line is usually obtained at each temperature. This can then be used to extrapolate time to failure for longer times. Changes in slope of the stress rupture line are due to structural changes in the material. The higher the temperature, the lower is the value of stress to rupture.

Creep and stress rupture are the two elevated temperature properties that are of great importance in sizing the high-temperature boiler PPs, such as the tubes, pipes and headers of SH and RH.

- Creep strength is the average stress to produce 0.01% creep rate in 1000 h as per ASME and is considered as such for design with a factor of safety of 1.
- Stress to rupture is considered for 100,000 h. It is the lower of 80% for minimum stress and 67% for average stress.

13.15.1.6 Fatigue Strength or Endurance Limit

Fatigue strength is of importance whenever a component is subject to cyclical or alternate loading inflicting a component with high number of deflections or stress reversals. A component can fail even though the stress levels are lower than yield, because of fatigue. The surface develops a large number of micro-cracks along the grain boundaries and the core structure taking the stresses reduces in size and fails by ordinary rupture.

The highest cyclical stress that can be applied for a certain number of cycles before the component fractures is the fatigue strength or endurance limit. The standard fatigue strength for steel is that reported for 100,000,000 cycles. *S–N* curves (stress–cycles) give the fatigue strength of a material for a certain temperature. A typical *S–N* curve is given Figure 13.8.

A stress level Se or lower in curve A is a level at which infinite cycles of stress can be impressed with no fear of failure of the component. Se is about 0.5 times the TS for steels, that is ~100,000 psi or 690 MPa.

Curve B gives the fatigue strength at the specified number of cycles. Fatigue strength for cyclical duty design is as important as the YS is to a static design. Endurance limit is adversely affected if there are changes in cross section and stress concentration notches. *Corrosion fatigue* is the fatigue strength in a corrosive environment and the values are lower.

13.15.2 Hardness and Hardenability

Hardness of metals, as it is usually understood, is a measure of their *resistance to permanent or plastic deformation*. It is the *resistance to indentation under static load.*

Hardness is a very significant characteristic as it is indicative of several other important properties of steel such as wear resistance, heat treatment, machinability, fabrication and so on.

There are, however, three types of hardness measurements:

- Scratch hardness as in Moh's hardness
- Rebound or dynamic hardness
- Indentation hardness

Scratch hardness as measured by *Moh's scale* is not quite well suited for metals (see Moh's hardness below). In dynamic-hardness measurements, an indenter is usually dropped on the metal surface, and the hardness is expressed as the energy of impact. The *Shore scleroscope*, which is the most common example of a dynamic-hardness tester, measures the hardness in terms of the height of rebound of the indenter.

Only indentation hardness is of major engineering interest for metals. The smaller the indentation, the harder is the material. Hardenability is the *measure of the depth of full hardness achieved* during heat treatment and it is not to be confused with hardness. It is related to the type and amount of alloying elements as well as the heat treatment process. Different alloys, which have the same amount of C content, will achieve the same amount of maximum hardness; however, the depth of full hardness will vary with different alloys.

Hardenability and Weldability Do Not Go Together

The reason to alloy the steels is not merely to increase their strength, but increase their hardenability as well in many cases—the ease with which full hardness can be achieved throughout the material.

Usually when hot steel is quenched, most of the cooling happens at the surface, as does the hardening. This propagates into the depth of the material. Alloying helps in hardening and by determining the right alloy one can achieve the desired properties for the particular application.

Such alloying also helps in reducing the need for a rapid quench cooling—thereby eliminating distortions and potential cracking. In addition, thick sections can be hardened fully.

There are three types of steels from hardenability point of view:

- Non-hardening
- Shallow hardening
- Deep hardening

C and other alloying elements decide the hardenability:

- Low C and very low ASs are non-hardenable. They are good for welding and cold working.
- Medium C and low ASs are shallow hardening with hard outer surface and soft inner core as required for gears, cam shafts and so on.
- High C and high ASs are deep hardening.

Figure 13.9 shows the solubility of various elements in ferrite and their influence on the hardenability of iron in the presence of carbon.

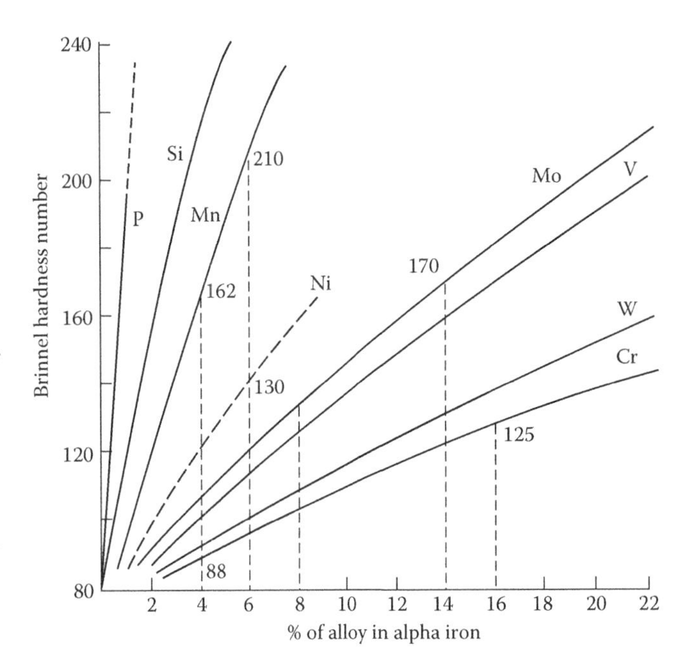

FIGURE 13.9
Solubility of elements in ferrite and the effects on hardenability.

With a slight increase in the C content, they respond markedly to heat treating, because C acts as a ferrite strengthener.

P improves the hardness of ferrite significantly even by a small addition while Cr will not, even with generous addition.

13.15.2.1 Moh's Hardness Scale

Moh's hardness refers to a material's *ability to resist abrasion or scratching.*

The Moh's (Mohs) scale of hardness is the most common method used to rank gemstones and minerals according to hardness. Devised by German mineralogist Friedrich Moh in 1812, this scale grades minerals on a scale from 1 (very soft) to 10 (very hard). In that relative scale talc is graded as one and diamond at 10. A mineral with a given hardness rating will scratch other minerals of the same hardness and all samples with lower hardness ratings.

Moh's scale is not very linear, in the sense that each step of increase does not correspond to the same increase in hardness, as the system was evolved over two centuries ago with the minerals known at that time.

Moh's hardness is compared with every day examples in Figure 13.10.

The Mohs' scale is not well suited for metals since the intervals are not widely spaced in the high-hardness range. Most hard metals fall in the Mohs' hardness range of 4–8.

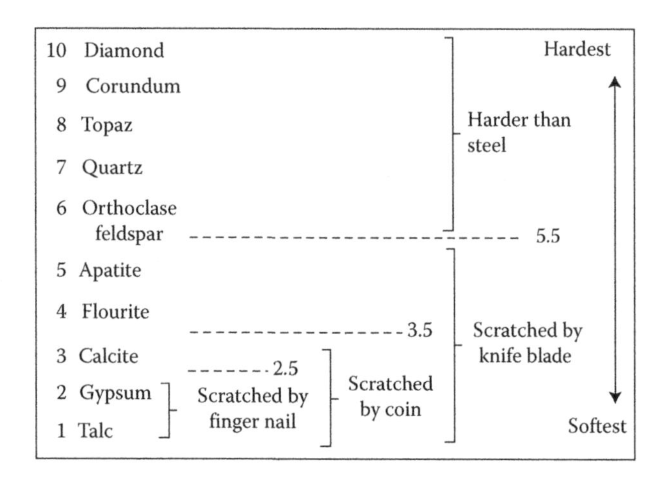

FIGURE 13.10
Moh's scale of hardness for various minerals.

13.16 Metallurgy

Fundamental metallurgical aspects of steels as needed by boiler engineers are covered here. They are elaborated under the sub-topics of

1. Iron carbon (Fe–C) diagram
2. Heat treatment
3. Heat treatment processes
4. Alloying elements of steel
5. Grain size
6. Deoxidation of steels

13.16.1 Iron Carbon Diagram

Fe–C equilibrium diagram or Fe–FeC phase diagram is a graph showing the transformation of Fe with respect to C content and temperature which is shown in Figure 13.11. This is one of the most important diagrams in metallurgy.

When C is added to Fe in small quantities, steel is obtained. The influence of C on mechanical properties of Fe is much larger than the other alloy elements. α *Iron or ferrite is a very pure form of Fe with C < 0.02%, and it is soft*. It flows plastically. Addition of C to Fe increases the mechanical strength but decreases ductility and weldability. Various forms of Fe–C alloys appearing in the Fe–C diagram are explained along with their microstructure in Figure 13.12.

Ferrite (α) is body-centred cubic crystal (BCC) in structure and is virtually pure Fe. It is stable up to 910°C. C solubility, which is temperature dependent, is minimum of 0.02 and maximum of 0.2% at 723°C.

Cementite is Fe_3C, Fe combined with C of 6.67% by weight, which is the maximum solubility of C in the form of Fe_3C. Addition of C beyond this would result in the formation of free C or graphite.

Pearlite is a fine mixture of ferrite and cementite arranged in lamellar form and stable at all temperatures <723°C.

Austenite (γ) is a phase-centred cubic structure (FCC) crystal. It is stable at temperatures >723°C depending on C content which it can dissolve up to 2%.

Generally, C content of structural steels is in the range of 0.12–0.25%. A structure of

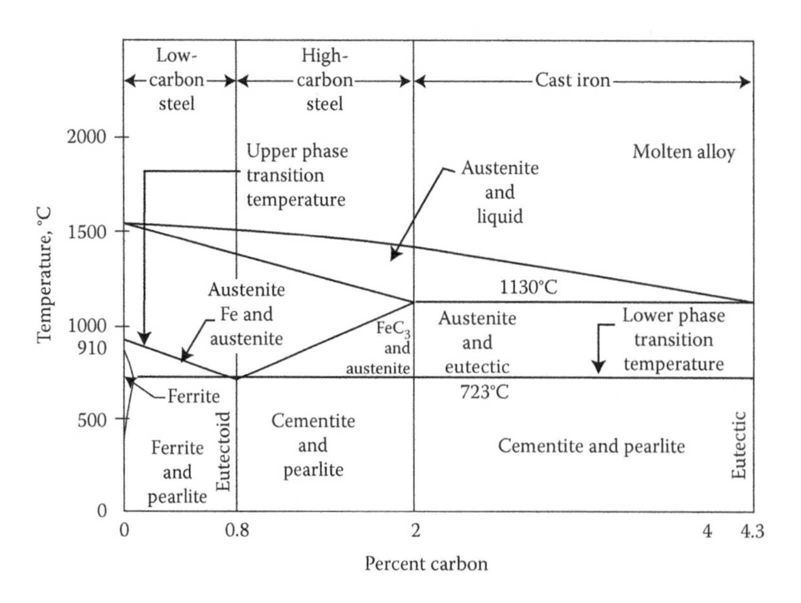

FIGURE 13.11
Iron–carbon (Fe–C) diagram for steel.

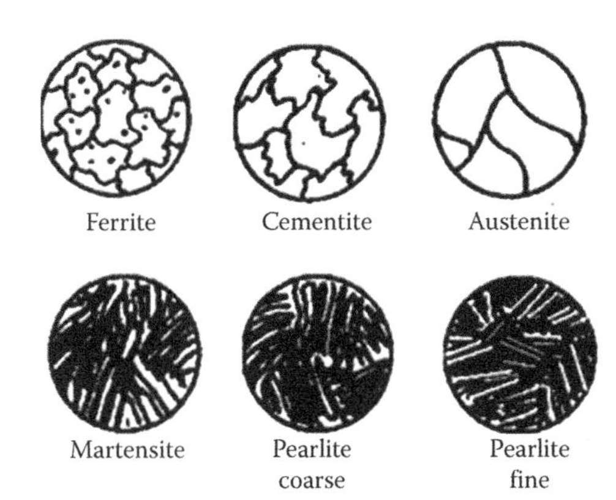

FIGURE 13.12
Microstructures of various alloys of Fe–C diagram.

- Ferrite + pearlite prevails up to 0.8% of C (hypoeutectoid steels)
- Pearlite + cementite from 0.8% to 2.0% (hypereutectoid steels)

Beyond 2% of C, CI is formed. The phase diagram is thus split into two parts at 0.8% of C which is called *Eutectoid point*.

Cooling rates of steel from austenite region to room temperature produce different microstructures with different mechanical properties.

- When cooled slowly in a furnace, called *annealing*, structural steels have a microstructure consisting of ferrite and pearlite.
- When cooling rate is increased as in *normalising* (cooling in still air), a small increase in the ferrite content is obtained along with fine lamellar pearlite. The decrease in grain size of pearlite improves mechanical strength. When hot rolling is done, with temperatures maintained in austenitic region, fine grains are obtained because of heavy mechanical deformation. When it is followed by normalising, even more number of grains are in small range, increasing mechanical strength further.
- If the steel is cooled very rapidly, like quenching in cold water, because of insufficient time for diffusion of C to form ferrite + pearlite structure, *martensitic* structure is formed. Slightly lower cooling rate would result in *bainite* structure depending on the composition of steel.
- Martensite is very hard and not ductile and therefore, undesirable for structural steels as welding also becomes difficult. Martensitic structure is preferred in some parts like high-strength bolts due to its hardness, which is a function of C content. When it is heated to a temperature of 600°C, martensite softens and toughness improves. This process of reheating martensite is called *tempering*. Quenching and tempering results in many varieties of steel with varying hardness, wear resistance, strength and toughness.

13.16.2 Heat Treatment

Heat treatment of steel is the *controlled heating and cooling of steel to alter the physical and mechanical properties without changing the product shape*. Sometimes it happens inadvertently in processes that either heat or cool the metal, such as in welding or forming.

Heat treatment helps in

- Increasing the strength of material
- Altering certain manufacturability objectives such as improving machining and formability, restoring ductility after a cold working operation and so on

Thus, it is a highly enabling process that helps other manufacturing processes and also improves product performance by increasing strength or other desirable characteristics.

Steels respond well to heat treatment. Steels are heat treated for one of the following reasons:

a. Softening
b. Hardening
c. Material modification

a. Softening: softening is done to
  - Reduce strength or hardness
  - Remove residual stresses
  - Improve toughness
  - Restore ductility
  - Refine grain size or change the electromagnetic properties

Restoring ductility or removing residual stresses is a necessary operation when a large amount of cold working is to be performed, such as in a cold-rolling operation or wiredrawing. Annealing—full process, spheroidising, normalising and tempering—austempering and martempering are the principal ways by which steel is softened.

b. Hardening: Hardening is done to increase the strength and wear properties. For hardening, there should be sufficient C and alloy content. With sufficient C, steel can be directly hardened.

Hardness depends on the C content of the steel. Steel is heated to austenitic region. When it is

- Suddenly quenched, martensite is formed which is a very strong and brittle structure.
- Slowly quenched, austenite and pearlite is formed which is a partly hard and partly soft structure.
- Very slowly cooled, pearlite is formed (mostly), which is extremely soft.

c. Material modification: Heat treatment is used to modify properties of materials in addition to hardening and softening. These processes modify the behaviour of the steels in a beneficial manner to maximise service life, for example, stress relieving, or strength properties.

13.16.3 Heat Treatment Processes

Only those heat treatment processes involved in boiler making are included here. Figure 13.13 explains these processes on a simplified Fe–C diagram.

a. *Full annealing* is the process of slowly raising the steel temperature ~50°C (90°F) above the *austenitic temperature* line A_3 in the case of hypo-eutectoid steels (steels with <0.77% C). It is held at this temperature for sufficient time for all the material to transform into austenite. It is then slowly cooled at the rate of about 20°C/h (36°F/h) in a furnace to about 50°C (90°F) into the ferrite–cementite range. At this point, it can be cooled in room temperature air with natural convection. The steel becomes soft and ductile.

b. *Normalising* is the process of raising the temperature to over 60°C (108°F), above line A_3 or line A_{CM} fully into the austenite range. It is held at this temperature to fully convert the structure into austenite, and then removed from the furnace and cooled at room temperature under natural convection. This results in a grain structure of fine pearlite with excess of ferrite or cementite. The resulting material is soft; the degree of softness depends on the actual ambient conditions of cooling. This process is considerably cheaper than full annealing since there is no added cost of controlled furnace cooling.

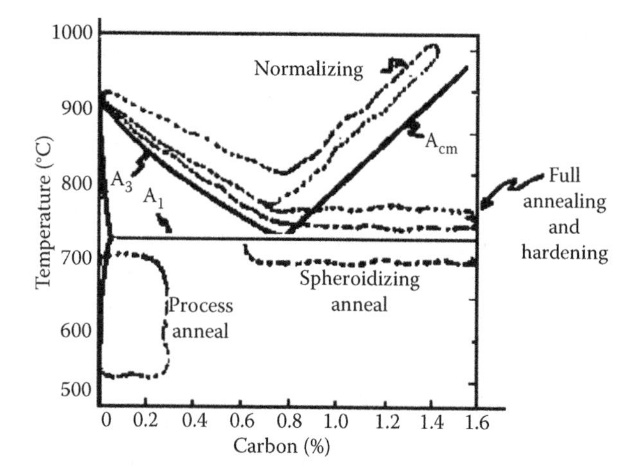

FIGURE 13.13
Heat treatment ranges plotted on Fe–C diagram.

The main difference between full annealing and normalising is that fully annealed parts are uniform in softness (and machinability) throughout the entire part; since the entire part is exposed to the controlled furnace cooling. In the case of the normalised part, depending on the part geometry, the cooling is uneven resulting in non-uniform material properties across the part. This may not be desirable if further machining is involved since it makes the machining job somewhat unpredictable. In such a case, it is necessary to do full annealing.

c. *Process annealing* is used to treat work-hardened parts made out of low CSs (<0.25% C). This allows the parts to be soft enough to undergo further cold working without fracturing. Process annealing is done by raising the temperature to just below the ferrite–austenite region, line A_1 on the diagram. This temperature is ~727°C (1341°F). So heating it to about 700°C (1292°F) should suffice. The part is held long enough to allow recrystallisation of the ferrite phase, and then cooled in still air. Since the material stays in the same phase throughout the process, the only change that occurs is the size, shape and distribution of the grain structure. This process is cheaper than either full annealing or normalising since the material is not heated to a very high temperature or cooled in a furnace.

d. *Stress relief anneal* is used to *reduce residual stresses* in large castings, welded parts and cold-formed parts. Such parts tend to have stresses because of thermal cycling or work hardening. The parts are heated to temperatures of up to 600–650°C (1112–1202°F), and held for an extended time (about 1 h or more) and then slowly cooled in still air.

e. *Spheroidisation* is an annealing process used for high CSs (C >0.6%) that will be machined or cold formed subsequently. This is done by one of the following ways:

f. *Tempering* is a process done subsequent to quench hardening to remove the brittleness of the parts (caused by martensite). Tempering results in a desired combination of hardness, ductility, toughness, strength and structural stability.

Tempering is done immediately after quench hardening. When the steel cools to about 40°C (104°F) after quenching, it is ready to be tempered. The part is reheated to a temperature of 150–400°C (302–752°F). In this region a softer and tougher structure troostite is formed. Alternatively, the steel can be heated to a temperature of 400–700°C (752–1292°F) that results in a softer structure known as sorbite. This has less strength than troostite but more ductility and toughness.

The heating for tempering is best done by immersing the parts in oil up to 350°C (662°F) and then heating the oil along with the parts to the appropriate temperature. Heating in a bath also ensures that the entire part has the same temperature and will undergo the same tempering. For temperatures above 350°C (662°F) it is best to use a bath of nitrate salts. The salt baths can be heated up to 625°C (1157°F). Regardless of the bath, gradual heating is important to avoid cracking of steel. After reaching the desired temperature, the parts are held at that temperature for about 2 h, then removed from the bath and cooled in still air.

g. *Austempering* is a quenching technique. The part is not quenched through the martensite transformation. Instead, the material is quenched above the temperature when martensite forms M_S, around 315°C (600°F). It is held at this temperature till the entire part reaches this temperature. As the part is held longer at this temperature, the austenite transforms into bainite. Bainite is tough enough so that further tempering is not necessary, and the tendency to crack is severely reduced.
h. *Martempering* is similar to austempering except that the part is slowly cooled through the martensite transformation. The structure is martensite, which needs to be tempered just as much as martensite that is formed through rapid quenching. The biggest advantage of austempering over rapid quenching is that there is less distortion and tendency to crack.
i. *Stress relieving:* Machining and cold working induces stresses in parts. The bigger and more complex the part, the more are the stresses. These stresses can cause distortions in the part over a long term. If such parts are clamped in service cracking can occur. Hole locations can change causing them to go out of tolerance. For these reasons, stress relieving is often necessary.

Typically, the parts that benefit from stress relieving are large and complex weldment, castings with a lot of machining, parts with tight dimensional tolerances and machined parts that have had a lot of stock removal performed.

Stress relieving is done by subjecting the parts to a temperature of about 75°C (165°F) below the transformation temperature, line A_1 on the diagram, which is about 727°C (1340°F) of steel—thus, stress relieving is done at about 650°C (1202°F) for about 1 h or till the whole part reaches the temperature. This removes more than 90% of the internal stresses. ASs are stress relieved at higher temperatures. After removing from the furnace, the parts are air cooled in still air.

13.16.4 Alloying Elements of Steel

Various alloying elements are used to improve the properties of steels to suit. The effects of only a few of them which matter in boiler manufacturing are explained in brief here. It is interesting to note that almost all of them have beneficial effects on steel except for S, P and Cu.

C has a major effect on steel properties, acting as the primary hardening element in steel. Hardness, hardenability and tensile strength increase as C increases up to about 0.85% broadly as captured in Figure 13.14. Ductility, impact, machinability and weldability decrease with increasing C.

1. *Aluminium (Al)* is widely used as a deoxidiser.

   Al can control austenite grain growth in reheated steels and is therefore added to control grain size. Al is the most effective alloy in controlling grain growth prior to quenching. Al improves resistance to ageing, scaling and heat resistance (as in **calorising**)

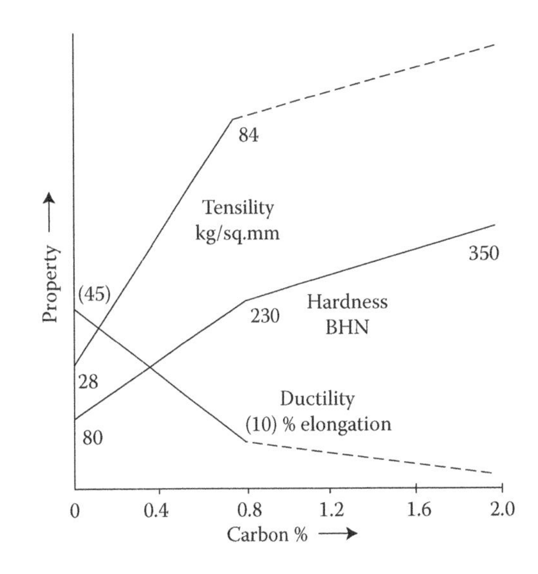

FIGURE 13.14
Effect of carbon on properties of steel.

2. *Boron (Bo)* is added to fully killed steel to improve hardenability in very small amounts of about 0.001%.

   Bo-treated steels are produced to a range of 0.0005–0.003%. Bo is most effective in lower C steels.

3. *Copper (Cu)* in significant amounts is detrimental to hot-working steels.

   Cu negatively affects forge welding, but does not seriously affect arc or oxyacetylene welding. Cu can be detrimental to surface quality. Cu is beneficial to atmospheric corrosion resistance when present in amounts >0.20% as in weathering steels.

4. *Chromium (Cr)* is commonly added to steel to increase corrosion and oxidation resistance, to increase hardenability and to improve high-temperature strength.

   As a hardening element, Cr is frequently used with a toughening element such as Ni to produce superior mechanical properties.

   At higher temperatures, Cr contributes increased strength. Cr is strong carbide former. Complex Cr-Fe carbides go into solution in austenite slowly; therefore, sufficient heating time must be allowed prior to quenching. Cr >13% improves atmospheric corrosion.

5. *Lead (Pb)* is virtually insoluble in liquid or solid steel. However, Pb is sometimes added to CS and AS by means of mechanical dispersion during pouring to improve machinability.

6. *Manganese (Mn)* is a deoxidiser.

   Mn is generally beneficial to surface quality especially in resulphurised steels. It has a significant effect on the hardenability. The effect of Mn to increase strength and hardness and decrease ductility and weldability of steel is similar to but less than C. The increase in strength is dependent upon the C content of steel.

7. *Molybdenum (Mo)* increases the hardenability of steel. Mo may produce secondary hardening during the tempering of quenched steels.

   It enhances the creep strength of low ASs at elevated temperatures. It improves UTS, YS, impact, wear resistance and high-temperature strength. It reduces resistance to scaling.

8. *Niobium (Columbium)* increases the YS and, to a lesser degree, the UTS of CS. The addition of small amounts of Nb can significantly increase the YS of steels.

   Nb can also have a moderate precipitation strengthening effect. Its main contributions are to form precipitates above the transformation temperature, and to retard the recrystallisation of austenite, thus promoting a fine-grain microstructure having improved strength and toughness.

9. *Nickel (Ni)* is a ferrite strengthener.

   Ni does not form carbides in steel but remains in solution in ferrite, strengthening and toughening the ferrite phase. Ni increases the hardenability and impact strength of steels. It improves UTS, YS and toughness.

10. *Phosphorus (P)* is impurity in steel. It increases strength and hardness and decreases ductility and notch impact toughness of steel. P is normally controlled to low levels.

    Higher P is specified only in low-C free-machining steels to improve machinability.

11. *Silicon (Si)* is one of the principal deoxidisers used in steel making.

    Si is less effective than Mn in increasing as-rolled strength and hardness. It improves TS, YS and resistance to oxidation and scaling. It reduces hot and cold workability. It also reduces electrical conductivity. In low CSs, Si is generally detrimental to surface quality. It is ideal for spring steels and electrical apparatus.

12. *Sulphur (S)*, like P, is an impurity in steel.

    It decreases ductility and notch impact toughness, especially in the transverse direction. Weldability decreases with increasing S content. Like P, S is also controlled to low levels with exception in free-machining steels, where it is added to improve machinability.

13. *Titanium (Ti)* is a deoxidant and is used to retard grain growth and thus improve toughness. It increases the strength at high temperatures.

    Ti is also used to achieve improvements in inclusion characteristics. Like Zr, Ti causes sulphide inclusions to be globular rather than elongated, thus improving toughness and ductility in transverse bending.

14. *Tungsten (W)* is primarily added for high-temperature strength.

    W improves TS, YS, toughness and resistance to high-temperature wear in a similar way as Mo. It reduces the resistance to oxidation.

15. *Vanadium (V)* is a deoxidant and a degasifier. It is a grain refiner.

    It significantly increases TS, YS and wear strengths of CS even in small amounts. V is one of the primary contributors to precipitation strengthening in micro-alloyed steels. When thermo-mechanical processing is properly controlled, the ferrite grain size is refined and there is a corresponding increase in toughness. The impact transition temperature also increases when V is added.

16. *Zirconium (Zr)* can be added to killed high-strength low ASs to achieve improvements in inclusion characteristics.

    Zr causes sulphide inclusions to be globular rather than elongated, thus improving toughness and ductility in transverse bending.

All *micro ASs* contain small concentrations of one or more strong carbide and nitride forming elements. V, Nb and Ti combine preferentially with C and/or N to form a fine dispersion of precipitated particles in the steel matrix.

13.16.5 Grain Size

In Fe–C diagram, above the upper critical or phase transformation temperature, all grains are austenitic and of the smallest size. Grain size increases as the temperature increases.

Steels are classified as

- Fine grained when grain size is from one to four
- Coarse grained when grain size is from five to eight

The number is derived from the formula

$N = 2^{n-1}$ where n refers to the number of grains/in^2 at a magnification of 100 diameters. Finer grains produce stronger steels with higher value of UTS and YS. Fine-grained steels give a better combination of strength and toughness whereas coarse steels have better machinability. The growth of grains sets in earlier for coarse grain than fine-grain steel. In rimmed steel, coarsening is rapid whereas in killed steels, fine grains are maintained for a much longer period at higher temperatures.

13.16.6 Deoxidation of Steels

Solubility of O_2 in molten steel is ~0.23% at 1700°C, which decreases sharply during cooling, dropping to ~0.003% in solidified steel. The O_2 so liberated forms blowholes (gas pores) and non-metallic inclusions entrapped within the ingot structure, both of which affect the quality of steel.

Deoxidation is the process of reducing the dissolved O_2 in molten steel to the acceptable level by

1. Deoxidisers
2. Vacuum
3. Diffusion

1. *Deoxidation by metallic oxidisers* is the most popular method where elements like Mn, Si and Al are commonly used to form strong and suitable oxides. According to the degree of deoxidation, CSs are divided into three groups.
   a. *Killed steels* which are fully deoxidised steels with no CO formed during solidification. They have homogenous structure and no gas porosity. They have more uniform properties of ingot. Most CS having <0.25% C and ASs are of killed variety. Killed steels are used for items required in forging, carbonising and heat treatment.
   b. *Semi-killed steels* are incompletely deoxidised steels containing some amounts of excess O_2 forming CO during the last stages of solidification. Semi-killed steels have 0.15–0.25% C and 0.05% Si. Semi-killed steels are used for plates, sheets and structurals.
   c. *Rimmed steels* are partially deoxidised or non-deoxidised low-C steels with sufficient amount of CO formed during solidification. The gases are trapped at the edge of the ingot with increasing amounts at the bottom. The thin outer rim is low in C and high in pure Fe. Rimmed steels are cheaper to make and widely used for making structural plates and steel strips for welding pipes. They are good for forming and deep drawing operations but unsuitable for forging or carburising.

   Figure 13.15 describes the three types of steels.

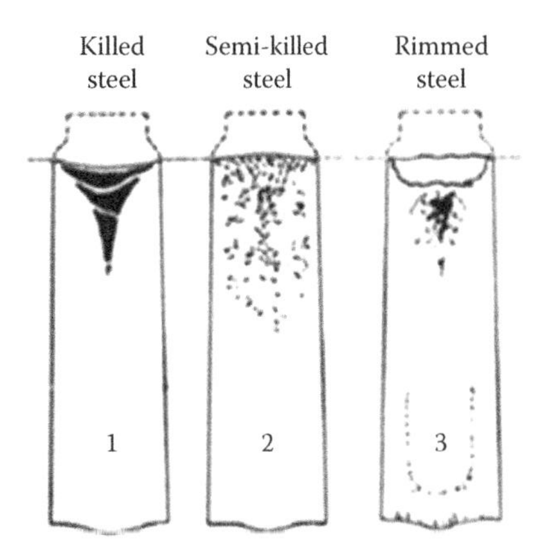

FIGURE 13.15
Classification based on deoxidisation of steels.

2. *Deoxidation in vacuum* utilises C dissolved in steel as deoxidiser which floats on the top, forming CO on reaction with O_2. Vacuum treatment of molten steel removes the floating CO as well as the H_2 dissolved in steel. Vacuum deoxidised steels have a very homogenous structure, low content of non-metallic inclusions and low gas porosity. Vacuum degassing is a popular method for deoxygenating steel and producing killed steels.

   Vacuum degassing is increasingly practised in the steel industry to

   a. Remove H_2
   b. Improve cleanliness by removing part of the O_2
   c. Produce steel of low C content (< 0.03%)
   d. Produce steels to close chemical composition ranges (including deoxidisers)
   e. Control pouring temperatures, especially for continuous casting operations

   Vacuum degassing processes, in the broadest sense, refer to the exposure of molten steel to an LP environment to remove gases (chiefly H_2 and O_2) from the steel. The effectiveness of any vacuum degassing operation depends upon the surface area of liquid steel that is exposed to LP. The mechanisms of H_2 and O_2 removal from liquid steel are related directly to surface area.

   Vacuum treatment is used in manufacturing large ingots required for making heavy plates and other high-quality steels. BQ plates are usually produced by vacuum degassing.

3. In *diffusion deoxidation*, the slag is deoxidised which makes the dissolved O_2 in steel diffuse into the slag to achieve equilibrium. Coke, Si, Al and other elements are used for slag deoxidation. Since the deoxidisers are not introduced directly into the steel melt oxide, non-metallic inclusions are not formed resulting in steel less contaminated by non-metallic inclusions.

13.17 Methane (CH_4) (*see* Gaseous Fuels)

13.18 Mills or Milling Plant (*see* Pulverisers)

13.19 Mineral Matter (*see* Coal Ash)

13.20 Mineral Wool (*see* Insulating Materials)

13.21 Moh's Hardness Scale (*see* Metal Properties)

13.22 Molecular Weight/Mole/Molar Volume (*see* Thermodynamic Properties)

13.23 Mollier Chart/Diagram (*see* Thermodynamic Properties)

13.24 Mono Wall (*see* Pressure Parts *and* Furnace in Heating Surfaces)

13.25 Moody's Friction Factor (*see* Fluid Flow)

13.26 Mountings and Fittings (*see* Auxiliaries)

13.27 Multiclones (*see* Dust Collection Equipment)

13.28 Municipal Solid Waste

Municipal solid waste (MSW), urban solid waste or household trash, refuse or garbage is a combination of solid and semi-solid wastes collected by a municipality in a designated area. It is predominantly household/domestic waste with some additional commercial wastes and generally excludes industrial and hazardous wastes.

Unlike the other waste fuels, MSW is a combination of several different types of wastes containing

- Biodegradable waste like food and kitchen waste, green waste, paper and so on
- Recyclable material like paper, glass, bottles, cans, metals, certain plastics, and so on
- Inert waste like construction and demolition waste, dirt, rocks and debris
- Composite wastes like waste clothing, Tetrapaks, and waste plastics such as toys
- Domestic hazardous waste and toxic waste like medication, e-waste, paints, chemicals, lightbulbs, fluorescent tubes, spray cans, fertilizer and pesticide containers, batteries and shoe polish

MSW, as fuel, can be described as follows:

- It can be compared with young fossil fuels like wood or bagasse in terms of its M and cv.
- But it has fairly high levels of corrosive elements, higher than in any coal or FO. It has Cl2, S and low-melting solids like Na, K Pb, V and Zn. However it is Cl2 that causes most corrosion.
- Its cv is derived mainly from its organic/cellulosic content.
- The composition and cv of refuse are highly variable depending on time, geography, weather, social status of a community and so on.
- Urban garbage has more paper and plastic while suburbs have more leaves and grass.
- Over the years, the hv of waste has been steadily going up contributed by paper, paper board and plastic.
- There is a large difference in the trash produced in various societies, both in quality and quantity. The hv of trash in developing countries is lower than in rich nations mainly because of the established practice of rag pickers who salvage the recyclable materials. Garbage in poor countries therefore has more food and inorganics.

A broad range of as-received analyses is presented in Tables 13.7 and 13.8 in PA and UA forms of analyses.

There are several stages of processing involved between generation of waste and its disposal. The stages depend upon the amount of waste, type of waste, distance to the disposal site and the final mode of disposal.

- It starts with separation of waste components in households and storage at the source ready for collection.
- Collection is the gathering of solid waste and recyclable materials and transportation to the location where the collection vehicle is emptied. This location may be a material processing facility, a transfer station or a landfill disposal site.
- Separation, processing, and transformation of solid wastes take place at this stage. In many societies, separation takes place before collection when recyclable waste is removed by the rag pickers.

TABLE 13.7
Proximate Analysis of MSW

Constituent in PA				Gcv		
VM w/w%	FC w/w%	M w/w%	Ash w/w%	MJ/kg	kcal/kg	Btu/lb
37–56	<1–15	20–32	9–27	7–15	1720–3610	3100–6500

TABLE 13.8
Ultimate Analysis of MSW

Constituent in UA							
C w/w%	H w/w%	N w/w%	Cl w/w%	S w/w%	O w/w%	M w/w%	Ash w/w%
23–43	3.4–6.4	0.2–0.4	0.1–0.9	0.1–0.4	15–32	20–32	9–27

- This is followed by transfer and transportation which involves the transfer of wastes from the smaller collection vehicle to the larger transport equipment and the subsequent transport of waste, usually over long distances, to a processing or disposal site.
- In a majority of the cases, the wastes go for land filling. Unlike in the past a modern sanitary landfill is an engineered facility for disposing of solid wastes on land without creating nuisance or hazards to public health or safety, such as the breeding of insects and the contamination of ground water.
- Energy generation is an alternative when landfill sites are either not easily available or are too far away and also there are monetary incentives to generate expensive power for sale.

The high cost of steam and power is due to

- The high cost of preparation of a bulky fuel
- Having a low cv
- Generating highly corrosive flue gases

all of which need large fuel handling facilities, expensive firing equipment, big boiler and sophisticated gas cleaning systems.

Despite the cost of power from WTE plants there are certain other advantages of incineration of waste, namely, big reductions in

- Total quantity of waste by nearly 60–90% depending upon the composition of waste and the technology adopted
- Demand for land, which is already scarce in cities, for land filling
- Cost of transportation of waste to far-away landfill sites which also helps in decongestion of cities with garbage trucks
- Environmental pollution

At present, <15% of MSW in the United States is used for power generation. In Europe, it is higher. Sweden leads the list of WTE plants installed worldwide. Denmark, Germany, France, the Netherlands and Spain are the other leading countries with a large number of installations. Lack of landfill sites, stringent environmental regulations and generous state subsidies have all contributed to such growth in Europe.

With steadily rising urbanisation, increasing consumerism and better prosperity, waste creation is bound to escalate despite efforts to the contrary. Rich countries produce more waste per person on a smaller population while poor nations generate less trash on larger urban population. Power generation from MSW is considered to be a good solution all over the world for both rich and poor societies. But such power from MSW is presently very expensive at ~2–3 times the normal cost.

Typically, a ton of trash in the rich societies can produce ~300 kWh of power. Or a 200 tpd plant can generate ~6 MWe and a 1000 tpd ~30 MWe.

In addition to the above, there are techniques like landfill gas capture, combustion, pyrolysis, gasification and plasma arc gasification to deal with MSW. These are all emerging techniques not fully proven.

As MSW is going to be a bigger menace in the future, prudence demands that all societies should

1. *Reduce*
2. *Reuse*
3. *Recycle*

waste that is technically and economically feasible to do so.

Energy from MSW Combustion

MSW is burnt in two ways. In unprocessed condition, it is burnt on pusher-type grates in MB mode, which is a European practice. It can also be processed into refuse-derived fuel (RDF) which is an American evolution. RDF is mainly burnt in suspension firing in SS. It can also be fired in BFBC, CFBC and PF boilers usually in combination with other fuels.

13.28.1 Mass Burning

This is the most common technology where MSW is burned in as received form with no fuel preparation other than the removal of non-combustible and oversized items. 70–80% installations worldwide employ MB. It is burnt on PG or RG system where feeding of fuel to the grate is from the top and combustion air is from the bottom. Fuel feeding typically includes a feed hopper and hydraulic ram that pushes fuel from the bottom of the grate. Long or multi-stage pusher grates are needed for MSW which are quite expensive. Small to big units can be built burning 50–1000 tpd of MSW.

Figure 13.16 shows the cross section of a typical MB installation for MSW. There are essentially five major sections to a WTE plant, namely

1. Fuel receiving section—where the garbage trucks enter a receiving hall and tip the MSW into the refuse bunker
2. Grate section—where the fuel is loaded by the grab buckets on to the feed chutes from where it is pushed over to the top of the grate for combustion

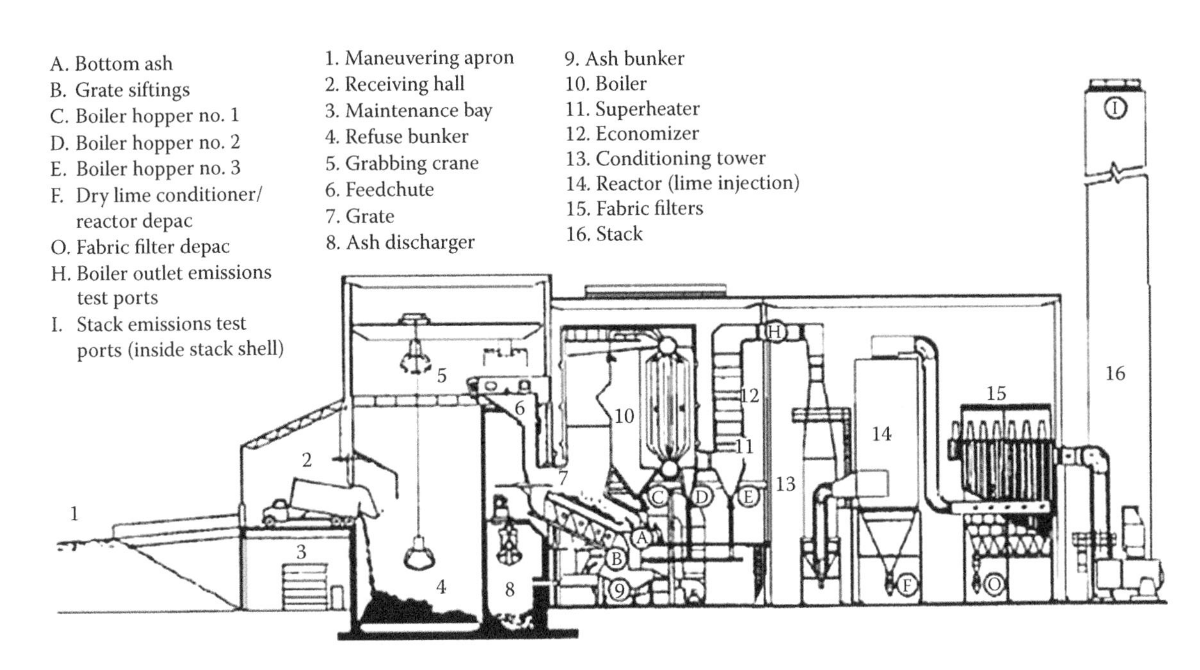

FIGURE 13.16
Cross section and details of a typical MSW fired boiler plant.

3. Boiler section—where heat from combustion is absorbed by various HSs to generate steam at the desired p & t
4. Gas cleaning system—where the gas is stripped of its PM in ESP or bag filter and conditioned in $DeNO_x$ and FGD plants
5. Ash disposal system

13.28.2 Refuse-Derived Fuel Burning

This technology of fuel enrichment involves various processes to improve physical and chemical properties of the solid waste. First of all, the combustible and non-combustible fractions are separated. The combustible material, further processed, is called RDF and is fired in boilers just like wood firing. The processing may typically involve primary shredding, magnetic separation, sieving through trommel screens, secondary shredding and storage in bins. The non-combustibles go directly for landfill unlike in mass burn, where only the residues after all the fuel is burnt are used for land filling.

The MSW receiving facility includes an enclosed tipping floor called municipal waste receiving area, with a storage capacity equal to about 2 days of typical waste deliveries. The sorted MSW is then fed to either of the two equal capacity processing lines. Each processing line typically includes primary and secondary trommel screens, three stages of magnetic separation, eddy current separation, a glass recovery system and a shredder (as shown in Figure 13.17). The resultant fuel from such a system is fuel which is 100% <150 mm and 50% <75 mm and other fuel properties, as per Table 13.9.

There are a lot of equipments in RDF processing and therefore it is more expensive to build and operate than mass-burn. There are also reports of shortfall in availability as the RDF processing plants get bogged down with O&M issues.

Owing to the reduction in fuel particle size and non-combustible material, the RDF fuels are more homogeneous, better sized and hence easier to burn than the MSW feedstock. RDF has higher cv and lower ash as the inorganics are removed. The boilers firing RDF are much smaller, simpler and cheaper. Figure 13.18 shows the schematic of an RDF power plant.

Typical analyses of the two types of wastes, MSW and RDF, are given in Table 13.9, which shows the effects of

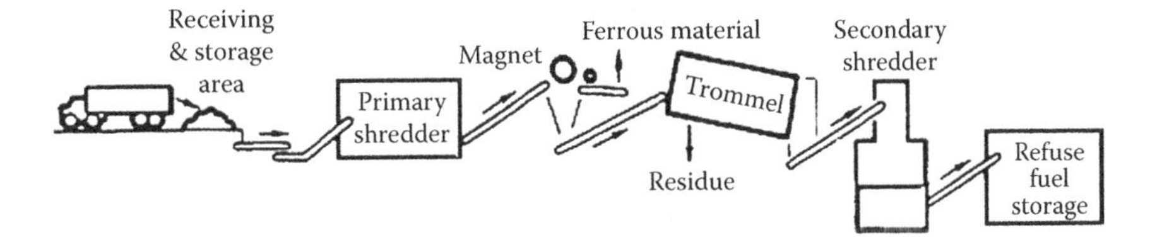

FIGURE 13.17
Processing of MSW.

TABLE 13.9

Properties of MSW and RDF Compared with Low-Ash Coal

	GCV			Burden in Fuel		
Fuel	**MJ/kg**	**kcal/kg**	**Btu/lb**	**%M**	**%A**	**% Total**
MSW	11,000–12,000	2630–2860	4730–5150	30–40	25–35	50–70
RDF	12,000–16,000	2860–3820	5150–6880	15–25	10–22	20–40
Coal	21,000–32,000	5010–7640	9020–13,750	3–10	5–10	10–20

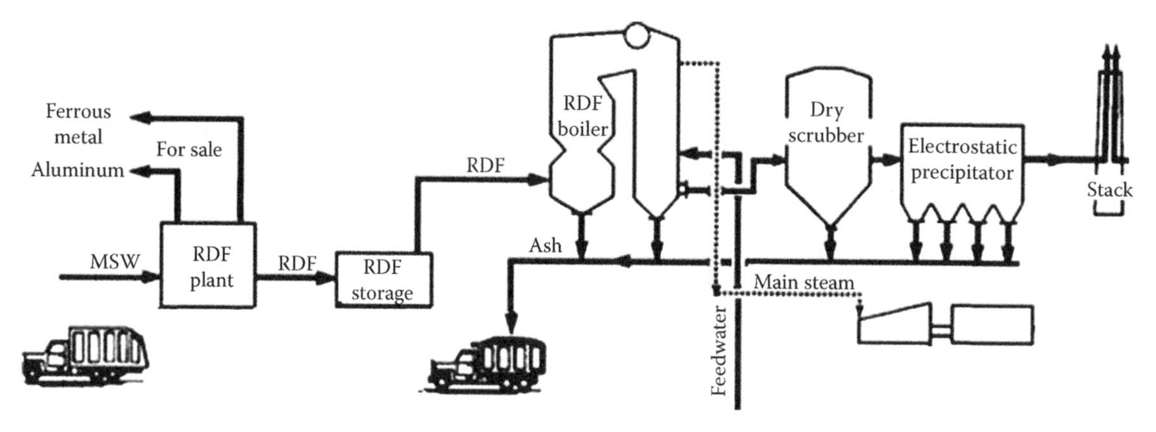

FIGURE 13.18
Cross section of a typical boiler plant firing RDF.

beneficiation. Properties of good quality of low-ash coal are also shown for comparison.

RDF has been successfully burnt in a variety of combustion equipments, such as SS, PG, PF, BFBC and CFBC boilers either alone or in combination with other fuels depending on the fuel sizing and other factors.

In RDF system, the fuel preparation at the front end of the scheme is more expensive and boiler plant at the rear/tail end is cheaper while in MB it is the reverse. RDF firing needs lower excess air than MSW and hence works at better η. Also, fuel handling is easier. RDF boilers operate typically at ~50% excess air and ~175°C (350°F) exit gas temperature while MSW units operate at ~100% excess air and ~200°C (400°F) exit gas temperature resulting in ηs of 75% and 70% respectively. 5% points difference in η would result in higher steam and power production from the same amount of waste. In a 550 tpd plant, this works out to a difference of ~25 tph of steam. MSW would generate ~70 tph and RDF on SS ~95 tph.

The unit sizes in RDF range from 300 to 1000 tpd. They produce better quality of ash and ash disposal is less problematic.

Some important sizing parameters for grate fired MSW and RDF boilers are tabulated in Table 13.10. It brings out the differences to show how large the boiler for MSW can be in comparison to RDF boiler.

13.29 MSW Boilers-Mass Burning (*see also* Grates: Reciprocating, Pulsating, Step or Pusher)

Unprocessed MSW is burnt in PG or RG-fired boilers. Burning an unfriendly fuel like MSW efficiently is a

TABLE 13.10

Parameters for Sizing the Grate and Furnace for Boilers Firing MSW and RDF

	Grate HRR			Furnace HRR			Max. FEGT		Gas Velocity	
	kW/m²·h	**kcal/m²·h**	**Btu/ft² h**	**kW/m³·h**	**kcal/m³·h**	**Btu/f³·h**	**°C**	**°F**	**mps**	**fps**
MSW	85,000–110,000	730,000–946,000	270,000–350,000	62,000–100,000	54,000–90,000	6000–10,000	815	1500	6.1–7.6	20–25
RDF	<235,000	<2 × 10⁶	<750,000	125,000–155,000	107,000–134,000	12,000–15,000	870	1600	7.6–10.7	25–35

challenge. The fuel characteristics of MSW can be summarised as follows:

- It has large volume with no sizing.
- It has low and variable hv of ~7–15 MJ/kg, ~1700–3600 kcal/kg or ~3000–6450 Btu/lb.
- It has highly variable physical and chemical properties.
- It has a high fuel burden with large M content of ~20–32% and fairly high A content of ~7–27%.
- It has high VM and low FC, needing a large furnace to burn the VM with good residence time ~3 s.
- It has relatively significant amounts of Cl_2 and S to cause severe cases of corrosion at both high and low gas temperatures.
- It has low-melting compounds of Na, K and Pb to cause both slagging and fouling.

MB on an overfeed PG is an ideal solution because

- The voluminous mass of fuel can be loaded on to the inlet hopper of the grate and introduced into the combustion chamber quite easily.
- The grate provides ample time for taking the high-M low-cv fuel through the classical stages of drying, ignition and combustion in three distinct overlapping stages as no other firing equipment can do.
- With multiple drives the grate can be set at three decreasing inclinations to suit the three stages of combustion.
- Only this grate has the ability to reciprocate the grate bars in a way that a voluminous mass can be dug into and turned around to expose fresh surface for combustion whereas the others provide an undisturbed quiescent combustion more suited for clinker forming coals.

Figure 13.19 portrays an isometric view of a typical PG or RG for MSW application along with the lower furnace.

- Combustion of such lean and moist fuel is possible only when there is radiation from the top and sweeping of hot gases on to the fuel bed. This is accomplished by the deep front and rear arches as well as the SA nozzles located in the neck.
- The excess air required is ~70–80% for complete combustion with <100 ppm of CO Of this 60–70% is provided as PA below the grate and 40–50% as SA through the front and rear walls.
- While PA is invariably hot, for Sa, cold air is preferred, as it has higher ρ and better penetration power. SA is admitted at ~4.5 m above the grate.
- All walls of the furnace are completely water cooled with suitable tube protection like SiC refractory layer on pin studding or weld overlays on tubes up to ~10 m as the lower furnace is highly susceptible to corrosion and fouling. Alternately, the walls are covered with SiC tiles.
- Conventional AHs are not feasible as they get plugged with dust in flue gases. At the same time, high M in fuel desperately demands hot air for better combustion. It is normal practice to have Scaph at the discharge of fans to provide hot air of 150–175°C.

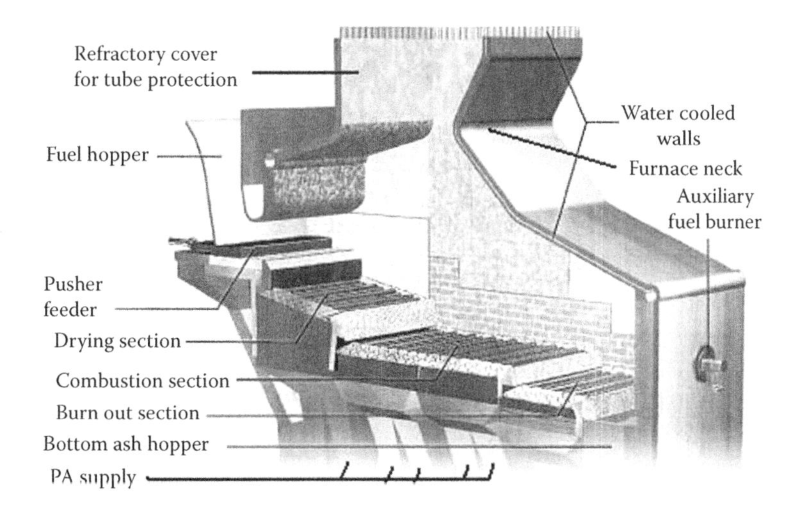

FIGURE 13.19
Isometric view of a three-stage pusher grate for burning MSW.

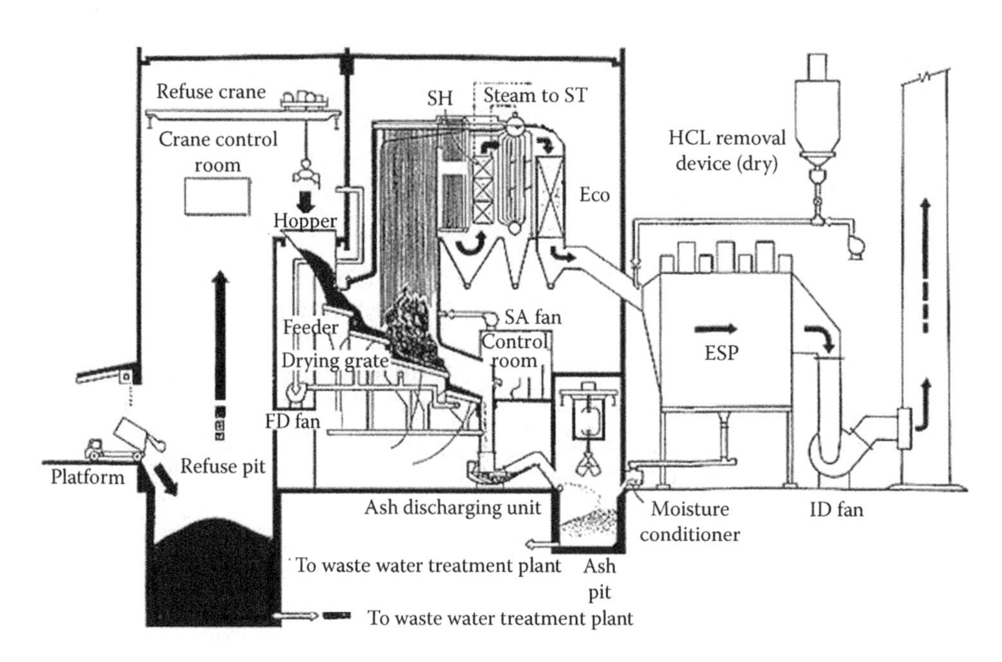

FIGURE 13.20
Cross section of a typical boiler plant for burning MSW on three-stage step pusher grate.

- SHs and Ecos are usually made of vertical tubes to reduce erosion. Many times, longitudinal gas flow is employed over tubes. Horizontal banks offer more compact arrangement but are used only when the gases are known to be not so aggressive.
- FEGT is usually <815°C (1500°F). The initial rows of SH tube banks are in parallel flow to keep the tube metal temperatures low and prevent corrosion. The spacing is quite wide apart to avoid bridging and erosion. Gas velocities are restricted to ~9 m/s and the actual velocities may be half that.

Grate dimensioning considerations are as follows for a typical MSW with ~11.7 MJ/kg, ~2800 kcal/kg or ~5000 Btu/lb of Gcv. As the fuel varies widely, sizing parameters also vary a good deal from market to market and the following guidelines are only indicative:

- Grate width is decided basically from the point of carrying the amount of fuel.

  It is taken at ~90 te/day/m (2.7 te/day/ft) width of grate.
- The area of grate is set at a burning rate of 320 kg/m^2 h (~65 lb/ft^2 h).
- The grate HRR is ~0.85–1.10 MW/m^2, ~730,000–946,000 kcal/m^2 h or 270,000–350,000 Btu/ft^2 h. The rates may be lowered for more most and inferior fuels. Low HRR not only ensures a better burn out because of long residence time but also provides for a long grate life.

PGs as wide as 15 m are built. Figure 13.20 shows a typical mass-fired boiler plant employing a step pusher type PG along with the draught plant.

The main concern in MSW boilers is the severe corrosion and fouling of surfaces. The entire boiler design is directed to address this issue.

- It is normal to provide empty second pass as shown here to permit cooling of gases by radiation and shedding the ash from flue gases before they come in contact with SH panels. ~60% of total heat is absorbed in the furnace.
- Steaming conditions are preferably restricted to 60 bar/450°C for trouble-free operation from the boilers as the tube metal temperatures are low and corrosion issues are under control. Corrosion is less when metal temperatures are <480°C.

As Cogen is more efficient at higher p & t boilers of higher steaming conditions are also made. Corrosion protection has to be intensified and more expensive tube materials are required to be used. Superior WT with minimal internal scales is another necessity as the tubes are required to be cooled more effectively.

- Longer residence times both on grate and in furnace are essential to ensure that the oxidation of fuel is complete and the emission of corrosion causing CO is minimised.
- This will also minimise the carryover of fly ash which can initiate a secondary corrosion

reaction. The furnace velocities are kept low at 5.5–6 mps (18–20 fps).

- The complex chemicals in fly ash catalyse corrosion reactions involving Cl_2 and S present in the flue gases. Sulphation reaction escalates with temperature and lower metal temperatures help retard the reaction.
- Eco exit gas temperature is kept ~200°C (400 °F) for safety of ESP internals and breechings from corrosion.

13.30 MSW Boilers: RDF Burning

Enriched MSW is RDF. The higher the enrichment, the higher the cv and lower the fuel burden, making it easier to burn in a boiler. It is also more expensive to increase the enrichment. The amount of RDF produced for a given weight of MSW is termed as the *yield of RDF* which is generally held at ~70–80%. It is a more uniform fuel than MSW with regard to fuel sizing and cv resulting in a more efficient combustion.

13.30.1 RDF on Spreader Stokers

SS firing is the most common way to burn RDF even as there are installations in Europe that employ PGs. The general norms for firing RDF on SS are:

- Grate HRRs are <350,000 kW_{th}/m^2, 2×10^6 kcal/m^2h or 750,000 Btu/ft^2h which is ~65–70% of the limit for bio-fuels like wood. This is to prevent clinker formation on the grate and slagging on the furnace walls.
- Grate HRR is twice as high as in mass firing as RDF is better than MSW and also the combustion is partly in suspension and partly on grate.
- HRR per unit width of grate is <53 MW/m, 12.6×10^6 kcal/m h or 15.3×10^6 Btu/ft h.
- Fuel sizing being better and more uniform, the requirement of excess air is lower at 40–50%. Under grate PA is ~50–70% with balance as overfire SA.

Boiler design basics for RDF are as follows:

- RDF burning bears a lot of similarity to a bio-fuel like wood. SS-fired RDF boiler can be patterned after wood- or bagasse-fired SS boilers with differences in burning of RDF accounted for. RDF has high VM and M and naturally, a large furnace and ample SA are needed for satisfactory combustion.
- Compared to wood firing, a much larger furnace volume of ~40% is required here because of low permissible furnace HRRs of 125,000–155,000 kW/m^3, 107,000–134,000 kcal/m^3 h or 12,000–15,000 Btu/ft^3 h.
- However, there are important dissimilarities with RDF burning. The ash is highly abrasive as there are fine particles of glass and metal in addition to silica. This requires that the flue gas velocities are kept lower and suitable protection to tubes is provided against erosion in line with requirements of boilers firing MSW.
- Also, RDF has highly corrosive components in the form of Cl2 and S. All precautions needed in the MB of MSW are also required here. Low FEGT of <870°C (1600°F), protection for lower furnace walls, wider spacing of SH tube panels, better WT and so on are needed.
- Construction of lower furnace with composite tubes of Inconel outer covering on CS core seems to be preferred over refractory covered studded tubes. This is because nearly half the amount of heat release is in suspension causing higher gas temperature which can lead to slagging in furnace with refractory covering of tubes.
- Presence of low-melting compounds of Na, K and Pb cause high degree of fouling. SH spacing has to be sufficiently wide apart.
- Parallel flow of steam in SH and use of Inconel in high-temperature sections of SH are the usual solutions common for both MSW and RDF.
- Compared to MSW, the ash burden is less in fuel and consequently the dust carryover is also less. This permits the inclusion of AH as backend equipment to heat the combustion air.
- Final gas exit temperatures can be lower at ~175°C (350°F) against 200°C (400°F) with MSW firing.

A typical RDF-fired spreader stoker boiler is shown in Figure 13.21.

13.30.2 RDF in Suspension Firing

Burning of RDF in suspension along with coal is possible in burners arranged alongside PF burners. The issues involved can be summarised as follows:

- High amount of M in RDF does not permit burning it alone.

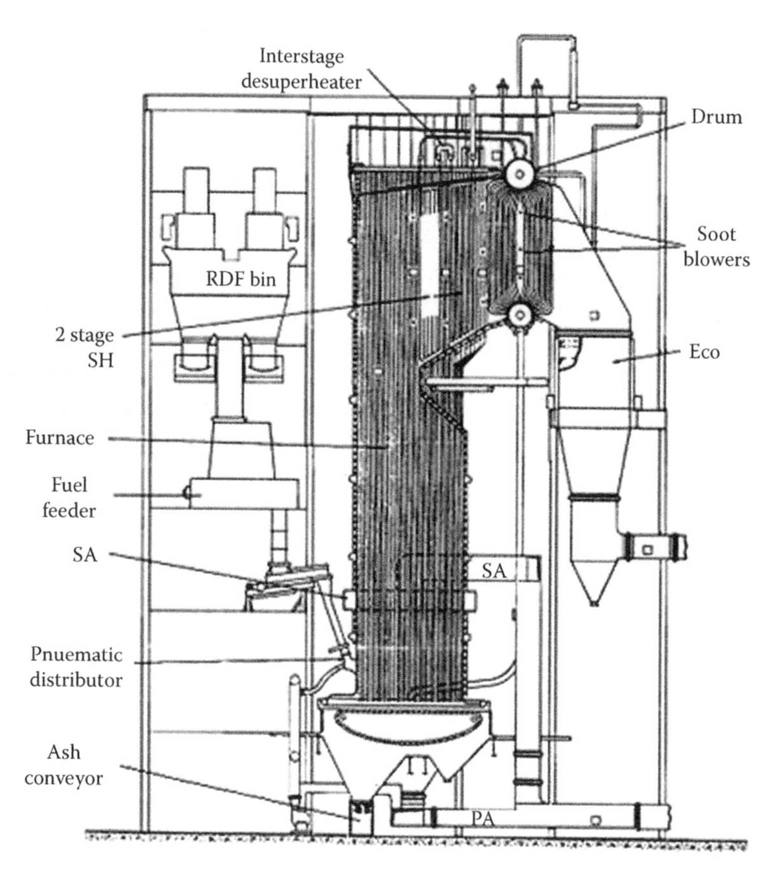

FIGURE 13.21
Cross section of a boiler with spreader stoker for firing RDF.

- The corroding elements in fuel do not make it safe to co-fire high amounts of RDF without seriously damaging the tubes. Flue gases from RDF are far more corrosive than most coals and FOs.
- 10–20% of heat input from RDF is found to be a good compromise.
- This limit is also set from the M content of the fuel which is constantly fluctuating. M in RDF is about three times that of coal for the same heat input, thereby increasing the flue gases. Not only does the large M reduce the η but also varies the SH and RH temperatures.

14

N

14.1 Naphtha (*see* Liquid Fuels)

14.2 National Fire Prevention Association Codes

These are American codes that deal with fire prevention in all walks of life.

ASME B&PV codes are basically concerned with the construction and operation aspects of the main boiler. They do not address the safety aspects of fuel and firing equipment, which is taken care by National Fire Prevention Association (NFPA) codes.

NFPA has more than 300 codes of which NFPA 85 namely, The Boiler and Combustion Systems Hazards along with five other referenced codes are relevant to boiler engineers and are listed in Table 14.1. This code numbered 85 is applicable to

- Single burner oil/gas-fired boilers
- Multiple burner oil/gas-fired boilers
- Stoker-fired boilers
- Atmospheric fluidised-bed boilers (AFBC) with a fuel input rating of 3.7 MWt (12.5 m Btu/h) or greater
- PF systems
- Fired or unfired steam HRSGs

14.3 Natural Circulation (*see* Circulation)

TABLE 14.1

NFPA Codes Relevant to Boilers

NFPA	Title	Edition
85	Boiler and Combustion Systems Hazards Code	
30	Flammable and Combustible Liquids Code	2003
31	Standard for the Installation of Oil-Burning Equipment	2001
54	National Fuel Gas Code	2002
69	Standard on Explosion Prevention Systems	2002
70	National Electrical Code®	2002

14.4 Natural Convection (*see* Convection in Heat Transfer)

14.5 Natural Gas (*see* Gaseous Fuels)

14.6 Net Calorific Value (*see* Calorific Value)

14.7 Newtonian Fluid (*see* Fluid in Fluid Characteristics)

14.8 Nitrogen Oxides in Flue Gas

NO_x is a generic term for mono-nitrogen oxides—NO and NO_2 (nitric oxide and nitrogen dioxide) and traces of other N_2-bearing gases produced during combustion. All fossil fuel combustion generates some level of NO_x due to high temperature and the presence of N_2 and O_2 both in fuel and air.

NO_x does not include nitrous oxide (N_2O), commonly known as *laughing gas*, which has many uses as an oxidiser, anaesthesia and food additive. However, it contributes to the destruction of stratospheric ozone.

This topic is divided into the following sub-topics:

1. NO_x generation
    1. Fuel NO_x
2. Thermal NO_x
3. Prompt NO_x
    2. Ill effects of NO_x
    3. NO_x reduction or denitrification
a. Pre-combustion modifications
    1. LNBs
    2. SOFA

3. Reburning
4. FGR
5. Operational modifications

b. Post-combustion processes
1. SNCR
2. Selective catalytic reduction (SCR)
3. Hybrid process

14.8.1 NO_x Generation

NO_x from combustion of coal typically contains ~95% NO, ~5% NO_2 and <1% of N_2O. With no control, NO_x levels can vary from 175 to ~3200 ppm based on N_2 in coal and the combustion intensity.

Once out of the stack, NO eventually oxidises to form NO_2 in the atmosphere. It is NO_2 which renders the plume brownish from stack.

NO_x in flue gas is the result of oxidation of otherwise inert N_2 under the influence of high temperature. Mainly two types of reactions promote NO_x formation with the third one playing a minor role.

1. *Fuel* NO_x production is due to the *N_2 in fuel*.
   - It is formed by pyrolysis and oxidation of organically bound N_2 compounds in fuels.
   - About 80% of total NO_x emission from coal firing and 50% from oil firing is fuel NO_x.
   - Conversion of fuel N_2 to fuel NO_x relation is complex and unclear. So, it cannot be generalised that fuels with higher N_2 will produce higher fuel NO_x all the time, though they generally tend to do so.
   - O_2 concentration plays an important role in fuel N_2 conversion into NO_x.

   Fuel NO_x generation is most effectively controlled by the reduction of O_2 in the combustion reaction.
2. *Thermal NO_x* production is due to the *N_2 in air.*
   - It is formed by the attack of O_2 atom on N_2 in combustion air.
   - Only about 20% of total NO_x emission from coal firing is thermal NO_x. However, this is dominant when fuel contains less or no N_2 as in NG.
   - It is mainly affected by flame temperature and concentration of O_2 in combustion air with the former being more important.

   Thermal NO_x generation is minimised by reducing the combustion temperature. There are three methods, namely
   a. Reducing the peak and average flame temperatures.
   b. Staged combustion, which is a progressive admission of air into the combustion zone, thereby protracting fuel–air reaction to prevent flame attaining high temperatures.
   c. FGR from low-temperature zone like Eco or AH exit which increases the flue gas quantity and lowers the furnace gas temperature.
3. *Prompt* NO_x is formed by the reaction of HC radicals with N_2 in air. First, HCN (hydrogen cyanide) is produced and then NO_x by a series of complex reactions.

No additional measures are needed to control prompt NO_x as the actions taken for fuel and thermal NO_x are adequate. The contribution of prompt NO_x is also small. It is <5% of the total NO_x in case of PF firing.

NO_x is produced from the reaction of N_2 and O_2 during combustion. NO_x production starts at temperatures of ~1200°C (2400°F) and escalates sharply at high temperatures >1540°C (2800°F), as shown in Figure 14.1. NO_x production is the highest at fuel-to-air combustion ratios of 5–7% O_2 (25–45% excess air). Lower excess air levels starve the reaction of O_2 while higher excess air levels cool down the flame temperature, slowing the rate of reaction.

It is recognised that NO_x generation is a very complex process and is dependent on localised temperatures, stoichiometry and chemical reactions, principally in the near-burner region. Currently, there are no simple means to accurately predict NO_x when only limited plant operating or design data is available. Boiler manufacturers continue to use in-house empirical correlations, not CFD models, to predict NO_x formation.

14.8.2 Ill Effects of NO_x (see *also* Acid Rain)

Of all the stack cleanup efforts, NO_x reduction has been the area of most concern for the past nearly four decades. Thermal NO_x is the main offender.

- NO_x combines with other pollutants in the atmosphere and creates ground-level O_3. When NO_x and VOCs enter the atmosphere, they react in the presence of sunlight to form ground-level O_3, which is a major ingredient of smog. The present limit for O_3 in most parts of the world is 0.08 ppm (8 h average). Automobile emission is the main cause while fossil-fired power plants are also substantial offenders.
- NO_x eventually forms HNO_3 when dissolved in atmospheric M, forming a component of acid rain. It has harmful effects on humans, plants,

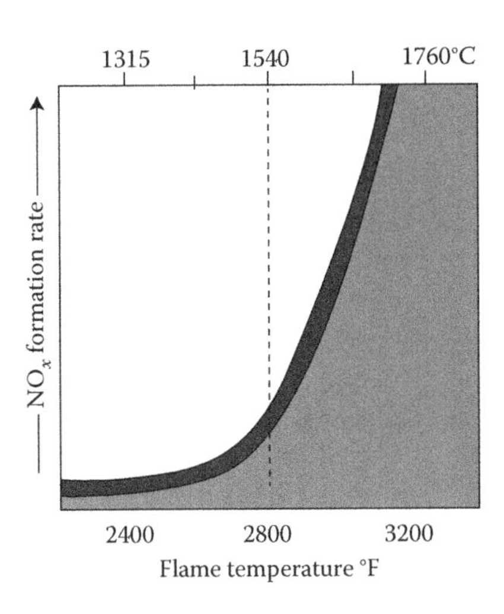

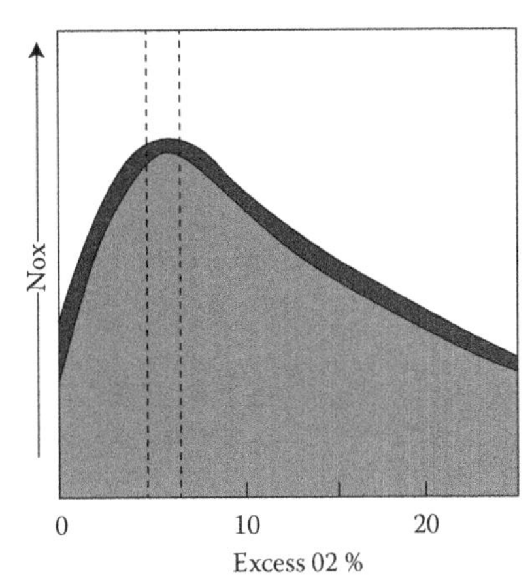

FIGURE 14.1
Relationship of NO_x generation with flame temperature and excess air.

aquatic animals and infrastructure through the process of wet deposition.

$$NO_2 + OH = HNO_3 \tag{14.1}$$

SO_2 contributes significantly more to acid rain formation than NO_x which contributes to about a third.

14.8.3 NO_x Reduction or Denitrification

NO_x in boiler burners can be reduced with either

a. Pre-combustion controls or
b. Post-combustion controls

Pre-combustion methods minimise NO_x formation in the first place. It is accomplished by either

a. Air staging or
b. FGR

Post-combustion control allows NO_x to form and then breaks it down in the exhaust gases (a process called *catalytic reduction*). This method is normally confined to larger utility-size equipment.

14.8.3.a Pre-Combustion Methods

1. *Low-NO_x burners (LNBs)* (*see also* Low-NO_x Burners)

   LNBs are modified axial flow burners designed to mix fuel and air stagewise so as to achieve protracted combustion, as shown in the schematic of a low-NO_x gas burner in Figure 14.2. This results in a lower peak flame temperature and reduced O_2 concentration during

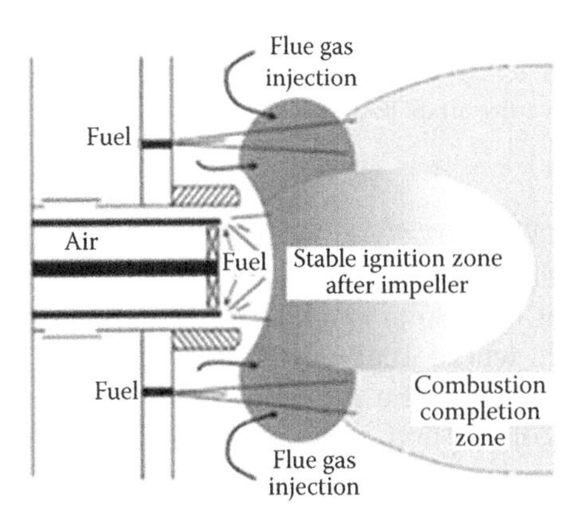

FIGURE 14.2
Typical LNB for gas firing with FGR.

   some phases of combustion, thus resulting in both lower thermal and fuel NO_x production. FGR also helps to lower NO_x as the cool gases depress the flame temperature without altering combustion air quantity.

2. *Separated overfire air (SOFA)* (*see also* Separated Overfire Air)

   SOFA, as the name suggests, is the diversion of a part of combustion air and injection in the upper furnace above the normal combustion zone. The main burners are consequently operated at air–fuel ratios less than normal, which promotes less NO_x formation. SOFA, which is added from above the main combustion zone (as shown in Figure 14.3), makes up for the starvation of air from below and helps to complete the combustion. SOFA is frequently used in conjunction with LNBs such that there is NO_x reduction beyond what is achieved by LNBs. Figure 14.3 also shows reburn fuel arrangement.

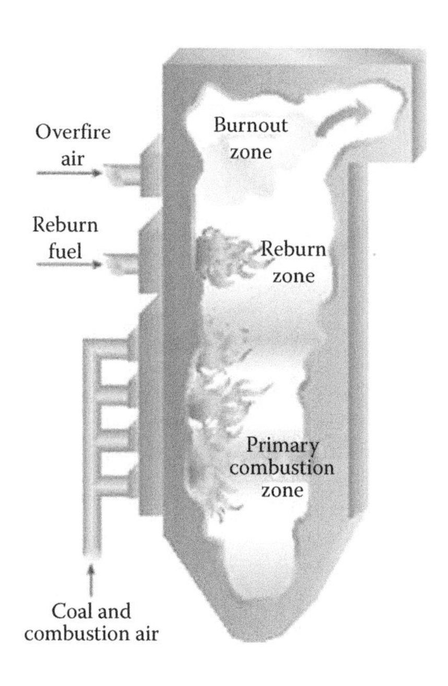

FIGURE 14.3
Separated overfire air (SOFA).

3. *Reburning* (*see also* Reburning)

   In reburning process a part of the boiler heat input (~10–30%) is added in a separate reburn zone, where fuel-rich conditions are created that lead to the reduction of NO_x formed in the combustion zone below. SOFA is injected above the reburn zone to complete combustion. Thus, with reburn, there are three zones in the furnace:

   1. Combustion zone with air fuel ratios at slightly below normal.
   2. Reburn zone, where the added fuel results in a fuel-rich reducing condition.
   3. Burnout zone, where OFA leads to completion of combustion.

   Coal, oil and gas can all be used as the reburn fuel.

4. *Flue gas recirculation (FGR) (see also* Gas Recirculation and Gas Tempering)

   FGR can be used to modify conditions in the combustion zone (lowering the temperature and reducing the O_2 concentration) to reduce NO_x formation. Another use of FGR is to act as a carrier to inject fuel into the reburn zone to increase penetration and mixing.

5. Operational modifications.

Certain boiler operational parameters can also be altered to create conditions in the furnace that lower NO_x production, such as

a. Burners-out-of-service (BOOS): In BOOS, selected burners are removed from service by stopping fuel flow, but air flow is maintained to create staged combustion effect in the furnace.

b. Low excess air (LEA): LEA involves operating at the lowest possible excess air level while maintaining good combustion.

c. Biased firing (BF): BF involves injecting more fuel to some burners (typically the lower burners) while reducing fuel to other burners (typically the upper burners) to create staged combustion conditions in the furnace.

14.8.3.b Post-Combustion Processes

1. *Selective non-catalytic reduction (SNCR) (see also* Selective Non-Catalytic Reduction)

   In SNCR, a reducing agent (NH_3 or urea) is injected into the upper furnace, as shown in Figure 14.4, where it reacts with NO_x to form N and H_2O, thus reducing NO_x emissions. For the success of SNCR there should be

   - Adequate residence time in the appropriate temperature range
   - Equal distribution and mixing of the reducing agent across the full furnace cross section

2. *Selective catalytic reduction (SCR) (see also* Selective Catalytic Reduction)

   In SCR, a catalyst vessel is installed downstream of the furnace (as shown in Figure 14.5). NH_3 is injected into the flue gas before it passes over the fixed-bed catalyst. The catalyst promotes a reaction between NO_x and NH_3 to form N_2 and H_2O. NO_x reductions as high as 90% are achievable, but careful design and operation are

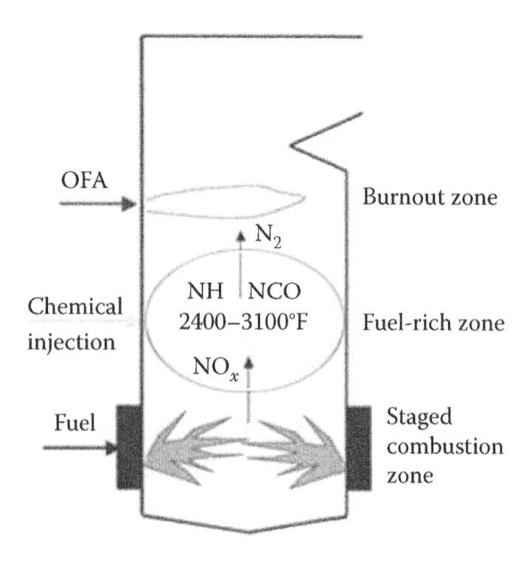

FIGURE 14.4
Injection of NH_3 above the burner zone for SNCR.

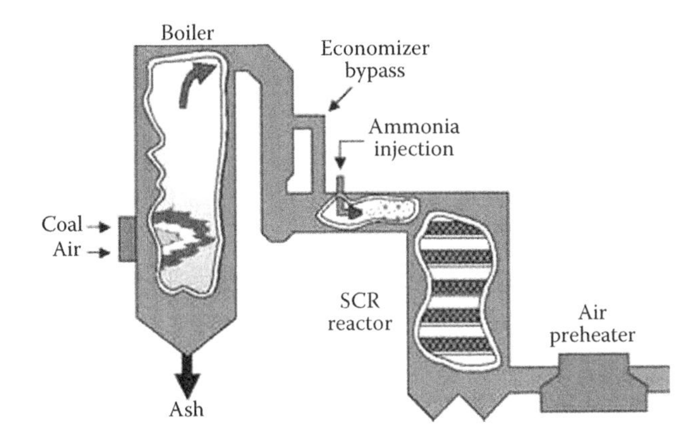

FIGURE 14.5
SCR reactor placed ahead of AH.

necessary to keep NH_3 emissions (referred to as NH_3 slip) to a concentration of only a few ppm.

3. Hybrid process

   SNCR and SCR can be used together with some synergistic benefits. Also, both processes can be used in conjunction with LNBs.

14.9 Noise Control

Noise is Unwanted Sound

Intensity ('loudness') of noise is measured in decibels (dB). As the decibel scale is logarithmic, a 3 dB rise in the scale represents a doubling of the noise intensity. While normal conversation may be ~65 dB and someone shouting can be ~80 dB. The difference is only 15 dB but the shouting is 30 times as intensive.

To take into account the fact that the human ear has different sensitivities to different frequencies, the strength or intensity of noise is usually measured in A-weighted decibels—dB(A).

It is both the intensity and duration of exposure that determine if noise is hazardous. To take this into account, time-weighted average sound levels are used. For workplace noise, this is usually based on an 8-h working day. Permissible noise exposure limits as per OSHA are given in Table 14.2.

Noise control is done by a combination of employing low-noise equipment and sound-attenuating devices. The main sound-generating equipments in a boiler plant are the rotating equipment and valves. Noise attenuation is needed despite installing low-noise machines. Two main devices for noise reduction are acoustic enclosures for equipments and silencers for equipment and vents of the valves.

TABLE 14.2
Noise Exposure Limits Permitted as per OSHA

Duration: Hours per Day	Sound Level dB(A) Slow Response
8	90
6	92
4	95
3	97
2	100
1.5	102
1	105
0.5	110
0.25 or less	115

Acoustic Enclosures

These are shrouds for the entire equipment with suitable insulation material attached to the inside of the housing for absorption of sound. Fans, blowers, compressors, BFPs and mills are the equipments which may be provided with acoustic enclosures in a boiler plant.

Silencers

These are provided for fans, blowers and vents of various valves such as SVs, start-up vent valve, blowdown valves and so on. The vent silencers are made out of concentric annular acoustic cylinders enclosed with a robust steel casing with dished end or flat end at one end. A typical vent silencer is shown lying flat in the manufacturing shop in Figure 14.6.

Silencers for fans and blowers can be both at suction and discharge sides either in circular or rectangular construction. The internals of a silencer comprise of a series of acoustic splitter elements arranged in parallel inside the steel casing to ensure

FIGURE 14.6
Safety valve vent silencer in manufacturer's shop.

FIGURE 14.7
Fan inlet silencer in manufacturer's shop.

the flow of air/gases through the rectangular air gap between acoustic splitters for sound absorption. Such construction also offers reduced pressure drop. The leading and trailing edge of the acoustic splitters will be designed aerodynamically to offer minimum resistance to the flow medium. A typical fan inlet silencer is shown in Figure 14.7 after its manufacture in the shop floor.

14.10 Non-Destructive Examination/Testing (*see* Testing of Materials)

14.11 Non-Luminous Radiation (*see* Radiation in Heat Transfer)

14.12 Normal Continuous Rating (*see* Boiler Ratings)

14.13 Nozzle

A nozzle is a short pipe for controlling the direction or characteristics of a fluid as it exits or enters an enclosed vessel or pipe. A nozzle can have constant or varying cross-sectional area. It is used to regulate the rate of flow, speed, direction, mass, shape and/or pressure of the stream that emerges from it.

14.14 Nucleate Boiling (*see* Boiling in Properties of Steam and Water)

14.15 Nusselt Number (*see* Dimensionless Numbers in Heat Transfer)

15

O

15.1 Operation and Maintenance Topics

The following topics are covered:

1. Banking of boilers
2. Burner management systems (BMSs)
3. Explosion and implosion
4. Filling of boilers with water
5. Foaming and priming
6. Furnace cleaning
7. Purging of furnace and boiler setting
8. Layup/preservation/storage of boilers

15.1.1 Banking of Boilers

Banking refers to a condition of readiness of a boiler to deliver steam almost immediately on demand.

A cold boiler requires considerable amount of time with controlled heating to reach its full pressure depending on the size and construction. The amount of refractory in a boiler decides as to how fast it can be heated or cooled. At the time of shutting down, if a boiler is boxed up at full p & t and it is cooling naturally by itself, the boiler can return to service usually in <2 h even after a day. When there is a process interruption or ST trip the boiler is boxed up and periodically fired to raise the pressure, so that it can be loaded immediately when the fault is set right. This *free floating hot standby condition is called the banking of boiler.*

The boilers that always keep operating in a battery are referred to as lead boilers, while the others are called lag boilers. The most efficient boiler is always started first and stopped last while the least efficient always started last and stopped first to conserve fuel.

Boiler banking keeps boilers in hot standby mode by intermittently firing unused boilers, thus maintaining the required pressure. Boiler banking acts as a warm-start facility, improving the plant's response to sudden load changes.

15.1.2 Burner Management Systems or Burner Management and Safety Systems or Flame Supervisor Safety System

Boilers equipped with **burner** systems, be they **FO**, **gas**, waste gas or **PF**, all are prone to explosions, particularly during the unsteady parameters that prevail at start-up and low-load conditions. To avert all adverse situations, which can lead to unstable firing that can cause explosions, the following precautions are necessary:

- Careful sequencing of operations
- Constant monitoring of parameters
- Continuous observation of flames

BMS or BMSS, also called FSSS, is the safety logic of the boiler, in which individual burner management is done under the supervision of central logic which ensures the safety of the entire boiler. The primary aims of safety logic are safe

- Start-up and shut-down of the burners
- Operation of all the burners in the boiler under all loads

The BMS is designed to perform the following functions:

1. Prevent firing unless a satisfactory **furnace purge** has first been completed
2. Prohibit start-up of the equipment unless certain permissive interlocks have first been completed
3. Monitor and control correct component sequencing during start-up and shut-down of the equipment
4. Conditionally allow continued operation of the equipment only while certain safety interlocks remain satisfied
5. Provide component condition feedback to the operator, plant control systems and/or data loggers
6. Provide automatic supervision when the equipment is in service and provide means to actuate a Master Fuel Trip (MFT) should certain unacceptable firing conditions occur
7. Execute an MFT upon certain adverse unit operating conditions

BMS at the burner level may be considered as the brain of the burner, consisting of indication, controls and perhaps some software, all of which reside in the local burner panel. The central control panel exercises overall control with local panels supporting it.

The complexity of the logic is directly proportional to the variables of the system, such as the

- Number of burners
- Variety of fuels
- Combination of firing equipment
- Operational flexibility desired
- Local or remote mounted BMS panel
- Extent of manual intervention

NFPA, who have done elaborate studies, have provided detailed guidelines and elaborate procedures to be followed for single and multiple burner installations covering all the above conditions. The guidelines refer to

a. Oil or gas burning
b. Oil and gas burning
c. Multiple fuel combinations
d. Single or multiple burners

As the variables increase, the logic gets more advanced. Reduced to the most essentials, the working of BMS can be described in the following manner:

Boiler start: For boiler starting sequence, the permissives to be satisfied and sequence to be followed are broadly as indicated below:

- A burner can be lighted only with its own igniter.
- The **igniter** cannot be started unless the furnace purging is complete.
- The purging can be initiated only on a satisfactory report on the five main boiler conditions namely:
  1. Water level in the drum ok
  2. FO pressure and FO temperature ok
  3. Atomising steam pressure and temperature (p&t) ok
  4. FD fan ok
  5. Igniter conditions ok

Purge: On finding these conditions proper, it is now possible to start the FD fan. The first step is to carry out a purge of the boiler setting as elaborated earlier.

Light up: Fuel can be turned on only after purge is completed and when the igniter can provide a spark/flame. Within a set time the main flame should be sighted or else the whole sequence has to be repeated.

Monitor: In an operating boiler, all the above conditions and flame health are monitored at all times. Any variation is alerted. Any deviation of main parameters beyond the limits results in either the burner or the boiler tripping depending on the cause.

Boiler master fuel trip: BMS should initiate the boiler trip if any of the following conditions should arise. MFT results in the shutdown of all fuel and requires another furnace purge cycle before any attempt at re-lighting:

1. Excessive steam pressure or water temperature
2. Low water level (LWL)
3. High or low fuel pressure
4. Low FO temperature (for FO)
5. Loss of combustion air supply
6. Loss of flame
7. Loss of control system power
8. Loss of atomising medium (for FO if used)

Interface with the combustion control system (CCS): The following list of signals, as a minimum, should be sent to the CCS:

1. Controls to purge position
2. Controls to light-off position
3. Normal run condition: release controls to modulation
4. Main NG block valve open: permissive to place gas CV in automatic.
5. Master fuel trip: run boiler load to zero and place combustion controls in manual.
6. FO recirculation signal

Alarms: The following is a list of recommended alarm conditions:

1. Any boiler or burner trip signal
2. High or low water level
3. High furnace pressure
4. Partial loss of flame (For the typical two scanner system, one indicates 'no flame')
5. Main fuel shutoff valves closed
6. Loss of control system power
7. Unsuccessful burner shutdown

This, very briefly, sums up the BMS. A sound BMS, needs for dependable operation, a

- Reliable flame monitor
- Good logic system
- Robust hardware

BMSs can be fairly simple to very elaborate for the same boiler depending on the customer's operating practices and the sophistication of the O&M staff.

15.1.3 Explosion and Implosion

a. Explosion

Explosion is violent release of energy with large increase in volume and pressure of the substance invariably accompanied by loud sound.

In boilers explosions can be of three types:

1. PP
2. Furnace setting
3. Chemical reaction namely smelt-water reaction

1. PP explosion

This is the result of over pressure built-up in the inside of tubes, headers and drums because of sudden and excessive steam formation. This is often because of run-away combustion conditions in furnace.

In **water tube boilers**, PP explosion these days is rare as the internal water volume is inherently low and combustion controls are reliable.

However, such explosions do occur, off and on, with **fire tube boilers** which have large shells containing lot of water and steam.

2. Explosions in furnace and boiler setting

These are likely to take place mostly during lighting up and low-load operation of boilers whenever

a. There are abrupt changes in the fuel-air ratios
b. Combustible gases are trapped which catch fire resumption of supply of air or
c. The accumulated pockets of fuel burn on sudden agitation

Burner-fired boilers are more susceptible to explosions as the fuel–air contact is more intimate than in **grate** or **FBC boilers**.

Purging of furnace before light up is the most effective way to prevent explosions. This is now compulsorily done with all burner-fired boilers and also FBC boilers in many cases, where light up is always preceded by blowing the furnace with fresh air to remove the unburnt gases or fuel accumulations. Furnace purging usually lasts about 20 min when furnace setting is rinsed by four to five volume changes. NFPA has laid down the procedures which are to be compulsorily built into the boiler start-up logics.

3. Explosion due to smelt-water reaction (see also Recovery Boilers)

Another type of explosion specific to **BLR boilers** is the result of **smelt-water reaction**. When water comes in contact with hot and molten smelt in the hearth, usually because of tube leakages, violent explosions have occurred leading to extensive damage to the boiler and surroundings. There are rules laid out by BLRBAC on how to deal with tube leakages to avert these disasters. A very detailed explanation is given under BLR boilers.

b. Implosion

Implosion is the opposite of explosion. Instead of positive pressure, a large negative pressure builds up in the boiler under certain conditions of maloperation leading to the inward collapse of draught plant and furnace in particular. In the large high-efficiency modern balanced-draught utility boilers, the suction heads of ID fans are very high because of large back end (Ecos + AHs) and gas cleaning equipments (ESP/bag filter + SCR equipment + scrubbers). The design draught of ID fan can exceed the structural design pressure of furnace and ducts.

If there is a sudden

- Loss of air from FD fans
- Flame by the burners or
- Control on ID fan inlet vanes

there can be a runaway increase in the draught causing an inward collapse. The resultant damage can be considerable needing a few months to rebuild.

Recognising this aspect the furnaces and flues and ducts in boilers are strengthened with sufficient stiffening. The buckstays and stiffeners are designed for both positive and negative pressure excursions matching the heads of ID fans. The furnace implosion protection system has to comply with the guidelines established by NFPA 85G. These guidelines include

- Redundant furnace pressure transmitters and transmitter monitoring system
- Fan limits or run-backs on large furnace draft error
- Feed-forward action initiated by a main fuel trip
- Operating speed requirements for final control elements, and interlock systems

15.1.4 Filling of Boilers with Water

The act of topping the water up to the normal water level (NWL) of steam drum or more initially in a near-empty boiler is called filling. Every time the boiler is drained, it is required to be filled up. In small boilers, water filling is done directly into the steam drum through the fw line using the BFP. Δp across the feed water regulator (FWR) station is very large, particularly for HP units, which can lead to wire drawing effect on the CV or its bypass, depending on which valve is pressed into service. A small fill line (25 or 38 nb), parallel to the fw line with a separate CV, is a good solution often provided in many industrial boilers. No separate fill pump is needed.

For large boilers, independent high-capacity LP fill pumps are provided which can fill the boilers quickly without the problems of damage to main CVs. These are placed on the ground level with the discharge connected to the bottom ring main with a flexible connection. The additional advantage with this arrangement of filling is the positive displacement of air as the water level rises from below.

15.1.5 Foaming and Priming

Foaming and priming in boilers are two very different phenomena both causing **carryover**.

1. *Foaming is due to saponification* (soap forming) of bw which results in a steady carryover while
2. Priming is due to the sudden rise in drum water level causing slugs of water to enter steam space

Foaming: When more steam bubbles reach the water surface than bubbles bursting, foam accumulates. This foaming in steam drum is due to increased saponification agents in water, such as

- Total and suspended solids
- Alkalinity
- Oil
- Phosphates

Erratic fw controls also cause foaming. Steady high carryover is the characteristic of foaming.

Priming: **Priming** is a result of the increased drum level and not due to sapnofication of water. The main reasons for priming are

- Increased steaming
- Sudden drop in drum pressure
- Erratic fw control

While foaming is substantially a water-related problem, priming is an operational issue.

15.1.6 Furnace Cleaning (*see also* SBs)

Furnace is a large enclosure lined with smooth walls. As such, there is no need for cleaning a furnace unless ash in fuel melts due to high combustion temperature and re-crystallises on the tubes in cooler zones of the furnace, forming thick deposits:

- It naturally follows that fuels with little ash, such as NG and light FOs, do not have furnace deposits and consequently need no cleaning.
- Likewise, FBC boilers, with the combustor temperatures limited to 850–900°C, also have clean furnaces. In fact, the CFBC boilers, with continuous downward ash flow along the tube walls, keep the tubes in a well polished condition.
- It is the high-ash fuels like coal and lignite or fuels with low-melting ash such as certain biofuels that foul the furnace. They need *on-line cleaning* by having furnace wall deslaggers or short refractory **SBs**.

Furnace cleaning can be either *on-line* or *off-line*. On-line cleaning on hot, running boiler once or twice a shift is by steam or compressed air with SBs while off-line cleaning on a cold, non-operating boiler every few months is with HP water.

- Off-line cleaning is strictly necessary only for furnaces which develop thick deposits
- Every furnace must have a provision for draining water after off-line water washing
- Most of the ash deposits are easily water soluble but it is unusual to encounter sticky deposits too

For highly slagging furnaces soot blowing by steam may not be adequate to dislodge the thick ash deposits. **Water lancing** may have to be adopted where a jet of HP water cools the deposit and makes it friable and then dislodges it from the wall. Water lancers are similar to wall deslaggers.

15.1.7 Purging of Furnace and Boiler Setting (*see also* Explosions)

Burner-fired boilers are susceptible to explosions. Most explosions occur during lighting up and low-load operation of boilers.

Before any fuel firing is permitted in burner-fired boilers, either initially or after a boiler trip, a satisfactory furnace purge cycle must be completed to expel combustible gases and unburnt fuel from furnace and boiler setting to prevent explosion. Experience has shown that in the furnace:

- Four to five volume changes are needed at a steady rate of about 30% of full load air volume for a period of minimum 20 min

- In the case of single burner units, eight volume changes with 70% air flow are required

Furnace purge cycles are integral to the boiler start-up or re-start procedures and it is well laid out by NFPA.

Prior to starting a furnace purge cycle in a single-burner boiler installation firing FO and gas, the operator must ensure that the following purge requirements are satisfied. For other configurations, including multiple burners and PF burners, similar procedures apply with suitable modifications:

1. Drum level is within operating range (neither high nor low)
2. Instrument air header pressure is within operating range
3. At least one fan set is in service
4. Purge airflow capable of a minimum of 70% of the full load airflow established through the unit
5. All flame scanners are reading 'No Flame'
6. NG and FO block valves are proven closed
7. Air dampers are in the fully open position
8. NG, FO and pilot gas header pressure upstream of block valves are satisfactory
9. Burner control system is energised
10. A 'No MFT condition' is established

As each of the above conditions is reached, their respective indicator lights illuminate as 'Purge Permissives' on the operator control console. Upon completing a satisfactory purge cycle the operator presses the 'Reset MFT' pushbutton. The boiler control system will then indicate that the MFT has been reset with a steady 'MFT Reset' light on. At this point, the boiler control system is ready to allow the main flame start-up sequence.

15.1.8 Layup/Preservation/Storage of Boilers

At some point in time it may be necessary to take a boiler off line and store it for an extended period of time. Rapid generalised corrosion and extensive pitting take place in water-steam circuitry in boilers that sit idle. When they resume operation, the corrosion products move to areas of high heat input and also into the steam turbine. The operational efficiency and overall health of both boiler and turbine are thus affected by the corrosion products. Prudent storage methods are essential to address the corrosion matter.

There are two ways to store a boiler:

- *Wet storage for short duration* and when the boiler is to be returned to service at a short notice
- *Dry storage for long duration* and when there is ample time for restarting

Wet storage is more popular than dry storage. It is typical for a boiler that needs a short or long repair. Wet storage permits a unit to be returned to service more quickly. The only precaution is that the water should not go below freezing point anytime during the storage. Wet storage with N_2 blanketing is often the most practical method and it offers an excellent protection as O_2 is eliminated from the environment. N_2 leakage is the only point to guard against.

Dry storage is for long-term storing, typically for a standby, a boiler which would not be returned to service for some months. This method also offers excellent protection against corrosion. Environmental concerns are also minimal here.

15.1.8.a Wet Storage Procedure Is as Follows

- As the boiler is being shut down and pressure subsides, but before the stopping of steaming, chemicals are to be added to the boiler to scavenge O_2 and to control pH as per the recommendations of the owner's WT consultant.
- Typically, this consists of treating water with ~500 ppm of N_2H_4 for O_2 removal and sufficient NH_3 to attain pH of 10.
- The unit is to be completely filled with treated water including SHs either through the drum overflow or by the SH drains. This is followed by closing of all vents.
- The MSSV should already be closed. N_2 should be added through the drum vents until the pressure is built up to ~0.35 to 0.7 barg (5–10 psig).
- All fittings and valves are to be checked for leakage.
- Frequent water samples should be taken and analysed by the WT consultant. If the analysis indicates a need for additional chemicals, the level in the boiler steam drum should be lowered to normal level and chemicals added. The boiler should then be steamed to circulate the solution, and the process of wet storage repeated as previously described.
- All gas side access doors are closed to prevent cold air from reaching the HSs by isolating the system. Periodic inspections of the external surfaces of the PPs are to be made to guard against condensation and subsequent corrosion.
- If the storage continues into winter with ambient temperature falling below the freezing point

15.1.8.b Dry Storage Procedure Is as Follows

- The boiler is first fired and pressure raised to ~3.5 barg (50 psig). The firing is stopped and the pressure is allowed to decay to about 1.5 barg (20 psig).
- The unit is completely drained with air vents fully open making sure that no pockets of water remain in the drum, piping, water column and so on. All vents are opened to assure that air circulates on hot surfaces and helps to dry the M. Any M left on boiler surfaces will eventually corrode
- Residual particulate from the gas side metal surfaces are mechanically cleaned and the systems thoroughly inspected.
- The system is dried thoroughly.
- Flat wooden trays of M absorbent, such as quicklime or silica gel, are placed inside the drums to absorb any M that will be trapped when the unit is closed up. The trays are placed on supports to allow air to circulate under them. For recommendations on quantity of M absorbent, the owner's water consultant should be contacted but usually 1 kg/te of water is adequate. The trays should not be more than 3/4 full of the dry absorbent to prevent overflow of the corrosive liquid that has been absorbed.
- After the entire system is dried the boiler should be pressurised with N_2 to approximately 0.35 barg or 5 psig though the drum vent. The steam outlet, drain valves and fw block valves are closed.
- To ensure that M cannot enter the boiler steam lines, fw lines and any points of entry for air are to be closed.
- Then, all man ways and hand holes are closed and the boiler is isolated to prevent M from reaching the HSs.
- A sign board should be placed on the boiler prominently stating that M absorbent material is inside the boiler so that nobody makes the mistake of firing it without removing the material. Moreover, no one should enter the drum without full exhaustion of N_2 as it can lead to death due to asphyxiation.
- Finally, the boiler should be inspected every 2 or 3 months and the absorbent materials replaced with new or regenerated materials.

Simple draining without any preservation is also done when maintenance and inspection needs to be carried out and full access is desired. Usually such stoppage is limited to ~3 weeks and offers no protection. The costs are minimal and concerns on safety and environment are also negligible.

15.2 Odourisation of Natural Gas (*see* Natural Gas in Gaseous Fuels)

15.3 Oil Ash

Ash in FO is quite low, rarely exceeding 0.2% by weight. Distillates are free of ash. It is the residuals where the ash in the original crude consisting mainly of metallic compounds and S which gets concentrated. It is not comparable to the ash in solid fuels but in terms of problems, it is nearly of the same importance.

Vanadium (V)
V is a metal that is present in all crude oils in an oil-soluble form. The levels found in residual FOs depend mainly on the crude oil source, with those from Venezuela and Mexico having the highest levels. The actual level is also related to the concentrating effect of the refinery processes used in the production of the residual FO. Majority of residual FOs have V levels of less than 150 ppm by weight. However, some FOs have a level >400 ppm. There is no economically viable process for removing V from either the crude oil or residue economically viable.

V is burnt to one or more of its oxide forms depending largely on the amount of excess air. V_2O_5 the highest oxidation state, is a relatively low-melting-point form which is adsorbed on refractories causing spalling. It also builds up on SH tubes causing **corrosion** and forming rockhard deposits that are extremely difficult to remove. Fluxing of the deposit will occur as Na gradually reacts with it to convert it to sodium vanadyl vanadate ($Na_2O\ V_2O_4$–$5V_2O_5$), which results in a high corrosion rate on the SH tubes. Another corrosion causing daughter product of V_2O_5 is orthovanadate ($3NiO \cdot V_2O_5$). Deposits in the SH area can build-up to a point where they bridge the screen and SH tubes and substantially cut down on boiler draught. If excess air can be kept to a minimum through improved atomisation V oxidation can be kept to the V_2O_3 and V_2O_4 forms with melting points of ~1970°C. They will not fuse and adhere to boiler metal even in the combustion zone.

Sodium (Na)
In general, crude oil as delivered contains a small amount of Na, and typically <50 ppm by weight. The presence of sea water increases this value by ~100 ppm for each % of sea water. If not removed in the fuel treatment process, it will build up to a high level of Na that will give

rise to post-combustion deposits which can be normally removed by water washing.

Na in the fuel reacts with SO_3 and the V oxides to form relatively low-melting-point salts such as Na vanadyl vanadate ($Na_2O\ V_2O_4–5V_2O_5$) and so on. These can cause corrosion in SH areas and at the cold end (sodium acid sulfates). Na can be removed from the FO to a large extent by water washing with ~5% of water followed by centrifuging to remove the M. The cost of such a procedure makes it uneconomical except for marine installations, although it is common practice in GT operations using high-Na–V fuel.

Sulphur (S)

Concentration of S in residuals varies from 0.3% to 0.5% in hydro-desulphurised blends; ~2% to 3.5% for Venezuelan, Persian Gulf, Texas and other fuels.

- S burns to SO_2 and SO_3. The amount of SO_2 formed is ~98% of the total S present.
- Although SO_2 is not generally detrimental to the boiler and passes harmlessly up the stack, it is an undesirable air pollutant.
- The remaining 2% of the S is converted to SO_3 depending on the amount of excess air used in combustion and the catalytic effect of coke formed at the burner and throughout the unit.
- The reaction of SO_3 with Na results in the formation of acid sulphate deposits.
- SO_3 raises the acid dew point of the flue gas and if the metal temperatures of Eco, AH or last pass tubes are below the dew point temperature condensation occurs with the formation of H_2SO_4 and/or acid sulphate salts which corrode the metal surfaces.
- Low excess O_2 (<2%) helps to minimize SO_3 formation and thereby keeps the dew point at a more normal level.

To sum up, the metallic constituents of ash, namely, Na and V, form low-melting compounds. The metals vary from traces to as much as 30% of ash by weight. S, the other troublesome element varies from 0.5% to 5.0%. Together with the metal ions, low-melting compounds ~250–680°C (~480–1255°F) are formed, coinciding with the tube metal temperature range.

In boilers they cause

1. **Slagging** and **corrosion** of SH and RH
2. **Low-temperature corrosion** of Eco and AH

Experience has shown that

a. **High-temperature corrosion** is avoided if the tube metal temperatures are kept under 600°C (~1110°F).

b. Low-temperature corrosion can be avoided if the Eco and AH tube metal temperatures are maintained well above the dew point corrosion range, which is dependent on the S content of the fuel.

The other precautions are

- Treatment of FO to remove V, Na and S
- Use of additives
- Use of corrosion resistant alloys and coatings
- Low excess air operation

Acid Smuts

When the flue gas temperatures fall below dew point, H_2SO_4 acid condenses on the ash particles which, then agglomerate, to form highly corrosive acid smuts. Acid smuts destroy public property by falling in the vicinity of the plant and corroding buildings, cars and so on. The only way to prevent this is to keep the flue gas temperatures above the dew point at all loads of operation.

15.4 Oil and Gas Burners (*see* 1. Oil and Gas Burners, 2. Pulverised Fuel Burners, 3. Lignite and Brown Coal Firing and 4. Low NO_x Burners)

A burner is a device in which fuel and air are brought together appropriately so that combustion ensues upon ***ignition****.* A burner is attached to the windbox, which in turn is bolted to the furnace. A burner facilitates the mixing of air and fuel in an optimum and safe fashion to enable combustion to actually take place inside the furnace chamber.

This section deals mainly with oil and gas burners, including duct burners and is covered under the following sub-topics:

1. Burner components
2. Burner types by construction
    - Circular burners
    - Tangential burners
    - Scroll burners
3. Burner sizes for circular burner
4. Burner performance
    - Burner turndown
    - Hot air
    - Excess air in burners
5. Duct burners

    In-line duct burners

    Side burners

15.4.1 Burner Components

Any boiler burner has the following four sub-assemblies:

1. *Air register* for dividing and supplying air with a proper swirl for optimum turbulence. A burner derives its name from the type of register.
2. *Fuel pipe* of one or the other design for bringing fuel to the combustion space in a suitable manner. For
   - FOs, it is the *atomiser* which divides the FO into fine spray.
   - Gases, it is multiple *gas spuds, gas ring* or a *central gas pipe* depending on gas cv for uniform distribution.
3. *Igniter* for providing ignition energy for lighting up the fuel.
4. *Flame monitor* for checking the health of flame whose input is vital for flame supervisor safety system (FSSS) or burner management system (BMS).

A typical industrial register-type circular oil and gas burner is shown in Figure 15.1. Here, FO is supplied through the oil gun placed centrally in the middle of the burner while gas is supplied through a set of gas spuds spaced around the swirler. PA and SA enter burner tangentially at different velocities so that scrubbing action on fuel is maximised. Flame monitor mounted on the front plate of the burner is not shown. The entire burner assembly is held in place by fixing the throat in the furnace wall and bolting the front plate of the burner to the windbox.

15.4.2 Burner Types by Construction

There are basically two types of boiler burners, namely, *circular* and *tangential*. The other category, namely the **duct burners**, is for mounting inside the ducts.

In addition to bringing fuel and air together, the circular burners also thoroughly mix the two and help in scrubbing the burning fuel with air streams so that the combustion is complete and rapid. Fuel and air mixing is associated with a lot of turbulence. Tangential burners, on the other hand, split and admit fuel and air streams in several horizontal layers from the four corners of the furnace tangential to an imaginary circle in the centre of furnace to create a flame ball, where a thorough scrubbing takes place to complete combustion. The air and fuel jets issue out of burners in streamline manner and thereafter create a violent turbulence in the flame ball.

15.4.2.1 Circular Burners

Circular burners come in a variety of internal designs in a wide capacity range. They can be placed in a boiler either

- Horizontally on any wall, on the opposite walls or even in corners
- Vertically firing downwards from the roof or vertically upwards from the floor or
- At any suitable inclination

Circular burners can be of two types, namely, register and parallel-flow types:

- *Register*-type burners are simpler and use higher excess air between 10% and 15%. The

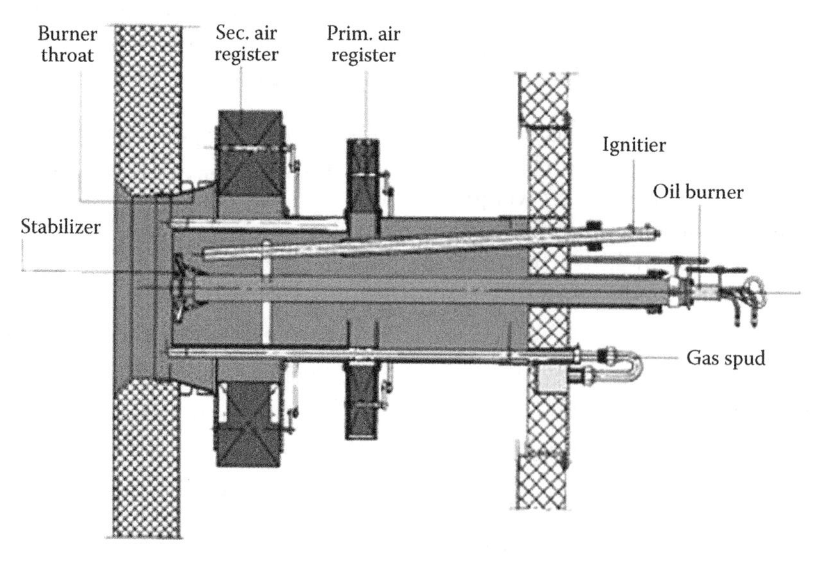

FIGURE 15.1
Register-type circular oil and gas burner.

airflow is provided tangentially to the flame. This results in a short, bushy and hot flame with higher NO_x production. They are used for compact furnaces. As the limits of NO_x are getting tightened, use of register-type burners is on the decline.

- *Parallel-flow* type of burners use lower excess air of 5–10% and produce long and lazy flames with attendant low NO_x generation. Parallel-flow burners are used in large package boilers and field-erected boilers as they have large furnace volumes.

Figure 15.2 shows a circular parallel-flow oil-fired burner for low excess air. Unlike in the register burner where the air and fuel flow directions are perpendicular to each other, they are set at parallel in these air registers. As there is no turbulence created in the burner, they need more space in the front to mix and burn. The flames are long and narrow and the furnace depth needed is more. As the combustion is protracted the maximum flame temperatures attained are lower. With venturi construction of the throat, excess air can be controlled to very low levels like 3%. See also the last paragraph in Burner Performance.

15.4.2.2 *Tangential Burners*

Tangential burners are placed only horizontally and for corner firing. They have to be always in sets of four burners to be placed at the four corners of a furnace. The flames are long and lazy as they come out of burners but turn highly turbulent when they meet tangentially in the flame ball. Tangential burners for oil and gas are similar in construction to those employed for PF shown in Figure 16.30.

15.4.2.3 *Scroll Burners* (see also *Whrbs for Steel Plants* under *Waste Heat Recovery Boilers*)

Scroll burners are especially for burning high-volume low-cv lean gases. **BFG** is one of the leanest gases and scroll burners are ideally suited there. Tangential inlet of fuel into the scroll provides a long and swirling path in such a way that the gas comes into intimate contact with combustion air. Also, the swirling fuel–air mixture gets to dwell in the furnace for a longer period so that combustion is more complete.

Figure 15.3 shows a typical scroll burner for BFG and FO firing for supporting the main fuel. Scroll burners can also have a provision for firing high-cv gas like **COG** centrally around the FO flame. In addition, a high-cv gas like **NG** can also be fired through gas spuds around the periphery.

15.4.3 Burner Sizes for Circular Burner

The size of a circular burner is always designated by the diameter of its *throat*.

Throat is the slightly narrowed channel in the air register through which all the air is directed into the combustion area.

For oil and gas burners, the throat sizes vary from 250 to 900 mm (10 to 36″).

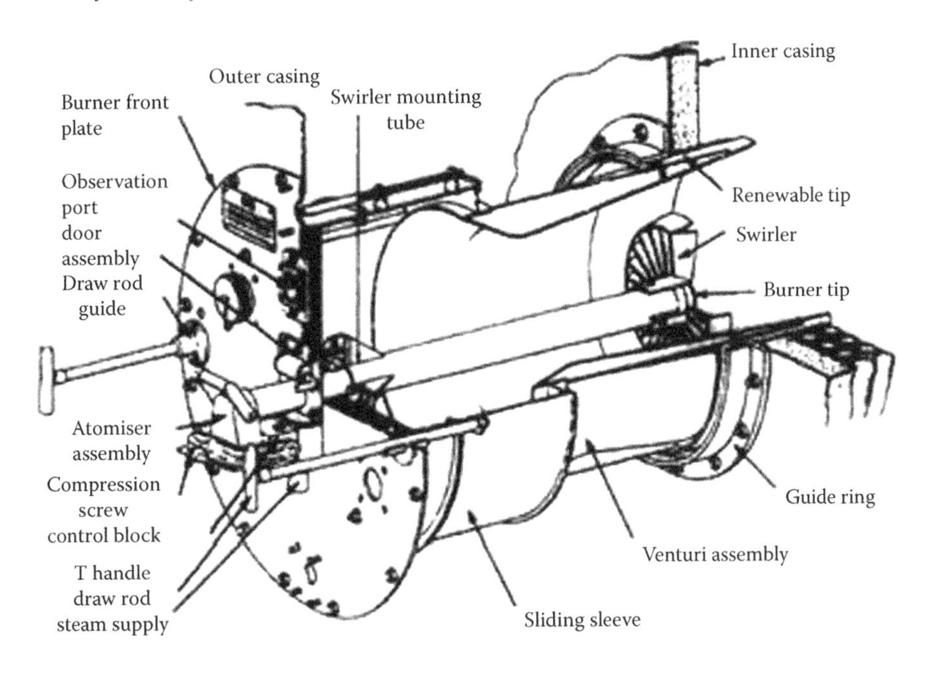

FIGURE 15.2
Parallel-flow manually operated venturi-type circular burner for oil firing.

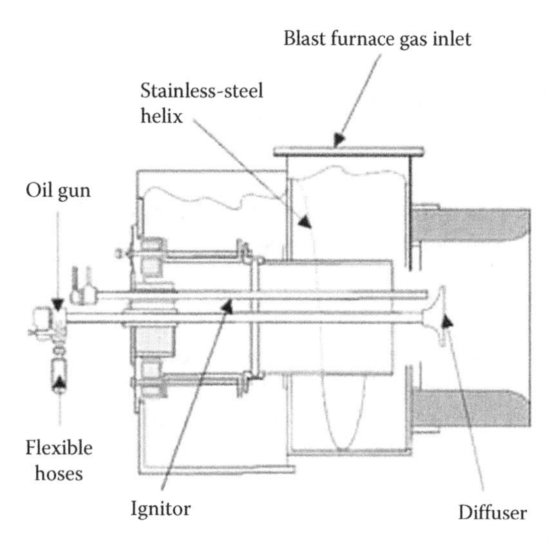

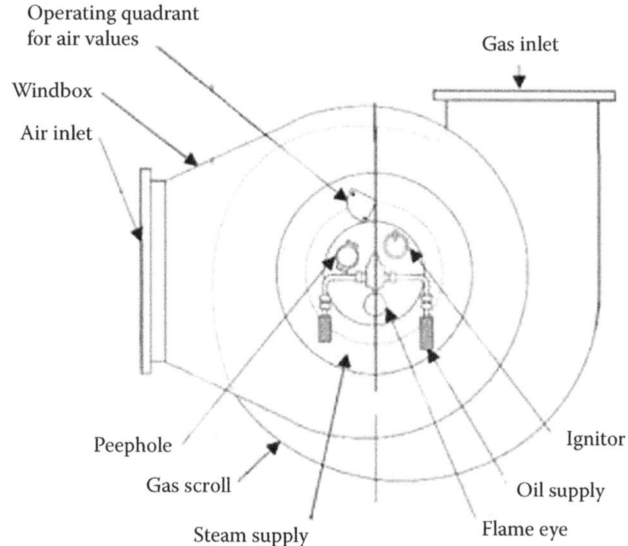

FIGURE 15.3
Scroll burner for BFG and FO.

15.4.4 Burner Performance

15.4.4.1 Burner Turndown

- Burner turndown for any fule is the ratio of maximum to minimum fuel that can be burnt in a burner in stable manner.
- A burner turndown is different from an atomiser turndown, which is essentially the range of satisfactory atomisation of FO.
- An atomiser independently can have a turndown of 1:10 while the burner as a whole has normally a turndown of 1:4.
- Higher turndown for burner of 1:6 for FO and 1:10 for gas is possible. The higher the turndown, the greater is the air and fuel pressure drops and hence the need for higher fan and fuel pump power.

15.4.4.2 Hot Air

- Hot air is preferred in all modes of combustion for enhancing its speed and η.
- Hot air is essential for PF and HFOs while it is also preferred for gases and LFOs.
- Air temperature of 200–300°C (400–600°F) is the normal range in burners.
- Most FOs and gases burn completely with no perceptible unburnt loss except for HFOs (~0.1%) and pulverised coals (<1%).
- Unburnt loss is best measured by the CO level in exit flue gas which should be <200 ppm with 3% O_2.

15.4.4.3 Excess Air in Burners

Low excess air for combustion is desirable as it

a. Increases the boiler η by lowering the stack loss
b. Reduces fan power

At the same time, the HS of all tube banks will be marginally higher due to the reduced gas flow.

- Typically, 5% excess air is normal with large package and field-erected boilers firing FO and gas. Parallel-flow or low-NO_x burners can be employed as the furnace volumes are sufficiently large. Tangential burners can also be used.
- High excess air of 10–15% is normal with smaller oil/gas package boilers. Small furnace volumes need short and bushy flames which are provided by circular register-type burners. Also, the HSs being less it is compensated by higher gas flows. The overall boiler η is marginally lower but the arrangement is economical. NO_x generation is also high.
- Very low excess air of ~3% is possible in venturi-type parallel-flow burners where the burner barrel is reduced to form a venturi to enable accurate air flow measurement. The Δp for air is magnified by the venturi throat. Individual control of air for each burner is necessary at such low levels of excess air. Such burners are more expensive and controls are more elaborate.

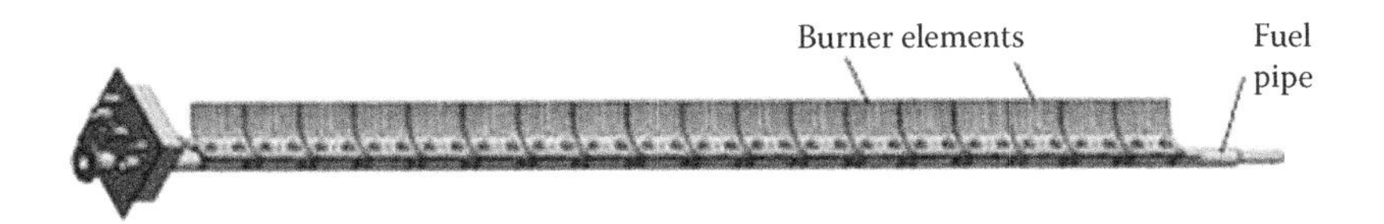

FIGURE 15.4
Duct burner elements.

15.4.5 Duct Burners

Duct burner is a burner arrangement inside an air duct for burning high-cv gases or light FOs. Unlike a furnace, a duct has extreme limitations of size and shape. To contain the flame without impinging on the duct, it is necessary that the combustion is completed in the shortest possible time with small size even with quick-burning fuels. Burning heavy FOs is almost not possible.

Duct burners are needed in boilers for two purposes:

- For providing hot gases for heating of fuel beds in **FBC boilers**
- For burning supplementary fuels in the **Hrsgs** to produce additional steam

Duct burners come in two varieties, namely

1. *In-line duct burners*

These are more popular and are arranged either horizontally or vertically within a duct with multiple small flames along the airflow. These are usually for burning fuel gases. The burner element takes the form of an SS 304 drilled manifold of 50–200 mm diameter with SS 309 wing like flame protectors (as shown in Figure 15.4). ss perforated screens are installed upstream for equalisation of airflow over the entire duct. Suitable upstream and downstream straight lengths are needed for proper flame development as gas Δp across the burner is very small.

Figure 15.5 shows the duct burner assembly consisting of several burner elements stacked one over the other to fill the inlet duct in a horizontal Hrsg. Horizontal elements are shown here but vertical arrangement is also possible. Elements vary in length from ~2 to 18 m. For lighting up of burners, HEA or GE igniters are installed.

The burner turndown is nominally 10:1 on gas and 5:1 on FO. Duct burners are usually shop assembled along with the valve skid with minimal site erection. Duct burners for large Hrsgs can be quite big with typical heat output of 300 MW_{th} or 250 × 10^6 kcal/h or 1000 × 10^6 Btu/h. Gas pressure drops vary from 3 to 25 mm wg across the duct burners.

2. *Side burners*

These are located perpendicular to the air flow stream in separate enclosures. These burners are actually circular burners with short flames usually for FO firing. Figure 15.6 shows such a side burner.

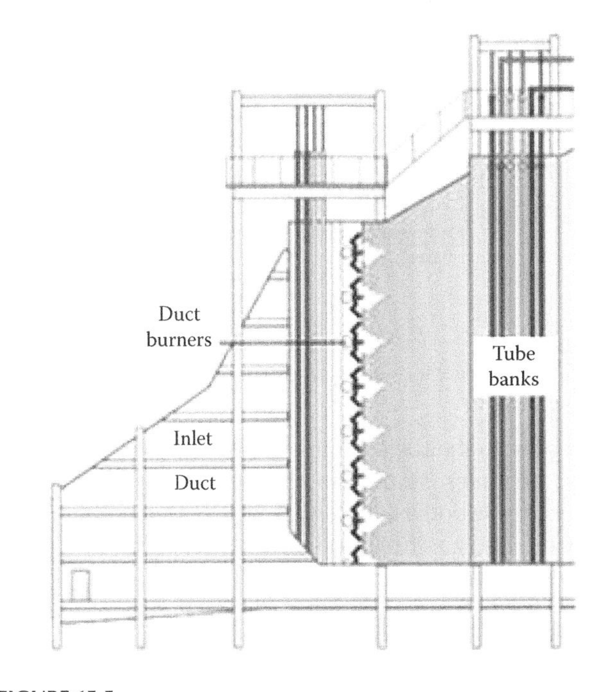

FIGURE 15.5
Horizontal duct burners for supplementary oil/gas firing in an Hrsg.

A duct burner system, in all, consists of the following sub-assemblies:

- Duct burner assembly
- Perforated plate for gas flow equalisation in in-line burners
- Igniter
- Scanner and burner management system
- Fuel valve trains
- Control system

15.5 Oil- and Gas-Fired Boilers

FO and **NG** are both very-quick-burning fuels. They are also very easy to handle and burn, provided adequate safety measures are followed. The burners can handle

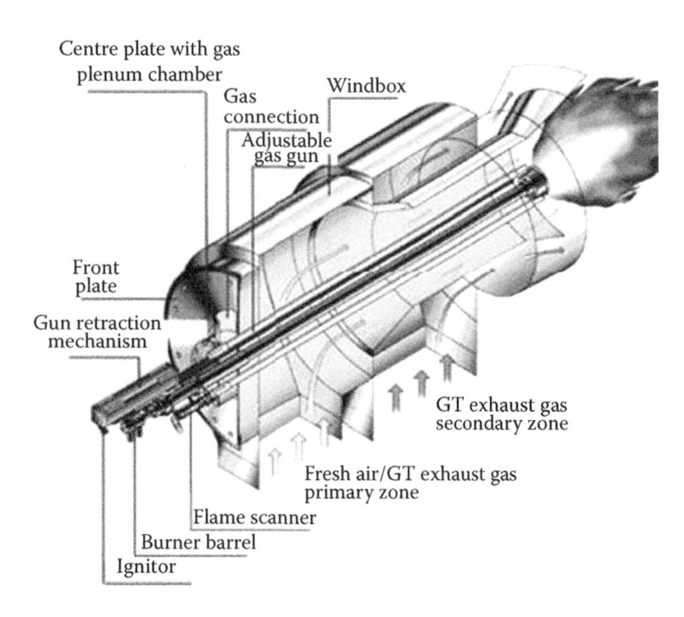

FIGURE 15.6
Circular duct burner for sidewise installation.

either or both the fuels in most cases. The two fuels share many attributes similar to the boilers. Similar to the burners, these boilers also can handle both or either fuels.

The oil- and gas-fired boilers can be divided into three groups, namely

1. Package boilers
   a. O-type boilers
   b. A-type boilers
   c. D-type boilers
2. Field-erected boilers
3. Utility boilers

which are described below.

15.5.1 Package Boilers

They are so called as they can be fully shop assembled and transported to site in a single package, at least in smaller sizes. The package boilers come in a wide range starting from 5 tph and going upwards of 300 tph.

Package boilers are mainly for process steam and Cogen in chemical and HC industries. Despite the wide capacity variations, all of them share some essential features such as listed below

- They are bottom supported.
- The flow of flue gases is mostly horizontal; travelling to the rear and exiting towards the front.
- They are fully or substantially shop assembled.
- Their designs are highly standardised.
- Practically no foundations are required other than a flat floor and a few stub columns.
- They are made very compact to suit the limits of road or barge dimensions.
- Naturally, this makes them the most highly rated of all boilers. The boiler life is shortened.

Package boilers in many developing countries are partly assembled in shops and completed at site due to the limitation and high cost in transportation.

Package boilers broadly fall into three categories depending on the shape of their cross section. They are, in the order of size progression, O-, A- and D-type boilers.

15.5.1.a O-Type Boilers

As shown in Figure 15.7, these are essentially **bi-drum BBs** pulled apart at the centre to make place for a small furnace. Usually one or at the most two burners can be installed.

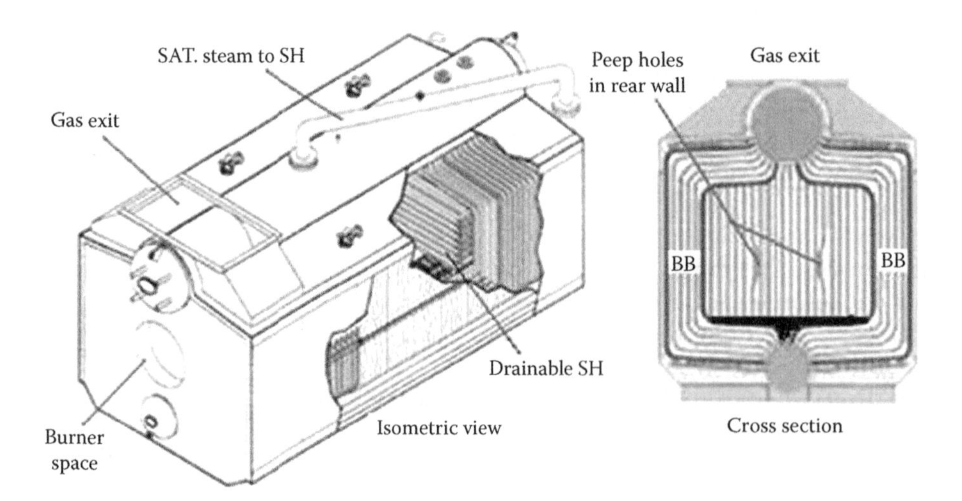

FIGURE 15.7
O-type oil- and gas-fired boiler for small and medium steaming.

SH is provided by removing some BB tubes at the far end. SHs are usually drainable and they can be placed either on one or on both sides depending on the SOT.

As the size of the furnace chamber is small, this design is suited for high-cv and quick-burning FOs and gases. The access for cleaning being limited and the flue gas flow being horizontal, these boilers are less suited for HFO with sludge and dust.

These boilers are normally available in sizes ranging from 5 to 35 tph, for design pressures of ~15–100 bar and SOT up to 500°C. With two burners and field erection, boilers as large as 115 tph can be built. Waste fuels that produce clean gases can also be used by burning them in an external furnace.

As the boiler is symmetrical about its axis, the handling and transportation is easier.

15.5.1.b A-Type Boilers

These are tri-drum boilers with the BB tubes from steam drum coming down into the two bottom drums, as shown in Figure 15.8. The space available for furnace is, thus, larger than in O-type boilers. As the weight is equally distributed, A- type boilers are easier to transport, a feature they share with O-type boilers.

A-type boilers can be built to much larger size of ~140 tph as there is scope to install one to four burners. SOP and SOT are about the same as in O type. As the furnace is larger, slightly inferior fuels can be accommodated in A-type design. The bottom being open and access better, even solid fuels can be burned in a separate furnace and flue gases cooled in the boiler.

15.5.1.c D-Type Boilers

Unlike in O- and A-type boilers, the BB of D-type boiler is kept intact and only the last tube is pulled apart to make provision for a large furnace (as shown in Figure 15.9). This enables D-type boilers to accommodate up to four burners and generate steam in excess of 300 tph. The maximum design pressure is also slightly higher at 115 bar. Typically, steam generation of a boiler with

- Single burner with 2.75 to 3.75 m (9–12′) drum centres is ~5–80 tph
- Two burners with 4.25–8 m (14–26′) drum centres is ~80–275 tph
- Three or four burners with >8 m (>26′) drum centres is >275 tph

The entire boiler is shop assembled and shipped depending upon the shop capability and transportation limits, at least in twin burner designs. Otherwise, the BB and the furnace can be shipped in two pieces with reasonable amount of site work to complete the boiler.

Figure 15.9 shows a small D-type boiler with single burner and no SH. A fabricated base frame supports the boiler. The following are the salient aspects of D-type boilers which are also applicable for A and O types as relevant.

Boiler:

- The smallest boiler is usually employing ~2.5 m in depth to accommodate the flame length of the smallest burner. These are **register-type burners** employing ~15% excess air or more to produce short flames.
- The longest boilers are ~9.5 m with about 100 rows of tubes in the BB. Usually **parallel-flow** type or **LNBs** are used for better efficiencies and lower $\mathbf{NO_x}$.
- With ~100 tubes deep BB, the Δp in the BB limits the permissible furnace pressure.

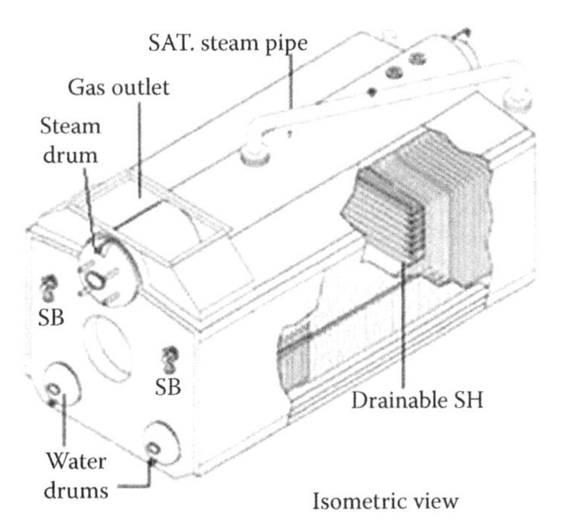

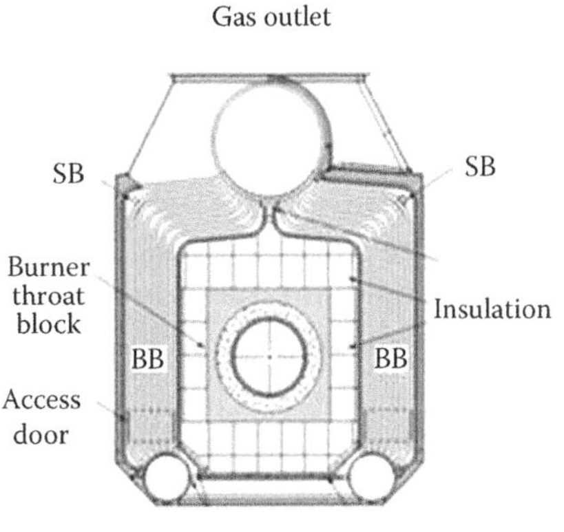

FIGURE 15.8
A-type three-drum oil- and gas-fired boiler for small and medium steaming.

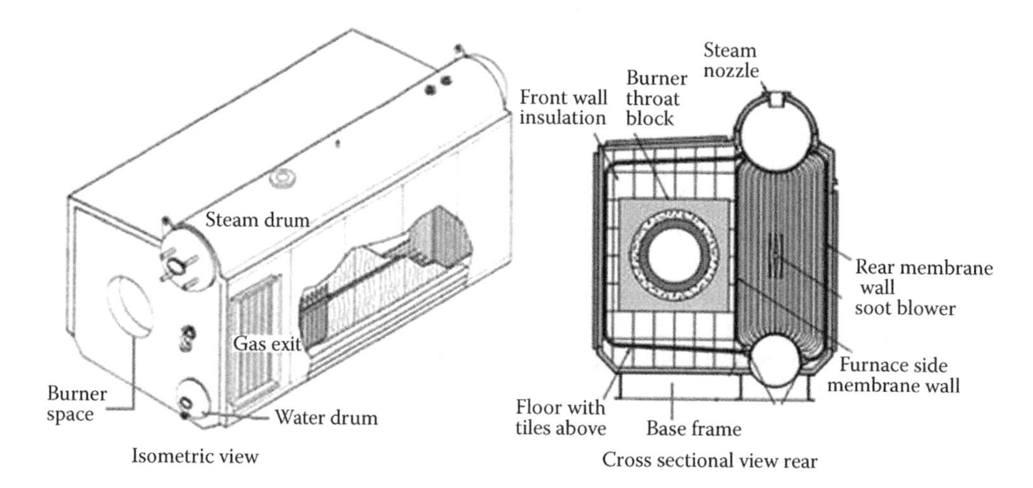

FIGURE 15.9
D-type bi-drum oil- and gas-fired boiler for medium and large steaming.

- Also, the Δp between the furnace and the BB exit makes sealing a difficult job.

Furnace
Furnaces in package boilers are highly rated:

- Volumetric HRR is limited to 1.45 MW/m^3, ~1.25 × 10^6 kcals/m^3/h or 14,000 Btu/ft^3/h.
- Maximum heat flux is up to 0.63 MW/m^2, ~54 × 10^6 kcal/m^2 h or 200,000 Btu/f^2 h.
- At these high heat ratings, the furnaces are required to be fully water cooled with minimum exposure of refractory to heat.
- FEGT is limited to 1400°C or 2550°F to prevent fouling of surfaces and overheating of SH coils.
- Furnace floors are invariably covered with refractory tiles to prevent their overheating.
- Furnace pressure is limited to 600 mm wg or ~24 wg.

BB

- The drums are spaced apart by a minimum of 2.75 m (~9 ft) from circulation considerations.
- Drum centres exceed 8 m (~26′) in large boilers.
- Boiler exit gas temperature (BEGT) is limited to 450°C (~850°F) to prevent a reverse circulation of water in the final bank tubes and also to avoid the need of insulating the exit duct of boiler from inside.
- The BB tubes are of in-line or staggered construction for clean or fouling fuels, respectively.
- The BB tubes across gas flow are limited to ~10 in number.
- High gas velocities of maximum 30 m/s (~100 fps) are adopted.

SH:

- Drainable SHs with single or double loops are installed. A typical vertical single loop drainable SH is shown in Figure 15.10.

D-type boilers are adapted for burning low-cv waste gases such as **COG** and **BFG**. The front portion of the furnace is refractory lined for reradiating the heat. The volumetric HRRs are considerably reduced to provide adequate residence time in furnace for these slow-burning fuels. Typically, the rates are 0.4 and 0.2 MW/m^3 for CO and BF gases, respectively.

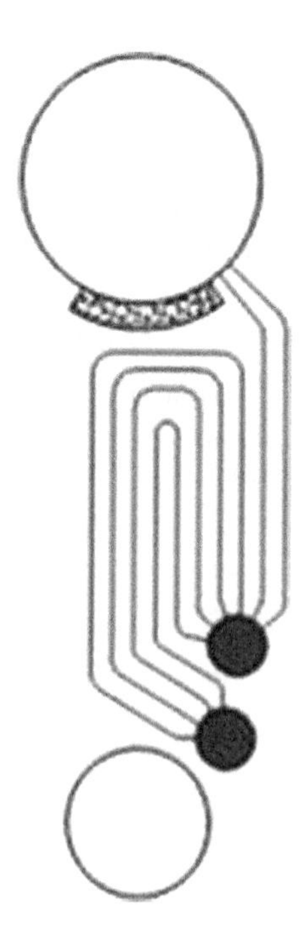

FIGURE 15.10
Drainable inverted loop SH in D-type boilers.

TABLE 15.1

Broad Steaming Capabilities of All Standard Package Boilers

Boiler	Capacity (tph)	Design Pressure (bar)	SOT (°C)
O type	5–115	15–100	~500
A type	5–140	15–100	~500
D type	5–>300	17–115	~500

Table 15.1 briefly captures the broad capabilities of all the three package boilers.

D type is the most popular among all package boilers as they meet the most usual requirements in the industry and are also substantially shop assembled. An isometric view of a relatively small, single-burner-fired boiler from B&W range (called FM boiler) is shown in Figure 15.11 which provides a lot of constructional details. Gas flow is entirely horizontal making this design suitable for relatively clean fuels which do not precipitate ash. The furnace floor is covered with refractory tiles to prevent tube overheating as the steam bubbles collected on the top side of the tubes cannot provide adequate tube cooling. Also, the floor is slightly slanted upwards by ~5 degrees to help steam bubbles flow upwards and keep the tubes cool. An inverted loop vertical drainable SH is provided here. The steam separators are horizontal to squeeze into the limited space of the compact steam drum. The top drum is also provided with lifting hooks. These boilers do not need much foundation work other than a flat RCC floor on which they can be seated with their bottom frame.

15.5.2 Field-Erected Boilers

Over the years, the range of package boilers has been greatly increased to cover most steam and Cogen applications of the process and HC industry. However, when

- Higher p & t is desired for power generation
- HFO and dirty fuels are to be fired
- Lean and slow burning gases are to be combusted
- Customisation is desired outside the design features of package boilers

field-erected boilers are chosen. Even here, the boilers are substantially standardised and the designs are modular. The boiler shown in Figure 15.12 is a typical bottom-supported standard pressurised bi-drum boiler very popular in refineries and petrochemical plants. Unlike in the smaller package boilers, the gas flow here is up and down enabling burning of HFO and residues. Ash and sludge get collected at the bottom of furnace and BB from where they can be even water washed quite easily. The furnace volume is large which can provide adequate time for combustion of slow-burning fuels. It is front wall firing in this design. Corner firing is also possible. SH can be inverted and made drainable. BB here is a two-pass arrangement with gas flow in longitudinal fashion and gas exiting at the top. It can also be made three-pass with bottom exit to suit layout. If there is no erosion fear, the BB can be in cross flow of a well to make it more effective. However, as they are extensively standardised to reduce on costs and delivery,

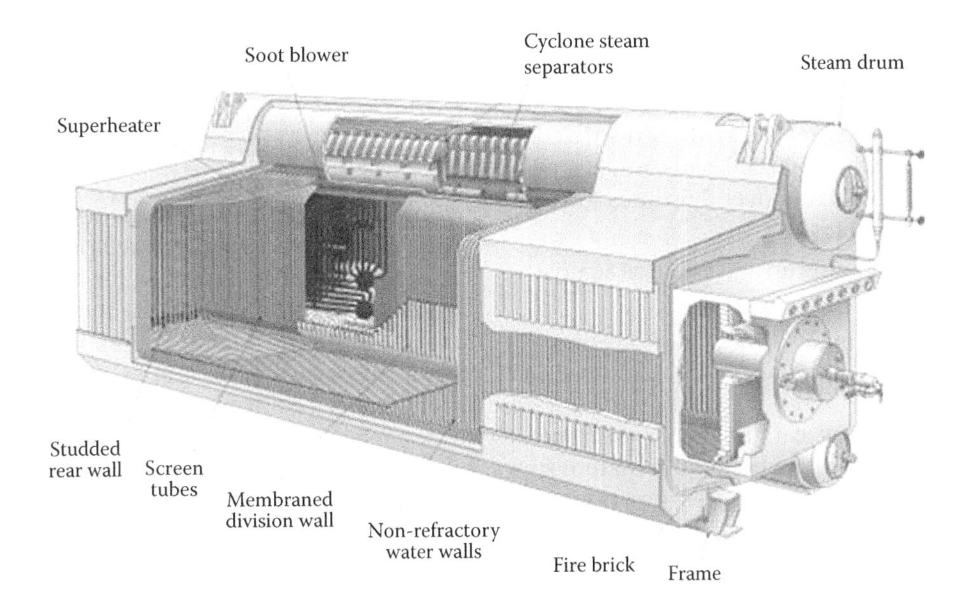

FIGURE 15.11
Small single-burner oil- and gas-fired FM boiler. (From The Babcock and Wilcox Company, USA. With permission.)

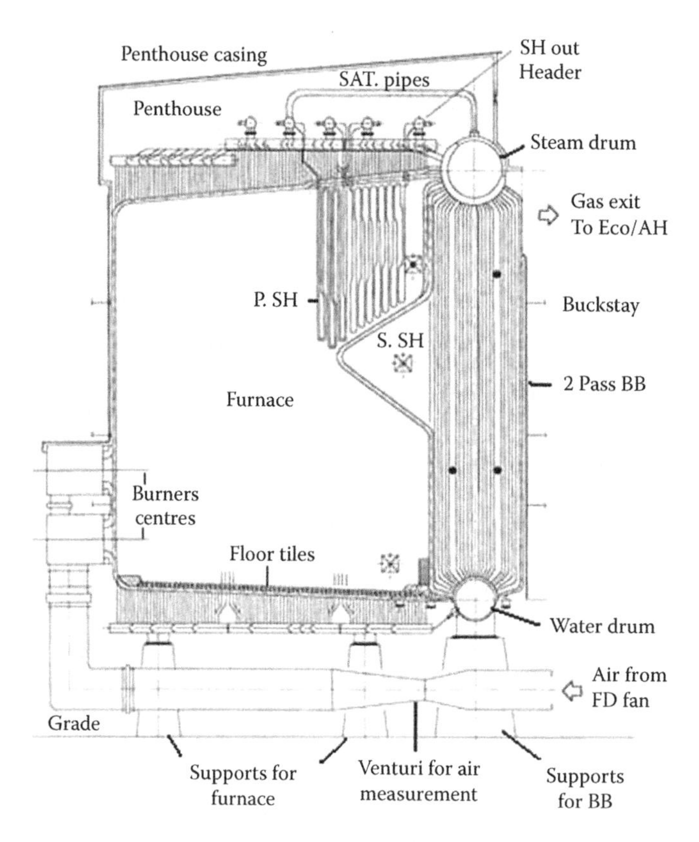

FIGURE 15.12
Bottom-supported pressurised field-erected bi-drum boiler for oil and gas.

each manufacturer offers a set of features which are exclusive.

Package versus field-erected boilers—A comparison
The principal differences between the package and field-erected boilers are as follows

- The heat ratings of the field-erected boilers are considerably lower than that of package boilers. For instance, volumetric HRR is restricted to ~0.58 MW/m^3, ~500,000 kcals/m^3/h or 56,000 Btu/ft^3/h which is over 2.5 times lower than in package boilers.
- Flue gas flow is normally up and down as against horizontal flow.
- As SHs are large many times, they are non-drainable.

The field-erected boilers can be either *bi-drum* or *single-drum* boilers. The bi-drum boilers are bottom supported and hence are generally more competitive. But the single-drum modularised boilers are suitable for higher capacities, pressures and temperatures which make them particularly attractive if power generation is normal requirement. The two designs are broadly compared in Table 15.2.

TABLE 15.2
Broad Capabilities of Bi-Drum and Single-Drum Boilers

Parameter	Bi-Drum Boilers	Single-Drum Boilers
Capacity (tph)	40–450	50–550
Design pressure (bar)	15–130	20–140
SOT max (°C)	540	570

Figure 15.13 shows a bottom-supported bi-drum front-wall-fired BFG fired boiler.

Single-drum boilers are generally preferred for their superior load following characteristics as thermal inertia is lower due to lower water-holding-capacity and lighter PPs. At the same time the single-drum boilers are less forgiving for the errors in operation as there is not ample water in the system. The flexibility of arrangement of burners in a single-drum boiler is far more as the burners can be mounted on the boiler either on the roof or at the bottom in addition to the front and the corners. The boiler would normally yield a smaller foot print.

Figure 15.14 shows a typical pressurised top-supported front-wall-fired 1½-pass boiler with horizontal drainable SHs which is one of the popular designs for high p&t.

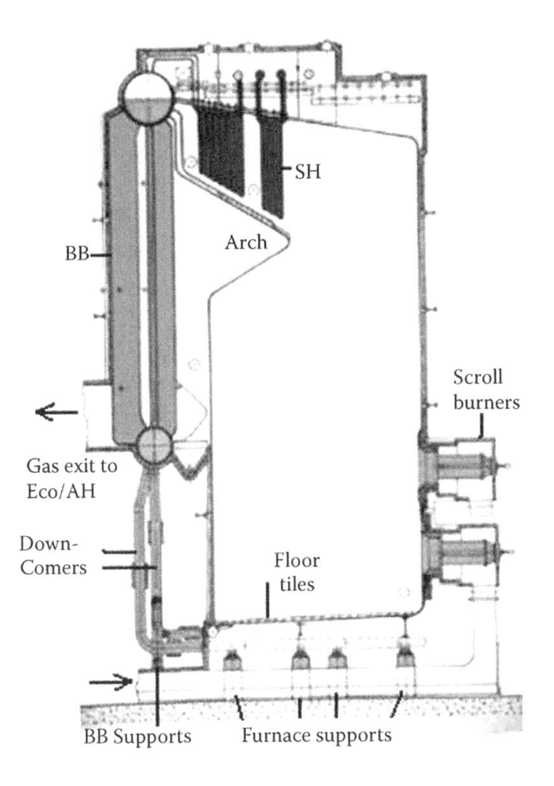

FIGURE 15.13
Field-erected bottom-supported bi-drum BFG boiler.

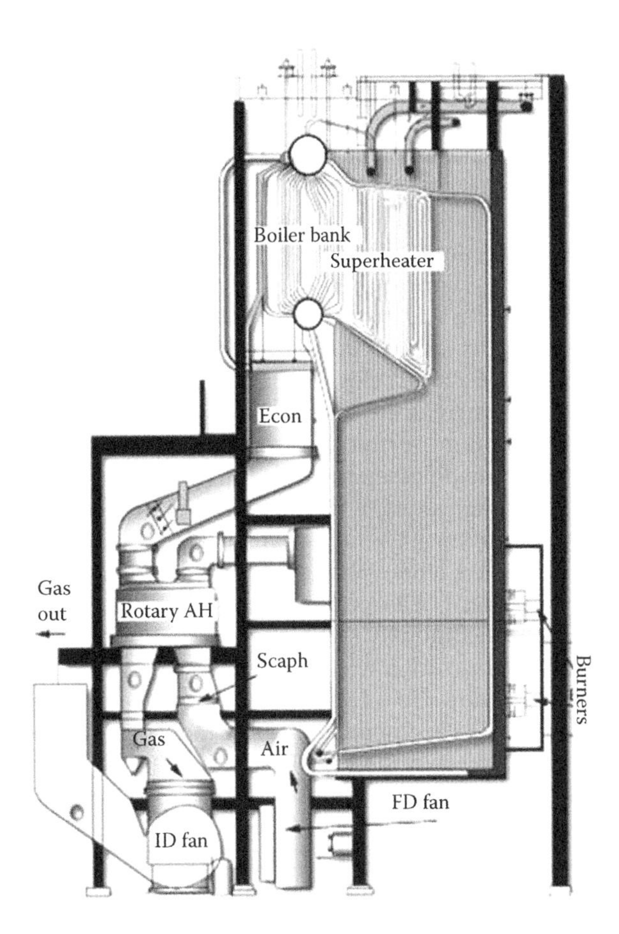

FIGURE 15.15
Large field-erected top-supported bi-drum boiler with pendant SH.

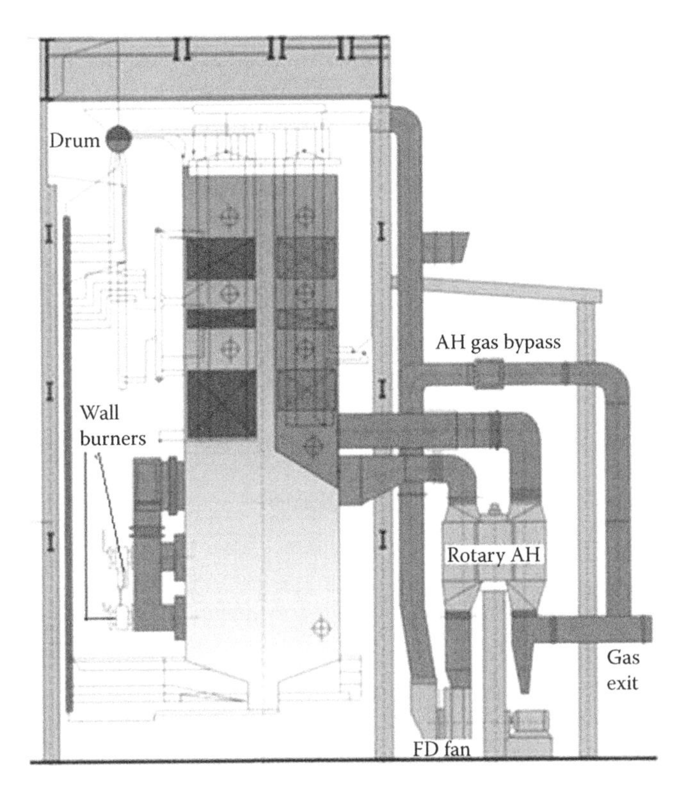

FIGURE 15.14
Large top-supported field-erected HP 1½-pass oil and gas boiler with drainable surfaces.

A typical top-supported bi-drum front-fired balanced-draught boiler with vertical pendant non-drainable SHs is shown in Figure 15.15.

It is seen that in general, single- and bi-drum preferences are displayed where European and American practices prevail, respectively.

15.5.3 Utility Boilers

Very large oil- and gas-fired utility boilers were built in the past before the scale-up of GTs took place. The introduction of large-capacity GTs and acceptance of large CCPPs has practically ended the era of oil- and gas-fired utility boilers. Occasionally, they are built in the refineries to burn heavy residues.

15.6 Oil Pumping and Heating Unit

Called the OPH unit or simply the P&H unit, this arrangement is needed in all the FO-fired units and also many solid-fuel-fired boilers that employ FO for initial start-up and low-load stabilisation.

FO has to be raised to the appropriate p & t depending upon the type of FO and the atomisation. Pressure atomisation needs higher pressure than steam atomisation. FO has to be heated to the temperature range that provides appropriate viscosity for the type of atomisation adopted in the burners (as shown in Figure 12.12). Pressure/mechanical atomisation needs lower viscosity and hence requires higher FO temperature.

Conditioning FO from the storage to firing status is the purpose of an OPH unit. Small units are skid mounted and placed close to the boiler along with the day tanks. Such units support the start-up duties of industrial boilers or small oil-fired boilers. Larger units are placed inside a separate building a little far away from the boiler along with storage tanks meeting the regulations of the local Explosives act. Such an arrangement is needed when a battery of boilers require start-up arrangement or large FO-fired boilers are to be fed with fuel continuously. Also, the PF-fired boilers need low-load stabilisation for extended periods when coals are wet or loads <50% MCR are to be generated. Large amounts of FO are needed and an independent large OPH unit placed in the yard is provided.

Shown in Figure 15.16 is the schematic of a basic OPH unit. FO pumps withdraw oil from the storage tanks through the suction filters. If it is HFO, the tanks are appropriately kept heated by steam coils or electric heating pads. Many times, outlet heaters are installed to raise FO temperature further to make it pumpable. Pressurised FO passes through the heaters to get heated, either by steam or electric power, to the required temperature and thereafter led to the boiler front in FO pipes of proper size. FO from the burner front is recirculated to the pump suction initially to set up circulation and heat the FO until it reaches the desired temperature. This return pipe brings back the excess FO from the boiler front during the FO firing. All HFO pipes are electrically or steam traced so that the HFO does not congeal and solidify inside the pipes when the FO firing is stopped.

Many times light diesel oil (LDO) is used for start-up and HFO for load carrying. LDO needs no heating. So LDO duly pressurised is fed to the boiler front FO header.

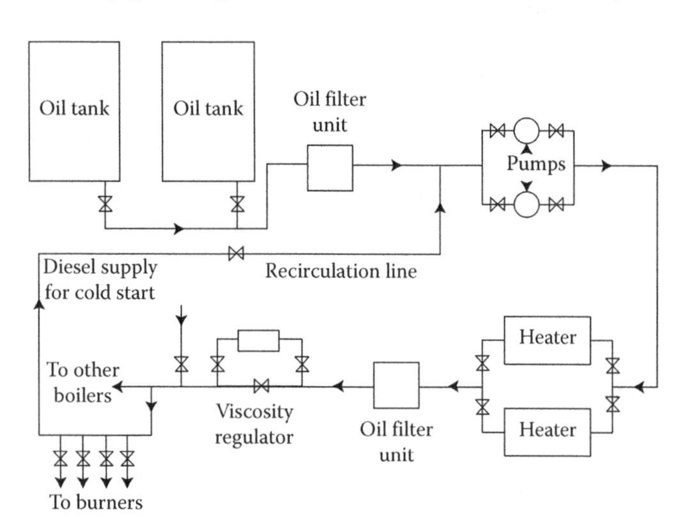

FIGURE 15.16
Schematic of oil pumping and heating unit.

15.7 Oil Shale (*see* Liquid Fuels)

15.8 Omega (Ω) Tubes

As depicted in Figure 15.17, these omega (Ω) tubes are specially extruded tubes, usually for use in heavily dust-laden gas passes, to withstand the erosive forces of dust moving past them at high speeds. The uneven disposition of metal helps to provide sacrificial thickness where needed and promotes far superior erosion withstand capability compared to plain circular tubes. Unlike the erosion shields mounted on tubes which are barely cooled, the integral flanges are cooled far better, promoting longer life.

For high-temperature SH and RH placed in very hot and dusty gases, omega tubes are used at times. In **CFBC** boilers, particularly in tower-type construction, high-temperature panels made of omega tubes are located in the furnace pass (as shown in Figure 3.13). They can also be used in PF and WH boilers also having dusty gases.

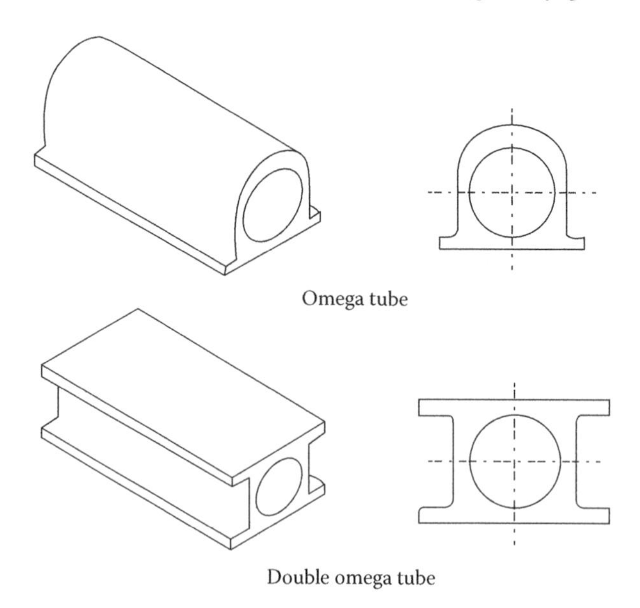

FIGURE 15.17
Specially extruded omega tubes.

The tubes are made in ~25–50 mm (1″ – 2″) bore with tube thicknesses varying from 4 to 8 mm. The flanges are even thicker. The small bore helps to increase steam velocities and keep the tubes cooler. Ω tubes are made in all metallurgies as needed by boilers, that is, CS, AS and ss. As they are specially extruded, the Ω tubes tend to be very expensive.

In plain single-flanged Ω tube, the flange faces the gas flow and deflects the dusty gas stream away from the tube behind. Dust in gas impinges the flat surface of the flange and falls off causing little harm. The tube stays protected. The width of the flange is to be chosen carefully to prevent creation of eddies or deflection of gases to another tube downstream.

Double-flanged Ω tubes are welded along the flanges to create flat surfaces which are swept by the gases. Gas-side **erosion** is substantially avoided. **Heat transfer** is reduced as the gases flow over flat surfaces and not on tubes. HS needed is more. As the flanges are welded end to end with no gaps in between, an expansion problem can show up.

15.9 Once-Through Boiler

OT boilers, as the name suggests, are *forced flow* boilers in which the fw turns to steam as it passes sequentially through the Eco, furnace walls and SH. There is little or no circulation of steam water mixture in the furnace. In La Mont Boilers, water equivalent to ~2–3 times the steam output is forced through the furnace tubes for keeping them cool. In Benson boilers, it is only one time the steam output and there is no circulation.

La Mont and *Benson* are the original inventors with slightly differing principles of working. In comparison to natural circulation boilers, in both designs

- Water in circulation is much lower, with no circulation in at all Benson boilers at higher loads.
- Tubes required in furnace are therefore much lesser and of smaller bore.
- Tubes are smaller and hence thinner.
- Downcomers and risers are lesser.
- Boilers are therefore less heavy.
- Boiler dynamics are superior.

On the other hand

- There is additional power requirement for pumping fw
- Controls are sophisticated
- Fw quality required is more stringent

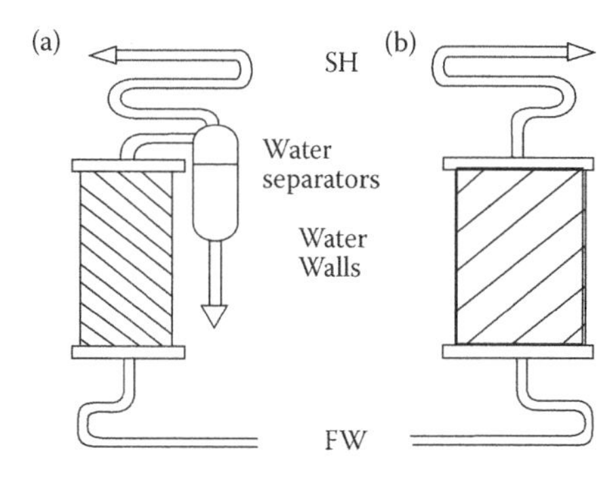

FIGURE 15.18
Classical design of (a) La Mont and (b) Benson type of OT boilers.

Power plants with SC boilers enjoy thermal η benefit because of higher cycle η, even though the boiler by itself is no more efficient than the sub-c boiler. In the case of OT boilers, in the sub-c range, the cycle efficiency advantage is not there, but the plant dynamics are superior and **variable-pressure or cyclic** operation is more easily possible.

La Mont design was purchased by Sulzer and improved upon. Several boilers are built and the system is very well suited for sub-c range. Most forced flow designs with furnace spiral tubing are modified La Mont Boilers; fw is pumped through the furnace walls to the water separators where steam is separated and pushed through SH (as shown in Figure 15.18a).

Mark Benson developed a design in 1922 that turned water into steam directly at SC pressure which was purchased by Siemens in 1924 and refined further. Here, the fw is pumped all the way from Eco to SH, as shown in Figure 15.18b, and there is no steam drum or steam separators. This type of design is very popular at SC pressures. In addition to the advantages of La Mont boiler, the evaporation end point in Benson boiler is variable. Benson boilers can have both spiral and vertical tubes in the furnace.

The current designs of OT boilers are substantially improved over the original classical designs which are elaborated in SC boilers under the topic of Utility Boilers.

15.10 Once-Through Steam Generator

Hrsgs mostly generate steam by natural or forced circulation principles for generating steam up to pressures of ~140 bar. The dynamics of a combined or Cogen cycle is mainly governed by the Hrsgs as the GTs have very high response rates. It is possible to improve this aspect if sub-c OT system can be employed in the Hrsg.

Once through steam generator (Otsg) is such an advanced version of Hrsg that employs sub-c OT principle of flow which is amenable to variable pressure operation. The advantages are that Otsg

- Requires no drums, risers and downcomers and blowdown system and so on that are associated with circulation
- Has no distinct Eco, Evap and SH sections; fw enters the coil and comes out as steam
- Dispenses with various tube sections and headers making them all into a single bank
- Works out much lighter and compact both for PPs and further in its supporting structure
- Space requirement is low
- Erection time required is naturally much lower
- Instrumentation and control is also lower as a single fw CV at inlet varies the output

Moreover

- It permits dry running of the unit dispensing with the bypass stack as Inconel tubes can withstand the exhaust gas temperatures of GT.
- Otsgs are slightly more efficient than Hrsgs as the preheater gas outlet temperature can be lowered without the fear of **low-temperature tube corrosion** as the tubes are made of Inconel. Also, the outward gas leakage through bypass stack is not present here.

Otsgs are made in vertical design to have the benefit of improved dynamics inherent in the vertical arrangement. As the cooling effect of steam is substantially less, the tubes are required to be made of very high-temperature-resistant materials like Inconel alloys. This makes the Otsg more expensive than Hrsg.

Where compactness and less weight are very important considerations like in a barge mounted plant or a marine application Otsg provides a unique solution. Also where it is highly beneficial to advance the boiler erection time Otsg can be considered favourably.

Otsgs have so far been confined to about 60 tph steam output behind smaller GTs like LM6000 or Frame 6 or Trent. In large conventional plants it is still the conventional Hrsgs that find favour. Figure 15.19 gives an idea about the compactness of the Otsg for an application behind LM 6000 GT.

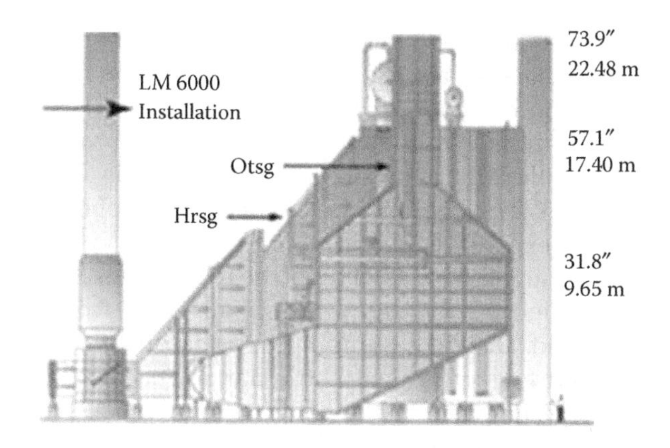

FIGURE 15.19
Size comparison of Otsg and Hrsg behind LM 6000 turbine. (From Innovative Steam Technologies, Canada. With permission.)

15.11 Opacity of Stack

Opacity is the *visible indication of completeness of combustion:*

- Dark plume represents incomplete combustion and can be set right by optimising the combustion parameters
- White plume is because of the presence of H_2SO_4 fumes and attention is needed in controlling excess air
- Light brown haze is the desired opacity

Opacity is also % reduction of light intensity. Smoke is commonly measured in terms of its apparent ρ in relation to a scale of known greyness. The most widely used scale is the one developed by Professor Maximilian Ringelmann in 1888. It has five levels of smoke ρ inferred from a grid of black lines on a white surface which, if viewed from a distance, merge into known shades of grey. There is no single definitive chart and several versions are used. The data obtained has limitations as the apparent darkness of smoke depends upon the concentration of the PM in the effluent, the size of the particulate, the depth of the smoke column being viewed and natural lighting conditions such as the direction of the sun relative to the observer while the accuracy of the chart itself depends on the whiteness of the paper and blackness of the ink used. Figure 15.20 shows Ringelmann charts in a series of blackening squares to ascertain the opacity while viewing the stack.

Figure 15.21 depicts the relationship between Ringelmann number and opacity or transmittance and optical ρ.

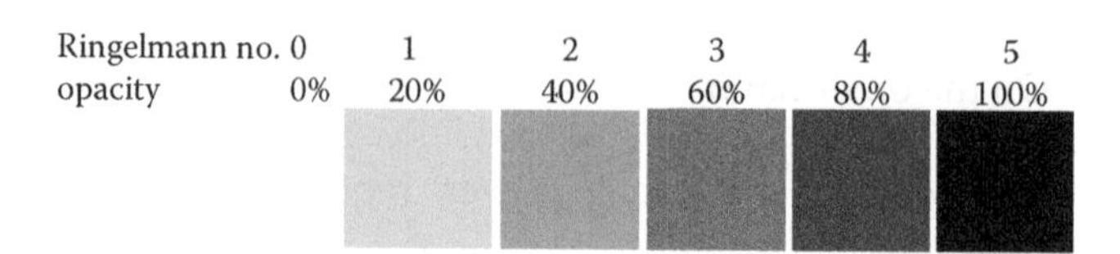

FIGURE 15.20
Ringlemann's chart in progressively darkening squares.

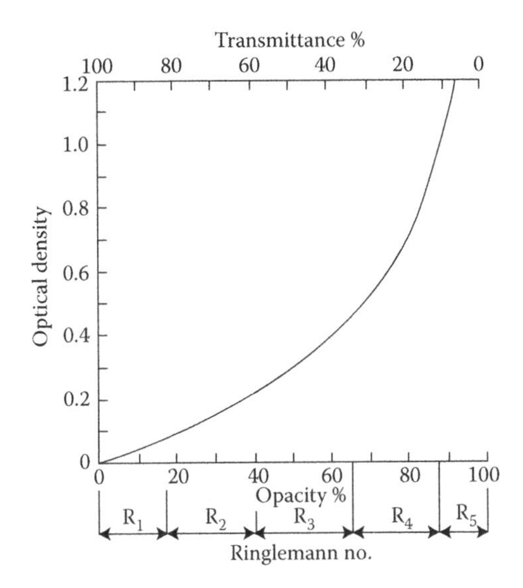

FIGURE 15.21
Opacity/Ringlemann number versus optical density.

Smoke or Opacity Meters

These meters detect and measure the amount of light blocked in smoke emitted in large stacks from industrial operations. The smoke meter readout displays the smoke ρ giving a measure of the efficiency of combustion. This makes the smoke meter an excellent diagnostic tool to ensure proper maintenance of combustion equipment in boilers for improved fuel economy and protection of the environment.

15.12 Open Cycle in Gas Turbines (*see* Thermodynamic Cycles)

15.13 Operating Pressure of Boiler (*see* Boiler Pressures)

15.14 Optical Pyrometer (*see* Temperature Measurement in Instruments)

15.15 Orbital Welding (*see* Welding)

15.16 Orifice

An orifice is any opening, mouth or hole from one vessel, pipe or tank into another vessel, pipe, vent or atmosphere.

15.17 Orifice Meter/Plate (*see* Flow Measurement in Instruments)

15.18 Orimulsion (*see* Liquid Fuels)

15.19 Orsat Analyser (*see* Flue Gas Analysis in Instruments for Measurements)

15.20 Oxy-Fuel Combustion

Oxy-fuel combustion is the process of *burning* ***fossil fuel*** *in nearly pure O_2 rather than in air.* The advantage is that it produces

- N_2-free flue gas with only water vapour
- A high concentration of CO_2

as its main products. It is then relatively easy to remove H_2O and concentrate the flue gas to an almost pure stream of CO_2 suitable for sequestration.

Coal is the cheapest and most widely spread of all fuels around the globe with coal-based power being the most competitive except in a few provinces blessed with cheap alternative fuels. But coal is also the dirtiest of all fuels. While the issue of **PM** emission is adequately resolved, there are limits to **NO_x** abatement and most importantly, there is no solution for major reduction of **GHG**. Oxy-fuel combustion followed by CCS can reduce the pollutants so effectively that coal firing can compete with **NG** in clean combustion. Both these processes are under active development and it is expected that a commercially viable solution may emerge by around 2020.

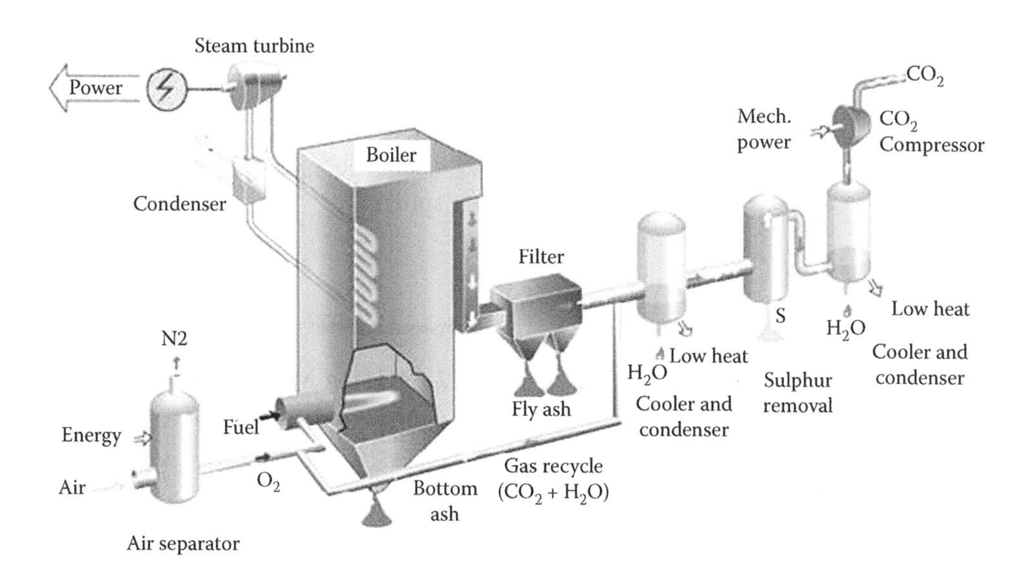

FIGURE 15.22
Schematic of oxy-fuel combustion.

The principle of oxy-fuel combustion is captured in the diagram shown in Figure 15.22.

O_2 is first separated from combustion air in an air separating plant and then admitted into a boiler along with fuel for combustion. As there is no N_2 there is no fear of NO_x formation even though the furnace temperatures are higher. However, flue gas recycling is continuously done to increase the gas quantity to maintain adequate mass flows over the convection tube banks of the boiler for effective utilisation of heat transfer surfaces. Flue gas containing mainly CO_2 and H_2O along with small amounts of SO_2 are then passed through cooler condensers and S removal units to remove H_2O and SO_2. Only CO_2 is now left in gas, which forms ~90% on volume basis, which is compressed and sent for sequestration.

The advantages claimed are

- Flue gas volume is reduced by ~75%.
    - Consequently, stack losses are proportionately reduced thereby increasing boiler η.
    - The size of the flue gas treatment equipment is reduced by 75%.
- Combustion η is slightly improved as pure O_2 is provided for combustion.
- NO_x production is greatly reduced.
- Concentration of pollutants in the flue gas is higher, making separation easier. Compression separation is possible as most of the flue gases are condensable.
- Heat of condensation can be captured and reused rather than lost in the flue gas.

At the present stage, oxy-combustion is not economical as air separation costs are quite high consuming ~15% of the generated power. Research is being conducted at several places to make this promising system economically viable.

15.21 Oxygen Scavenging (*see* Water)

15.22 Ozone (O_3) at Ground Level (*see* Air Pollution)

16

P

16.1 Package Boilers (*see* Oil- and Gas-Fired Boilers)

16.2 Parallel Flow of Fluids (*see* Fluid Flow Patterns in Fluid Flow)

16.3 Partial Pressure of Gas (*see* Properties of Substances)

16.4 Particulate Matter in Flue Gas (*see* Air Pollution)

16.5 Peak Rating (*see* Boiler Ratings)

16.6 Peat

Peat may be considered as the earliest stage of metamorphism of vegetable matter into coal. When plant remains in marsh lands decompose in the absence of air, peat is formed. *Peat is the youngest of coals.* See Figure 3.23 in Coalification.

Peat is a complex mixture of C, H_2, O_2, S, N_2 and ash. It has high levels of M in undrained bogs, as high as 80–95%. It reduces to about 70–90% on draining and 35–55% on air drying. Properties of peat vary considerably from place to place. Peat can be burnt in PF and FBC boilers. As fired, M must be <50% for stable firing.

Peat is found mostly in the higher latitudes of Northern countries. There are extensive deposits next only to coal. Low cv of 8.4–12.5 MJ/kg, 2000–3000 kcal/kg or 3600–5400 Btu/lb, even on reducing the M to 50–55%, makes peat very uneconomical to transport.

Sweden, Finland, Russia and Ireland are some of the countries which have a number of peat-based power stations.

16.7 Pensky Marten Flash Point Testing Apparatus (*see* Liquid Fuels)

16.8 Percentage Saturated Water Head (*see* Circulation)

16.9 Performance Test Codes

PTCs of ASME are the codes laid out for establishing the performance of most main and auxiliary equipment in a power plant for the both for the purpose of testing and acceptance of the new plant and also for routine testing of the running plant. Boiler, Hrsg, AH, mills and so on are the equipments covered by PTC. In all, there are 21 PTCs which are of relevance to boiler engineers. These are listed in Table 16.1.

16.10 Performance Testing (*see* Efficiency Testing of Boilers)

16.11 Petroleum/Crude Oil (*see* Liquid Fuels)

TABLE 16.1

List of Performance Test Codes of Relevance to Boilers

S. No.	PTC	Year	Title
1	4	1998	Fired Steam Generators
2	4.2	1969 (R1997)	Coal Pulverisers
3	4.3	1968 (R1991)	Air Heaters
4	4.4	1981 (R1992)	Gas Turbine Heat Recovery Steam Generators
5	6	1996	Steam Turbines
	6A	2001	Test Code for Steam Turbines—Appendix to PTC 6
	6-S		Procedures for Routine Performance Test of Steam Turbines
	8.2	1990	Centrifugal Pumps
6	11	1984 (R1995)	Fans
7	12.3	1997	Deaerators
8	19.1	1998	Measurement Uncertainty
9	19.2	1987 (R1998)	Pressure Measurement
10	19.3	1974 (R1998)	Temperature Measurement
11	19.7	1980 (R1988)	Measurement of Shaft Power
12	19.8	1970 (R1985)	Measurement of Indicated Power
13	19.10	1981	Flue and Exhaust Gas Analyses
14	19.11	1997	Steam and Water Sampling, Conditioning, and Analysis in the Power Cycle
15	21	1991	Particulate Matter Collection Equipment
16	22	1997	Performance Test Code on Gas Turbines
17	25	1994	Pressure Relief Devices
18	36	1998	Measurement of Industrial Sound
19	38	1980 (R1985)	Determining the Concentration of Particulate Matter in a Gas Stream
20	46	1997	Overall Plant Performance
21	PM	1993	Performance Monitoring Guidelines for Steam Power Plants

16.12 Petroleum Coke or Petcoke

Petroleum cracking results in separation of light, medium and heavy products leaving behind carbonaceous residues such as delayed coke, fluid coke and petroleum pitch depending on the type of process adopted. Petroleum coke is thus the end product of the refineries today.

Coker units are of two types—delayed coker and fluid coker.

16.12.1 Delayed Coking

This is the most widely used (over 90%) process having been developed in 1930 by Standard Oil, producing *shot coke* or *sponge coke*, so named after its appearance. Delayed coke has the appearance of run-of-mine coal with dull black colour. See Figure 19.7.

- Delayed coking is an endothermic process. Batches of reduced crude are rapidly heated in a furnace and are then confined to a reaction zone or coke drum under suitable p & t conditions, until the unvapourised part of the furnace effluent is converted to vapour and coke.
- Delayed cokes are generally available in large pieces that must be ground for proper utilisation. It is relatively soft compared to the other forms.
- Delayed coke has low ash content and higher VM making it a better fuel for boilers.

16.12.2 Fluid Coke

This accounts for <10% production, and is produced in a fluidised-bed reactor operating at 540°C, with a second stage of the process taking place in a fluidised-bed burner.

- The resultant fluid withdrawn from the burner consists of small solid spherical particles of around 200 μm resembling black sand that are very abrasive and hard with grindability index (HGI) at times as low as 17.
- Fluid cokers are more efficient in terms of conversion than delayed cokers. Fluid coke generally contains higher S and metal levels. It usually has lower VM and higher FC.

Petroleum coke has several uses depending on its composition and structure but the main use is in the form of fuel.

As all the distillates are extracted, petcoke naturally has low VM (3–20%) and high FC (70–95%). Gcv is very high, ranging from ~31.4 to 37.7 MJ/kg, ~7500 to 9000 kcal/kg or ~13500–16200 Btu/lb. Ash and M are very low. These properties are similar to anthracite.

The impurities of the original crude oil are of concentrated in petcoke. S is usually very high ranging from 3% to 8% depending upon the S of the original crude. Also heavy metals in ash such as V and Ni are also high. These impurities severely lower the fuel value of petcoke.

The grindability of petcokes varies considerably from 35 to 80 HGI averaging 50–60.

High cv of petcoke combined with low A and M is a definite advantage. Due to high FC, the ignition temperature is very high at >600°C. But, once ignited, the fuel burns very well provided there is adequate residence time in furnace.

TABLE 16.2

Properties of Delayed Petroleum Coke

Analysis	Delayed Coke % by Wt
M	3–12
Ash	0.2–3.0
VM	3–20
FC	70–95
S	3–8
GCV MJ/kg	31.4–37.7
kcal/kg	7500–9000
Btu/lb	13,500–16,200
HGI	35–80
Ignition temperature	>600°C

The real problem is with very high S that causes **low-temperature corrosion** and high V that causes **high-temperature corrosion**. The properties of delayed petroleum coke are shown in Table 16.2.

Until 20 years ago, **downshot firing** in PF boilers was the only way to burn petcoke. U or W type of firing was adopted to provide long flame path and high residence time. FGD units were installed ahead of AH to protect from low temperature corrosion. High-temperature SH materials were also selected with great care to avoid corrosion. Due to the high FC in fuel the pulverisers were required to grind petcoke to higher fineness together with much higher temperature of PA.

With the advent of FBC, petcoke burning has become easier, mainly due to the facility of in-bed de-sulphurisation. In CFBC, with de-sulphurisation ranging from 90% to 95%, the flue gases are rendered substantially harmless with no corrosion of back end equipment. Large input of limestone into and large output of gypsum from the furnace are the new issues to deal.

In both PF and CFBC, co-firing of petcoke is quite easy and many plants adopt it. Petcoke is largely available in the United States, Mexico, Venezuela, China, the Middle East and to an extent in India. Being energy intensive and produced mostly on the sea coasts, petcoke is a fuel that is amenable to fuel trading. At the same time, as the pollution norms get tightened, its attractiveness is reducing. This is mainly due to its high S which contaminates the atmosphere and the ground on which it is stored, particularly in rainy countries.

The pricing of petcoke is also debatable. For the refineries, petcoke is a by-product which they want to dispose off at any cost. At the same time, there is a temptation to link the prices with the prevailing oil prices. With many refineries adding coker units over the years, the prices have only tended to soften.

16.13 pH (*see* Properties of Substances)

16.14 Piezometer (*see* Pressure Measurement in Instruments for Measurement)

16.15 Pile Burning

As the name suggests, *pile burning is burning of certain solid fuels in piles or heaps*. This is one of the earliest practices in burning of wood and various bio-fuels.

Pile burning can be in thick piles like in horse shoe-type furnaces or in thin piles like in PHG, which are described in grate firing. By pile burning is normally meant the thick pile.

Pile burning is carried out in refractory cells built on refractory floor. These cells are called horse shoes because of their shape. Such boilers are called horse shoe boilers and they were very popular in the sugar industry for burning bagasse in the former times when **bagasse** used to be in excess and **Cogen** was not in use. The cross section of a typical horse shoe furnace is shown in Figure 16.1.

In thick pile burning, fuel is dropped from a height on a stationary grate or floor on which it piles up in a conical form to a height of 1.5–2 m and a diameter of ~2 m. The pile is surrounded by a refractory cell of ~2 m diameter in which air nozzles are embedded. Fully developed pile touches the cell walls at the bottom. Hot or cold air is blown all around the pile at usually 2 and at times 3 levels at a gentle pressure of ~50 mm wg to let its surface burn vigorously. More air is directed at the lower end of the pile for char burning. Air is not to be given from the bottom as it would tend to escape from the periphery of pile without participating in the combustion due to less resistance to flow.

Combustion takes place in three overlapping stages starting from the top of the pile. The green fuel gets substantially dried during the travel from feeder to the top of the pile, where the remaining M and VM, are driven off by the heat of the pile. The volatiles catch ignition and burn just above the surface along the slope of the pile. Char then starts burning and reaches to the bottom of the pile where ash is left behind. Ash is periodically removed manually from the man-doors at the front.

Sometimes, SA is also provided in the chamber above for a more complete combustion. Excess air levels are quite high as control of tramp air is difficult.

The cells can be placed independently like shown in Figure 16.2 or can be integrated at the bottom of the main furnace. The cells can be 2, 3 or 4 in number.

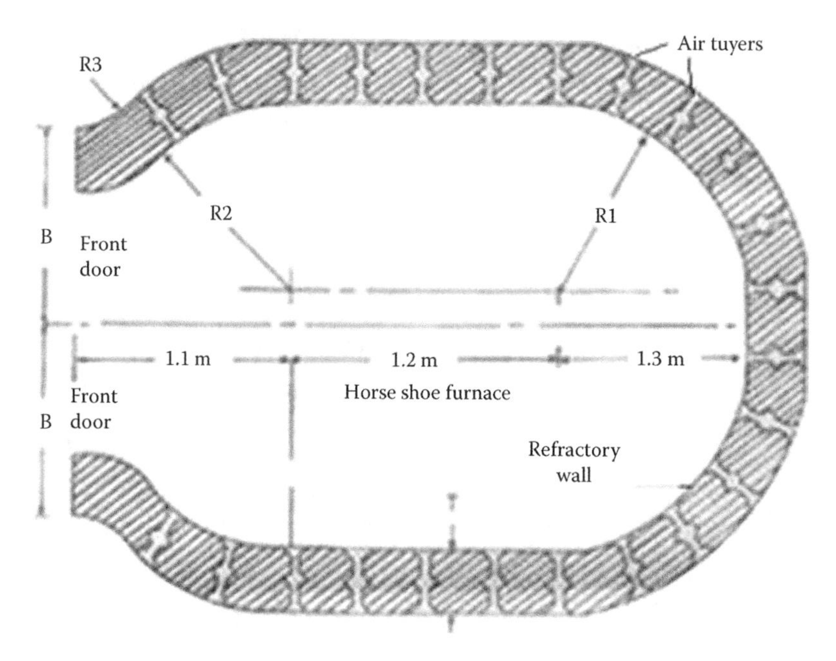

FIGURE 16.1
Plan view of a cell of a horse shoe furnace used in pile burning of bio-fuels.

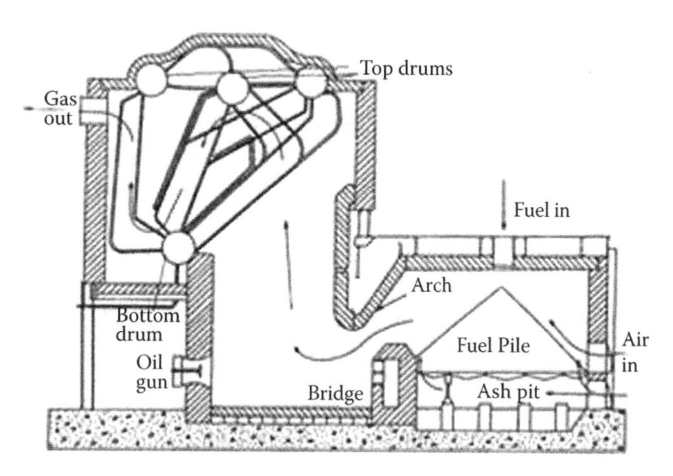

FIGURE 16.2
Cross section of a LP process boiler employing pile burning.

The advantages of pile burning are as given below.

- It is characterised by its utmost simplicity.
- The burning is vigorous and HRR as high as 0.4 kg/m^2 s (~300 lbs/ft^2 h) can be realised with the cold air with bagasse. With hot air temperatures ranging from 150°C to 300°C, the HRR can be increased by another 20%.
- Fuel sizing is not a concern as large blocks of even 100 mm can be occasionally be burnt.
- Fuel M can be as high as 60%.
- A large stepless turn down of 5:1 can be realised by varying the air.

The drawbacks of pile burning are many.

- Combustion is inefficient.
- Periodic manual ash removal is needed.
- De-ashing removes certain amount of green fuel and gives rise to large amount of tramp air ingress. During that time, the adjacent cells need to be over-fired. All these lower the boiler η.
- The boiler operation is not steady.
- Due to high heat, the refractory maintenance is very heavy.
- Clinker formation is also high. The proximity of refractory makes the clinker adhere to the brick work and its removal almost always results in the destruction of the refractory cells.

Pile burning is practiced in remote areas where simplicity of operations is the only concern due to the shortage of skilled operators. Fuel in such places is also likely to be quite cheap. For all practical purposes pile burning is phased out even though boilers as large as 100 tph have been built with bagasse firing.

16.16 Pinch Point (in Hrsg) (*see* Thermal and Mechanical Design Aspects of Hrsg *under* Heat Recovery Steam Generators)

16.17 Pin Hole Grate (*see* Grates)

16.18 Pipes (*see* Materials of Construction *and see also* Tubes)

16.19 Piping or Ducting Losses (*see* Fluid Flow)

16.20 Pitot Tube (*see* Pressure Measurement in Instruments for Measurements)

16.21 Pitot-Static or Prandtl Tube (*see* Pressure Measurement in Instruments for Measurements)

16.22 Pitting or Pitting Corrosion (*see* Corrosion)

16.23 Plastic Chrome Ore Refractory (*see* Refractories)

16.24 Platen Superheater (*see* Superheater in Heating Surfaces)

16.25 Pour Point (*see* Liquid Fuels)

16.26 Power Plant Cycles (*see* Thermodynamic Cycles)

16.27 Prandtl Number (Pr) (*see* Dimensionless Numbers in Heat Transfer)

16.28 Pressure (*see* Properties of Substances)

16.29 Pressure Parts or Scantlings (*see also* Heating Surfaces)

Parts subjected to internal pressure of water and steam are called the boiler PPs. Scantlings is another term for PPs.

Heat transfer from hot flue gases takes place in almost all PPs except in the

- Drums and headers when placed outside the flow path of gases
- Integral piping

The PPs are

1. Drums
2. Headers
3. Panels for furnace and second pass
   a. Membrane or monowall panels
   b. Tangent tube walls
4. Division walls
5. Coils for SH, RH, Eco and in-bed
6. Platens for SH and RH
7. BB tubes
8. Risers and downcomers
9. Integral piping

Each of these parts is elaborated.

16.29.1 Drums

Drums are present in all sub-c natural circulation boilers, that is, where drum pressure is <210 bar, while they are absent in SC and OT boilers. Water or steam separator vessels take the place of steam drum instead in sub-c OT and SC boilers, respectively.

In natural circulation boilers, steam drum is invariably present, be it a radiant (single drum) or bi-drum design. In most drum-type boilers there is only one water drum. Occasionally, in the likes of **A-type package boiler** (see oil- and gas-fired boilers), there are two water drums. In the decades gone by, it was not unusual to have 2, 3 or even 4 water drums when the boilers were of low p & t. Big **water-holding capacity** in boiler banks was an insurance against the deficiencies of the firing, instrumentation and operation. See Figure 8.30.

Drums are the most important of all PPs in a boiler because they are not only the largest and heaviest pieces of equipment but also require the maximum fabrication time. In fact, it is the drum manufacturing cycle in most cases that dictates the delivery time of a boiler. Steam drum is the heaviest lift needing biggest crane and drum lifting is a major milestone in boiler erection.

Boiler drums contain saturated steam and water and hence do not experience temperatures >370°C on the wetted side. For this temperature, drums do not need to be made of high AS. By and large, drums for industrial boiler applications are usually made of CS such as SA515, SA516 and SA 299 for generally thickness up to ~40, <100 and >100 mm for the three steels, respectively. When the pressures are high and the drum thickness is to be restricted, low AS such SA302 grade B is employed.

Drums up to the size of about 1100 mm OD are known to be forged in pipe making shops. Normally they are made from **BQ plates** ordered to specific dimensions. Drums are fabricated in two ways:

a. BQ plates in either hot or cold condition are rolled to form shells varying between 1.5 to 4 m in length. The shells are then welded circumferentially and finally closed with the dish ends.
b. BQ plates, in either hot or cold condition, are pressed to form bottom or top halves and then are longitudinally to form shells. The bottom and top plates are of different thicknesses as per their ligament efficiencies.

Hot or cold rolling/pressing of plates depends on the thickness, length and material of shell with a nominal dividing line as ~100 mm.

16.29.1.a Steam Drum

Steam or top drum is very important equipment in the water circuit. It is a surge tank where

- fw pumped through the Eco joins the bw which is in **circulation**.
- This combined water circulates through the various evaporating circuits and re-enters the drum for **separation** of steam from water. For this a steam drum contains **baffle plates, steam separators, steam dryers** and a **dry box**.
- **Chemical dosing** is carried out through the dish ends of the drum with perforated pipes along the drum axis.
- Collection of drum water below the **drum internals** for **CBD** is also carried out with perforated pipes running lengthwise along the drum.

See also Drum Internals and Figure 4.18.

Drum sizing is governed by the following factors:

(a) Drum Internals

The steam drum is primarily sized to contain all the drum internals. Steam drum diameters normally vary from 900 mm (~36″) to 2200 mm (~87″) and thickness from 25 to 250 mm.

As the drum thickness increases with diameter, it is crucial that the drum ID is kept to a minimum by having compact internals.

(b) Swell Volume

The drum size is also decided by the *swell volume* when the BB is made of horizontal serpentine tubes as in the case of vertical Hrsgs and similar construction. When load changes abruptly steam formed in all the legs of bank tubes tends to get cumulative. As this accumulated steam bubbles enter the drum, the water level is pushed up until all the steam goes out of the drum. This abnormal rise in water level is called the *swell effect*.

During this period, it is vital that the steam dryers should stay above the swollen water level lest the water should enter SH tubes. Drums have to be often increased in size to take care of such water swells.

(c) Water-Holding Capacity

A steam drum is also required to offer certain amount of water storage to provide for inherent sluggishness of instrumentation and operator errors. The *water-holding capacity* between the NWL and the top of the downcomers is the water available for quickly boiling and releasing steam during short overloads for maintaining the steam flow. Depending on the operational practices, certain water-holding capacity (or hold up time) may be desired by customer which often increases the drum diameter.

Larger the drum diameter, thicker is the drum. Inherently single drum construction with large bore downcomers enjoys higher **ligament η** as compared to bi-drum arrangement with several closely spaced rows of tubes. Higher the ligament η lower is the drum thickness. Thinner the drum the better it is in following the load fluctuations without the fear of high stresses in the drum plates. For drums >100 mm thickness it becomes necessary to monitor the drum metal temperatures on the inside and outside over its entire length. The load ramping get severely curtailed with higher drum thicknesses.

A steam drum can be either

- Stubbed drum or
- Expanded drum

depending upon whether the riser and downcomer tubes/stubs are welded to the shell or expanded in it.

16.29.1.b Water Drum

Water drum is also called as bottom drum and mud drum. Mud drum is the name from former times when there used to be a large amount of sludge or mud for removal.

The purpose of water drum is to

a. Merely distribute the water received from steam drum to various downcomer circuits and also to the riser tubes of the BB. It is practically a large header.
b. Provide a place from where **sludge** can be removed for **IBD**.

Water drums vary in sizes from 600 mm (approximately 24″) to 1070 mm (approximately 42″).

16.29.2 Headers

Headers are pipes of large diameter connected to the tubes in panels or coils. Their main function is to *either collect or distribute steam and/or water.* Usually they are circular but square headers were also popular in the olden times. Square headers with their flat bearing surfaces on all sides were well suited in bottom-supported boilers to prop up water walls. As far as possible headers are placed outside the flue gas flow to avoid undue thermal stresses. All bottom headers have provision for draining and top most headers for venting but all of them have provision for internal inspection.

Header sizes range from 150 to 610 mm (6 to 24″) in most boilers. CS (SA106) and AS (SA 335 P11, P22 and P91) headers are both used in the construction of sub-c boilers.

It is essential to reduce the thickness of headers to improve the heating and cooling rates of the boiler as header thicknesses are next only to drums. Often higher metallurgy is adopted as the first step. **Swaging** of tubes is another solution as it improves the **ligament η** and reduces the header thickness.

16.29.3 Panels for Furnace and Second Pass

16.29.3.a Membrane or Monowall Panels (see also *Furnace Wall Construction* under *Heating Surfaces)*

Modern boilers employ fully water-cooled membrane panels for walls of furnaces and other enclosures as well as division walls. Membrane walls, also called monowalls practically dispense with refractory that calls for costly down time and maintenance, thereby greatly improving the boiler availability.

In the history of boiler making it has been a long journey of several decades from full brick wall to full water wall construction for furnace, going through several stages of partial cooling, namely, spaced, tube, tube and tile, bailey wall and so on paralleling the advancements in combustion and circulation. The tangent walls, with tubes touching each other, was a major advancement in the early 1950s with practically no refractory facing the gases. Big boilers were built with tangent construction but due to their deficiencies they were superseded in favour of membrane walls.

Membrane panels, have a steel strip/bar inserted between the tubes and shop welded in **SA** or tungsten inert gas (welding) **(TIG) welding** machines (to give adequate weld penetration between the bar and the tube) and finish fabricated in transportable panels. With the end headers fitted and **heat treatment** carried out in the fabrication shops itself the site work is now considerably simplified. The advantages of the membrane walls are

- Reduced time and easier erection making it possible to build very large boilers
- Much simpler tube repair in a working boiler
- Far superior fit and finish of shop made panels compared to the individually erected tubes
- Furnace puffs withstand capacity far better as the membrane panels behave like plates
- Less number of furnace tubes for the same EPRS
- Smaller tube diameters in membrane panel save the PP weight

The membrane strip is usually 6–8 mm in thickness and varies from 12 to 60 mm depending on the heat flux. Only high heat flux like in PF and package boilers for oil and gas call for 8 mm thick strips.

The maximum temperature attained by the strip is at its midpoint and the stress levels generated govern the strip width which, in turn, is dependent on the heat flux.

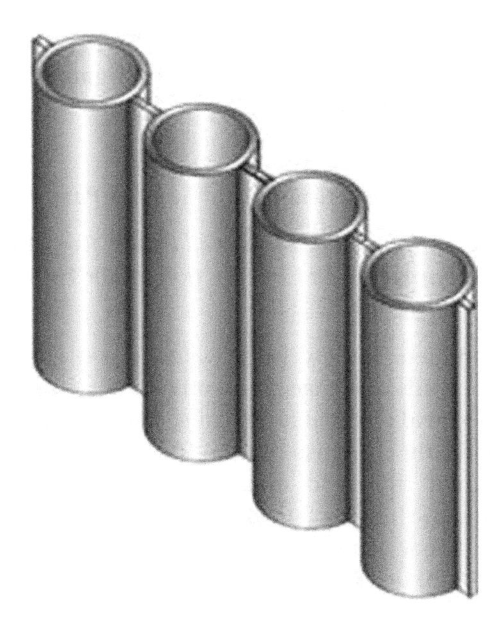

FIGURE 16.3
Membrane or monowall panel.

Panel welding and bending along with the other required facilities and cranes are expensive and occupy considerable shop space. Figure 16.3 shows the cut view of a membrane panel.

16.29.3.b *Tangent Tube Walls* (see also *Furnace Wall Construction* under *Heating Surfaces)*

In the early days furnace construction was entirely of refractory brick work. Maintenance was high, availability low and operational problems were more. Tubes began to be placed in front of refractory to improve the cooling. There were several improvements in this direction of enhancing the water cooling until finally tangent tube walls were introduced by the early 1950s where furnaces were fully water cooled and refractory was totally dispensed with. It was a major breakthrough as the reliability of the boilers began greatly increasing with no refractory problems coming in the way. Also it was the beginning for building bigger and bigger boilers.

In tangent tube wall construction, the tubes are placed side by side with a minute gap of 1–1.5 mm (merely to accommodate the tolerance on the tube od) and covered on the outer side with a thin steel sheet, called the inner casing, for gas tightness. Tubes are welded for better tightness only in case of oil-fired boilers as the sulphurous gases would otherwise cause corrosion of the casing. Suitable thickness of block insulation would be held in place by means of studs and wires before covering with veneer or remote casing, which can be either galvanised or corrugated Al.

Aside of dispensing with refractory the other big advantage was that the tangent tubes needed no special

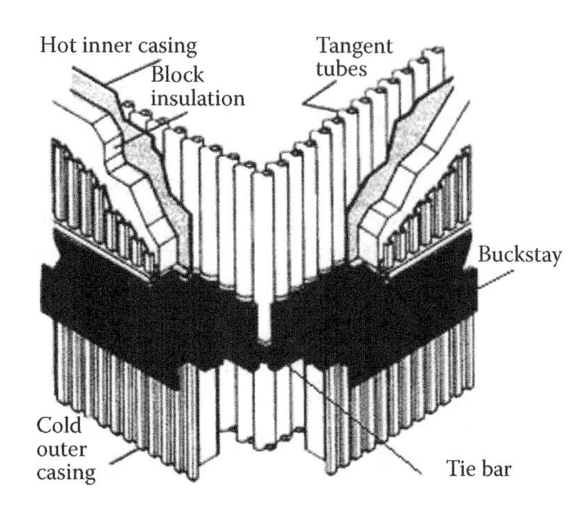

FIGURE 16.4
Tangent tube wall construction.

machinery in the shops or field. Really tall and big utility boilers were built with tangent walls. Construction speeded up as there was no refractory to be erected. As more experience was gathered and field erection costs began increasing, the deficiencies of the tangent tube walls began to show up.

Two major issues that needed to be addressed were

- The construction time, though lower than with the others, was still high as each tube was to be lifted into position.
- The tube repair was difficult as the affected tubes were required to be jacked out of position.

Membrane panels, with their distinct advantages, have practically replaced tangent construction worldwide. However tangent tube is still adopted selectively in certain parts of boilers for arrangement reasons.

Figure 16.4 shows a typical tangent tube wall construction in furnace.

16.29.4 Division Walls

Furnace is an enclosure surrounding the firing equipment in a conventional boiler to contain the flue gases and cool them appropriately. As boiler becomes larger the furnace volume increases much more than the cooling surfaces placed on all sides. This is because the volume increases in a cubic manner while the areas increase only as square. To set this volume to area proportion in a right manner in most utility boilers division walls are introduced in the furnace.

Division walls are tangent or closely spaced membrane panels placed usually from bottom to top of the furnace, covering the entire depth, parallel to the firing direction. As they are subjected to heat from both sides ample supply of water from the bottom and adequate evacuation of steam–water mixture from the top is very important to save the tubes from overheating. At times they are studded and covered with refractory to reduce over-cooling of flue gases.

16.29.5 Coils for SH, RH, Economiser and In-Bed (see *also* Superheaters *under* Heating Surfaces)

Coil is serpentine loop formation of one or several tubes connecting a pair of headers. A stack of drainable horizontal Eco coils in the manufacturing shop is shown in Figure 16.5. It has one tube per loop construction.

Coil work forms the bulk of HS in most boilers. After the flue gases are cooled to less than the IADT and there is no serious fear of bridging of ash SH, RH and Eco coils are placed serially in an ever decreasing spacing to extract heat from gases as depicted in Figure 8.18.

In conventional boilers coils are made of plain tubes. In Hrsgs the tubes are **helically finned**. At times Eco coils are also provided with extended surfaces to save on space and draught loss. Plate or rectangular fins are used. For high ash fuels gilled tubes are employed. See Economisers and Extended Surfaces for more details.

When tubes are to be protected from heavy dust loads, such as those in CFBC boilers, heavy pin studding of initial rows of tubes, as shown in Figure 16.6, is undertaken.

Bed coils, immersed in the bed of BFBC boilers to keep the bed temperature to the desired limit, do not extract as much heat from flue gases as they take away the heat from the bed material. Bed coils are always placed horizontally. Usually at an upward inclination of ~5 degees to help in **circulation**. Erosion forces on the bed coils are very heavy as the tubes are immersed in a violently swirling bed of inert material. Widely spaced studding, as shown in Figure 16.7, is often found to be a good protection.

Tube coils can be either *vertical* or *horizontal*. Vertical coils are *non- drainable* unless they are *inverted*. Inverted vertical coils are usually for modest superheating. *See* Figures 8.26 and 15.10 showing **drainable SH** in package boilers. Non-drainable vertical coils are also called ***pendants***.

FIGURE 16.5
Eco coils stacked in the shop floor complete with supporting arrangement.

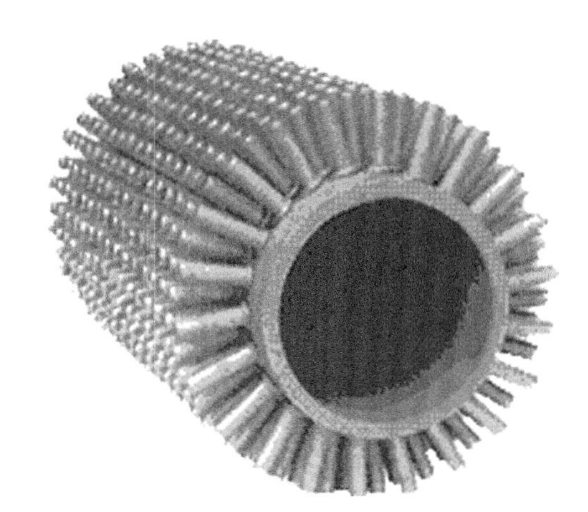

FIGURE 16.6
Closely spaced pin studding.

Supporting vertical non-drainable coils is easy as they are hung from their headers. However sealing through roof is not always perfect. Lugs are needed for alignment of tubes along and across gas flow and not much for load sharing. In the case of horizontal coils supporting is more complex. *See* Figure 8.25. Load of the coils is to be finally transferred to the vertical walls of the enclosure.

For low-temperature elements like economisers loads can be transferred from upper to the lower coil as shown in Figures 8.25 and 16.5 and thereafter finally transferred to the beams that rest on structure. In high-temperature elements like horizontal SH and RH coils rest on support tubes that are hung from the top (as shown in Figure 16.8).

FIGURE 16.7
Widely spaced studding of tubes.

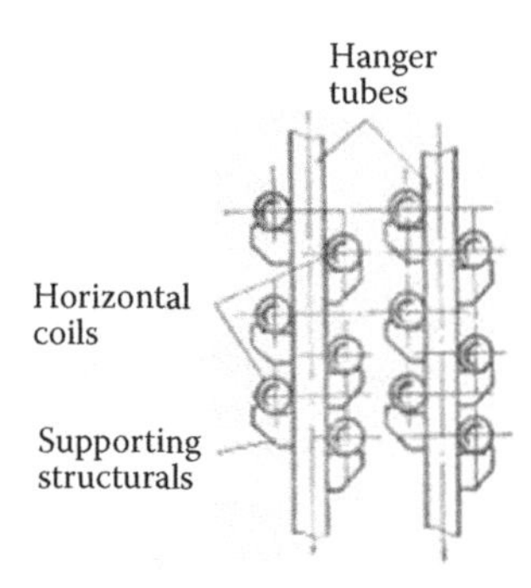

FIGURE 16.8
Supporting of horizontal coils from the top.

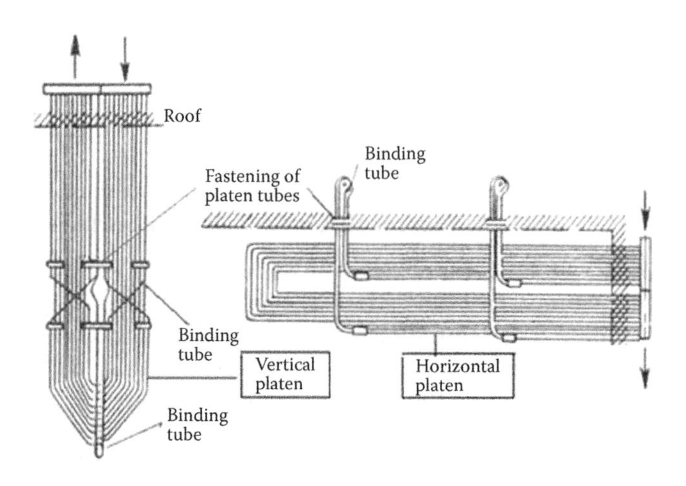

FIGURE 16.9
Vertical and horizontal platen SHs.

Depending on the metal temperature coils are made of C, C–Mo, low AS and high ASs including ss. Tube diameters vary from 31.8 to 63.5 mm od (1–2.5 in).

Spacers/lugs for the tubes, either for alignment in vertical coils or for support in horizontal coils, are always made of high-alloy high-temperature castings or ss possessing high hardness and wear resistance. Sometimes steam-cooled tubes are also ingeniously used as spacers.

16.29.6 Platens for SH and RH

When high temperatures like 500°C or higher are desired from boiler, it is normal to place some portion of SH inside the furnace in two-pass boilers to obtain high radiation rates of furnace so that the SH surface is compact. Widely spaced across the gas flow at >400 mm these platens 'see' the furnace radiation. They are called **platens** as the tubes are placed practically touching each other (with only 1.5 mm gap) so that molten ash does not lodge itself between the tubes. The wide spacing is also for preventing the bridging of gap by slagging ash deposits.

Heat transfer is very high. All heat is received as radiant heat and, therefore, the platen exit steam temperature drops as the boiler load is increased. But the overall SH characteristic tends to be fairly flat as the pendant SH absorbs heat mostly by convection and has rising characteristic with load as shown later in Figure 8.22.

Due to wide spacing of platens the number of tubes in platens is quite high. The resulting appearance is of a plate and hence the name platen.

As ash is in molten stage where platens are placed the cleaning requirements are heavy. Heavy-duty retractable SBs are placed adjacent to the platen SHs. Due to heavy soot blowing the platens in two-pass boilers swing gently which helps them further to shed the ash deposits. See Figure 18.15. Platens can be both vertical type as in two-pass boilers or horizontal type as in tower-type boilers.

Platens are not positioned as final SHs as the heat absorption is very high and there will be problems of excessive metal temperatures. As intermediate SHs they experience lower temperatures and can be made of T11 or T22 tubes. Usually platens are made of 50.8 or 76.2 mm (2 or 3 in) od tubes.

Figure 16.9 captures the details of both vertical and horizontal platen SHs.

16.29.7 Boiler Bank Tubes (*see also* Boiler Bank, Convection Bank *or* Evaporator Bank)

BBs employ tubes ranging from 38.1 to 63.5 mm od (1½" to 2½"), smaller tubes being popular with smaller package boilers. Smaller tubes have better heat transfer and larger tubes have higher strength. Oil- and gas-fired package boilers have to be very compact from transportation point and hence employ 38.1 mm od (1½") tubes in most cases. Field-erected boilers, particularly the bottom-supported units, need the tubes to act as struts as the BBs can be as tall as ~12 m where 50.8 mm (2") od is the most common size. Occasionally 63.5 mm od (2½") is also used. Tubes are scarcely larger than these sizes in both large package and field-erected boilers.

BB tubes are invariably made of low CS with high ductility to facilitate the easy expansion in the drums. SA 192, 178A or 210A-1 are the usual specifications. In most cases, the tube ends are also **annealed**. Both **ERW** and **seamless** tubes are employed. If the fear of opening up of weld joint is not present ERW tubes are preferred up to 70 bar pressure, as the perfect concentricity of id and od eases the tube expansion. Higher pressures demand thicker tubes which fall outside the ERW range and needing HFS tubes.

BB heights vary from

- As low as ~3 m (10 ft) in small **D-type boilers** and other small package boilers with high heat flux to generate adequate thermo-syphonic head to
- As high as 12 m (40 ft) in large **field assembled package** boilers in refineries

Tubes are to be placed quite closely to obtain good heat transfer rates. At the same time, access is needed for tube repair. To have a clear gap of 50–60 mm between tubes is the normal practice.

Swaging of bank tubes is often done to improve the ligament η of the drums and reduce their thicknesses. **Bi-drum boilers** are made up to drum pressures of 120 bar pressure limited by the feasibility of tube expansion. Close cluster of bank tubes demands on-line soot blowing to keep tubes clean in all but gas and light oil-fired boilers. Rotary SBs are adequate as the gas temperatures are low.

16.29.8 Risers and Downcomers

16.29.8.1 Risers

Risers/releasers and downcomers/supplies are part of circulating system of a boiler. In downcomers or supplies, almost steam-free water from the top drum is taken to the bottom headers of the furnace in downward direction. On the contrary, in risers, the steam-water mixture is taken from the bottom headers to the top drum in an upward direction.

They are termed *riser tubes* when heat is received by risers, as in the convection bank or furnace. Outside the gas path they are called *riser pipes* collecting the steam–water mixture from the furnace top headers and delivering to the steam drum.

Generous sizing of supply and riser pipes ensures better water circulation under all loads but physically accommodating them is a problem that becomes progressively difficult as boiler pressures increase.

Riser pipes are usually 76.2–114 mm od. The total area of risers is ⅓ – ½ the area of water walls for low and medium pressures. The area is higher for higher pressures. Riser pipes longer than 6 m are usually supported by sling rods to prevent sagging and vibration.

As only saturated steam–water mixture flows through the riser/releaser pipes, they are made of CS. For reducing thickness, low-alloy steels such as SA213 P1 and P11 are also occasionally used.

16.29.8.2 Downcomers or Supplies

When placed inside the gas flow they are *downcomer/supply tubes* as in BBs and when placed outside they are *downcomer or supply pipes,* distributing water to various furnace headers.

Large-bore downcomers are usually adopted in bigger boilers from layout considerations. They are attached to the top and bottom drums in single- and bi-drum boilers, respectively. In bottom-supported boilers, they are employed in preference to the small supply pipes as they are additionally required to carry the boiler loads. They vary in size from 193 to 457 mm od (7¾″ to 18″). Two large-bore downcomers, on either side of the drum, is provided as a minimum, and they increase with the size of boiler.

Package boilers do not have large bore downcomers but only the field-erected boilers of medium and large sizes have them. As only saturated water flows through the downcomer/supply pipes, they are made of CS. For reducing thickness low-alloy steels such as SA213 P1 and P11 are also occasionally used.

16.29.9 Integral Piping

Integral piping is the piping within the boiler envelop between the FW check valve and MSSV. Integral piping usually consists of piping for

- SBs
- Blowdown—both CBD and IBD
- Drain, vent and sampling
- Terminal piping—fw inlet, main steam outlet and RH inlet—RH outlet

16.30 Pressure Measurement (*see* Instruments for Measurement)

16.31 Pressure Part Design (*see also* Stresses, Allowable or Permissible in Pressure Parts)

Establishing thickness requirement of all parts subjected to either internal or external pressure of steam and water is the aim of the PP design in a boiler. It is no way connected with the sizing of the equipment which is done by the thermal design. Boiler codes deal elaborately and exclusively with PPs as they are concerned with safety of operators and equipment.

PPs constitute 30–40% cost of a total boiler including auxiliaries and take the maximum time to manufacture. As the p & t of the boiler increase the PP materials and manufacturing becomes more sophisticated. Even though the PP design is relatively simple in concept it becomes very elaborate and involved as it has to meet the conflicting requirements of safety, reliability and cost competitiveness. When related aspects like materials, welding, fabrication quality, attachments, strength compensation and so on are all taken into account, PP design becomes an elaborate and structured subject.

The PPs of a boiler consist essentially of rounds of three types and can be divided into

- Drums
- Headers
- Tubes
- Pipes

The PP design of the boiler parts is primarily built around the thin shell design of the component where the thickness is derived from the formula

$$t = \frac{pd}{2f} + c \tag{16.1}$$

where

p = calculation pressure of the component
d = mean diameter
f = allowable stress at the design temperature
c = design allowance to cover for effects such as corrosion, erosion and so on

16.31.1 Pressure

The *highest SVLP* is usually the *design* pressure of the boiler. Some codes

- Demand that all the components be designed uniformly to a single pressure, namely, the boiler *design* pressure. In such a case the design and calculation pressures are the same.
- Permit adjustments to the design pressure to take into account the pressure drops in the steam pipes and tubes and also the effects of the static head. The design pressure, in such case, will be adjusted to arrive at the *calculation* pressure which will be slightly different for each component.

The pressure that a component can safely withstand without exceeding the safe permissible limits at the specified design temperature is the *maximum working* pressure of the component.

16.31.2 Temperature

The maximum temperature to which a component can be subjected during the upset conditions is the design temperature. These upsets can occur typically when there are

- Load ramp ups
- Peak loads
- Unbalance in steam and gas flows and temperatures
- Operation with slagging and fouling and so on

For different components the margins to be added for arriving at the design temperature over the maximum operating temperature are different and the codes specify the values.

16.31.2.1 Mean Metal Temperature

This is the arithmetic mean of the inside and outside metal temperature of a tube, pipe or plate subject to heat across its thickness. Whenever the metal temperature is mentioned it refers to this mean temperature unless specifically indicated as inside or outside.

Codes indicate limiting operating temperatures for all materials, which invariably means the mean temperature, unless qualified specifically otherwise.

For thick-walled components, the temperature drop across the thickness can be quite large, particularly in SH tubes. The inside, outside and mean temperatures can vary significantly and the design limits have to be applied suitably lest there should be failures due to overheating.

16.31.3 Allowable Stresses

The allowable stress values to be used in the design are temperature dependent. The following is based on ASME Section I:

- UTS governs the stress values up to a temperature of about 300–350°C (~570–660°F). Drums and all headers, except for SH and RH, are therefore decided by tensile strength considerations. Eco and Evap tubes also fall in this category in most cases. ASME Section I specifies, since 1998, a safety factor of 3.5 over room temperature tensile strength. European codes are less conservative as they adopt a lower factor of 2.5–2.7 in this range.
- YS governs the stress values between ~300°C and 500°C (~570–930°F) depending upon the metallurgy. The safety factor adopted by ASME is 1.5 which is the same as European practice. Bulk of the lower end SH and RH tubes and headers fall in this range.
- Creep strength or stress to rupture, whichever is lower, governs the stress values to be adopted for high end tubes and headers of the SH and RH. Stress to rupture is considered for 100,000 h. It is the lower of 80% for minimum stress and 67% for average stress
- Creep strength is the average stress to produce 0.01% creep rate in 1000 h as per ASME and is considered as such for design with a safety factor of 1.

Many other codes do not specify the allowable stresses but provide the ultimate strength values and the formulae from where the stress values are to be derived.

As PP designing is extensively codified, the real complexity of the subject is kept out when a designer deals with the coded calculations. There are a set of design rules to be followed and a set of manufacturing norms to be observed to achieve good results in PP making. This serves well, because the PP design affects the safety.

16.32 Pressure Reducing and Desuperheating (*see* Steam Conditioning)

16.33 Pressurised Firing (*see* Draught)

16.34 Pressurised Fluidised Bed Combustion (*see* Fluidised Bed Combustion)

16.35 Pressurised Milling (*see* Pulverisers)

16.36 Prime/Principal Fuel (*see* Fuel Firing Modes)

16.37 Priming (in Steam Drums) (*see* O&M [Operation and Maintenance] Topics)

16.38 Producer Gas (*see* Gaseous Fuels)

16.39 Projected Radiant Surface (*see* Furnace in Heating Surfaces)

16.40 Propeller Fans (*see* Axial Fans)

16.41 Properties of Air and Flue Gas

16.41.1 Air (*see also* Combustion Air *under* Combustion)

Dry air is a mixture of gases, 78% N_2, 21% O_2 and 1% inerts (with traces of CO_2, Ar and various other components) by volume. By treating inerts as part of N_2 the composition of dry air can be considered as

	% vol.	% wt
N_2	79	76.85
O_2	21	23.15

Air is treated as a uniform gas with properties that are averaged from all the individual components. Table 16.3 provides the important properties of air at STP conditions, namely, 15°C and 760 mm wg in mks and fps systems.

Table 16.4 gives a range of properties of air at various temperatures.

16.41.2 Flue Gas

In **combustion** reaction gas and ash are the end products. Combustibles in fuel and O_2 in air chemical react to produce various products of combustion, consisting mainly CO_2, N_2 and H_2O the sum of which is the flue gas. Excess O_2, unburnt CO and SO_2 from S in fuel are also present. Boiler gases are called flue gases because they are let out into the atmosphere by flues.

Unit and total flue gas are the two stages in combustion calculations. Unit gas quantity is the gas generated on burning a unit quantity of fuel while the total gas quantity is calculated by multiplying the unit quantity with total *fuel burnt* (fuel-fired less the unburnt fuel).

Dry flue gas is the calculated theoretical flue gas weight which is the sum of combustible weights and corresponding air required. This is an intermediate step in further calculations.

Wet flue gas is the actual amount of flue gas after adding (1) humidity in air, (2) M in fuel and (3) M due to the combustion of H_2 in fuel.

TABLE 16.3

Notable Properties of Air at STP Conditions (15°C and 760 mm wg)

Property	Symbol	Dimensions	Value	
			Metric	British
Density	ρ	Mass/volume	1.229 kg/cum	0.00237 slugs/cft
Specific volume	v	Volume/mass	0.814 cum/kg	422 cft/slug
Pressure	p/P	Force/area	101.3 kN/m²	14.7 lb/ft²
Temperature[a]	t/T	Degrees	15°C 288°K	59°F 519°R
Viscosity	μ	Force-time/area	1.73×10^{-5} N-s/m²	3.62×10^{-7} lb-s/ft²
Gas constant	R	Energy/mass/deg	0.286 J/g/K	53.5 ft-lb/lb/°R
Specific heat at constant volume	c_v	Energy/mass/deg	0.715 J/g/K	53.5 ft-lb/lb/°R
Specific heat ratio	γ		1.4	1.4

[a] Standard temperature at MSL (STP).

TABLE 16.4

Properties of Air at Elevated Temperatures

Temperature t in °C	Density ρ in kg/m³	Specific Heat c_p in kJ/kg K	Thermal Conductivity k in W/m K	Kinematic Viscosity v in 10^{-6} m²/s	Coefficient of Expansion b in 1/1000
0	1.293	1.005	0.0243	13.30	3.67
20	1.205	1.005	0.0257	15.11	3.43
40	1.127	1.005	0.0271	16.97	3.20
60	1.067	1.009	0.0285	18.90	3.00
80	1.000	1.009	0.0299	20.94	2.83
100	0.946	1.009	0.0314	23.06	2.68
120	0.898	1.013	0.0328	25.23	2.55
140	0.854	1.013	0.0343	27.55	2.43
160	0.815	1.017	0.0358	29.85	2.32
180	0.779	1.022	0.0372	32.29	2.21
200	0.746	1.026	0.0386	34.63	2.11
250	0.675	1.034	0.0421	41.17	1.91
300	0.616	1.047	0.0454	47.85	1.75
350	0.566	1.055	0.0485	55.05	1.61
400	0.524	1.068	0.0515	62.53	1.49

16.42 Properties of Steam and Water

The following sub-topics are covered here:

1. Water
2. Steam
3. Phase change of water
4. Latent heat, heat of evaporation or heat of transformation
5. Boiling or evaporation
   a. Nucleate boiling
   b. Film Boiling
6. Superheat (SH)
7. Reheat (RH)
8. Density (ρ) of steam and water
9. Specific volume of steam and water
10. SBW and volume (SBV)
11. Throttling

16.42.1 Water

Water is a chemical containing two H_2 and one O_2 atoms in its molecule. Water is a liquid at ambient conditions, but often co-exists with its solid state, ice, and gaseous state, water vapour or steam. It is the most abundant compound on earth, covering about 70% of the planet's surface. It is nearly colourless with a hint of blue at ambient conditions, tasteless, and odourless. It is commonly referred to as *the universal solvent* as many substances dissolve in water. Naturally, water in nature and in use are rarely pure.

Water can be *sub-cooled* or at *saturated* conditions. Water at temperatures lower than its saturation at the specified pressure is called sub-cooled water. When it is heated to a point when its temperature no longer increases it is the saturation temperature at that pressure. It is the same as boiling temperature. Water heated to saturation temperature is called saturated water.

When water is cooled to <0°C it solidifies and turns into ice.

16.42.2 Steam

Steam is *water in vapour state*. It has inherently a lot of considerable heat energy as steam is produced only when considerable heat is added to water. Steam, therefore, forms an ideal working fluid for heat engines like steam engines and STs. Not only is water available widely but it is also non toxic. The main uses of steam are for process heating and power generation.

Steam can be in *wet*, *saturated* or *superheated* conditions.

Wet Steam

When heat is added to *sub-cooled water* (water at temperature lower than its saturation temperature at that pressure) at certain pressure it reaches the saturation temperature and boils at that constant temperature while absorbing *latent heat of evaporation*. There is a progressive conversion of water into steam. During this process of *boiling* the vapour contains both steam and water and it is known as *wet steam*. *Dryness fraction* of the wet steam is the% of steam in the steam–water mixture.

Saturated Steam

When all the water is just converted to steam it is called *saturated steam* and it has no water vapour. It is 100% dry. Steam is at its highest ρ when it is saturated. Also the specific heat is at its highest. Process industry needs mostly saturated steam for these attributes.

Superheated Steam

As more heat is added to the saturated steam the temperature increases and steam is now said to have become SH steam. The difference between the prevailing temperature and the saturation temperature is the *degree of SH*. Higher the temperature greater is the heat content lower the ρ. Heat engines, like the STs, need SH steam due to the high heat content.

16.42.3 Phase Change of Water

There are three states for any substance, namely solid, liquid and gaseous states. Those states for water are ice, water and steam, respectively, with latent heats of fusion and evaporation for phase conversions from ice to water and water to steam. This is captured in Figure 16.10 for water at atmospheric pressure.

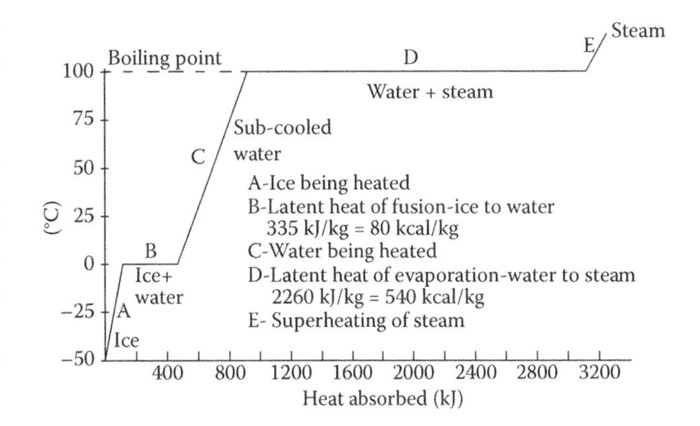

FIGURE 16.10
Phase change diagram of water at atmospheric pressure.

As the pressure increases, the latent heat of evaporation shrinks until it becomes zero at the critical pressure. This is described in detail in Critical Pressure Point.

16.42.4 Latent Heat, Heat of Evaporation or Heat of Transformation

Latent heat is the quantity of heat absorbed or released by a substance undergoing a change of state, such as ice changing to water or water to steam, at constant p & t. It is also called *heat of transformation*.

Heat is absorbed by the substance for loosening of the atomic particles while changing from lower to higher state. Hence the temperature of the substance remains constant. Heat is given off when the reverse takes place.

In changing from water to steam the heat absorbed is the *latent heat of evaporation*. At room pressure it is 2260 kJ/kg, 540 kcal/kg or 970 btu/lb at a constant temperature of 100°C or 212°F.

16.42.5 Boiling or Evaporation

Boiling or evaporation can be defined as the *addition of latent heat* to a liquid *at constant temperature*. It is phase transition. When heated a liquid reaches its boiling point and further heating goes only towards the increase of its latent heat with no further rise in temperature. The liquid progressively evaporates to form vapour.

Boiling point is the temperature at which the vapour pressure of the liquid equals the pressure exerted on liquid by the surrounding pressure.

Depending on the heat flux boiling occurs in two stages, namely *nucleate* and *film boiling*. See also Flow Types in vertical and horizontal tubes. Figure 16.11 describes the phenomenon very well set in four regions.

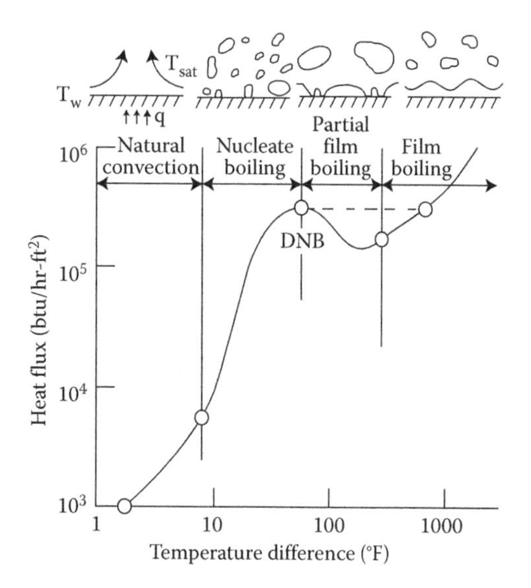

FIGURE 16.11
Nucleate and film boiling shown on *H–t* curve.

16.42.5.a Nucleate Boiling

The first and second regions show that as heat flux increases, the Δt between the surface and fluid does not change very much. Better heat transfer occurs during nucleate boiling than during natural convection.

Nucleate boiling is characterised by the formation of bubbles on the heated surface. This happens with relatively lower heat flux and low temperature of the surface. Bubbles form at the bottom and rise to the top, while growing in size, carrying heat with them. The number of nucleation sites increase with the increased roughness of the heated surface. The heated surface is always covered by the liquid film which keeps the surface close to the liquid temperature. In heated tubes there is a film of water always adhering to the tube wall during this phase and hence it is also known as *wet wall flow*.

16.42.5.b Film Boiling

As the heat flux increases, steam bubbles become numerous enough that partial film boiling (part of the surface being blanketed with bubbles) sets in. This region is characterised by an increase in Δt and a decrease in heat flux. Further increase in Δt thus causes total film boiling, in which steam completely blankets the heat transfer surface. Since the steam film has much lower conductance compared to the water film, the hot surface is not effectively cooled raising its surface temperature alarmingly.

In boiler tubes subjected to heat, the efforts are always to maintain the boiling process within the nucleate boiling regime and avoid excursions to film boiling. This way the tube temperatures are kept close to the boiling point of the water inside the tubes thus preventing overheating.

16.42.5.c Departure from Nucleate Boiling and Critical Heat Flux

In practice, if the heat flux is increased, the transition from nucleate boiling to film boiling occurs suddenly, and the temperature difference between the fluid and the inside of tube increases rapidly, as shown by the dashed line in Figure 16.11. The point of transition from nucleate boiling to film boiling is called the point of departure from nucleate boiling, commonly written as DNB. The heat flux associated with DNB is commonly called the critical heat flux. DNB is largely a governed by heat flux and tube metal temperature and both are controlled to prevent the onset of DNB.

As the nucleation points can be increased with higher surface roughness or more irregular surface evaporation, tubes in the areas of the highest gas temperatures and heat flux are made of *ribbed/rifled* tubes. The inside of the tube has flutes similar to the barrel of a gun, as shown in Figure 16.12. Rifled tubes are available in all metallurgies as the conventional tubes in single or multi-fluted ribs with differing shapes of the flutes.

16.42.6 Superheat

When steam is heated beyond the boiling temperature it enters SH condition. *SH is the differential temperature between the prevailing steam temperature and corresponding boiling temperatures.*

SH steam and water cannot co-exist. For heating applications, saturated steam with its higher specific heat is preferable whilst for power production in ST SH steam is better with its higher energy content.

In a boiler, dry steam emanates from the steam drum after its initial separation from water and subsequent purification. The HS that imparts SH to steam is called SH. The higher the SH, the higher is the η of steam cycle. The metallurgy of the ST components and the size of the ST determine the SOT. In the smaller power up to ~30 MWe, 485, 510 and 530°C (905°F, 950°F and 985°F) are the common SOTs. For larger STs like 60 MWe and above, 540°C (1005°F) is most common. In the large Hrsgs and sub-c utility boilers 565°C (~1050°F) is gaining popularity with the acceptance of T91 tube material. In SC boilers, temperatures up to 620°C (~1150°F) and even higher are employed.

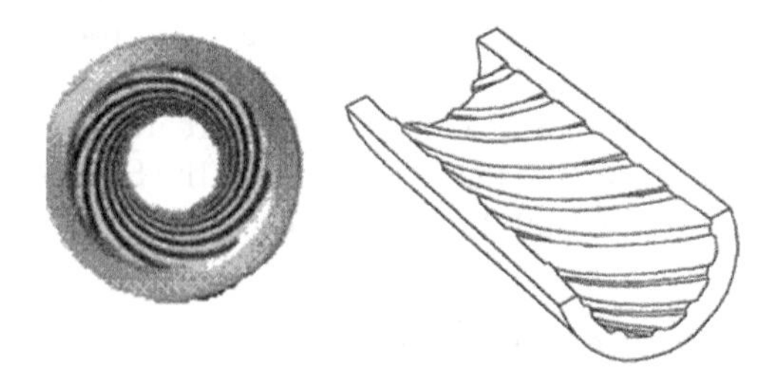

FIGURE 16.12
Ribbed or rifled tube.

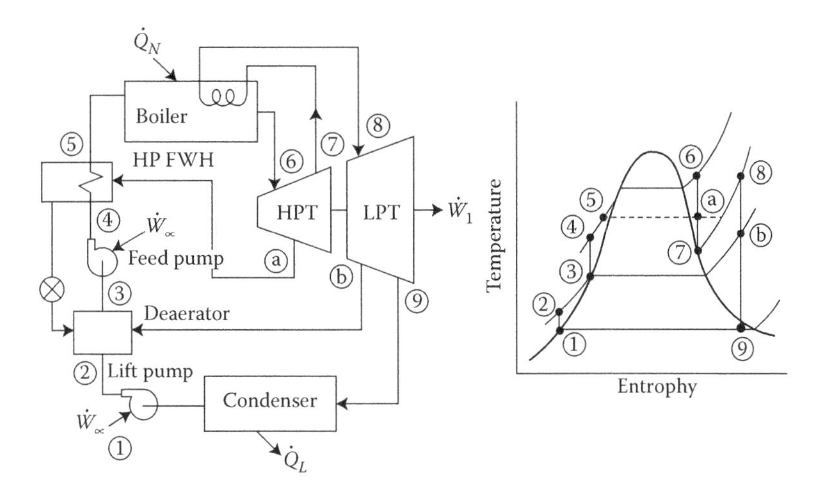

FIGURE 16.13
Reheat steam cycle and its depiction on T-s chart.

16.42.7 Reheat

As the name suggests, reheating is, *heating steam again*. In a steam cycle after its expansion in ST to 25–35% of initial pressure, the steam is withdrawn and reheated to the original or higher temperature. Again it is expanded in the ST to the desired vacuum. In doing this, the cycle η is improved because the additional work done due to reheating is more than the additional heat dissipated in condenser. This is shown in Figure 16.13.

Cycle η can be further improved by double RH. This is presently feasible in plants' sizes ~800 MWe, while single RH is economical at >100 MWe.

16.42.8 Density (ρ) of Steam and Water

ρ of a fluid is its *mass per unit volume* in kg/cum. At 4°C, pure water has ρ of 1 g/cc, 1 kg/L, 1000 kg/cum or 62.4 lb/cuft. ρ of pure water is constant at a particular temperature and varies with temperature and impurities. ρ is the inverse of *specific volume*.

Density Differential Between Steam and Water

ρ of saturated water decreases with increase in pressure. As the saturation temperature increases, water expands and gets lighter. Typically, the water ρ which is 958 kg/cum (59.8 lb/cuft) decreases progressively, as shown in Figure 16.14, to 315.5 kg/cum (19.7 lb/cuft) at the critical pressure of 225.5 bar.

The ρ of steam, on the other hand, increases with pressure. This is because the steam is compressible fluid and occupies smaller volume when pressurised.

There is considerable ρ differential at lower pressures between saturated water and steam. This is the main driving force for circulation in Evap circuits in boiler. This ρ differential progressively narrows with the increase in pressure until it becomes 0 at the critical point.

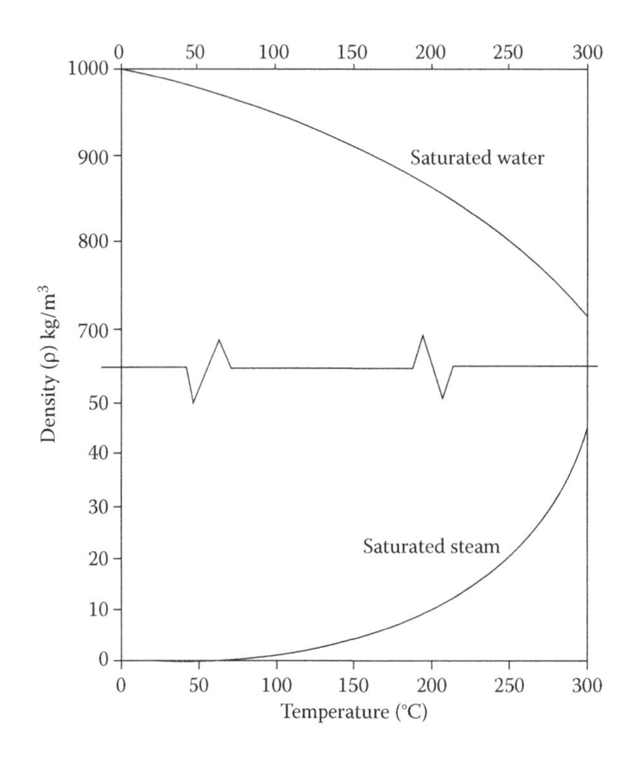

FIGURE 16.14
Density variations of saturated water and steam with pressure.

At 70 bar, water is ~20 times heavier than steam which narrows to as low as ~3 at 200 bar at which is just adequate for having natural circulation in a boiler. Figure 16.14 brings out this aspect very clearly.

16.42.9 Specific Volume of Steam and Water

Specific volume is the volume occupied (in cum/cft) by unit mass (1 kg/1 lb) expressed in cum/kg or cft/lb. Specific volume is in inverse of ρ.

16.42.10 Steam by Weight and Volume

Water, contained in an enclosure like a vertical tube, when heated over the whole length from outside, generates bubbles of steam which keep increasing with height. As pure water begins to turn into a mixture of steam and water, single-phase flow changes to two-phase flow. The amount of steam formed depends on the heat flux, length of tube, water pressure, velocity of mixture and so on. The percentage of steam in the mixture at any place can be expressed by either the weight or volume, that is, SBW or SBV.

In natural circulation boilers, the steam at the top of circuits generally varies from 55% to 80% by volume to ensure nucleate boiling or wet wall flow over the entire height of tubes. This depends on the drum pressure, typically, 80% at 20 bar and 55% at 180 bar. However, this would correspond to 5–35% by weight for the same pressures. At higher pressure due to higher ρ of steam % SBW is higher than at lower pressures. This would mean a circulation ratio of 20 and 3 for the circuit at 20 and 180 bar, respectively. This relationship between % SBW and % SBV can be seen in Figure 16.15.

16.42.11 Throttling

When steam gently flows through an orifice or an equivalent pressure reducing device from higher to lower pressure, with no change in velocity or static head or any radiation loss, it expands adiabatically without doing any work. The enthalpies of steam upstream and downstream of the orifice, that is, the higher and lower pressures are the same. This expansion is called throttling. The downstream temperature adjusts accordingly.

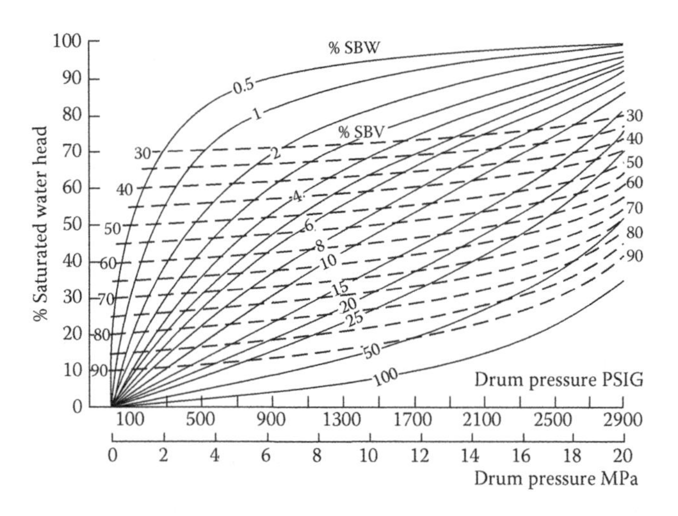

FIGURE 16.15
SBV and SBW at various drum pressures.

16.43 Properties of Substances

The sub-topics covered here are listed below.

1. Angle of repose
2. Bulk and true densities of solids
3. Partial pressure
4. pH
5. Relative ρ
6. Relative humidity
7. SG of fluid
8. Specific weight (Y)
9. Specific volume (υ)

16.43.1 Angle of Repose

Angle of repose is the *angle of rest* of bulk materials. This is the angle which the surface of a material of a *normal freely formed pile makes with horizontal.* This is of much importance in the design of bunkers, bins and hoppers. Angle of repose is greatly influenced by the grain size of material, its shape, M content, ρ, coefficient of friction and so on. Hence it is not a fixed figure but a range. Table 16.5 provides typical values of angles of repose for many common materials.

16.43.2 Bulk and True Densities of Solids

Bulk ρ is the weight per unit volume of divided solids (bulk material) such as powders, PM, granules or pellets. The volume here includes particle volume, voidage and internal pore volume. *Bulk ρ is not an intrinsic property of a material* as it can change depending on how the

TABLE 16.5
Angles of Repose for Various Bulk Materials

Angle of Repose	Characteristics
0–19°	Uniform size, very small rounded particles, either very wet or very dry, such as dry silica sand, cement, wet concrete and so on
20–29°	Rounded, dry polished particles, of medium weight, such as whole grain and beans
30–34°	Irregular, granular or lumpy materials of medium weight, such as anthracite coal, cottonseed meal, clay, and so on. Ash has an angle of ~35°
35–39°	Typical common materials such as bituminous coal, stone, most ores and so on
40° upwards	Irregular, stringy, fibrous, interlocking material, such as wood chips, bagasse, bark, shredded coconut, fresh coffee beans, tempered foundry sand and so on

material is handled. Freely stored material has lower bulk ρ than compressed material.

In contrast, the particle ρ or true ρ of a particulate solid or powder is the actual ρ of the particles. It is relatively well defined as it is not dependent on the degree of compaction of the solid. However, definitions of particle ρ differ based on whether pores and voids are included in the particle volume. ASTM D 1895 B is usually followed for measuring the bulk ρ of solids.

Bulk ρ is important in the design of bins, bunkers, hoppers and silos. For storage of coal it is normal to consider a bulk ρ of 800 kg/cum (50 lbs/cft) and 40° angle of repose. In case of well compacted open coal piles bulk ρ is typically 1050–1150 kg/cum (~65–72 lbs/cft).

16.43.3 Partial Pressure

John Dalton (1766–1844), the English physicist who gave us Dalton's atomic theory, enunciated in Dalton's law of partial pressures that the *total pressure of a mixture of gases equals the sum of the pressures that each would exert if it were present alone*. In a mixture of *ideal gases*, each gas has a *partial pressure* which is the pressure which the gas would have if it alone occupied the volume.

If P_t is the total pressure of a sample which contains a mixture of gases and P_1, P_2, P_3 and so on are the partial pressures of the gases in the mixture then $P_t = P_1 + P_2 + P_3 + \cdots$

16.43.4 pH

When a neutral solution, say pure water, dissociates an equal number of hydrogen (H^+) and hydroxyl (OH^-) ions are formed. *pH is the negative logarithm of H ion.*

$$pH = \log \frac{1}{\langle H^+ \rangle} \tag{16.2}$$

Neutral solution has a pH of 7. A solution with pH < 7 is acidic and corrosive while a solution with pH > 7 is alkaline and protective. Very high pH can cause scaling and deposition. Most natural waters have pH in the range of 6–8.

16.43.5 Relative ρ (see Specific Gravity)

16.43.6 Relative Humidity (*see also* Psychometric Chart)

Humidity is the amount of watervapour in a mixture of air and water vapour stated commonly in g/m³. Relative humidity is the fraction of water vapour held by air compared what it can hold under saturated condition at that temperature expressed in %. It is also the partial pressure of water vapour in the air–water mixture, given as % of the saturated vapour pressure under those conditions.

The relative humidity of air thus changes not only with respect to absolute humidity but also p & t. If the actual vapor ρ is 10 g/m³ at 20°C compared to the **saturation vapor ρ at** that temperature of 17.3 g/m³, the relative humidity is 10/17.3 = 57.8%.

16.43.7 Specific Gravity of Fluid

SG of any substance is the *ratio of ρ of the fluid to the ρ of reference fluid*. It is a dimensionless ratio. Water is used as reference fluid for solids and liquids and air for gases. SG of liquids is usually stated, for example, as 15/15°C where the upper temperature refers to the liquid and lower to the water. If no temperatures are stated, the reference is of water at its maximum ρ (1000 kg/cum or 62.4 lbs/cft) which occurs at 4°C under atmospheric pressure. For gases, it is common practice to use the ratio of MW of gas to that of air (28.96).

16.43.8 Specific Weight

Specific weight of a fluid is its *weight per unit volume* (not mass as in ρ) in lbf/cuft, N/cum:

$$\text{Specific weight } Y = \rho g,$$

where ρ is ρ and g is the acceleration of gravity.

16.43.9 Specific Volume

Specific volume of a fluid is its volume per unit mass in cum/kg-mass or cuft/lb-mass. Specific volume is the inverse of ρ.

16.44 Proximate Analysis (*see* Coal)

16.45 Psychrometric Chart

A typical psychrometric chart, shown in Figure 16.16, displays the thermodynamic properties of moist air at a constant pressure (often equated to an elevation relative to sea level). It depicts and relates comprehensively the

- Dry bulb temperature
- Wet bulb temperatures
- Humidity

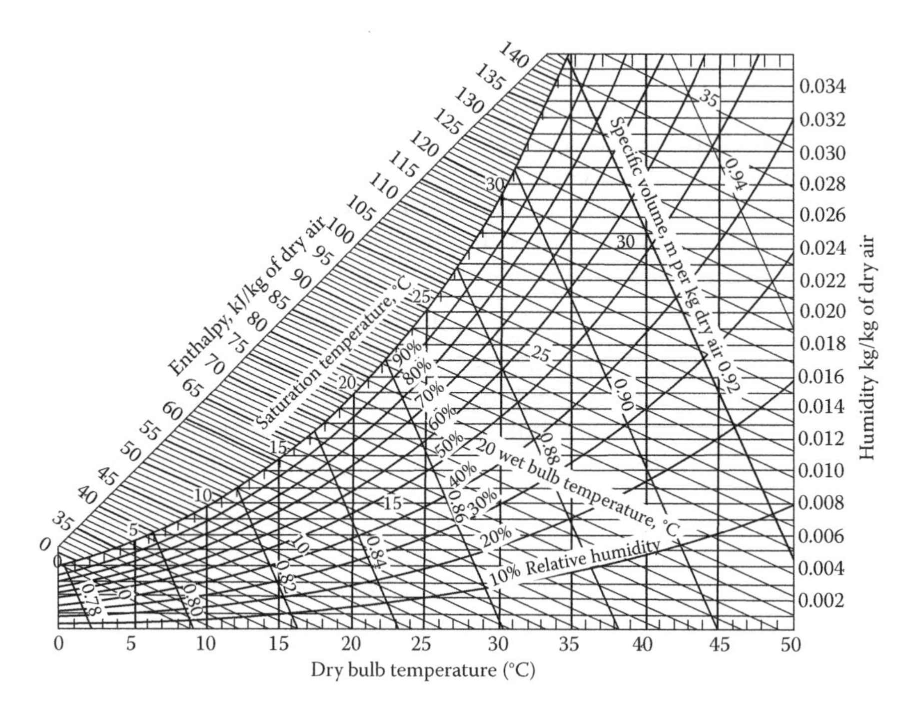

FIGURE 16.16
Psychometric chart for air.

- Relative humidity
- Saturation temperature
- Specific volume
  - Psychrometric chart is of great importance to boiler engineers as air is one of the inputs into the boiler and its properties are needed in combustion calculations.
  - Dry bulb temperature (T_{db}) is the temperature of the surrounding air measured by a thermometer. It is the ambient air temperature.
  - Wet bulb temperature (T_{wb}) is the temperature indicated by a thermometer when its bulb is moistened, that is, wrapped in wet muslin. Adiabatic evaporation of water from the thermometer and its resulting cooling effect is indicated by wet bulb temperature.
  - T_{wb} is always lower than the T_{db} unless the relative humidity is 100% when they are both equal.
  - Dew point temperature (T_{dp}) is the temperature at which water vapour starts to condense out of the air when air is completely saturated.
  - Humidity is the quantity of water vapour present in air. It can be expressed as an absolute, specific or relative value.
  - Relative humidity is the ratio of vapour ρ of air to the saturation vapour ρ at the dry bulb temperature. Relative humidity is usually expressed in %.
  - Specific volume of moist air is the total volume of humid air per mass unit of dry air (not dry air + M).
  - Enthalpy of wet (moist and humid) air: In atmospheric air, water vapour is present from 0–3% by mass. The enthalpy of wet air is the sum of sensible heat of dry air and latent heat of evaporated water per unit of wet/humid air.

16.46 Pulsating Grate (*see* Grate Types)

16.47 Pulse Jet Filter (*see* Bag Filters)

16.48 Pulverising or Grinding

Pulverising or grinding is the *action of reducing any solid fuel to fine powder to increase the surface area greatly* for enhancing both the speed and completeness of

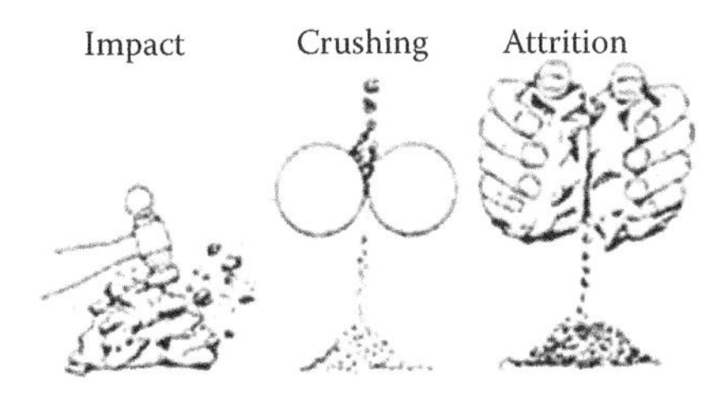

FIGURE 16.17
Actions of impact, crushing and attrition.

combustion. *Any solid that is not fibrous or stringy can be pulverised.*

Action of Pulveriser

In coal firing large pieces of coal are crushed to <50 mm (2″) size in the coal-yard by the crushers. The crushed fuel is transferred to the fuel bunkers or silos from where it is extracted and fed into pulverisers for grinding. The pulverising action inside any mill is a *combination of impact, crushing and attrition*. These actions are as shown in Figure 16.17. In horizontal mills impact + attrition while in vertical mills crushing + attrition predominate.

Self-Pulverisation

Low-rank coals produce 'self-pulverisation' on combustion because of sudden evaporation of the large amount of inherent M. This M is held trapped inside the fine pores of coal structure. On application of heat it evaporates instantly. As steam occupies nearly 1600 times the volume of water at atmospheric pressure it is nothing short of an explosive eruption, tearing apart the surrounding coal into fine pieces. This phenomenon is termed as self-pulverisation. That is the reason for only sizing and not pulverising of the low-grade coals like **lignite** and **peat**.

16.49 Pulverisers (*see also* Beater Mills in Lignite and Firing)

Pulverisers or mills are the equipments that grind fuel to very high fineness—talcum powder consistency in case of coal—for firing in the **PF boilers**.

Any pulveriser has to perform the four essential tasks of

a. Grinding the fuel
b. Evaporating the fuel M
c. Separating the coarse particles for regrinding
d. Discharging ground particles of proper size

Mill air at high p & t enters the pulveriser and helps to perform the above tasks of b, c and d. The sub-topics under pulverisers are listed below.

1. Pulveriser types
   a. Verticals mills/pulverisers
   b. Horizontal mills/pulverisers
   c. Beater mills
   d. Atrita mills
2. Pulveriser capacity
3. Pulveriser power
4. Fires in pulverisers and mill inerting
5. Pulveriser requirements
6. Milling plant

16.49.1 Pulveriser Types

Based on the *method of operation*, pulverisers can be divided as *suction or pressure* types. In *suction mills*, the mill operates under suction with an exhauster fan is located downstream of the mill (between mill and burner). This sucks the air and fuel mixture from mill and discharges to the burners.

- The big advantage of the suction mill is that there are no chances of fuel leakage and consequent fires in the milling plant. The entire area is clean and safe.
- As the exhauster fan operates at mill outlet temperature of ~90°C, it is compact and simple.
- But the fan ηs are low, as the impeller has to be radial type to deal with dusty air. The resulting auxiliary power consumption of the milling plant works out quite high in comparison to the pressurised system.
- Also this system is workable only if the fuel is not erosive.

Modern plants employ mostly *pressurised milling* where the hot PA fan pushes the hot air all the way up to the burners carrying the ground fuel powder with it from the mill. Mills are under pressure but good sealing arrangement in mills is now well established. Leakage of fuel is a thing of the past. Even with abrasive fuels, this system works well as fan deals with clean air and does not experience wear.

Based on *construction*, pulverisers can be divided as

a. Vertical mills
b. Horizontal mills

with the former being highly popular. These mills are explained under separate headings.

16.49.1.a Verticals Mills/Pulverisers

Vertical coal mills were initially developed in the steel industry and later in cement and finally in power. PF boilers even during the 1950s were built with separate milling and boiler sections with *indirect firing*. With the advent of vertical mills *direct firing* gained popularity and the milling and boiler sections were integrated.

Salient features of vertical mills

Vertical mills are basically air swept roller mills. This type of mill combines in a single chamber the (a) grinding (b) drying and (c) classification sections in a compact manner.

- These mills can be made in very wide ranging capacities from ~5 to 200 tph.
- It is possible to provide certain redundancy in a boiler by having a standby mill which is not the case with horizontal ball mills due to their large sizes.
- With low weights and less inertia, these mills are very dynamic and load responsive. Mills are compact.
- There is also a greater flexibility in laying out the mills.
- Vertical mills are also much less noisy.

All these advantages helped the development of direct firing of PF boilers with progressive marginalisation of the horizontal ball and tube mills.

Types of Mills

Based on the speed of rotation, vertical mills can be classified as

1. a. High-*speed mills* operating between 50 and 90 rpm

   These mills are for the lower capacities usually ranging between 5 and 25 tph. Mill drive is through an integral worm or bevel gear.

   b. Low-speed mills operating between 20 and 40 rpm

   These mills are for medium and large capacities ranging from ~15 to 200 tph. Here, the drive is through a planetary gear box which can be detached from the mill for maintenance purpose.

   By way of construction, vertical mills can be divided as follows:

2. a. Ball and racer mill

   Here, the grinding elements are 7–12 nos. of steel balls rotating on a lower race that is attached to the rotating table and upper stationary race. Typical ball and racer mill is shown in Figure 16.18.

   b. Roller and bowl mill

   Here, 2–6 rollers attached on a wheel connected to or through the housing, rotate on a bowl fixed to the rotating table. Arrangement of a typical vertical bowl mill is shown in Figure 16.19 which is proprietary, called MPS mill.

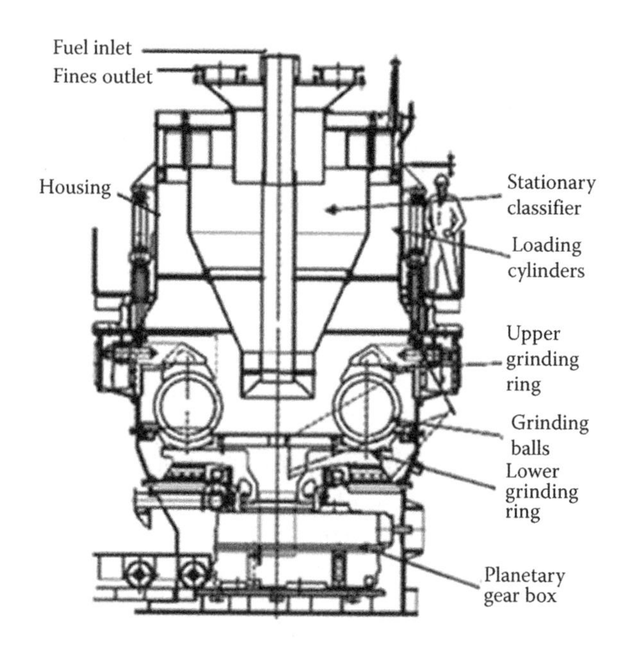

FIGURE 16.18
Ball and racer mill.

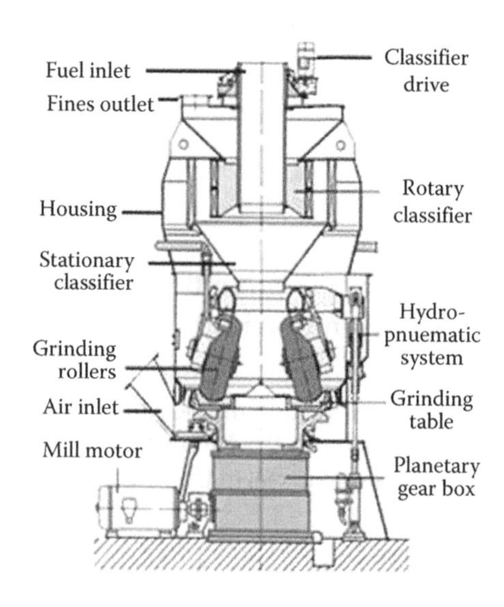

FIGURE 16.19
Roller- and bowl-type MPS mill. (From Riley Power Inc, USA, with permission.)

In all the vertical mills, discharge from the fuel feeder, which is crushed fuel of <50 mm, enters the mill through a central or a side pipe on to the rotating table. The coal pieces are hurled to the periphery of the table due to centrifugal action. The grinding elements, namely the balls or rollers, trap and pulverise the fuel pieces. Hydraulic or hydro-pneumatic pressure is applied on balls or rollers for optimum grinding. Hot air, up to a maximum temperature of 450°C, entering the mill through the annular ring, located between the housing and table, picks up the ground particles and starts travelling upwards. Around the roller area, a fluidised bed is created with an intense swirling of air and fuel particles, where maximum drying takes place. Drying is instantaneous and the grinding zone temperatures are between 60°C and 120°C. Air and fuel travel further upwards towards the classifier vanes. Primary and secondary stages of separation take place by gravity before entry into the classifier, where an abrupt change of direction precipitates the heavier particles. They return to the grinding zone by rolling down on the inside of the classifier. The fine particles reach the outlet chamber and go through the coal pipes.

Ball and racer mills are also called *E-mills*. These were developed originally by Claudius Peters of Germany. Slow-speed mills for coal grinding were licensed to Babcock of the United Kingdom who developed them further for larger sizes up to 120 tph base coal grinding capacity.

Even though the roller (and bowl) and ball (and racer mills) are both vertical mills and share many features there are some notable differences too in construction which contributes to minor differences in performance.

- In roller mills, the roller shafts are fixed to the casing permitting rollers to rotate on the bowl which is fixed to the rotating table. Upward hydraulic force is applied to the roller shafts from outside the casing to exert necessary pressure for coal grinding.
- In ball mills, the balls rotate between the rotating bottom racer and fixed top ring. Downward force is applied on the top ring for grinding coal.
- In roller mills, the position of rollers is fixed and they rotate around their axes. In ball mills, balls rotate in all directions as well as move on the racer.
- Rollers have bearings on their shafts which need periodic greasing which is not a requirement with balls.
- Rollers wear out after months of operation and need periodic metal build up. In case of balls they reduce in size and filler balls of reduced diameter are inserted in track.
- Over time product fineness reduces with roller wear in bowl mills but improves with ball wear in ball mills. Fineness is proportional to the racer diameter. As the racer also wears (along with balls) the track diameter increases prolonging the grinding action.
- Mill noise is claimed to be a little higher in ball mills. There is tendency for balls to touch each other while in motion and the top loading ring and balls have metal-to-metal contact both of which are absent in roller mills.

Figure 16.20 shows the various flows and their measurements in a vertical mill. There are two most important controls in any mill during its operation, namely air flow and outlet temperature.

- Mill output is varied by the variation of PA flow either by control damper or fan speed which picks up the ground fuel in a set proportion. This is fixed at the time of boiler commissioning by mill calibration when the differential pressures of mill and air flow measuring device are set in a desired ratio.

  There is a sympathetic variation in the speed of the fuel feeder which follows the PA variation.
- Mill outlet temperature is held constant by varying the tempering air quantity to raise or lower the PA temperature entering the mill.

Product Fineness

Product fineness is expressed as the % of the ground fuel passing through the specified standard mesh, which is usually the 200 mesh for coal with a nominal particle size of 75 μm.

- The fineness is set at the beginning by adjusting the vanes of the classifier.
- Fineness required, in general, for various coals through 200 mesh is as per Table 16.6. About <2% is to be retained on 50 mesh (297 microns). However, the final fineness setting is decided in actual operation depending upon the unburnt C in ash for a particular coal.
- For instant ignition and rapid burn out very fine particles in abundance are needed with coarse particles held to minimum to obtain the highest combustion η and also to avoid slagging.
- Fineness is an indirect indication of surface area generated. When coal is ground typically to a fineness of 80% through 200 mesh, the surface generated is about 1500 cm²/g.
- Unless the coarse particles are also controlled to less than the specified amount, the intermediate sizes also become coarser which reduces the

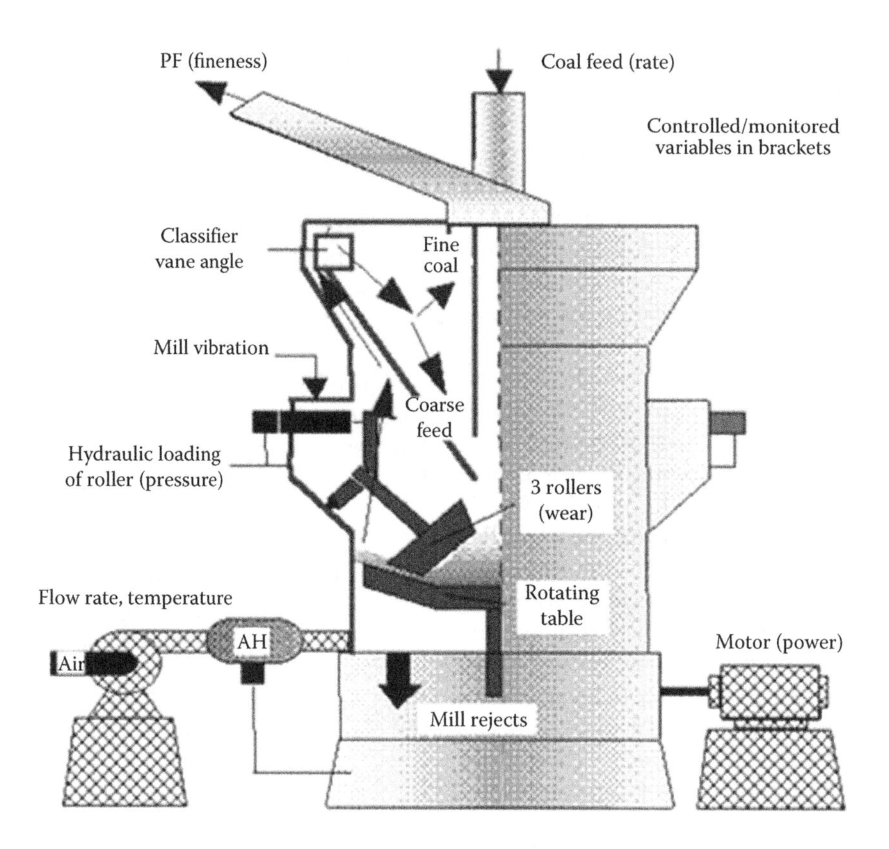

FIGURE 16.20
Flow streams in vertical coal mill and their measurements.

TABLE 16.6
Fineness Requirements from the Mills for Various Types of Coals

Type of Fuel	Fineness <75 μm
Sub-bituminous coals and lignites with high VM	60%
Bituminous and sub-bituminous coals with normal to high VM	70%
High-GCV low-volatile bituminous coals with FC of 78–86%	80%
Anthracites, petcoke and so on with FC of ≤98%	95%

combustion η. Both grinding and classification should be in step to achieve the desired fineness over the whole range.

- The higher the fineness, the greater is the power consumption and the lower is the output of the mill and the better is the C burn up.
- With *LNBs* and also for better C burn up rates the fineness requirement is higher by about 5–10% higher than normal.

Higher rank of coal demands finer powder for good combustion, as shown in Table 16.6.

Reproduced in Table 16.7 is the standard U.S. sieve sizes as per ASTM. It gives the sieve number size in microns and inches and also equivalent BS and Tyler standards. The mesh number system is a measure of how many openings there are per linear inch in a screen. Sizes vary by a factor of √2.

Mill Sizes

Each manufacturer follows proprietary construction and range of sizes of mills evolved over decades. Nominal diameter (DN) of the ring, racer or wheel attached to the bottom table is the prime dimension of any vertical mill. Slow-speed mill sizes range from ~1750 to 3500 mm (70–140″). This corresponds to a mill output of 20–200 tph. Drive motor sizes vary from ~120 to 1000 kW. The mill weights, including gear box, range from 60 to 300 tons.

Mill Power

Mill power consumption for vertical mills varies from 7 to 8 kWh/t at full load based on 70% fineness <75 μm (through 200mesh) for coal of 50 HGI which is nearly a third of what the ball and tube mills consume.

Mill Improvements

Mill improvements are basically aimed at

1. Reducing auxiliary power
2. Increasing combustion η

TABLE 16.7

US Sieve Sizes and Their Equivalents

Microns	Inches	ASTM E 11-61	BSS: 410-1989	Tyler
3360	0.1320	6	5	6
2830	0.1110	7	6	7
2380	0.0937	8	7	8
2000	0.0787	10	8	9
1680	0.0661	12	10	12
1410	0.5550	14	12	12
1190	0.4690	16	14	14
1000	0.0394	18	16	16
840	0.0331	20	18	20
707	0.2800	25	22	24
595	0.0232	30	25	28
500	0.0197	35	30	32
420	0.0165	40	36	35
354	0.0138	45	44	42
297	**0.0117**	**50**	**52**	**48**
250	0.0098	60	60	60
210	0.0083	70	72	65
177	0.0070	80	85	80
149	0.0059	100	100	100
125	0.0049	120	120	115
105	0.0041	140	150	150
88	0.0035	170	170	170
74	**0.0029**	**200**	**200**	**200**
63	0.0025	230	240	250
53	0.0021	270	300	270
44	0.0017	325	350	325
37	0.0015	400	400	400
32	0.0013	450	440	450
25	0.0010	500	–	500
20	0.0008	635	–	635
13	0.0005	1000	–	1000
10	0.0004	–	–	1250
5	0.0002	–	–	5000
1	0.00004	1	–	10,000

3. Reducing emissions
4. Improving reliability

1. Net η of boiler improves with the reduction in the power consumption of milling plant. This is possible if the overall weight of the rotating components is reduced. This conflicts with the need for enhancing boiler availability for which the wear parts must be quite heavy (to provide for adequate sacrificial material). The way to economise is to identify the wear pattern and redistribute the weight suitably.

   Rotating throat rings introduced in the 1980s enhanced air distribution and reduced air resistance, both of which contribute for lowering the power consumption. More importantly, erosion in the throat ring is reduced thereby increasing the reliability of the mills.
2. Increasing combustion η is possible by improving the coal fineness. But this would consume more power. *Dynamic/rotating classifier* is a means for reducing the coarse particles in the final product and thereby increasing the content of fines. Dynamic classifiers do not consume any appreciable power.
3. Reduction of emission follows when the fineness is improved as the fuel burns more completely and evenly without excessive flame temperature which leads to the lowering of NO_x generation.
4. Reliability improvement is achieved by introducing advancement in the metallurgy and mechanical design. Notable improvements achieved over the decades are by way of lining the mill for abrasion resistance with appropriate wear-resistant tiles, introduction of planetary gear box, improve mill loading systems and so on.

Dynamic Classifier or Rotating Classifier

Mill classifier is placed at the top end of the mill housing in the flow stream of air and coal. It separates coarse from fine coal by allowing the fine coal to pass through and precipitating the coarse particles for regrinding. This separation in static classifier is by change of direction or inertial.

A dynamic classifier has an additional inner rotating cage besides the outer stationary vanes. This can be seen in Figure 16.19. Together, they provide centrifugal classification, with output being finer than in standalone static classifier.

Static classifier of a pulveriser is being replaced with dynamic classifier for improving performance of the boiler unit in the following two critical areas:

- The finer powder reduces the level of unburnt C in the combustion of coal.
- The NO_x levels are reduced as the burning is more uniform.

Shown in Figure 16.21 is the large-capacity slow-speed vertical pulveriser of B&W, with a brand name of Roll Wheel pulveriser, fitted with rotating classifier as a standard. They are made nominally in six sizes with outputs ranging from ~17 to 104 te/h. These mills always have three grinding wheels, rollers or tyres with large amount wearing material to provide long life. Ceramic lining is provided on all surfaces where the coal fines slide to improve the life of wear parts.

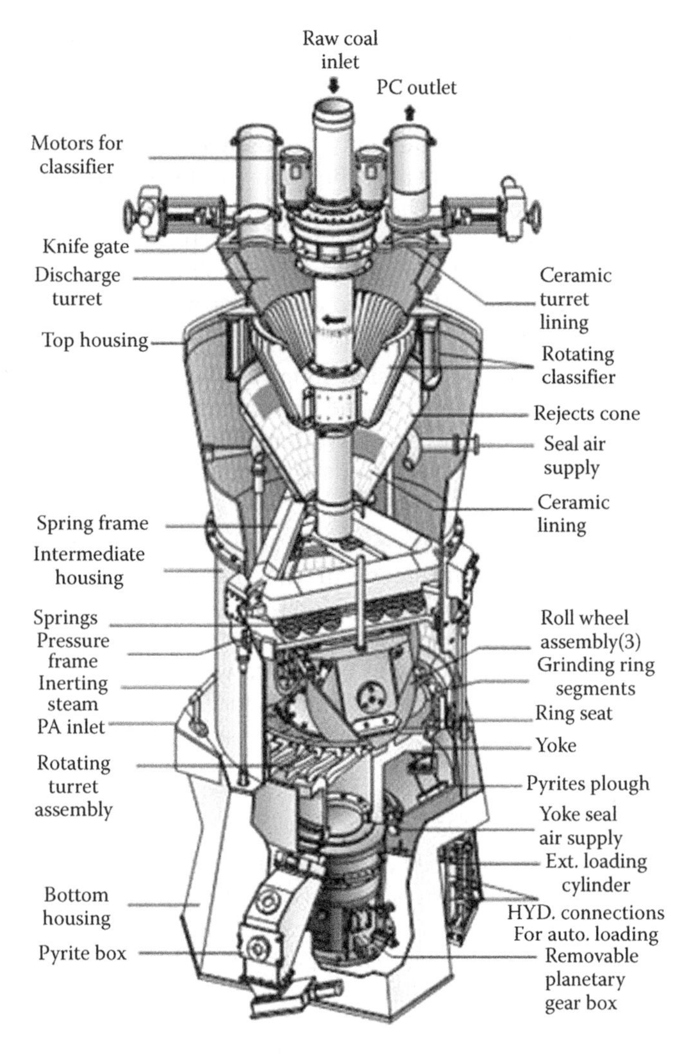

FIGURE 16.21
Large roll wheel pulveriser. (From The Babcock and Wilcox Company, USA, with permission.)

16.49.1.b Horizontal Mills/Pulverisers

Horizontal mills predate vertical mills by a few decades going back to those early days of PF with *indirect firing*. Big mills were built as the early PF boilers had large centralized mill and storage system.

There are two types of horizontal mills based on their speeds of rotation:

- Low-speed mills (15–35 rpm) for heavy grinding as required for nutty fuels like coals, anthracite and so on.

  They are also called *ball mills, tube mills or ball and tube mills* and steel balls are employed as crushing medium in horizontal shells or tubes.

- High-speed mills (400–750 rpm) for grinding of soft and high M lignites and brown coals. They are also called *beater mills*. Grinding action is by the impact of beater fan impeller on the soft coal. These are described separately under the topic of lignite.
- High-speed mills for grinding small to medium quantities of coal have also been made for long time in the United States under the trade name of Atrita mills by Riley Power Inc, described later. No steel balls are employed but a high-speed grinding wheel reduces the size by attrition after the initial crushing action.

Ball Mill Operation

Ball mills can operate either under *suction* or *pressure*. They are built in medium to large sizes ranging from 25 to 250 tph. They are essentially long shells with wear-resistant liners on the inside and having steel balls of various sizes for grinding. The shells are closed on both sides and coal enters centrally from one or both the ends. Double entry mills are primarily two independent mills joined together to economise on space and electrical drives. Coal classification is always carried out externally. In case of double entry mills, failure of one of the two feeders is not a matter of worry, as the speed of the running feeder can be doubled up and the mill output can be maintained substantially at both ends. In the same way, half mill operation is also possible.

The grinding media consists of forged steel balls varying in size from 25 to 100 mm with nearly one fourth of the shell filled with them. Grinding action is a combination of impact and attrition. Mill speed is kept ~0.8 times gravity such that the balls climb up the drum to some height and fall back to the bottom and in the process impacting the coal pieces heavily. Slow grinding of balls produces very fine coal powder. Balls can be charged in a running mill without the mill stoppage. Naturally the wear of the balls and liners is quite high. The concept of a ball mill is shown in Figure 16.22.

External classifiers of double cone construction with wear-resistant liners separate the heavier particles from the fine powder and feed them back to the trunions where the impacting coal is also fed to reach the grinding zone.

- The mill outlet temperature usually is between 90°C and 140°C.
- The fuel reserve inside a ball mill is typically ~15 min as opposed to ~5 min in a vertical mill.
- As there is no fluidization of coal inside the pulveriser as in vertical mills, the maximum M in a ball mill is limited to 20% despite a longer residence time.
- Power consumption of ball mills is significantly higher than vertical mills as the power is mainly for rotating the huge mass of shell and balls, practically independent of the coal charge.

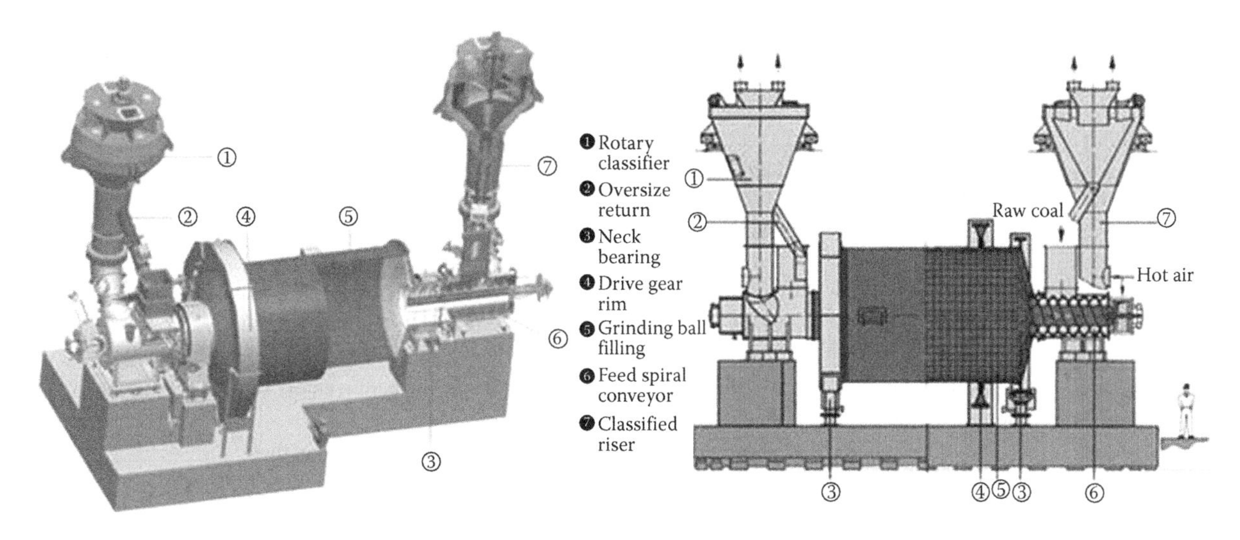

FIGURE 16.22
Low-speed horizontal ball mills—isometric and cross-sectional views. (From Hitachi Power Europe GmbH, Germany, with permission.)

- Further, the specific power consumption rises sharply at lower loads. Table 16.8 indicates the differences in the specific power between the two types of mills for some particular coal. The trend is worth noting here than the absolute figures.

Advantages and Limitations of Ball Mills

Advantages include

- High availability
- Ease of maintenance. Only periodic ball change needed. Bolted Mn-steel liners need replacement only after several years. On the whole, the maintenance cost would work out nearly the same as in vertical mills
- Tolerance to a lot of foreign material in fuel
- Large reserve capacity and hence quick response to varying loads
- Good fineness of the product, which increases at low loads, making the flames extremely steady
- High suitability for hard and abrasive fuels

The limitations are

- Significantly higher power consumption particularly at low loads
- M in fuel limited to 20%

TABLE 16.8

Approximate Specific Power Consumption of Mills in kWh/te of Coal

Mill Load	100%	80%	60%	40%
Ball mill	24	32	44	68
Vertical mill	10	10	12	16

- Mill equipment and foundation costs are both high
- Space required is high
- Mill redundancy is usually not possible due to high cost
- High noise levels needing insulated enclosures
- Not particularly suited for intermittent operation

Areas of Application

In present-day context, when very large vertical mills have been built and perfected, the horizontal ball mills have only certain niche areas of application, such as for grinding of

- Slow-burning densely structured high FC fuels like *anthracite, petroleum coke* and so on that require much higher fineness of ~80–85% <75 µm (through 200 mesh). They are preferred some times for low volatile coals as well.
- Very abrasive coals where the mills can only have long intervals between shut-downs.

Comparison of Vertical and Horizontal Mills

Table 16.9 provides a comparison between the two types of pulverisers in a very succinct manner.

16.49.1.c Beater Mills (see Lignite and Firing)

Pulverisers are usually for hard fuels like various types of **coals, petroleum cokes, anthracites** and so on. They are used even for lignites containing lower M. But for younger lignites with fibrous content and high M of 40% and above, or brown coals *beater mills* are required. For

TABLE 16.9

Comparison of Vertical and Horizontal Mills

Parameter	Vertical Mill	Ball Mill	Comments for Ball Mill
Space	Less	More	
Milling plant weight	Low	Much higher	
Foundation costs	High	Higher	
Lay out flexibility	High	Low	Bulky equipment
Spare mill provision	Usual	Not possible	Too expensive and space consuming
Power	Less	more	Even more at part loads
Product fineness	Fine	Finer	Fineness increases at lower loads. Stable firing
Load response	Dynamic	Slightly lower	Despite higher inventory due to mill weight
Inventory of coal inside	5–6 min	15–20 min	
Total M in coal	~40%	~20%	Fluid bed drying is not there
Intermittent operation	No problem	Problematic	High amount of fines pose fire hazard
Foreign material in coal	Problematic	No problem	
Abrasive coals	Problematic	No problem	This is the real advantage
Availability	High	Higher	
Maintenance	Less	Slightly more	Over long period almost same
Noise level	Medium	High	Enclosures often needed
Drive motors	Heavy duty	Heavier duty	Higher starting torque due to casing and ball load
Overall mill cost	Low	Much higher	
Best suited for	Normal coals, low-M lignites	Abrasive coals, anthracites, petcoke and so on	

the removal of high amount of M very high air temperature is required. For breaking down the lignites into fine particles, there is no grinding action involved but the dry fuel needs to be lightly beaten by the fan impeller to make the dry mass disintegrate. Beater mills are high-speed horizontal mills which are separately explained under the topic of lignites.

16.49.1.d Atrita Mills

These are high-speed horizontal mills made by Riley Power Inc of the United States for several decades under the trade name of Atrita mills for small capacities of 3–25 tph of coal grinding. It is a light, compact mill with small plan area and height suited for industrial PF boilers. All the four sections of the mill consisting of

a. Crusher dryer
b. Pulveriser
c. Classifier
d. PA fan

are mounted on a single shaft and arranged inside the same housing (as shown in Figure 16.23). The figure depicts the duplex pulveriser for larger sizes. Simplex pulveriser is also made.

The coal feed into the mill is first reduced in size to ~6 mm in the crushing section for primary size reduction and drying. Screened by the grid segments under the hammers, the reduced-size coal is introduced to the grinding section for pulverisation. The conveying or primary air, developed by an integral fan in the mill's fan section, transports the PC from the grinding section to the burners.

- Hammers and a breaker plate perform crushing function in the crusher-dryer section. Below the breaker plate, there is an adjustable crusher block to vary the gap between hammer tips and crusher block. This gap controls the size of crushed coal entering the grinding section.
- In the grinding section, major grinding components are stationary pegs and moving clips that are attached to the high-speed rotating wheel. The turbulent flow and impact momentum on coal particles by the moving clips create an intense particle-to-particle attrition or a pulverising effect to grind the crushed coal particles.
- To control coal fineness, there is a whizzer type rotating classifier or rejecter arm assembly between the grinding and the fan sections. The V-shaped rejecter arms magnify the intensity of centrifugal forces within the grinding section, which retains coarser coal particles in the grinding zone for additional pulverisation.

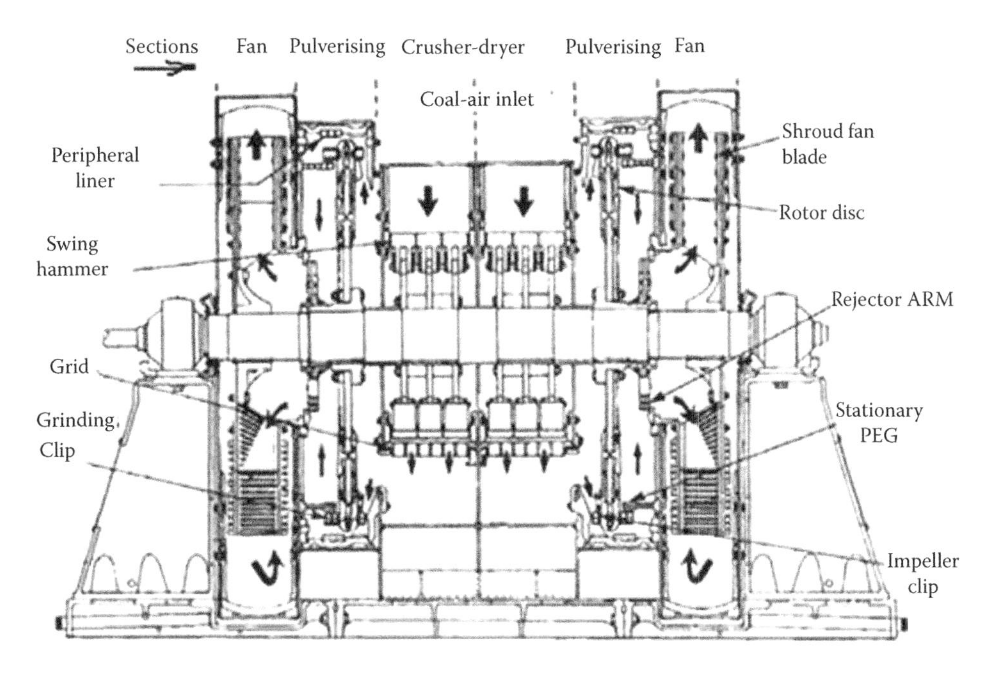

FIGURE 16.23
High-speed horizontal Atrita mills. (From Riley Power Inc, USA, with permission.)

- The finer coal particles, subjected to less centrifugal forces due to reduced mass and sectional area, pass through the rejecter arms along with PA into the fan section and are delivered to the burners through coal pipes.

In this mill there is no metal-to-metal contact in the crushing or grinding zones. Wear-resistant materials or liners are provided in areas of wear. Load response has to be good in such a compact mill with low thermal inertia.

Compactness and cost competitiveness are the main advantages of this type of mill. Being a high-speed mill its ability to deal with abrasive coals is facilitated by the use of abrasive-resistant materials.

16.49.2 Pulveriser Capacity

Pulveriser capacity is expressed in different ways.

a. *Base capacity* is the nominal capacity of a pulveriser to grind a fuel of 50 HGI having 10% total M to fineness of 70% <75 μm (through 200 mesh).
b. *Maximum capacity* is the base capacity of the mill adjusted for actual HGI, M and fineness.
c. *Actual capacity* is the maximum capacity reduced by a factor of 0.85–9.0. This 10–15% margin is to take care of the possible deterioration of fuel and mill in future.

Mill *capacity* increases with the HGI, as the fuel is softer. The capacity reduces if the fineness requirement is higher, as more grinding is involved. With the increase in fuel M, mill inlet air temperature has to be proportionally higher.

16.49.3 Pulveriser Power

Power consumption is by far the most important factor in the selection of a pulveriser as the milling constitutes the highest power consumption in the whole PF boiler. The net output from the boiler plant is optimum when the milling power is the least. This is the reason for an overwhelming popularity of the vertical pressurised mills which consume the least power.

16.49.4 Fires in Pulverisers and Mill Inerting

Fires and explosions may occur during upset conditions within the mills, namely, during start-up, shut-down and significant load changes which result in situations when fuel has the possibility of self-heating faster than heat dissipation to its environment causing smouldering to take place. *Smouldering is the slow, flame less, low temperature and smoky burning.* During smouldering high amount of CO is emitted as a result of incomplete combustion. Under favourable circumstances this can result in an explosion.

The areas prone to fire are

- Mill inlet air box which has the highest air temperature and where coal particle have a tendency to precipitate
- Mill outlet chamber where the dried PC particles along with PA are present

Mill fires and explosions can be prevented by CO detection and inerting. Three elements are needed for a fire to start and sustain, namely, heat source, fuel and air. Inerting is to remove at least one of the elements. Mill-inerting systems protect the pulveriser by injecting sufficient amounts of inert gas to reduce the O_2 content in the equipment to such low concentration that explosion is impossible.

Coals have different tendencies to self-ignite and form explosive mixtures. An empirical measure for assessing the safety of milling coals has been developed, based on the fuel ratio (the ratio of FC to VM) which varies from ~0.5 for lignites to ~20 for anthracites. Most coals with a fuel ratio >1.5 can be processed safely. In the range 1–1.5, caution is required, while at <1, some form of intermittent or continuous inerting is advised.

Mill inerting needs a steam inerting system along with a water-based fire-extinguishing system, and possible modifications to the pulveriser inlet ducts. Steam is taken from the ST bleed steam system. Any coal left in a tripped pulveriser or during a normal shut-down can be safely removed under the inert atmosphere to reduce the risk of fire and explosion. The mill-inerting system is interlocked with the mill controls to activate during a mill start-up or shut-down. An existing local warning system, also incorporated into the design, warns plant personnel that an explosive condition exists and to clear the area. Mill temperatures are monitored and in the event of abnormally high rise, or rate of rise, the system automatically shuts down and isolates the mill besides activating the water-based extinguishing system.

16.49.5 Pulveriser Requirements

The expectations from pulverisers are the

- Capability to handle a wide variety of coals
- Accommodation of load swings
- Fineness being optimum throughout the mill operating range
- Low wear rates of the grinding elements
- Low noise levels
- Ease of maintenance
- Low O&M costs
- Low power consumption
- Minimum volume of building

16.49.6 Milling Plant

Milling plant in any PF boiler is a sub-system consisting of

- Fuel feeders
- Feeder discharge/mill inlet pipes

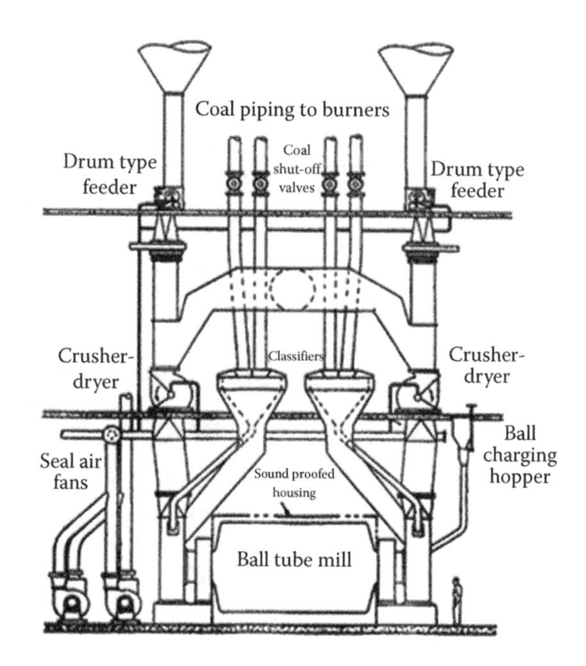

FIGURE 16.24
Milling plant in a boiler with horizontal boiler. (From Riley Power Inc, USA, with permission.)

- Pulverisers
- Fuel pipes up to furnace

In other words, it is the equipment package in the fuel circuit, starting from the bunker discharge and ending at the inlets of burners. Typical milling plant around a ball mill is depicted in Figure 16.24.

16.50 Pulverised Fuel Firing

Burners bring the fuel, air and ignition energy together to initiate combustion and it is in the furnace chamber that the combustion is completed in PF burning. For optimum combustion, therefore, not only should the burner be of the best design but also the furnace should be of appropriate shape and volume to provide the necessary radiant heat and residence time.

PA + PF enter the furnace at a low velocity through the central pipe of burner while SA enters at a higher velocity to scrub away the ash layer formed on the fuel particles on combustion. An igniter provides the initial ignition energy but later on it is the burning of VM and radiation from furnace that keep the fuel at high enough temperature for the sustenance of combustion.

The manner of burning of PF particles is illustrated in Figure 16.25. PF, substantially dried in the pulveriser, enters the furnace chamber. With the addition of further heat in furnace, there is an instant flash drying and devolatilisation. VM burns out in ~0.1 s adding to

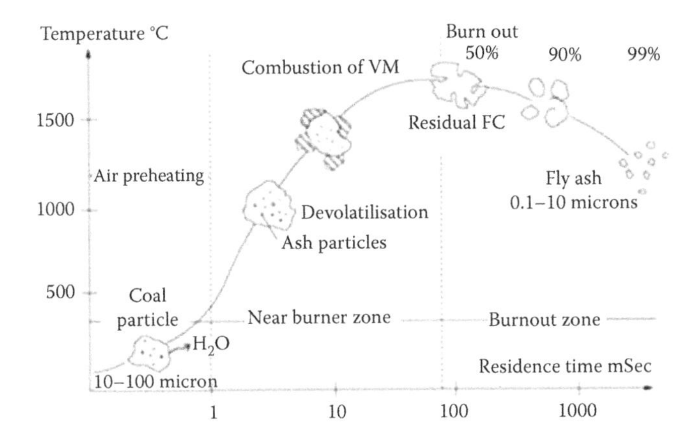

FIGURE 16.25
Stages and manner of combustion of PF particles.

the heat of flame. Minimum 19% daf VM is required in circular burners for the flame temperature to be maintained above the ignition temperature. The devolatilised fuel particles or char (FC) need long time (1–2 s) to complete the burning but coarser particles >100 microns need even longer time.

With daf VM <19% downshot or arch burners, on one or both sides of furnace, with 'U' or 'W' shaped flames, respectively, are needed to provide the required heat energy from the burning char and hot furnace refractory. See PF burners for more details.

With fuel M of more than 40% combustion cannot be sustained with circular PF burners, unless fuel has high VM. For instance, brown coals have high M levels of even 70% and combustion is easily sustained because of the high daf VM of ~60%.

Velocities of SA are significantly higher than PA to provide scrubbing action. In circular burners, SA is at near right angles to PA flow while in tangential burners it flows in parallel streams. But it is the higher SA velocity and longer traverse that helps to scrub the fuel particles of ash layer.

16.50.1 Direct and Indirect PF Firing

Any PF boiler has two main sections, namely

1. Milling/pulverising section, in which the fuel is ground to fine powder.
2. Burning section, in which the fuel is burnt.

In direct-fired PF boilers, the milling and burning sections work together in unison while in indirect firing, they worked independently; the milling plant would work for part of the day to fill up the bunkers and the burning system would work all the time drawing the PF to burn in the boiler.

Until the 1950s, indirect system was quite popular. It was the days of large horizontal mills. Several mills would work for 10–12 h a day to fill up a set of fine coal bunkers. For the remaining time, the mills would be under daily maintenance. The fine coal bunkers may even be common for several boilers. It was also the time when the boilers rarely exceeded 50 MW capacities.

- The system was untidy and relatively complicated.
- The whole area was dusty.
- The powdered coal is inherently disposed towards spontaneous combustion.
- The boilers were prone to explosions due to minor discontinuities in flow pattern between the bunkers and the boiler. This was unavoidable because the PC being hygroscopic, would turn to small lumps here and there in the bunker by absorbing humidity from the atmosphere and interfere in the smooth flow of fuel.
- Despite full loading of mills the power consumption was high.

The development of vertical mills dealt a death knell to this system. Vertical mills made it possible to eliminate the intermediate fine coal bunkers. Several mills share the boiler load and improve the operational flexibility.

- Vertical mills were small and very responsive.
- Power consumption was significantly lower.
- They could be integrated with the boiler operation effortlessly making the milling and firing activities seamless.
- Most importantly fires in bunkers and boilers could be practically eliminated making PF firing a safe practice.
- Boiler surroundings were much cleaner.

Indirect firing in PF boilers is of historical significance today. The concept of a modern direct-fired PF boiler is illustrated in Figure 16.26.

16.50.2 Limitations of PF Firing

- Even though PF firing for solid fuels is undoubtedly the most versatile, it tends to be fuel specific, that is, a boiler designed for a particular heat duty and fuel with a certain configuration of mills and burners can address a very limited variation in fuel. It is quite unlike a CFBC boiler which has much wider range of fuel acceptance.
- Slagging and fouling are the annoying features of PF firing due to the melting of ash during combustion and its recrystallisation during the cooling process. Boilers designed for normal fuels cannot accept slagging or fouling fuels.

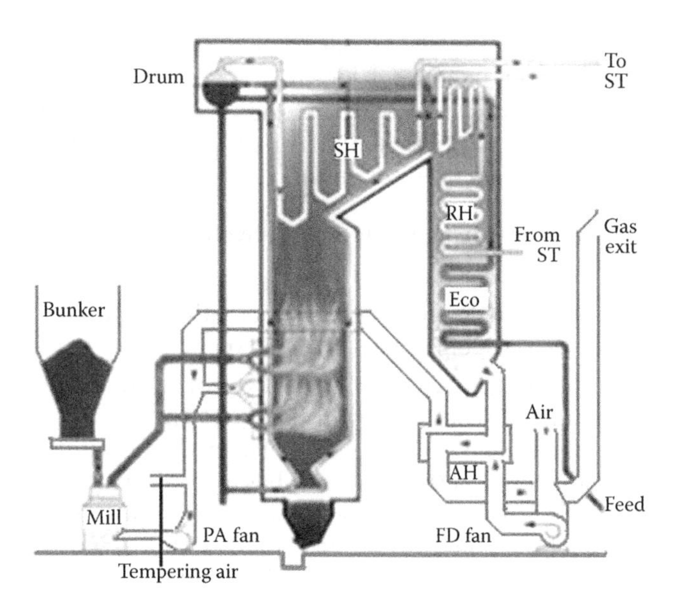

FIGURE 16.26
Direct-fired PF boiler.

- Fuel flexibility by way of multi-fuel firing is highly limited.
- Fuels with Gcv <8.4 MJ/kg, ~2000 kcal/kg or ~3600 Btu/lb and fuel burden (A + M) >62%, can be burnt only with auxiliary fuel support.
- Stepless turndown is limited to only 70–100% load. At lower loads, mills have to be progressively cut out.

16.50.3 PF Firing versus CFBC

PF has been an undisputed leader for over eight decades. Boilers as small as 40 tph to as large as 4000 tph have been built. Almost every type of solid fuel has been burnt but in the last couple of decades, there is a steady progression of FBC boilers because of certain advantages. CFBC in particular has practically displaced PF from the industrial and captive power market. The largest CFB boiler in operation is 462 MWe (1300 tph) and there are plans to make them as large as 1000 MWe, which seems quite possible.

- CFBC has an enormous multi-fuel firing capability. This is best captured by Figure 16.27 where CFBC straddles all types of PF firing systems. Additionally, it covers a lot of area which used to be traditionally serviced by stokers.
- All the debilities in firing of fuels are addressed by CFBC.
- In-furnace desulphurisation and low-temperature combustion of ~850°C take care of environmental compliance quite effortlessly.

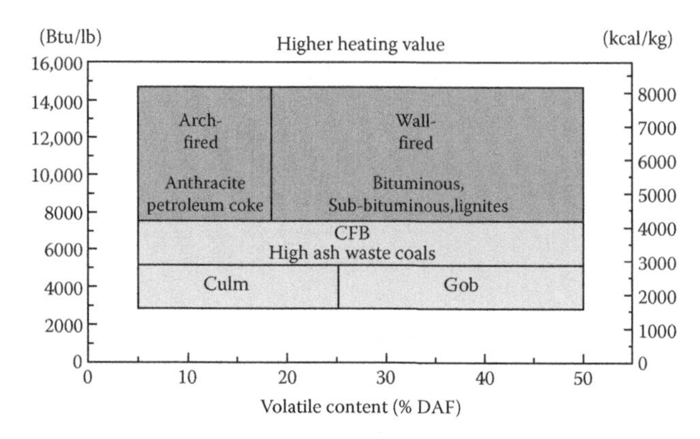

FIGURE 16.27
Combustion versatility of CFBC compared to PF firing. (From Foster Wheeler Corporation, USA, with permission.)

- CFBC boilers are simpler to operate and maintain and O & M costs are lower.
- They present little danger of explosion.
- The combustion efficiencies are nearly the same as PF, while the auxiliary power consumption may be slightly higher.

A more exhaustive comparison is given in Table 16.10.

16.51 Pulverised Fuel Burners (*see also* 1. Burners 2. Oil and Gas Burners 3. Lignite and Brown Coal Firing *and* 4. Low NO_x Burners)

A burner is a device in which fuel and air are brought together appropriately so that combustion ensues upon ignition. Fuels should be in fluid state like gas, FO, PF or biomass waste. In PF firing, PA or mill carrier air brings the dry hot fuel powder to the PA nozzle for admission into the furnace. Hot SA at higher velocities is admitted around the PA of the stream in such a way that the fuel particles are scrubbed of the thin layers of ash formed on combustion to continuously expose fresh surface.

There are three types of PF burners for coal firing:

1. Circular horizontal burners for wall mounting
2. Tangential horizontal burners for corner mounting
3. Downshot vertical burners for arch mounting

16.51.1 Circular Horizontal Burners

A typical circular burner is shown in Figure 16.28 which consists of a central PA pipe in which PA + PF are brought from the mill at temperatures of 60–100°C and velocities of ~18–23 m/s (~60–75 fps) for bituminous

TABLE 16.10

Comparison of PF Firing and CFBC

Parameter	PF	CFBC	Comments for PF
Fuel preparation	Ground to <75 μ	Crushed to <6–8 mm	Aux. power for fuel grinding is heavy
Fuel flexibility	Limited	Very wide range	
Limit of burden (A + M)	~62%	~75%	Cannot burn brown coals, peat and washery rejects easily. Need lot of changes
Abrasive fuels	Heavy wear of milling plant	No special wear problems	
High S fuels	FGD plant required	Lime dosing with fuel	
Oil firing requirement	Start-up and up to 50%	Only for start-up	Oil bill significant
Thermal η	~88%	~88%	Thermal η marginally better
Aux. power	Lower despite mills	High due to fluidisation	Aux. power ~2–4% lower
Nox ppm dv (typical)	500–1000	<100	Needs special
with low NO_x burners	100–300		deNO_x measures
Sox mg/NM^3 (typical)	~2600 without deSox	<100	Need deSox plant
Moving equipments	Mills and feeders	Only feeders	
O&M costs	High	Low	Higher O&M costs generally
SBs	Many and all over	Only in second pass	
Ash handling	Wet or dry, complicated	Dry, simple	
Boiler controls	Elaborate	Relatively simple	Many mills and burners and hence more I&C
Burner controls	Complicated	Very simple	
Boiler dynamics	Very fast	Nearly as fast	No fuel reservoir makes it very highly responsive
Cold start after 48 h	Normal time ~6 h	~2 h as bed	
	Nearly normal time	~500C	
Warm start after 8 h	~1 h	~1 h	
Hot start		~20–30 min	
Explosion risk	High	None	Needs more skilled operation

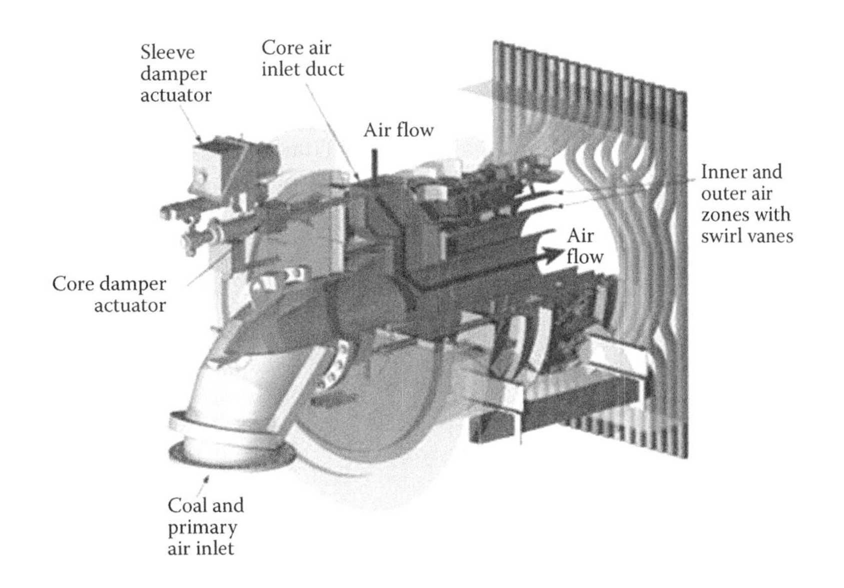

FIGURE 16.28
Horizontal circular PF burner.

coals. There is an oil gun or gas pipe located centrally inside the PA pipe for burning oil or gas for ignition and low load support. An impeller is placed centrally at the exit of the PA pipe over which the coal–air mixture passes into the furnace. The impeller reduces the speed of mixture and acts as a bluff body to help flame stabilise on it. The empty space immediately in front of the impeller provides a small negative pressure due to the swirling of the coal-air streams which helps the flame to anchor to the impeller.

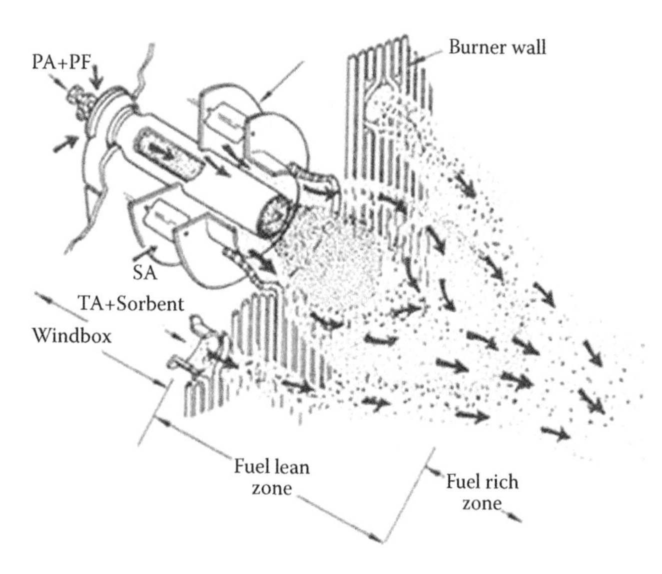

FIGURE 16.29
Splitting of air streams for the reduction of NO_x in circular PF burners.

The SA is provided through the vanes in the wind box at higher velocities of ~35–50 mps (~115–165 fps) nearly at right angle tangentially around the PA stream to provide the desired scrubbing action. The resulting flame is short, bushy and turbulent with high flame temperature. To reduce the flame temperature and attendant NO_x formation the SA is split to elongate the combustion. Further dividing of SA around the combustion zone increases the staged combustion reducing the NO_x even further. Figure 16.29 shows such arrangement.

Circular PF burners are made from ~15–90 MW_{th} (~13–77 × 10^6 kcal/h or ~51–307 × 10^6 btu/h).

16.51.2 Tangential Burners

Tangential burners are placed at the corners of an approximately rectangular furnace. A typical tangential burner is shown in Figure 16.30. Here, all the burners are placed one above the other and thus there are alternate streams of PA and SA. TA is also included in case of **LNBs**. Typical ratios of PA, SA and TA are 20%, 65% and 15%.

The PA + PF are admitted in the PA nozzle in the same way as in circular burners. The SA, however, is not swirling around the PA stream but flowing parallel to it at a higher speed with no appreciable intermixing of streams. At this stage the combustion is slow and even with no hot spots. The four streams from the four corners of the furnace are directed tangentially around an imaginary circle in such a way that a turbulent swirling ball of flame is formed at the centre of the furnace (as shown in Figure 16.31).

The combustion thus takes place in two stages:

- In front of the burner
- In the flame ball

Unlike in wall burners with individual bushy flames, in corner firing, the four sets of flames from four corners merge to form a single flame in the middle of the furnace. During the travel from burners to flame ball coal burns in an air deficient atmosphere resulting in a lower NO_x generation by as much as 50% compared to wall firing. NO_x is further reduced when 15–20% of total air is provided as OFA.

Capabilities of tangential burners are very similar to circular burners in terms of burning various types of coals. Low-VM coals can be better handled due to longer travel and residence time. Capacity of burners is similar to circular burners.

In most PF boilers, the burners are provided with tilting mechanism so that the ball of flame could be moved upwards or downwards by ±15 degrees maximum to

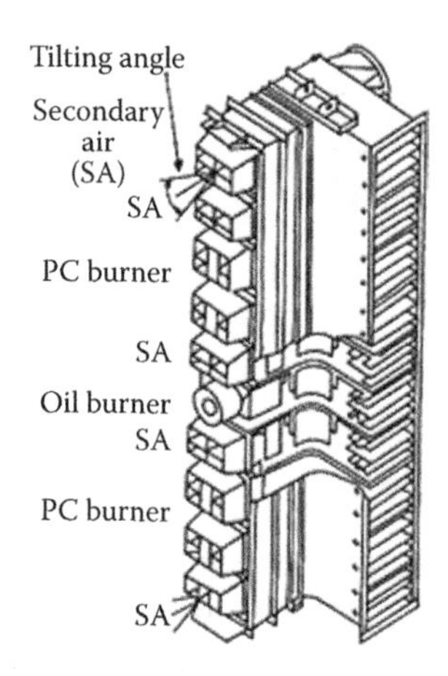

FIGURE 16.30
Tangential PF burner placed at the four corners of furnace.

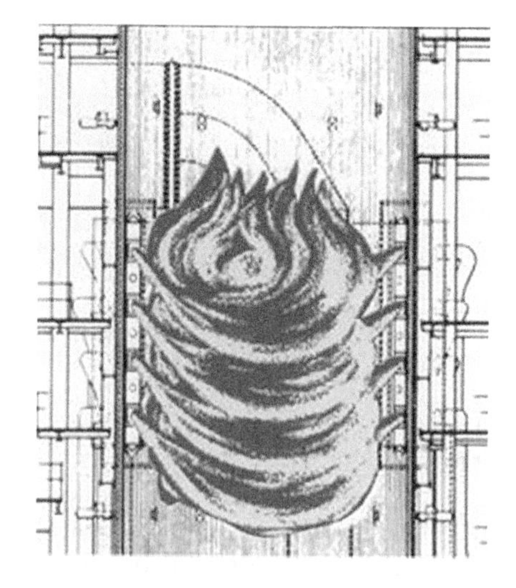

FIGURE 16.31
Tiltable flame ball created in the middle of furnace by the four tangential burners.

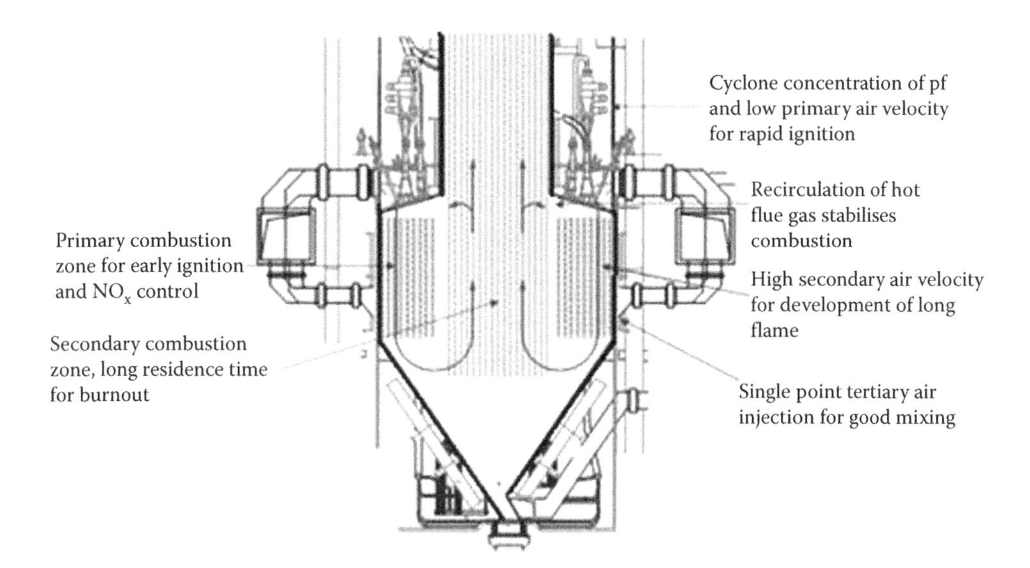

FIGURE 16.32
Lower furnace in a downshot-fired boiler.

raise or lower the FEGT. This is also helpful in controlling the SH and RH temperatures and their metallurgy. This is shown in Figure 16.31.

16.51.3 Arch Mounted/Downshot/Vertical Burners

Downshot firing is adopted mainly for low-VM fuels (<19% on daf basis) such as low-volatile coals, **anthracites, petcoke** and so on, where the high FC in the fuels requires a long residence time and high heat in furnace. The fuel is also ground to higher fineness (100% <75 μm or through 200 mesh). The lower part of the furnace of a downshot boiler is shown in Figure 16.32.

The lower furnace is altered by having arches in such a way that downshot burners can be installed. These non-turbulent burners are actually small jets fixed between the arch tubes. Fuel is introduced at high mill outlet temperatures of ~100°C. The firing is vertical and the flames are directed downwards. Refractory lined slopes of furnace bottom guides the flames while the hot refractory radiates heat back to flames with little absorption. Ignition takes place primarily due to the radiation of the furnace heat.

The flue gases take a 'U' turn and travel upwards and this long journey provides the increased residence time of ~3 s. Depending on the boiler size, the flames can be 'U' or 'W' type for which the downshot burners are located on one or both walls of the furnace. C burn out can scarcely exceed 95% despite these measures.

Figure 16.33 shows the manner in which the long flames are developed in w-type firing. TA is provided near the slopes to achieve staged combustion.

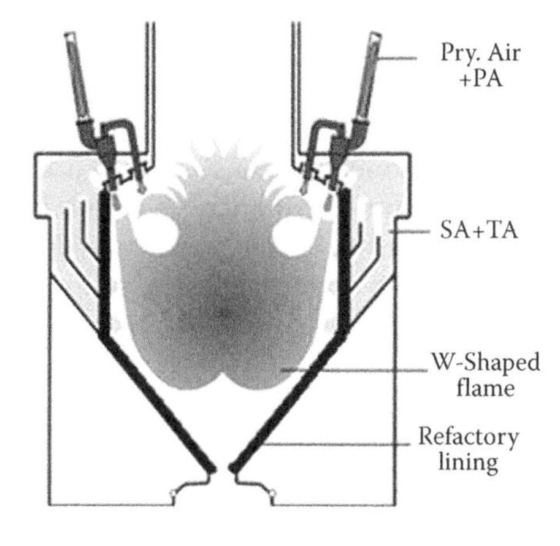

FIGURE 16.33
Long flames in the downshot firing. (From Foster Wheeler Corporation, USA, with permission.)

16.52 Pulverised Fuel Burner Arrangement

Based on the type of burner and size of boiler, the burners are arranged in some definite pattern. Often the boiler derives its name from the burner arrangement. Figure 16.34 shows the burner arrangements in PF furnace.

16.52.1 Front/Rear Wall Firing

When horizontal wall-mounted circular burners are used for steaming capacities up to ~400 tph or 100 MWe, it is economical to arrange the burners in three or four rows on either the front or rear furnace wall. The furnace is rectangular with depth of ~7–8 m to accommodate the

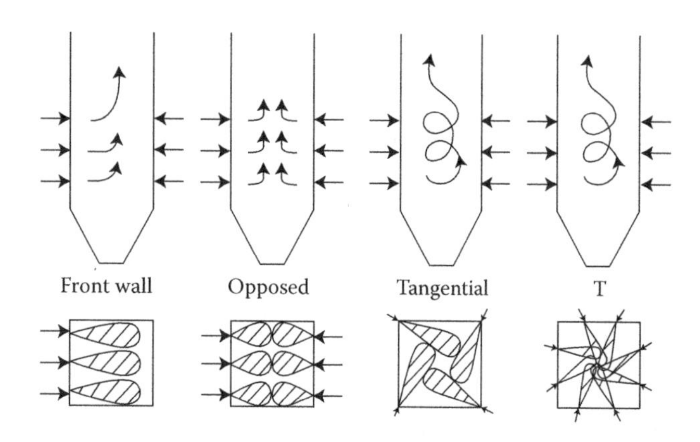

FIGURE 16.34
Different modes of PF firing.

flame length. Usually the mills are all located in one line and the coal piping is short.

16.52.2 Opposed/Dual/Twin Wall Firing

When circular burners are used for steaming capacities >400 tph, it is economical and efficient to adopt opposed wall or twin wall firing where the burners are mounted on both front and rear wall. The furnace is squarer than in front wall-fired boilers. Furnace depths are ~13–14 m or more to provide for 2 flame lengths from both ends. Coal piping from mill to furnace can be reduced if the mills are arranged in two rows on either side of the boiler.

16.52.3 Tangential Firing

The arrangement of burners in a furnace is always fixed. They are always located at the four corners so that a flame ball is created at the centre of the furnace. This type of firing inherently produces less NO_x due to the long residence time.

16.52.4 'T' Firing

When circular LNBs with their characteristic long flames are located at the furnace edges or furnace corners, it is possible to a form a flame ball somewhat similar to that in tangential firing. This increases the residence time and lowers NO_x generation. Figure 16.35 shows the arrangement with burners placed at the corners of furnace with OFA being supplied from furnace walls.

16.52.5 Corner Firing

In the case of brown coal firing, one, two or three rows of burners are located on each wall in such a way that the long flames create a fireball in the centre of the furnace. Reduction of NO_x is of a less consideration here as the high M in fuel any way depresses the flame

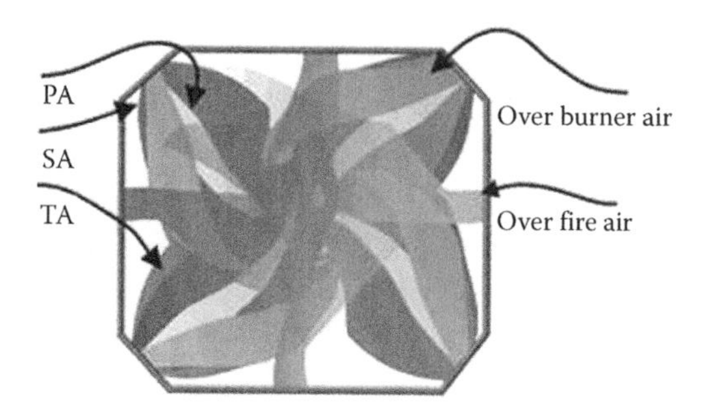

FIGURE 16.35
T-type PF firing.

temperature. The enormous amount of flue gas rotates inside the furnace before going up in a spiral fashion which enhances the residence time. The massive burners are also not tiltable as in the case of tangential firing.

16.52.6 Downshot or Vertical Firing

This method is exclusively for low-volatile coals, anthracites and petcoke. The furnace shape is altered to provide arches. Downshot firing is described adequately under the heading of PF burners.

16.53 Pulverised Fuel-Fired Boilers (*see also* Utility Boilers)

PF-fired boilers are the most dominant as they can burn all types fossil solid fuels, namely, **coals, lignite** and **peat**. They are the most versatile, efficient, scalable and reliable. They have come to dominate the boiler scene for the last eight decades.

Fuel is ground to very fine powder, typically 70% <75 μm (200 mesh) for most bituminous coals and fired through burners in large combustion chambers that provide 1 or 2 s residence time and burn the fuel completely. Typical C burn out is ~99%for most coals and can be as high as 99.7% for most lignites in large boilers.

PF boilers were introduced around the 1920s. The following salient developments propelled the PF boilers to the forefront and there was dramatic growth in size, η and reliability.

- **Radiant boiler** design which came into acceptance in the 1940s helped the growth of PF boilers immensely. Those times the boiler construction method, based on stoker firing of good coals, was to have a small furnace and big BBs with several drums with heat transfer substantially by convection. In radiant boilers, the furnace was enlarged and the convection banks were shrunk with radiation becoming the dominant heat transfer mode. Inferior fuels could be burnt because the large furnace could handle the slagging and fouling problems effectively.

- Development of **vertical mills** helped the integration of milling and firing sections of the boiler making the PF firing very reliable.
- *Direct* **firing system** of PF boiler could be introduced and developed further because of the vertical mills. The severe drawbacks of the **indirect firing** could be put behind.

These major developments along with big improvements in auxiliary plants, advances in **metallurgy**, developments in I&C—all helped rapid development of PF firing. In the 1950s and 1960s during the post-war reconstruction period, there was a frenzy of power development activity in the West, culminating in the commissioning of the largest boiler of ~4000 tph or 1300 MWe as far back as 1973. Since then, the boiler sizes have not increased but the η, reliability, flexibility and even some fundamental design aspects have been steadily improved in the last three decades.

The salient aspects of PF firing can be broadly summarized as follows:

- PF firing is by far the most versatile. Any non-fibrous solid fuel with Gcv > 8.4 MJ/kg, ~2000 kcal/kg or ~3600 Btu/lb and adequate VM to provide auto ignition can be burnt in PF mode. All types of coals starting from anthracite to peat have been successfully burnt.
- As PF is burnt in burners similar in construction to oil and gas burners, a variety of FOs and gases can be co-fired with very high η.
  - Modern PF boilers are environmentally highly compliant.
  - LNBs with SOFA arrangement (if needed) are necessary to reduce the flame temperatures by delayed and protracted combustion, thereby lowering the NO_x to the desired levels.
  - Denitrification or $DeNO_x$ plant is added at the back end for the removal of traces of NO_x if the efforts of LNBs are not adequate.
  - FGD or $deSO_x$ plants in the form of gas scrubbers are installed behind the boilers to capture the S in flue gases while firing high S coals.
  - The issue of particulate emission is addressed by having either an ESP or bag filter as appropriate.
- The two main departments of any PF boiler are the pulverisation and firing.
- Majority of the modern PF boilers are equipped with vertical mills because of their low power consumption, better flexibility, superior dynamics and other advantages. It is normal to adopt one spare mill to improve the availability. Sometimes, even two spare mills are provided when fuels are known to be exceptionally aggressive. In some advanced countries on the other hand, no redundancy is kept on better η and layout considerations.
- Several types of burners are adopted. Normal coals are burnt either by wall firing in a number of circular burners or tangential firing. Downshot or arch firing is adopted in case of low-VM coals and anthracites. Brown coals are burnt in either corner firing or tangential firing.
- Fuels are transported from bunker to the pulverisers either by gravimetric belt feeders or by volumetric drag chain feeders in case of coals and by plate feeders in case of lignite.
- Load response of PF boilers is exceptionally high, nearly matching that of oil and gas fired boilers. This is possible as the fuel in powder form burns instantaneously and fully. 5 min of fuel storage in the pulverisers further comes to the rescue during rapid load increases.

Typical designs of PF boilers are elaborated under the heading of utility boilers.

16.54 Pulverised Fuel Piping

PF piping or coal piping is the piping between mills and burners. Coal pipes are usually made of CS pipes of 12 mm thickness. As the bends are prone to erosion, cast **basalt** is the normal lining material. For more abrasive coals special ceramics are used. Fittings like isolating valves at mill outlets and burner ends, diverters and splitters are employed for channelising the fuel laden PA flow. All these parts are subject to heavy erosion. Wear-resistant construction is necessary for improved availability.

Coal pipes are sized for velocities of 15–25 or even 35 m/s.

- 15 m/s (50 fps) is the minimum velocity to avoid separation of coal dust from the PA in the horizontal runs.

- 25 m/s (82 fps) is the normal limit for erosion prevention for abrasive coals.
- It can be relaxed to 35 m/s (115 fps) when coals are known to be soft.

Coal pipe sizing is to balance between air pressure loss and pipe flexibility. The lengths of piping should be approximately the same to prevent unequal flow between burners. Balancing orifices may be needed occasionally to equalise the flow.

16.55 Pumps

Boiler feed (BFP) and boiler circulating pumps (BCP) are the two main centrifugal pumps falling within the boiler scope. Both are hot water pumps and are required to run all the time, which demands a great care in the design and construction of pumps. BFP is multi-stage pump while the BCP is usually of single stage.

The other pumps are the boiler fill pumps and chemical injection/dosing pumps. Boiler fill pumps are medium head cold water centrifugal pumps used only for filling the boilers at the starting. Dosing pumps are small reciprocating pumps to inject the hp and lp dosing chemicals into the drum and BFP suction piping, respectively.

Selection parameters for centrifugal pumps are the

- Pump design data (flow rate or capacity Q, discharge head H, speed of rotation n and NPSH)
- Properties of fluid to be pumped
- Application, place of installation
- The applicable specifications, laws and codes

The main design features on which centrifugal pumps are classified are

- The number of stages (single stage/multi-stage)
- The orientation of the shaft (horizontal/vertical)
- The type of pump casing (radial/volute or axial/tubular casing)
- The number of impeller entries (single entry/double entry)
- The type of motor (dry motor/dry rotor motor, as in submerged motor or wet rotor motor, as in canned motor)

Some of the important design features of centrifugal pumps are briefly explained below.

- Capacity or flow rate (Q)
  - It is the *useful volume of fluid delivered to the pump discharge nozzle in a unit time* in m³/s, L/s, m³/h or gpm.
  - This flow rate is the *net flow of fluid from pump* not considering leakage flow as well as the internal clearance flow.
  - The flow rate varies as the pump speed N.

$$Q \alpha N \quad (16.3)$$

- Developed head or pressure (H)
  - It is the *useful mechanical energy in Nm transferred by the pump to the flow*, per weight of fluid in N, expressed in Nm/N = m (also called 'meters of fluid').
  - A given centrifugal pump will impart the same head H to various fluids (with the same kinematic viscosity ν) regardless of their ρ.
  - The head develops in proportion to the square of the impeller's speed and is independent of the ρ of the fluid.

$$H \alpha N^2 \quad (16.4)$$

- Input power (P) or bhp
  - It is the *mechanical power in kW or W consumed by the shaft or coupling.*
  - It is proportional to the third power of the pump speed.

$$P \alpha N^3 \quad (16.5)$$

  - It is also linearly proportional to the fluid ρ

$$\rho \alpha N \quad (16.6)$$

- No load pump pressure (NLPP)
  - Head–capacity (H–Q) curve of any centrifugal pump has drooping characteristic. Pump performance is stable when the curve has a steady droop. A hump at the start can make the pump to hunt. *NLPP is the pressure developed by pump with the discharge valve closed.* In other words, it is the pump's zero discharge pressure. The desirable NLPP is ~110% of the design discharge head.
- Net positive suction head (NPSH)
  - The NPSHa (NPSH available) is the difference between the *total pressure at the centre of the pump inlet and the vapour pressure* p_v,

expressed as a head difference in m. It is a measure of the probability of vaporisation at that location and it is determined only by the operating data of the system and the type of fluid.

- NPSHr (NPSH required) is the actual NPSH required in a system at the centre of pump inlet to avoid cavitation.
- NPSHa should be always greater than NPSHr by a good margin for safety.

16.55.1 Boiler Feed Pumps

BFPs form part of the ST island in the power plants as they are rotating machines like STs and are physically located in the ST hall. However, in the industrial boilers they are usually supplied by the boiler makers and are placed on the ground in front of the boilers.

BFP is truly the heart of the boiler as it sets into motion the water and steam circuit. Also *it is the highest power consumer in the whole power plant* consuming between 3% and 5% of the plant output. It is natural, therefore, that a BFP should have the best η besides the highest reliability. It should also have the finest build quality for sustained availability.

Centrifugals are the pumps employed in all boilers save the very small. Reciprocating pumps are used for applications of ~5 tph and are not considered for elaboration here.

The following guidelines are important in the sizing and selection of centrifugal BFPs:

a. Number of BFPs
b. BFP drives
c. System resistance and BFP sizing
d. Parallel and series operation of pumps
e. BFP characteristics
f. BFP control
g. BFP protection
h. Types of BFP

a. Number of BFPs

- BFPs can feed a single or a battery of boilers.
- A minimum of two BFPs is required with *independent sources of power.*
- However, certain codes permit a single BFP operation provided the following points are taken care without any failure.
  - When there is failure of the source of energy leading to BFP stoppage the firing in boiler is stopped immediately.
  - Steam pressure and fw supply is continuously and automatically controlled. Reliability of FWR is very important and evidence of its reliable operation has to be demonstrated.
- There is a reliable controlling of the firing in the boiler. After any interruption of fire the heat stored in the furnace and boiler passes does not produce excess evaporation that can cause damage. This is particularly important in boilers with large amounts of refractory, such as the CFBCs with hot cyclones or boilers with refractory furnaces.
- Boiler is equipped with automatic trip system to cut off the fuel and air when the water level in the steam drum goes below the trip level.

b. BFP drives

- BFPs can be driven by either electric motors or ST.
- In process plants where steam availability is assured and there is use for LP steam exhausted by the drive STs, the turbo drives are overwhelmingly preferred to motors.
- If the electric motors alone are used care should be taken to employ independent feeders to assure power all the time.
- In utilities it is normal to have ST drives for BFPs from cycle η considerations. Besides, they are more amenable to speed variation which can make sliding pressure operation of plant possible.
- From cycle and plant η considerations STs are preferred, as the energy conversion from steam to power is not involved.
- With two BFPs there should be a reliable mechanism to instantly start the standby pump when the running pump fails.

  The convenience of installation and lower O&M of motors against the expense, complexity and higher O&M of STs has to be carefully evaluated when a particular system permits both.

c. System resistance and BFP sizing

BFPs consume a lot of power and have to be sized carefully to save on the auxiliary power. Sizing for a typical sub-c boiler plant is broadly as follows:

- The MCR capacity of a BFP should equal its share of boiler capacity + allowance for CBD + any added user like the deSH spray water.

- A safety margin of usually 10% is added to the pump capacity to arrive at the design or test block margins in industrial plants. This depends on the class of pump which defines the tolerances on *H, Q* and *P*. This can also be dictated by the client or consultant based on satisfactory practice. The margin is lower in utility boilers.

The operating pressure of a BFP is the sum of the following losses added to the highest SVLP

- Loss through fully open FCV
- Loss through Eco and connections
- Loss through FWH
- Loss through fw piping and valves
- Static head loss

- A safety margin of 2.5% over the operating pressure should be sufficient to arrive at the design condition.

Shown in Figure 16.36 is a typical characteristic of any boiler fw system which shows the 'system resistance' at various loads on an *H–Q* curve.

On this system resistance curve MCR and design conditions of pump are marked. BFP selected should have its *H–Q* characteristic curve passing through the design point. This is shown in Figure 16.37.

Two pumps are shown in Figure 16.38 to be operating in parallel to meet the system demand which is often the case in a boiler plant.

d. Parallel and series operation of pumps

Parallel operation is planned when more capacity is needed than the capacity of a single pump. Pumps should be identical for a parallel operation with steadily drooping curves with each pump sharing exactly half the load. If the curves are not continuously drooping but decline at the front end they are unsuitable for parallel operation as they tend to hunt. One pump will share more load than the other and suddenly move horizontally to the other point with lower flow. This repeated shifting of load is termed as *hunting*.

When more head is to be delivered pumps are put in series operation. Pumps need not be identical for series operation. As the discharge pressure of the first pump is the inlet pressure of the second pump leakage through the seals can be a problem unless the gland sealing is strengthened.

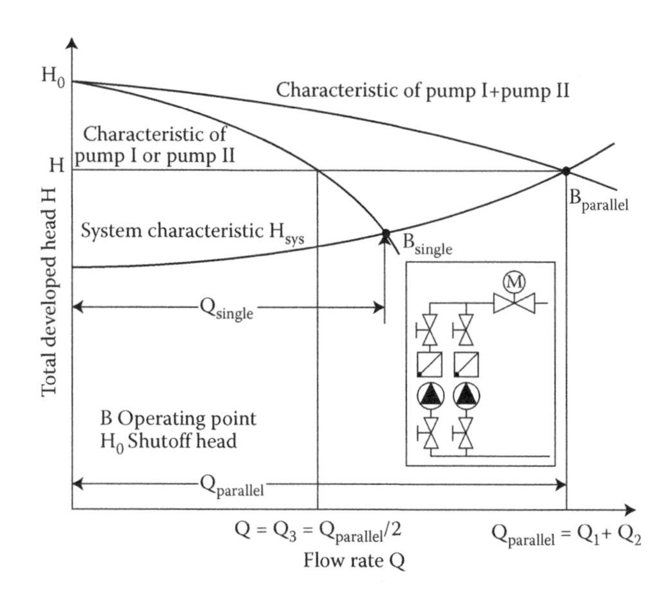

FIGURE 16.37
Pump characteristics imposed on the system resistance curve.

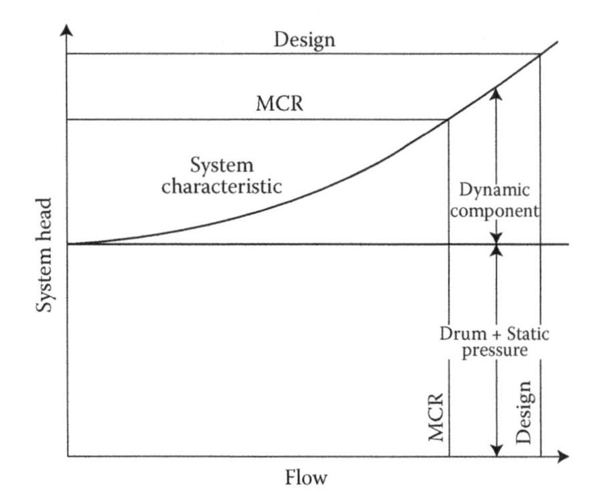

FIGURE 16.36
Water side system resistance in a boiler.

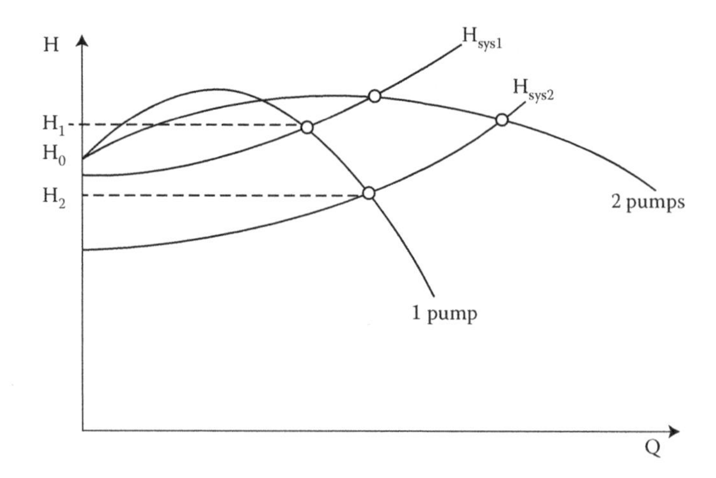

FIGURE 16.38
Parallel operation of two BFPs to meet the load.

e. BFP characteristics

- The operating and design temperature of a BFP is the same as that of the upstream deaerator.
- NPSH.
  - NPSH is a very important consideration as the BFPs operate at relatively high temperature and speed.
  - The minimum required NPSH is a function of pump design at the operating speed and has to be furnished by the pump maker.
  - The suction losses are to be added and there must be adequate margin in the available head.
  - Flooded suction is necessary.
- Inadequate NPSH causes *cavitation* in a pump leading to the *pitting* of the impeller.
- A continuously drooping characteristic is to be opted so that the parallel running of pumps can take place without the fear of hunting.

Figure 16.39 shows a typical set of characteristic curves of a centrifugal BFP.

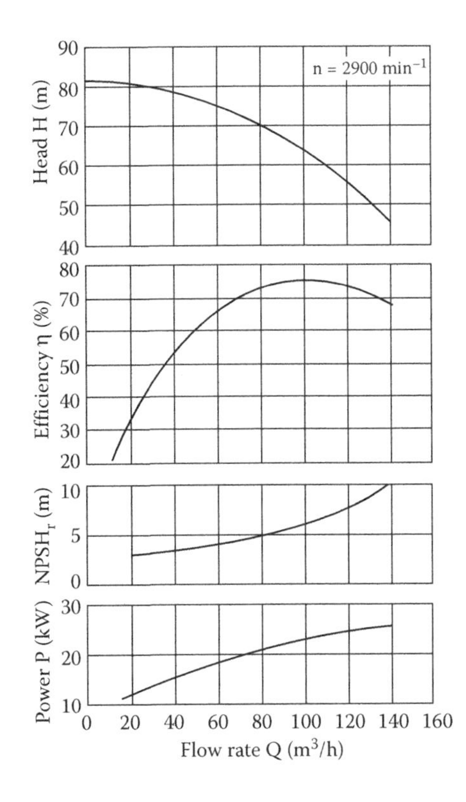

FIGURE 16.39
Characteristic curves of BFPs.

f. BFP control

Q, N and P are inter-related, as shown here.

$$Q_2/Q_1 = N_2/N_1 \quad (16.7)$$

$$H_2/H_1 = (N_2/N_1)^2 \quad (16.8)$$

$$P_2/P_1 = (N_2/N_1)^3 \quad (16.9)$$

If the speed is lowered by half head is lowered by a fourth and pressure by eighth. As compared to simple throttling control by valve at the pump discharge, speed control of pump is highly power saving as seen in Figure 16.40 where A_1 and A_2 are two system curves. P is the power absorbed by throttle control while P_1 and P_2 are the power by speed variation. Power saving at 50% load, namely, $P_1 - P_2$ (marked on the left), is quite sizable.

g. BFP protection

During start-ups and low loads

- An automatic recirculating control (ARC) valve is absolutely necessary to protect the pump from working against closed discharge valve required during start-ups and low loads. It is to prevent the overheating and possible seizure of the pump internals due to the idle churning of water. ARC valve works by automatically opening the valve below a predetermined flow of about 10–15% full load flow and discharging the

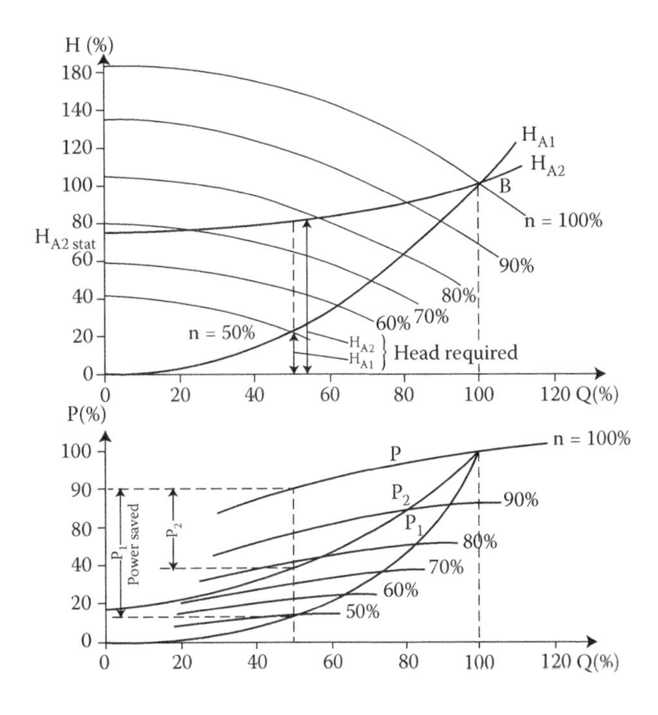

FIGURE 16.40
Pump control by throttling and speed variation.

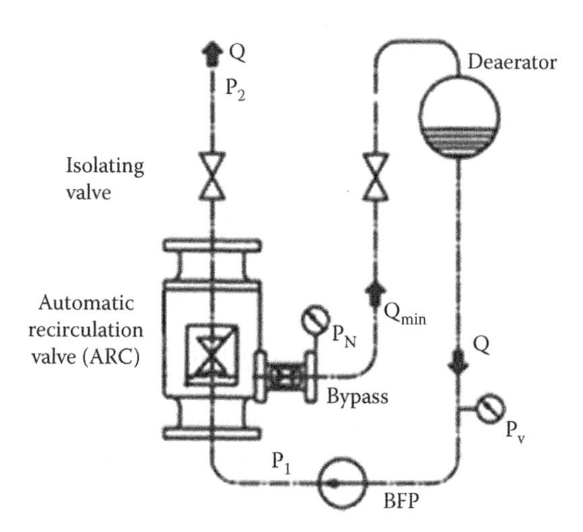

FIGURE 16.41
Working of automatic pump recirculation valve.

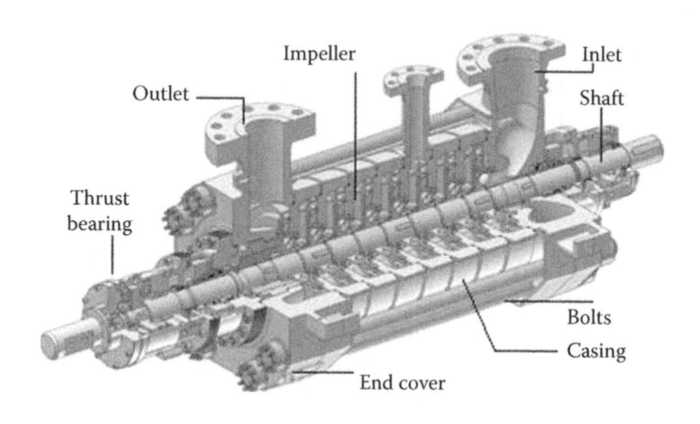

FIGURE 16.42
Radial or ring section centrifugal pump.

water either mostly to the deaerator or sometimes the pump suction. This way a minimum flow through the pump is established which prevents the overheating of internals. There are suitable pressure break down orifices to match the downstream LP system to prevent flashing. The scheme is illustrated in Figure 16.41.

- The convenience, reliability and simple construction of a special ARC valve has more or less made the earlier system obsolete, where a conventional CV in bypass line used to open when the main flow was measured to be less than minimum acceptable.

Pump balancing

When water flows along the pump shaft towards the discharge end axial thrust is created at the discharge end on the bearing. This thrust has to be counter balanced, which is usually done by a balance disc attached to the shaft or other means. HP water from discharge side is supplied from the opposite side of the balance disk in the balancing chamber. The leak off from the balance chamber is usually piped to the pump suction.

Another way to balance is to have pump suction from either end with discharge at the middle when the axial forces work in opposition.

h. Types of BFP
- There are mainly two designs in the centrifugal BFP:
  1. Radial split or ring section pumps
  2. Axially split pumps
- Radial pumps come with
  1. Individual ring sections, consisting of impeller and casing sections, which are bolted together and hence called *radial ring section pumps*
  2. The impeller and casing are inserted from one end in an external barrel and called *barrel casing pumps.*
- *Radial or ring section* pumps are popular due to simplicity and low cost with process and Cogen boilers. For maintenance and access purposes of pump the suction and discharge pipes are to be removed in ring sections, which is its main disadvantage. Figure 16.42 shows a typical multi-stage centreline mounted ring section centrifugal pump.
- *Barrel casing pumps* overcome this drawback of disturbing the piping by placing all the pump parts within a shell, called barrel. Only the nozzles protrude from the barrel, which are connected rigidly to the piping. The pump parts can be fully withdrawn for maintenance and inspection from one end of the barrel without disconnecting the piping.
- Being expensive, this construction is usually reserved for HP and large capacity pumps. Pressures >400 bar is possible, as the inside of the barrel, which has a gap between the impeller ring and barrel, is pressurised with water. This way the thicknesses of both components, namely the casing and barrel, are lowered. Also, HP pumps with impellers exceeding 8–10 in number are arranged in opposed manner to balance the thrust. Figure 16.43 depicts the cross section of a typical barrel casing pump. Note the thrust bearing and balancing leak off connection installed to counter the axial thrust.

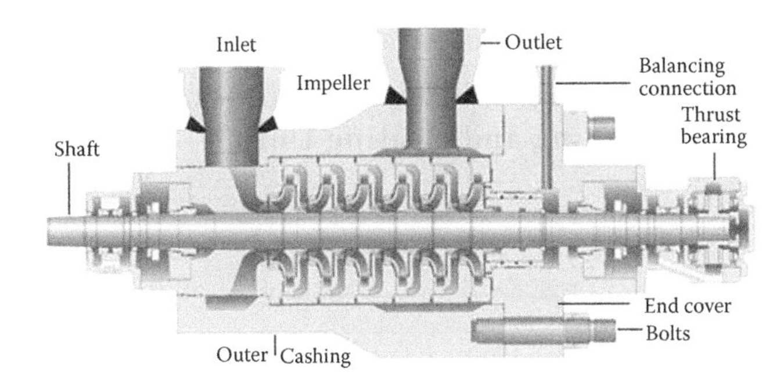

FIGURE 16.43
Barrel casing centrifugal pump for higher pressures.

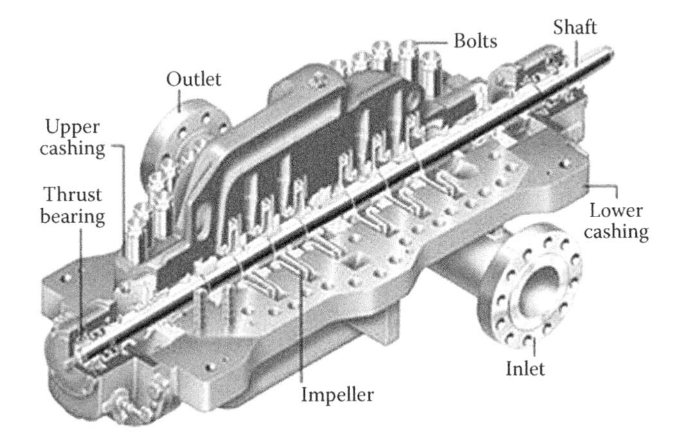

FIGURE 16.44
Axially split centrifugal pump.

- *Axially split pumps* shown in Figure 16.44 offer even better advantage of full access to pump internals without disturbance to the connected piping, as the pump casing is made in upper and lower halves and bolted together. The disadvantages are that the casings are thick and heavy. Each pump requires a pair of castings, making the pumps expensive. Large HP pumps, comparable to barrel casing pumps, are not possible. These pumps are not as popular because of their high cost.
- The pump η is decided by the impellers, which are common for all designs of casings.

16.55.2 Boiler Circulating Pumps

Natural circulation boilers do not need any pumps for helping the bw to circulate. They are needed in forced and assisted circulation boilers to pump the hot bw. They have to pump circulating water of high volume and temperature but at low heads. Single-stage pumps are usually adequate. Sealing the pumps is critical as the pumps operate at very high pressure. Being high-temperature pumps, the NPSH requirement is high.

There are two types of pumps available for such duty, namely

- Vertical glandless pumps with water-cooled motors
- Horizontal single-stage centrifugal pumps

Vertical glandless pumps are hung from downcomers and need no separate foundations. In these zero-leak designs, the motor is integral to the pump and both are placed in the same casing. Pumps are glandless with no stuffing boxes or mechanical seals. Motors are water cooled with process water which in turn, is cooled in an external HX. Such glandless pumps are much specialised and are available in large sizes up to ~2500 cum/h, 70 m head and 350°C. A typical glandless pump is shown in Figure 16.45. Such pumps are usually provided in large PF boilers where assisted circulation is offered.

There are two variations in glandless pumps.

1. *Canned or dry stator pump* in which the process water passes through the gap between the stator and rotor. The dry stator winding is completely isolated from the HP water a 'can' lining the stator bore.
2. *Submerged motor or wet stator pump* in which the water passes through the stator coils also.

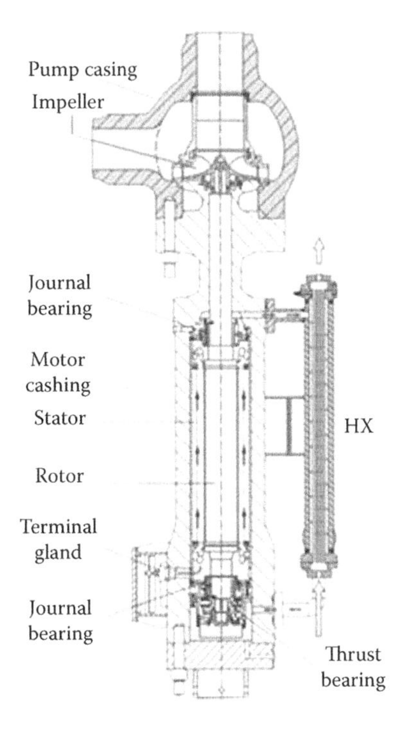

FIGURE 16.45
Vertical glandless wet stator pump for bw recirculation.

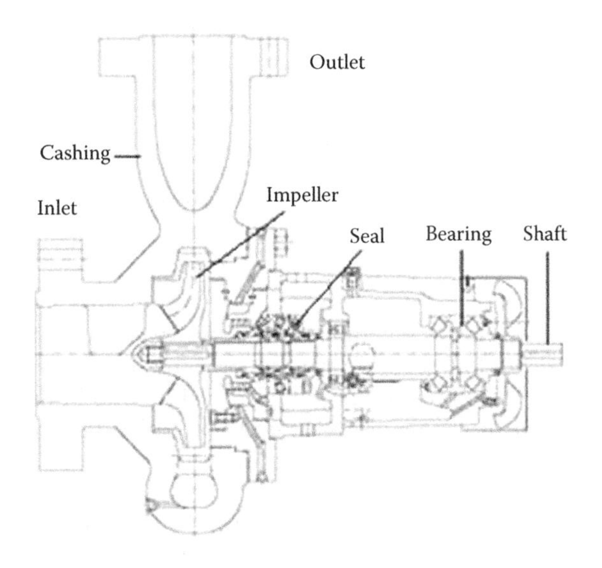

FIGURE 16.46
Horizontal back pull out single-stage hot water centrifugal pump.

The other design is the more conventional horizontal back pull out single-stage pump, very similar in construction to chemical pump, which is shown in Figure 16.46. These pumps are also available to similar capacities as vertical pumps. These are popular with Hrsgs and other forced circulation boilers where the number of pumps is less and there is adequate space to locate the pumps.

16.56 Pumping and Heating Unit (*see* Oil Pumping and Heating Unit)

16.57 Purge Systems (*see* Operation and Maintenance Topics)

16.58 Pyrometer (*see* Temperature Measurement *under* Instruments for Measurement)

16.59 Pyrometric Cones (*see* Refractory Properties *under* Refractories)

17

R

17.1 Radiant Boilers (*see* Industrial Boilers)

17.2 Radiation Heat (*see* Heat Transfer)

17.3 Radiant Superheater (*see* Superheater Classification in Heating Surfaces)

17.4 Ramming Mix (*see* Refractories)

17.5 Rankine Cycle (*see* Thermodynamic Cycle)

17.6 Rayleigh Number (*see* Dimensionless Numbers in Heat Transfer)

17.7 Refuse Derived Fuel-Fired Boilers (*see* MSW Boilers: RDF Burning)

17.8 Reburn (*see also* Reburning in Nitrogen Oxides in Flue Gas)

Reburn is a process where NO_x is reduced by reaction with HC fuel fragments as illustrated in Figure 17.1 for a front-wall-fired boiler.

In some cases, **LNBs** alone may not be able to achieve the NO_x emissions to the desired levels in a boiler mainly due to

- The non-availability of LNBs (such as in cyclone firing)
- The performance problems with LNBs (such as in C loss or tubewall wastage) or
- Insufficient NO_x reduction

Reburn can be applied to any unit firing any fuel to achieve NO_x reductions, typically in the range of 50%–70%. Moreover, it can be integrated with N_2-agent injection and other technologies for even greater NO_x reduction.

No changes to the main burners or **cyclone furnaces** are required. It is only that they are operated with the lowest excess air commensurate with acceptable lowered furnace performance and slightly reduced load to compensate for the additional reburn fuel.

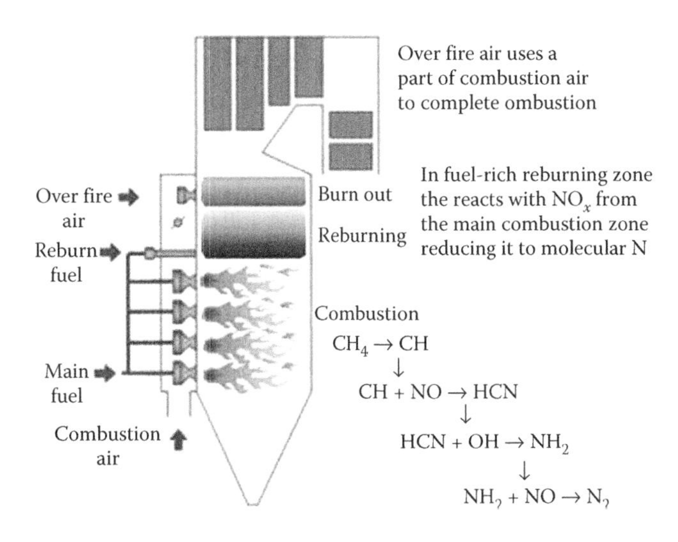

FIGURE 17.1
Reburning process.

A slightly fuel-rich reburn zone (with ~90% theoretical air) is created above the main combustion zone where reburn fuel is injected and where most of the NO_x reduction occurs.

OFA is injected above the reburn zone to complete combustion.

Reburn is always custom designed to match unit-specific factors. It can be applied to any system including boilers with **wall-**, **tangential-**, and **cyclone-firing** configurations as it requires no modifications to the main combustion system

NG is the preferred reburn fuel as it produces the greatest NO_x reduction per unit of reburn fuel injected, has no A or S and requires no pulverisation or atomisation. The only issues are its availability and cost. On coal-fired units, the use of PC as the reburn fuel avoids any cost penalty of the reburn fuel over the main fuel. Other fuels of interest include **FO** (with oil-fired units) and even **Orimulsion**. Lately, **agro-fuels** are gaining favour as reburn fuels.

For implementing reburn, the requirements are that there must be

- Sufficient space above the burners or cyclones to install the reburn components
- Adequate residence time in the reburn and burnout zones

By designing the reburn fuel and OFA injectors for rapid mixing, space requirements are in the range typically available on fullscale utility boilers. This is usually sufficient to provide the needed residence time.

NO_x reduction as high as 70% in a cyclone furnace application has been achieved with an effective reburn-zone *residence time* of only 0.25 s. Although such applications are feasible, longer residence times reduce the amount of reburn fuel required to achieve a specific NO_x control goal while improving C loss, particularly with coal reburn.

The reburn fuel injectors should be located close to the upper firing elevation, leaving enough space above the burners to achieve essentially a complete combustion in the main burner flames. For best results, injection of the reburn fuel should

- Penetrate across the furnace depth
- Mix rapidly with the furnace gases

As the amount of reburn fuel is small compared to the furnace gas flow, achieving penetration and rapid mixing is a challenge, especially with large furnaces. Increasing the momentum of the injected stream via a carrier gas or by high-velocity injection are means to achieve penetration.

Coal as the reburn fuel requires a gaseous carrier for pneumatic transport. Flue gas is preferred over air to minimise the reburn fuel quantity.

Char oxidation for most primary fuel occurs in the O_2-rich combustion zone. The burnout zone completes combustion of the reburn fuel. For gas reburn, CO is the primary combustible. For fuels that contain FC such as coal, the combustibles are CO as well as the C in fly ash.

OFA port location should allow for balancing the NO_x reduction of the reburn zone with the combustion η of the burnout zone. This is done by locating the ports substantially higher in the furnace than for conventional boilers but well below the furnace exit.

Reburn system can be integrated with normal boiler controls for fully automated operation. Depending on the NO_x control goal, the reburn fuel injection can be fixed or varied in response to boiler operating conditions and/or NO_x emissions.

17.9 Recovery Boiler (*see also* Kraft Process)

As the name indicates, a recovery boiler is for recovering the heat in fuel. In the case of BLR boiler, it has to additionally recover the chemicals. BLR boiler is an integral part of the **Kraft pulping** system and needs to be planned and executed along with pulp mill unlike many other WH boilers which can be added later.

This topic is covered under the following sub-topics:

1. Black liquor (BL)
2. BLR boiler size
3. BLR boiler process
4. Chemical recovery in BLR boiler
5. BLR boiler construction
6. BLR boiler auxiliaries
7. Smelt–water reaction

17.9.1 Black Liquor (see *also* Kraft Pulping Process)

BL from digester in Kraft pulping process is the *spent cooking (white) liquor along with the dissolved lignin of wood and some organic matter.* It is also called 'weak liquor' with solids content of only 15–18%, consisting of both organic and inorganic matter, when it comes out of the digester. Essentially, the

- Organic matter of BL is burnt for heat recovery for steam generation
- Inorganic chemicals of BL are reduced and recovered in the form of smelt for further use

BL concentration should be 55–58% solids for a self-sustaining combustion. As weak liquor is excessively dilute, it is first concentrated to *strong BL* in MEE to ~60 or even as high as 80% solids if liquor viscosity permits.

The *strong BL* is turned into *heavy BL* after adding Na_2SO_4 in the *mix tank* for making up the unavoidable losses in the pulping process.

This liquor is then heated in steam heaters for easy pumpability before it is sent to recovery boiler at 70–75% concentration.

If BL is concentrated to only ~55–60% in the evaporators, further concentration is done in *cyclone or cascade-type direct-contact evaporators* using the heat of flue gases. This practice has practically gone out of fashion as the flue gases pick up malodour of the liquor and spoil the environment.

In smaller boilers, decades ago, venturi scrubbers were employed for concentrating the BL, where weak liquor of ~45–50% concentration from MEE was sprayed into high-velocity flue gases in a venturi to scrub and cool the gases as illustrated in Figure 4.35. The chemicals of the flue gas and the M of the BL would exchange places. Chemicals are separated from the flue gas in a cyclone separator. The system was very effective in cooling the gases as well as recovering the chemicals. But the malodour in flue gas was very high. Such a wet scrubber is described in Wet Scrubbers under the topic of Dust Collection Equipment.

Cooking chemicals undergo a change in the digester. NaOH and Na_2S of white liquor turn principally into Na_2CO_3 and Na_2SO_4, respectively, in BL. Chemically, BL is a mixture of several elements where the largest fractions are C, O, Na, H and S. Analysis and cv is naturally dependent on the raw material and the variables in the pulping process. Typical analyses of dry BL are given in Table 17.1 for various raw materials used in the pulping process.

Points to be noted about the BL analysis and Gcv are as follows:

- BL analysis is always given in its elemental form and not as **UA**.
- UA gives a large value for A which needs to be distributed into its various constituents, particularly to Na and O_2 to obtain elemental analysis.
- **Bomb calorimeter** operates in oxidising atmosphere while the BLR boiler furnace is under reducing conditions. Bulk of Na_2SO_4 reduces to Na_2S and releases SO_2 into furnace, contributing to reduction of heat due to endothermic reactions. cv given by bomb calorimeter has to be reduced to the extent of 'heat of reaction' correction which is the difference between the endothermic and exothermic reactions.

17.9.2 BLR Boiler Size

The pulping capacity of a mill determines the size of a BLR boiler.

- For each 1000 kg of pulp, about 1500 kg of dry solids are generated.
- Of this ~650 kg or ~45% are recoverable chemicals.
- BL has nominal Gcv of ~15.35 MJ/kg, ~3660 kcal/kg or ~6600 Btu/lb.

A typical small BLR boiler will process 350 tpd dry solids and produce ~60 tph steam at a SOP of 20 bar.

TABLE 17.1

Analysis of Dry BL for Various Raw Materials Used in Pulping Process

Raw Material	Wood	Wood	Bagasse	Bagasse	70% Eucalyptus + 30% Bamboo	40% Hard Wood + 60% Bamboo	30% Hard Wood + 70% Bamboo
Element	% w/w	% w/w	% w/w	% w/w	% w/w	% w/w	% w/w
C	36.4	43.7	40.0	38.1	33.7	40.0	37.7
Na	18.6	17.3	13.0	22.3	26.3	19.0	15.6
S	4.8	3.4	3.0	4.4	5.2	3.8	1.4
H	3.5	3.7	4.0	3.4	3.2	3.5	4.5
K	2.0		1.0				2.2
Cl	0.2		—				0.6
N	0.2						0.4
O_2 by diff.	34.3	31.7	37.0	31.6	31.4	31.5	34.9
SiO_2/inerts		0.2	2.0	0.2	0.2	2.7	2.7
Total	100.0	100.0	100.0	100.0	100.0	100.0	100.0
GCV MJ/kg	15.35	15.60	12.98	14.15	12.98	14.65	15.26
kcal/kg	3665	3725	3100	3380	3100	3500	3645
Btu/lb	6600	6705	5580	6085	5580	6300	6560

While a medium-size BLR boiler will process 2700 tpd dry solids and produce ~380 tph at a SOP of 105 bar. Very large boilers processing solids >5000 tpd are built and even larger sizes of 8000 tpd are being planned.

17.9.3 BLR Boiler Process

The purpose of the BLR boiler is to

1. Burn the organic matter to capture the cv
2. Reclaim the spent pulping chemicals

Steam is produced to generate electricity and supply process steam demands. The design of the BLR boiler is attributed to G.H. Tomlinson and the first unit having water walls was constructed in 1934 in the United States.

A BLR boiler forms an integral part of Kraft pulping system. The various flow streams around the boiler are shown in the schematic diagram shown in Figure 17.2.

Typical modern BLR boiler has no direct-contact Evap in the form of a *cyclone or cascade* Evap to make the plant truly odourless by precluding any contact of flue gas and liquor.

- Strong BL from the MEE is pumped into the *mix tank* where fresh Na_2S from bin along with Na_2SO_4 collected in Eco and ESP hoppers is added to make up for the loss of chemicals.
- The liquor, now termed concentrated BL is heated in direct-contact steam heaters to reduce its viscosity (which was increased due to the addition of Na_2SO_4) and pumped to the spray guns for spraying inside the hot boiler furnace.
- In the furnace, the liquor gets dried as it falls to the bottom where combustion and chemical reduction takes place to convert the mass into smelt.
- Red hot molten smelt is tapped into the dissolving tank where it is mixed with ample weak white liquor from causticising plant to form green liquor from where it is taken to causticising plant.
- Green liquor is very dilute unlike BL. Green liquor mainly contains Na_2CO_3 and Na_2S as active chemicals besides C and inorganic impurities from smelt (mostly compounds of Ca and Fe).
- Flue gases from combustion are cooled in various heat transfer banks like SH, boiler and Eco like in a conventional boiler before dedusting in ESPs and discharging into the atmosphere.
- BLR boilers are always balanced draught units with PA, SA and ID fans. In medium- and large-sized units, PA and SA fans are separated as the pressures are different. In some units, TA fan is also given. Combustion air has to be necessarily heated, which is done with Scaphs with MP steam.

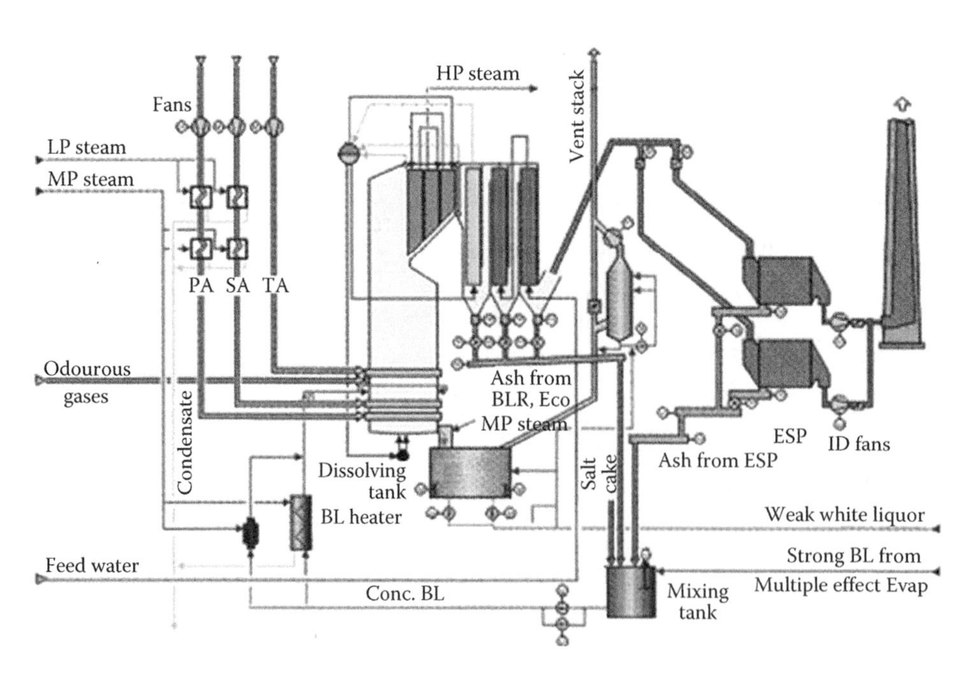

FIGURE 17.2
Schematic showing the flow streams around a BLR boiler.

17.9.4 Chemical Recovery in BLR Boiler

The lower part of the furnace is practically a chemical retort in which the combustion and chemical recovery reactions take place simultaneously. The combustion is a combination of suspension firing in furnace and pile burning on the floor. Air is provided at three levels as PA, SA and TA. The furnace, up to a certain height above the level of sprayers, has to be made of **composite tube** (CS covered with typically **Incoloy 825**), weld overlay of Inconel 625 on CS tubes or refractory lining on CS tubes (in LP and MP boilers) to prevent corrosion due to the prevailing reducing conditions (required for reducing Na_2SO_4 to Na_2S). Figure 17.3 shows the lower furnace where all the reactions take place.

The chemical recovery process, called *smelting*, is essentially the reduction of Na_2SO_4 to Na_2S by red hot C in the fuel bed. Na_2CO_3 remains practically unchanged and is turned into NaOH in the causticising plant that follows the BLR boiler in the Kraft process.

Concentrated BL is introduced into the furnace either horizontally by spraying on to the walls or vertically by spouting in the furnace. Both ways it is dehydrated by the time the droplets fall on the hearth below. Part of the liquor will flow down the furnace walls.

a. Smelting takes place in the bottom pile, called the char bed, where only ~40–50% of air is supplied at an LP of 75–100 mm wg to create a reducing atmosphere conducive to the reduction of Na_2SO_4 by C, as shown by the chemical reactions below. PA is supplied at ~1 m above the floor from all four sides of the furnace at LP and velocity for minimum penetration of the bed. The combustion takes place on the surface of the pile.

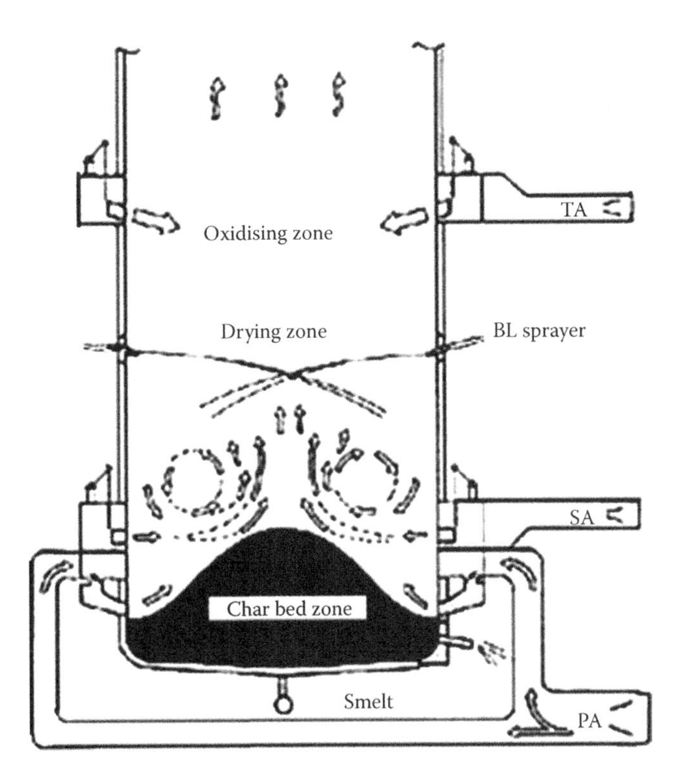

FIGURE 17.3
Lower furnace of a BLR boiler.

$$Na_2SO_4 + 2C = Na_2S + 2CO_2 \quad (17.1)$$

$$Na_2S + CO_2 + H_2O = Na_2CO_3 + H_2S \quad (17.2)$$

$$Na_2CO_3 = Na_2O + CO_2 \quad (17.3)$$

$$Na_2O + C = Na_2\text{ (gas)} + CO \quad (17.4)$$

Temperatures of the char bed and furnace should be as low as possible to minimise

- The loss of Na by volatilisation
- Fouling of HSs

It is important to maintain as small a bed as possible to contain the explosion forces should water leak into the bed inadvertently.

The floor of the furnace can be, with suitable differences in construction, either

- *Sloping* towards the dissolving tank or
- Flat decanting

b. Another 40–50% of air is given as SA at ~2.5 m above the floor through large ports in the side walls at HP of ~200–250 mm wg so that there is adequate penetration to burn off the VM from the bed and maintain the height of the pile. The burning of VM creates a high-temperature zone below the liquor spray level. The important reaction at this level is the oxidation of Na vapour into Na_2O vapour.

$$Na_2 + \tfrac{1}{2}O_2 = Na_2O \quad (17.5)$$

c. Above the liquor spray in the oxidising zone, TA nozzles are provided on front and rear walls where 5–10% air is at HP is provided for good penetration into the furnace to create turbulence and thorough mixing so that the combustion is completed. As oxidation of S and H_2S of BL takes place, as shown below, odorous discharge of flue gases ensues from the furnace.

$$SO_2 + \tfrac{1}{2}O_2 = SO_3 \quad (17.6)$$

$$2H_2S + 3O_2 = 2H_2O + 2SO_2 \quad (17.7)$$

d. In the upper furnace, CO_2 and SO_2 react with Na_2O fumes to form Na_2SO_4 and Na_2CO_3, which upon cooling, turn into dust and get captured in various hoppers.

$$Na_2O + CO_2 = Na_2CO_3 \quad (17.8)$$

$$Na_2O + SO_2 + ½O_2 = Na_2SO_4 \quad (17.9)$$

Preheated air at ~150°C (300°F) is needed to help in combusting a wet fuel like BL. With higher inorganic content in liquors of bamboo and grasses, even higher air temperature of ~200°C (400°F) is desired. As no gas–air AH can be installed in BLR boilers for fear of fouling, **Scaphs** are used at the discharge of FD fans with MP steam provided for air heating.

Furnace gas velocities in hearth and liquor spray zones are maintained quite low at 3.5–5 mps (11.5–16.5 fps) to minimise the carryover of Na. This translates to furnace plan HRR of ~2.84 MJ/m², 2.44 × 10^6 kcal/m² h or 900,000 btu/ft² h.

Furnace volume is generous so that the FEGT <925°C (1700°F) to enable the low-velocity flue gases to precipitate the sticky dust to the maximum extent so that the fouling of HSs is low.

17.9.5 BLR Boiler Construction

Na and S are the dominant elements of BL. Fouling and corrosion are the resultant problems in the boilers.

Na with its low melting point causes high degree of fouling and sulphurous gases contribute to corrosion. The obsession of the BLR boiler designer is to find appropriate solutions for these twin problems of fouling and corrosion.

A typical large-HP high-capacity boiler is shown in Figure 17.4.

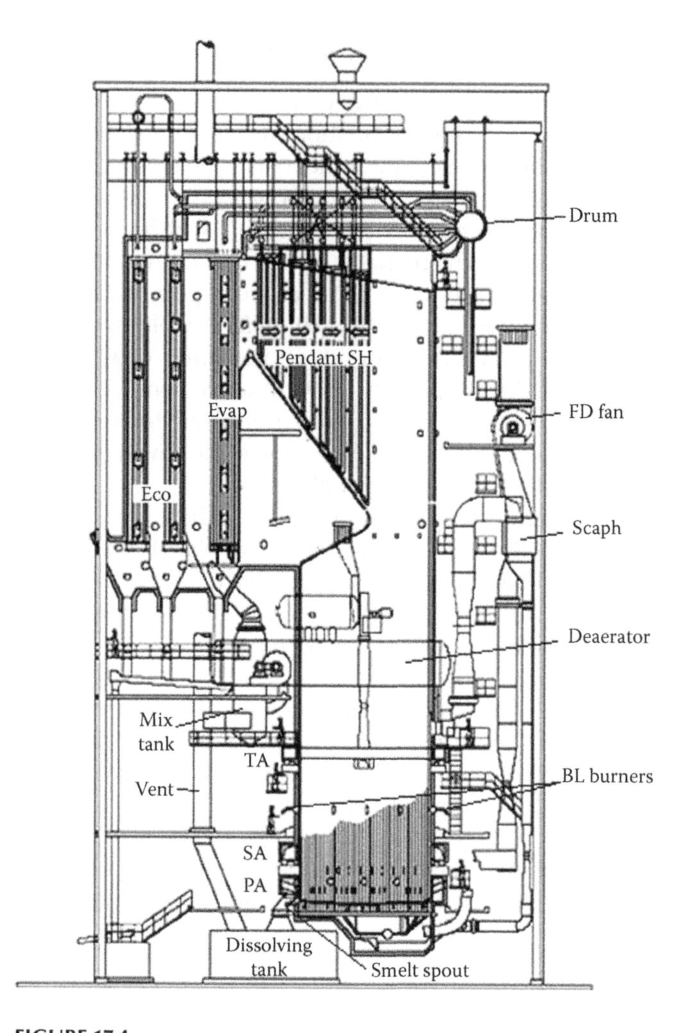

FIGURE 17.4
Cross section of a large-capacity HP single-drum BLR boiler with vertical Eco.

Furnace:

BLR boiler, therefore, is characterised by

- Very tall, generously proportioned and cool furnace aiming to minimise carryover that causes fouling
- Lower furnace of suitable construction for corrosion and erosion resistance
- Widely spaced tube panels in SH, BB and Eco banks with least ash collection
- Liberal provisioning of SBs and poke holes

Over the years, the boilers have become more generously proportioned with better materials of construction with improved build quality aimed at enhancing the availability and reliability. In terms of η there may have been no major advancements worth stating. The furnace bottom here has decanting floor.

Single- and bi-drum boilers:

Boilers are generally of either bi-drum design suitable for operating pressures <62 barg/900 psig or single-drum suitable for higher pressures. HP boilers are usually built for ~105 barg and 490°C (~1500 psig and 915°F) even though some units are operating at ~130 bar and 515°C (1900 psig and 950°F). It took many years to move from MP to HP as the corrosion of water walls sharply escalates when the tube metal temperature rises above 300°C. It is the advent of composite tubes that has mainly helped to overcome this hurdle.

Water wall construction:

Furnace construction is quite sophisticated here. Furnace floor is made of **ribbed/rifled** CS tubes for improved circulation and pin studded with refractory covering to hold out to the erosive forces of the flowing smelt. Furnace walls up to PA and SA ports are of Inconel tubes of 825 specification or 625 weld overlay to withstand the highly corrosive environment. Between SA and TA ports it is usually composite tubes of CS with outer sheath of SA 304L to contain costs. This may extend 1–5 m above the TA. Above that CS tubing is adequate. A typical composite tube wall is shown in Figure 13.4.

Screen and SH:

At the exit of the furnace there is no screen, which is provided many times to reduce carryover, if the surfacing needs so demand. SH tubes here are completely of pendant construction. Platens are also used when high SOT is needed. SH construction is conventional except for a wide spacing of >300 mm across the gas flow entirely. Also, the first bank at the furnace exit is PSH to keep the metal temperatures down to minimise fouling. The SH floor is much steeper ~50 degrees to horizontal and pendant to floor gap much higher to help the precipitated dust to easily flow back to furnace and not build up.

BB:

Evap tubes are made into vertical panels and placed quite wide apart to help easy cleaning.

Eco:

Eco is of vertical construction with either plain tubes or tubes having flat longitudinal fins. Fins impart strength to the long tubes and also reduce the HS. These Ecos are invariably top supported. Flow direction is always water up and gas down, with gas flowing longitudinally over the tube banks for minimum fouling and improved heat transfer. The tube panels are placed a little closer than in SH as the dust is now distinctly less sticky. Gases are cooled to within 50°C (90°F) of Eco inlet fwt in odourless units. If a direct-contact Evap is used, ~300°C (570°F) from Eco is the exit gas temperature desired. Figure 17.5 shows a typical vertical tube Eco.

Sootblowers:

Heavy-duty SBs are required to be generously provided in all convection passes for an effective removal of A. It is normal to cover the entire SH zone by retractable SBs despite the wide spacing of SH elements as high-temperature fouling is very severe with many BLs.

Often air is preferred to steam as the blowing medium from safety considerations. Water leakage from SBs may initiate smelt–water explosion. Another reason for choosing air for soot blowing is the relatively large amount of expensive steam consumed for blowing instead of being available for process or power generation. Since the steam output is modest, the soot blowing steam as % of evaporation works out quite high, unlike in a utility boiler.

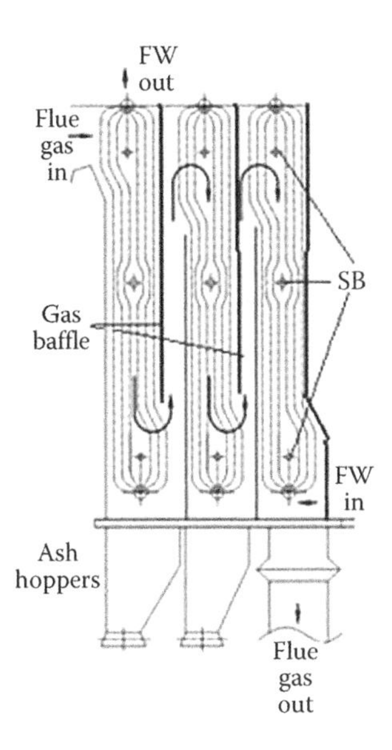

FIGURE 17.5
Vertical economiser in a BLR boiler.

Mix tank:

This is the place where Na_2SO_4 lost in the pulping process is made up by adding to the BL from the salt cake bin.

- Ash from all the hoppers, including ESP, is taken to the mix tank by scraper conveyors in dry form. Ash contains inerts and Na_2SO_4.
- In some units, strong BL from MEE is pumped to the boiler and Eco hoppers before being sluiced back to the mix tank in wet form. In this system, the scrapers are avoided. This is not favoured these days as BL-flue gas produces malodour.
- The mix tank is located at good elevation to provide adequate NPSH to BL firing pumps.
- Na_2SO_4 makeup is done in the mix tank from the salt cake bin kept at a higher elevation.
- Mix tank is provided with a small vent of ~200–250 mm to remove any gases formed.
- The liquor inside is continuously stirred by vertical agitator to help dissolve the solids in the strong BL.
- The tank is indirectly heated by steam to lower the viscosity to keep the BL pumpable. A steam coil is wrapped around the body of the tank for heating without diluting the BL.

BL firing pumps:

They raise the BL to the desired pressure and pump it through direct-contact BL heaters where steam injection into BL dilutes the liquor, increases the temperature and lowers the viscosity for improved spraying.

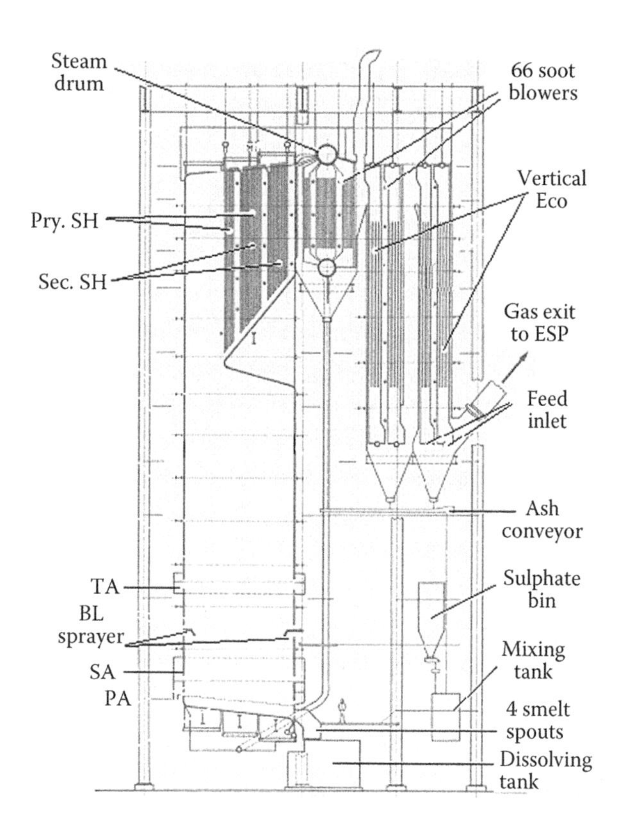

FIGURE 17.6
Bidrum BLR boiler with vertical economizer.

Figure 17.6 shows a typical MP odourless boiler for processing 800 tpd of dry BL solids and producing 185 tph of steam at 66 bar and 450°C (~408,000 pph at 940 psig and 840°F). The main difference here is the presence of a conventional BB and furnace of substantially CS construction. The LP units are also of nearly the same construction. The furnace has a sloping floor in this design.

17.9.6 BLR Boiler Auxiliaries

BLR boiler has some typical auxiliaries, in addition to the normal, as it is also a process equipment forming part of the pulping cycle besides being a steam generator.

- Mix and dissolving tanks are exclusive to BLR boilers. Mix tank is briefly covered under boiler description above and the elaboration of the dissolving tank follows.
    - Dissolving tank, always located on the ground floor in front of the smelt spouts, is comparatively a much larger tank. It is usually circular in shape. In large units it is made oval to take additional spouts.
    - It is made of heavy construction to withstand the minor explosions continuously taking place as the hot smelt falls into water in tank.
    - One or two horizontal agitators located at the lowest point help in hastening the dissolving action.
    - Smelt spouts, numbering one to four, depending on unit size, transfer the hot smelt from the furnace hearth to the dissolving tank. MP steam jets shatter the smelt flow into much smaller streams to avert possible explosion from smelt–water reaction.
    - Spout hoods at the furnace discharge protect the personnel and equipment from smelt spatter.
    - A large vent taken to the top of boiler relieves any pressure build-up. The fumes need to be washed to prevent spreading of any odour from the vent stack, for which weak white liquor is used.
    - Weak white liquor taken from causticising plant is mainly used for mixing with smelt to form green liquor, which is basically an aqueous solution of $Na_2S + Na_2CO_3$. Green liquor tank is always kept nearly full to help dissolve the smelt and more importantly provide adequate NPSH for green liquor pumps.
    - Green liquor in dissolving tank is at ~95°C having an SG of ~1.16 and viscosity <30 SSU.
- Shown in Figure 17.7 is an oscillating BL sprayer assembly unit. The two rotary couplings help in simultaneous tilting and rotating movements to provide a uniform flow throughout the furnace. Large droplets help to minimise carryover. Typical spraying pressure is 1.5 barg at 105°C. It is normal to provide one oscillating sprayer up to ~500 tpd in the front/rear wall, two for 500–1000 tpd on both front and rear walls and four for >1000 tpd.
- Figure 17.8 shows a typical start-up gun for firing FO and/or NG. Start-up guns are introduced through the SA ports. These burners are for

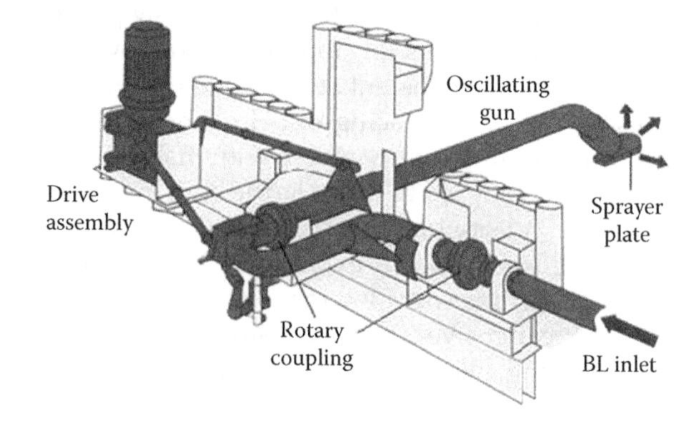

FIGURE 17.7
Oscillating BL sprayer assembly unit.

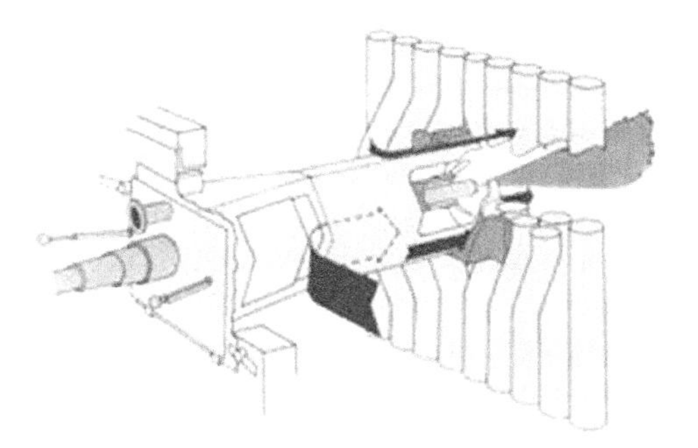

FIGURE 17.8
Start-up gun in a BLR boiler.

- Start-up duty to raise steam at the beginning
- Load stabilisation during upset condition
- Sustaining ignition while building up char bed
- Burning the char bed when shutting down
- Shaping and regulating the bed
- Carry on certain amount of load

As they are located in a zone where highly reducing conditions prevail, special provisions are necessary to prevent ash build-up leading to corrosion.

- In some mills, auxiliary firing is employed for full load operation on oil or gas for Cogen. For this register burners are employed at TA level or higher, as shown in Figure 17.9. As the burners are in oxidising zone, there is no effect of the lower furnace except that they are likely to accumulate a lot of dust. Special provisions are needed to keep them clean.
- The solid concentration of strong BL coming from MEEs at ~50% needs to be raised to 65–68%. This is done in direct-contact evaporators, either in cyclone or cascade type.

Cyclone Evap: In *cyclone Evap* shown in Figure 17.10 strong BL is sprayed in the inlet duct where flue gas from Eco exit is admitted at ~300°C. The gases and liquor droplets have an intimate

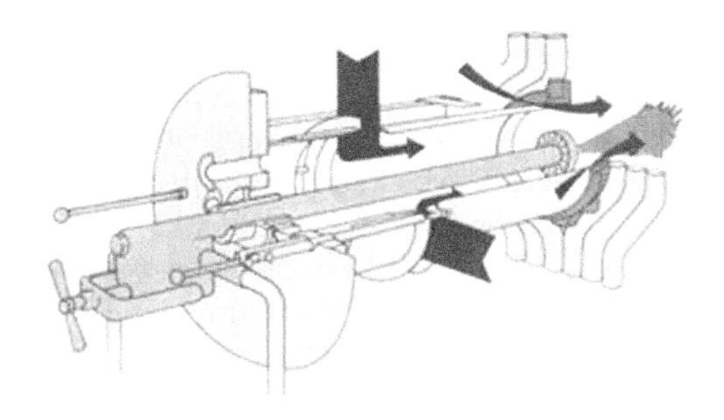

FIGURE 17.9
Load-carrying gun in a BLR boiler.

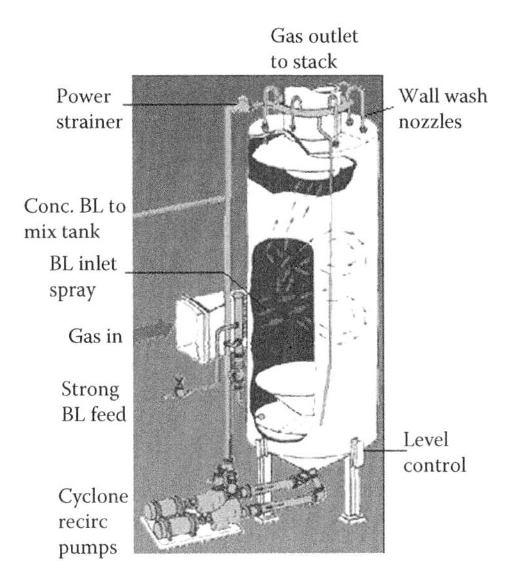

FIGURE 17.10
Cyclone evaporator for BL concentration.

contact and as they rotate within the Evap body, the dried liquor droplets are thrown to the wall by centrifugal force. Wall wash flow captures the dried matter and brings it into the bottom sump. BL recirculating pumps lift the liquor to the top of the tower through a mechanical strainer. The strained liquor is sprayed as wall wash to keep the Evap wetted and capture the dust. The concentrated unstrained BL is sent to the mix tank. Cyclone exit gas temperature is ~150–160°C for the protection of ESP from low-temperature S corrosion.

Cascade Evap: Cascade Evap shown in Figure 17.11 works on an entirely different principle. Strong BL is pumped into a steel trough in which a rotor or a cascade wheel is moving slowly. There can be a single or a pair of rotors depending on

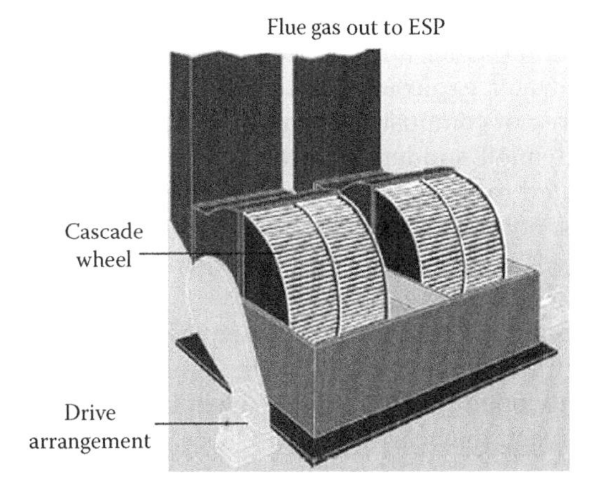

FIGURE 17.11
Cascade evaporator for BL concentration.

the size of boiler. The rotor assembly consists of a pair of end plates which carry a number of tubes between them. BL sticks to the tubes as the rotor comes out of liquor and comes in contact with the flue gases admitted from the exit of Eco at ~300°C. The gases dry the liquor film and take its M as they exit towards the ESP. The dried liquor falling into the BL in the trough increases the overall concentration. The exit gas temperature is ~150–160°C.

- Dust collection in BLR boiler is both for the recovery of expensive chemicals from flue gases and stack cleaning. The dust precipitated is carried to the mixing tank by scraper conveyors. ~150°C (300°F) is considered as a safe limiting temperature for ESP operation to avoid corrosion of internals. A two-stream ESP with each stream of 60% capacity is normally chosen for having high availability of boiler plant.

17.9.7 Smelt–Water Reaction (*see also* Emergency Drain Valve in Control Valves *under* the Topic of Valves)

It is the pool of molten smelt at the bottom of the furnace, at temperatures of 925–980°C (1700–1800°F), that presents a unique BLR boiler explosion hazard. As with any high-temperature molten material, mixing water can result in an expanding vapour explosion caused by an extremely high heat transfer rate between molten smelt and water. The rapid vapourisation of a kg of water can release as much energy as half-kg of TNT. A smelt–water explosion generates a shock wave that travels at supersonic velocity and causes severe local damage when the wave contacts the furnace. It is nearly 10 times as severe as the other furnace explosions that occur with FO, NG or PF, develop much lower peak pressure and travel at sonic velocity. These conventional furnace explosions take place when an explosive concentration of fuel–air mixture accumulates and ignites. Smelt–water explosions are entirely different, because they are not combustible, that is, nothing ignites. When molten smelt suddenly contacts liquid water, steam is generated extremely rapidly. The seriousness of explosion is not related to the amount of smelt or water involved. A teaspoon of molten smelt in a cup of water can produce a big explosion.

Interestingly, the smelt–water explosions do not occur every time water touches molten smelt but may happen only one-third of the time when large quantities of water reach the hearth of the furnace. The chance of an explosion is less with smaller quantities of water.

Although much study has been done, the only known method to avoid a smelt–water explosion is to prevent smelt–water contact which is a challenging task in a water-cooled vessel like a boiler. In addition to bw, several other water sources exist. If the water content of the BL exceeds 42% (reduces to <58% solids), an explosion can result. The condensate may enter through a faulty SB system or a faulty Scaph. Water can be introduced from external water wash hoses, from the BL piping water wash system or from faulty liquor HXs.

Tube leaks are the main cause of smelt–water explosions. The probability and intensity of an explosion increases with the amount of liquid water reaching the bed and so depends on leak size and location. There are two ways to avoid explosions from tube leaks:

1. Eliminating tube leaks by *identifying causes* and *taking proactive action* during operation. Better build quality also contributes towards this cause. Superior quality tested tubes, better welding materials and procedures, and more rigorous testing of all welds are some of the ways of improving the integrity of construction.
2. Minimising water input to the furnace if a leak occurs by initiating an *emergency shutdown procedure (ESP)* that includes *rapid-draining* the boiler. This consists of initiating the following steps on suspicion of leak:
   - Stop firing BL and other fuels
   - Shut the air supply to PA ports by tripping the emergency damper in the PA duct
   - Keep SA and TA on
   - Keep running ID fan to maintain 5–10 mm wg draught in the furnace
   - Shut off fw supply
   - Drain the boiler to 2.5 m (8 ft) from the floor. By draining the boiler to such a low level, the furnace is nearly made bereft of water but prevented from overheating from the heat of the hearth.

All personnel should be cleared from the area. No one should go near to locate the leak as the explosions can occur from 5 min to 5 h of smelt after contact.

Explosions from tube leaks are primarily caused by large leaks, but floor tubes are an exception. Experience shows that the most dangerous type of tube leak is a large leak in the furnace cavity. A small floor tube leak also carries a significant risk. Small leaks in wall tubes near the hearth can result in explosions, but the intensity will likely be low. Large tube failures in the BB have caused explosions, but the relatively remote location decreases the likelihood. There have been no explosions caused by an Eco tube failure of any size, nor by a small leak in the BB or upper furnace.

There has been a significant reduction in the frequency of large leaks in the furnace cavity in the last few years mainly due to

1. *Low drum level trips*
2. *Better WT*
3. *Effective inspection and maintenance procedures*

The Black Liquor Recovery Boiler Advisory Committee (BLRBAC) has done a great deal of work over many years and provides recommended procedures and guidelines for safe operation of BLR boilers.

17.10 Redwood Seconds of Viscosity (*see* Liquid Fuels *and* Instruments for Measurement)

17.11 Refinery Gas (*see* Gaseous Fuels)

17.12 Refractories

Refractories are *non-metallic inorganic materials which substantially retain their strength, chemical and physical properties at high temperatures* that make them suitable for construction or lining of high-temperature furnaces, kilns and reactors. As per ASTM C71 refractories are expected to withstand hot conditions >1000°F (~540°C). They should be able to sustain in hot condition

- Pressure from self-weight or the weight of furnace parts or contents
- Thermal shock resulting from rapid heating and cooling
- Chemical attack by heated solids, liquids, gases and fumes
- Mechanical wear

The salient features are that the

- Refractories are made of a combination of minerals
- Properties are dependent on what combination has gone into the making
- Properties are also heavily dependent on the type of local minerals used
- Very small change in the ingredients can affect the properties in a big way

Refractories are made mostly from Al_2O_3, SiO_2 and MgO; CaO (lime) is another oxide usually found in refractories. Fireclays are also widely used.

17.12.1 Refractory Types

a. On the basis of properties or the composition, the refractories can be classified as
   1. Fireclay
   2. High alumina (Al_2O_3)
   3. Silica (SiO_2)
   4. Basic and
   5. Insulating refractories
b. There are special types like
   1. Silicon carbide (SiC)
   2. Graphite (C)
   3. Zirconia (ZrO_2) and so on

   The bulk of the refractories are made from silica and alumina in various combinations.
c. Refractory materials are supplied as
   1. Preformed shapes
   2. Special-purpose clays
   3. High-temperature cements
   4. Bonding mortars
   5. Plastic refractories
   6. Ramming mixes
   7. Hydraulic setting castables
   8. Gunning mixes
   9. Granular materials
   10. Ceramic fibres

17.12.2 Refractory Properties

Properties of refractories which are of practical importance can be listed as below:

1. Refractoriness pyrometric cone equivalent (PCE)
2. Refractoriness under load (RUL)
3. Thermal expansion
4. Bulk ρ
5. Volume stability
6. Thermal shock resistance

Thermal properties of importance are

1. Thermal conductivity (*k*)
2. Specific heat (c_p)
3. Thermal diffusivity

17.12.2.1 Refractoriness

Refractoriness is the *temperature withstanding ability* of the refractory.

Owing to the heterogeneity of their composition and structure, ceramic refractories do not exhibit a uniform melting point. The refractoriness is characterised by the optical determination of the pyrometric cone equivalent (PCE according to Seger), that is, the temperature at which the tip of a cone made of the sample material softens to the point that it touches the base plate. The cone is of specified size (usually 12 mm sides and 38 mm tall with one edge perpendicular to the base) and heated slowly at the rate of 5°C/min.

Pyrometric cones are made in series, the temperature interval between the successive cones usually being 20°C. The best known series are Seger Cones (Germany), Orton Cones (the USA) and Staffordshire Cones (the UK).

RUL-refractoriness under load (2 kg/cm^2) is a more valuable data as it represents the practical situation better.

PCE and its equivalent temperature is given in Table 3.3 under the topic of refractory brick.

17.12.3 Refractory Selection

Besides the ability to survive high gas temperatures, refractory selection is based on the following considerations:

- They have to withstand specific type of chemical attack from the ash of flue gas.
- The selection of the type of refractory is also guided by the gas velocity.
- Ceramic fibres can be used up to a gas velocity of 7.5 m/s while
  - Monolithics <60 m/s
  - Bricks >60 m/s
- Monolithics have no joints and are in a way better as there is no risk of erosion of the mortar joints in brick construction, provided the application can accept lower ρ and wear resistance.

Some applications require special refractory materials. ZrO_2 is used when the material must withstand extremely high temperatures. SiC and C (graphite) are also used for very high temperatures.

17.12.4 Various Types of Refractories

17.12.4.1 Fireclay Refractories

A group of refractory clays which can stand temperatures >PCE 19 are called fireclay. In nature it is usually found to contain 24–32% Al_2O_3 and 50–60% SiO_2. Impurities like oxides of Ca, Fe, Ti and Mg and alkalies are invariably present, making it white, grey and black in colour.

Refractory fireclays consist essentially of hydrated aluminium silicates (generally $Al_2O_3 \cdot 2SiO_2 \cdot 2H_2O$) along with smaller quantities of other minerals.

- *Hard clays*, deriving their name from their extreme hardness, are the flint and semi-flint clays, which form the principal component of the high- and super-duty fire bricks having PCE of 32–35.
- *Soft clays*, are plastic and semi-plastic refractory clays, which vary considerably in refractoriness, bonding strength and plasticity having a PCE value ranging from 29 to 33.

A good fireclay should have 24–26% plasticity and shrinkage after firing should be within maximum 6–8%. It should also not contain more than 25% Fe_2O_3.

Because of the abundant supply of fireclay and its comparative cheapness, the refractory bricks made out of it are the most common and extensively used in all places of heat generation.

Refractoriness and plasticity are the two main properties needed in fireclay for its suitability in the manufacture of refractory bricks. A good fireclay should have a high fusion point and good plasticity. Depending upon their capacity to withstand high temperatures before melting, the fireclays are graded into the following:

- Low duty—withstand temperatures between 1515°C and 1615°C (PCE 19 to 28)
- Intermediate duty—1650°C (PCE 30)
- High duty—1700°C (PCE 32)
- Super duty—1775°C (PCE 35)

17.12.4.2 High-Alumina Refractories

These can be produced using many minerals but mostly are made from bauxite ($Al_2O_3 \cdot H_2O + Al_2O_3 \cdot 3H_2O$) or diaspore ($Al_2O_3 \cdot H_2O$) or a mixture of the two blended with flint and plastic clay.

Alumina refractories are more refractory than fireclay, approximately in proportion to their content of alumina. They are highly resistant to chemical attack by various slags and fumes and in general, have greater resistance to pressure at high temperature than fireclay refractory.

High-alumina bricks have Al_2O_3% ranging from 50 to 99, Fe_2O_3% from 0.2 to 3.0, and apparent porosity from 15% to 23%.

17.12.4.3 Silica Refractories

These are made from high-purity crystalline mineral quartz and silica gravel deposits with low alumina and alkali contents. They are chemically bonded with 3–3.5% lime.

The characteristic of this refractory is that the thermal expansion is high at low temperatures and negligible beyond 550°C. They possess high refractoriness, high abrasion resistance, strength and rigidity at temperatures close to its melting point. They are particularly suited to containing acidic slag.

However, they are susceptible to thermal spalling (cracking) at 650°C but at higher temperatures they are free from spalling.

17.12.4.4 Basic Refractories

Basic refractory is any heat-resistant material used for basic linings. The raw material used here includes magnesite ($MgCO_3$), dolomite ($CaMg(CO_3)_2$), chrome ore and so on.

17.12.4.5 Insulating Refractories

These are lighter and porous as they trap a lot of air and possess much lower conductivity and heat storage capacity. They can withstand high service temperatures in cold and hot facing duties. Typical properties of insulating bricks are displayed in Table 17.2. Higher grades are also in vogue.

They are normally used as backing material. If the furnace conditions are clean, with flue gases free from ash, they can be used as the facing materials also as in oil- and gas-fired boilers.

17.12.5 Various Types of Bricks

17.12.5.1 Refractory/Fire Bricks and Tiles

Bricks or tiles are preformed shapes. They are obtained by pressing the green mass to the required shape and size and firing at the specified temperature until the refractory bond is formed by chemical action under heat.

Standard refractory bricks are made in two sizes 227 × 116 × 76 mm (9″ × 4½″ × 3″) and 227 × 116 × 63 mm (9″ × 4½″ × 2½″). Tiles are flattened bricks usually made in 50–76 mm thickness with overlapping edges and arrangement to hold them. Tiles can also be made in different shapes and in more advanced composition than bricks.

Refractory or fire brick construction today is confined mainly to the following areas in modern boilers:

- Around the fire in pile burning like in horse shoe or ward furnaces in firing bagasse or similar fuels
- Shaped refractory arches in gravity-fed grates for radiating the heat on to the bed
- Boiler enclosure in brick set boilers
- Brick lining for underground brick flues
- Stack lining
- Brick lining in cyclones and external HXs in CFBC boilers

Refractory tiles are used in the following areas in modern boilers.

- On top of the floor tubes to protect the tubes from overheating as in the case of package boilers
- In burner quarls to give the right shape to the flame
- Between tubes to form gas baffles as in cross or longitudinal BBs
- Lining of hot cyclones
- Lining of hot and dust-laden gas ducts and hoppers

In comparison to the former times, the use of refractory bricks in modern boilers is negligible. As bricks are pressed in hydraulic presses, they are strong, dense and heavy which makes them ideally suited to face the fire and dust-laden gases.

Several types of refractory bricks are manufactured to suit the duties of furnaces, of which bricks for boiler duty are, mainly

TABLE 17.2

Properties of Insulating Bricks

Item	Service Temperature (°C max)	Bulk ρ (g/cc max)	Apparent Porosity (% min)	CCS (kg/cm² min)	PLC[a] at ST/3 Hrs% max	PCE Orton Cone min	Thermal Conductivity (*k*) (kcal/m/h/°C max)
Cold face insulation bricks	1050	0.8	70	12–15	±1.5		0.20
Hot face insulation bricks	1250	0.9	65	20–25	±1.5	27	0.24

[a] PLC, permanent linear change at standard temperature.

- Fireclay
- High alumina
- Insulating bricks

17.12.5.1.1 Fireclay Bricks

These are made of a blended mixture of flint and plastic clays. Some or all of the flint may be replaced by highly burnt or calcined clay called grog. The dried bricks are burnt at 1200–1500°C. Fire bricks can withstand spalling and many slagging conditions but are not suitable for lime or ash slag.

17.12.5.1.2 High-Alumina Bricks

These are graded as per their alumina content as 50%, 60%, 70%, 80% and 90% and are used for unusually severe temperature and load conditions. They are more expensive than fire bricks.

17.12.5.1.3 Insulating Firebricks

These are made from porous fireclay or kaolin. They are lighter, 1/2–1/6 of the brick weight, low in thermal conductivity and can withstand high temperatures. They are graded as per their thermal withstand temperatures as 870, 1100, 1260, 1430 and 1540°C (1600, 2000, 2300, 2600, and 2800°F). They are not slag resistant.

High-burned kaolin (alumina-silica china) refractories can withstand high temperatures and heavy loads or severe spalling conditions as in oil-fired boilers.

The salient properties of some of the types of bricks are given in Tables 17.3 and 17.4.

17.12.5.2 Refractory Castables

These are refractory materials that are not preformed but cast *in situ* to desired shapes. Because of this flexibility, coupled with advances made both in the materials and binders, castable refractories have gained greatly in popularity at the expense of the formed refractory in many applications. They come in special mixes or blends of dry granular or stiffly plastic refractory materials with which practically any type of joint-free linings (monolithic) can be made or masonry repaired. The transportation and handling are also easier as they

TABLE 17.3

Refractoriness of Bricks Used in Boilers

Brick Type		FB 23–30%	FB 30–35%	FB 35–40%	Al_2O_3 40–45%	SiO_2 Brick
Refractoriness	PCE	27–29	28–31	29–33	32–35	32–33
	°C	1610–1630	1630–1690	1650–1730	1710–1770	1710–1730
Refractoriness under load	PCE	13–16	16–19	19–26	26–28	29–32
	°C	1380–1460	1460–1520	1520–1580	1580–1630	1650–1710
True SG		2.5–2.6			3.1–3.4	2.3–2.4
Bulk ρ		1.9–2.1			1.8–2.1	1.7–1.8

Note: % represents Al_2O_3; FB = fire brick.

TABLE 17.4

General Physical and Chemical Properties of Shaped Refractories Used in Boilers

Type of Brick	Fire Brick	Fire Brick	High Al_2O_3 Brick	Extra High Al_2O_3 Brick	SiO_2 Brick	SiC Tiles	Insulating Bricks
Nominal composition	40% Al_2O_3	42% Al_2O_3	50–85% Al_2O_3	90–99% Al_2O_3	95% SiO_2	80–90% SiC	
Fusion temperature (°C)	1720	1745	1760–1870	1650–2010	1700	2300	
Apparent porosity%	18	15	20	23	21	15	65–85
Permeability	Moderate	High	Low	Low	High	Very low	High
Hot strength	Fair	Fair	Good	Excellent	Excellent	Excellent	Poor
Thermal shock resistance	Fair	Good	Good	Good	Poor	Excellent	Excellent
Resistance to acids	Good	Good	Good but for HF	Good but for HF	Good	Good but for HF	Poor
Resistance to alkalis	Good at low temperature	Good at low temperature	Slight attack at high temperature	Slight attack at high temperature	Good at low temperature	Attack at high temperature	Poor

Note: HF: hydrofluoric acid.

come packed in bags. The application procedure is also made very easy with little or no preparation. There are four types of monolithics:

1. Plastic refractories
2. Ramming mixes
3. Gun mixes
4. Castables

Monolithics develop their strength either by *air setting* or by *hydraulic setting*. The entire thickness becomes hard and strong at room temperatures. At higher temperatures, it becomes even stronger on the development of the ceramic bond. Heat setting monolithic refractories have very low strength at low temperatures and develop their full strength only on attaining the full temperature. Linings on furnace wall tubes require that the water walls are fully drained before the application of refractory layer, lest the water-cooled walls hinder the lining from attaining the necessary temperature. Usually, castable linings need some anchor material to hold.

17.12.5.2.1 Plastic Refractories

These are mixtures of refractory materials prepared in stiff plastic conditions of proper consistency for ramming into place with pneumatic hammers or mallets. Plastics are similar to castables in formulation as both use calcined aggregates and binder. However, the plastics, which are premixed at the factory, use phosphates or other heat setting agents to develop a bond when fired. Castables use hydraulic cements which form a permanent bond when mixed with water.

High-alumina phosphate-bonded plastics are used in hot cyclones of CFB boilers to withstand high erosion of the heavily dust-laden flue gases flowing at high velocity.

Plastic chrome ore

Typically, PCO is made from an admixture of sized chrome ore, sodium silicate, hectorite, and sufficient water to form a plastic mass. It may contain a plasticiser such as plastic kaolin clay in addition to the hectorite. This is a proven lining material for furnaces of BL recovery boilers which experience reducing conditions and have to continuously face the abrading flow of high-temperature molten smelt. This air setting plastic compound is rammed into position on the studded walls to develop a dense monolithic layer which displays high resistance to spalling and erosion of smelt.

17.12.5.2.2 Ramming Mixes

These are ground refractory materials with minor amounts of other materials added to make the mixes workable. Most ramming mixes are supplied dry and while some are wetready to apply. Ramming mixes are required to be mixed with water and rammed into place followed by drying and heating when they form, by self-bonding, a dense and strong monolithic refractory structure.

17.12.5.2.3 Gun Mixes

These are granular refractory materials prepared for spraying at high velocity and pressure by guns. The resulting lining is homogenous and dense and is free from lamination cracks. The spray can be either by dry mix or by wet mix. When the dry mix is used, the guns have a water nozzle to moisten the mixture. The gun mixes can be either air setting or heat setting.

Refractory lining of steel stacks is often done by gunning.

17.12.5.2.4 Castables

These are granular refractory materials combined with suitable hydraulic setting bonding agent. They are supplied in dry form to be mixed at site with water and poured or cast in place when they develop a strong hydraulic set. They are rammed or trowelled or tamped into position and occasionally applied with air guns. These castables have negligible shrinkage in service and low coefficients of thermal expansion. They are resistant to spalling. Some are capable of withstanding severe erosion. Some are good insulators while some are good conductors.

a. High-alumina dense castables

   These conventional castables with refractory cement content in excess of 2.5% and with varying alumina contents in excess of 90% are used up to an operating temperature of 1800°C to withstand high erosion. Burner quarls and furnace linings are some examples of usage of this castable.

b. Low-cement, ultra-low-cement and no-cement castables

   These are relatively recent developments where refractory cement used as binder is progressively brought down to withstand high temperatures without weakening the lining. Less cement means reduced lime content and reduced water requirement for setting. Less water makes the refractory less porous.

   CaO content in low-, ultra-low- and no-cement castables is in the region of 1–2.5%, 0.2–1.0% and <0.2%, respectively. The porosity here is reduced by more than 50% resulting in very dense structure capable of standing up to very high erosion.

   Low-cement castable is used quite successfully at the inlet of cyclones in CFB boilers which experience very high erosive forces due to the combination of gas velocity and dust content.

c. Insulating castables

These serve as hot face lining in clean (such as in petrochemical) and as back-up lining for dusty (such as in process industry) applications. They are light, strong and low in conductivity making the whole liner thin and cheap. They are made in a range of densities from about 400 to 1600 kg/m^3.

d. Silicon carbide (SiC)

SiC is extremely hard and highly heat-conducting ceramic. Grinding wheels in machine tool industry are made of SiC for the same reason. SiC lining of FBC and other furnaces is done with a view to withstand extreme levels of erosion and have good heat transfer to the water walls.

A bluish-black crystalline compound, one of the hardest known substances, SiC is the only chemical compound of C and Si. It is found very rarely in nature. However, it is manufactured by high-temperature electro-chemical reaction of SiO_2 and C and is used extensively as abrasive. In the recent past, the material has been developed into a high-quality technical-grade ceramic with very good mechanical properties. It is characterised by

- Low ρ and thermal expansion
- High strength, thermal conductivity, hardness and elastic modulus
- Excellent thermal shock resistance with superior chemical inertness

17.13 Refuse Derived Fuel (*see* Municipal Solid Waste)

17.14 Reheat (*see* Properties of Steam and Water)

17.15 Reheater (*see* Heating Surfaces)

17.16 Reheater Inlet and Outlet Pressure (*see* Boiler Pressures)

17.17 Reheater Outlet Temperature (*see* Boiler Temperatures)

17.18 Relative Density (*see* Specific Gravity in Properties of Substances)

17.19 Relative Humidity of Air (*see* in Properties of Substances)

17.20 Releasers or Risers (*see* Pressure Parts)

17.21 Remote Water Level Indicator (*see* Water Level Indicators *and* Direct Water Level Indicators)

When the drum level is higher than ~20 m, viewing the DWLIs from the operating floor is not easy.

A telescopic arrangement reflecting the image of the level gauge through a pair of mirrors, a practice of the former times on industrial boilers, is also not quite satisfactory. RWLI then becomes indispensable. Also, RWLIs are required for viewing the drum level in the control room. In a modern boiler of medium or large size, both DWLIs and RWLIs are employed for viewing the water level at the drum and firing floors as well as control room.

There are basically two types of RWLIs:

1. Manometric
2. Electronic

17.21.1 Manometric Gauges

Manometric gauges are simple, reliable and inexpensive. There are two types of manometric gauges—one in which a heavy fluid is employed (Figure 17.12) and used for short distances and the other a diaphragm type (Figure 17.13) which can be used for long distances.

In diaphragm-type manometric gauge, the differential head between the bottom tapping of the steam drum and a reference point is magnified by a diaphragm to give an indication of water level in the drum. Figure 17.13 illustrates the principle along with the details of the diaphragm-type manometer. This is suitable for remote indication. Further, the signal can be carried to a farther location by having an electronic repeater station.

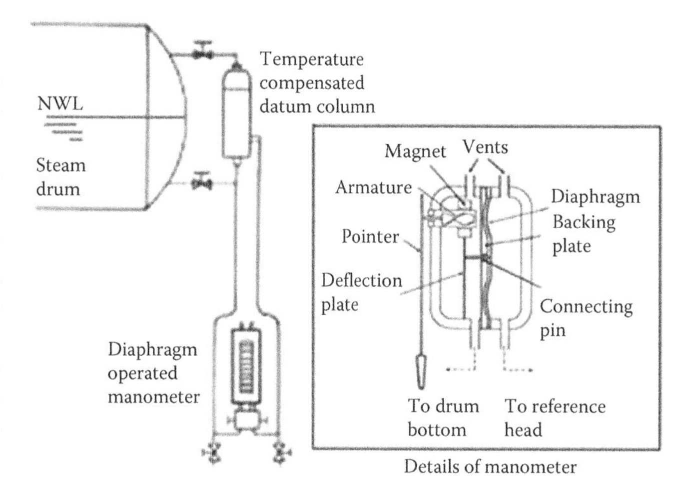

FIGURE 17.13
Diaphragm-type manometric remote water level indication.

17.21.2 Electronic Drum Level Indicator of Conductivity Type

Electronic drum level indicator (EDLI) of conductivity type is illustrated in Figure 17.14. Depending on the number of electrodes submerged by water in the probe vessel, the indications are given in both the panel mountings and local indicators. As fibre optic cable is used for transmission of the picture, distance is no consideration. Also, at the predetermined water levels, interlocks and alarms can be actuated to energise solenoid valves or pumps as required. EDLIs have established themselves as a more versatile, reliable and competitive option to manometric gauges.

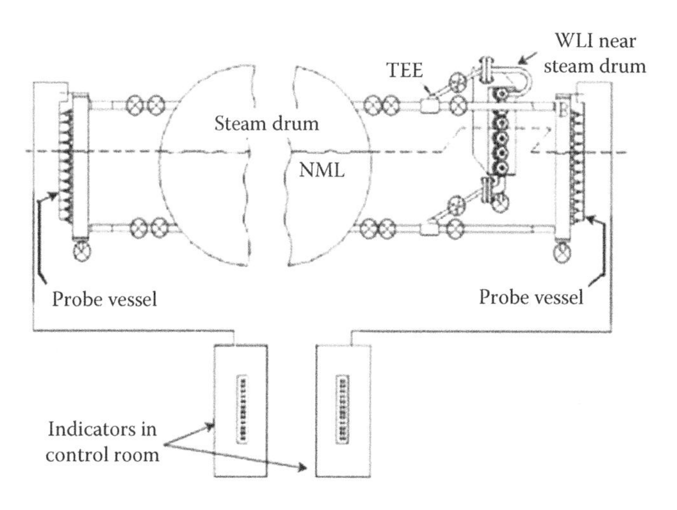

FIGURE 17.14
Electronic drum level measurement.

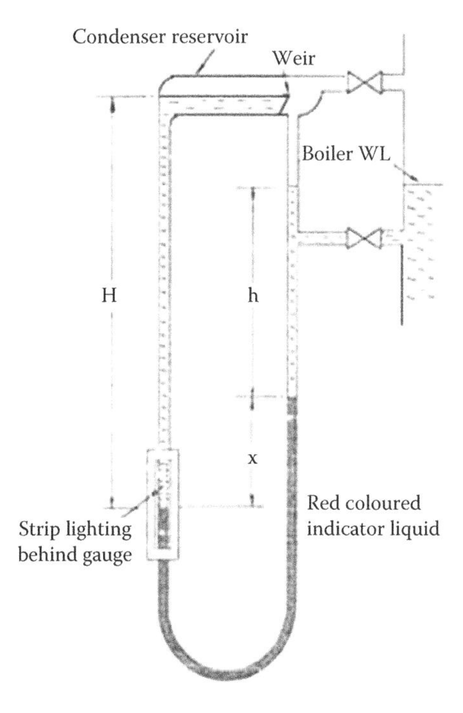

FIGURE 17.12
Liquid-filled manometric gauge for remote drum level indication.

17.22 Residence Time in Furnace (*see* Furnace Size or Volume in Heating Surfaces)

17.23 Reverse Gas Filters (*see* Dust Collection Equipment)

17.24 Reynold's Number (Re) (*see* Dimensionless Numbers in Heat Transfer)

17.25 Ribbed Tubes (*see* Boiling or Evaporation in Properties of Steam and Water)

17.26 Rice Husk/Rice Hull and Firing (*see* Agro-Fuels)

17.27 Rimmed Steel (*see* Deoxidation of Steels in Metallurgy)

17.28 Rifled Tubes (*see* Boiling or Evaporation in Properties of Steam and Water)

17.29 Ringlemann Number (*see* Opacity of Stack)

17.30 Risers (*see* Pressure Parts)

17.31 Residual Life Assessment (*see* Life Extension)

17.32 Rockwell Tests for Steel Hardness (*see* Testing of Materials)

17.33 Rockwool (*see* Mineral Wool *or* Rockwool in Insulating Materials)

17.34 Roughness Factor (*see* Moody's Friction Factor in Fluid Flow)

18

S

18.1 Safety Valves (*see* Valves)

18.2 Safety Valve Floating (*see* Commissioning)

18.3 Saybolt Seconds (*see* Liquid Fuels *and* Instruments for Measurement)

18.4 Saturated Water (*see* Water in Properties of Steam and Water)

18.5 Scantlings (*see* Pressure Parts)

18.6 Schedule (of Pipes)

Pipes are specified by their nominal bore (NB), nominal pipe size (NPS) for sizes up to 12″ (304.8 mm) and by their od for higher sizes. Thickness is specified by the schedule, which is a number indicating the approximate value of

Sch. = 1000 P/S, where P = service pressure (psi), S = allowable stress (psi).

The higher the schedule number, the thicker the pipe and the greater is the pressure withstand capability. Since the od of the pipe is standardised, a particular nominal pipe size (NPS) will have different id depending on the schedule specified. In the same way, for the same schedule, the thickness increases with the pipe size.

Nominal wall thickness for seamless and welded steel pipes according to ANSI B36.10 is indicated in Table 18.1. Pipes of 2–36″ sizes (60.3–965 mm od) are covered in this abridged table, which is of greater relevance to boilers, while standards cover a much wider range of 1/8–42″ (10.3–1219 mm od).

The schedules standardized are 5 (light), 10, 20, 30, 40, strong (STD), 60 extra strong (XS), 80, 100, 120, 140, 160 double extra strong (XXS). Not all thicknesses are rolled in each size of pipe. Many of the schedules are identical in certain sizes.

18.7 Scrubber

In boiler plants, the word scrubber has several meanings:

- A scrubber is an air pollution control device to remove particulates and/or gases from industrial exhaust streams. Usually a scrubber is a device that employs liquids to wash unwanted **pollutants** from a gas stream. It is also a system that injects a dry reagent or slurry into a dirty exhaust stream to remove acid gases. Scrubber is the most important equipment in an **FGD** system.
- Venturi scrubber was extensively employed till the 1970s for cooling and dedusting the flue gases in the small BLR boilers in pulp mills. It was for both heating and chemical recovery by flue gas condensation.
- Steam scrubber is a device to separate traces of water from steam after steam separation as explained in drum internals.

18.8 Seal Pot (*see* Loop Seal in Circulating Fluidised Bed Combustion)

TABLE 18.1

Nominal Wall Thickness for Seamless and Welded Steel Pipes According to ANSI B36.10

Pipe Size	O.D. (in.)	5		10S		10		20		30		40S/STD		40	
2	2.375	0.065	1.651	0.109	2.769	0.109	2.769					0.154	3.912	0.154	3.912
2-1/2	2.875	0.083	2.108	0.120	3.048	0.120	3.048					0.203	5.156	0.203	5.156
3	3.500	0.083	2.108	0.120	3.048	0.120	3.048					0.216	5.486	0.216	5.486
3-1/2	4.000	0.083	2.108	0.120	3.048	0.120	3.048					0.226	5.740	0.226	5.740
4	4.500	0.083	2.108	0.120	3.048	0.120	3.048					0.237	6.020	0.237	6.020
5	5.563	0.109	2.769	0.130	3.302	0.134	3.404	0.203	5.156			0.258	6.553	0.258	6.553
6	6.625	0.109	2.769	0.134	3.404	0.134	3.404	0.203	5.156			0.280	7.112	0.280	7.112
8	8.625	0.109	2.769	0.148	3.759	0.148	3.759	0.250	6.350	0.277	7.036	0.322	8.179	0.322	8.179
10	10.750	0.134	3.404	0.165	4.191	0.165	4.191	0.250	6.350	0.307	7.798	0.365	9.271	0.365	9.271
12	12.750	0.165	4.191	0.180	4.572	0.180	4.572	0.250	6.350	0.330	8.382	0.375	9.525	0.406	10.312
14	14.000			0.188	4.775	0.250	6.350	0.312	7.925	0.375	9.525	0.375	9.525	0.438	11.125
16	16.000			0.188	4.775	0.250	6.350	0.312	7.925	0.375	9.525	0.375	9.525	0.500	12.700
18	18.000			0.188	4.775	0.250	6.350	0.312	7.925	0.438	11.125	0.375	9.525	0.562	14.275
20	20.000			0.219	5.563	0.250	6.350	0.375	9.525	0.500	12.700	0.375	9.525	0.594	15.088
22	22.000			0.219	5.563	0.250	6.350	0.375	9.525	0.500	12.700	0.375	9.525		
24	24.000			0.250	6.350	0.250	6.350	0.375	9.525	0.562	14.275	0.375	9.525	0.688	17.475
26	26.000					0.312	7.925	0.500	12.700			0.375	9.525		
28	28.000					0.312	7.925	0.500	12.700	0.625	15.875	0.375	9.525		
30	30.000			0.312	7.925	0.312	7.925	0.500	12.700	0.625	15.875	0.375	9.525		
32	32.000					0.312	7.925	0.500	12.700	0.625	15.875	0.375	9.525	0.688	17.475
34	34.000					0.312	7.925	0.500	12.700	0.625	15.875	0.375	9.525	0.688	17.475
36	36.000					0.312	7.925	0.500	12.700	0.625	15.875	0.375	9.525	0.750	19.050

Pipe Size	O.D. (in.)	60		80S & XS		80		100		120		140		160		XXS	
2	2.375			0.218	5.537	0.218	5.537							0.344	8.738	0.436	11.074
2-1/2	2.875			0.276	7.010	0.276	7.010							0.375	9.525	0.552	14.021
3	3.500			0.300	7.620	0.300	7.620							0.438	11.125	0.600	15.240
3-1/2	4.000			0.318	8.077	0.318	8.077									0.636	16.154
4	4.500	0.281	7.137	0.337	8.560	0.337	8.560			0.438	11.125			0.531	13.487	0.674	17.120
5	5.563			0.375	9.525	0.375	9.525			0.500	12.700			0.625	15.875	0.750	19.050
6	6.625			0.432	10.973	0.432	10.973			0.562	14.275			0.719	18.263	0.864	21.946
8	8.625	0.406	10.312	0.500	12.700	0.500	12.700	0.594	15.088	0.719	18.263	0.812	20.625	0.906	23.012	0.875	22.225
10	10.750	0.500	12.700	0.500	12.700	0.594	15.088	0.719	18.263	0.844	21.438	1.000	25.400	1.125	28.575	1.000	25.400
12	12.750	0.562	14.275	0.500	12.700	0.688	17.475	0.844	21.438	1.000	25.400	1.125	28.575	1.312	33.325	1.000	25.400
14	14.000	0.594	15.088	0.500	12.700	0.750	19.050	0.938	23.825	1.094	27.788	1.250	31.750	1.406	35.712		
16	16.000	0.656	16.662	0.500	12.700	0.844	21.438	1.031	26.187	1.219	30.963	1.428	36.271	1.594	40.488		
18	18.000	0.750	19.050	0.500	12.700	0.938	23.825	1.156	29.362	1.375	34.925	1.562	39.675	1.781	45.237		
20	20.000	0.812	20.625	0.500	12.700	1.031	26.187	1.281	32.537	1.500	38.100	1.750	44.450	1.969	50.013		
22	22.000	0.875	22.225	0.500	12.700	1.125	28.575	1.375	34.925	1.625	41.275	1.875	47.625	2.125	53.975		
24	24.000	0.969	24.613	0.500	12.700	1.219	30.963	1.531	38.887	1.812	46.025	2.062	52.375	2.344	59.538		
26	26.000			0.500	12.700												
28	28.000																
30	30.000			0.500	12.700												
32	32.000			0.500	12.700												
34	34.000			0.500	12.700												
36	36.000			0.500	12.700												

18.9 Selective Catalytic Reduction (*see also* Post-Combustion Processes in Nitrogen Oxides in Flue Gases [NO_x])

Selective catalytic reduction (SCR) systems are the most widely used post-combustion NO_x control technology for achieving significant reductions in NO_x emissions. Operating usually between 300 and 400°C which is normally the flue gas temperature at the Eco outlet, NO_x reduction of 80–90% of the incoming level is achieved. NH_3 vapour is the reducing agent and is injected into the flue gas stream, passing over a catalyst. The NO_x and NH_3 reagent react to form N_2 and H_2O. The reaction mechanisms are very efficient with a reagent stoichiometry of approximately 1.05 (on a NO_x reduction basis) with very low NH_3 slip. A simplified schematic diagram of a typical SCR reactor is illustrated in Figure 18.1. However, most modern SCR systems are built without bypass systems and *sonic horns* are used in place of steam or air *SBs*.

The catalyst and its support are housed in SCR reactor which is basically a widened section of duct work, modified by the addition of gas flow distribution devices, access doors and sonic horns/SBs. There are three typical layout arrangements of SCR systems applied to coal-fired power stations:

- High dust position is the most widely used, especially with dry bottom boilers, as it does not require hot **ESP** upstream of the reactor.
- Low dust positioning has the advantage of less catalyst degradation caused by fly ash erosion but requires a more costly hot-side ESP.
- Tail-end position SCR has been used primarily with wet bottom boilers with ash recirculation to avoid catalyst degradation caused by As poisoning. The configuration is also favoured with retrofit installations (because of SCR space requirements) between the Eco outlet and ESP.

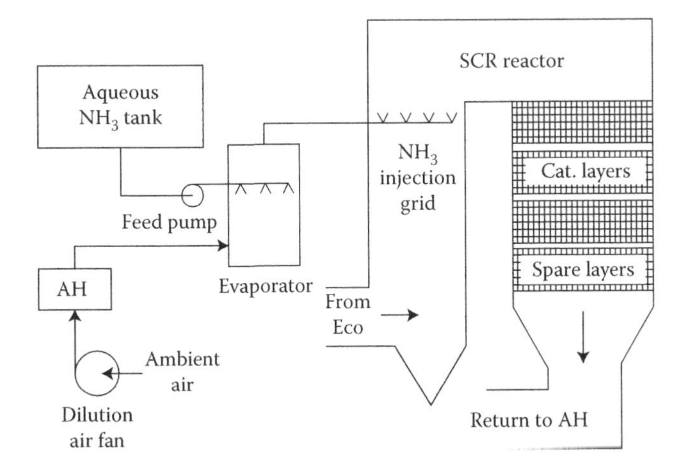

FIGURE 18.1
Scheme of selective catalytic reduction (SCR).

Flue gas inlet temperature to the SCR reactor is controlled by mixing the flue gas exiting the Eco with the flue gas from the ECO bypass.

The NH_3 injection grid is located in the inlet ductwork to the SCR catalyst sufficiently upstream to ensure optimum gas and reagent distribution across the cross section of the catalyst. The NH_3 reagent can be supplied by

a. Anhydrous NH_3
b. Aqueous NH_3 or
c. Conversion of $(NH_2)_2CO$ (urea) to NH_3

As NH_3 is used in vapourised form, its type does not influence the performance of catalyst. However, it does affect other subsystem components, including reagent storage, vapourisation, injection control and balance-of-plant requirements. The vast majority of worldwide installations use anhydrous NH_3.

Different compositions for catalysts based on TiO_2, zeolite, iron oxide or activated C are available. Most catalysts in use in coal-fired plants consist of V (active catalyst) and Ti (used to disperse and support V) mixture. However, the final catalyst composition can consist of many active metals and support materials to meet specific requirements in each SCR installation. Catalyst geometry may typically be a flat plate or honeycomb. A moving bed is used for granular activated C. Some experience shows that plate types generally have a higher resistance to deposition and erosion than honeycombs. In this case, catalytic converters are used in an AH.

Draft losses can range from 100 to 250 mm wg based on arrangement and performance requirements needing additional ID booster fans. Ductwork and boiler casing reinforcement may also be required. An expansion of the auxiliary power system might be necessary. Auxiliary power modifications may also be necessary for NH_3 supply system requirements

AH is where there are two areas of concern:

1. Deposition of $(NH_4)HSO_4$ on the AH surface causes an increase in the pressure drop of AH and degrade its performance and decrease plant η.
2. High SO_3 in flue gas can, if the acid dew point temperature has been increased to more than the exhaust temperature, lead to the condensation of acid gases and cause pluggage and corrosion. This can be corrected by the right composition of the catalyst to minimise the SO_2 to SO_3 conversion.

SCR technology has been used commercially in Japan since 1980, in Germany since 1986 and in the United States since the 1990s. Their successful commercial use has followed the introduction of stringent limits to regulate NO_x emissions in each country.

18.10 Selective Non-Catalytic Reduction (*see also* Post-Combustion Processes in Nitrogen Oxides in Flue Gases [NO_x])

The aim of SNCR is to reduce NO_x emissions in boilers that burn biomass, waste and coal.

NH_3 or $(NH_2)_2CO$ (urea) is injected into the boiler at multiple levels as illustrated in Figure 18.2 where the flue gas is at 815–1205°C (1500–2200°F) to react with the NO formed in the combustion process and reduce it to elemental N_2, CO_2 and H_2O.

$(NH_2)_2CO$ is easier to handle and store than the more dangerous NH_3 even though it reacts like NH_3.

$$(NH_2)_2CO + H_2O \rightarrow 2NH_3 + CO_2 \quad (18.1)$$

The reduction happens according to (simplified)

$$4NO + 4NH_3 + O_2 \rightarrow 4N_2 + 6H_2O \quad (18.2)$$

The reaction to be effective requires a certain temperature range, typically 815–1205°C (1500–2200°F), NO and NH_3 do not effectively react. NH_3 that has or else reacted is called *ammonia slip* and is undesirable, as it can react with other combustion products like SO_3, to form ammonium salts. $(NH_4)HSO_4$ will precipitate at AH operating temperatures and can ultimately lead to AH fouling and plugging.

SNCR systems rely solely on reagent than a catalyst. Essential for the success of the process are

- Suitable reagent injection temperature
- Good reagent/gas mixing by multiple levels of injection
- Adequate reaction time to achieve NO_x reductions

SNCR systems can achieve NO_x reduction of 50–60% under optimum conditions stated above with high baseline NO_x with NH_3 slips of 10–50 ppmvd (volume dry). Lower NH_3 slip is possible with lower NO_x reduction. As the optimum conditions are difficult, the resulting emission reduction levels are only 20–40%.

The actual performance is very site specific and varies with fuel type, boiler size, allowable NH_3 slip, furnace CO concentrations, and boiler heat transfer characteristics. SNCR systems reduce NO_x emissions using the same reduction mechanism as SCR systems. The SNCR reagent storage and handling systems are highly similar to those for SCR. However, SNCR processes require three or four times as much reagent as SCR systems to achieve similar NO_x reductions.

Most of the undesirable chemical reactions occur when reagent is injected at temperatures above or below the optimum range:

- When reaction occurs at >1000°C, NO_x removal decreases because of thermal decomposition of NH_3.
- On the other hand, the NO_x reduction rate decreases below 1000°C and NH_3 slip may increase.

This complicates the application of SNCR for boilers larger than 100 MWe, but this is not an issue for biomass boilers since they are typically no larger than 100 MWe.

- Injection lances are usually located between the boiler SBs in the pendant SH section.
- To accommodate SNCR reaction temperature and boiler turndown requirements, several levels of injection lances are normally installed. Typically, four to five levels of multiple lance nozzles are installed if sufficient boiler height and residence time are available.
- Optimum injector location is mainly a function of temperature and residence time.

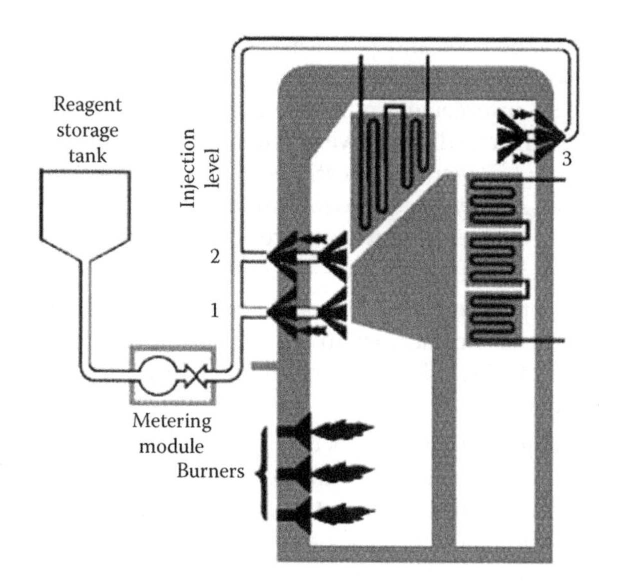

FIGURE 18.2
Scheme of selective non-catalytic reduction (SNCR).

- A flue gas residence time of at least 0.3 s in the optimum temperature range is desired to assure adequate SNCR performance. Residence times in excess of 1 s yield high NO_x reduction levels even under less than ideal mixing conditions.
- CFD and chemical kinetic modelling can be done to establish the optimum NH_3 injection locations and flow patterns.
- For an existing boiler, minor water wall modifications are necessary to accommodate installation of SNCR injector lances. Steam piping modifications would probably be required to achieve optimum performance.

SNCR came to be used on oil- or gas-fired boilers in Japan in the mid-1970s. They spread to Western Europe in the 1980s and the United States in the early 1990s.

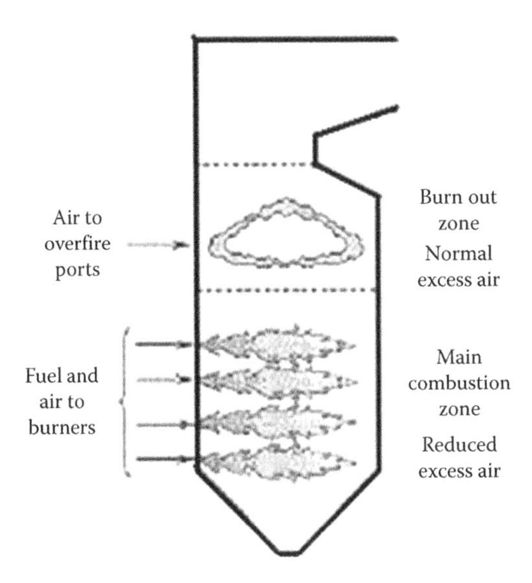

FIGURE 18.3
Separated overfire air (SOFA) system for wall-fired boilers.

18.11 Self-Pulverisation (*see* Pulverising)

18.12 Separated Overfire Air (*see also* Nitrogen Oxides [NO_x] in Flue Gas)

SOFA, also known as *furnace air staging* or *two-stage combustion*, is a well-established technology for NO_x reduction in fossil fuel-fired furnaces. The process is equally applicable to both **wall** and **tangential firing**. In fact, air staging is an inherent part of tangentially fired combustion systems.

In the context of wall firing, SOFA involves the diversion of a portion of the main combustion air from the burners to SOFA injectors located at a discrete elevation above the main burners.

Early applications of this technology were undertaken in conjunction with conventional burners; more recently, SOFA has been combined with **LNBs**.

The application of SOFA restricts the availability of O_2 to the fuel in the area of most intense combustion, reducing both fuel NO_x and thermal NO_x. The introduction of the balance of combustion air through the SOFA ports allows the burnout of the fuel to take place in a less-intense, lower-temperature combustion zone. OFA is one of the most established NO_x reduction technologies available, having been applied in the power industry for over 25 years.

Figure 18.3 depicts a typical SOFA system for wall-fired boilers.

18.13 Shale Gas (*see* Gaseous Fuels)

18.14 Shell-Type Boiler (*see* Smoke Tube Boilers)

18.15 Silencers (*see* Noise Control)

18.16 Silica (SiO_2) in Steam (*see* Steam Purity)

18.17 Silicon Carbide (*see* Castables *under* Refractories)

18.18 Simple Cycle Operation (of GTs) (*see* Thermodynamic Cycles)

18.19 Single Drum Boilers (*see* Industrial Boilers)

18.20 Single-Pass Boilers (*see* Utility Boilers)

18.21 Siphon (*see* Loop Seal in Fluidised Bed Combustion, Circulating)

18.22 Slag Wool (*see* Insulating Materials)

18.23 Slagging and Fouling (*see* Ash)

18.24 Sliding Pressure Operation (*see* Variable Pressure Operation)

18.25 Sling Rods (*see* Structure, Boiler)

18.26 Sludge (*see* 1. Water *and* 2. Liquid Fuels)

18.27 Slug Flow (*see* Flow Types in Vertical and Horizontal Tubes)

18.28 Smelt (*see* Recovery Boilers)

18.29 Smoke Tube Boilers also Fire/Flue Tube and Shell/Shell and Tube Boilers

These boilers are called by several names. Essentially, here, *flue gases pass through flue tubes while water surrounds them*, unlike in water tube boilers where water flows through the tubes and gases surround them.

- As smoke flows through the tubes they are called *smoke tube* boilers.
- Since combustion gases pass through flue tubes they are termed as *flue tube boilers*.
- They are *fire tube* boilers because combustion is performed within the central large bore fire tube.
- They are also called *shell* boilers as the whole boiler is contained within a cylindrical shell.
- They are *shell tube* boilers as the boiler construction is primarily a shell and tube HX.

These boilers are very compact as combustion and heat transfer are both contained within the shell. The large-diameter fire tube, located towards the bottom of the main shell, is fitted with an oil/gas burner. Combustion has to be substantially completed within the confines of the fire tube. Hot gases then flow through the flue tubes either in single or two passes (called two- or three-pass boilers as the first pass is always the fire tube) depending upon the required cooling of gases. Steam is generated within the shell from the pressurised water. Water pressure is limited by the plate thicknesses of shell and ends. A small SH is accommodated at the end of fire tube where gases take a turn. Limits for pressure, temperature and output are all quite low.

Smoke tube boilers are suitable basically for

- Clean fuels like LFO and gas as dusty fuels cause fouling and erosion as they travel inside the tubes
- Smaller sizes up to 35 tph limited by the size of shell
- Small SH of ~50°C as there are arrangement problems for accommodating large HS
- LPs of usually <25 ata as the pressure-holding capacity of the shell is limited

They are popular in the process industry for small steam requirement despite

- A poorer safety record inherent to the design
- Lower thermal efficiencies
- Poorer environmental compliance

mainly due to the low cost, compact size and quick deliveries. The minimum capacity of water tube boiler is usually considered as 5 tph. Fire tube boilers are vertical in smaller sizes and horizontal in larger sizes.

Thermal η of smoke tube boilers is much lower as the flue gas exit temperatures are ~250°C or higher. Fuel cost therefore is a major deciding factor.

With good-quality coals, smoke tube boilers of ~15 tph have been built where the burners are replaced with narrow gravity-fed CGs that can fit in the fire tube. However, for slow-burning and high ash coals, the volume available in the fire tube is the real constraint. Also, the fouling of smoke tubes needing a constant scraping is another problem. Hybrid boilers, explained below, can offer a reasonably satisfactory and economical solution for such cases.

Figure 18.4 shows a typical horizontal skid-mounted/gas-fired fully automatic three-pass dry back saturated smoke tube boiler.

Figure 18.5 is a schematic diagram of a typical horizontal skid-mounted FO-fired fully automatic two-pass wet back saturated smoke tube boiler along with its auxiliaries.

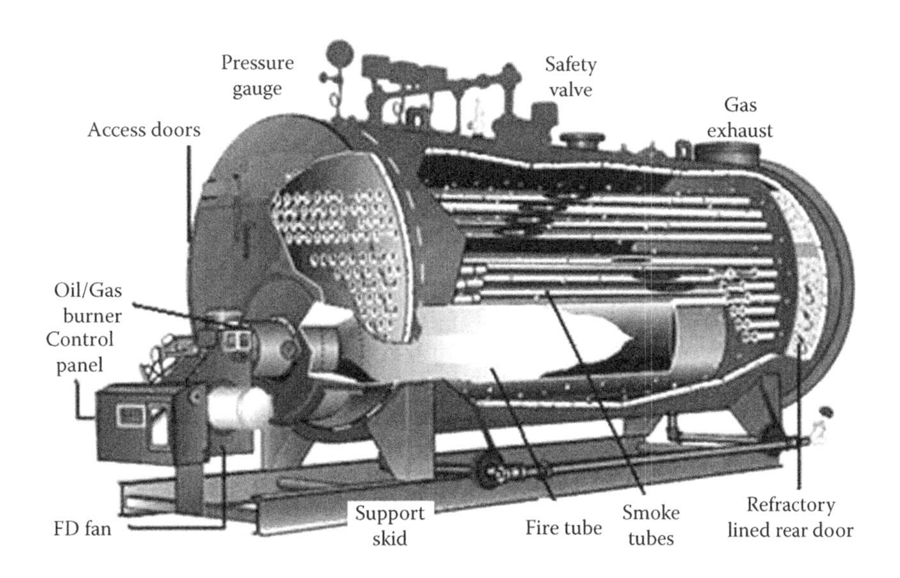

FIGURE 18.4
Three-pass oil/gas-fired dry back saturated smoke tube boiler.

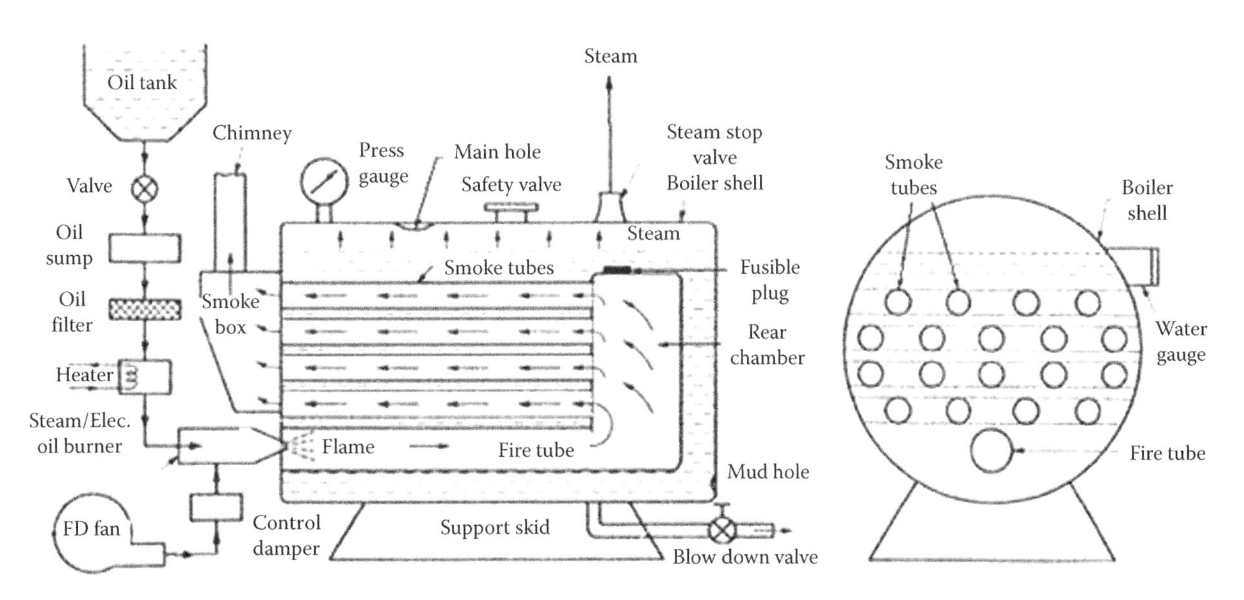

FIGURE 18.5
Schematic diagram of two-pass wet back saturated smoke tube boiler along with its accessories.

18.30 Hybrid Combination Boilers

Hybrid or combination boilers are process boilers which combine the water tube and fire tube constructions for arriving at very competitive configurations in low pressure and low temperature range of usually <25 bar and 400°C. Even though fire tube boilers enjoy the advantages of being very compact and cost effective they are only good for quick burning fuels like oil and gas because of very low furnace volume available in fire tube. This limitation is overcome by building a suitably large external furnace with refractory and/or water wall construction and retaining the shell and tube construction for boiler duty. Hybrid boilers are suitable for burning many types of solid fuels, including biomasses by fitting suitable grates and making appropriate changes to the furnace construction.

Shown in Figure 18.6 is a typical dual fuel biomass and FO-fired combination boiler with a refractory lower furnace and water-cooled upper furnace. A CG burns solid fuel in MB mode. Gases are cooled in the furnace as well as the boiler shell which has two gas passes. SA injection is done for the efficient combustion of VM and GR is employed for improving the combustion η. SH can be fitted if needed in the chamber at the rear of boiler shell.

18.31 Sodium Flame Photometry (*see* Steam Purity in Measurements in Boilers)

18.32 Softening Temperature of Ash (*see* Ash Fusibility)

18.33 Solid Fuels

Solid fuel is a term given to various types of *solid materials that release energy on combustion.* Solid fuel is also used in referring to a type of rocket propellant.

Solid fuels can be

- Fossil fuels which include coal (bituminous and sub-bituminous), lignite, peat and anthracite
- Agro-fuels like wood, bagasse, rice husk, bark and many other minor agro-fuels
- Waste fuels like petroleum coke and MSW
- Manufactured fuels like RDF and Orimulsion

The immediate use of all these fuels is for power generation, industrial use and domestic heating. Coal was the fuel of the Industrial Revolution, from firing furnaces, to running steam locomotives on railways. Wood was also extensively used to run steam locomotives. Lignite, peat and coal are widely used for power generation today. In fact, the share of the solid fuels in power generation worldwide ranks the highest of all types of fuels. Figure 18.7 shows typical solid fuels covering fossil, agro- and waste fuels. In Figure 22.1, these fuels are arranged by their cv and their relative ease of firing. All the fuels mentioned above are covered in the book at appropriate places.

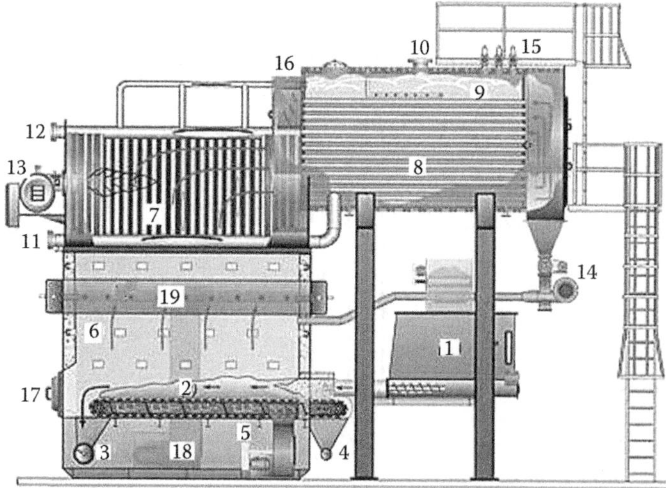

FIGURE 18.6
Biomass- and oil-fired hybrid boiler.

FIGURE 18.7
Some common solid fuels.

18.34 Sonic Horns

Sound waves are generated at low frequencies in acoustic blowers which produce rapid pressure fluctuations causing vibrations in the ash deposits loosening them from the tubes and making them fall off into the gas stream.

The sound waves carry energy levels exceeding the forces that tend to make particles suspended in a gas flow to adhere to each other and the surrounding surfaces, that is, preventing a build-up by breaking up the particles before they can form a hard layer. In practice, this is achieved by activating one or more sound emitters for a short period and thereafter repeating this cycle continuously with certain intervals, more frequently for harsher conditions and with longer intervals between insonations for lighter conditions.

For optimum results

- Ash build-up has to be dry and powdery; the lower the M in flue gases, the better the cleaning result.
- The sound pressure level has to be high enough throughout the whole volume that is supposed to be cleaned, that is, make certain that correct number of sound emitters are engaged.
- The time interval between insonations must be short enough to ensure that particles do not become firmly adhered to each other (a normal cycle is 10–15 s of insonation every 5–10 min).
- In installations with vertical downward flue gas flow, such as in boilers and cyclones, gravity alone will transport the loosened parts; in other cases, a minimum gas velocity of 5 m/s is required.
- It is important to start with clean surfaces, as sonic cleaning is a method of preventing ash build-ups to form.

The ash deposits have to be necessarily loosely sitting on tubes or gently attached and should not be too sticky or hard. The vibrations are not strong enough to dislodge slagging or sintered deposits.

Thus, the sonic horns are more suited for the second pass of the conventional boiler for cleaning the dust from the low-temperature SHs, Ecos and tubular AH in solid fuel-fired boilers.

An acoustic blower consists of a wave generator and resonance section or amplifier. A standard plant air compressor produces the required compressed air to generate powerful sound waves as it enters the wave generator. This air forces the only moving part of the system, namely, a titanium diaphragm to create a powerful base sound wave. Different bell sections transmit this wave to produce a range of sounds of selected fundamental frequencies.

The advantages of these blowers are many:

- There is practically no maintenance as there are no moving items other than the diaphragm.

FIGURE 18.8
Sonic horn.

- The operation and installation costs are also low.
- Unlike the steam or air SBs which operate with short bursts and high energy, the acoustic blowers operate with comparatively low energy levels but steadily for a long time as it takes quite a while to resonate the ash deposits and dislodge them. This is better for installations with ESPs as the outlet concentration can be maintained the same at all times with no objectionable spikes during soot blowing.
- Sonic horns are popular with CFBs as the cleaning is required, if at all, only in the second pass and the ash is usually in loose deposit form.
- Besides the boilers the sonic blowers can be employed in **bag filters**, **ESPs**, silos and bins.

They may not be suitable for high-M lignite-fired or biomass-fired boilers where the M makes the ash attach to the tubes. A typical sonic horn is shown in Figure 18.8.

18.35 Soot Blowers (*see also* Furnace Cleaning in Operation and Maintenance Topics)

As the name suggests, *SBs are for blowing off the soot and ash from the boiler HSs*. Tubes are kept free of ash and the gas lanes between them are free of bridging deposits. Heat transfer is impaired when a layer of ash settles on the tube. By removing this layer, heat absorption is restored and η improved. The dustier the flue gases, the more SBs are needed.

Boilers firing clean fuels such as NG or distillate oil do not need SBs. Even FBC boilers do not need them as the **low-temperature combustion** at about 850°C does not melt the ash particles to deposit them on the tubes when gases cool down. At times they are provided in second pass over the SH, Eco and AH to dust away the dry friable ash. Fuels with high ash or fouling characteristics, such as BL, require an elaborate combination of SBs.

By construction, SBs can be classified as

1. Wall
2. Rotary
3. Retractable
4. Rake types

18.35.1 Wall Blowers, Short Retractable Blowers or Furnace Deslaggers

These are SBs installed in the furnace for the removal of *slag* deposits from the walls. They are invariably installed in PF boilers in the burner zones where **slagging** takes place due to high gas temperatures. They are not required in any other boiler unless the fuel is heavily slagging type. In stoker-fired boilers they are installed sometimes when the combination fuels such as FO and coal/biomass can produce slagging deposits. Figure 18.9 shows a typical motorised wall blower.

The action of wall blower on the furnace wall is depicted in Figure 18.10. The travel of the wall blower is short (40–50 mm into furnace), the operation swift (<2 min) and steam flow heavy (3.5 tph) so that the blower is saved from overheating due to furnace heat. Nominal bowing radius is 1.5 m which can reduce to 0.75 m with heavily fouling fuels.

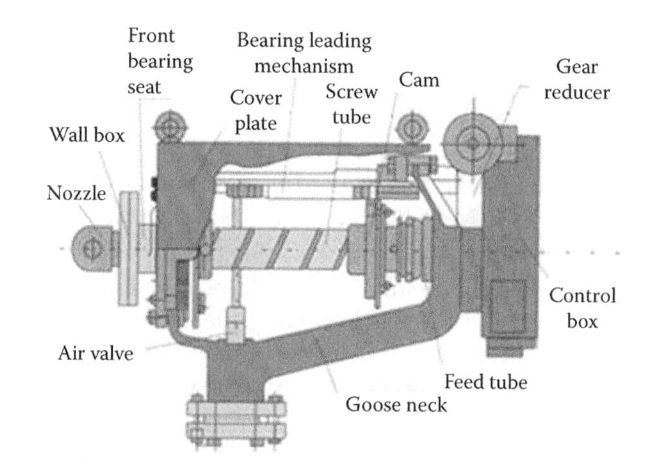

FIGURE 18.9
Motorised wall blower.

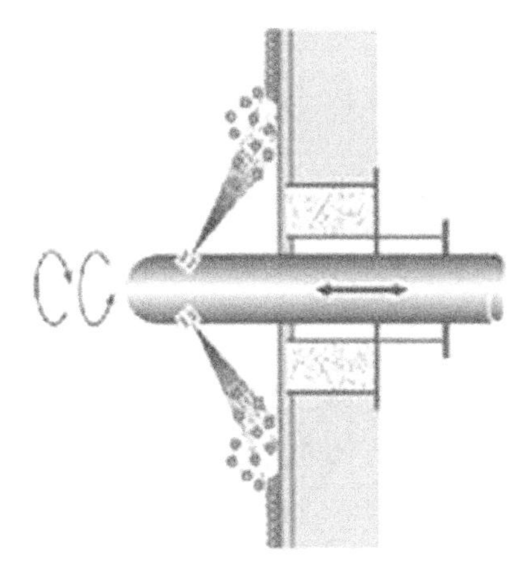

FIGURE 18.10
Action of a wall blower, short retractable or furnace deslagger.

18.35.2 Rotary SB

This is an axially stationary SB which is used in the *cooler part* of the boiler where an uncooled lance can survive inside the boiler passes without damage. At gas temperatures <~1100°C, ash deposits on tube banks are *light and fluffy* which can be blown away with light force by means of small jets. A rotary blower (~40–80 mm diameter) has a long lance with several small nozzles (6–10 mm diameter) and is located between the tube banks blowing steam either directly on the tubes or through the lanes (mass or lane blowing, respectively). The blowing radius is limited to ~1.5 m. A typical motorised rotary blower is shown in Figure 18.11. The maximum length of rotary blower is limited to about 7 m. At higher lengths, steam reaching the last nozzle is at too small a pressure to provide effective blowing. This is the only SB which comes with the option of manual or motorised operation. Smaller process boilers adopt manual SBs for reasons of economy. Naturally, sophistication like automatic sequential operation of SBs is not possible with manual blowers.

The lance is made of high AS or ss pipe to withstand the gas temperature and corrosion. It is supported at its ends and suitable intervals from the tube banks to prevent sagging. The action of rotary SB is depicted in Figure 18.12.

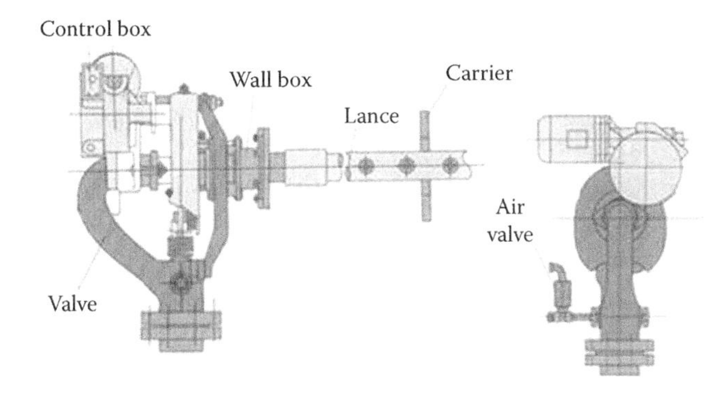

FIGURE 18.11
Motorised rotary SB.

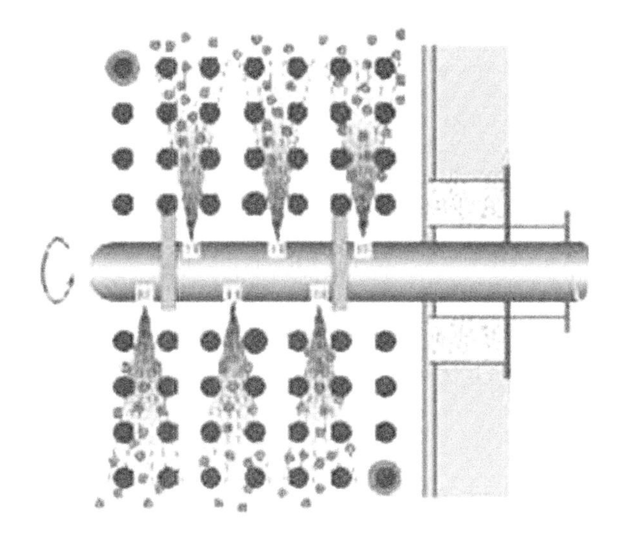

FIGURE 18.12
Action of a rotary SB.

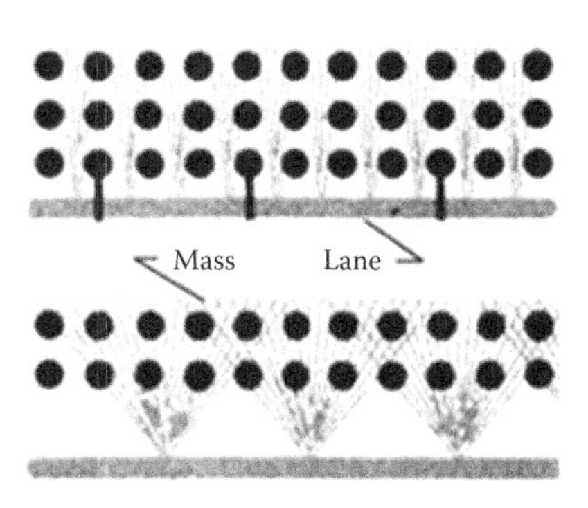

FIGURE 18.13
Mass and lane blowing.

These rotary blowers can be either

- Manually operated as in small boilers or motorised as in large boilers or for automatic sequential operation.
- Mass or lane blowing depending on the way the tube banks are cleaned, as shown in Figure 18.13. Lane blowers need more steam holes and less cavity space and mass blowers need the opposite.

18.35.3 Long Retractable SB

Long retractable or simply retractable SB is the heaviest and most complicated of all blowers both in design and operation. Retractables are provided in the upper part of the furnace to clean the platen, wing walls, final SH and RH elements where the gas temperatures are usually in excess of 900°C and the ash deposits are large. In these

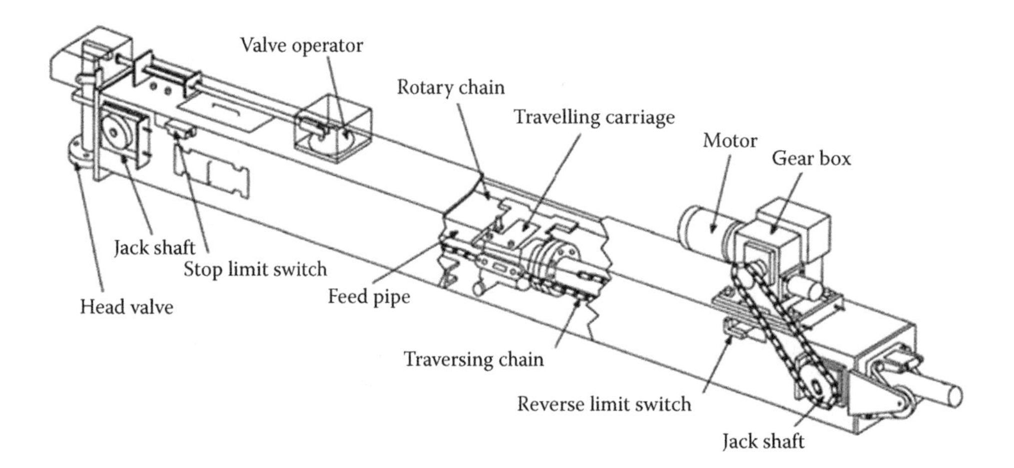

FIGURE 18.14
Isometric view of a long retractable SB.

regions, the ash deposits require heavy impact to dislodge. Also, the blowers have to be in a retracted position, outside the gas pass, lest they are damaged by overheating. The *retractables are always motorised.* There is always a manual back-up for emergency retraction of lance in case of failure of motor while blowing, to save the lance from damage due to overheating. During normal operation the lance can be inside the boiler for as long as ~30 min in hot gas. Figure 18.14 shows a typical retractable blower.

Retractable SBs with just a pair of large nozzles (16–25 mm diameter) on either side of the lance pipe (60–80 mm diameter) with high steam flow (4–20 tph) alone can perform the task of dislodging the heavy deposits or washing the slag off the platens with their massive blows. Blowing pressures are 4–12.5 barg (~60–180 psig) for air and 5–25 barg. Slanted nozzles give a better dwell of steam on platen assemblies. Also, the heavy impact delivered on the vertically hanging elements sets them into a gentle sway helping them to shed the deposits better. The action of retractable SBs is depicted in Figure 18.15.

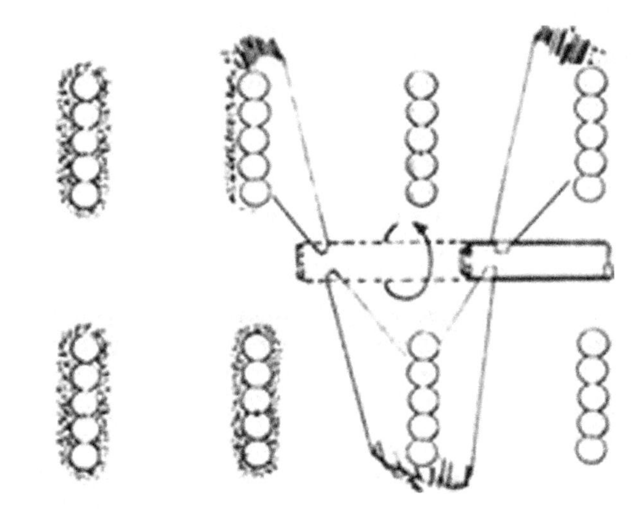

FIGURE 18.15
Action of retractable SB.

Fully retractable SBs are employed only in the high-gas-temperature zones. They are usually placed on the sides of the boiler. The retraction length of such SBs equals the full or half width of the boiler pass depending on whether they are on one or both sides. Obviously, adequate space outside the boiler setting should be available.

Part-retractable blowers

In slightly lower-gas-temperature zones, such as in the second pass of two-pass boilers, where

- Gas temperatures are low enough for the lance to stay inside the gas pass but the deposits are too heavy for rotary blowers to remove with their small jets
- Enough gallery space outside the gas pass is not available for the fully retractable SB to be placed

part-retractable SBs are employed. Depending on the space for lance withdrawal either a half or one-third retractable blower is chosen.

These blowers are like the conventional full retractable SBs in construction but with three main points of difference:

- They are provided with two or three pairs of large nozzles along the lance.
- Either half or two thirds of the lance is permanently located inside the gas pass like a rotary blower.
- The retraction of the lance is accordingly half or a third of the width of the pass.

Because of the multiple nozzles, the entire bank width is covered in one traverse of the lance.

FIGURE 18.16
Rake-type sootblower for cleaning gilled tube economisers.

18.35.4 Rake-Type Blowers

These are particularly effective for Ecos with **extended surfaces** and **vertical tubular** *AHs*. A rake blower usually has a maximum length and width of ~5 × 3 m. A central lance has several cross lances attached to it with multiple small nozzles. The steam blows through several hundred nozzles and the resultant cleaning is much better than by rotary blowers. This is because of better distribution of nozzles despite the small individual steam flows. A typical rake blower on top of a fin tube Eco bank with rectangular fins described in Figure 5.3 is shown in Figure 18.16.

Table 18.2 gives a summary of all types of soot blowers employed in boilers.

Steam versus air blowing:

A majority of SBs employ steam for blowing as it is readily available from the boiler. Steam is required to be dry and preferably with SH of 10–20°C to assure no condensation in the SB piping, as the water droplets can cause severe damage to the PPs as they impinge along with steam at high velocity. Saturated steam with its high ρ renders good impact over tubes compared to SH steam which is lighter.

Air blowing is also equally effective. In BLR boilers, air is preferred as condensate from steam SBs can have explosive reaction with smelt. In large utility boilers, it is cheaper to use air than consume the expensively prepared water.

For air blowing, an elaborate arrangement consisting of air compressor station to deliver HP air between 25 and 35 bar (300–500 psi) with redundancies, controls, safeties and so on is required, which is quite expensive. But air piping is simpler in comparison to steam piping.

Sequence of blowing

SBs are always blown sequentially along the direction of gas flow, starting with furnace and ending with Eco or AH, so that the dust is effectively pushed towards the exit. In all but the smallest boilers a PLC-based controller sequences the SBs. Depending upon the extent of

TABLE 18.2
Summary of SBs

Type of SB	Wall	Rotary	Retractable	Rake
Location	Furnace	BB, economiser, AH	SH & RH	Fin tube eco
Gas temperatures	Very high	Medium and low	Very high and high	Low
Lance temperature limit	NA	1100°C	NA	530°C
Lance material		CS ≤ 500°C Calorised (500–900°C) Ferritic ss ≥ 900°C	CS 1¼%Cr1%Mo Ferritic ss	CS, 1¼%Cr1%Mo
Nozzles in lance Nozzle dia	1 or 2 nos Average 25 mm Ø	Many 6–10 mm Ø Average 8 mm	2 at 180° opposed or angled	Many sets
Blowing rotation	1–3 rpm	2–3 rpm	4–12 rpm	NA
Angle	360°	<360°	360°	NA
Travel			1–5 m/min	~1.5 m/min
Steam consumption	~60 kg/min	~60 kg/min	~60 kg/min or more 120 for heavy duty	60–75 kg/min
Blowing time	40–60 s	40–60 s	10–30 min	2–10 min
~Blowing radius	1.5 m nom.	1.5–2 m	1.2–2.7 m	1.5–2 m
~Blowing arc	200–300 mm travel ~10 m² cleaning area	Up to 7 m lance Manual and motorised options available	0.6 to 17 m travel	~$L \times W = 5 \times 3$ m
Motorised	Always	Manual in small boilers	Always	Manual in small boilers
Application	PF boilers. Rarely for grate boilers	All	All	Only TAH, finned tube eco

Notes: NA, not applicable.

fouling soot, blowing is carried out once a day to once or even twice a shift. Soot blowing is an expensive process as steam made from treated water is wasted away. This has to be balanced against the fuel saved due to the better cooling of gases.

Intelligent soot blowing

In many large utility boilers, automatic sequencing is further refined by employing intelligent blowing to derive better life of PPs and greater fuel savings. It is common knowledge that not all SBs are fully effective while, at the same time, certain locations need more cleaning, both of which cannot be addressed in a standard sequential blowing that goes through a set cycle of operations. By running SBs repeatedly where there are no deposits or failing to run them where there are, or even waiting too long to run them, SB performance can continuously fail to meet expectations. Hence, a system which processes the necessary inputs and interprets them intelligently to tell which SB to run when, where and even how, is a critical step in achieving the full potential of a boiler cleaning system.

Intelligent soot blowing is quite sophisticated and expensive to be deployed in all but large boilers. It employs a PLC which accomplishes these requirements by processing feedback from heat flux sensors and strain gauges, besides the usual boiler parameters such as steam and gas temperatures, spray water flow and draft pressures. Besides fuel savings because of lower exit gas temperatures and better overall boiler cleanliness there are secondary benefits also claimed, namely reductions in

- Tube erosion
- Use of water
- Use of SBs
- Spray water rates (extension of tube lifetime)

Water lancing

With severely slagging furnaces, both in coal and FO firing, wall blowers are ineffective beyond a certain point. At that time water lancing is employed in furnace walls. In place of steam HP water jet is employed which delivers a much higher and concentrated impact. Moreover, the action of water on ash deposits is very different from that of steam. As water is much cooler and denser, the mass of ash on being quenched with water shrivels and becomes friable, leading to easy dislodgement under the HP of water.

Water lancing is a highly skilled application needing a good understanding of boiler and its operation. The water jet should be carefully directed and not allowed to impinge on the tubes or cause disturbance to the furnace pressure or water circulation.

Ash deposits from furnace, division and wing walls are cleaned which are in the line of sight. Thermal sensors are installed both to alert the operators about the ash build-up as well as when they are cleaned up, so that the water lancing operation can be timed properly.

Water lancers are also tied in with the intelligent soot blowing systems.

18.36 Sour Gas (*see* Natural Gas in Gaseous Fuels)

18.37 SO_x or Sulphur Oxides (*see* Air Pollution)

18.38 Specific Gravity (*see* Properties of Substances)

18.39 Specific Gravity of Fuel Oils (*see* Fuel Oils in Liquid Fuels)

18.40 Specific Volume (υ) (*see* Properties of Substances)

18.41 Specific Weight (Y) (*see* Properties of Substances)

18.42 Spontaneous Combustion

It is *combustion without external ignition source*. The process is slow and takes many hours of decomposition/oxidation to build up the required temperature, as the heat is mostly unable to escape. Heat of generation exceeds the heat of dissipation. The temperature slowly rises up to the point of **ignition** and **combustion** gets triggered when conditions turn favourable in the presence of air.

The temperature at which the oxidation reaction of coal becomes self-sustaining and spontaneous combustion results varies a great deal depending on the type (nature and rank) of coal and surrounding conditions of heat dissipation. In low-rank coals and where the heat retention is high, the carbonaceous material in coal may start burning at temperatures as low as 30–40°C. Also, the low-rank coals are more susceptible to spontaneous combustion due to their high level of fuel O_2:

- As three factors, viz. carbonaceous material (fuel), O_2 and heat are needed for spontaneous combustion, all fire prevention plans are based on the elimination of one or more of these components in the fire triangle.
- Spontaneous combustion occurs, not infrequently, in fuel yards and fuel bunkers. It is usual to practice compaction of fuel piles in the yard to prevent the ingress of O_2. Fuel bunkers are normally provided with accelerated discharging through the fuel feeders to remove burning coal.
- In case of **lignite**, the bunkers are designed for a mere 4–5 h storage in place of the 10–16 h storage with coal to limit the damage.

18.43 Spreader Burning

Spreader burning is a combination of

- Suspension firing in furnace above
- Mass burning on the grate below

when sized solid fuel is evenly metered, distributed and projected into the furnace. The grate can be stationary **PHG, DG, CG/TG, VG** or **pusher grate**. **Coal, lignite, rice husk, bagasse, biomass** and **RDF** are some of the fuels ideally suited for spreader burning. For un-sized and very high-M fuels like **MSW**, mass burning is needed and suspension firing does not help. Spreader burning is the most common type of firing for a majority of industrial applications using a variety of solid fuels.

18.44 Spreader Stoker

SS is a combination of **fuel spreaders** and **grate** of any type. It is primarily an expression to indicate a combination firing consisting of suspension and mass burning. It is further classified depending upon the type of grate employed, such as SS with CG and so on.

18.45 Stack or Chimney

The purpose of a stack in a modern boiler is merely to vent out the flue gases into the atmosphere at a stipulated safe height. This is unlike in the olden times of **natural draught** boilers when the stack had a very important function of creating adequate draught to provide sufficient flow of air and gas in the boiler setting as needed for combustion of fuel.

As the stack has no other productive function it is best restricted to minimum height consistent with the local **pollution norms**. The taller the stack, the larger is the diameter from stability considerations and deeper are the foundations. Thus, the total cost of chimney is largely proportional to its height.

The stack height should be no lower than the surrounding buildings for proper diffusion of flue gases. It should be tall enough to spread the plume quite far and wide to avoid unacceptable ground-level precipitation in the immediate surroundings. For this the diameter is so chosen to give a sufficiently high exit velocity. Many times, the discharge end is slightly tapered inwards and made conical in steel chimneys to increase the exit velocity. At any rate, there should be no inversion of plume (downward bending of exit gases due to low exit velocity) at any load, which can happen particularly during winter months when the **ambient** air is heavy. Besides high precipitation around the stack, the inversion of plume can cause corrosion of the stack when it is made of steel, particularly with high S in fuel.

To distribute the pollutants far and wide, the dispersion height of gases should be at a good elevation. Dispersion height of gas plume is the sum of actual stack height + apparent stack or plume height. To improve the dispersion besides a tall stack, it is necessary to increase the gas exit velocity.

Stack height is made a function of fuel S in many countries. Stacks of Hrsgs and gas-fired boilers are, therefore, among the shortest. Stack height can be reduced for FBC boilers with their inherent in-bed desulphurisation.

Stacks of **Hrsgs** are particularly short and stout as the GT exhaust gases are very voluminous. It is normal to have steel stacks for most **industrial boilers**. For large industrial and **utility boilers**, concrete stacks are economical as the stacks are quite tall. The choice of steel versus concrete stack is dependent on the costs of material, fabrication and erection prevailing locally.

Steel stacks can be

1. Self-supporting
2. Guyed
3. Structurally supported

For heights of maximum ~100 m with diameters of ~10 m, self-supporting stacks are quite popular. Typically, they are made of 10–25 mm plates reducing in thickness in stages, with approximately the bottom 1/3 height made into a conical shape with brick lining for proper stability. The lining may be extended to full height when corrosion protection is desired. This is usually done by guniting the chimney with refractory concrete, that is, spraying the castable refractory under pressure with the help of special guns. In self-supporting chimneys, the lateral forces (wind or seismic forces) are transmitted to the foundation by the cantilever action. The chimney, together with the foundation, remains stable under all working conditions without any additional support. Figure 18.17a shows a typical self-supported chimney under erection.

In guyed steel chimneys, which can be quite tall, steel wire ropes or guys are attached to support the stack and transmit the lateral forces. All the externally applied loads (wind, seismic force, etc.) are shared by the chimney shell and the attached guys which ensure the stability of the guyed steel chimney. There may be one, two or three sets of guys. Main advantage of the guyed chimney is that it is much slimmer with no enlarged conical bottom portion, thus needing less steel. The serious disadvantage is that the guys come in the way or movement of men and machinery, severely reducing their appeal. If the guys are short and they can be terminated on top of buildings this disadvantage can be overcome. This is shown in Figure 18.17b.

Structurally supported steel stacks can be much taller where the chimney is supported by a lattice of structural members all around the stack (as shown in Figure 18.17c).

A concrete stack can be made much taller, as high as 420 m, as the world's tallest chimney in Kazakhstan. Usually, they are multi-flue chimneys serving several boilers.

Stacks are required to be lined very often. Flue gas entering the stack is already cooled to the optimum in the boiler. Further cooling of gas takes place in stack as heat flows out to the ambient which is at a much lower temperature. This cooling of gases warrants application of a suitable lining on the inside to prevent sulphurous gases from condensing and attacking the steel. Internal lining can be either of brick for small boilers or castable refractory. External lining with insulation mattresses is done in case of very cool stacks of Hrsgs to avoid both condensation of stack gases and cooling of Hrsgs in banked condition.

Flow of gas from bottom to top of chimney is due to the stack effect. The tendency of hot air or gas to rise in a shaft or other vertical passage, owing to its lower ρ compared with that of the surrounding air or gas, is termed as stack or chimney effect. Accordingly, the bottom of the chimney displays negative draught or suction. The taller the stack and higher the average inside gas temperature, the greater is the stack effect and higher the suction. Draught created by stack effect should exceed the skin friction loss experience by the flue gases as they flow up the chimney. In principle, a stack is not expected to add its draught requirements to the load of fans but be self-reliant.

Gas velocities for stack are in the range of 15–30 m/s. Tall stacks with higher draughts can sustain higher velocities.

Tall steel chimneys are susceptible to vibrations producing large deflections normal to the wind direction at steady wind velocities that are lower than the design speeds. This is due to the von Karman vortices that form around the stack at higher levels whenever steady winds are present. The vortex shedding sets up periodic pressure pulsations. Whenever their frequency matches the natural frequency of the stack, resonant conditions are set up causing stack deflections of large magnitude. It is normal to add vortex shedders in the top portion of

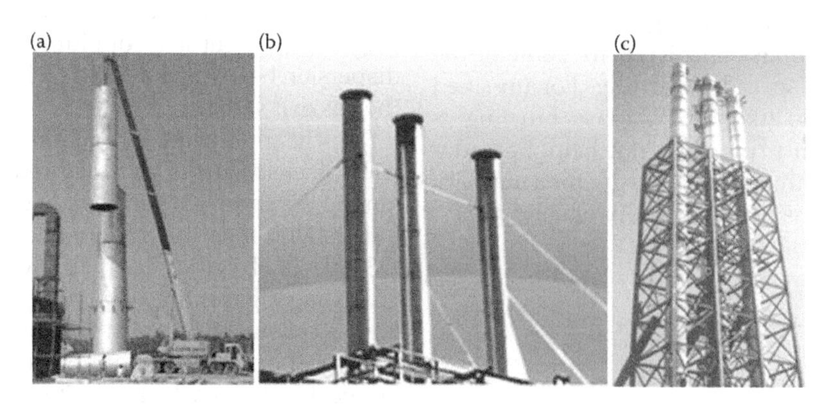

FIGURE 18.17
(a) Self supporting, (b) guyed and (c) structurally supported stacks.

the stack in fabrication stage itself which helps to prevent the formation of vortices by deflecting the winds. Vortex shedders are helical plates around the body of the stack guiding the winds away.

18.45.1 Stack Instrumentation (*see also* Continuous Emission Monitoring System in Acid Rain).

As part of the **acid rain** prevention programme, it is normal these days to install comprehensive continuous emission monitoring system (CEMS) to measure CO_2, O_2, CO, SO_2, NO_x, SO_2, particulates, **opacity** and other gases as desired by the local authorities.

18.46 Stack Effect

Stack or chimney effect is the suction effect created at the bottom of a vertical duct or chimney containing hot air or gas entirely due to the difference in densities between the inside hot and outside cold fluids. At the discharge end, the pressure is atmospheric. As the hot fluid would like to go up due to its lower ρ, the cold air would try to enter at the bottom creating thereby a suction effect, which is termed the stack effect.

It is very important to calculate this effect and add to the pressure losses in the loss calculations of the flues and ducts, as the stack effect is quite considerable with tall boilers/stacks and high gas temperatures.

As the stack effect produces suction at the bottom of the tower, it aids the upward gas flow and acts against the downward flow of gases. Typically, in a two-pass boiler, the suction effect created at the bottom of the furnace is large as the furnace temperature is high. It helps to reduce the head requirement of fan. In the second pass where the gas flow is downward, the suction effect works against the system and increases the head requirement of fan. In the third pass, which is the stack, again it is helpful to the system. Even though the heights may be similar, the gas temperatures being very different the stack effects vary considerably and need to be assessed at every gas turn. The actual draught readings at different places would be affected differently. That is the reason why in tall balanced draught boilers, while there is suction pressure at the bottoms of furnace and second pass, the top of the pass has positive pressure and it is not advisable to open the man doors there. Stack effect affects the vertical passes but not the horizontal duct runs.

18.47 Stainless Steels (*see* Metals)

18.48 Standard Wire Gauge (*see* Wire Gauge)

18.49 Stanton Number (*see* Dimensionless Numbers in Heat Transfer)

18.50 Start up of Boiler (*see* Commissioning)

18.51 Start up Vent (*see* Vent Piping)

18.52 Start up Fan (*see* Boiler Fans *under* Fans)

18.53 Steam (*see also* Properties of Steam and Water)

Steam is vaporised water. As water is abundantly available, steam is a relatively inexpensive working fluid. Steam can pack considerable amount of heat energy making it an ideal working fluid. It has two main uses in the industry. In process plants, steam is used for heating, cooking and drying. In power plants, it is used for driving the STs to generate power.

18.54 Steam and Water Sampling System

In any boiler, the quality of

- Water (fw and bw)
- Steam (saturated and SH/RH)

are to be monitored continuously for variations which can lead to deposits in drums, SHs, RHs and STs. SH and RH deposits cause overheating leading to failure of tubes while those in STs cause loss of cycle η. Both deposits are detrimental to the plant η and availability and therefore SWAS is very important in boiler operation.

The steam and water monitoring can be either

a. On-line as in automatic systems or
b. At regular intervals as in manual systems

In both cases, the sampling methodology remains the same. The main object of sampling followed by testing is to help control the station steam-water chemistry and improve plant availability. The sampling system continuously collects and transports the required small quantities of water and steam to the central sampling area, reduces the pressure and cools them to a temperature of ~25°C as specified in codes like Steam and Water Sampling, Conditioning, and Analysis in the Power Cycle PTC 19.11-2008. A typical sample cooler is shown in Figure 18.18.

The sampling system consists, for each point, of an isolating valve, sampling piping and sample cooler. It is essential that the sampling piping and coolers are made of ss material, such as ANSI 304 or 316 to ensure that the sample is not contaminated by any corrosion products on account of the high temperature of steam and water in the system.

The samples so collected are analysed mainly for conductivity to ascertain the levels of solids. In steam, SiO_2 is also monitored. The other parameters measured are pH, dissolved O_2, dissolved SiO_2, Na and PO_4 as per the monitoring philosophy of the plant.

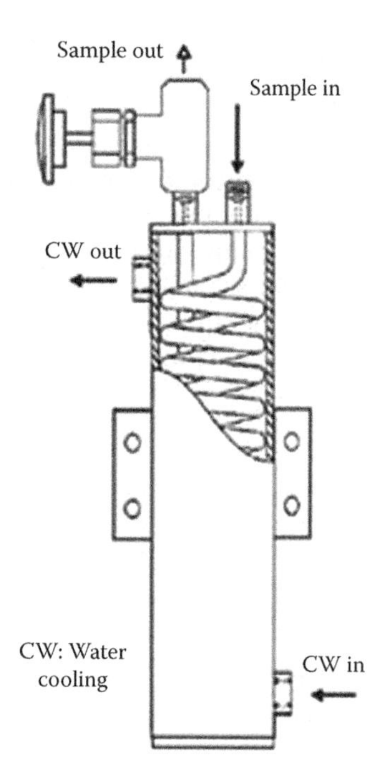

FIGURE 18.18
Sample cooler.

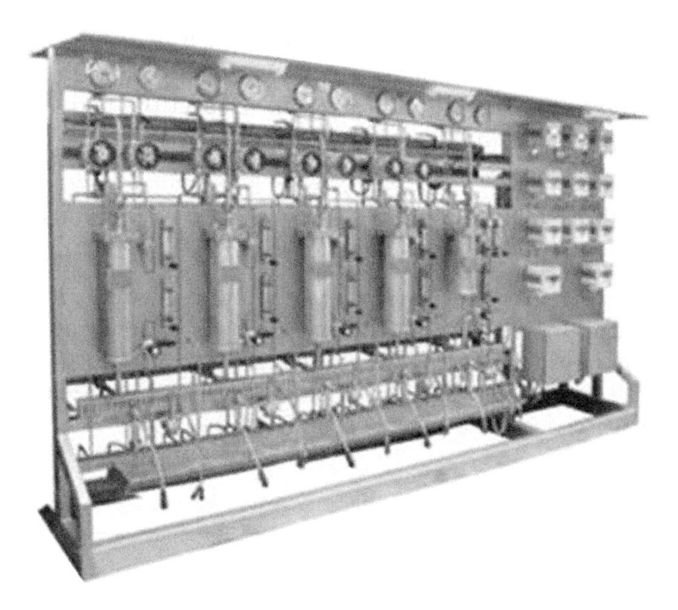

FIGURE 18.19
SWAS panel.

Sample coolers associated instruments and analysers are arranged in a rack as shown in Figure 18.19 and placed close to the boiler control room.

18.55 Steam Blowing (*see* Commissioning)

18.56 Steam Coil Air Preheater (*see* Airheaters)

18.57 Steam Conditioning (*see also* Turbine Bypass Systems)

Steam conditioning is the *combination of pressure reduction followed by temperature reduction of SH steam*. It is also commonly called as PRDS (pressure reducing and desuperheating station). This is to

- Meet process needs
- Protect downstream equipment
- Allow the use of less expensive materials or schedules for downstream piping

This is done with a pressure reducing valve and a spray water additional section either as separate units or

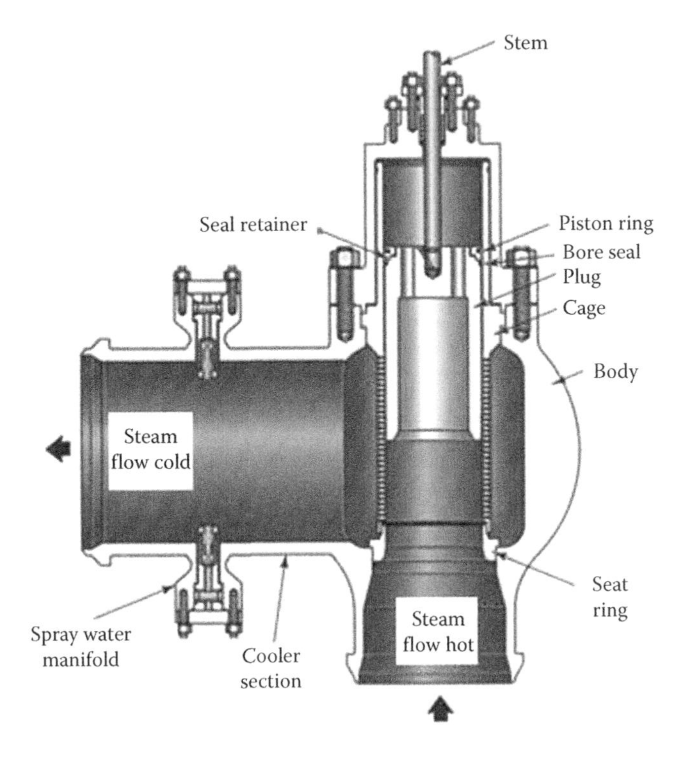

FIGURE 18.20
Steam conditioning valve or PRDS.

a single device. The latter requires less space and is more cost effective. Cooling water is injected in proportion to the steam flow volume. The multi-stage, low-noise designs are available both as a globe or angle style valve.

ST bypass system is the best example of the use of steam conditioning valves. In plain deSHs and attemperators, there is no meaningful pressure reduction while there is only temperature reduction.

Shown in Figure 18.20 is a typical steam conditioning valve PRDS of angle construction which integrates, in the same assembly, both the sections consisting of the steam pressure reduction followed by water spraying. In place of angle, globe-type valve arrangement is also possible. In more compact planning, the spray water section is also placed inside the main body. Several types of trims and materials of construction are possible. Noise attenuation and cavitation elimination are points of concern.

18.58 Steam by Weight and Volume (*see* Properties of Steam and Water)

18.59 Steam Cycle (*see* Thermodynamic Cycles)

18.60 Steam Drum (*see* Pressure Parts)

18.61 Steam Generator (*see* Boiler)

18.62 Steaming Economiser (*see* Economiser in Heating Surfaces)

18.63 Steam Mass Velocity (*see* Superheater in Heating Surfaces)

18.64 Steam Purification (*see also* Drum Internals)

The process of steam purification inside steam drum consists of two steps, namely, to separate the

- Bulk of *steam from water* in steam separators, called *steam separation*. The action is centrifugal or inertial.
- Minute amount of *water from steam* in scrubbers, called the *drying or scrubbing* of steam. Here, the action is adsorption of water either on corrugated plates or on wire mesh.

Steam purification is complete when there is

- Water—Free steam to SH
- Steam—Free water to downcomers

18.65 Steam Purity

Steam *purity* is the amount of *contaminants* in steam. It is measured in ppm or ppb.

Steam purity is different from *steam quality* which is the amount of *M* in the steam expressed as % dryness of steam.

The contaminants can be solid, liquid or gaseous. However, solids are the most troublesome of them all as they form scales in subsequent equipment. *By steam purity is meant the carryover of solids.* Carryover cannot be eliminated even in the best designed and operated boilers, but can be minimised to a very low level of 0.001–0.01 ppm or 1–10 ppb solids. Experience shows that when carryover is limited to 30 ppb the chances of troublesome deposits are minimum.

Deposits are formed in SH, MSSV and NRV, piping, ST CVs and blades, all of which are bad for the respective equipment. Corrosion and overheating take place in SHs. η of ST can drop by as much as 5% due to the scales on blades.

Steam purity levels of <1 ppm (1000 ppb) solids was the industry standard for a long time. With mean time intervals of ST, stoppage for overhauling has gone up substantially, the purity standards have also gone up sympathetically. HP ST manufacturers usually accept a figure of 0.01–0.02 ppm or 10–20 ppb of SiO_2 in steam today, together with ~0.03 ppm or 30 ppb total solids in steam.

Steam purity is reduced or carryover is increased by the following factors:

1. Mechanical—Minute inefficiency of steam separators, high drum water level, leakage through separators and so on.
2. Chemical—High TDS, excessive alkalinity, oil and other organic impurities
3. Design—Higher loading of separators
4. Operational—High water levels in the drum

Maintaining fw and bw quality as recommended by boiler manufacturer is most essential. These are evolved from the general guidelines provided by bodies like ASME. Table 23.9 provides ASME guidelines for fw and bw quality in sub-c drum-type boilers up to ~140 bar.

In a majority of the modern bw conditioning practices, the soluble chemical constituents of fw, namely, Ca and Mg salts, are reduced to Na form with Ca, Mg, Fe and Cu being precipitated. The solid and liquid impurities carried over, therefore, will be dust and water droplets containing mostly Na salts with minor amounts of Ca, Mg, Fe and Cu. The gaseous impurities are NH_3, CO_2, N_2, SO_2, N_2H_4, amines and SiO_2. Of all these, only dissolved SiO_2 is most troublesome as it forms thin scales downstream while the others are harmless and only interfere in the measurements of purity.

Silica (SiO_2) in steam

SiO_2 in bw has a tendency to dissolve in steam and escape the separation efforts of the most highly efficient steam separators in the drum. This tendency is observed even at pressures as low as 30 bar but is more pronounced at HP of 70 bar and above.

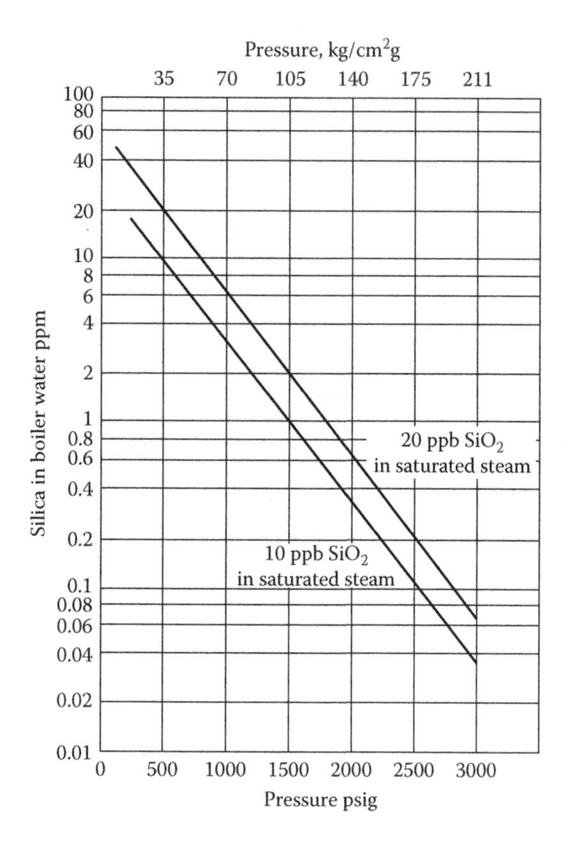

FIGURE 18.21
Permissible SiO_2 limits in drum water at various drum pressures.

Dissolved SiO_2 in steam is a highly dreaded contaminant as it tends to precipitate out of the steam at lower pressures to form scales on the blades of ST. SiO_2 scales are thin, dense and impervious like porcelain. As these scales are difficult to dislodge, the ST η suffers permanently.

There is a general consensus among the utility owners and ST makers that if SiO_2 in steam is maintained between 10 and 20 ppb (0.010–0.02 ppm) at pH > 10, this problem of scaling might not occur. This is achieved only by reducing the SiO_2 in bw. Figure 18.21 indicates the permissible SiO_2 levels in bw at various drum pressure to attain low SiO_2 levels in steam. This graph gives slightly higher values than the recommended levels as per ASME 1975. But the point to note is that the SiO_2 permitted in the bw reduces sharply over 70 bar drum pressure. The higher the pH value maintained in the drum, the higher the permissible SiO_2 content.

18.66 Steam Purity Measurement (*see* Instruments for Measurements)

18.67 Steam Quality or Dryness of Steam

While steam purity is a measure of impurities carried by steam quality is the M in steam. It is measured in % dryness which is given by

$$\frac{\text{Wt of M}}{\text{Wt of steam}} \times 100\% \qquad (18.3)$$

In the early times, dryness of steam was used as a measure of the quality of steam. It is still used for very small boilers. Modern boilers have long replaced quality by purity. Dryness has relevance in the process systems and piping.

18.68 Steam Quality Measurement (*see* Instruments for Measurements)

18.69 Steam Sampling (*see* Steam Purity Measurement *under* Instruments for Measurement)

18.70 Steam Scrubber (*see* Drum Internals)

18.71 Steam Separator (*see* Drum Internals)

18.72 Steam Temperature Control

A boiler is required to provide a constant SOT to the process or ST from certain part load to its peak load to derive optimum operational η. This range is called the STC of a boiler.

In most cases, there is certain decay in the SOT at part loads unless the boiler has a full radiant SH. As this is not ususally practicable, the SH is increased in size and the excess steam temperature is cooled in **attemperators**. While there may be no loss of heat energy the cost of boiler increases due to the large SH.

An STC of 70–100% is normal. At times even 50–100% is demanded. As the HS in SH grows larger, there is a greater expense in the materials of construction, larger attemperators, more instrumentation and even a larger boiler as the corresponding penalties.

18.73 Steam Turbine (*see* Turbine, Steam)

18.74 Steel (*see* Metals)

18.75 Steel Alloying Elements (*see* Metallurgy)

18.76 Steel Classification (*see* Metallurgy)

18.77 Steel Properties (*see* Metal Properties)

18.78 Stefan–Boltzmann's Law of Radiation (*see* Radiation Heat *under* Heat Transfer)

18.79 Stellite Alloys

Stellite alloys, a trademarked name of the Deloro Stellite Company, are principally Co–Ni alloys designed for primarily wear resistance. In boilers they are Co-based alloys commonly applied as coating to valve trims to provide high resistance to wear, corrosion and high temperatures.

There are a large number of stellite alloys composed of various amounts of Co, Ni, Fe, Al, B, C, Cr, Mn, Mo, P, S, Si and Ti, in various proportions with most alloys containing four to six of these elements. Because of the combination of exotic metals Stellite alloys are very expensive.

The salient properties of Stellite alloys can be summarised as follows:

- Non-magnetic and corrosion resistant
- Extremely high melting points due to their Co and Cr contents
- Very high hardness and toughness
- So hard that they are very difficult to machine Usually a Stellite part is precisely cast so that only minimal machining is necessary. They are more often ground than machined or cut.

There are a number of Stellite alloys, with various compositions optimised for different uses. Some alloys are formulated to maximise combinations of wear resistance, **corrosion** resistance, or ability to withstand extreme temperatures.

Typical applications include saw teeth, hardfacing and acid-resistant machine parts.

18.80 Steam Injected Gas Turbine Cycle (*see* Turbines, GT *under* GT Performance)

18.81 Stoichiometric Air (*see* Combustion)

18.82 Stoker Firing (*see* Grate Firing)

18.83 Stress

Stress is the force experienced by a member on its unit cross section.

$$\text{Stress} = \frac{\text{Force}}{\text{Cross-sectional area}} \quad (18.4)$$

See metal properties for more details.

In boilers there are a number of cylindrical PPs like the drums, headers, tubes, pipes and so on which are subject to internal pressure, which produces longitudinal and circumferential/hoop stresses. *Longitudinal stress* is the tensile stress produced along the axis of the cylindrical part while the *circumferential/hoop stress* is the stress produced along the circumference of the cylindrical part.

18.84 Stress Corrosion (*see* Corrosion)

18.85 Stress Rupture (*see* Metal Properties)

18.86 Stresses, Allowable or Permissible in Pressure Parts (*see also* Pressure Part Design)

The basis of arriving at the allowable or permissible stress values for the design of PPs is explained under the topic of PP design. Based on those rules laid down by ASME B&PV code, all PP steels have steadily reducing stress values with rising metal temperature. This is illustrated in Figure 18.22 for various steels for plates, tubes and pipes based on B&PV code of 2007. Different codes would have slightly differing considerations in arriving at the allowable stress values but they do not vary very much. More importantly, the pattern of the high-temperature stress values remains the same.

Figure 18.22 provides the graphs of high-temperature stresses for the most commonly used PP steels of boilers. Table 18.3 is the basis for the graph where the stress values are derived from the B&PV code of 2007.

18.87 Boiler Structure

The purpose of boiler structure is to support and hold in place the PPs, NPPs, AHs firing equipment and other related items so that they can function safely in an integrated manner to deliver steam. A boiler structure contains several parts.

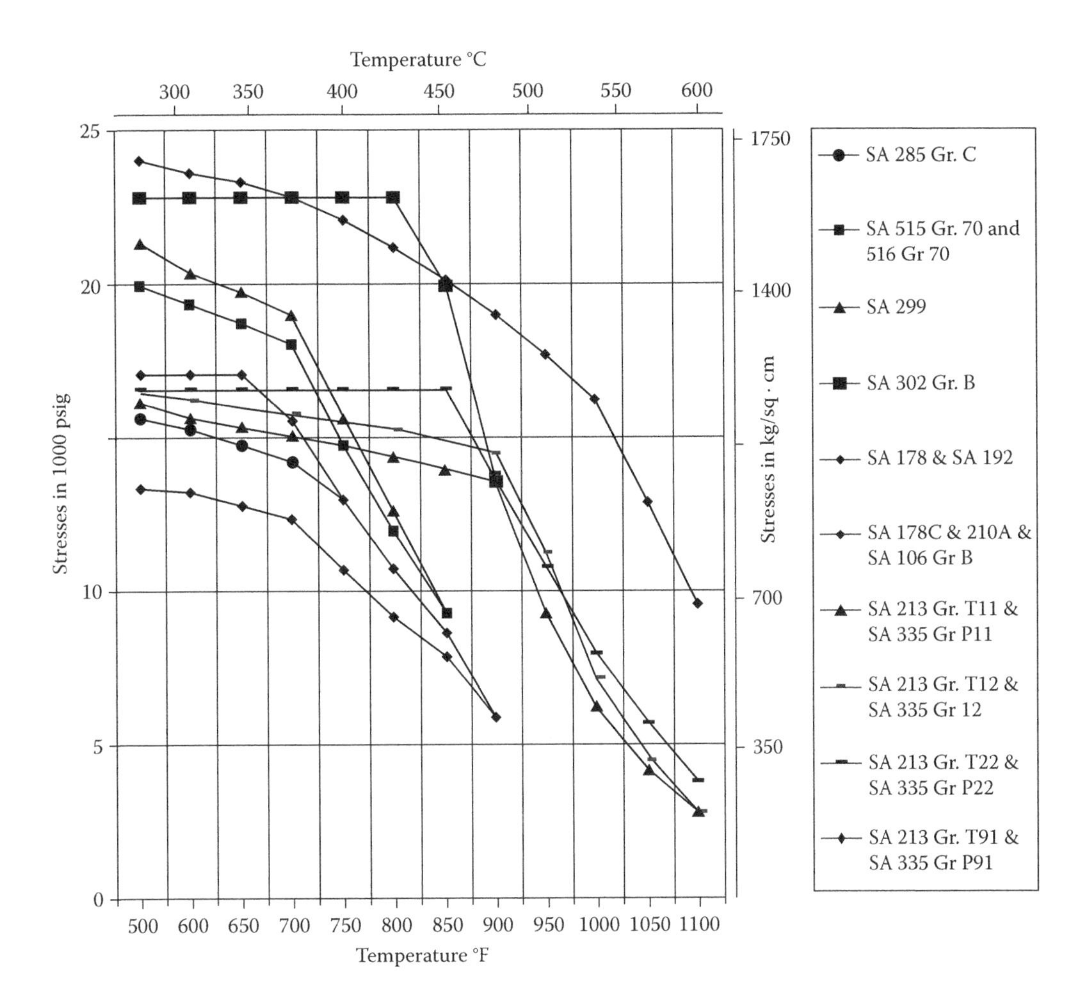

FIGURE 18.22
Allowable high-temperature stresses for common PP steels.

TABLE 18.3
Allowable High Temperature Stresses for Common PP Steels (ASME B&PV Code 2007)

Material/Temperature in °F	500	600	650	700	750	800	850	900	950	1000	1050	1100	Material
Plates													
SA 285 Gr. C	15.7	15.3	14.8	14.3	13.0	10.8	8.7						C-steel
SA 515 Gr. 70 & 516 Gr 70	20.0	19.4	18.8	18.1	14.8	12.0	9.3						C–Mn
SA 299	21.4	20.4	19.8	19.1	15.7	12.6	9.3						C–Mn–Si
SA 302 Gr. B	22.9	22.9	22.9	22.9	22.9	22.9	20.0	13.7					alloy
Tubes and Pipes													
SA 178 & SA 192	13.4	13.3	12.8	12.4	10.7	9.2	7.9	5.9					C-steel
SA 178C & 210 A & SA 106 Gr B	17.1	17.1	17.1	15.6	13.0	10.8	8.7	5.9					C-steel
SA 213 Gr.T11 & SA 335 Gr P11	16.2	15.7	15.4	15.1	14.8	14.4	14.0	13.6	9.3	6.3	4.2	2.8	1¼Cr½Mo
SA 213 Gr T 12 & SA 335 Gr 12	16.5	16.3	16.0	15.8	15.5	15.3	14.9	14.5	11.3	7.2	4.5	2.8	1Cr½Mo
SA 213 Gr.T22 & SA 335 Gr P22	16.6	16.6	16.6	16.6	16.6	16.6	16.6	13.6	10.8	8.0	5.7	3.8	2¼Cr1Mo
SA 213 Gr T91 & SA 335 Gr P91	24.1	23.7	23.4	22.9	22.2	21.3	20.3	19.1	17.8	16.3	12.9	9.6	9Cr1MoV

Structure and parts:

- *Top suspension girders* are the framework of big fabricated beams to which all the loads are transferred in a top supported boiler by means of *sling rods.*
- This framework rests on the *main boiler columns* which, in turn, transfer the entire load to the ground.
- The columns are held in position with a lattice work of *main beams and bracings.*

- Along the height of the boiler, there are *platforms* for accessing various equipments, manholes and peepholes which are held in place by *floor beams.*
- The various floors are interconnected by *galleries and ladders.*
- There are *guides and restraints* for controlling the movement of the pressure parts in a planned manner.

Cold and hot structure:

All these above-mentioned parts put together are called the *cold structure.* The structural members experience only ambient temperature as they are in no way connected with the hot pressure parts.

On the other hand, *hot structure,* which is comparatively lighter, is attached to the PPs. *Hot structure* essentially consists of

a. Top suspension slings/rods
b. Buckstays and tie bars

Loads on structure:

Supporting structure constitutes a significant weight in all boilers excepting in the package boilers which sit on the ground and the structure is only for supporting galleries and ladders. It is nearly the same in the case of bottom supported boilers also.

Boiler structure is required to support the loads of the following items:

- PPs filled with water
- Buckstays, tie bars and allied attachments to the PPs
- Refractory insulation, casing and penthouse
- Firing equipment supported on the PPs
- All auxiliary loads on the platforms
- Weights of slag, ash and bed material in case of FBC boilers

In addition to supporting various boiler loads, the structure has to withstand the external loads imposed by the local conditions, namely

- Wind
- Earthquake

Structural design of a boiler is required to strictly conform to the local building codes.

Indoor, outdoor and semi-outdoor structure:

Boiler structure can be either *indoor or outdoor* structure depending on whether there is a boiler house or not on top of the boiler. Since an indoor structure is not exposed to the wind loads, it turns out to be a very light structure. Indoor structure is necessary for plant and personnel protection in boiler plants located in cold countries where there is a steady fall of rain or snow for most months. Outdoor structure is popular in places like the Middle East where there is not much rain. In other places, mostly *semi-outdoor structure* is employed with a canopy at the top of the structure.

Bolted and welded structure: Bolted or welded refers to the type of joints employed in the structure. Bolted construction takes longer to manufacture, is a bit more heavy and expensive due to the

- Additional jointing material (support stools and gusset plates) and fasteners
- Shop fabrication and trial assembly
- Heavier construction as the beams are designed as simply supported

But bolted construction renders erection very easy, accurate and fast. Where erection costs are high, the overall cost of bolted construction works out nearly the same as the welded design with the added advantage of faster and error-free erection despite higher material costs.

Welded construction is more popular as

- There are no additional jointing material
- It is lighter being part of an integrated structural framework
- It is easy to correct at site

despite taking more erection time, welding and man power.

Boiler maker's practice or customer's preference usually dictates the choice between the welded or bolted structure.

The tower-type front-wall-fired PF boiler is presented in Figure 18.23 which shows the outdoor structure with top suspension girders resting on main columns that are strengthened by bracings. Only vertical tubular AH and SCR units are in the second pass and are bottom supported. The main columns are starting from the firing floor which is independently supported from the ground. The picture is interesting as it shows the arrangement of various sub-assemblies of a large boiler in an isometric view.

18.87.1 Supporting Methods

Boilers can be supported from the (1) bottom, (2) middle or (3) top.

In bottom-supported boilers, the PPs act as load-carrying struts and stand on themselves transferring the imposed loads to the ground. The thermal expansion

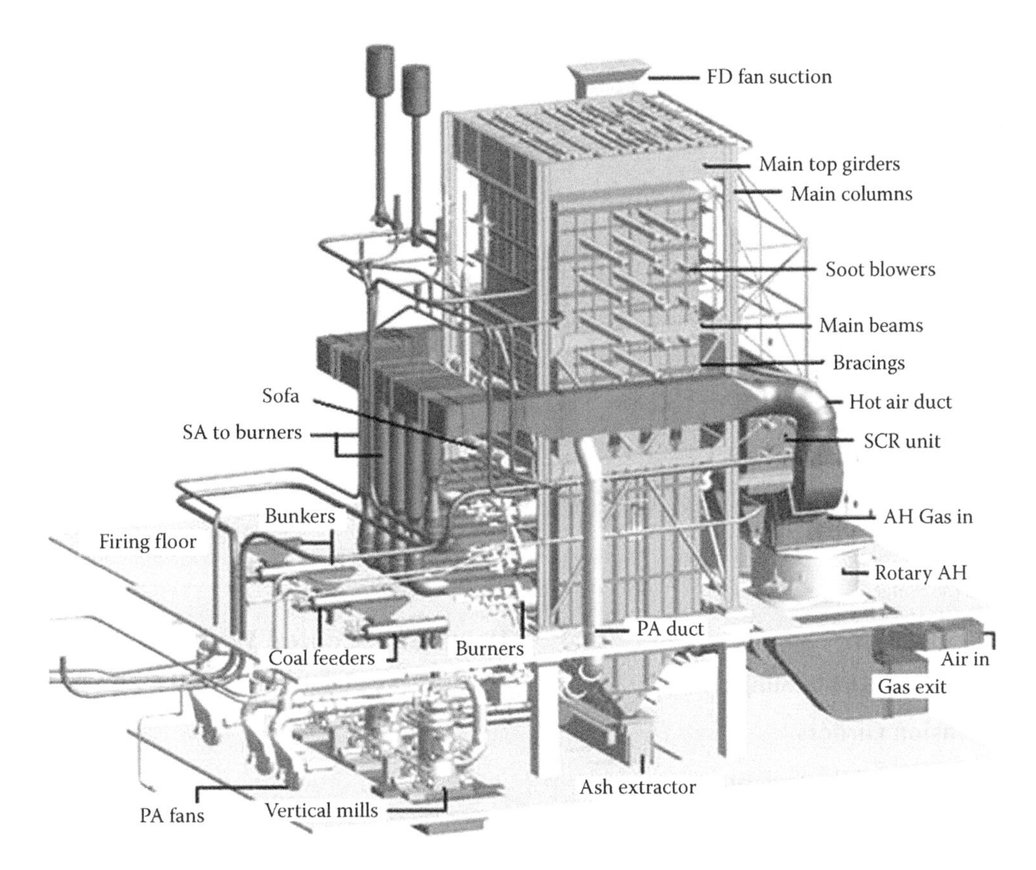

FIGURE 18.23
Isometric view of the boiler structure for a large top-supported utility boiler. (From Hitachi Power Europe GmbH, Germany, with permission.)

of the PPs is upwards. The structure is lighter and is required practically for providing access. Package boilers, pre-engineered oil- and gas-fired boilers of even large size and solid fuel-fired boilers up to ~100 tph are amenable to bottom supporting. Boiler costs can be lowered by adopting bottom supporting, therefore, it is the first choice of designers.

Beyond a certain size as the boilers become tall the differential expansion increases and the weights become heavier, making it beneficial to adopt top support. Top support design can be likened to a church bell. Just like the weight of the bell is suspended from the casing of the bell, the entire PP assembly of the boiler is suspended from the structure, which encloses and transfers the weight to the ground. The PP assembly, now slung from the top girders, is free to expand downwards and move sideways. Naturally, the top support design is more expensive albeit more easy to design.

There are certain boiler configurations which are amenable to a middle support, also called *girdle support*, where the boiler is supported at some mid-point allowing thermal expansion to take place both upwards and downwards. Grate-fired boilers fall in this category.

Regardless of top or bottom support a heavy item like the AH (both tubular AH and rotary AH) are supported from bottom. The Eco is also supported from the bottom in most bi-drum designs as it is placed in the second pass above the AH. However, it is slung from the top in single drum radiant or tower-type boiler designs.

18.87.2 Drum Supporting (*see* Drum Supporting)

18.87.3 Sling Rods/Top Suspension Hangers/ Suspension Slings

Sling rods, top suspension rods or suspension slings are the supporting rods connecting the load of the PPs located at the top of the boiler to the top suspension framework. The PPs are usually the furnace, SH, RH, Eco and other headers besides the riser and saturation tubes.

The suspension rods are made of high tensile CS or C–Mo steels in most cases. However, with boilers prone to puffing, like FBC and PF, the penthouse has a tendency to get filled with hot fly ash creating differential expansion problems in the sling rods between the inside and outside of the penthouse. The rods are usually made of Cr–Mo ASs and designed for higher temperatures. Figure 18.24 shows the completed sling rod assembly during construction stage supporting various top headers.

FIGURE 18.24
Sling rods at the top of the boiler.

18.87.4 Drum Slings (see Drum Slings)

18.87.5 Top Suspension Girders

In top-supported boilers, all the weights are hung from the top. The weights of all the components of boiler are finally transferred to the grid of girders placed on top of the main columns by means of suspension slings or sling rods. The main suspension girders and most intermediate girders are fabricated from steel plates as rolled girders readily available in the market are too small and hence inadequate to take the loads. The main girders are easily the deepest fabricated sections in most boilers.

The top beams finally transfer their loads to the main girders which are placed on top of the main columns. The main girders are not welded to the columns but are simply located and supported.

18.87.6 Buckstays (see Furnace Pressure in Heating Surfaces)

18.88 Stubbed Drum (*see* Steam Drum *under* Pressure Parts)

18.89 Sub-Bituminous Coal

Sub-bituminous coals are generally dark brown to black in colour. They are intermediate in rank between **lignite** and bituminous coal. Geologically, they are younger than bituminous coals. They are also widely spread across the globe.

Sub-bituminous coal contains 40–50% FC on daf basis and has Gcv ranging from ~19 to 26 MJ/kg (~4470–6100 kcal/kg or 8200–11,200 Btu/lb). S content is often low at <1% making it a cleaner coal to burn. Less water makes it easier to transport, store and use.

Combustion of sub-bituminous coals is very easy. Ignition temperature is low (~400°C) and burning is quicker compared to bituminous coals. Many times they are also easier to grind. Sub-bituminous coals do not contain enough tar-like substances to produce coking coals.

18.90 Sub-Cooled Water (*see* Properties of Steam and Water)

18.91 Sub-Critical Pressure (*see* Critical Pressure Point of Steam and Water)

18.92 Suction Milling (*see* Pulveriser Types *under* Pulverisers)

18.93 Sulphation Reaction (*see* Desulphurisation)

18.94 Sulphur Oxides (*see* Air Pollution)

18.95 Sulphur (in Coal) (*see* Sulphur in Ultimate Analysis in Coal)

18.96 Supercritical Boilers (*see* Utility Boilers)

18.97 Super-Critical Pressure (*see* Critical Point of Steam and Water)

18.98 Superheat (*see* Properties of Steam and Water)

18.99 Superheater (*see* Heating Surfaces [HSs] *and* Pressure Parts)

18.100 Superheater Outlet Pressure (*see* Boiler Pressures)

18.101 Superheater Outlet Temperature Boiler Steam Temperatures

18.102 Supplies (*see* Downcomers)

18.103 Surface Tension (*see* Fluid Properties *under* Fluid Characteristics)

18.104 Suspension Hangers/Slings/Rods (*see* Sling Rods)

18.105 Swaging (*see* Tube Swaging *under* Manufacturing Topics)

18.106 Sweet Gas (*see* Natural Gas)

18.107 Swell Effect in Steam Drums (*see* Swell Volume in Steam Drum *under* Pressure Parts)

18.108 System Resistance of Boiler

Air and gas are the working fluids passing through a boiler overcoming the resistance offered by the

- Flues and ducts
- Firing and milling equipment
- Various tube banks

This resistance to flow is termed as system resistance. The system resistance varies nearly as the square of the flow, as shown in Figure 18.25.

Modern boilers need the assistance of fans to overcome the resistance offered. Only in clean boilers the system resistance remains substantially constant. When dusty fuels like coal are fired, the gas side resistance varies over a band depending upon the extent of fouling.

Fan margins are provided because of the practical difficulties of estimation of system losses and also to account for its variability due to fouling. Fan selection should be such that the fan curve intercepts the system resistance curve at the design point after applying proper margins to the MCR point (as shown in Figure 18.25).

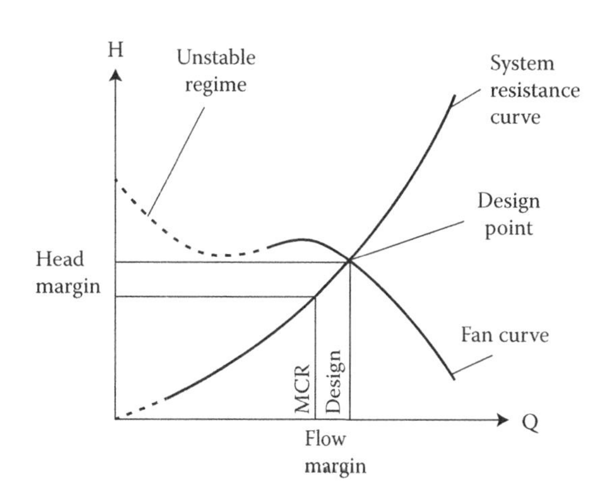

FIGURE 18.25
System resistance and fan selection.

19

T

19.1 T Firing (*see* Pulverised Fuel Burner Arrangement)

19.2 Tangent Tube Walls (*see* Heating Surfaces Pressure Parts)

19.3 Tangential Burners (*see* Pulverised Fuel Burners)

19.4 Temperature (*T*, *t*) (*see* Thermodynamic Properties)

19.5 Temperature of Gas (High-Velocity and Multiple-Shield High-Velocity Thermocouples) (*see also* High-Velocity and Multiple-Shield High Velocity Thermocouples *under* Instruments for Measurement)

Flue gas temperature measurement is desirable for both boiler designers and operators. Accurate measurement, while essential, is particularly difficult at high temperatures. Optical and radiation pyrometers are not designed for measuring gas temperatures. Ordinary **TCs** indicate lower temperatures and give large error as can be seen in Figure 19.1.

Temperature-sensitive element receives heat from gas primarily by convection, as it is immersed in the gas currents. It also receives heat by radiation from its surroundings. A small amount is also provided by **conduction** through its own body. True gas temperature is the one when heat by all the three modes by the element is integrated.

Usually the flue gas measurement device is at a lower temperature than the surrounding gas and therefore, particularly at higher gas temperatures, the **radiation** heat is not adequately picked up by the element. This leads to lower temperature indication than the actual. This difference between the observed and true temperatures can be narrowed when true flow conditions of flue gas are tried to be simulated in HVT and MHVT.

HVT and *MHVT* are described in Chapter 9. MHVT is unsuitable for dusty flue gases for fear of clogging. Calculated temperatures correspond to MHVT and measured temperatures to HVT or TC and are required to be indicated as such, so that suitable correction can be incorporated. TC starts giving lower than true temperatures from ~400°C/750°F while MHVTs from ~870°C/1600°F.

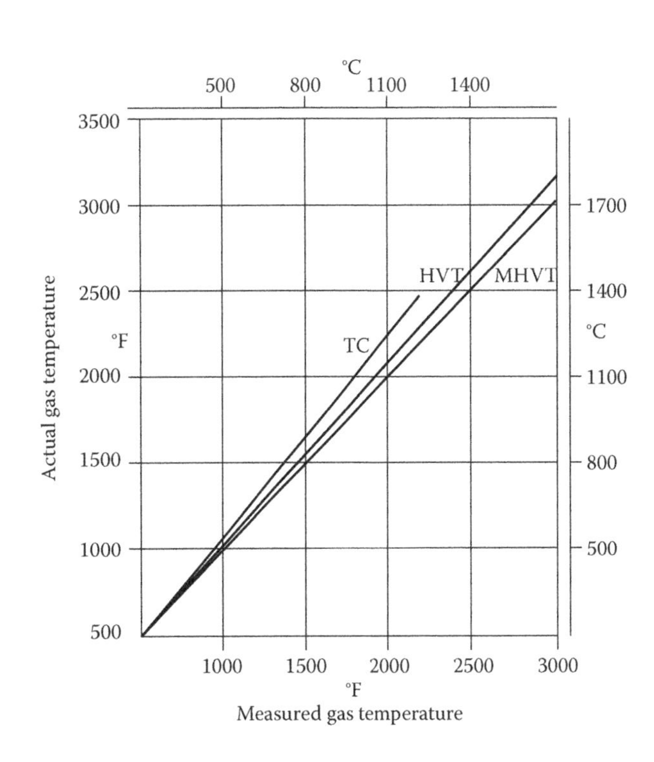

FIGURE 19.1
Measured and actual gas temperatures by different measuring techniques.

19.6 Temperature–Entropy Diagram (*see* Critical Pressure Point in Steam and Water)

19.7 Temperature Measurement (*see* Instruments for Measurements)

19.8 Tempering Air for Mill

Tempering air is the air admitted to the suction of the **coal mill** to regulate the temperature of fuel–air mixture (PF + PA) leaving the mill.

Certain mill outlet temperature is needed for optimum burning of PF. Hot PA is admitted to the mill for drying the fuel and transporting it to the burners. PA is at the AH discharge temperature, which is high enough to dry the most moist coals expected in the plant. When such high M fuel is not to be used, the PA temperature is too high and needs to be cooled to the required level. Tempering air is mixed with PA to do this cooling.

Tempering air can be taken from the atmosphere or from the discharge of the FD fans before the AHs.

This can be seen in Figure 16.26 given under the topic of Pulverised fuel firing.

19.9 Tensile Strength (*see* Metal Properties)

19.10 Tensile Test (*see* Testing of Materials)

19.11 Testing of Materials (Metals)

This topic is elaborated in the following sub-topics.

a. Destructive testing (DT) of metals
   1. Hardness testing
   2. Impact testing (Charpy and Izod)
   3. Tensile testing
b. Non-destructive testing/examination (NDT/NDE)
   1. Liquid penetrant test (LPT)
   2. Magnetic particle test (MPT)
   3. Ultrasonic test (UT)
   4. Radiographic test (RT)

Material testing is two types—DT and NDT.

- In the DT, tests are carried out *until the specimen fails*, to understand performance or behaviour under different loads. DTs are usually much easier to perform and interpret than NDT. They also provide more information, and DT is carried out using specimens or actual work pieces if they are mass produced.
- NDT is analysis-based technique to know the properties of a material without causing damage to the work piece or the weld.

19.11.1 Destructive Testing

19.11.1.1 Hardness Testing of Metals (Brinnel, Vickers, Rockwell)

There are several accepted methods of measurements for *indentation* **hardness** of which the following are the most popular.

a. Brinnel hardness

This test consists of indenting the metal surface with a 10-mm-diameter steel ball at a load of 3000 kg mass (~29,400 N) for a standard time, usually 30 s. For soft metals the load is 500 kg and for very hard metals a tungsten carbide (WC) ball is used. The diameter of the indentation is measured with a low-power microscope after removal of the load. The average of two readings of the diameter of the impression at right angles is made.

Brinell hardness number (BHN) is load P divided by the surface area of the indentation, given by the formula

$$\text{BHN} = \frac{P}{(\pi D/2)\sqrt{(D - D2 - d2)}} = \frac{P}{\pi Dt} \tag{19.1}$$

where
P—applied load, N
D—diameter of ball, mm
d—diameter of indentation, mm
t—depth of the impression, mm
It will be noticed that the units of the BHN are MPa.

b. Vickers hardness

The Vickers hardness test is similar to Brinnel but uses a square-base diamond pyramid as the indenter and hence, is frequently called the *diamond–pyramid hardness (DPH) test*. This test has received fairly wide acceptance for research work because it provides a continuous scale of hardness, for a given load, from very soft metals with a DPH of 5 to extremely hard materials with a DPH of 1500. The Vickers hardness test is described in ASTM Standard E92-72.

c. Rockwell hardness

This test utilises the depth of indentation, under constant load, as a measure of hardness. The wider acceptance of this test is because of its

- Speed of testing
- Freedom from personal error
- Ability to distinguish small hardness differences in hardened steel
- The small size of the indentation

so that finished or heat-treated parts can be tested practically without damage.

Relationship between the various hardness measurements is depicted in Figure 19.2. Vickers–Brinnel relation is nearly linear up to 400 with Vickers ~5% higher than Brinnel. Since the tests measure slightly different parameters the comparisons are not strictly accurate.

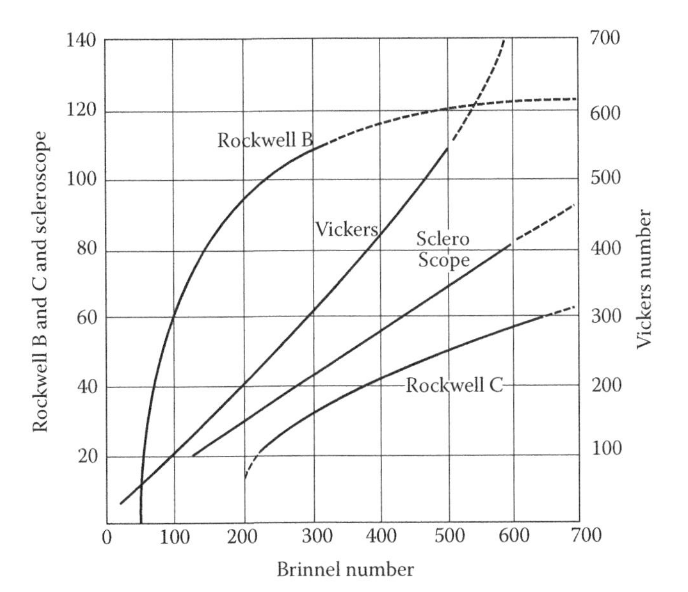

FIGURE 19.2
Comparison of material hardness scales.

19.11.1.2 Impact Testing of Metals (Charpy and Izod)

Impact tests provide information on the *resistance of a material to sudden fracture where a sharp stress riser or flaw is present*. These tests are quick and inexpensive and data obtained from such tests is frequently employed for engineering purposes.

Impact tests involve breaking of notched specimens by a swinging pendulum and noting the energy expended. *Charpy V-notch* and the *Izod tests* are the most popular and are described in ASTM E23 Standard Test Methods. Both tests are similar but the specimens are slightly different. In Charpy test the notched specimen is held at both the ends in a simply supported and in Izod it is held at one end in a cantilever manner. A swinging pendulum from a specified height is allowed a free swing. The extent of rebound is measured directly in kg-m or ft-lbs as the energy absorbed. Figure 19.3 depicts the two specimens.

Many materials, including metals, exhibit marked changes in impact energy with temperature. It is known that there tends to be a region of temperatures over which the impact energy increases rapidly from a lower level that may be relatively constant to an upper level that may also be relatively constant. Such temperature-transition behaviour is common for metals.

The mode of failure in the impact test changes from ductile (shear) to brittle (cleavage) as the temperature is lowered. The temperature range in which this change takes place is known as the transition temperature. Material within this range and lower can crack extensively under impact load.

Fortunately, boiler PP steels have their transition temperatures considerably below their operating temperatures. Heat-treatable steels in quenched and tempered conditions have superior transition temperature properties. High Ni steels exhibit low transition temperatures. There is no abrupt change in toughness properties of austenitic ss and other pure metals such as Cu and Al making them suitable for cryogenic duty conditions.

These simple impact tests have given way to testing methods that make use of fracture mechanics. Fracture mechanics methods allow more sophisticated analysis of materials containing cracks and sharp notches. However, the advantages of fracture mechanics are achieved at the sacrifice of simplicity and economy. Impact tests such as the Charpy and Izod, have thus remained popular despite their shortcomings, as they serve a useful purpose in quickly comparing materials and obtaining general information on their behaviour.

19.11.1.3 Tensile Testing of Metals (see also *Tensile and Yield Strengths in Metal Properties*)

Tensile testing, also known as *tension testing*, is a fundamental test of materials science, the purpose of which

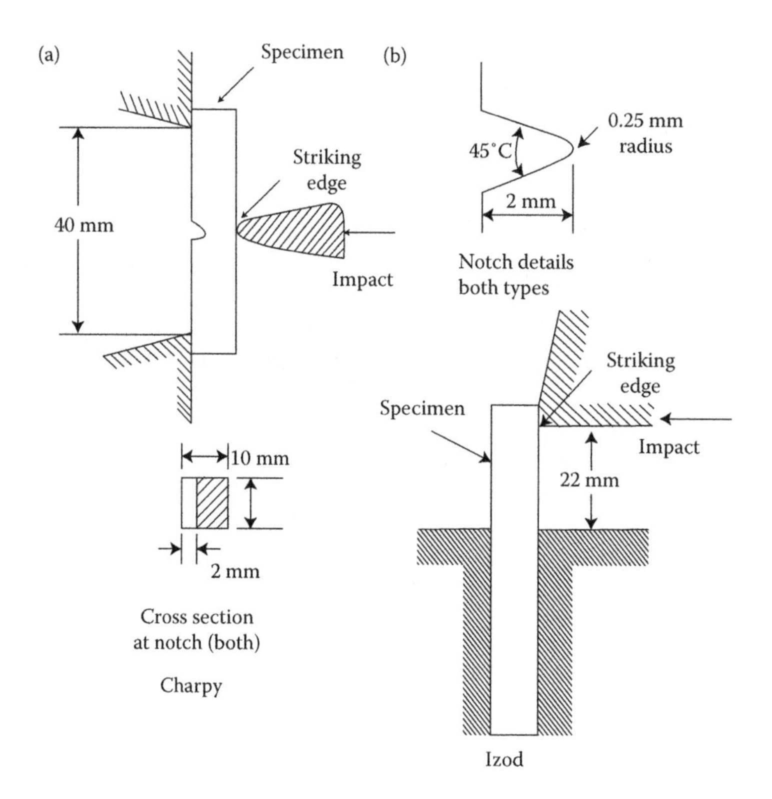

FIGURE 19.3
Specimen in (a) Charpy and (b) Izod impact testing.

is to provide data for use in design or verify compliance with spec. It may therefore be either a quantitative or a qualitative test. Relevant codes are ASTM E8 Standard Test Methods for Tension Testing of Metallic Materials or ISO 6892 Metallic materials—Tensile testing at ambient temperature

The test comprises of holding the ends of a suitably prepared standardised test piece in a tensile test machine and applying a continually increasing uni-axial load until failure occurs. Test pieces are standardised in order that results are reproducible and comparable, as shown in Figure 19.4.

Specimens are said to be *proportional* when the *gauge length*, L_0 is related to the original cross-sectional area, A_0, expressed as

$$L_0 = k\sqrt{A_0} \tag{19.2}$$

The constant k is 5.65 in EN specifications and 5 in the ASME codes. These give gauge lengths of ~5× and 4× specimen diameter, respectively. This difference may not be technically significant but it assumes is importance when claiming compliance with specifications.

Both the load (stress) and the test piece extension (strain) are measured and from this data an *engineering stress/strain curve* is constructed as in Figure 19.5. From this curve the following can be determined:

a. The *tensile strength (TS)* or the *ultimate tensile strength (UTS) is the load at failure Pmax divided by the original cross-sectional area* A_0.

$$\text{UTS} = P_{\text{max}}/A_0 \tag{19.3}$$

In EN specifications this parameter is also identified as 'R_m'.

b. The *yield point* (YP), the stress at which *deformation changes from elastic to plastic behaviour*, Below the YP unloading the specimen it returns to its original length, while above the YP permanent plastic deformation has set in.

$$\text{YP} = P_{\text{yp}}/A_0 \tag{19.4}$$

where P_{yp} = load at the yield point.

In EN specifications this parameter is also identified as 'R_e'.

c. By reassembling the broken specimen we can also measure the *percentage elongation*, El% shows how much the test piece had stretched at failure where

$$\text{El\%} = (L_f - L_0/L_0) \times 100 \tag{19.5}$$

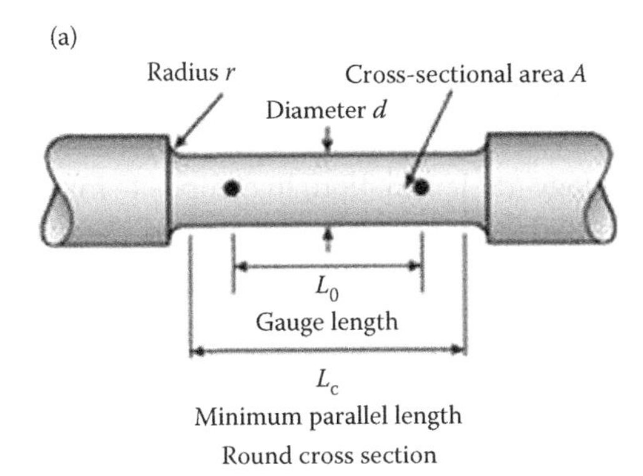

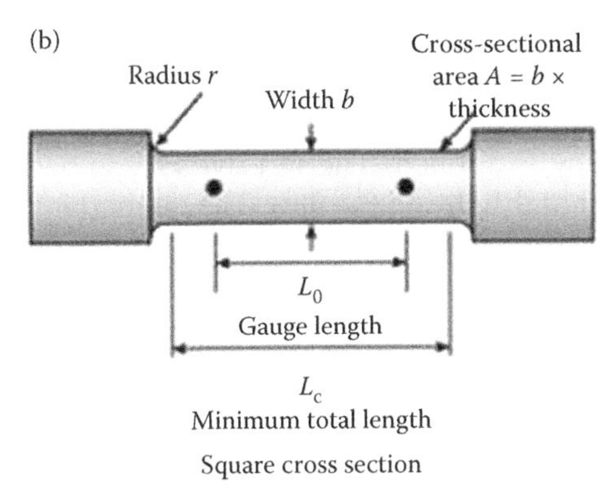

FIGURE 19.4
(a) Round and (b) square test specimen in tensile testing.

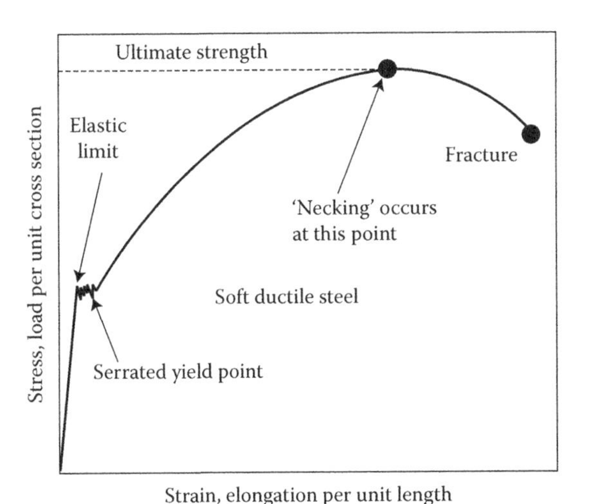

FIGURE 19.5
Stress–strain curve from tensile testing.

where L_f = gauge length at fracture and L_0 = original gauge length.

In EN specifications this parameter is also identified as 'A' (Figure 19.6a)

d. The *percentage reduction of area*, shows how much the specimen has necked or reduced in diameter at the point of failure where

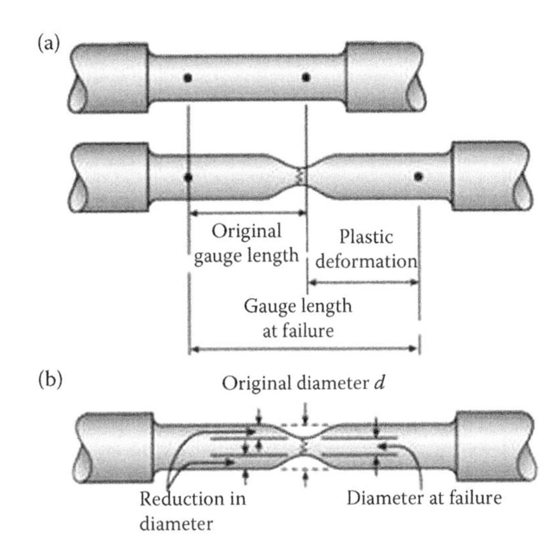

FIGURE 19.6
(a) Elongation (b) fracture of specimen in tensile test.

$$\text{Reduction of } A\% = (A_0 - A_f/A_0) \times 100 \qquad (19.6)$$

where A_f = cross-sectional area at site of the fracture.

In EN specifications this parameter is also identified as 'Z', (Figure 19.6b). (a) and (b) are measures of the strength of the material whereas (c) and (d) indicate the *ductility* or ability of the material to deform without fracture.

The slope of the elastic portion of the curve, essentially a straight line, will give *Young's Modulus of Elasticity* (E), a measure of how much a structure will elastically deform when loaded.

A low modulus indicates flexibility of structure whereas high modulus its stiffness.

The stress strain curve in Figure 19.7 shows a material that has a well pronounced YP but only annealed CS exhibits this sort of behaviour. Metals that are strengthened by alloying, by heat treatment or by cold working do not have a pronounced YP. In such cases YPs determined by measuring the *proof stress* (*offset yield strength* in American terminology), the stress required to produce a small specified amount of plastic deformation in the test piece.

The proof stress is measured by drawing a line parallel to the elastic portion of the stress/strain curve at a specified strain, this strain being a% of the original gauge length, hence *0.2% proof, 1% proof* (Figure 19.6).

For example, 0.2% proof strength would be measured using 0.2 mm of permanent deformation in a specimen with a gauge length of 100 mm. Proof strength is therefore not a fixed material characteristic, such as the YP, but will depend on how much plastic deformation is specified. It is essential therefore when considering proof strengths that the % figure is always quoted. Most

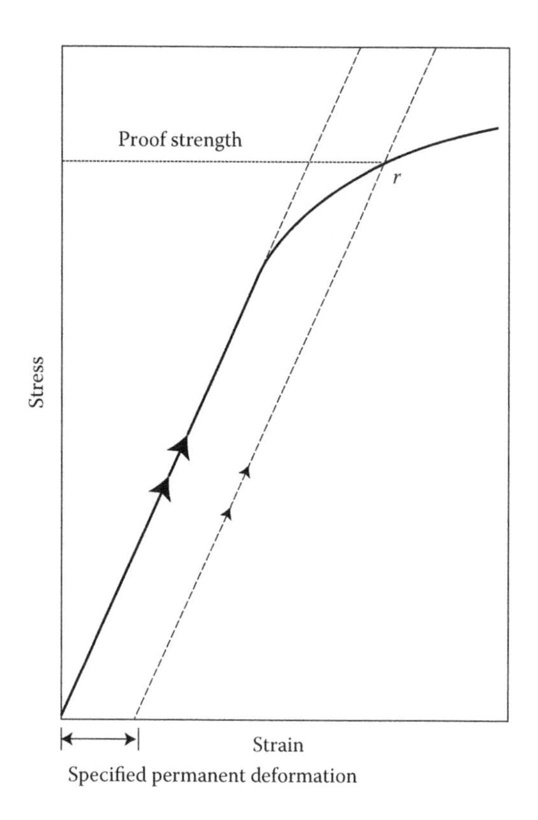

FIGURE 19.7
Proof stress.

steel specifications use 0.2% deformation, $R_{P0.2}$ in the EN specifications.

Some materials such as annealed copper, grey iron and plastics do not have a straight line elastic portion on the stress/strain curve. Then the usual practice, analogous to the method of determining proof strength, is to define the 'yield strength' as the stress to produce a specified amount of permanent deformation.

19.11.2 Non-Destructive Testing/Examination

NDE/NDT is a method of *inspecting without in any way destroying the utility of the item.*

In boiler manufacture, erection and maintenance NDT techniques are constantly in use for the inspection of the integrity of

- Raw materials, namely PPs, castings, forgings and structures
- Welds

There are basically four types of NDT techniques for inspection of various different aspects.

It can be generalised that NDT can helps identifying the following:

- **Casting** defects
- **Welding** defects
- Crack detection (internal or emergent)
- Damage (de-lamination) in composite materials
- Material heterogeneity

19.11.2.1 Liquid Penetrant Testing

LPT is used for *locating defects of non-porous materials which are open to the surface.*

In LPT a highly fluid, red dye is sprayed on the surface of the joint and allowed to soak in any open surface defect by gravity and capillary action. The surface is then wiped clean and a white developer with a powder consistency is applied. The red dye bleeds back out of the defect highlighting the flaw. The method is typically used on completed welds.

LPT and MPT are somewhat interchangeable tests in certain circumstances. Owing to the problems associated with additional surface preparations and the time involved with LPT, it is recommended that MPT is applied whenever possible. However, there may be situations where, because of geometry or restricted access, MPT cannot be performed. LPT is an allowable option, keeping in mind, that additional surface preparation may be necessary.

The main advantages of LPT are as follows:

- It is extremely simple and versatile. Both equipment and testing are simple.
- Testing can be done on metal, glass and ceramics.
- Its main use, however, is on austenitic ss material and welds which are non-magnetic and not amenable to magnetic particle testing.

The limitation is that the imperfections of only the surface can be detected and not beneath.

Figure 19.8 shows the detection of a surface crack in liquid penetrant test (LPT).

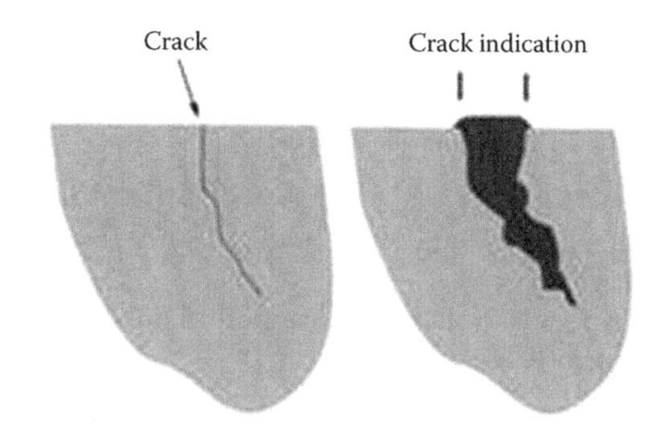

FIGURE 19.8
Liquid penetrant test.

19.11.2.2 Magnetic Particle Testing

MPT can be used *for surface and near-surface linear flaw detection* limited to a depth of <3 mm.

The instrument consists of an electro-magnetic yoke which sets up a magnetic flux field around a weld. A very fine magnetic powder dust is applied to the area being tested. As the flux lines cross a linear defect, the field is interrupted and the powder aligns with the defect. Spurious indications are sometimes encountered along the areas of poor weld bead contour, undercut or overlap. The use of a white background paint to improve contrast can improve the reliability of this method.

A key use of this method is during air-arc gouging to determine if a crack has been totally removed. Root-pass testing is also commonly done with MPT.

The main advantage of MPT is the simplicity of equipment, its portability and ease of use. However, it cannot be used on a metal that cannot be magnetised such as austenitic ss. It is used in boiler manufacturing commonly in defect location in

- Welds of fusion-welded panels
- Butt joints in plates of <30 mm
- Nozzle welds
- Internal and external attachments
- Drum welds

Figure 19.9 shows the manner in which the magnetic flux is disrupted by a crack in the material.

19.11.2.3 Ultrasonic Testing

UT is employed for

- Detecting *surface and subsurface imperfections*
- Also for *measuring thickness*

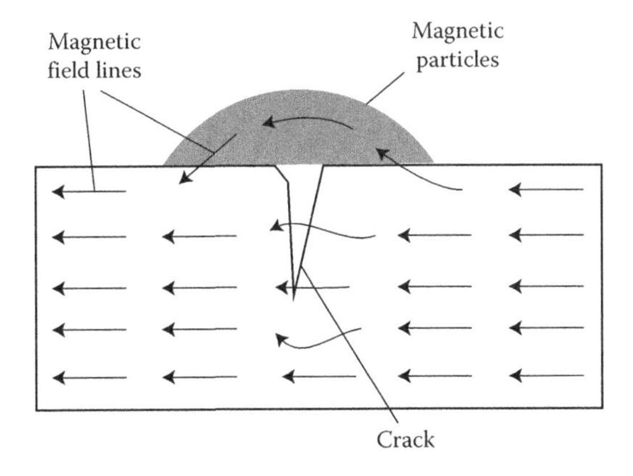

FIGURE 19.9
Magnetic particle testing.

UT involves sending sound waves in ultrasonic frequency into a weldment or parent material. Any reflector within the weld or parent metal sends back a reflected signal to the instrument. The sent and received signals are presented on an oscilloscope for interpretation. Unlike in RT, MPT and LPT, the interpretation of the received signal in UT is highly dependent on the skill and training of the technician. The location and depth of the flaw can be accurately determined. The shape and type can also be interpreted to some degree by competent operators. The scanning surfaces must be clean and free from fireproofing, upset metal and weld spatter for proper transducer contact.

Code requires that the entire area to be scanned by shear wave for weld flaw detection be first scanned by longitudinal wave to detect any lamellar defects. These defects can mask indications from the weld areas, if present, and are not favourably oriented for shear wave testing. UT is highly sensitive to planar defects if they are favourably oriented to the sound beam.

- UT can be done on metals, ceramics and plastics.
- Permanent records can also be generated.
- The equipment has moderately high initial cost.
- Unlike in other tests it needs highly trained and certified operators for operating the equipment and interpreting the results.

UT is routinely done on plates, particularly of high thicknesses, to verify compliance with the stated physical properties from the mill.

Figure 19.10 depicts the principle of UT.

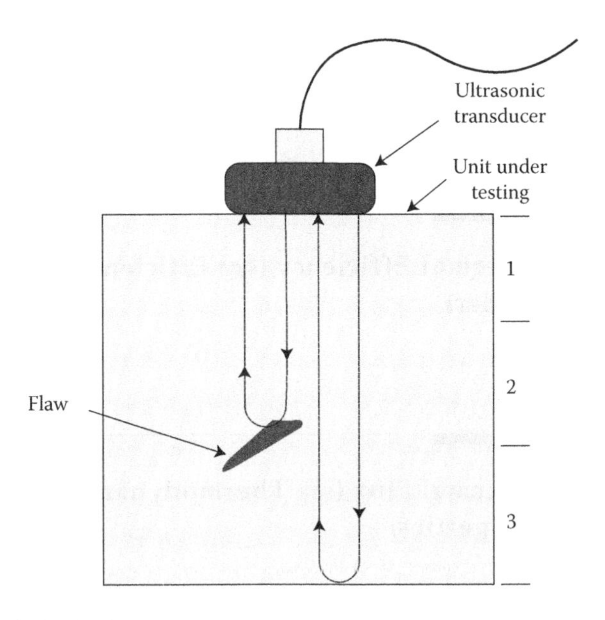

FIGURE 19.10
Ultrasonic testing.

19.11.2.4 Radiographic Testing

RT is specified for internal flaw detection. By definition *radiography is the shadow captured on a film produced by passing x-rays or γ-rays through an object.*

RT procedure consists of using an x-ray or γ-ray source to expose a film similar to that used in medical applications. The most common shop and field technique uses an *Iridium 192* source for γ-rays on one side of the member being inspected and a film cassette on the opposite side. An exposure is made and the film developed much the same way as photographic negatives are produced. Areas of different film density relate to flaws in the weldment.

RT is sensitive to cracks, lack of fusion, lack of penetration, slag inclusions and porosity defects. It is rather insensitive to lamellar-type defects perpendicular to the path of radiation. It does produce a permanent film record. Owing to its two-dimensional capability, it gives limited information about the depth of the defect or the angular orientation of a crack.

The big advantages of RT are

- Its versatility in being usable on metals and variety of other surfaces
- The permanent record of the discontinuities the test produces

On the contrary,

- The initial and running costs are high because of power source.
- Radiation hazards to personnel are always present.
- RT needs a separate area and at times a different timing for conducting the tests from radiation consideration.

19.12 Thermal Efficiency (*see* Efficiency of Boiler)

19.13 Thermal Flux (*see* Thermodynamic Properties)

19.14 Thermal Resistivity (*see* Thermal Conductivity *under* Thermodynamic Properties)

19.15 Thermodynamic Cycles

Thermodynamic cycle is *a process in which a series of operations on a working fluid altering its pressure, temperature and volume are performed by supplying heat. Work output results and certain heat rejection takes place, after which the working fluid returns to its original state.*

There are five practical thermodynamic cycles in operation.

1. Steam power or Rankine cycle
2. GT cycle or Brayton cycle
3. Internal combustion engine or Otto cycle
4. Compression ignition or diesel cycle
5. Refrigeration cycle

Of these, the first two, namely the steam and GT cycles, are of relevance to boiler engineers in steam and gas-based power plants. There are additionally the following two variations.

a. Cogen cycle
b. Combined cycle

Cogen cycle, which has the highest thermal η of ~80–85%, is increasingly employed where there is use of steam or heat which is produced along with power.

CC, which is a dual cycle of gas and steam (Brayton and Rankine) cycles, is employed for high η gas-based power generation.

Carnot cycle, which finds mention often, is only an ideal cycle.

Carnot cycle or ideal vapour cycle.

Carnot's principle, enunciated by Nicolas Leonard Sadi Carnot, states that heat transfer from a heat reservoir is proportional to its temperature.

Carnot cycle is an idealised gas cycle which results in the highest amount of work and maximum η from an engine working in a thermodynamically reverse manner. It is also called the *ideal vapour cycle.*

There are four reversible processes in this cycle, two isothermal and two adiabatic (isentropic), between the source and sink, as shown in Figure 19.11, on T–s and p–V diagrams.

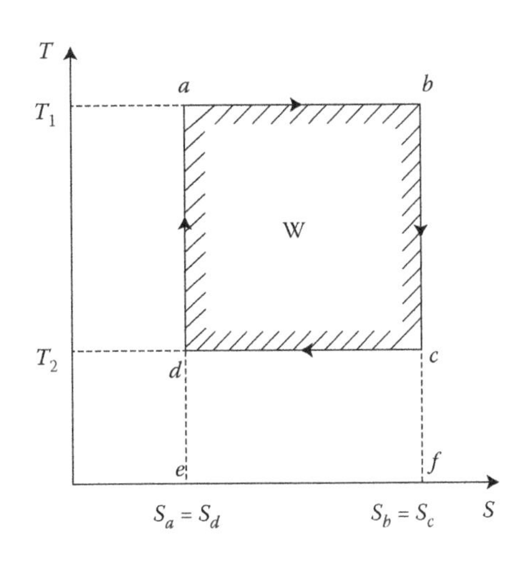

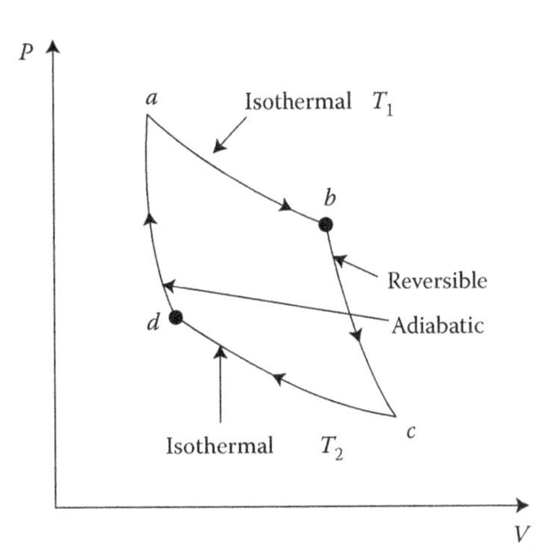

FIGURE 19.11
Carnot cycle on T–s and p–V diagrams.

No engine operating between temperatures T_1 and T_2 can be more efficient than Carnot heat engine. Carnot heat engine is only hypothetical as it is not possible to build real engines with thermodynamically reversible fluids.

$$\text{Carnot } \eta = \frac{T_1 - T_2}{T_1} \times 100\% \tag{19.7}$$

$$= \left(1 - \frac{T_2}{T_1}\right) \times 100\% \tag{19.8}$$

where T_1 and T_2 are the absolute temperatures of the source and sink, respectively.

The importance of Carnot cycle is that it establishes the maximum possible η of an engine cycle between two temperatures, which can be used as a bench mark for comparing the other cycles.

19.15.1 Rankine or Steam Cycle

Rankine or steam cycle is the practical Carnot cycle that uses water/steam as the working fluid to produce work from the heat of fuel. A total of 80% of all power produced world over is by using steam cycle.

The two isentropic (adiabatic) processes of Carnot cycle are retained and instead of the two isobaric (constant pressure) processes Rankine cycle adopts isothermal processes, as shown in the T–s chart in Figure 19.12. BFP pressurises water instead of gas.

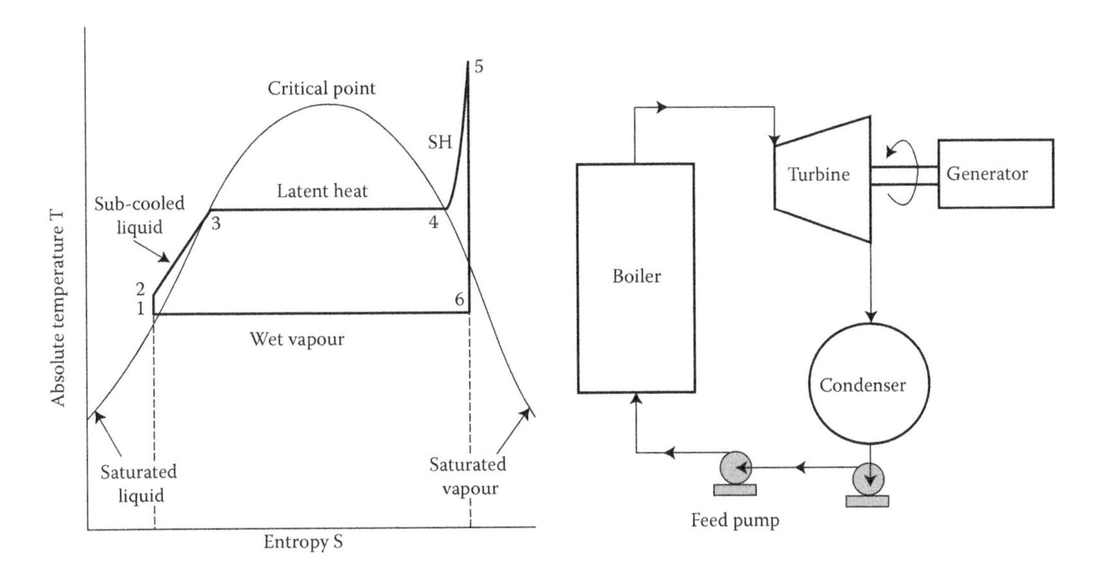

FIGURE 19.12
Rankine cycle presentation on T–s chart and the cycle components.

In a power plant boiler converts the heat of fuel to produce steam at a certain pressure from condensed water. The steam so produced is expanded to vacuum levels in an **ST** to produce mechanical power. Condenser then condenses the exhaust steam to water which is pumped to the required pressure by the BFP. Common heat source is the combustion of fossil fuel, heat of waste fuel, heat from hot gases and nuclear fission. This, essentially, is the steam/Rankine cycle in a power plant.

$$\text{Cycle } \eta = \text{total heat of steam} - \text{heat loss in condenser} \quad (19.9)$$

Total heat of steam = **area under** *1234561* (19.10)

area under 12345

19.15.2 Brayton, Joule, Open or Simple Gas Turbine Cycle

Named after the American developer George Brayton (1830–1892), the Brayton or Joule cycle works on air/gas as the working fluid and captures the working of a GT.

It consists of two constant pressure (isobaric) processes interspersed with two reversible adiabatic (isentropic) processes. *Heat rejection is to the atmosphere and not to condenser* as in Rankine cycle.

The thermal η is solely dependent on the compression ratio.

GT essentially has three components:

- Gas compressor
- Combustion chamber
- Expansion turbine

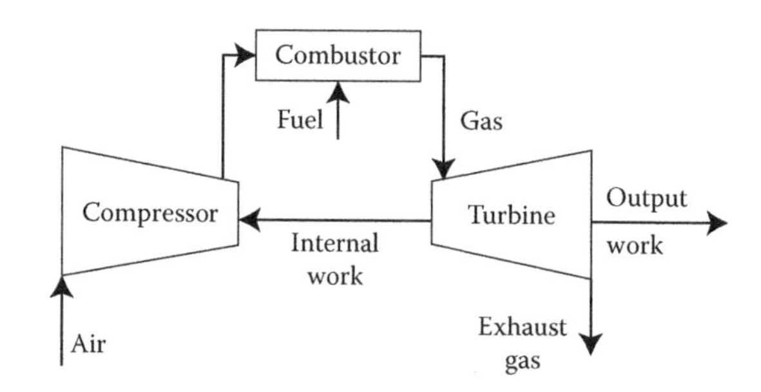

FIGURE 19.13
Main components and working streams of a GT, Brayton or open cycle.

This is shown in Figure 19.13.

Ideal Brayton or GT cycle is depicted in Figure 19.14 both on p–V and T–s charts.

Brayton cycle is an open cycle with hot gases exhausted to atmosphere. The cycle η is quite low. Open cycle GT operation is only carried on if the gas prices are very low in comparison to installing a Hrsg and adopting CC mode.

19.15.2.1 Cogeneration Cycle

Cogeneration, as the name indicates, is the *generation of two forms of energy* from a thermodynamic cycle, usually power and steam/heat.

In many industrial processes, instead of condensing the exhaust steam of ST in the condenser, it is possible to use the LP steam for process heating, thereby reducing the heat loss to minimum. Cogen is therefore very efficient as the heat lost to condenser, which forms ~50% of fuel heat, is substantially utilised. From ~30–35% thermal η of steam cycle it improves to ~80–85% in a cogen cycle.

Cogen is practical only when there is need for process steam or heat. Whereas many processes such as sugar

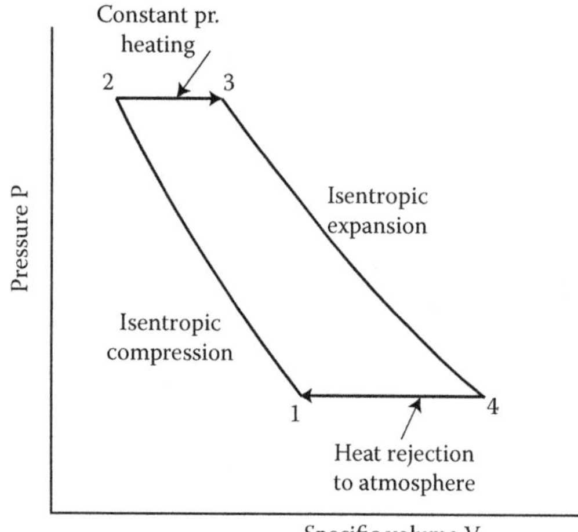

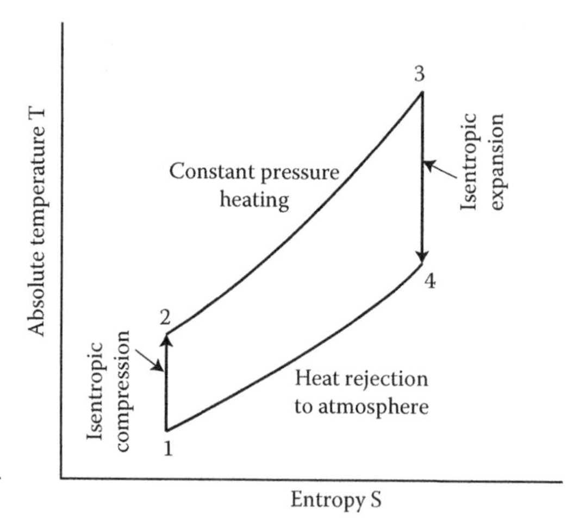

FIGURE 19.14
Ideal Brayton or GT cycle on both p–V and T–s charts.

and paper making have been historically benefitting from cogen, many other plants like cement and mining simply cannot adopt cogen as there is no need for steam in the plant.

GTs used in the process and heating are capable of cogen.

19.15.2.2 Combined Cycle

When more than one thermodynamic cycle is at work, it is termed as CC.

Heat engines generally discard more than 50% of input heat to the condenser or atmosphere. When this heat, which is otherwise wasted, is used by another engine, it is termed as CC operation.

GT cycles

GTs are used in the following three modes:

1. Simple/open cycle (OC)
2. Cogen cycle
3. Combined cycle

1. Simple/open cycle

It is the same as Brayton cycle. When *GT* is operated on its own with exhaust gases discharge to atmosphere without any heat recovery, it is termed simple or OC. The exhaust gases carry away 35–40% of the fuel heat. This mode of operation is highly inefficient with only 28–39% gross η. OC operation makes sense only if the fuel costs are very low as it is the case in the Middle East. Figure 20.13 shows a typical simple cycle.

2. Cogen cycle

In cogen two useful forms of energy are produced—usually power and steam.

The exhaust gases from the **GT** are cooled in *Hrsg* to produce steam. This steam can be for process or for power. Many times as the steam generated from only the exhaust gases may be insufficient for power or process, additional fuel is fired as supplementary fuel, usually to the extent of 100%, such that the plant is never short of power or steam. The η levels in a cogen cycle can exceed 80% with steam generated at two or three pressure levels in the Hrsg. A cogen cycle is shown in Figure 19.15 where heat from fuel is extracted to produce electrical power, steam and hot water.

3. Combined cycle

Power is produced both by GT and ST. In this mode GT/Brayton cycle rides over steam/Rankine cycle to deliver overall efficiencies of 45–60% on Ncv. Figure 19.16 shows the CC in a diagrammatic manner with the two cycles plotted on T–s diagrams.

19.16 Thermodynamic Processes

A thermodynamic process is any process which *changes the state of the working fluid*. It is the *energetic evolution of a thermodynamic system from an initial state to final state.* These processes can be classified by the nature of the state change that takes place. Common types of thermodynamic processes include the following:

- A *reversible process* is an ideal process where the working fluid returns to its original state by conducting the original process in the reverse direction. For a process to be reversible, it must be able to occur in precisely the reverse order. All energy that was transformed or distributed during the original process must be capable of being returned to its exact original form, amount and location. Reversible processes do not occur in real life.
- An *irreversible process* is any process which is not reversible. All real-life processes, such as the basic steam cycle, are irreversible.

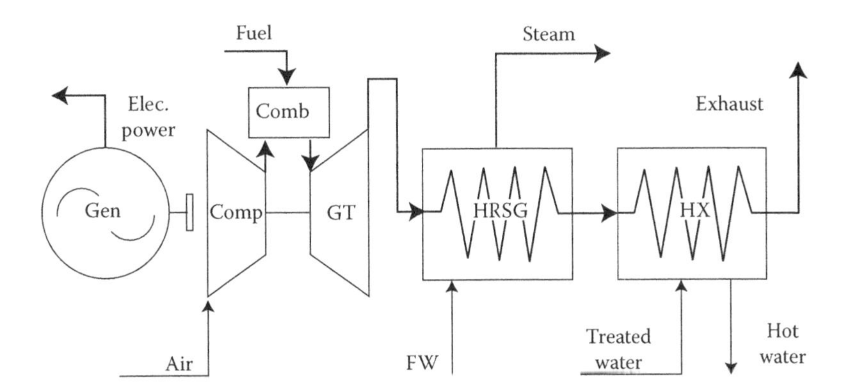

FIGURE 19.15
GT-based cogen cycle generating power, steam and hot water.

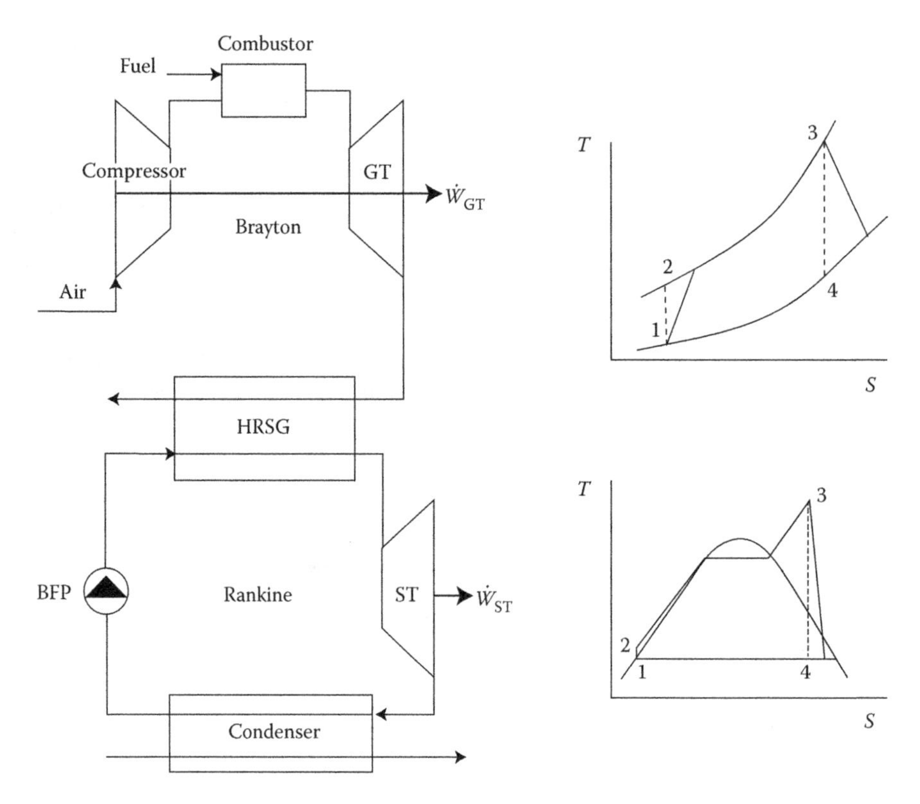

FIGURE 19.16
Combined cycle of GT and ST.

- An *adiabatic or isentropic or constant entropy process* is a state change where there is no transfer of heat to or from the system during the process. Because heat transfer is relatively slow, any rapidly performed process can approach to being adiabatic. Compression and expansion of working fluids are frequently achieved adiabatically with pumps and STs.
- An *isothermal or constant temperature process* is a state change in which no temperature change occurs. Note that heat transfer can occur without causing a change in temperature of the working fluid.
- An *isobaric or constant pressure process* is a state change in which the pressure of the working fluid is constant throughout the change. An isobaric state change occurs in the boiler SH, as the heat of the exiting steam is increased without increasing its associated pressure.
- An *isochoric or constant volume process* is a state change in which volume of working fluid remains constant throughout the change. As there is no change in volume no work is performed.

These processes can be visualised in the diagrams of Carnot and Brayton cycles provided in thermodynamic cycles.

19.17 Thermodynamic Properties

Stated in the simplest terms *thermodynamics is the study of* the *inter-relation between energy, heat and work of a system.* Conversion of energy from one form to another, the direction in which heat will flow and the availability of energy to do work are all examined.

The following thermodynamic properties of interest to boiler engineers are covered.

1. Calorific value (cv) or heating value (hv) of fuel
2. Energy (E)
3. Enthalpy (H)
4. Entropy (s or φ)
5. Heat or heat energy or thermal energy
6. Heat flow (q) equation
7. Heat flux or thermal flux
8. Heat of evaporation or latent heat
9. Heat rate
10. Heat of transformation or latent heat
11. Laws of steam and water
12. Laws of thermodynamics
13. Mass velocity

14. Molecular weight/mole/molar volume
15. Pressure (p, P)
16. Temperature (t, T)
17. Thermal conductivity (k)

19.17.1 Calorific Value or Heating Value of Fuel

cv or hv is the *amount of heat released by unit fuel*. It is the heat content of fuel and expressed in MJ/kg, kcal/kg or Btu/lb.

Values of cv vary depending on whether the fuel is

a. As delivered
b. As fired
c. Dry or
d. Dry ash-free (daf)

There are two values for the cv of any fuel:

- Gross cv (Gcv) or Higher cv (Hcv)
- Net cv (Ncv) or Lower cv (Lcv)

The former is popular in the USA and the UK while the latter in Europe.

Gcv is the total *heat released* during combustion of a unit fuel, whereas Ncv is the net *heat available* which is the total heat (Gcv) less than the heat in water vapour.

Gcv is determined by the bomb calorimeter while Ncv is calculated from the Gcv by the following formula:

$$\text{Ncv} = \text{Gcv} - (m + 9\text{H}_2) \times \text{latent heat} \qquad (19.11)$$

where latent heat is used variously from ~2.42–2.45 MJ/kg, 577–586 kcal/kg or 1045–1055 Btu/lb.

Formation of water vapour is inevitable during any combustion because of the

a. Fuel H_2
b. Fuel M

Water vapour so formed stays in flue gas in SH condition at a partial pressure of 0.07–0.08 kg/cm^2 and leaves a boiler without transferring its latent heat. This heat in water vapour is, therefore, inevitably lost to the atmosphere and not available for use. This is captured by the Ncv/Lcv.

Ncv is needed in boiler calculations while Gcv for assessing fuel required to be fired.

Gcv is determined in the laboratory for solid fuels by burning unit quantity of ground fuel in a submerged bomb calorimeter and noting the rise in the temperature of the fixed quantity of water. Gcv of the fuel is the heat picked up by this water. See Bomb Calorimeter for full details.

19.17.2 Energy

Energy is the *capacity of a physical system to perform work*.

Energy exists in several forms such as heat, light, kinetic or mechanical energy, potential energy, electrical energy, or other forms.

According to the law of conservation of energy, the total energy of a system remains constant, though energy may transform into another form. Energy cannot be created or destroyed.

The SI unit of energy is the Joule (J) or Newton-meter (Nm). The joule is also the SI unit of work.

In mks system the unit is kg force m and in Imperial units it is ft-lb force.

19.17.3 Enthalpy

Enthalpy *is the sum of the internal energy of the system and the system's volume multiplied by the pressure exerted by the system on its surroundings.*

$$H = E + p \times v \qquad (19.12)$$

Its value is determined by the temperature, pressure and composition of the system at any given time.

Enthalpy was referred to earlier as total heat or heat content, but this definition is inadequate.

According to the law of conservation of energy the change in internal energy is equal to the heat transferred to the system minus the work done by the system. If the only work done is a change of volume at constant pressure, the enthalpy change is exactly equal to the heat transferred to the system.

If heat is given off during a transformation from one state to another, then the final state will have lower heat content than the initial state, the enthalpy change ΔH will be negative, and the process is said to be *exothermic*.

If heat is absorbed during the transformation, then the final state will have a higher heat content, ΔH will be positive, and the process is said to be *endothermic*.

SI units for enthalpy are kJ/kg while in mks units it s kcal/kg and in fps units it is Btu/lb.

19.17.4 Entropy

Entropy is a property of a substance, such as p, t, v and so on. Changes in entropy can be determined by knowing the initial and final conditions of a substance. It quantifies the *energy of a substance that is no longer available to perform useful work*. Entropy is sometimes referred to as a *measure of the inability to do work for a given heat transferred*.

Specific entropy given in steam tables is entropy per unit mass, S/m.

Mollier chart/diagram, named after Richard Mollier (1863–1935) is a graphical representation of the thermodynamic properties and states of materials involving enthalpy on one of the coordinates. They are routinely used in the designing work associated with power plants, STs, compressors and so on, to visualise the working cycles of thermodynamic systems. A typical water–steam enthalpy–entropy (H–s) diagram or Mollier chart is depicted in Figures 19.17 and 19.18.

19.17.5 Heat or Heat Energy or Thermal Energy

Heat is energy unlike temperature. *Energy is the capacity to work.*

It is the total amount of energy possessed by the molecules in matter comprising both potential and kinetic energies.

When a substance is heated, there is a

- Rise in temperature without change of state or
- Change of state without change of temperature

Unit of heat is joule (J), kilocalorie (kcal) or British thermal unit (Btu).

19.17.6 Heat Flow Equation

The general equation for heat flow is

$$q = UA\Delta t \tag{19.13}$$

$$\text{Or} = \frac{A\Delta t}{R} \tag{19.14}$$

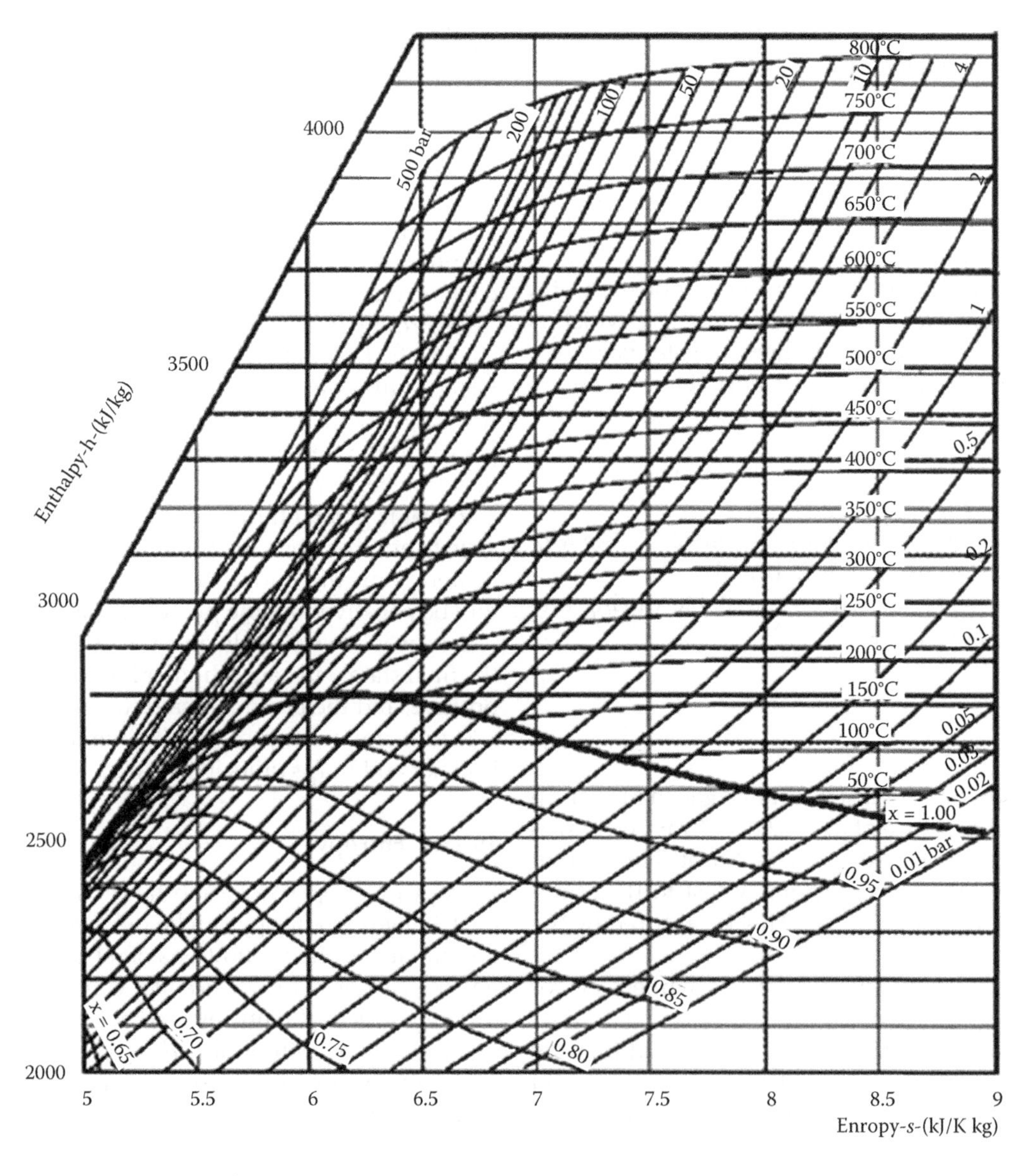

FIGURE 19.17
Mollier or enthalpy–entropy diagram for steam in SI units.

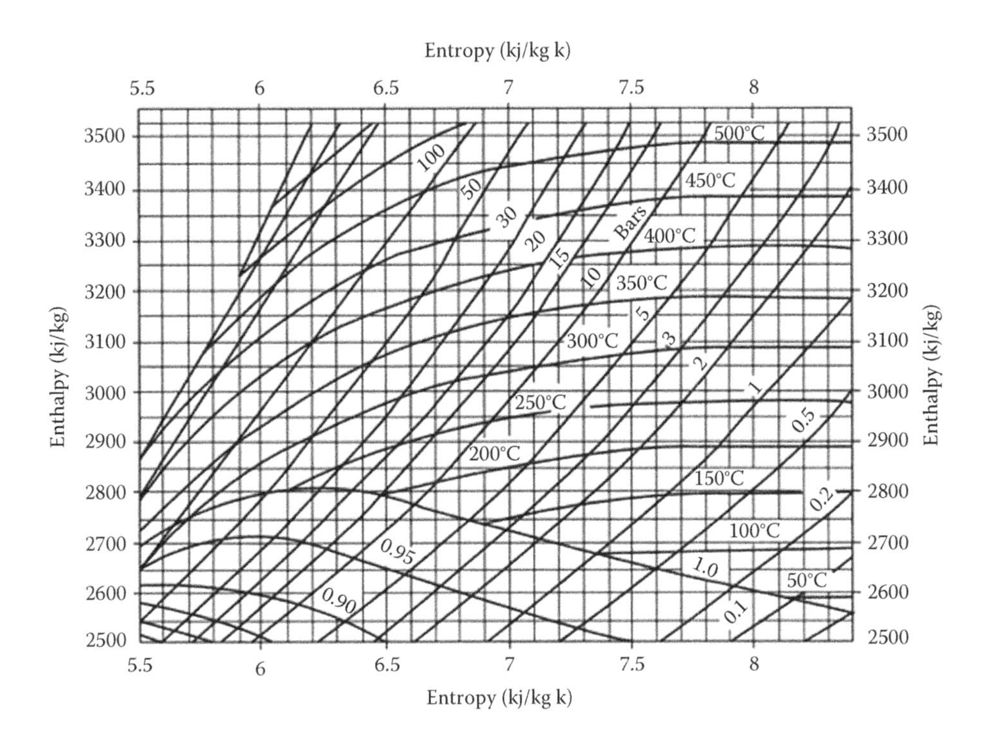

FIGURE 19.18
Part of Mollier diagram which is of relevance to boiler engineers.

where

- q = rate of heat flow in watts or kilocalories per hour or Btu/h
- U = Over all heat transfer coefficient in W/m² °C, kcal/m² h °C, or Btu/ft² h °F
- A = Surface area in m² or ft²
- Δt = Temperature difference causing flow in °C or °F
- R = 1/U, the overall combined resistance

19.17.7 Heat Flux or Thermal Flux

Heat flux is the flow of energy per unit time per unit area expressed in w/m², kcal/m² s or Btu/ft² h

$$\text{Heat flux} = q/A \tag{19.15}$$

where

- q = rate of heat flow in W or kcal/h or Btu/h
- A = Surface area in m² or ft²

19.17.8 Heat of Evaporation, or Latent Heat (see Properties of Steam and Water)

19.17.9 Heat Rate

Heat rate is the *conversion* η *from heat to electrical energy.* It is applicable to ST, GT and even a total power plant. Its unit is kJ/kwh, kcal/kwh or Btu/kwh. It is computed by dividing the total heat input (fuel fired/h × cv) by the electrical power output in kwh.

It is of profound significance as the ηs of different power generation schemes can be compared or η of power plant at different conditions can be studied.

Heat rate can be stated in terms of gross or net cv. Likewise it can be evaluated on gross basis or net of auxiliary power consumption.

19.17.10 Heat of Transformation or Latent Heat (see Properties of Steam and Water)

19.17.11 Laws of Steam and Water

Water is incompressible and obeys all laws of liquids.

Dry steam is a compressible vapour and obeys all the laws of gases.

Wet steam is a mixture of water and steam and follows the laws of compressible fluids in two phase flow.

19.17.12 Laws of Thermodynamics

There are three main laws of thermodynamics which are profound in their influence. Stating them in a very simple manner

1. The first law states that whenever energy is converted from one form to another, the total *quantity of energy remains the same.* This is the *law of conservation of energy.*

   Work output cannot be higher than the energy input.

2. The second law states that, in a closed system, the entropy of the system does not decrease.

   In the transformation between energy and heat there is always some loss.

3. The third law states that as a system approaches absolute zero, further extraction of energy gets more and more difficult, eventually becoming theoretically impossible.

Although one can try to approach absolute zero as closely as one desires, this limit cannot actually be reached.

19.17.13 Mass Velocity

Mass velocity is the weight flow rate of a fluid divided by the cross-sectional area of the enclosing chamber or conduit expressed in kg/s m^2 or lb/(h ft^2).

Instead of volume flow mass flow rate is considered. This is very useful in sizing of pipes and tubes in heat transfer banks and calculating the pressure drops.

19.17.14 Molecular Weight/Mole/Molar Volume

MW is the *sum of the atomic masses of all the atoms in a molecule.* It is also called *formula weight.* It is actually the molecular mass.

For a combustion engineer the mass of a substance in kg/lb equal to its MW is the kg-mole/lb-mole of the substance. kg-mole is referred to as kmole while lb-mole is simply mole.

Molar volume is the volume occupied by one mole of ideal gas. For all compositions of gas this volume is constant for a given p & t.

1 kg mole occupies 22.4 m^3 at NTP (0°C and 760 mm wg/1.01 bar) while 1 lb-mole occupies 394 cft at 80°F and 30″ wg/14.7 psia or 359 *t* at NTP (32°F).

19.17.15 Pressure

Pressure (p) is the force per unit area applied in a direction perpendicular to the surface of an object. It is the resultant effect on a surface when a *force* is applied.

$$p = \frac{F}{A} \tag{19.16}$$

The SI unit for pressure is the Pascal (Pa), equal to one Newton per m^2 (N/m^2 or kg/m s^2).

In mks and fps systems, the units are kgf/m^2 and lbf/in^2.

It is also expressed in bars which is 10^5 Pa.

Pressure is measured either by

a. Taking local atmospheric pressure as zero or
b. Taking absolute vacuum as zero

Atmospheric or barometric pressure is the pressure exerted by the atmospheric air on the earth surface because of its own weight. It is the height of a column of mercury (Hg) in a barometer that can be balanced at sea level. It equals 760 mm Hg, 1.01 bar, 14.7 psia.

1 atm = 101.325 kN/m^2 = 101.325 kPa = 14.7 psia = 0 psig = 29.92 in Hg = 760 torr/mm Hg = 33.95 ft. water = 10.33 m of water = 407.2 in wg = 2116.8 lb/ft^2.

Gauge pressure is the pressure of fluid measured with respect to the local atmospheric pressure. If the pressure is above atmospheric pressure, it is called +ve pressure and if it is below, it is vacuum or –ve pressure.

Absolute pressure is the pressure of a fluid measured from complete vacuum or –760 mm Hg.

Figure 19.19 explains the various pressures.

Fluid pressure can be static, dynamic and total/stagnation pressures.

Static pressure is the pressure experienced uniformly by fluid in motion or rest.

Dynamic pressure is the additional pressure experienced when the surface faces the direction of fluid flow as the moving or kinetic energy of the fluid is brought to rest.

Total pressure or stagnation pressure is the sum of static and dynamic pressures and comes into reckoning only when the surface faces the flowing fluid

19.17.16 Temperature

- Temperature is not energy.
- It is a number related to the average kinetic energy of the molecules of a substance.

Unit of temperatures is °K, °C or °F in SI, mks and fps systems respectively.

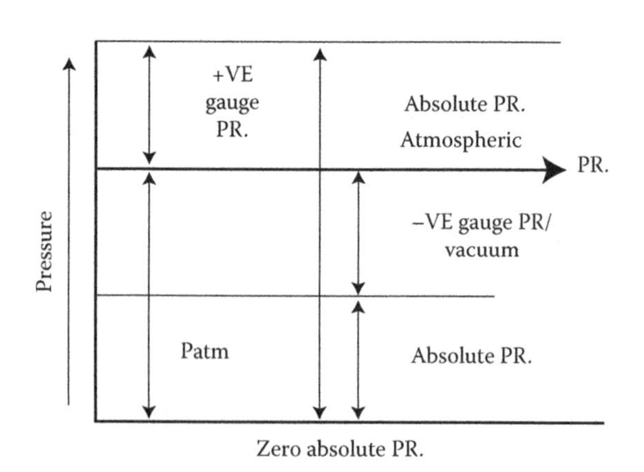

FIGURE 19.19
Designating pressures.

19.17.17 Thermal Conductivity

k is the rate of flow of heat flux caused by unit temperature difference through unit thickness.

It is the intrinsic property of a material to conduct heat.

It is the quantity of heat conducted by a body of unit surface area normal to the direction of heat flow through unit thickness due to unit differential temperature under steady-state conditions.

$$k = \frac{q/a}{t_1 - t_2/l} \quad (19.17)$$

where

q = Rate of heat flow in watts or kcal per hour or Btu/h
A = Surface area perpendicular to the heat flow in m^2 or ft^2
l = Thickness of body in m/cm or ft/in
$t_1 - t_2$ = Temperature gradient in °C or °F
k = Thermal conductivity in W/m/°C, kcal/m^2/°C/cm, Btu/ft^2/in or Btu/ft^2 h/°F/ft

Thermal resistivity is the inverse of thermal conductivity.

Interesting aspects of thermal conductivity (k) are

- Metals have high conductivity and are good conductors.
- Refractory and insulating metals have low conductivity and are bad conductors.
- k for pure metals decreases with the rise in impurities.
- With rise in temperature, k of
    - Gases increase the most
    - Liquids reduce except for water
    - Insulating material increases
    - Alloys may increase or decrease
    - Most pure metals decrease except aluminium and uranium
- Variation of k with temperature is almost linear for most materials

19.18 Thermal Energy (*see* Thermodynamic Properties)

19.19 Thermocouple (*see* Instruments for Measurements)

19.20 Thermometer (*see* Instruments for Measurements)

19.21 Throttling (*see* Properties of Steam and Water)

19.22 Throttling Calorimeter (*see* Steam Quality in Instruments for Measurement)

19.23 Tie Bars/Tie Channels (*see also* Furnace Pressure in Furnace *under* Heating Surfaces)

The rectangular furnace chamber is held in position every 2-4 m in its height with the help of **buckstay** beams which run on all four sides. Buckstays help in retaining the furnace shape by containing the internal forces of the draught and also minor pulsations. In modern boilers, they act against the furnace implosions also. Buckstays are exposed to the ambient and are 'cold'. But furnace is hot and insulated.

The strip or channel rigidly connected to the furnace tubes is the tie bar/channel. Tie bar is hot and expands and contracts along with furnace. But the buckstay beam does not have an appreciable thermal expansion or contraction. The two are connected by means of several buckstay clips placed along their lengths which permit the differential expansion and transfer the outward forces from furnace to buckstay beams (see Figure 8.7).

19.24 Top-Supported Boiler (*see* Boiler-Supporting Systems in Structure)

19.25 Top-Suspension Girders (*see* Structure)

19.26 Top Suspension Hangers (*see* Structure)

19.27 Total Dissolved Solids in Water (TDS)

TDS is the combined content of all inorganic and organic substances contained in water that goes through a sieve of 2 μ in filtration process. Depending on the source of water the common chemical constituents of TDS are Ca, K, Na and their phosphates, nitrates and chlorides, which are found in nutrient runoff, general stormwater runoff and runoff from snowy climates where road de-icing salts are applied. More exotic and harmful elements of TDS are pesticides arising from surface runoff. Certain naturally occurring TDS arise from the weathering and dissolution of rocks and soils.

19.28 Toughness (*see* Metal Properties)

19.29 Tower-Type Boilers (*see* Utility Boilers)

19.30 Transition Flow (*see* Fluid Flow Types)

19.31 Transmissivity (*see* Radiation Heat *under* Heat Transfer)

19.32 Travelling Grate (*see* Grates, Travelling)

19.33 Tube Expansion (*see* Manufacturing Topics)

19.34 Tube Swaging (*see* Manufacturing Topics)

19.35 Tubes (*see* Materials of Construction)

19.36 Turbines, Gas

Gas or combustion turbines are complete power plants and not mere prime movers such as *STs*.

The topic of GT is covered under the following sub-topics:

1. Main components of a GT
2. Types of GTs
   a. Industrial/land turbines or frame machines
   b. Aero derivatives
3. Definitions of GT terms
4. GT and ST
5. GT performance
   a. Steam and water injection
   b. Evaporative cooling
   c. Regenerative/recuperative cooling

19.36.1 Main Components of a GT

GT consists of four sub-assemblies, namely

- *Air compressor* to produce HP air and feed to the combustor
- *Combustor* to burn the fuel with hot air to produce hot gases and feed the turbine
- *Turbine* to expand hot gases to generate mechanical power
- *Gen* to produce electrical power from the mechanical energy

The entire GT power plant is attractively skid mounted, at least in small and medium sizes. Air compressor in the GT consumes ~2/3 of the generated power. Aside of the main GT, the following auxiliary or BOP are needed to make the GT power plant complete.

1. Inlet air cleaning system
2. FO or gas conditioning system
3. Lubricating system
4. Start-up system
5. Control system

GTs are of interest to boiler engineers because the heat in the exhaust gases of GTs is recovered in **Hrsgs** to produce steam for power or process. GT in **simple/open *cycle*** mode is not particularly efficient as can be seen from Table 19.1.

TABLE 19.1

Performance Parameters of GT on Open/Simple Cycle Mode

Parameter	Value
OC η	28–39% in land turbines
	<45% in aero derivatives
GT exit gas	425–610°C
Temperature	800–1130°F
Excess air	220–300%
O_2 in flue gas	15–16% by volume

TABLE 19.2

Comparison of Aero Derivative and Land Turbines/Frame Machines

Parameter	Aeroderivative GT	Land GT
H	35–45%	28–39%
Compression ratio	18–30	10–18
Firing temperature	Higher	Lower
Gas exhaust temperature	415–540°C	490–615°C
Fuels	Gas, LFOs, Naphtha	Gas, LFOs, HFOs and crude oils
Maximum size MWe	~100	~350
Speed	Higher	
Suitability	Cogen and OC	Cogen and CC
Build quality	Slightly delicate	Rugged
Start-up time	Quicker	Quick
Weight	Less	More
Plan area	Smaller	
Noise	Lower	
Cost	Higher	

19.36.2 Types of GTs

There are basically two types of GTs—those that have been developed primarily for industrial duty and GTs of jet planes that are adapted for land duties, which are called, respectively, as

- Industrial/land turbines or frame machines
- Aero derivatives

19.36.2.1 Industrial/Land Turbines

Industrial and land turbines are built mainly for power and cogen applications. They are heavier, more rugged and have longer life. They can use LFOs as well as HFOs and even crude oils. The materials of construction are less exotic and less expensive. The size to which they can be built is much higher to ~350 MWe. They are comparatively lower in η and hence exhaust flue gases at higher temperatures. Naturally, the Hrsgs behind land machines produce more steam. In other words, compared with the aero derivatives land GT extracts less heat from fuel and leaving Hrsgs to extract more heat.

19.36.2.2 Aero Derivative Gas Turbines

In the early 1960s, when the GTs were used in industry and power plants mainly for peaking and emergency duties, it is the GTs used in the aerospace industry that were suitably modified for land application. In the last 50 years, there have been improvements on all counts including making large aero derivative GTs up to 100 MWe capacities. But essentially they continue to remain efficient and expensive machines made of special high-temperature materials. Because they operate at higher speeds they are lighter, more compact and somewhat delicate. They derive their higher η by employing higher firing temperatures and compression ratios. They fire NG other manufactured gases and light FOs. As they are more efficient, they extract more from the fuel as mechanical energy and hence are better suited when power is to be maximised. Aero derivatives extract more heat and discharge cooler gases that produce less from Hrsgs in sharp contrast to the land machines which perform the reverse. Table 19.2 compares both the types of GTs.

19.36.3 Definitions of Gas Turbine Terms

Various terms used in GTs and GT-based power generation are defined below.

- It is the industry practice to state GT performance for base load operation.
- *Base rating* of a GT is the power output at the base load under ISO conditions of 15°C (59°F) and 0 m MSL (sea level) site conditions firing NG excluding parasitic load of auxiliaries and gear box.
- Power output is (a) gross plant output for OC and (b) net plant output for CC
- Net output accounts for
    - Parasitic auxiliary plant loads
    - Losses in gear box
    - Intake losses
    - Outlet losses
- *Heat rate* is the fuel heat input *on NCV* basis for unit power output expressed in kcal/kwh, Btu/kwh or kJ/kwh.
- In GTs it is the Ncv that is used as a standard for heat rates unlike in conventional power plants where both Gcv and Ncv are prevalent.
- Power and heat rate are measured *across Gen* terminals for power plants.

- *H* is the ratio of power output to heat input expressed in%.
- *Compression ratio or pressure ratio* is the ratio of the discharge pressure of the GT compressor to the atmospheric pressure of air at the GT intake.
- *Flow* is the air flow through the GT expressed in tph, kg/s or lb/s.
- *Turbine speed* is the design speed of GT output shaft. 3000 and 3600 rpm are the speeds for power generation at 50 and 60 Hz, respectively.
- *Exhaust temperature* is the gas temperature at the outlet of the GT exhaust duct.

19.36.4 Gas Turbine and Steam Turbine

Although GT and ST are both big rotating machines sharing many mechanical design features, the GT is a power plant while the ST is a prime mover and they are not of the same family.

- Just as the ST expands the high p & t steam from higher to lower levels of pressure to produce electrical energy, GT expands high p & t gases to do the same.
- The input to ST is ultra-clean steam at temperatures <620°C while GT input is clean or slightly fouling gases at very high temperatures of up to 1450°C.
- Very important practical differences between the two are that the
- STs are custom built while the GTs come in certain standard sizes.
- GTs are first manufactured and then sold while STs are first sold and then manufactured.
- Lead times for procurement are thus much lower for GTs than either ST or Hrsg.

Shown in Figure 19.20 is the cross section of a typical small industrial GT. HP side of GT drives the compressor and the LP side of GT drives the electrical Gen or the drive load.

A number of GT models are in the market covering a wide range of <1 to ~340 MWe. The lower range up to 20 MWe is usually for small power generation and big drives for mainly oil and gas industry as well as for large ships. For gas-based power generation up to 20 MWe a battery of gas engines in many cases makes an economical choice. Tables 19.3 and 19.4 give the salient data of the aero derivatives and frame machines used in CC plants and hence are of relevance to Hrsg.

It is to be noted that

- These are ISO ratings (at 15°C, 60% RH and MSL)
- The data is with gas firing
- Without steam and water injection which are usually adopted for performance enhancement
- The heat rate is based on Ncv
- An inlet and outlet gas pressure loss of 100 mm wg is considered

The models listed are for 50 Hz. The figures are taken from *Gas Turbine World* 2009 GTW Handbook (2009). The latest figures may be at a slight variance from Tables 19.3 and 19.4 as the GT models are continuously under up gradation. It is worth noting the difference in the compression ratios and exhaust temperatures between the aero derivatives and frame machines. GT26 is the only frame machine which has compression ratios matching those of aero derivatives. Dry low NO_x (DLN) is the short form of dry low NO_x combustor. It is the normal performance of the GT with no additional measures. Water injection in the case of aero derivative GTs is a method of performance boosting and lowering of NO_x as explained below.

19.36.5 Gas Turbine Performance

GT is a volumetric machine quite similar to a fan. Its performance is highly influenced by the ambient

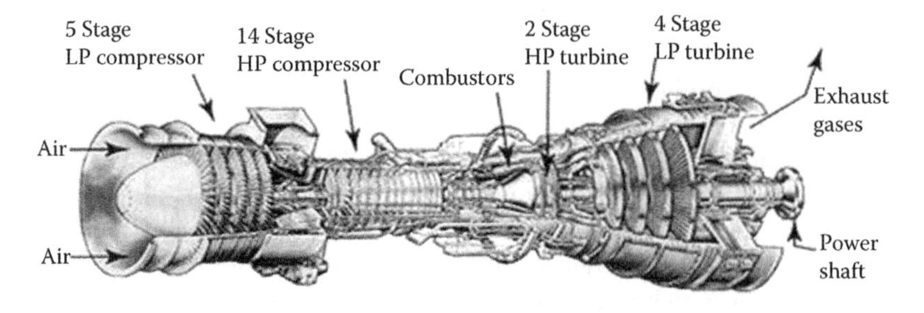

FIGURE 19.20
Cross-sectional view of a small industrial-type GT.

TABLE 19.3

Salient Data of Notable Aero Derivative GTs of 50 Hz

Make Model	Power, MWe	GT Eff, η%	Heat Rate kcal/kwh	Heat Rate Btu/kwh	Compression Ratio	Exhaust Gas Flow lb/s	Exhaust Gas Flow tph	Exhaust Gas Temperature F	Exhaust Gas Temperature C	Comments
GE										
LM2500 PE	23.09	34.0	2527	10027	18.0	157.8	257.5	963	518	Water injection
LM2500 PK	29.316	35.4	2426	9629	18	197.0	321.5	911	489	Water injection
LM 2500 + RC	36.024	36.8	2334	9263	23.0	213.0	347.6	945	507	Water injection
LM 6000 PC	43.339	39.8	2160	8571	29.2	284.7	464.6	803	429	
LM6000 PC Sprint	50.836	40.2	2136	8478	32.3	300.1	489.7	835	446	Water injection
LMS 100 PA	102.995	43.2	1981	7894	41.0	469.9	766.9	765	408	Water injection
RR										
RB211-GT61DLE	32.12	39.3	2187	8680	21.5	208	339.5	938	503	4850 rpm
Trent-60 WLE	64.0	41.1	2093	8299	38.5	365	595.7	767	408	Water injection
P &W										
FT8Pwr pack	25.49	38.1	2255	8950	19.3	187	305.2	855	457	

TABLE 19.4

Salient Features of Some of the Current Land Turbines of 50 Hz

Make Model	Power, MWe	GT η%	Heat Rate kcal/kwh	Heat Rate Btu/kwh	Compression Ratio	Exhaust Gas Flow lb/s	Exhaust Gas Flow tph	Exhaust Gas Temperature F	Exhaust Gas Temperature C	Comments and Former Model no
Alstom										
GT8 C2	56.3	33.9	2540	10080	17.6	434	708	946	508	
GT11 N2	113.6	33.3	2582	10247	16.0	882	1440	977	525	DLN combustor
GT13 E2	179.9	36.9	2330	9247	16.4	1243	2029	950	510	DLN combustor
GT26	288.3	38.1	2257	8956	33.9	1433	2339	11,411	616	
GE										
PG 6591C	45.4	36.50	2347	9315	19.6	269	439	1078	581	
PG 6111FA	77.1	35.5	2459	9760	15.6	467	762	1117	602	
PG 9171E	126.1	33.8	2545	10,100	12.6	922	1505	1009	542	
PG9351F	255.6	36.9	2331	9250	17.0	1413	2306	1116	602	
PG 9371FB	279.2	37.9	2272	9016	18.3	1444	2357	1164	630	For CC only
Siemens										
SGT-900	49.5	32.7	2633	10450	15.3	386	630	957	514	50/60 Hz, W251B11/12
SGT5-2000E	168	34.7	2476	9825	11.4	1170	1909	998	537	V 94.2
SGT6-5000F	202	38.1	2181	8955	17.4	1120	1828	1073	579	W501F
SGT5-4000F	292	39.8	2159	8567	18.2	1526	2490	1071	518	V94.3A
SGT5-8000H	340				19.2	1808	2951	1157	625	
MHI										
M701DA	144.1	34.8	2472	9810	14.0	972	1586	1008	542	
M701F4	312.1	39.9	2188	8683	18.0	1549	2528	1106	597	
M701G2	334	39.5	2175	8630	21.0	1625	2652	1089	587	

conditions, namely the surrounding temperature and altitude. It is also affected by the inlet and outlet pressure losses. Table 19.5 summarises the effects.

There are several ways of enhancing GT performance of which the following are popular. A good understanding of these techniques is needed as they have a direct relation to the Hrsg design.

TABLE 19.5
Effects of Ambient Conditions on the Performance of GTs

Parameter	SI/Metric Units		British Units	
Ambient temperature rise by	1°C	Lowers power by 0.5–0.9%	1°F	Lowers power by 0.3–0.5%
Site altitude rises by	100 m	Lowers power by 1–1.7%	1000 ft	Lowers power by 3–5%
Inlet loss increases by	25 mm wg/245 Pa	Lowers power by 0.5%	1″ wg	Lowers power by 0.5%
		Increases heat rate by 0.1%		Increases heat rate by 0.1%
Outlet loss increases by	25 mm wg/245 Pa	Lowers power by 0.15%	1″ wg	Lowers power by 0.15%

19.36.5.1 *Steam and Water Injection*

Steam or water is injected into the compressed air in combustion chamber of GT prior to the combustion so that the

- GT output is increased
- Fuel NO_x generation is lowered

The added gas flow increases the power output from turbine with the compressor input power remaining the same. It is important to note that with

- Water injection the heat rate increases while
- Steam injection the heat rate reduces

This is because the steam adds to the heat while water takes away the heat for its evaporation.

Steam for injection is drawn from the Hrsg. This is known as *steam-injected GT (STIG)* cycle or *Cheng* cycle. There is no other equipment involved in this cycle other than GT and Hrsg. This is popular in the cogen arrangement where the desired power can be produced during low-steam demand and power enhanced for peaking. NO_x is also reduced as the combustion temperature is lowered.

When more steam or water is added beyond what is required for NO_x control it is called *power boost* mode.

Steam injection is done up to 2 kg/kg of fuel. Aero derivatives are more used for power augmentation by steam injection. Figure 19.21 describes the STIG cycle.

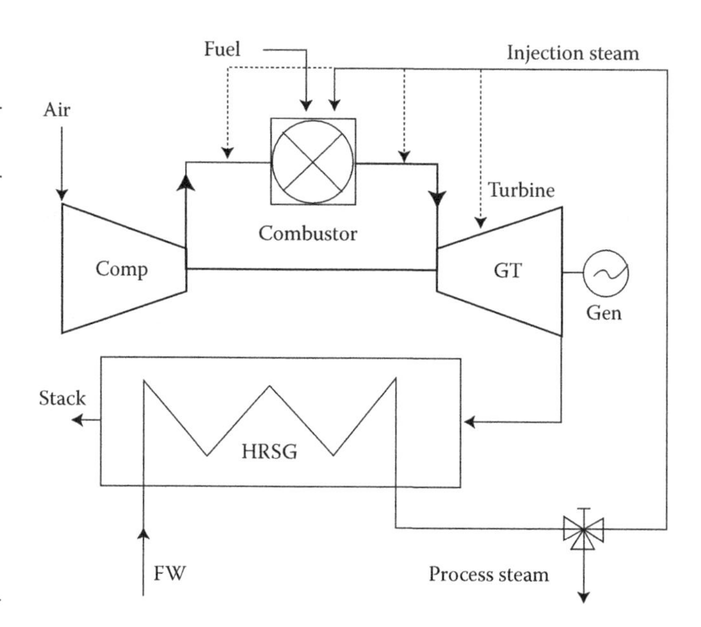

FIGURE 19.21
Steam-injected GT (STIG) cycle.

19.36.5.2 *Evaporative Cooling*

In hot and dry places it is possible to cool and humidify the inlet air to the extent of 90% of difference between the dry and wet bulb temperatures. Increased weight of entering air, because of the reduced temperature and enhanced humidity, increases the power output of the GT. The water quality needs to be high as any contamination affects the performance of the GT badly. Evaporative cooling cycle is shown in Figure 19.22.

19.36.5.3 *Regenerative/Recuperative Cooling*

To improve the η of OC, the exhaust gases are cooled by the combustion air in a counterflow heat exchanger, thereby transferring part of the heat to the air entering the combustion chamber. Between 20% and 25% of the exhaust heat can be recovered this way. Figure 19.23 portrays the regenerative cooling of exhaust gases.

19.37 Turbines, Steam

ST is a rotating machine and a *prime mover* that extracts thermal energy from pressurised steam by the gradual change of momentum and converts it into rotary motion. It is a heat engine in which steam expands theoretically in an isentropic or constant entropy process. Depending on the application and construction the isentropic η varies from 20% to 90% where isentropic or internal η is defined as the actual heat drop over theoretical isentropic heat drop.

This topic of STs is covered under the following sub-topics:

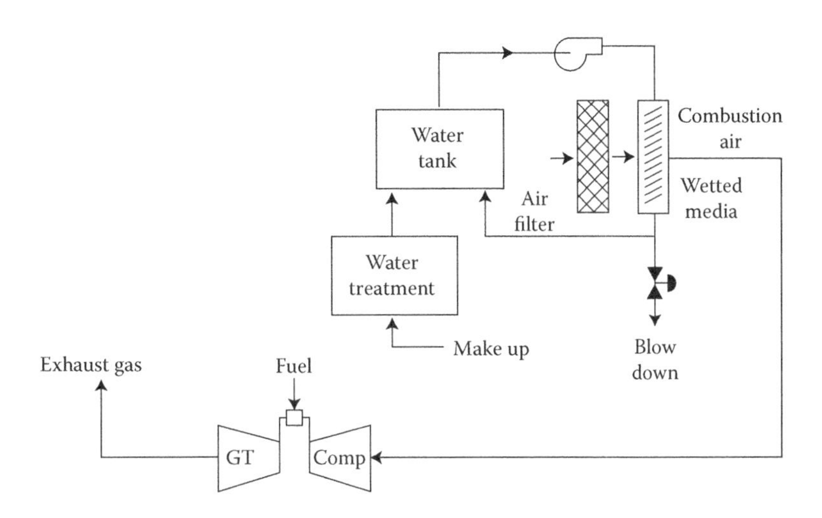

FIGURE 19.22
Evaporative cooling cycle along with regenerative heating of air.

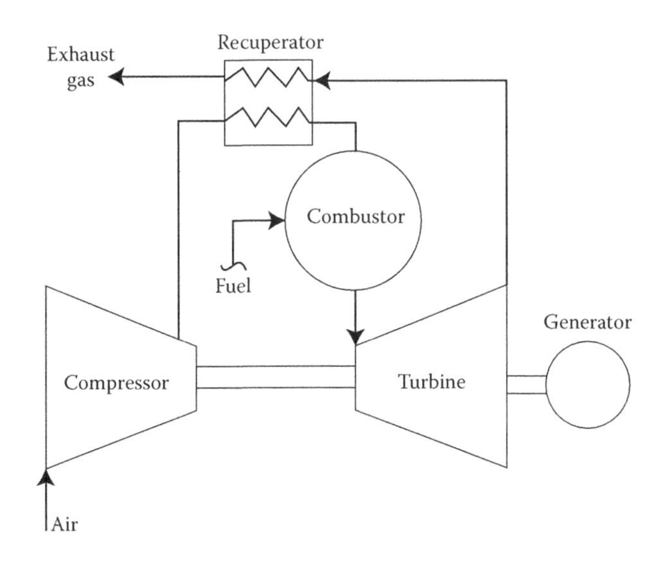

FIGURE 19.23
Regenerative or recuperative cooling of GT exhaust gases.

1. ST classification
2. ST versus GT
3. Range of ST

19.37.1 Steam Turbine Classification

STs can be classified in several ways. On the basis of

- Final use ST can be
  - Drive turbines
  - Power turbines

Drive turbines are single casing with back pressure while the power turbines are single or multi-casing with back pressure or condensing mode.

Figure 19.24 depicts a single-casing back pressure or non-condensing ST for prime mover or small power application while Figure 19.25 shows a single-casing straight condensing ST for industrial power generation.

- Supply and exhaust steam conditions of ST can be
  - Straight condensing for small power generation
  - Back pressure or non-condensing as in drive or process application such as refineries, paper mills, district heating and so on

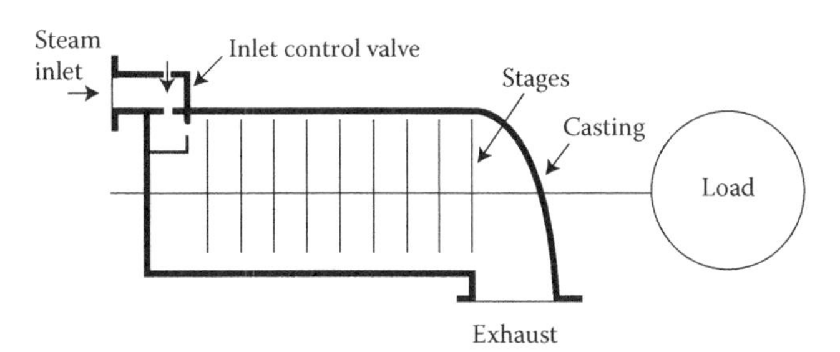

FIGURE 19.24
Single-casing back pressure turbine.

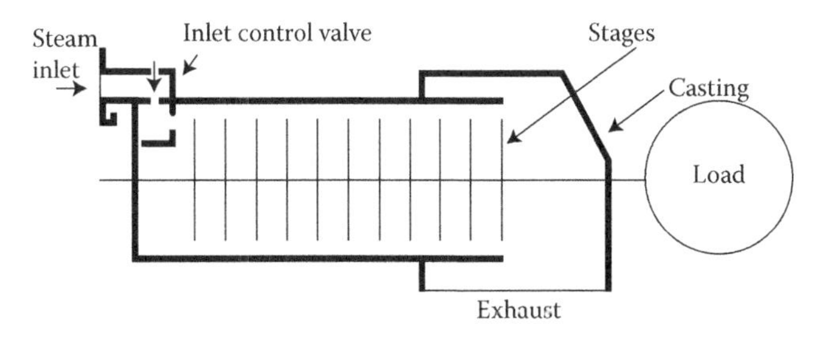

FIGURE 19.25
Single-casing straight condensing turbine.

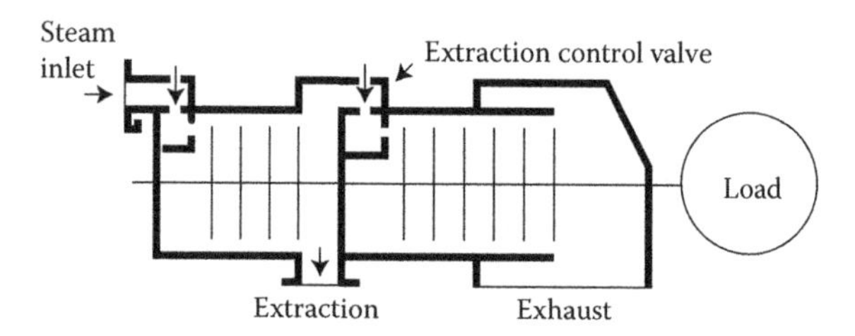

FIGURE 19.26
Extraction condensing process turbine.

- Extraction condensing turbines in power stations
- Injection STs, as in some process plants, where additional steam at a lower pressure is introduced to get more power
- RH STs exclusively for power generation usually for sizes >100 MWe

Figure 19.26 shows a typical extraction condensing process ST for medium-power generation. It is an automatic extraction with controlled pressure maintained by an extraction CV. In plain extraction arrangement there is no CV and the extraction pressure varies with load. Single extraction is shown here but multiple extractions are possible as in the case of large power turbines. Sufficient length of casing should be available to provide the nozzles for extraction.

- Casing and shaft arrangement STs can be
  - Single casing as in most drive, process and small power turbines
  - Tandem compound where two or more casings are connected to a Gen, as shown in Figure 19.27.
  - Cross compound where two or more casings are connected to two or more generators, as shown in Figure 19.28.
- The way steam is expanded STs can be
  - *Impulse* type where steam is fully expanded in the inlet nozzles and passed through the various stages as in most drive turbines. Theoretically, there is no pressure drop in the moving blades and hence there can be large internal clearances and no balance piston needed as there is no axial force.
  - *Reaction* type where steam is expanded throughout the ST in all the moving and fixed rotors. The internal clearances are tight as steam expands even on the moving blades and balancing piston is needed to take up the axial reaction forces generated. Reaction STs are more efficient and also more expensive.

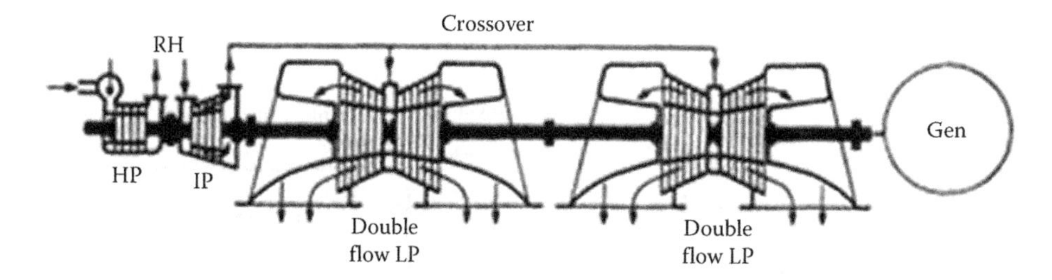

FIGURE 19.27
Tandem compound steam turbine.

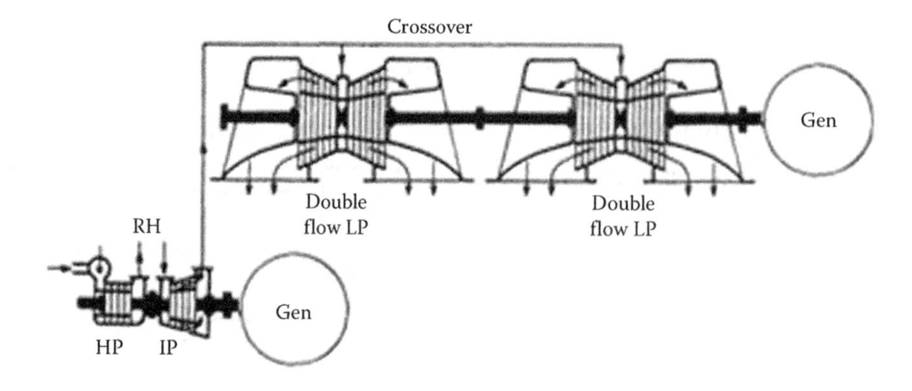

FIGURE 19.28
Cross compound steam turbine.

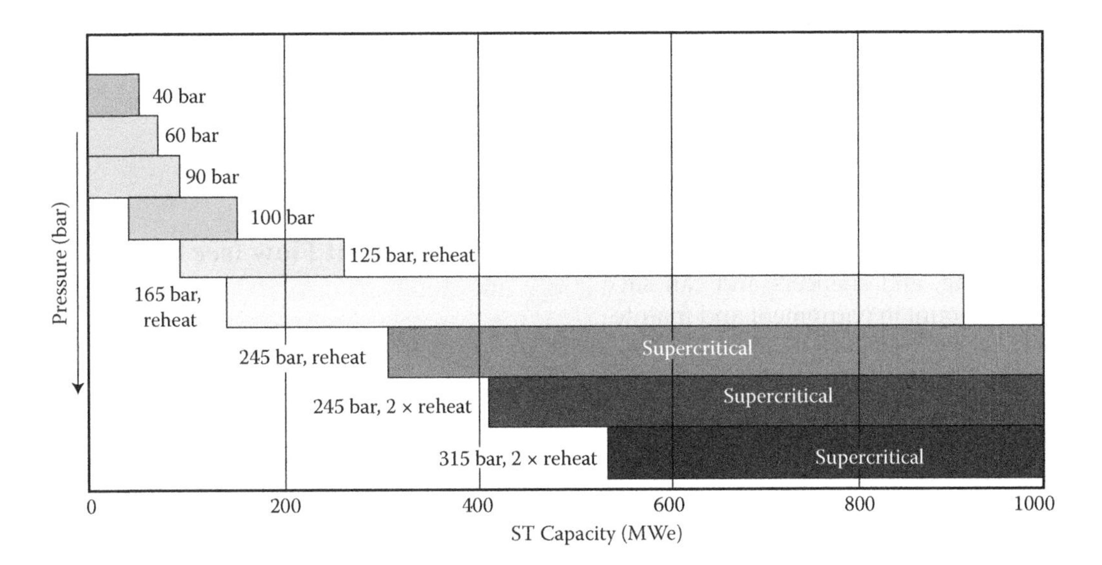

FIGURE 19.29
Typical standard sizes of STs for cogen and utility application.

    - *Impulse–reaction* STs which are most popular and combine the two principles.

- The way steam travels in the casing
    - *Axial flow*—most common arrangement
    - *Radial* flow—for smaller turbines in some countries
    - *Tangential* flow—not popular

19.37.2 Steam Turbine versus Gas Turbine

STs and GTs are very different types of heat engines; the former is a prime mover while the latter is a complete power plant. A more exhaustive comparison is presented in GT vs ST under the topic of GTs.

19.37.3 Range of Steam Turbine

STs can be made in a wide range of sizes from as low as <1 kw to as large as ~1500 MWe. Being fully or substantially factory assembled units, STs are amenable to substantial standardisation. For cogen and utility applications typical sizes of STs for power generation is presented in Figure 19.29. It can be seen that the inlet p & t rise as the ST sizes increase.

19.38 Turbine Bypass Systems

Turbine bypass system essentially decouples the boiler and ST operations permitting the working of boiler independent of ST. This permits ST start-up with zero temperature mismatches. Aside of the flexibility, the other advantages of the bypass are

- Faster start-up and minimised thermal stress

  A good ST bypass system reduces start-up time under cold, warm and hot conditions. It provides a continuous steam flow through the SH and the RH, and allows for higher firing rates which result in quicker boiler warm-up. It also controls SH and RH pressures during the entire start up, keeping thermal transients in the boiler to a minimum. Operating experience shows that power plants equipped with ST bypass systems experience reduced start-up times and much less solid particle erosion of the ST blades, reducing the need for expensive repair and replacement.
- Temperature matching

  ST bypass system allows optimum steam to metal temperature matching for all start-up modes.

  The boiler load can be selected to reach the desired SH and RH conditions for ST start. This results in reduced start-up time and extended life for main components of ST.
- Avoid boiler trip after load rejections

  A fast acting ST bypass system allows boiler operation to continue at an optimal standby load while demand for ST load is re-established after a load rejection. The ST can cover house load requirements. The p & t transients invariably associated with boiler trip and restarts are avoided.

- Eliminate HP SVs

  HP bypass valve sized for 100% MCR capacity can serve as an HP SV when equipped with necessary safe opening devices. This eliminates the need for separate spring-loaded HP SVs, associated piping, and silencers and can save considerable amount in equipment and maintenance costs.

The disadvantages are

- Increased investment
- More complex boiler control
- More maintenance due to additional valves
- Increased plant heat rate due to the higher condenser losses when ST is bypassed

Although ST bypass systems increase the operational flexibility in sub-c boilers it is mandatory for SC systems to prevent furnace tubes from getting overheated during start up and low loads.

Figure 19.30 shows a typical ST bypass system for coal-fired boiler.

Figure 19.31 shows the ST bypass system for CCPP with 3P Hrsg.

19.39 Turbulent Flow (*see* Fluid Flow Types)

19.40 Turndown

Turndown is the ratio of maximum to minimum operating condition in a stable manner. Usually this refers to the flow. Turndown can be for boiler or firing system, such as the burner or mill. It can be a stepless/seamless turndown or stepped turndown.

19.41 Two-Pass Boilers (*see* Utility Boilers)

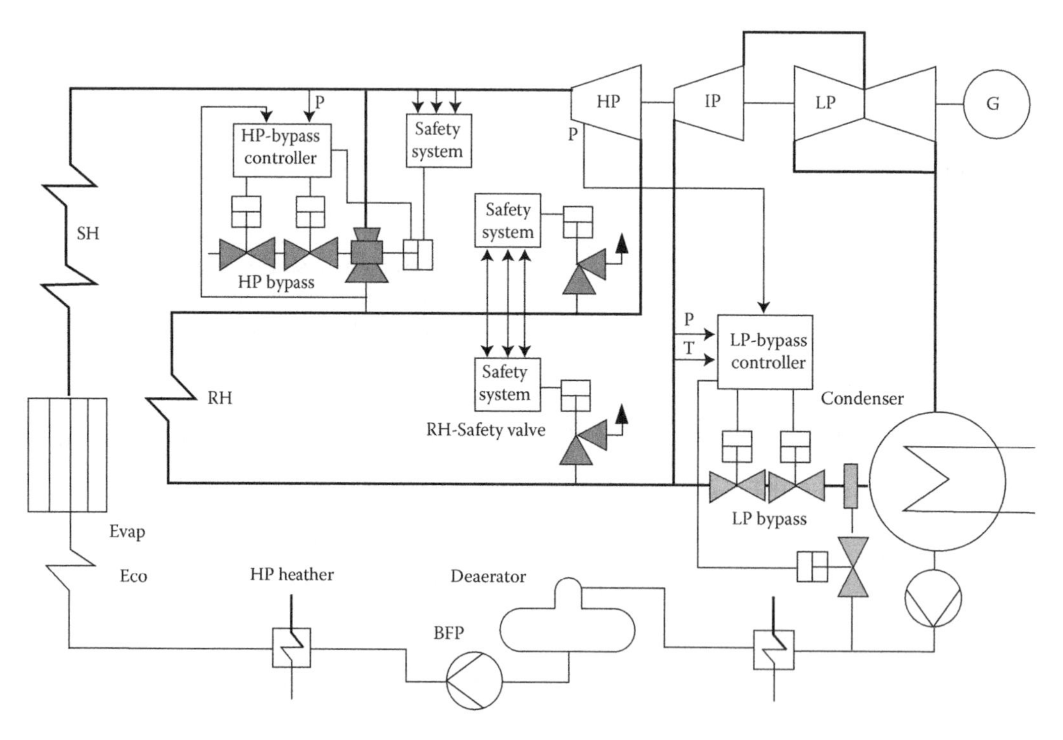

FIGURE 19.30
Turbine bypass system in a utility plant.

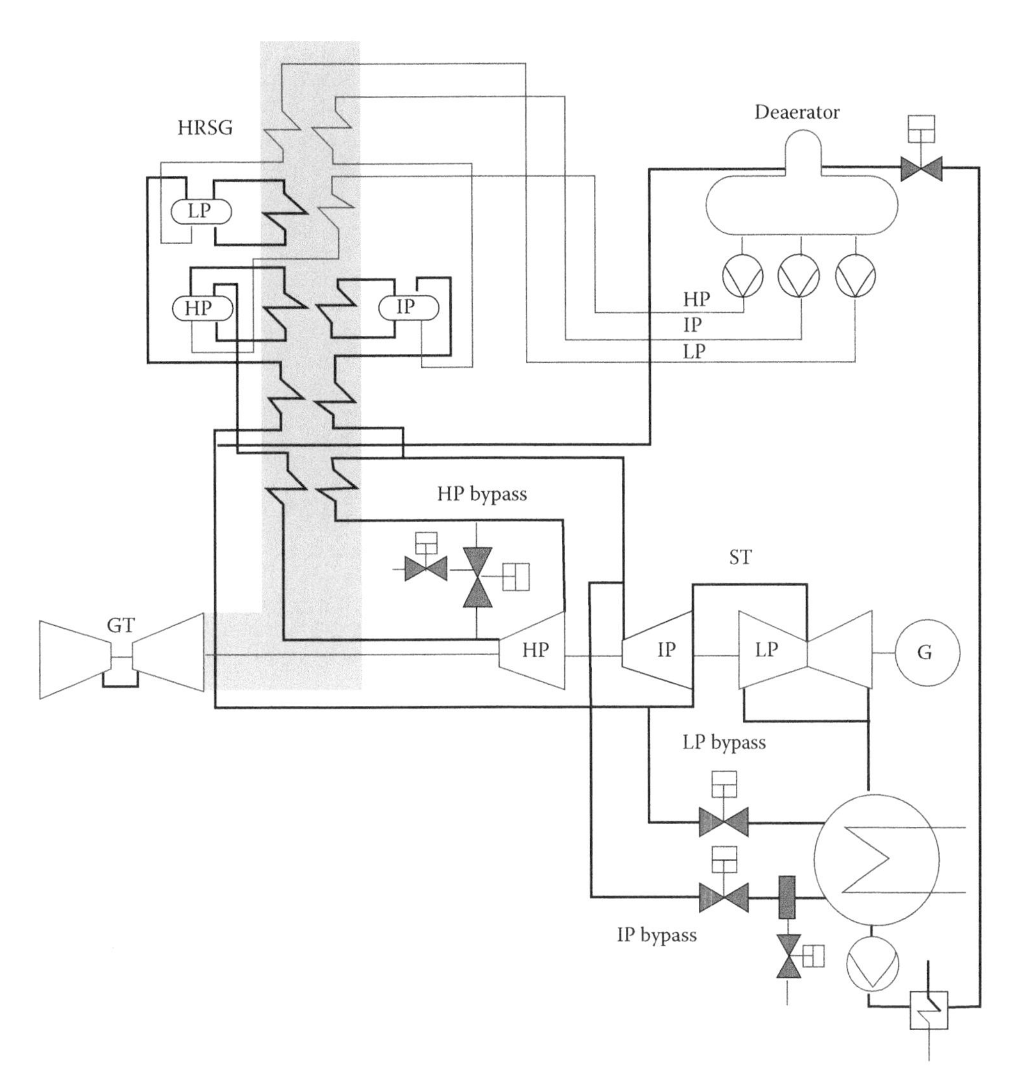

FIGURE 19.31
Turbine bypass systems in a CC power plant.

19.42 Two-Phase Flow (*see also* Circulation and Flow Types in Vertical *and* Horizontal Tubes)

Two phase flow in boilers refers to the flow of steam and water mainly through furnace tubes. Two phase flow takes place only in sub-c boilers in the circulating system where there is a gradual-phase transformation from water to steam upon absorption of latent heat. In other words, water walls, downcomers and risers are the boiler components experiencing the two phase flow. As water walls are subjected to heat, flow through them is more complicated than in the other pipes where there is no incidence of heat.

The challenging requirement is to prevent the onset of DNB in water walls particularly in HP boilers.

20

U

20.1 Ultrasonic Testing (*see* Non-Destructive Testing/Examination *under* Testing of Materials [Metals])

20.2 Utility Boiler (*see also* 1. Boilers, 2. Boiler Classification *and* 3. Industrial Boilers)

Utility boilers are so called as they are employed in power-generating public utilities. Large sizes with high operating p & t and use of prime fuels characterise these boilers.

This topic is elaborated under the following headings:

a. Utility versus industrial boilers
b. Small and large utility boilers
c. Single-pass or tower-type boilers

1. Utility PF boiler designs based on gas flow
   - 1a. Single-pass or tower-type boilers
   - 1b. Two-pass boilers
   - 1c. Down shot- or arch-fired boilers
2. Utility boiler designs based on steam pressure
   - 2a. Sub-critical (sub-c) boilers
   - 2b. SC boilers
     1. Multi-pass high mass flow arrangement (universal pressure boiler)
     2. Spiral tubing with high mass flow
     3. Vertical tubes with low mass flow
   - 2c. Advantages of SC over drum-type boilers
   - 2d. PP materials in SC boilers
   - 2e. Water conditioning for SC boilers
   - 2f. SC Boiler start-up and bypass systems
   - 2g. SC boiler arrangements
3. Utility CFBC boilers

20.2.1 Utility versus Industrial Boilers

The main objective of a utility boiler is to drive a turbo-Gen to feed power to the grid. As the power costs have to be most competitive the utility boilers are built for the largest sizes (as large as 4500 tph and increasing), best η using the highest steaming conditions—**SOPs** (~320 bar) **SOTs/RHOTs** (600/620°C). For the same reason, the **fossil fuels** they use are also the most economical available in the market. However, fuel flexibility is unlikely to be a prime requirement. The build quality and reliability have to be the highest as they have to power the STs practically all year around without interruption but for mandatory shutdowns. In terms of configurations utility boilers come in relatively less variety as compared to the industrial boilers.

Utility boilers have come to mean large coal-fired **PF** and **CFBC** boilers driving STs in Rankine cycle mode as FO and gas are used for generating power in **CCPPs** using GTs. The term 'coal' is used here in its broad sense to include **anthracites, lignite** and **peat** in addition to various **bituminous** and **sub-bituminous coals**.

Utility boilers represent the maximum technical advancement in boilers. The working conditions are the most arduous. The performance expectations are also high and can be summarised as the

- Highest η and the lowest fuel consumption with minimum auxiliary power consumption
- Ability to burn a range of coals with differing fouling characteristics
- Highest availability with minimum down time
- Short start-up time with minimum consumption of auxiliary fuel
- Quick load following capability to keep pace with grid fluctuations
- Least auxiliary fuel consumption for low load stabilisation

Power cycle design is optimised to increase the η of the total plant which reflects in the boiler as

- Increased p & t in SH and RH
- Double RH usually in sizes >800 MWe
- Minimum optimised flue gas exit temperature
- Increased FWT

Industrial boilers have very different conditions to satisfy. They are required to

- Meet the demands of the process industry of which they are a part, by producing both process steam and captive or Cogen power.
- Use fossil as well as waste fuels. Wide fuel flexibility is a very important requirement in most instances.
- Work at optimum η which may not be necessarily the best.
- Generate steam at p & t that is normally lower than the utility boilers; rarely exceeding ~140 bar and 565°C with capacities limited to ~400 tph with no RH.
- Be generally not as rugged and heavy duty as the utility boilers since they do get outages in line with the process duties to avail their maintenance needs.

The separating line between the industrial and utility boilers presently at ~400 tph is arbitrary. As the boiler sizes are increasing, what was a utility boiler some years ago has now become an industrial boiler. One of the main reasons for the relatively smaller size of industrial boilers is the requirement of high availability of the boiler house over fractionally better η or lower cost of each boiler. A battery of smaller boilers is preferred to single large boilers to keep the availability high and also meet reduced loads of process plants effectively. Table 20.1 succinctly captures the difference between the utility and industrial boilers.

20.2.2 Small and Large Utility Boilers

The utility or power boilers can be divide based on size, as

a. Small utility boilers ranging from 100 MWe (400 tph) to 300 MWe (~1000 tph)
b. Medium utility boilers from 300 MWe (~1000 tph) to 600 Mwe (1800 tph)
c. Large utility boilers >600 MWe (1800 tph)

These are arbitrarily dividing lines between small, medium and large utility boilers just like 400 tph is for industrial and utility boilers.

Small utility boilers are usually PF-fired boilers with

- Mostly conventional drum-type design
- Single RH
- Sub-c p & t
- Mostly natural circulation

Large utility boilers, by their size, can accommodate advancements like

- OT flow
- SC steam parameters
- Variable pressure operation
- Double reheat at >800 MWe

Large utility boilers can also be conventional drum-type natural or assisted circulation boilers when they are used for base load operation.

20.2.3 Single-Pass or Tower-Type Boilers

20.2.3.1 Utility PF Boiler Designs Based on Gas Flow

There are basically three types of PF boilers and their variants which are popular in the utility market for coal firing. These are, based on the gas flow

a. Single-pass or tower-type boilers
b. Two-pass boilers
c. Down shot- or arch-fired boilers

1a. Single-pass or tower-type boilers

Tower-type boilers were originally developed to suit the cold northern climate where anti-freezing facility of water in the fully drainable tube banks represented a big convenience, particularly for big boilers.

TABLE 20.1

Comparison of Utility and Industrial Boilers

Utility Boilers	Industrial Boilers
For grid power	For process steam, captive power, Cogen, heating
Generally up to 4500 tph with RH, up to 320 bar/620°C	Generally <400 tph, <150 bar/540°C, No RH,
For prime fuels. Occasionally waste gas, petcoke, etc.	For all types-prime, by-product, waste, manufactured and agro-fuels
Very rarely multi-fuel flexibility is required	Frequently multi-fuel flexibility is required
PF, cyclone and CFBC firing	All types—burner, stoker, pile burn, BFBC, CFBC,PF firing
Single drum and no-drum for once through	Bi-drum and single drum
Top supported	Bottom- and top-supported package and field-erected boilers
Only field-erected boilers	
Natural, assisted, SC and sub-c OT	Mostly natural. Sometimes forced circulation
Boiler dynamics to suit the grid fluctuations	Dynamics to suit process or captive or cogen power plant needs

Drainable SHs are highly beneficial as they

- Can be fully drained and therefore impose no limitation on the inlet gas temperature during start-up; very quick start-up is possible.

  On the other hand, in the non-drainable SHs, the condensed steam residing in the tubes during starting requires gentle heating so that the steam formed finds its way out and not create vapour locking leading to tube overheating. This increases the start-up time.
- Are easy to preserve during long layoffs as **dry preservation** can be adopted. In cold climates, this is beneficial as there is no risk of water freezing with **wet preservation**. **Platen** and **pendant** *SHs* cannot be drained and need wet preservation which is cumbersome.

In tower-type boilers, all the tube banks are

- Horizontal and fully drainable with all of them stacked one above the other
- Having no gas turns over them with minimal gas side erosion possibility

Cooled gases at the Eco exit come out of the first pass to get further cooled in AH (and sometimes in the first stage Eco) in second pass. Tower-type boilers are ~25% taller but occupy ~20% less floor area.

Further, they have the following advantages in addition to the above, namely

- Besides anti-freezing advantage the drainable surfaces permit faster start-up.
- Tower-type boilers are supported on four main columns making the arrangement very clean and elegant with good accessibility all around.

The drawbacks of the tower-type boilers are

- Being taller they are a bit more time consuming and difficult to erect, despite increased accessibility all around.
- **PP** erection generally takes longer as the pressure welds are ~30% more. Also only a single front is available for PP erection unlike two fronts in two-pass boilers.
- The supporting tubes and the support elements are in the gas path exposing them to the erosive forces and high temperatures. The start-up time is limited by the temperature attained by the support tubes which are dry initially.
- The **radiant** heat available to the SH and RH is limited thereby needing more HS for achieving the same turn down.

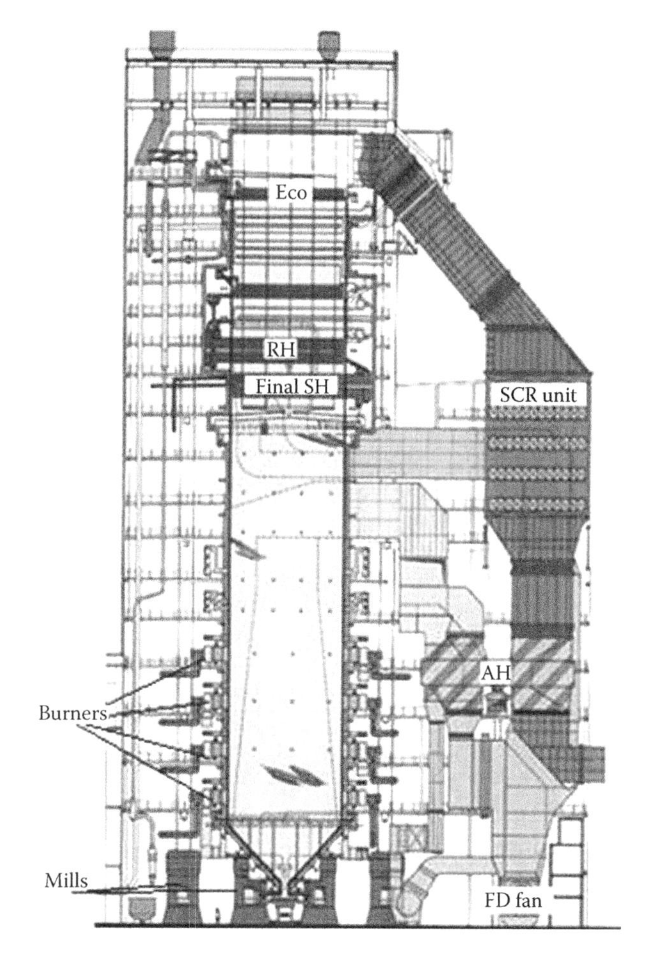

FIGURE 20.1
Tower-type SC boiler with opposed firing, SOFA and SCR arrangements. (From Hitachi Power Europe GmbH, Germany, with permission.)

The disadvantages of the two-pass design are the advantages of the tower design and vice versa.

Figure 20.1 shows a typical modern 550 MWe coal-fired opposed wall-fired SC tower-type boiler having fully drainable horizontal tube banks equipped with **SOFA** and **SCR for NO_x** control. The boiler layout does not follow the classical pattern of mill/bunker bay in front of the boiler. The mills are located on either side of the furnace, with bunkers above them. This type of layout is often preferred for opposed-fired boilers as the fuel piping is short. Coal handling and bunkering tends to be a bit more elaborate.

Table 20.2 compares the two-pass and tower-type boilers and Figure 20.2 brings out the differences.

The choice between two-pass and tower-type boilers is not easy to make as each has its advantages. Tower-type boilers are preferred for

- Lignite firing as they can stand up to gas side erosion better. Lignite has less ash but silica in lignite ash is more. Also, the large volume of

TABLE 20.2
Comparison of Two-Pass and Tower-Type Boilers

Two-Pass Boiler	Tower-Type Boiler
Gas turns on tube banks and hence erosion prone	No gas turns on tube banks and hence no erosion
Second pass erosion on mostly Eco top tubes needing erosion shields	Second pass in most cases has only AH
Vertical + horizontal SHs non-drainable	Horizontal SH only. Drainable. Boiler dynamics better.
SH and RH support from roof	SH and RH support from walls
Support elements outside gas path.	Support elements inside gas path.
Start-up time slower due to water in SH	Start-up time faster as SH is dry.
More floor space (H × W)	More height (1.25 H × 0.8 W)
Boiler short but all 4 sides not available for erection	Boiler tall but full access for erection
Boiler support on 6 or more shorter columns	Boiler support on tall 4 columns
Less pressure welds	More pressure welds
Two fronts available for PP erection	Only one front available for PP erection

flue gases with higher gas velocities makes it quite erosive.

- Frequent start-ups even for coal firing
- SC boiler with spiral tubes as greater furnace height is available

For normal cases of coal firing two-pass boilers with suitable erosion protection are quite satisfactory in most cases, besides being generally cost effective. Customer preference and their prevailing O&M practices also influence the choice, besides pure technicalities.

A modified tower-type boiler is a two-pass boiler with completely drainable horizontal surfaces and is shown

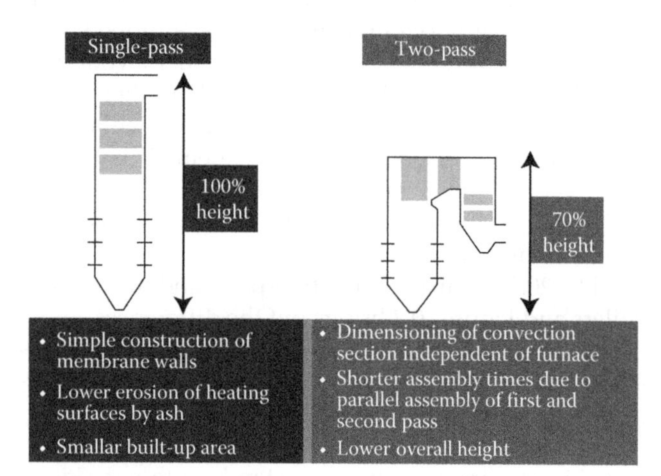

FIGURE 20.2
Differences between tower and two-pass boilers.

in Figure 20.3 which is taken from B&W. It is a design that combines the advantages of both the single- and two-pass designs; fully drainable surfaces and is yet not so tall. Naturally, it can be built at slightly lower cost than tower design. Ash in coal and its erosive index determines if such a compromise can be beneficially adopted. For brown coal firing conventional tower boilers are preferred due to large gas volumes. If they are fully cooled in the first pass before turning to the rear pass erosion potential would have been substantially mitigated.

1b. Two-pass boilers

In these boilers

- The first pass has the furnace, division walls and platens while
- The second or parallel pass has some RH and most of the backend equipment—Eco and sometimes AH
- With the connecting or inter-pass duct between the two passes containing most of SH and RH

Two-pass boilers are most popular as they have progressed from bi-drum boilers of smaller sizes. The main advantages are as follows:

- As the HSs are fairly evenly split between two passes the boilers are not so tall. Structure is short and erection is easier.
- Aside from shorter structure the boilers are usually supported on six or more columns and hence the structure can be a little lighter.
- The suspension hangers for the vertical HSs are located above the furnace roof outside the gas flow and are thus protected from the erosive forces of the gases.
- Vertical SHs and RHs are easy to support as they are held from the roof and are free to expand downwards.
- The roof is flat and usually made of SH tubes which minimises the relative expansion issues for the SH and RH tubes as they penetrate the roof. Hence the gas leakage issues are also minimal.
- Widely spaced radiant sections in platens and pendants as well as steam cooled walls in furnace have high heat absorption which increases the temperature control range.
- As a lot of radiant HS can be packed in the furnace this design needs lesser HS.

The main drawbacks of the two-pass design are that

- SH and at times the RH are not drainable. This requires careful monitoring of gas temperatures

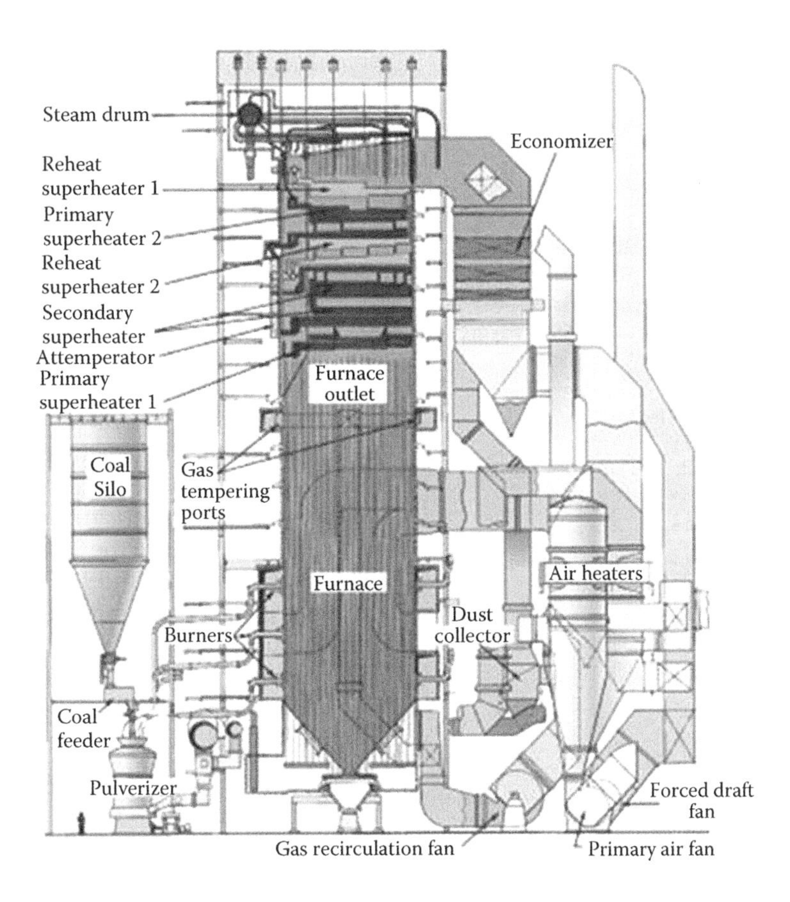

FIGURE 20.3
Modified tower-type sub-c boiler with drainable Eco in second pass. (From The Babcock and Wilcox Company, USA, with permission.)

at furnace exit leading to longer start-up times besides greater care during start-up.

- Flue gases turn over the convection banks as they go to the second pass from the first making them vulnerable to tube erosion both in the inter-pass area and the first set of tubes in the second pass. Suitable erosion protection measures are required to be installed.

There are two types of two-pass boiler designs with double and single rear passes. A double rear pass design is also called the divided pass.

a. In a *double or divided rear pass* boiler the SH and RH are arranged in separate passes with individual gas control dampers at the bottom of each pass to split the flow of flue gases for the control of steam temperatures. A drainable Eco is located at the bottom of both SH and RH banks. Usually a double-pass design is employed in furnaces with wall firing.

b. In a *single rear pass* boiler only part of the SH and all of the Eco are placed in the rear pass. RH temperature is controlled substantially by altering the heat absorption over the vertical RH pendant sections located in the inter-pass duct. This is possible because of the tilting burner arrangement in corner firing. This variation of centre of heat input can also be achieved by cutting the burner rows in and out suitably in wall-fired units.

Shown in Figure 20.4 is a typical top-supported dry bottom two-pass sub-c drum-type boiler with four levels of circular burners set in opposed firing. The second pass is arranged in single rear pass, as the entire RH is accommodated in the inter-pass duct between the first and second passes. PSH and Eco arranged in the rear pass are in horizontal pendant disposition and hence drainable. SSH is in **vertical platen** while final SH and the entire RH are in pendant construction and hence non-drainable. Vertical **rotary AH** is the last heat trap. **Coal mills** are vertical and placed in a single row in front of the boiler. It is a classical layout of the boiler plant with mill bay located in the front side of the boiler between the bunker bay and the boiler. While this layout is very popular, some plant owners do not favour the arrangement as the dirty parts, namely the bunkers and mills, are placed close to the clean part of the plant, namely the TG bay.

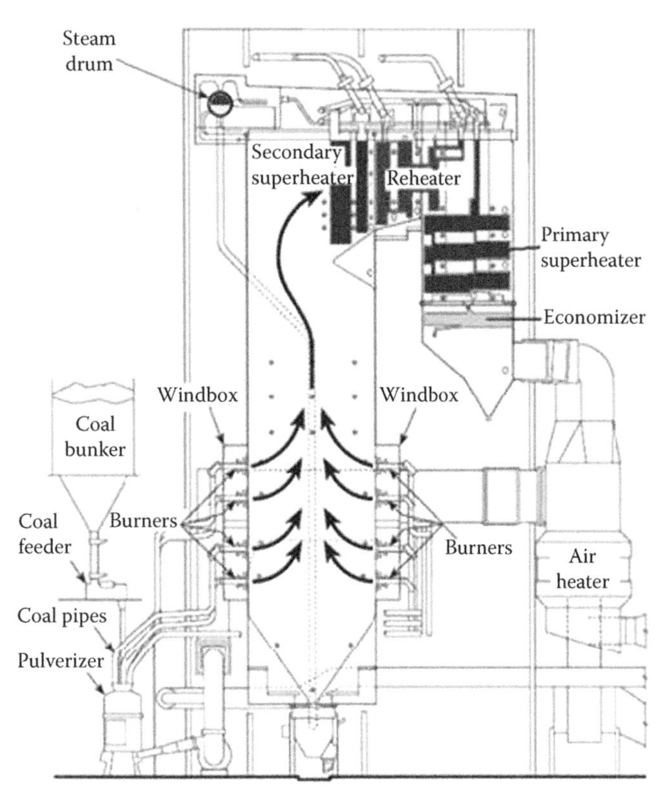

FIGURE 20.4
Two-pass drum-type opposed-fired boiler with single back pass.

In the single rear pass arrangement, with no division in the second pass, the RHOT is controlled by the burner tilt in the corner firing or cutting the burner rows in and out in wall firing. SOT is regulated by controlling the spray water in attemperators.

Figure 20.5 depicts a two-pass boiler with single rear pass very nearly similar in construction to the boiler shown in Figure 20.4 except that it is equipped with **corner firing** instead of **wall firing**. Also there are **wing wall evaporators** in the upper furnace.

Figure 20.6 is a picture of typical two-pass boiler with **opposed wall firing** at four levels and divided gas flow in the two parallel rear passes. The difference between the present and the earlier arrangements is the construction of the second pass. While it is divided flow here there is only a single pass in the former.

1c. Down shot- or arch-fired boilers

See also Arch Mounted Burners under Pulverised Fuel (PF) Burners

Down shot- or arch-fired boilers are employed where the **VM** in fuel is quite low at 19% or lower on daf basis. This is the design suited for **anthracites, petroleum coke** and low-volatile coals. These boilers are very large and hence expensive. They are more suited to base load operation. The **ignition temperature** of such low volatile fuels is quite high (~600°C) and the burning of high amount of char is slow, which requires

- Refractory lining of the lower part of **water walls**
- A large volume of furnace (longer residence times of ~3 s is desirable) respectively

to complete the combustion efficiently. **Refractory** in the lower part of furnace inhibits heat absorption and re-radiates heat to the flames to help ignition.

As shown in Figure 20.7, the firing is with multiple burners in a non-turbulent fashion, with the burners facing downwards such that the flames are long and there is a lot of time to burn the coal as the flames travel downwards and take a u-turn. Ignition takes place at some distance from the burners after receiving the heat from refractory and the burning of char is accomplished in the return path of the flue gases. Flue gas gets ample time in going down and coming up to complete the burning of char. Refractory keeps the lower furnace quite hot. Furnace can be designed for both dry and wet ash discharge. However the current NO_x regulations do not permit running the furnace hot enough to suit wet ash discharge.

The upper furnace and the rest of the boiler is a conventional two-pass arrangement as seen in Figure 20.7.

Down shot firing can be U type with a single set of burners on one side of the furnace or W type with burners on either side depending on the size of the boiler. For quite some time now only W-type firing is adopted, as the boilers built have been large.

For proper burnout of char it is essential to grind the low VM coals to a higher fineness, typically 100% through 75 microns. Ball mills are popular due to higher HGI of fuel.

Down shot boilers, by their nature of construction, tend to be big and hence expensive. Together with the costly NO_x reduction measures, the expenditure is too high to justify the resulting power tariff. As a result, down shot-fired boilers have not been built in the advanced parts of the world for quite some years now. A typical 700 MWe boiler firing anthracite will be nearly as large as a 1300 MWe coal-fired boiler! CFBC boilers are becoming popular in this segment.

20.2.3.2 Utility Boiler Designs Based on Steam Pressure

Based on drum pressure boilers can be divided as

a. Sub-c
b. SC

The dividing point is the critical point of steam, which as per ASME steam tables, is at

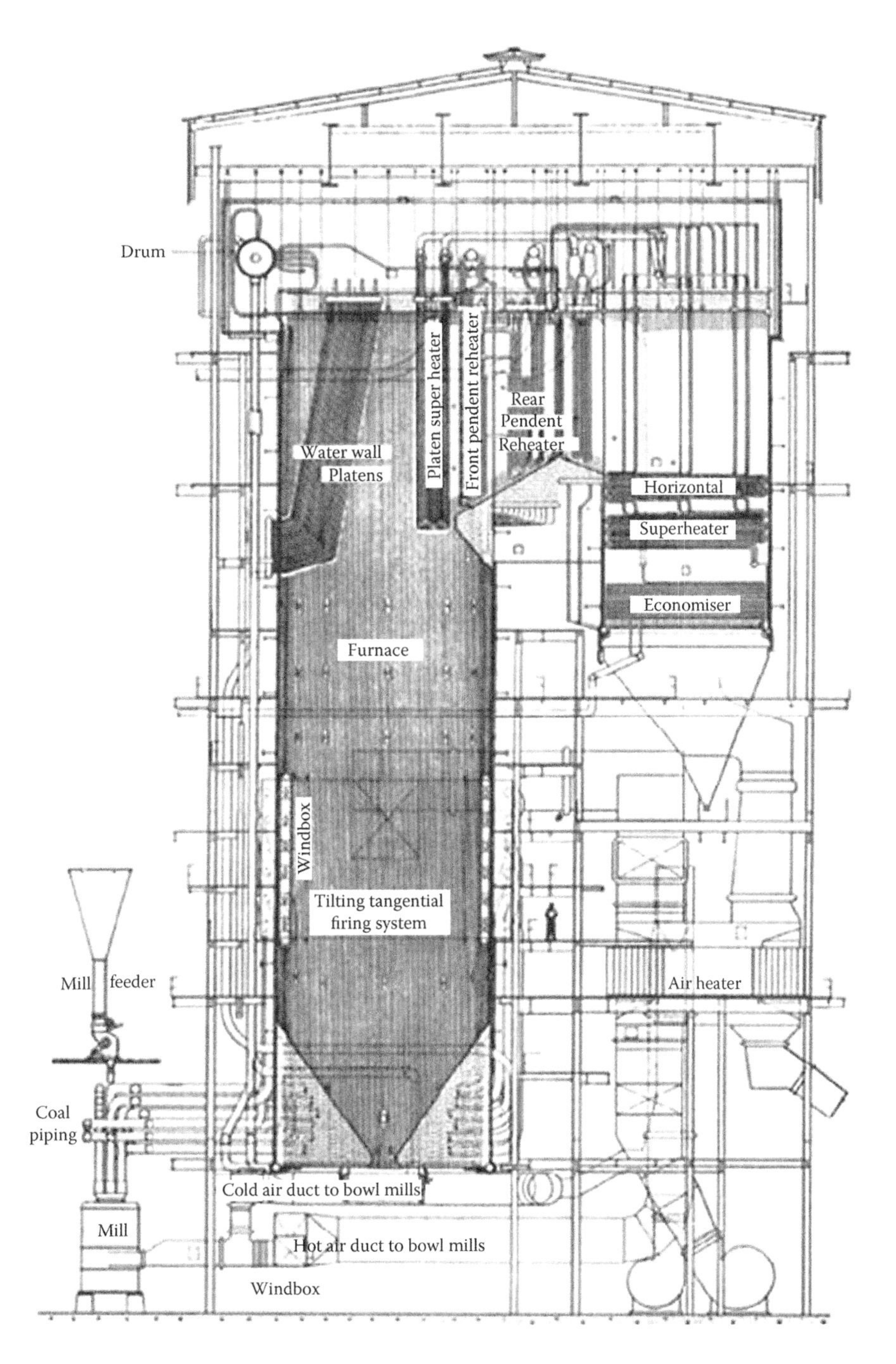

FIGURE 20.5
Two-pass corner-fired boiler with single rear pass and part vertical SH and RH.

220.7 bar and 374.1°C (3200.1 psia and 705.1°F) when water progressively converts directly to steam without the familiar intermediate stage of boiling. There is latent heat addition. These figures can vary fractionally depending on the steam tables followed, but that is merely of theoretical interest with little practical use. In a broad sense

- SC boilers are termed as those operating up to *~240 bar (3500 psi* nominal)
- USC are termed as those operating at *~310 bar (4500 psi* nominal) or higher

The operation of sub-c and SC steam cycles is plotted on the T-s diagram in Figure 20.8. It is evident that for nearly the same condensation loss work done is more in SC cycle, leading to higher cycle η. While the boiler ηs are practically the same in both, cycle η being higher, the overall fuel consumption is lower.

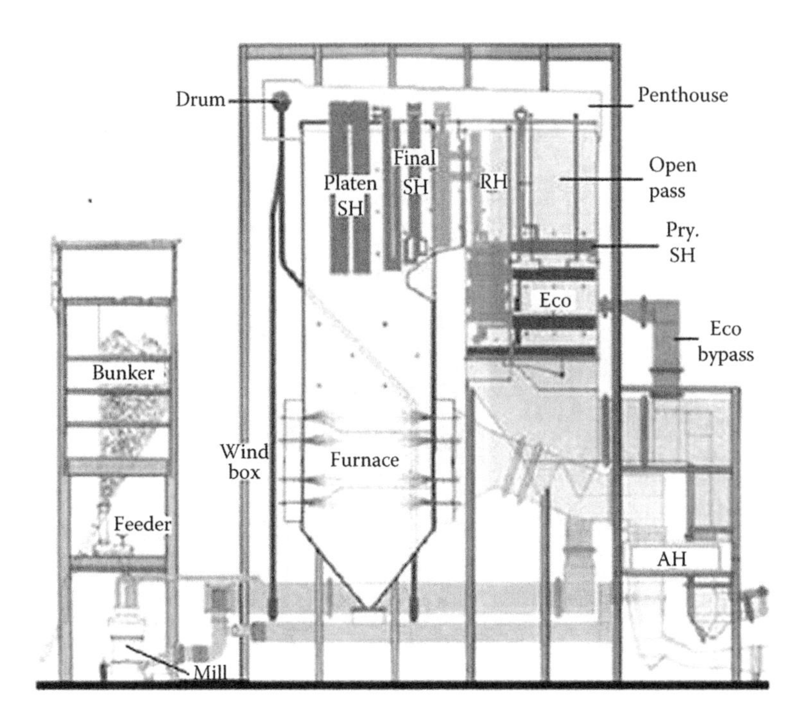

FIGURE 20.6
Two-pass opposed wall-fired boiler with divided back pass and part vertical SH and RH.

2a. Sub-c boilers

Over 90% of utility boilers are sub-c. Maximum drum pressure possible is ~200 bar with temperatures of 540/565°C. Very large boilers of nearly 1000 MWe capacity are in satisfactory operation. Their main advantages are that

- They do not demand ultra-pure quality of water as SC boilers
- They are conventional and slightly less stringent in their O&M
- As efficient as SC boilers

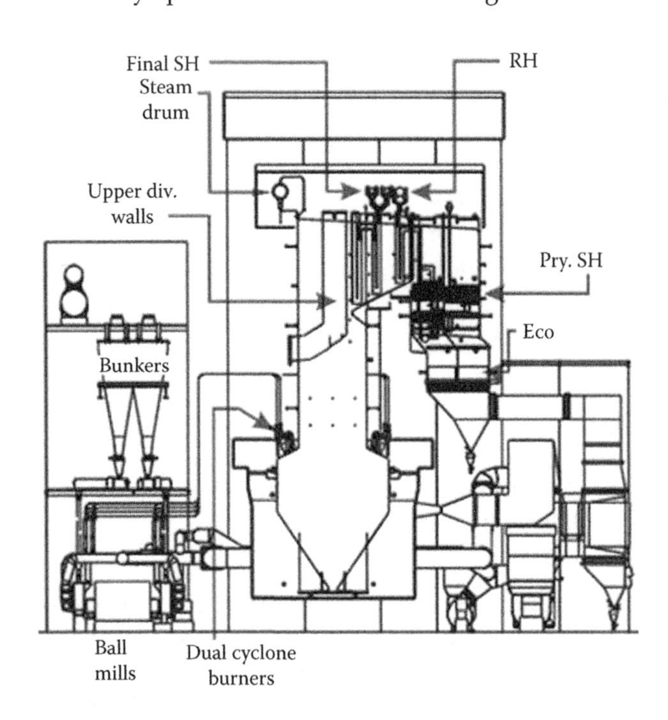

FIGURE 20.7
W-type down shot-fired boiler with divided back pass and ball mills. (From Foster Wheeler Corporation, USA, with permission.)

However, they have limited capabilities of **variable-pressure operation**. Load chasing ability is also slightly slower. There is also no further possibility of improving the boiler η. They are cheaper to install and operate. For base loaded power plant at pit head where load cycling is not needed they may in many cases work out to be more cost effective.

Power plants based on sub-c boilers are about 5–7.5% less in η than those with SC/USC boilers depending upon the pressure level. From ~40% in sub-c boilers

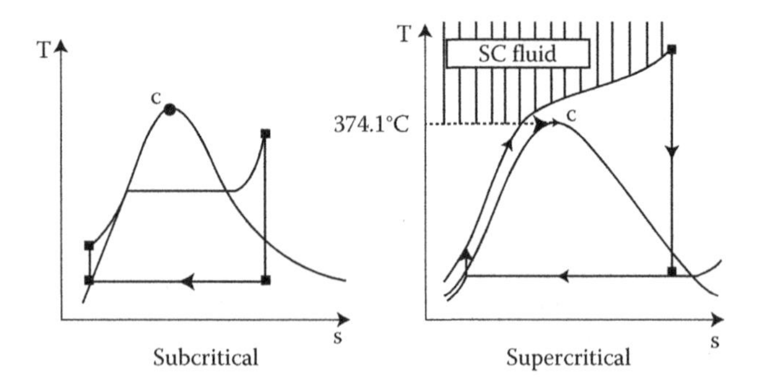

FIGURE 20.8
Sub- and supercritical steam cycles on T-s chart for comparison.

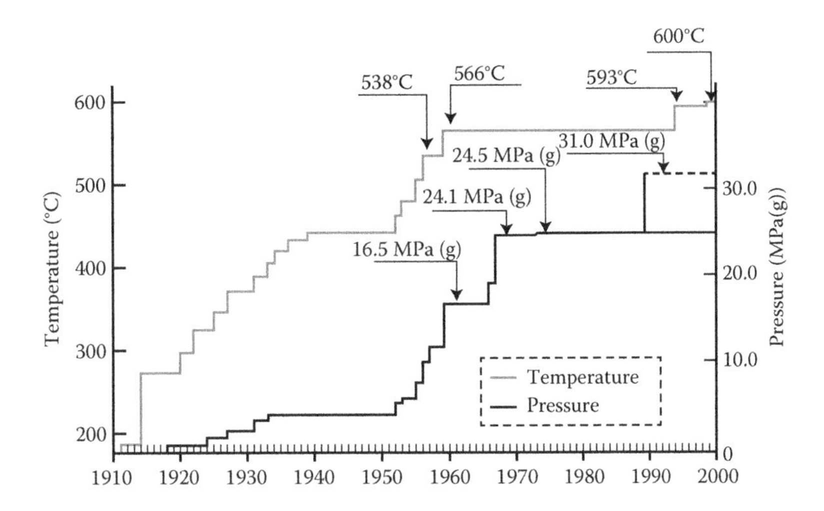

FIGURE 20.9
Rise of steaming conditions over time in Europe and Japan.

cycle η rises to ~42% at SC level. This lowers the GHG emissions by ~5% as the fuel input is lower by the same amount. With larger power plants coming up all over the world, this incremental benefit is viewed very positively. With added benefits of cycling capability the SC boilers are presently finding favour world over for capacities >600 MWe.

Both sub-c and SC boilers are built in all configurations namely, tower type, two pass and down shot firing.

2b. SC boilers

SC boilers have been around from the 1920s when Mark Benson built the first unit of 4 tph. Over the years they have grown in size and reliability as they possess certain unique advantages over sub-c boilers. Their growth over time is captured in Figure 20.9 which depicts the growth of p & t conditions in boilers in general in Europe and Japan.

In SC boilers there is *no circulation* of steam–water mixture in the water walls but a straight conversion of fw to steam in OT passage from Eco to SH. **Circulation** is substituted by forced flow where circulation ratio is unity. This is the essential difference between sub-c and SC boilers which results in sympathetic changes in furnace and related items connected with circulation. The most notable omission in SC boilers is the steam drum; but there are separating vessels to remove M when operating lower loads.

Figure 20.10 depicts the difference between sub-c and SC boilers very well. Note the absence of a two-phase flow in SC boiler.

Even at the highest drum pressure of 200 bar in drum-type boilers, there is a circulation of ~3. When this becomes unity in SC boilers the flow through the furnace tubes is too small to provide adequate cooling unless remedial measures are taken.

In a drum-type unit large bore tubes are used in the furnace to minimise flow resistance so that sufficient amount of steam and water can flow through the furnace tubing by natural circulation at all heights under all load conditions. Water passing through the furnace tubing never completely evaporates to steam and a liquid film is always maintained on the tube wall so that **DNB** and/or dry out do not occur. High heat transfer coefficient from nucleate boiling keeps all the Evap tubes at nearly saturation temperature.

With imbalances in heat absorption variations in tube temperatures occur. This is due to

- The geometric tube position (corner versus wall firing)
- Uneven burner heat release pattern
- Variation in furnace cleanliness
- Variations in flow rate (due to hydraulic resistance) differences from tube-to-tube.

If the imbalance in temperature is not limited, high thermal stresses will result which can lead to tube failure. The Evap of an SC boiler must provide for

- Means to accommodate differences in heat absorption of tubes so that the resulting temperature difference between adjacent tubes is limited
- Good tube cooling to avoid DNB and suppress dry out so that peak tube metal temperatures are minimised

There are three methods to achieve this, as shown in Figure 20.11.

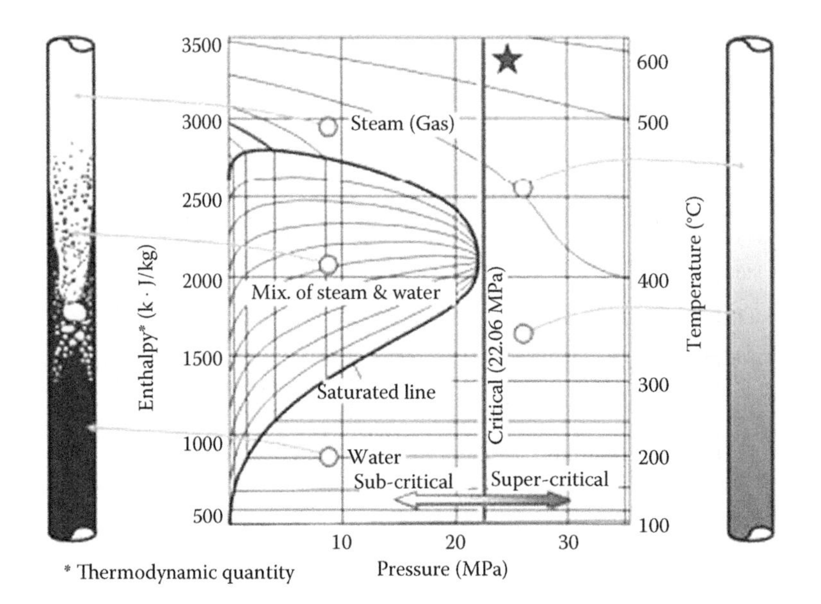

FIGURE 20.10
Differences between sub- and SC boilers plotted on H-p diagram.

1. Multi-pass high mass flow arrangement (universal pressure boiler)

   The earliest method was to have multiple passes with hotter water retuned from the upper parts of the circuits to the bottom headers to have adequate mass flow (as shown in Figure 20.11a). The tubes in furnace were vertical with conventional construction, somewhat in the same way as in sub-c boilers. Hence it is called universal pressure (UP) boiler by B&W Co. This arrangement operates at constant fluid pressure in furnace which is found suitable for base load and cycling loads and has been popular in the US. Several boilers have been built on this principle including the world's largest unit of 1300 MWe.

   The lower portion of the furnace, up to nearly mid-way between the upper level of burners and the furnace nose, is made of two sequential water flow paths. These are physically arranged in parallel around the furnace circumference to obtain the high mass flow needed for cooling furnace tubes. See Figure 20.12

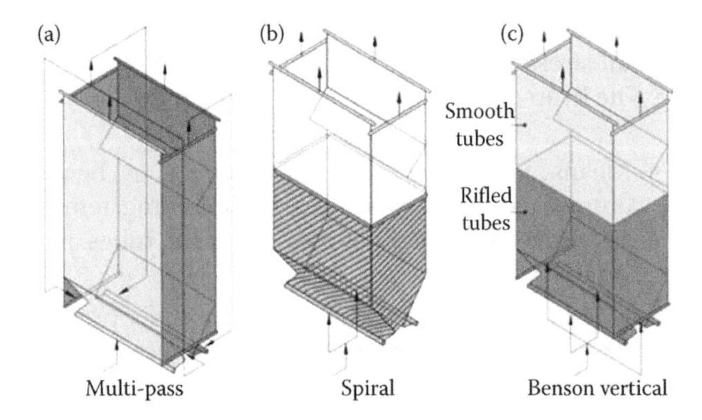

FIGURE 20.11
Three types of SC boilers namely (a) Multipass, (b) Spiral, and (c) Benson vertical.

- Water flows up first to first pass outlet headers.
- From there down a downcomer to the second pass inlet header at the bottom of the furnace.
- From there up the second pass tubes exiting again in second pass outlet headers.
- From there water then flows through enthalpy equalisation mix headers and back into the third pass tubes with the third pass making up the entire perimeter of the upper furnace.

Thus, in the lower heat flux zone of the upper furnace, the number of tubes is reduced to nearly half that of the lower furnace for economy of pressure loss.

Furnace pressure of the fluid is held constant in this design at SC level. At low loads it would have a tendency to go to sub-c levels which is unsuitable as the second pass tubes will have segregation of steam and water in the tubes. As steam is not an effective coolant those tubes will suffer from failures due to overheating.

For variable pressure ST operation during load turndown, pressure control division valves are located within the boiler between the PSH and SSHs. These valves maintain the furnace fluid pressure above the critical point while letting the ST follow its most economical pressure versus load operating profiles. With partial

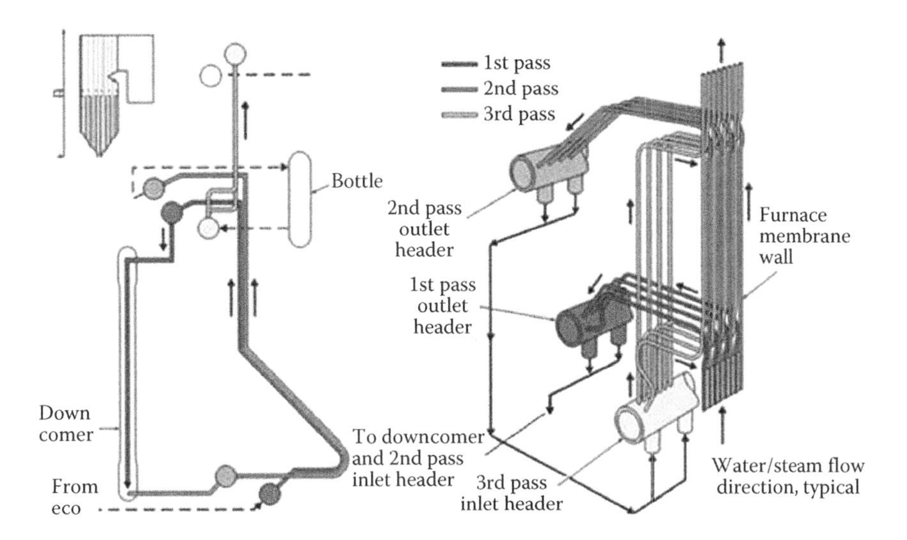

FIGURE 20.12
Furnace tube arrangement in UP boiler. (From The Babcock and Wilcox Company, USA, with permission.)

arc admission ST designs, the most economical operating condition is consistent pressure operation to the second valve point typically at between 60% and 70% load.

2. Spiral tubing with high mass flow

   As shown in Figure 20.11b, this arrangement, which is popular in Europe and Japan, has a few sets of tubes that are upwardly wound around the furnace at inclinations of 10° to 30° up to the furnace exit or nose before turning vertical. A boiler with spiral wall furnace is shown in Figure 20.13. At the elevation of furnace nose the spiral tubes are turned vertical.

   There are several advantages in this construction:

   - Smaller diameter tubes are used to obtain high mass velocities for keeping furnace tubes cooler, which also makes them thinner. The tubes vary from 31.8 to 42 mm od with strips of 19 mm width as in PF boilers.
   - The number of tubes in spiral arrangement reduces by nearly half, as shown in Figure 20.13. They can be even 1/3 at even lower angle.
   - All tubes experience the same amount of heat pick up as they wind around the furnace, as shown in Figure 20.14. This minimises the difference in tube to tube heat absorptions making the entire wall expand and contract like a single tube.

The main thermodynamic advantage of spiral tube arrangement is that this system allows for the fluid pressure to vary with changing loads

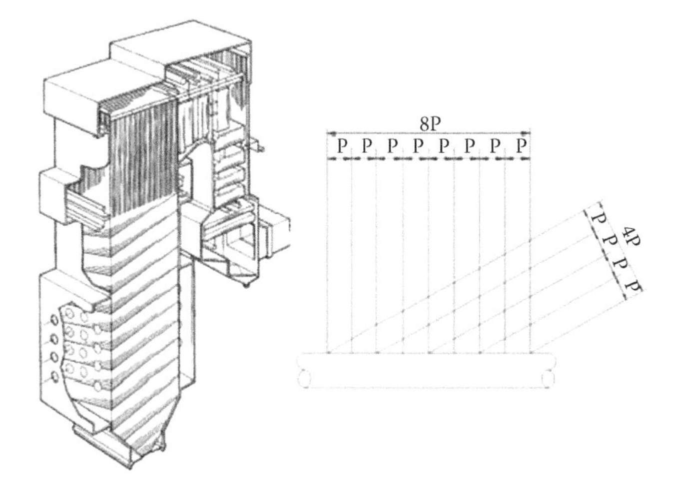

FIGURE 20.13
Furnace with spiral tubes and the reduction in the number of tubes in spiral wall construction.

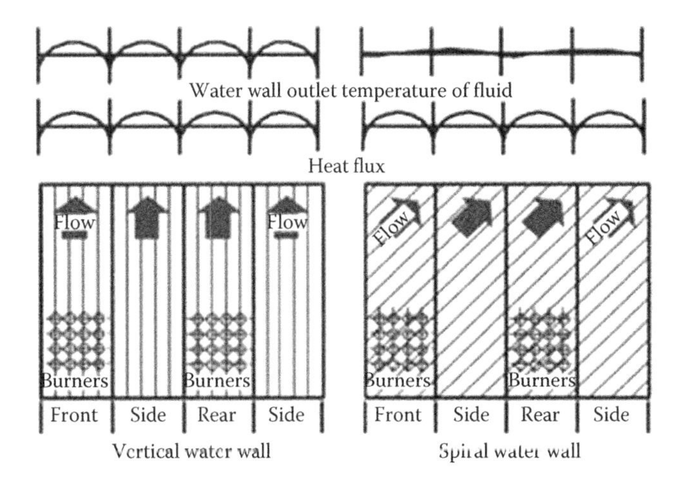

FIGURE 20.14
Heat absorbed by vertical and spiral wall tubes in a furnace.

making it conducive to variable pressure operation. Also the inertia of the system is low due to lower fluid content and lesser weight of the PPs.

However, in mechanical construction there are complications.

- As mass flows are high at ~2200–2400 kg/m² s (1.6–1.76 × 10⁶ lbs/ft² h) Δp in the furnace tubes is also high. This adds to the discharge head of BFP.
- Spiral furnace construction is more complicated and expensive.
- The sloping tube panels are more difficult to build and support. The load must be transferred through the fins by means of weld attachments, as shown in Figure 20.15.
- The number of tube welds is more.
- The transition to vertical tubes from spiral arrangement is done at a height where the heat flux is low with the help of a ring of transition forged pieces with suitable headers, as shown in Figure 20.16. This adds further to the cost, welds and the fabrication time.

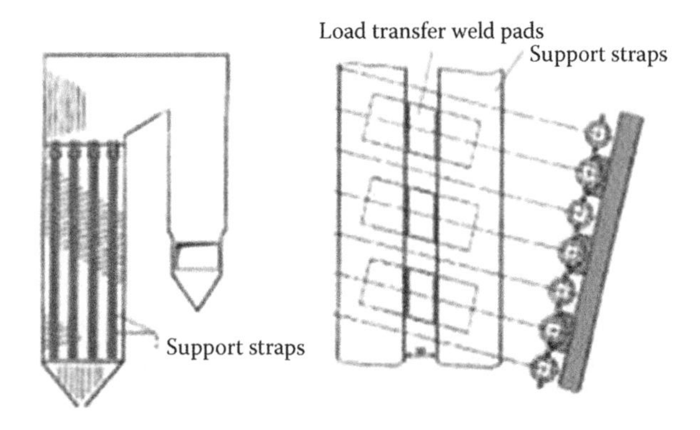

FIGURE 20.15
Supporting of spiral wall furnace in SC boilers.

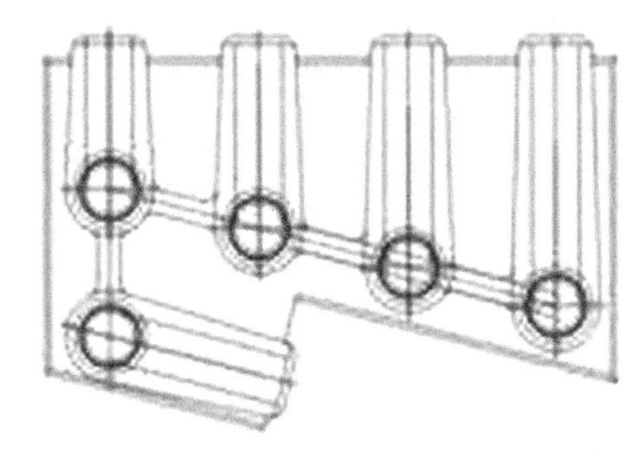

FIGURE 20.16
Forgings to connect spiral to vertical tubes in upper furnace of SC boiler.

3. Vertical tubes with low mass flow

   The latest design is to adopt vertical tubes, like in conventional boilers, with **ribbed/rifled tubes** in the lower furnace and plain tubes above. The jagged interiors of ribbed tube provide a place where water film can stick and protect against **DNB**, despite low mass flows. Tube diameter and spacing is chosen to give a typical low mass flow of ~1000 kg/m² s (738,000 lbs/ft² h) less than half of what is used in spiral tube arrangement. At such low mass flows the frictional drop in furnace tubes, compared to gravitational head, is low. As a result, when a tube is heated more it absorbs more heat and draws more flow similar to a **natural circulation** tube. With more flow the tube is cooled better, reducing the difference with the adjacent tube. In a high mass flow design the tube fails to increase the flow and the temperature rises creating an opposite effect. This design combines the advantages of spiral tubing, namely

   - Variable pressure operation
   - On/off cycling
   - Rapid load changing

   with the self-supporting feature of vertical tubing. Vertical tubes are easy and cheaper to fabricate and erect. Pumping pressure is less as the low mass flow results in lowering the pressure loss through the furnace tubing by over half. The margin against tube overheating is also more.

   Central to the success of this concept is the use of ribbed or rifled tubes which can provide a natural circulation characteristic. After considerable work Siemens have come out with optimised multi-lead ribbed (OMLR) tube, as shown in Figure 20.17, with increased rib height and flatter lead angle to create turbulence and mixing near the tube wall to increase cooling.

   Advantages of vertical water wall over spiral water wall can be summarised as follows:

   1. Simple structure
      - Easy manufacturing and construction
      - Shorter installation period
      - Less thermal stress
   2. LP drop in water wall
      - Low auxiliary power consumption
      - Low temperature imbalance
   3. Easy removal of slag
      - Less slag accumulation
      - No local concentration of slag

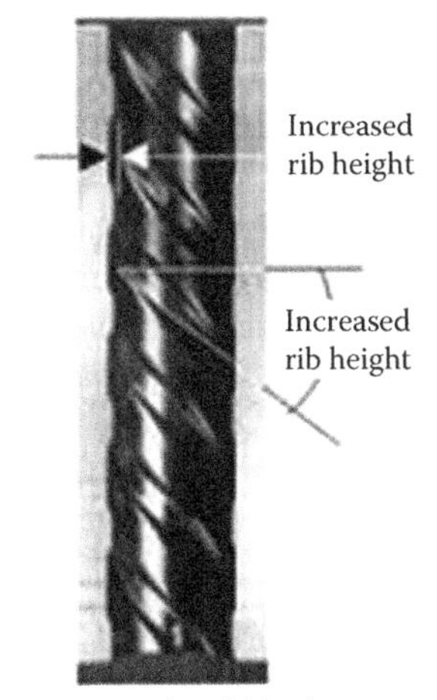

FIGURE 20.17
Optimised multi-lead ribbed tubes (OMLR) for vertical Benson furnaces.

2c. Advantages of SC over drum-type boilers

The main improvements of SC boilers over natural circulation boilers can be summarised as

- Fuel savings due to higher cycle η and hence reduced operating costs
- Reduction of CO_2 emissions due to lower fuel input for the same power
- Superior load dynamics as the thick walled components are avoided and there is no water in circulation
- Part load performance even better with **variable/sliding pressure operation**
- No blow down practically

Fuel savings depend on the p & t levels chosen for the operation of SC boiler. η improvement is higher when the operating p & t is more. Enough experience is available with SC boilers operating at ~250 bar (~3500 psig) level. There are a few reference plants for the next pressure level. Metallurgical problems have to be solved to reach 350 bar (5000 psig) level along with 700°C. The bar graph in Figure 20.18 taken from Siemens published data captures this aspect along with the reduction in emissions. A line on materials is added at the top of Figure 20.18 to show the metallurgical improvements needed at each stage.

The superior dynamics of SC plant are well illustrated by the start-up time comparison between sub-c and SC plants given in Table 20.3. There is a big reduction of time in cold starts for delivering first steam to ST. There is a reduction in time in reaching the full loads as well. The load change rate of the SC plant is ~3–5% per min which is much faster than the sub-c plant load change rate of 1–2% per min. This results in a much shorter start-up time in SC plants.

For boiler start-up a minimum fw flow of ~35% is established before starting of firing, which is called the 'minimum once-through load', required to provide adequate cooling of the furnace wall tubes. The excess water is dumped from the start-up separator vessel by a level control loop to condenser. Once the steam

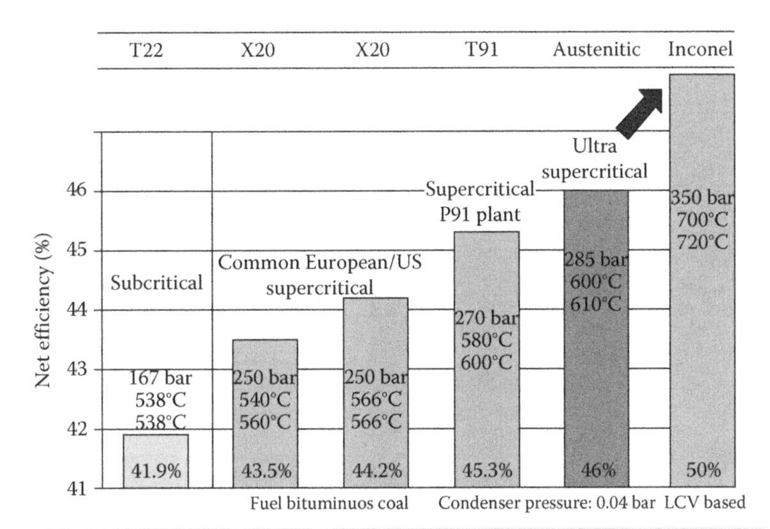

FIGURE 20.18
Enhanced cycle efficiencies, fuel savings and emissions with rise in steam parameters.

TABLE 20.3
Comparison of Start-Up Times between Sub-c and SC Boilers

Shut Down in hours	Sub-c Boiler 167 bar, 538/538 °C		SC Benson Boiler 250 bar, 540/560°C	
	Response Time in Min from Ignition to			
	First Steam	Full Load	First Steam	Full Load
<1	20–30	60–80	20–30	30–40
8	40–60	80–100	30–40	50–60
48	150–210	300–350	60–80	150–200
>48	150–210	450–600	60–80	400–600

generation matches the minimum fw flow the separating vessel will run dry and the true once-though mode is then established. The fw flow is then increased to match the increase in steam generation.

The dumping of water and heat on start-up can be avoided by the installation of a start-up boiler recirculation pump. Without a BCP it would not be economical to operate for any period of time below the minimum OT load.

2d. PP materials in SC boilers

The combination of high p & t demands superior materials of construction for various PPs

- To yield reasonably low thicknesses for reduced thermal inertia
- Better corrosion resistance for parts facing flue gases particularly at high metal temperatures.

The currently available commercial steels allow the construction of boiler plant for steam conditions as high as 300 bar/600°C/620°C, for a wide range of coals, even those producing an aggressively corrosive flue gas. The list and scrap view of upper boiler in Figure 20.19 depict the use of various materials in different parts of the boiler which is taken from the Best Practice brochure of DTI-UK (Department of Trade and Industry, 2006).

Advanced ferritic and ss, not required to be used in large sub-c boilers but needed for building USC boilers, are described below, with the exception of T12 and T22.

- *T12 and T22*: Furnace tubes are subject to the highest heat flux in the furnace. Typical tube materials are 1 ½Cr ½Mo (T12), 2 ¼Cr1Mo (T22) and a European material $15Mo_3$. These low ASs have excellent mechanical properties suitable for easy fabrication of the water wall panels. For non-corrosive coals for FEGT <1250°C (2280°F) T12 can be applied in the water walls of tower type boilers up to the water wall outlet temperature of ~ 470°C (880°F) in 280 bar (4060 psig) and 600°C (1112°F) cycle. T12 is used for lower temperatures in panel designs. For higher steam conditions and higher water wall outlet temperatures respectively, different materials are required. For example, T22 has slightly higher creep rupture strength and higher oxidation limit and is frequently used in place of T12. Its use, however, does not enable any significant increase in cycle parameters.
- *T23 and T24:* For components at even higher temperatures considering other manufacturing restrictions there is need for more advanced creep-strength enhanced ferritic (CSEF) steels. Two specific alloys within this family, both based on the much used T22 or 2.25% Cr steel are

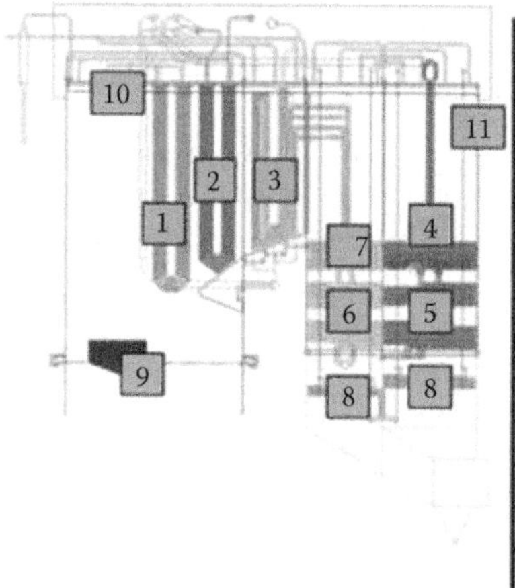

Ref	Heating surface	Material	Design temperature (°C)
1	Platen SH	SA 213 TP347H/ TP310HCbN/T92*	640
2	Final SH	SA 213 TP347H/ TP310HCbN/T92*	650
3	Final RH	T22/ T91/ TP347H/ TP310HCbN/ T92*	680
4	Prim SH	SA 213 T91	550
5	Prim SH inlet	SA 213 T23	530
6	RH inlet	SA 210C	470
7	RH	SA 213 T12	540
8	Econ	SA 210C	375
9	Water wall	SA 213 T12/ T23	515
10	Furnace roof	SA 213 T23	510
11	Rear cage	SA 213 T23	510

* SA213 T92 used for outlet tube stub connections

FIGURE 20.19
High temperature materials for various sections in a SC boiler of 300 bar/600/620°C.

- T23 (ASME code case 2199 and 2.25Cr 1.6W V steel) modified by the addition of 1.6% W and small amounts of Nb, V and B and reduction of Mo and C.
- T24 (ASTM A213 and 7CrMoVTiBo1010 steel) modified by the reduction of C and addition of V, Ti and B.

- These variations produce steels with very high creep strength comparable to T91 but with another compelling advantage of low C content needing no post-weld heat treatment (PWHT). This very attractive property makes these steels particularly suitable for membrane walls with no problematic PWHT.

  Other, non-membraned items such as the inlet end of the SH can also utilise these alloys simply due to their excellent creep strength.
- *T 92*: When selecting SH and RH tube materials, the creep strength of the selected alloys must be high enough to provide adequate margin of safety. In addition, the material's corrosion resistance, both on the flue gas side and on the steam side, must be considered. Oxides will always form on the inside and outside surface of tube. On some older boiler designs, exfoliating metal-oxide scale from the internal surfaces of SH and RH tubes, headers and piping have been a principal source of solid particle erosion in ST and valves. The exfoliated oxides are mostly ferritic types. Low-Cr ferritic alloys are predominant materials throughout the RH system and are more prone to exfoliation than austenitic alloys.

  CSEF steel which has been used successfully over 20 years is T/P 91 for tubes, headers and pipes. This has excellent creep strength but its temperature is limited by steam side oxidation. The internal oxide inside the heat transferring tube acts as a thermal barrier which increases both the metal temperature and the oxidation rate. The same oxidation controlled temperature limit applies to T92, which has an addition of 2% W and reduction in Mo so as to adjust the balance of ferritic-austenitic elements, and the addition of micro-alloying amounts of B. *T92 is the creep strongest* of all the so-called CSEF 9–12% Cr steels and is suitable for tube outside temperature up to 650°C.
- *HR3C*: In the case of SH and RH the choice of material is more likely to be dictated by flue gas corrosiveness than simple stress rupture characteristics. They employ lower ASs at the inlet and higher ASs at the outlet. TP347 is a frequently used 18% Cr 11% Ni austenitic alloy for which there is considerable experience, but for aggressive atmospheres higher Cr grades such as A213 TP310CNbN (HR3C) with 25% Cr and 20% Ni would be required.

  The curves in Figure 20.20 shows the influence of Cr in the tube materials on the corrosion rate at each test temperature. Tube materials with Cr of >25% are effective for management of corrosion rates for high S fuels and high steam temperatures. An advanced high strength austenitic ss with nominal Cr of 25%, of composition 25Cr20NiCbV (A213 TP310HCbN), has been employed since 1990 in various types of plants including utility boilers. By using this material in new coal-fired boilers to be designed for high S coals, the tube life of high temperature SHs and RHs can be well managed even for boilers with steam temperatures >1100°F (593°C).
- *P91 and P92:* There is an inherently difference in the coefficient of expansion and the thermal conductivity of austenitic and ferritic materials. The higher expansion and lower conductivity of austenitic materials induces twice the stress in a temperature transient than in a ferritic component of the same dimension. Hence ferritic materials like P91 and P92 are preferred for pipes and headers up to 620°C which are large and thick walled.

There are other steels used for higher p & t. Different boiler makers have developed their designs and

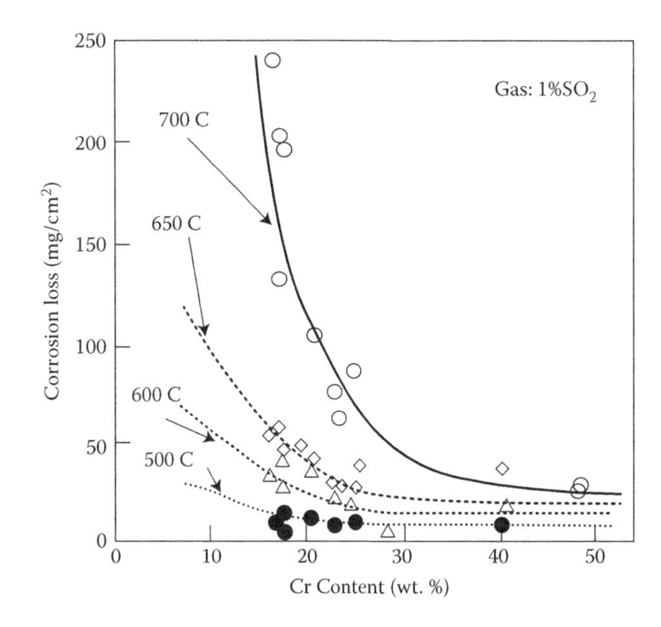

FIGURE 20.20
Influence of Cr on corrosion resistance of steels at elevated temperatures.

TABLE 20.4A

Nominal Composition of Ferritic Steels for Use at High Temperature in Boilers

		Specification		Chemical Composition											
	Steels	**ASME**	**Japanese Industrial Standards (JIS)**	**C**	**Si**	**Mn**	**Cr**	**Mo**	**W**	**Co**	**V**	**Nb**	**B**	**N**	**Others**
1-1/4 Cr	T11	T11		0.15	0.5	0.45	1.25	0.5	—	—	—	—	—	—	—
	MFIH			0.12	—	—	1.25	1.0	—	—	0.20	0.07	—	—	—
2Cr	F22	T22	STBA24	0.12	0.3	0.45	2.25	1.0	—	—	—	—	—	—	—
	SCM2S	T23	STBA24J1	0.06	0.2	0.45	2.25	0.1	1.6	—	0.25	0.05	0.003	—	—
	Tempaloy F-2W						2.0	0.6	1.0		0.25	0.05		—	
9Cr	T9	T9	STBA26	0.12	0.6	0.45	9.0	1.0	—	—	—	—	—	—	
	HCM9M	—	STBA27	0.07	0.3	0.45	3.0	2.0	—	—	—	—	—	—	—
	T91	T91	STBA28	0.10	0.4	0.45	3.0	1.0			0.20	0.08		0.05	0.8Ni
	E911		0.12	0.2	0.51	3.0	0.94	0.9	—	0.20	0.06		0.06	0.25Ni	
12Cr	HT91	(DIN × 20CrMoV121)	0.20	0.4	0.60	12.0	1.0	—	—	0.25	—	—	—	0.5Ni	
	HT9	(DIN × 20CrMoWV121)	0.20	0.4	0.60	12.0	1.0	0.5	—	0.25	—	—	—	0.5Ni	
	Tempaloy F12M						12.0	0.7	0.7	—					
	HCM12		SUS410J2T B	0.10	0.3	0.55	12.0	1.0	1.0	—	0.25	0.05		0.03	—
	TB12	—	—	0.08	0.05	0.50	12.0	0.5	1.8	—	0.20	0.05	0.30	0.05	0.1Ni
	HCM12A	T122	SUS410J3T B	0.11	0.1	0.60	12.0	0.4	2.0	—	0.20	0.05	0.003	0.06	1.0Cu
	NF12	—	—	0.08	0.2	0.50	11.0	0.2	2.6	2.5	0.20	0.07	0.004	0.05	—
	SAVE 12	—	—	0.10	0.3	0.20	11.0	—	3.0	3.0	0.20	0.07	—	0.04	0.07Ta, 0.04Nd

TABLE 20.4B

Nominal Composition of Austenitic Steels for Use at High Temperature in Boilers

	Specifications		Chemical Composition										
	ASME	JIS	C	Si	Mn	Ni	Cr	Mo	W	V	Nb	Ti	B
18Cr-8Ni	TP304H	SUS304HTB	0.08	0.6	1.6	8.0	18.0	—	—	—	—	—	—
	Super 304H	SUS304JIHTB	0.10	0.2	0.8	9.0	18.0	—	—	—	0.40	—	—
	TP321H	SUS321HTB	0.08	0.6	1.6	10.0	18.0	—	—	—	—	0.5	—
	Tempaloy A-1	SUS321JIHTB	0.12	0.6	1.6	10.0	18.0	—	—	—	0.10	0.05	—
	TP316H	SUS316HTB	0.08	0.6	1.6	12.0	16.0	2.5	—	—	—	—	—
	TP347H	SUSTP347HTB	0.08	0.6	1.6	10.0	18.0	—	—	—	0.8	—	—
	TP347HFG		0.08	0.6	1.6	10.0	18.0	—	—	—	0.8	—	—
15Cr-15Ni	17-14CuMo		0.12	0.5	0.7	14.0	16.0	2.0	—	—	0.4	0.3	0.006
	Esshete 1250		0.12	0.5	6.0	10.0	15.0	1.0	0.2	1.0	—	0.06	—
	Tempaloy A-2		0.12	0.6	1.6	14.0	15.0	1.6	—	—	0.24	0.10	—
20-25Cr	TP310	SUS310TB	0.08	0.6	1.6	20.0	25.0	—	—	—	—	—	—
	TP310NbN	SUS310J1TB	0.06	0.4	1.2	20.0	25.0	—	—	—	0.45	—	—
	NF707[a]		0.08	0.8	1.0	35.0	21.0	1.5	—	—	0.2	0.1	—
	Alloy 800H	NCF800HTB	0.03	0.8	1.2	32.0	21.0	—	—	—	—	0.5	—
	Tempaloy A-3[a]	SUS309J4HTB	0.05	0.4	1.5	15.0	22.0	—	—	—	0.7	—	0.002
	NF709[a]	SUS310J2TB	0.15	0.5	1.0	25.0	20.0	1.5	—	—	0.2	0.1	—
	SAVE25[a]		0.10	0.1	1.0	18.0	23.0	—	1.5	—	0.45	—	—
High Cr-High Ni	CR30A[a]		0.06	0.3	0.2	50.0	30.0	2.0	—	—	—	0.2	—
	HR6W[a]		0.08	0.4	1.2	43.0	23.0	—	6.0	—	0.18	0.18	0.003
	Inconel 617			0.40	0.4	54.0	22.0	8.5	—	—	—	—	—
	Inconel 617[b]		0.05	—	—	51.5	43.0	—	—	—	—	—	—

[a] Not ASME code approved.

[b] Low-strength material for use in co-extruded tubing. For weld overlays. IN72 (44%Cr-balNi) is the matching weld wire.

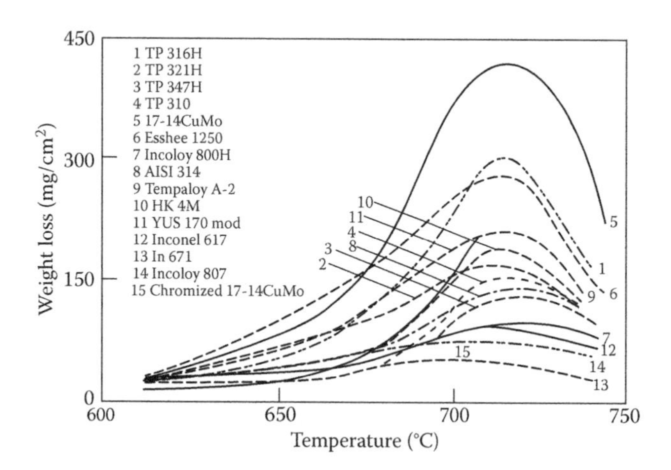

FIGURE 20.21
Reduction in corrosion at high metal temperatures of ~700°C with higher nickel in steel.

manufacturing methods on slightly differing materials. Table 20.4 provides a summary of ferritic and austenitic steels used in sub-c boilers of high p & t and SC boilers. These are taken from Central Research Institute of Electric Power Industry, Japan.

For the next level of SC boilers for 375 bar/700/720°C Ni-based alloys are required to be employed. These are still at various stages of development. Figure 20.21 sourced from EPRI shows the corrosion resistance to molten alkali sulphates of various materials exposed to hot flue gases from coal firing. It is evident that high Ni steels are needed as the temperature goes up.

2e. Water conditioning for SC boilers

The main objective of water conditioning is to ensure the integrity and safety of PP materials over a long time, while enabling a successful running of boiler-ST cycle. The following main differences between the sub-c and SC boilers demand that the contamination in water be kept as low as possible consistent with the requirement of the ST.

- There are no phase separating devices in SC boilers. There is no steam drum.
- The same fluid changes from water to steam as it passes from Eco to SH. fw, bw and steam are all the same fluid.
- Both sub-c and SC conditions prevail in a **sliding pressure** operation.

Corrosion products transported by fw and condensate should be contained to the lowest to minimise deposits in tubes and ST. This way the potential for damage to ST and fall in η are reduced. In SC boilers there is no way of removal of the contaminants unlike in sub-c boiler where there is blowdown from drum.

AVT and *oxygenated treatment* are the two methods of water conditioning for SC boilers to minimise general corrosion.

- AVT is the exclusive use of volatile conditioning agents. Volatile chemicals evaporate from the water and enter into the steam in gaseous form. When steam condenses, the chemicals dissolve into the condensate. As there are no solids, no scale formation takes place inside tubes. The common chemicals are NH_3, amines and N_2H_4 (or hydrazine substitutes).
- Oxygenated treatment provides a high electrochemical potential that promotes the formation of haematite (Fe_2O_3) layers, which are less soluble and hence more stable than the magnetite (Fe_3O_4) or mixed magnetite/haematite layers of AVT. As a result, the level of iron oxide in the fw is much lower which gives an excellent protective margin against flow assisted corrosion and minimises orifice fouling. The main issue with oxygenated treatment is that O_2 mixing with anions will be corrosive. A stricter control of fw impurities from condensate polishers is needed.

Table 20.5 provides general limits of contaminants for water at various stages in the steam cycle. Exact figures will be provided in each case by the consultant or boiler maker.

2f. SC Boiler start-up and bypass systems

In all boilers fluid flow through furnace tubes must be maintained at a level sufficient to cool the tube wall and prevent overheating. In SC boilers, where there is no circulation, during start-ups and lower loads the flow is too small to cool the tubes. Certain circulation has to be established by BFP by continuously pumping fw to the system to cool the water walls and SH surfaces prior to full firing. It is usual to design OT systems to ensure a flow rate of ~30% of rated flow at all times with the help of a ST bypass system to direct the steam formed in the boiler to flash tanks or condensers The start-up for OT boilers is, thus, different from and somewhat more complicated than drum-type boilers. OT boilers always come equipped with integral start-up systems, which vary in detail from manufacturer to manufacturer. The dumping of water and heat on start-up can be avoided by the installation of a start-up BCP. Without a BCP it would not be economical to operate for any period of time below the minimum once-though load.

The unit is initially fired, warmed up and brought to partial load on the bypass system. The BFP establishes a flow through the Eco and water wall tubes. The flow continues through the flash tanks (or separators) and on

TABLE 20.5

Acceptable Water Quality Limits for SC Boilers at Various Stages of Steam Cycle

		DM Water	Condensate	Fw (AVT Treatment)	Fw (Oxygenated Treatment)	SH and RH Steam
Parameter	Unit	at DM Plant Outlet	at Hotwell Outlet	At Eco Inlet	At Eco Inlet	At Boiler Outlet
Sp. conductivity	μS/cm	<0.10	2.5–11	4.0–11	2.5–7.0	2.5–7.0
	μS/cm	<1[a]	<0.20[b]	<0.20[b]	<0.20[b]	<0.20[b]
Silica as SiO_2	ppb	<10		<20		<20
Na + K	ppb	<5		<5	<5	<5
TOC	ppb	<300				
Dissolved O_2	–	Sat		<10	50–100[e]	
pH value			9.0–9.6[c]	9.2–9.6	8.5–9.0	9.0–9.6/9.2–9.6[c]
Iron as Fe	ppb	<20	<20	<20		<20
Hydrazine				[d]		

[a] At storage tank outlet including CO_2.
[b] After cation exchanger.
[c] For oxygenated treatment.
[d] NH_4 or equivalent should not be used if O_2 <10 ppb.
[e] Target for O_2 injection.

Note: 1. Values are consistent with long-term system reliability. A safety margin has been provided to avoid concentration of contaminants at surfaces.
2. All conductivities are referred to 25°C. Possible contributions from CO_2 may be excluded.
3. Operation is desirable at the lowest achievable impurity levels, with the shortest and least frequent excursions.
4. The specification is related to the following conditions:
- There are no Cu alloys in the system
- Conditioning is done with NH_3 and O_2 injection

to the condenser, then through the condensate polishers and back to the steam Gen.

A typical start-up schematic for a spiral wall variable pressure boiler is shown in Figure 20.22 taken from B&W. The main components are the vertical steam separators, water collecting tank and BCP. The steam separator acts similar to a drum in a natural circulation boiler. The fluid from this separator is recycled to the Eco inlet where it mixes with fw to maintain the minimum flow through the boiler furnace. Once the minimum load is achieved, the start-up system is put in warm standby for quick use when required. The steam from the separator passes through to the SH and then on to the ST. This system provides the maximum amount of heat recovery back to the boiler during start-up allowing for rapid start-ups with the least amount of thermal stress on the components.

2g. SC boiler arrangements

Figure 20.23 shows the arrangement of a typical large two-pass coal-fired SC boiler of ~660 MWe with spiral wall furnace and corner firing. There are 10 **mills** arranged 5 each on either side of the boiler. There are equal numbers of levels of **tangential tilting burners** in each corner of the furnace. There is a single rear pass with Eco and RH coils.SH is accommodated in the inter-pass and furnace. There is only a single **RH. Rotary AHs** are placed in the third pass separately.

20.2.3.3 Utility CFBC Boilers (see also Circulating Fluidised Bed Combustion Boilers)

Out of the various designs only **hot cyclone CFBC boilers**, either conventional **or compact**, have matured into utility segment. The others, namely, the **cold cyclone** and **internal recirculation CFBs** are expanded bed CFBs which were developed a few years after the hot cyclones were introduced. The gap is yet to be made up although these designs also appear to be suitable for larger sizes.

In the conventional hot cyclone design the largest boiler built at present is 300 MWe with four cyclones. The first of this class was the 250 MWe commissioned in the mid 1990s in Gardanne in France by former Stein Industrie, now part of Alstom, whose picture is given in Figure 20.24. Here the cyclones are placed two on either side of the furnace while in the next smaller size of boilers in 150 MWe class they were placed between the first and the second passes (as shown in Figure 3.13). **FBHEs** are used in this boiler. Furnace is split into two parts into a pant leg design with two fluidisation chambers while there was a single chamber in the smaller size. The boiler was designed to generate 700 tph of steam at 163 bar and 565°C with high A (~30%) and high S (~3.7%) coal.

In compact design of Foster Wheeler the largest CFBC boiler in operation since 2009 is a 460 MWe unit in Poland. This boiler is currently the largest CFBC boiler in the world and also is the first CFBC unit with SC

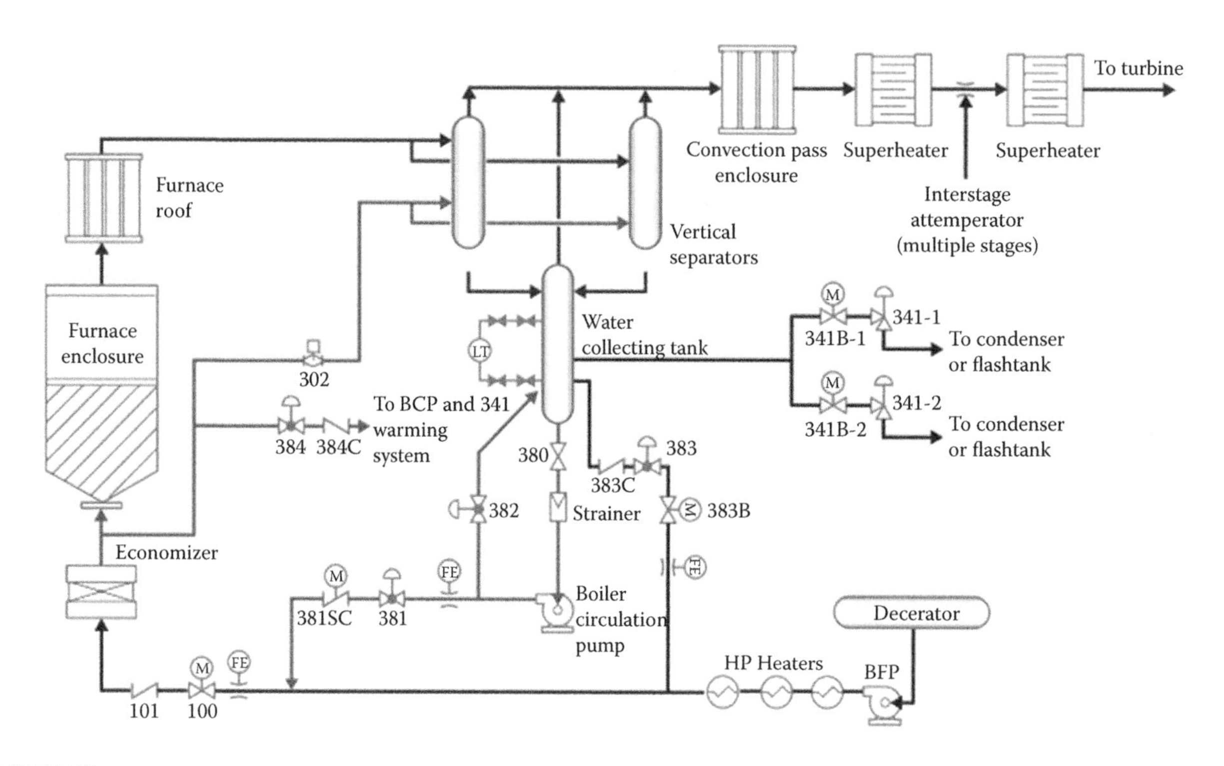

FIGURE 20.22
Start-up schematic of spiral walled SC boiler with variable operation. (From The Babcock and Wilcox Company, USA, with permission.)

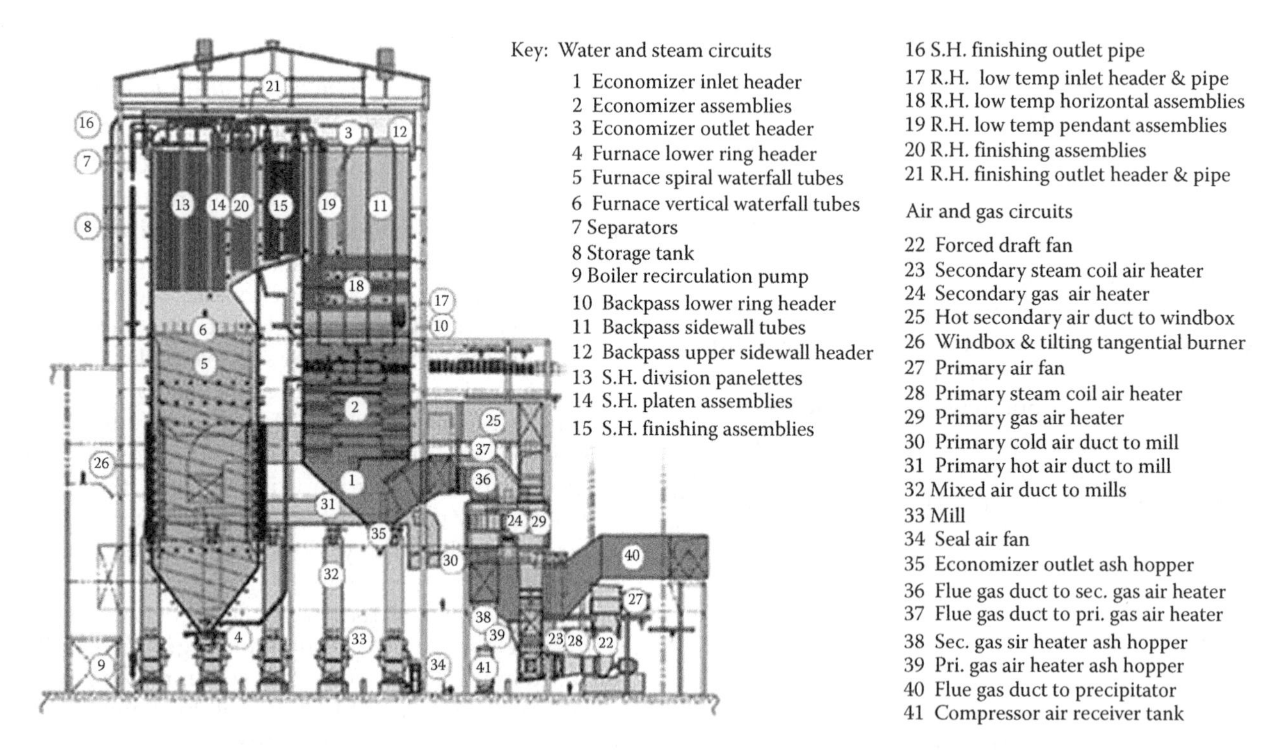

FIGURE 20.23
Two pass, coal fired, spiral furnace, corner fired SC boiler with a single rear pass and single RH.

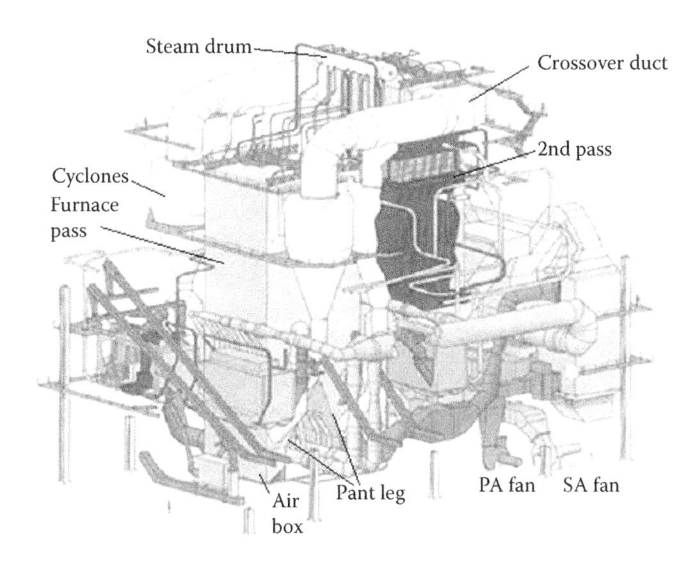

FIGURE 20.24
The first 250 Me boiler with four cyclones and pant leg arrangement of bottom combustor.

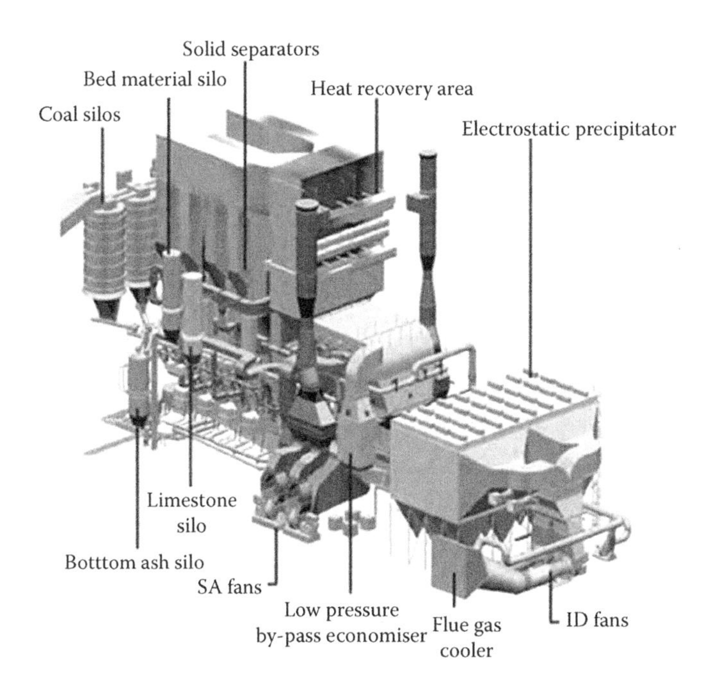

FIGURE 20.25
460 MWe SC RH boiler of compact design in operation in Poland. (From Foster Wheeler Corporation, USA, with permission.)

steam parameters. Its SH and RH steaming conditions are as follows:

- Flow SH/RH tph 1300/1100 kpph 2860/2423
- SOP/RHOP bar 275/55 psia 3990/800
- SOT/RHOT °C 560/580°F 1040/1076
- FWT °C 290°F 554
- Exit gas temperature °C 122°F 252,

The main fuel is coal with the flexibility of supplementary burning of coal sludge up to 30%, coal washery rejects up to 50% and bio-fuels in future up to 10%. Coal has M of 6–23%, A of 10–25% and S of 0.6–1.4% with LCV of 19–22 MJ/kg, ~4540–5255 kcal/kg or 8170–9460 Btu/lb. The emission limits are 200 mg/NM^3 for NO_x, SO_x and CO and 30 mg/NM^3 for particulates as per the current EU norms.

There are four compact dust separators on either side of the combustor with eight fuel feeding points. INTREX type FBHEs are placed in the return legs of all the separators for obtaining the necessary high steam temperatures Figure 20.25 gives an isometric view of boiler layout.

20.3 Ultimate Analysis (*see* Coal)

20.4 Ultrasonic Testing (*see* Non-Destructive Examination)

20.5 Ultra Supercritical Boilers (*see* SC Boilers)

20.6 Universal Pressure Boiler (*see* Supercritical Boilers)

21

V

21.1 Vacuum Degassing of Steel (*see* Metallurgy)

21.2 Valve Wide Open Condition (*see* Boiler Rating)

21.3 Valves

Valves are the *equipments in a fluid flow system that regulate the flow, pressure or both of the fluid.* This duty may involve stopping and starting flow, controlling flow rate, diverting flow, preventing back flow, controlling pressure, or relieving pressure. These duties are performed by adjusting the position of the closure member in the valve. This may be done either manually or automatically. Sealing, performance and flow characteristics are important aspects in valve selection.

21.3.1 Boiler Valves

These are pressure valves on steam and water lines whose construction is guided by codes like ASME B16.4. Valves are supplied by the boiler maker and are integrated with the boiler PPs. Valves on oil, gas and other fuel lines are quite different in construction, guided by a different set of codes and are usually supplied by the burner makers.

Valves on a boiler are mainly for start–stop and draining services of steam and water. CVs are only for regulating the fw and attemperator spray water. There is no direct regulation of steam in any boiler for output control, which is carried out only by regulating the inlet fw. However, CVs are at times employed in steam circuit in applications like drum attemperator where three-way CVs are used for dividing steam into the attemperator and bypass lines. Butterfly CVs are employed for larger sizes.

A bulk of the boiler valves is made up of *GGC* category, a trade acronym for *gate, globe* and *check* valves. Additionally, there are SVs and CVs.

21.3.2 Valve Terminology

The terms used in valve construction are first defined. A manually operated LP gate valve with flanged ends and bolted bonnet having inside screw and rising stem is shown in Figure 21.1 where the various parts are labelled.

Actuator: A device that operates a valve by utilising electricity, pneumatics, hydraulics, or a combination of one or more of these energies. Sometimes, actuators are referred to as *operators*. A hand wheel is in fact a manually operated actuator.

Bonnet: A valve body closure component that contains an opening for the stem.

Bore: The diameter of the smallest opening through a valve. It is also called *port*.

Bubble tight: A valve is termed *bubble tight* when the upstream side of the valve is pressurised with air and the downstream side is filled with water and no air bubbles are detected on the downstream side with the valve in fully closed position.

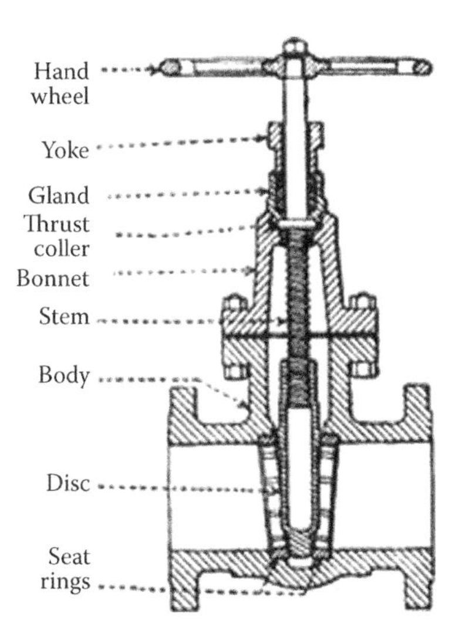

FIGURE 21.1
Manually operated LP gate valve with flanged ends.

Disc: The part of the valve which is positioned in the flow stream to permit or to obstruct flow, depending on closure position. In specific designs, it may also be called a *wedge, plug, ball, gate*, or similar.

Double-seated valve: A valve with two separate seating surfaces that come in contact with two separate seating surfaces of a disc or a double disc.

Full bore: When valve *bore (port)* is approximately of the same size as the inside diameter of the connecting pipe, it is called *full bore (full port).*

Fully stellited: A valve is termed fully stellited when seating surfaces of the valve seat(s) and the disc(s) are hard faced with wear- and corrosion-resistant material (***stellite*** or other such material).

Half stellited: A valve is termed half stellited when only the seating surfaces of the valve seat(s) are hard faced.

Hard facing: It is the application of hard, wear- and corrosion-resistant material on soft surfaces subject to wear.

Inside screw–non-rising stem (ISNRS): Threads on the stem are inside the valve body. The valve disc travels up and down the stem threads when the stem is rotated. The stem does not rise.

Inside screw–rising stem (ISRS): Threads on the stem are inside the valve body and exposed to the flow medium. The stem rises when it is rotated, thus opening the valve. Position of the stem indicates the position of the valve disc see Figure 21.1.

Leak-off connection: A pipe or tube connected to a hole in the stuffing box at the level of lantern ring. It is used to collect leakage past a lower set of lantern ring–type of packing or to inject lubricant into the stuffing box.

Non-rising stem: Refer to Inside screw, non-rising stem.

Outside-screw-and-yoke (OS&Y): A valve design in which the threaded portion of the stem is outside the pressure boundary of the valve. The valve bonnet has a yoke, which holds a nut through which the rotating stem rises as the valve is opened. The stem part inside the valve is smooth and is sealed so that stem threads are isolated from the flow medium.

Port: See Bore.

Quarter-turn valve: A valve whose closure member rotates approximately a quarter turn (90°) to move from full-open to full-closed position.

Reduced port: Valve port smaller than the inside diameter of the end-connecting pipe. It is approximately equal to the inside diameter of the one pipe size smaller than the end size for gate valves, and 60% of full bore on ball valves.

Regular port: A valve port smaller than the full bore, approximately 75–90% of full bore on ball valves and 60–70% on plug valves.

Rotary motion valve: A valve that involves a quarter-turn motion to open or close the valve closure element.

Seat: The portion of the valve against which the closure member presses to effect shut-off.

Seat ring: A separate piece inserted in the valve body to form a seat against which the valve-closure member engages to effect shut-off.

Short pattern valve: A valve that has face-to-face or end-to-end dimension for a short pattern design according to standard ASME B16.10.

Stroke: The amount of travel the valve-closure member is capable of from a fully closed position to a fully open position or vice versa. In linear-motion valves, it is expressed in mm (inches) and 0–90° for rotary motion valves.

Trim: Functional parts of a valve which are exposed to the line fluid. Usually refers to the stem, closure member and seating surfaces. The removable or replaceable valve metal internal parts that come in contact with the flow medium are collectively known as valve trim. Valve parts such as body, bonnet, yoke and similar items are not considered trim.

Venturi port: A valve bore or port that is substantially smaller than the full port, ~40–50% of full port. It is normally found in plug valves.

Wafer body: A valve body that has a short face-to-face dimension in relation to pipeline diameter and is designed to be installed between two flanges using special-length studs and nuts.

Wedge: A gate valve–closure member with inclined sealing surfaces which come in contact with valve-seating surfaces that are inclined to the stem centre line. Wedge is available in solid, split and flex designs.

Yoke: That part of the valve assembly used to position the stem nut or to mount the valve actuator.

Yoke bushing, yoke nut: Yoke nut, yoke bushing, or *stem nut* is the valve part that is held in a recess at the top of the yoke through which the stem passes. It converts rotary-actuating effort into thrust on the valve stem.

TABLE 21.1

Materials Used in the Construction of Cast and Forged Steel Valves

Nominal Composition	Forgings		Castings		Bars		Tubulars	
	Spec.	Grade	Spec.	Grade	Spec.	Grade	Spec.	Grade
CS	A105		A216	WCB	A675	70	A672	B70
					A105			
C–Si–31/2Ni	A350	LF3	A216	WCC	A350	LF3	A106	C
C–1/2Mo			A217	WC5	A182	F2	A691	CM-75
1 1/4Cr–1/2Mo	A182	F11 Cl.2	A217	WC6	A182	F11 Cl.2		
2 1/4Cr–1Mo	A182	F22 Cl.3	A217	WC9	A182	F22 Cl.3		
5Cr–1/2Mo	A182	F5	A217	C5	A182	F5		
9Cr–1Mo	A182	F9	A217	C12	A182	F9		
18Cr–8Ni	A182	F304	A351	CF3	A182	F304	A312	TP304
		F304H		CF8		F304H		TP304H
9Cr–1Mo–V	A182	F91	A217	C12A	A182	F91	A335	P91

21.3.3 Classification Based on Valve Size

Valve size is denoted by the nominal pipe size (NPS), which is equal to the size of valve-connecting ends or the flange-end size. In the metric system, valve size is designated by the nominal diameter (DN) of connecting pipe or the connecting flange ends. When a valve is installed with reducers on each end, the size of the valve will be equal to the size of the reducer-connecting ends attached to the valve. The valve size is not necessarily equal to the inside diameter of the valve. It is a normal industry practice to categorise valves, based upon size, into two classifications—small and large, the dividing line being 2/2½″ (50/65 NB) as there is no industry standard for this.

21.3.4 Materials of Construction of Valves

Table 21.1 gives a typical list of materials of construction for boiler valves.

21.4 Valves in Boiler

The main valves in a drum-type boiler are

- MSSV
- NRV
- Start-up vent isolating valve
- fw stop and feed check valves
- SVs on drum, SH and RH
- CBD and IBD valves
- Main, low load and start-up fw CVs and their bypass and isolation valves
- Attemperator control and isolating valves
- Boiler circulating pump's isolating valves on either side of pump
- Drain valves

Many are conventional globe, gate and check valves. Valves specific to boilers are described below.

21.4.1 Blowdown Valves: Continuous and Intermittent (CBD and IBD)

A blowdown valve is used to discharge the contents of a pressure vessel or piping. In a boiler, they are for

- Continuously draining the drum water for controlling **silica** in steam drum (**CBD**)
- Periodically discharging (blowing off) the **sludge** from bottom drum/headers (**IBD**)

respectively, and therefore, the construction of the two valves is entirely different.

CBD valve is a micrometer valve, that is, a globe valve with an elongated plug, amenable to fine control (as shown in Figure 21.2). It is invariably motorised, unless the boiler is quite small and CBD valve is close to control room.

The IBD valves, on the other hand, are usually quarter-turn or equivalent quick opening valves with high lift to permit effective flushing out of sludge. This valve is required to be operated about once in a shift and motorisation is for operator convenience. See Figure 21.3.

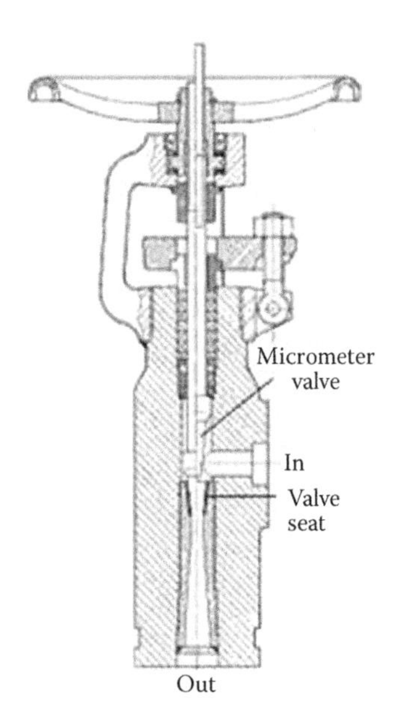

FIGURE 21.2
Continuous blowdown (CBD) valve with welded ends in forged steel.

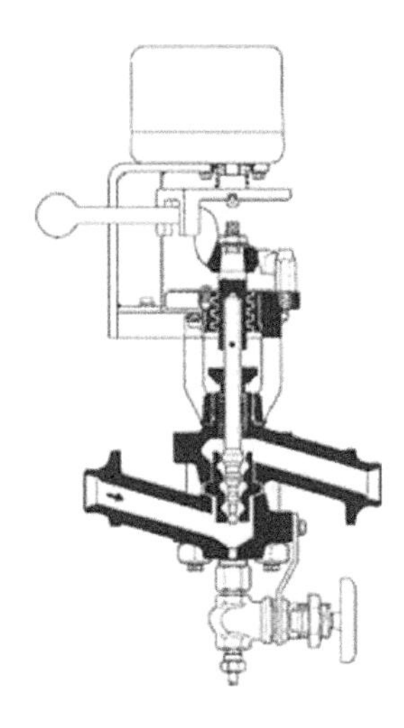

FIGURE 21.3
Motorised intermittent blowdown (IBD) valve.

Both CBD and IBD valves operate at the saturation temperature and are mostly made of CS.

21.4.2 Control Valves

A CV is the *final control element* in a process control loop. It manipulates a flowing fluid to compensate for the load disturbance and keeps the process variable as close as possible to the desired set point. Usually, by CV, one is referring to the entire CV assembly which typically consists of

- Valve body
- Internal trim parts

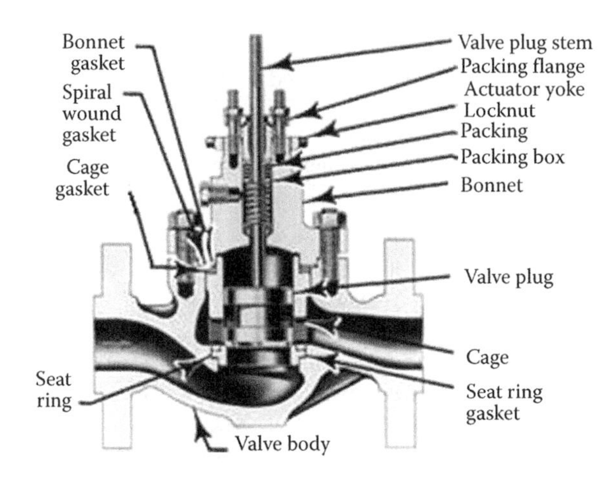

FIGURE 21.4
A typical sliding stem, single-seat globe-type CV assembly.

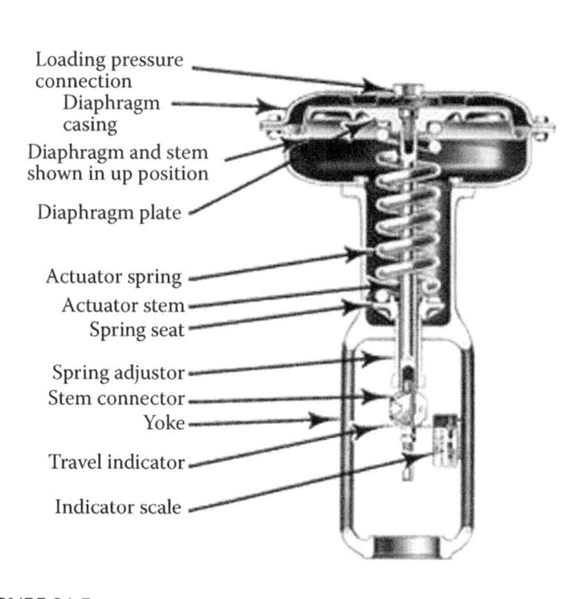

FIGURE 21.5
Diaphragm actuator.

- Actuator to provide the motive power to operate the valve
- Additional valve accessories which include positioners, transducers, supply pressure regulators, manual operators, snubbers or limit switches

Figure 21.4 shows a typical sliding stem, single-seat globe-type CV assembly with all major parts identified.

Figure 21.5 is the cross section of a typical direct acting diaphragm actuator which is the most popular type. For bigger valves and HPs needing greater force, a piston-type actuator is employed.

Some common CV terms of importance are listed below.

Cage is part of a valve trim that surrounds the closure member. Cage provides flow characterisation and/or a seating surface. It also provides stability, guiding, balance and alignment, and facilitates assembly of other parts of the valve trim. The walls of the cage contain openings that usually determine the flow characteristic of the CV.

Plug is the closure member.

Port is the flow control orifice.

Seat: The area of contact between the closure member and its mating surface that establishes valve shut-off.

Trim are the internal wetted components of a valve that modulate the flow of the controlled fluid. In a globe valve body trim would typically include closure member, seat ring, cage, stem and stem pin.

Some common process and performance terms are given below.

Hunting is an undesirable oscillation of appreciable magnitude, prolonged after external stimuli disappear. Sometimes called cycling or limit cycle, hunting is an evidence of operation at or near the stability limit.

Rangeability is the ratio of the maximum to minimum controllable flow rates.

Rated flow coefficient (Cv) is of the flow coefficient of valve at the rated travel.

Rated travel is the distance of movement of the closure member from the closed position to the rated full-open position. The rated full-open position is the maximum opening recommended by the manufacturers.

Seat leakage is the quantity of fluid passing through a valve when it is in the fully closed position with pressure differential and temperature as specified.

Vena contracta is the portion of a flow stream where fluid velocity is at its maximum, and fluid static pressure and the cross-sectional area are at their minimum. In a CV, the vena contracta normally occurs just downstream of the actual physical restriction.

Some salient features of boiler CVs are

- Large-size CVs in a boiler are only in the fw circuit. They are low-temperature valves requiring no more than CS execution. AS is to reduce the thickness and weight.
- Both SH and RH attemperator spray CVs, dealing with water quantities of not more than 10% of fw, are much smaller than FCVs and of similar construction. Some codes insist on attemperator and its isolating valves and piping designed for the same conditions as the steam line on which it is mounted, assuming the worst case of steam flowing back to a closed attemperator line. AS valves of appropriate rating are then required.
- Single-seated CVs with low leakage are the normal choice. For better control and economy, it is normal to adopt a CV of one size smaller than the line size.

Normal boiler CVs can be considered as improvised globe valves for water duty with definite flow versus Δp characteristics made possible by the special shape of the trim. Flow characteristics may be linear, equal percentage and modified parabolic where unit change in lift produces flow change of constant unit, constant % age or a combination of the two, respectively. This is depicted in Figure 21.6.

The trim is made of hard AS materials of hardness levels of Rc 43–47 to withstand erosion because of wire drawing, the cutting effect, produced by high-velocity fluid stream resulting from high Δp in the CV. FCVs are one of the most noisy equipments in a boiler house and it is required to reduce their noise level. This is possible by reducing the fluid velocities inside the cage and plug of CV by various ingenious ways. Several valve trims are possible, some of which are illustrated in Figure 21.7. Depending on the specific need, a suitable trim is selected.

Cavitation occurs when the fluid pressure at the *vena contracta* falls below the vapour pressure, followed by pressure recovery above the vapour pressure (as shown in Figure 21.8). The excursion of pressure below vapour pressure causes vapour bubbles to form, which then collapse as the pressure recovers. Collapsing bubbles can cause erosion of valve and downstream pipe metal

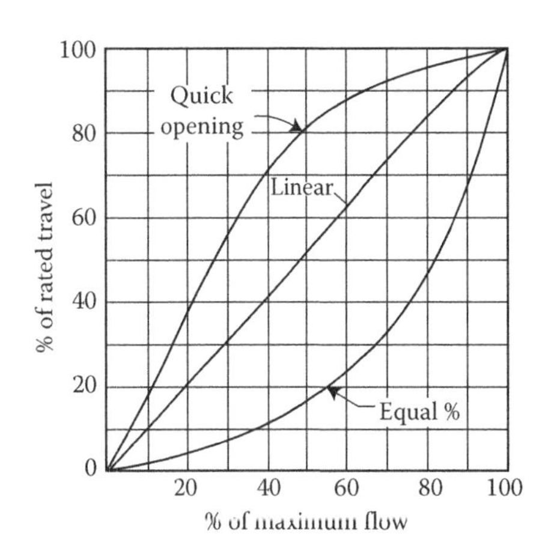

FIGURE 21.6
Flow versus lift for various types of trims.

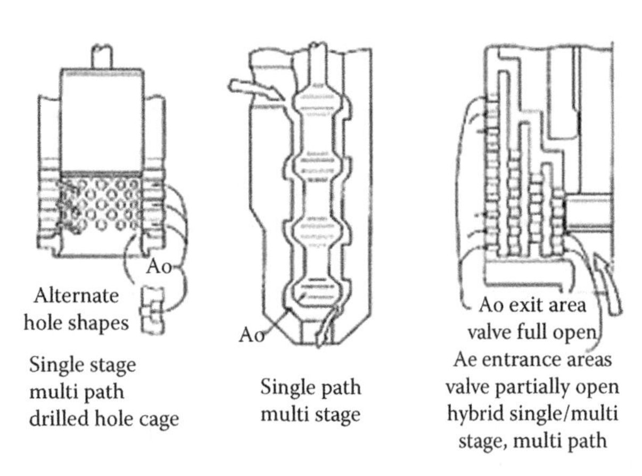

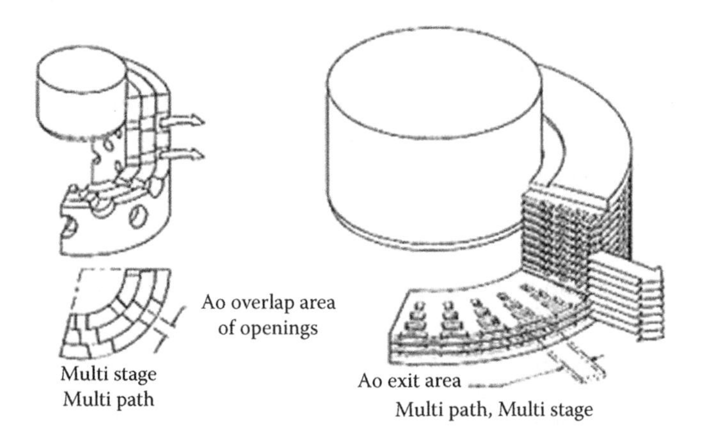

FIGURE 21.7
Examples of valve trims.

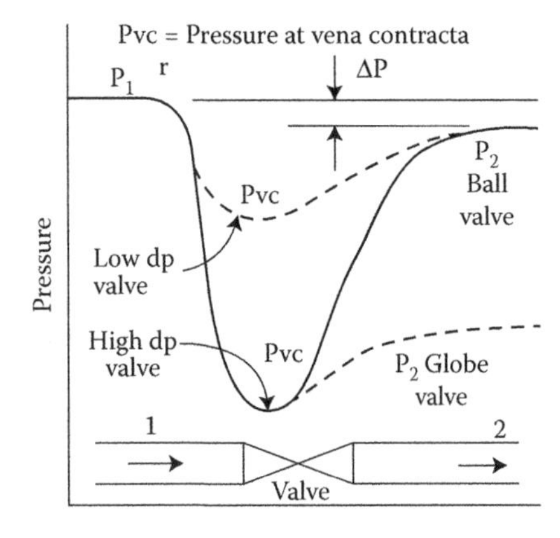

FIGURE 21.8
Pressure reduction and recovery in flow through valve.

surfaces. The phenomenon of *fluid pressure falling below and recovering above vapour pressure combined with forming and collapsing of bubbles is termed cavitation* which is to be avoided at all times.

Cavitation is avoided by a progressive reduction of pressure in place of abrupt reduction and preventing low pressures at vena contracta that can induce formation of bubbles. Pressure recovery depends on valve design. Ball and butterfly valves have a higher pressure recovery than globe valves.

Seat tightness is an important feature of CVs and is classified according to ANSI B16.104 where there are six classes of seat leakages ranging from class 1 to class 6 with increasing levels of seat tightness.

Actuation of CVs: For attemperator CVs, diaphragm actuation is adequate. FCVs also are usually diaphragm actuated but for larger sizes, piston-operated valves are employed. The main FCV is usually pneumatically actuated, with an identical electrically actuated CV as a standby, ready to be pressed into service if the compressed air fails.

FC station: In small boilers, a start-up FCV of 10–15% capacity is provided in parallel to the main FCV for boiler filling and start-up duty. In medium and larger boilers, it is customary to have a separate boiler fill pump along with valves. Also, it is a practice to have the low load running up to 30% or 40% MCR with low-load CVs which cut in and out automatically based on the flow.

CV size: CVs are usually chosen one size lower than the line in which they are located for better controllability. There are gate valves for isolation on both up- and downstream for on load valve removal for maintenance.

Three-Way CV: In drum-type attemperator, control steam flow is required to be divided between flow through drum and bypass which is done by three-way CVs or a pair of butterfly CVs when the flows are larger. Three-way valves can be for diverting or mixing duties which is illustrated in Figure 21.9.

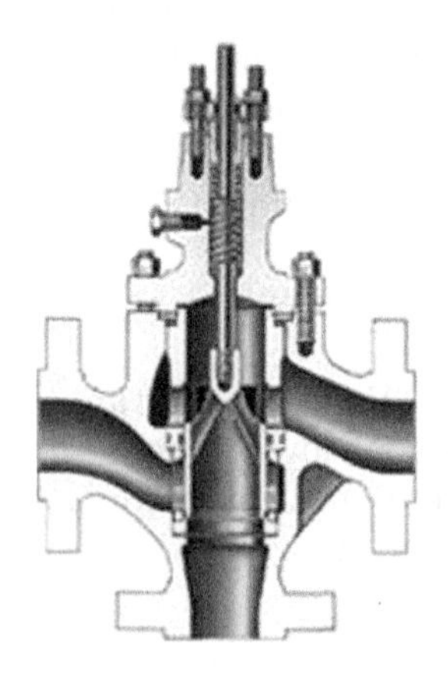

FIGURE 21.9
Three-way mixing CV.

FIGURE 21.10
Butterfly-type CV.

A suitable trim is selected depending on the flow requirement. Figure 21.10 shows a typical butterfly valve.

Combined spray and desuperheating (CSDH) valve:

CSDH valve, combining the functions of the spray nozzle and CV, are used in the boilers many times these days because of some definite advantages. A typical CSDH valve is shown in Figure 21.11. Control of spray water is by the actuation of the central spindle which uncovers more nozzles as the load is increased. Normal CV is replaced by this special arrangement:

- The spray quality remains fine over the whole range of operation because of multiple nozzles which are brought into action one by one as the load increases.
- As the water particles are very fine in size they are absorbed by the flowing steam with no fear of water impingement on the thick pipelines. The need of the sleeve on the downstream of the attemperator is eliminated.
- The straight distance requirement of the steam pipe is also relaxed as the water mist evaporation is faster.
- The entire arrangement can be more compact and the instrumentation can also be simpler.
- The turndown on the steam side can be 10:1 and on the water side 300:1 which gives a much wider range for operation.

21.4.3 Main Steam Stop Valve

This is perhaps the heaviest and most critical of all valves in a boiler and the one that experiences in full

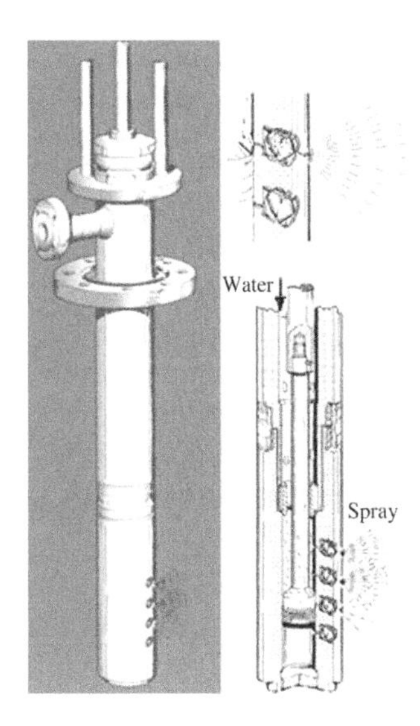

FIGURE 21.11
Combined spray and desuperheating (CSDH) valve.

the SH steam p & t. single MSSV is employed for most industrial boilers and boilers up to ~200 MWe size in utilities. For larger boilers, a pair of valves is used in parallel with two SH outlets.

In a variable-pressure boiler, many times the MSSV is merged with main **ST** inlet stop valve and located at the ST end.

MSSV is a heavy-duty isolating gate-type valve of the same size as the main steam line or a size smaller. It is required to have minimum steam Δp. Simple gate valves are used in lower pressures. But wedge gates are superior from leak tightness and maintenance considerations. An MSSV can be a *wedge gate, parallel slide or venturi* type:

- *Wedge gates* are preferred for LP and MP and also medium sizes. The tapered faces of the disc provide a positive sealing which is a big advantage. In larger sizes, however, the fluid forces on the disc tend to bend it, making parallel slide valves preferable.
- *Parallel slides* have two discs forcing themselves on the two faces by a set of springs to provide sealing. But the real sealing force is exerted by the upstream steam pressure on the disc. The thin discs are more flexible and can withstand more p & t than gates but are costlier and demand more maintenance.
- *Venturi* construction of the valve body has a taper at inlet and outlet so that the actual sealing disc has smaller diameter than in parallel slide valves. Venturi is preferred for very large sizes even though they are more expensive to make and maintain, mainly to reduce the line losses and maintenance.

The salient constructional aspects of MSSVs are as follows:

- Usually MSSVs are motorised and butt welded but in smaller LP process boilers they can be hand operated and flanged.
- It is normal to have integral body bypass valve of 15–25 (½–1″) NB for pressure equalisation of upstream and downstream sides to permit easy opening of MSSV for sizes of 150 (6″) NB and larger. Motorising of bypass valve and interlocking it with the main valve are common for large HP boilers.
- MSSVs vary a great deal in size, from as small as about 80 (3″) NB in small process boilers to as large as 600 (24″) NB or more in large utility boilers.
- The metallurgy varies from CS for steam temperatures up to about 450°C (842°F) to 1¼%, 2¼% and 9% Cr alloy steels for temperatures up to about 600°C (1112°F). Table 21.1 gives material selection criteria.
- The sealing of a valve against leakage of fluid from inside is achieved in two ways, either by bolting of the bonnet or by pressure sealing from inside, as shown in the wedge gate and parallel slide valve pictures in Figures 21.12 and 21.13, respectively. Pressure seal construction provides a more positive sealing over bolted construction besides being smaller and lighter as the internal pressure is what is used for sealing. But the valve is sealed for life unlike in bolted construction which permits inspection and access of valve internals whenever required. Pressure sealing is adopted for higher pressures.

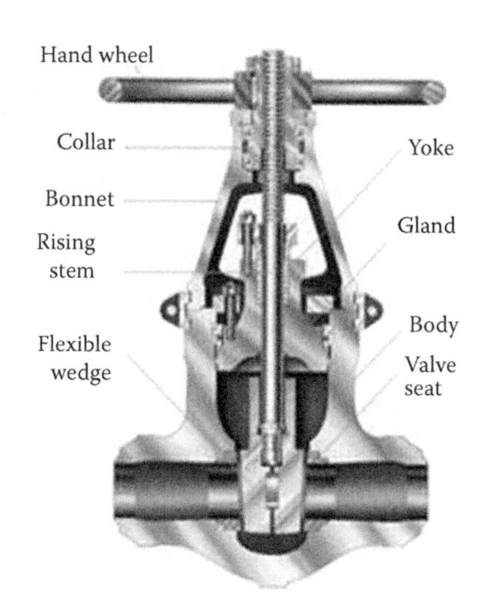

FIGURE 21.12
HP wedge gate-type MSSV with rising stem and pressure seal bonnet.

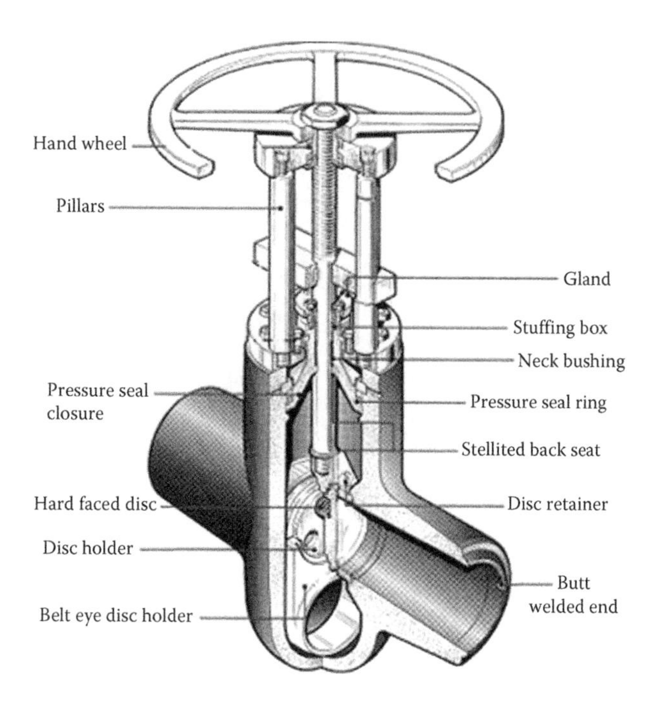

FIGURE 21.13
HP parallel slide-type MSSV with non-rising stem and pressure seal bonnet.

21.4.4 Non-Return Valves for Steam and Water

An NRV on steam line is employed when more than one boiler is connected to a common steam header to prevent back flow of steam into the idle boiler. This valve has to be of *swing check* design to limit the Δp of steam. It is of the same size as MSSV with the same material of construction. Figure 21.14 shows a typical swing check valve.

Swing check valves are self-actuated and require no external means to actuate the valve either to open or to close. They are also fast acting. As all moving parts are enclosed, it is difficult to determine whether the valve is open or closed. Furthermore, the condition of internal parts cannot be assessed. At times, valve disc can stick in open position.

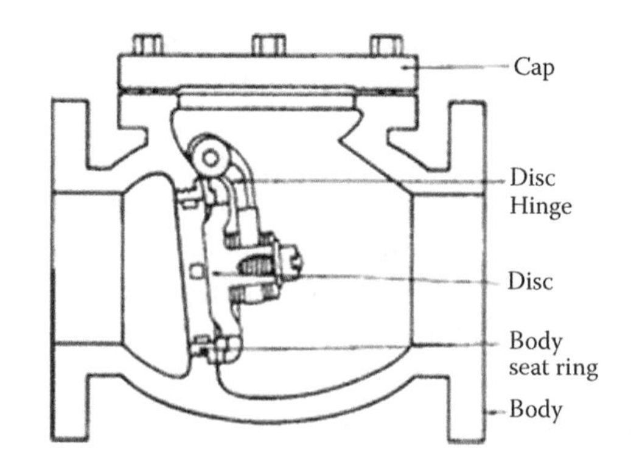

FIGURE 21.14
CS swing check valve for LP and MP application with flanged ends.

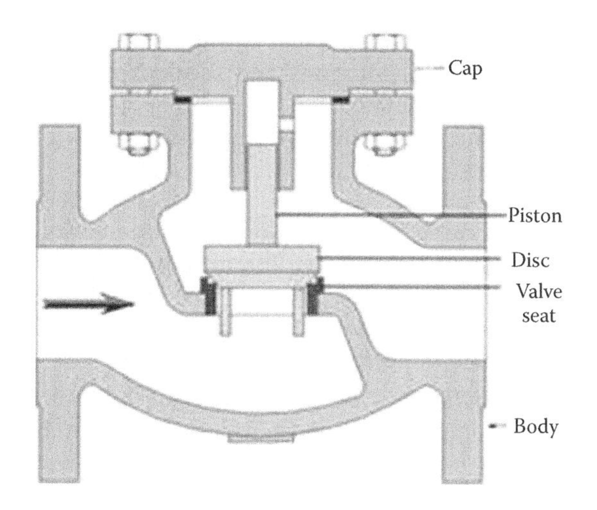

FIGURE 21.15
CS piston check valve for LP and MP application with flanged ends.

In the fw line, however, the compact and economical *lift check* valve is more commonly used as more Δp is permissible. Figure 21.15 depicts a typical LPlift check valve with flanged ends. Metallurgy of NRV in the fw lines is normally CS.

Sometimes the fw stop and check valves are combined into a *screw down non-return valve (SDNR)*. They are also called *stop check* valves. A typical manual piston check SDNR valve with flanged ends is shown in Figure 21.16.

NRVs are also required in water lines supplying spray water to the attemperators. Being small, valve ball-type construction may be adequate.

21.4.5 Safety Valves

SVs are the only glandless valves in boiler use. Since they are the valves that ultimately protect the boiler from over pressure, codes lay down detailed norms for their design, construction, sizing, selection and approval. ASME Boiler code, paragraphs 67.1-PG deals with SVs.

Valve selection and sizing: Sizing of the SVs in drum-type boilers is largely governed by the following principles. Boiler codes provide more exhaustive guidelines:

- The combined relieving capacity of all SVs (excluding RH SVs) should exceed the boiler evaporation.
- SVs for SH should be 20–30% of boiler evaporation to provide sufficient cooling to the SH tubes during blowing.
- Drums should have at least two SVs.
- The relieving capacity of RH SVs should exceed RH flow.
- Minimum number of SVs should be placed on the RH outlet header to provide ~15% RH flow during blowing from cooling considerations of RH tubes. The rest of the SVs should be on the inlet header to lower their metallurgy, size and cost.
- Drum SVs are normally made of CS, while SH and RH SVs can be either CS or AS depending on the steam temperature. Relieving capacity reduces with higher SH as steam attains greater specific volume.

SV terminology:

- *Relieving capacity* is the amount of steam that an SV can discharge when the valve is fully open.
- *Rated or certified discharge capacity* is the discharge capacity of a theoretically perfect nozzle having a cross-sectional flow area equal to the flow area of the valve multiplied by the rated or certified coefficient of discharge.
- *Actual discharge capacity* is the rated capacity which is multiplied by the actual instead of rated coefficient of discharge.
- *Coefficient of discharge* is the ratio of the measured to the theoretical relieving capacity of a pressure relief valve.
- *Coefficient of flow* is the flow rate that passes through the fully open valve at unit Δp. It is measured in gpm (3.8 L/min) of 60°F (16°C) water with 1 psi (6.9 kPa) Δp. It is also referred to as *flow coefficient* or *valve coefficient*.
- *Set pressure* is the pressure at which SV begins to lift to relieve the internal pressure.
- *Over pressure* is the maximum pressure increase attained when the SV is blowing expressed as % of set pressure. ASME limits it to 3%.

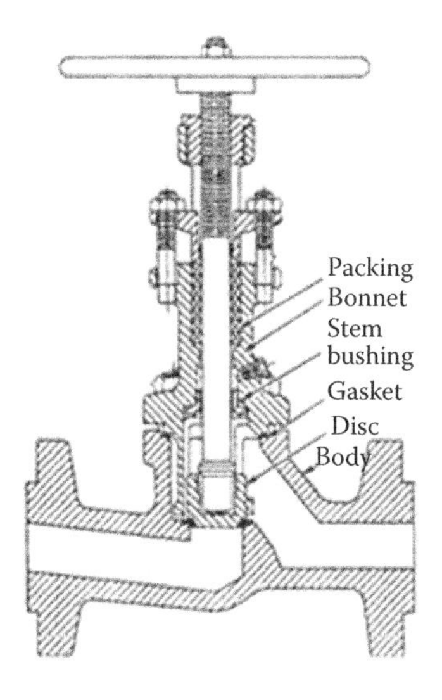

FIGURE 21.16
Screw down non-return/stop check valve.

- *Relieving pressure* is set pressure + over pressure.
- *Reseat pressure* is the pressure at which SV stops blowing and the valve seat establishes contact with the nozzle.
- *Back pressure* is the pressure that exists at the outlet flange as a result of the pressure in the discharge system. It is the sum of built-up and superimposed back pressures, expressed as a % of the set pressure or in pressure units.
- *Blowdown* is the difference between the set pressure and the disc-reseating pressure expressed as a % of the set pressure. ASME limits it to 2%.
- *Chatter* is the abnormal rapid reciprocating motion of the movable parts of a pressure relief valve in which the disc contacts the nozzle.
- *Flutter* is chatter in which the disc does not contact the nozzle.
- *Crawl* is the gradual decreasing of the set pressure of spring-loaded pressure relief valves from below to normal after the temperature of the spring has been raised by the fluid just discharged.
- *Simmer* is the audible or visible escape of compressible fluid between the seat and the disc at no measurable capacity as the set pressure is approached

Classification of SVs:

- SVs are classified as *high lift* and *full lift* depending on how much the valve lifts under full pressure. Full-lift valves open more than high-lift valves and hence have higher discharge.
- *Spring, torsion bar* and *pilot-operated* valves are the three types of SVs with increasing precision and cost. Spring-type SVs are the most popular and economical.
- *Safety, relief* and *safety relief valves* are the three classes of valves.
    a. *Safety valves* are characterised by full and instant lift of seat (pop lift) and full discharge on attaining the set pressure. They are for compressible fluids. Besides boilers SVs are also used for vapour and gas service.
    b. *Relief valves* have proportionate lift and discharge as the pressure increases above the set pressure. Such valves are for non-compressible fluids and are used for liquid service.
    c. *Safety relief valves* can be used for both water and vapour/gas services.

Figure 21.17 shows a typical spring-operated SV. The blowdown adjustment rings and the locking pins are shown in Figure 21.18.

Some of the important layout-related aspects of SVs are as follows:

- Great care in installation and operation is needed in protecting delicate equipment like SVs. Seat and disc are never to be subjected to scouring action of foreign particles lest the valve should leak.
- *Hydrostatic plugs* are to be compulsorily inserted at the entry to the inlet nozzles during any **HT** or **boil out** operations to prevent ingress of water or chemicals which can damage valve internals.
- Frequent lifting of SV is sure to lead to leakages calling for boiler stoppage for seat lapping and

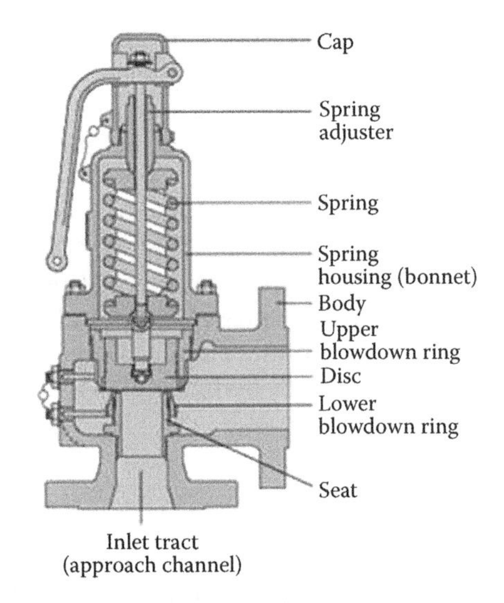

FIGURE 21.17
Spring-operated safety valve.

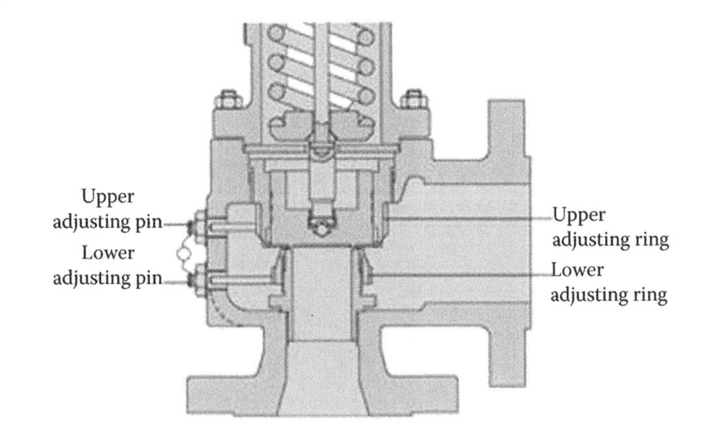

FIGURE 21.18
Blowdown adjustment by upper and lower rings in a spring operated SV.

reassembly. Often, this can be more expensive than the valve itself. This can be prevented by the installation of electromatic safety valves, explained later.

- SV is a glandless valve and all the internals are made from ss with mirror finish to provide high-quality alignment that stays for years. It will also ensure accurate and perfect pop action while lifting and reseating. The SV blowing (and reseating pressures) and blowdown are set at the manufacturer's end in the presence of an inspector and the rings are locked in position requiring no further adjustment at site. However, a provision exists for adjusting the blowdown at site by means of adjusting rings.
- SV chatter inevitably leads to valve leakage needing lapping of the seating surfaces to restore the tightness. Chatter is caused mainly by faults in installation or a long inlet pipe or faulty discharge pipe system.
- Spacing of SVs on headers and drums is to be done correctly lest the discharge forces should interfere with each other.
- Likewise, the discharge/escape piping is required to be laid out properly to lead the discharge steam safely to a suitable safe location, which is many times the boiler roof. SV escape pipes are rigidly tied to any nearby boiler column while the SVs are located on drum or piping having their own thermal expansions. A suitable expansion chamber between the SV discharge elbow and the escape pipe is essential to eliminate stresses because of differential expansion between the fixed escape piping and moving SV. The expansion chamber should be provided with suitable drain.
- The inlet nozzle length of the SV should be as short as possible to minimise the Δp which can lead to chatter. The SVs are usually, therefore, directly mounted on drums or pipes. The Δp in the inlet pipe has to be less than 50% of the SV blowdown. There is an upward force created during blowing as steam which takes a 90 degree turn inside the valve imposing downward stress on the drum and piping. To avoid this extra stress on PPs, the SVs in such cases can be located at some distance and rigidly supported on structural members. The inlet piping in such cases has to be carefully laid out and the set pressure of SV has to be selected taking into account the pressure loss in the inlet pipe for proper operation of the valves.

21.4.5.a Electromatic SV

After a few lifts, damage to internals which lead to SV seat leakages are inevitable. Therefore, it is normal these days to install an additional electrically assisted SV on SH header which operates first and saves the main SVs from lifting at all. Popularly known as the electromatic SV by its trade name, the discharge from this valve is not counted in the overall relieving capacity and is not recognised by the boiler codes. This valve is provided with an isolating knife gate valve which is kept open at all times except when the electromatic SV is required to be taken for its maintenance.

An additional advantage of this valve is that it can be lifted at any pressure from the control room to purge the pipeline.

21.4.6 Start-Up Vent Valves (see Vent Piping)

21.4.7 Air Vent Valves

The purpose of air vent valves is to help expel air from the inside of boiler PPs. Naturally, they are positioned at the highest points of the boiler on steam drum, SH headers, RH headers, Eco outlet pipes and any other headers located higher than these.

- **HT** at site to full pressure cannot be performed unless all air is expelled from PPs and the vent valves are closed.
- Air vents have to be opened during shutting the boiler down when pressure falls to ~0.3 atg to let the air in and avoid vacuum inside the boiler PPs.
- When raising pressure, they have to be closed at the same pressure after the vent steam flow is established to enable further pressurisation.

Air vent valves are small globe valves of usually 20–40 (¾–1½) NB size made of CS except for SH and RH vents which are of AS, usually of the same metallurgy as the header to which they are connected.

Vent valves are always motorised in medium and large boilers because of the remoteness of their location.

21.4.8 Drain Valves

Drain valves are meant for draining the water and condensate from various water walls and tube banks. Naturally, they are connected to the bottom headers.

Drain and vent valves together form nearly 80% of the total number of valves.

- Valves employed for draining the water walls are usually of CS execution. SH and RH drain can be of CS or AS depending on the temperature.

Usually the valves are made of the same material as the header to which they are attached.

- 40, 50, 63.5 and at times 80 (1½, 2, 2½ and 3″) NB globe or angle valves are used in drains. The drain valves in Eco and water wall headers are mostly of 40 or 50 (1½ or 2″) NB. But the SH drains, the final SH drain in particular, is larger.
- Drain valves of the water walls and Eco are required to be firmly closed during the boiler operation and opened only for draining during shutdowns. This group of valves is therefore mostly hand operated.
- Drain valves of the SH, RH and SBs are operated during the running of the boiler and are normally motorised for operator convenience except in very small boilers.
- Drain valves experience high Δp and velocities making them vulnerable to wear because of wire drawing effect. For frequently operated valves, it is normal to provide a martyr valve of the same size and construction ahead of the main valve. A martyr valve is normally used while the main valve stays open so that the martyr valve wears out and gets sacrificed while the main valve stays intact. When the martyr valve starts to leak, the main valve is operated until the next shutdown and thus the frequent boiler shutdowns are avoided.
- During start-ups, the final SH drain is used for inducing a flow through the SH to keep the tubes cool. Typically, they are sized for a flow of 10–15% MCR flow.

Emergency drain valve (see also Smelt Water Reaction under Recovery Boiler)

In BLR boilers, there are occasions when the boiler needs to be drained very fast to avert a possible explosion because of smelt water contact. The success of this emergency shutdown procedure (ESP) depends on the reliable performance of this rapid drain valve. This is a separate motorised emergency drain valve having high discharge coefficient which is connected directly to the IBD vessel and operation triggered by a specific command from control room. This valve is routinely tested once a month. Valve sizes range from 40 to 63 (1.5 to 2.5″) NB with high C_V of ~40–120.

21.5 Vapour

Vapour is a gaseous substance at a temperature below its critical temperature.

Vapour can be condensed liquid solid by increasing its pressure without change of temperature.

Vapour may co-exist with a liquid or solid. Then the two phases will be in equilibrium and gas pressure will be equal to the vapour pressure of the liquid.

21.6 Vapour Pressure (*see* Fluid Properties *under* Fluid Characteristics)

21.7 Vanadium in Oil Ash (*see* Oil Ash Corrosion)

21.8 Variable/Sliding Pressure Operation of Boilers

The output of an ST is varied by varying the steam pressure at the inlet of the steam nozzles.

In a constant pressure operation, the main steam pressure is held constant and the inlet nozzle pressure is varied by dropping the pressure across the ST inlet control/governing valves.

On the contrary, in variable/sliding pressure operation, the pressure of steam from the boiler is varied to suit the ST nozzle inlet pressure. Thus:

- The pressure variation is effected by controlling outlet pressure of the BFP and not by dropping it in the steam chest of the ST. In other words, the feature of steam pressure variation is shifted from ST inlet/governing valves to the BFP.
- Instead of raising full steam pressure and killing a part of it in governing valves, only the required amount of pressure is raised by the BFP.
- The boiler does not generate a constant outlet pressure but a variable pressure as per the ST load needs.

As the steam pressure at the inlet of ST is not held constant as in a conventional operation but continuously varied over the entire load range, this type of operation of the power plant is known as *variable or sliding pressure operation.*

The sliding pressure operation

- Is obviously more efficient thermodynamically as there is no loss of power in the inlet CVs/governing valves
- Consumes less power as the BFP losses are lower; even more so at lower loads
- Subjects all the pressure-bearing components to lower stress levels because of lower discharge pressure of the BFP

In a sub-c boiler, the sliding pressure operation needs to be done more carefully, with a reduced ramp rate. During variable pressure operation

- When the ST output is to be reduced, the boiler pressure is correspondingly reduced. This fall in pressure leads to the formation of steam bubbles in the drum and down comers, impairing the circulation.
- Bubbles cause disturbance to water level in the drum which, in turn, affects the controls.
- Δt between the top and bottom of the steam drum can induce thermal stresses in thick drums.

This situation is somewhat improved by adopting

a. an assisted circulation boiler or
b. modified sliding pressure operation where SOP is held constant up to a certain load and reduced below that load

SC boilers and even sub-c OT boilers are better suited for sliding pressure operation as there is no circulation and no thick drums. Usually, SC boilers employ sliding pressure operation to enhance the η differential over drum-type boilers.

21.9 Vegetable Fuels (*see* Agro-Fuels)

21.10 Velocity Coefficient

The ratio of the actual velocity of gas emerging from a nozzle to the velocity calculated under ideal conditions is less than 1 because of friction losses. It is also known as coefficient of velocity.

21.11 Velocities of Fluids (*see* Fluid Velocities)

21.12 Vena Contracta (*see* Fluid Flow)

21.13 Vent Piping (*see also* Integral Piping in Pressure Parts)

There are several vents in a boiler namely

- Start-up vent
- Drum, SH and RH vents
- SV vents
- Blowdown tank vents

The purpose of any vent is to let out steam at a suitable safe elevation ensuring no injury and damage to property and personnel. However, the purpose of each vent is different while operating the boiler.

Start-up vent

Start-up vent line, leading to atmosphere, is located just ahead of MSSV. The vent line is opened at the time of connecting the boiler to ST either from start-up or from banking conditions.

SOT has to be raised adequately as demanded by the metallurgy of ST to avoid any quenching effect. For this, a certain steam flow has to be created in SH without feeding the cold steam to the downstream ST until the specified acceptable minimum SOT is attained. Vent line comes to the rescue by artificially creating a steam flow through the SH which is vented and wasted. It meets the twin objectives of

- Inducing necessary steam flow (25–50% of MCR) in the SH so that the tubes can stay cool
- Raising the SOT rapidly to the desired level for admission into ST

The vent line is closed once the minimum SOT is attained and steam is admitted to the ST.

Start-up vent is provided even in process boilers to help raise the SOT quickly.

The start-up vent assembly consists of an isolating valve, motorised vent valve, silencer and vent piping to atmosphere. Start-up vent valve is a globe valve that can permit a large Δp across itself. The valves and piping up to the silencer are required to withstand the high temperature of SH outlet. The material of construction of the start-up valve is usually the same as that of MSSV.

21.14 Ventilation of Boiler Room

Outdoor and semi-outdoor installations are very well ventilated, needing no additional efforts. An indoor installation, with a proper boiler house surrounding it, needs ventilation to create a comfortable working environment. This is done by aspirating and expelling the warm air surrounding the boiler envelop by means of exhaust fans located at the ceiling of the boiler house. These are low-head axial fans called boiler ventilation fans.

21.15 Venturi Meter/Tube (*see* Instruments for Measurement)

21.16 Vertical or Down Shot Burners (*see* Pulverised Fuel Burners)

21.17 Vertical Economisers (*see* Economiser in BLR Boiler Construction *under* Recovery Boilers)

21.18 Vertical Mills (*see* Pulverisers)

21.19 Vicker's Test (*see* Hardness Testing of Metals in Testing)

21.20 Viscosity, Absolute/Dynamic and Kinematic (*see* Fluid Characteristics)

21.21 Volatile Matter (*see* Coal)

21.22 Volatile Organic Compounds

Volatile organic compounds (VOCs) or HCs are organic chemical compounds or *HCs in flue gases as a result of incomplete combustion*. They can affect the environment and human health, particularly the audible and visible senses. Their boiling points are low and vary from 50°C to 250°C. VOCs in boiler emissions are best controlled by having prudent combustion practices with adequate time, temperature and turbulence (3Ts). NO_x-controlling measures like LNBs, staged combustion, FGR and so on tend to increase VOC formation because of lower combustion temperature.

21.23 Volumetric Machines (*see* Ambient Conditions)

22

W

22.1 Wall Blowers (*see* Soot Blowers)

22.2 Waste Fuels

Waste fuels are essentially industrial by-product fuels and **MSWs**. They can be solid, liquid or gaseous. Many solid waste fuels are bio-wastes. Purpose-grown trees and other similar fuels all fall under the category of bio-fuels or agro-fuels. Typical industrial wastes with significant cv are listed in Table 22.1.

Waste fuels are being burnt in boilers in increasing quantities for one or more of the following reasons:

- Waste fuels are often cheaper than prime fuels.
- Many times disposal being problematic there is a compulsion to utilise waste fuel generated by the process.
- Simple flaring, incineration or land filling is increasingly getting difficult in most parts of the world due to environmental restrictions.
- Co-firing becomes attractive as most waste fuels contain less S which helps in meeting the limits.

Conventional firing equipments such as **burners, grates** and **FBC combustors** are adequate for burning waste fuels as well. Usually **bi-drum boilers** are employed even though **single-drum boilers** are also occasionally used. Both **natural** and **forced circulation boilers** are popular. As waste fuel availability cannot be steady the boilers are invariably built with multi fuel firing flexibility to improve the availability. Oil or gas are extensively used for

- Supporting waste fuels and to stabilise boiler at low loads
- Pilot flames for continuous support of fuels with low cv or
- Carrying full load during off season when the main waste fuel is not available

Figure 22.1 shows a number of conventional and waste solid fuels arranged by their cv and relative difficulty in burning. Many biomasses are not difficult to handle both from combustion and pollution aspects. Burning MSW and RDF and some wood and plastic wastes is problematic. The diagram taken from FWC is really insightful.

22.3 Waste Heat

For a boiler engineer, WH represents the heat contained in hot waste or exhaust gases. Hot gases are usually from the

- Various industrial processes or
- Exhaust gases from engines like GTs or diesel engines

The hot gases may contain

- Residual fuel and heat as in the case of BFG, CO gas and COG or
- Only high temperature as in the case of kiln, GT or diesel engine exhaust gases

In the former, the residual fuel has to be burnt in the boiler while in the latter only the heat from gases has to be extracted. The gases can be either

- Clean or dusty
- Passive or erosive
- Neutral or corrosive

The waste gases can be at high, medium and low temperatures of ~1000°C, 500°C and 300°C (~1830°F, 930°F and 570°F). Heat extraction gets progressively harder as the waste gases get cooler. Correspondingly the boilers become bigger.

GT exhaust gases are clean, passive and at medium temperature with no residual fuel. GTs employ very high excess air to produce short flames and relatively cooler gases in their combustion chambers. Consequently, the exhaust gases generally have 15–16% O_2 requiring no additional air to carry on supplementary fuel firing if required.

Boilers producing steam from the process waste gases are **Whrbs**. Cement, metallurgical, non-metallurgical, C black plants and so on are the usual processes

TABLE 22.1

Common Industrial and Agro Wastes and Their Temperatures

	Range of Gcv as Fired		
Waste	**MJ/kg**	**kcal/kg**	**Btu/lb**
Solids			
Bagasse	8.37–15.12	2000–3600	3600–6500
Bark	10.47–12.10	2500–2890	4500–5200
Wood wastes	10.47–15.12	2500–3600	4500–6500
Saw dust and shavings	10.47–17.45	2500–4170	4500–7500
Coffee grounds	11.40–15.12	2720–3600	4900–6500
Nut hulls	16.28–17.91	3890–4280	7000–7700
Rice hulls	12.10–15.12	2890–3600	5200–6500
Corn cobs	18.61–19.31	4450–4610	8000–8300
Liquids			
Industrial sludge	8.60–9.77	2050–2330	3700–4200
BL	10.23	2440	4400
Sulphite liquor	9.77	2330	4200
Dirty solvents	23.26–37.21	5555–8890	10000–16,000
Spent lubricants	23.26–32.56	5555–7780	10,000–14,000
Paints and resins	13.96–23.26	3330–5555	6000–10,000
Oily waste and residue	23.26	10,000	18,000
Gases			
COG	45.82	10,940	19,700
BFG	2.65	630	1140
CO gas	1.38	320	575
RG	50.71	12,110	21,800

Note: All values are rounded off.

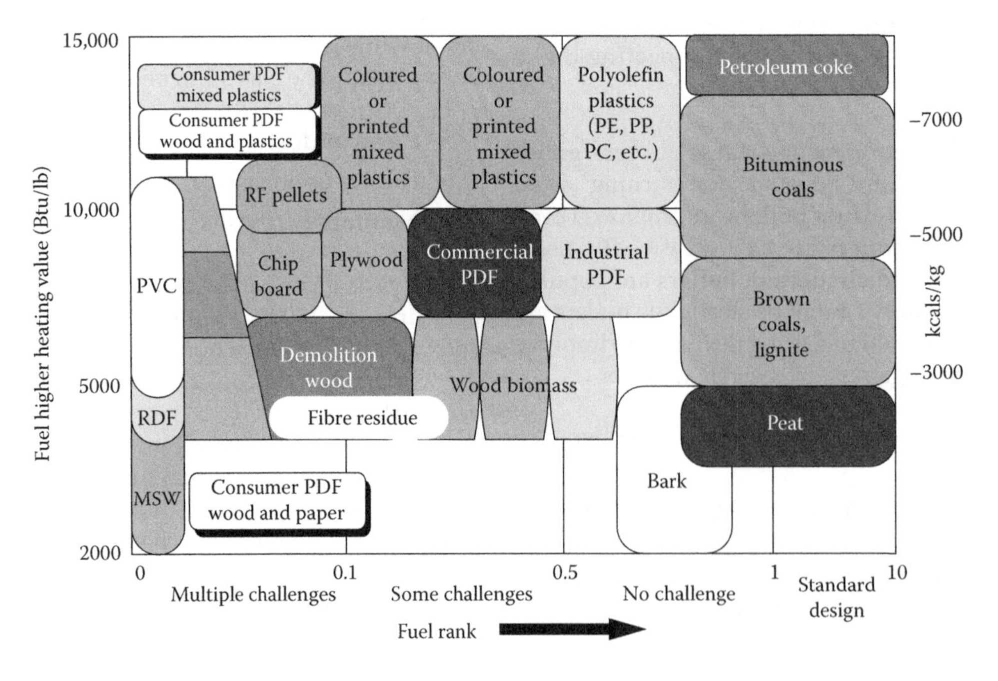

FIGURE 22.1
Solid wastes arranged by their hv and ease of burning. (From Foster Wheeler Corporation, USA, with permission.)

TABLE 22.2

Various Process Waste Gases and Their Temperatures

Source	°C	°F
Nickel refining furnace	1370–1650	2500–3000
Aluminium refining furnace	650–760	1200–1400
Zinc refining furnace	760–1100	1400–2030
Copper refining furnace	760–815	1400–1500
Steel heating furnaces	925–1050	1700–1920
Copper reverberatory furnace	900–1100	1650–2030
Open hearth furnace	650–700	1200–1290
Cement kiln (dry process)	620–730	1150–1345
Glass melting furnace	1000–1550	1830–2820
Hydrogen plants	650–1000	1200–1830
Solid waste incinerators	650–1000	1200–1830
Fume incinerators	650–1450	1200–2640

producing large amount of waste gases. Whrb for each plant is different. Table 22.2 lists various waste gases from process plants.

Boilers producing steam from the GT exhaust gases are **Hrsgs**. As the GT exhaust gases are clean and at medium heat, Hrsgs are made entirely of closely spaced finned tubes with tight spacing of fins to cool the gases efficiently. Hrsgs are the only WH boilers in the power plant market.

22.4 Waste Heat Recovery Boilers (*see also* Industrial Boilers *and* Heat Recovery Steam Generators)

Whrb is a boiler that recovers major part of heat from a WH stream in a process plant, which is otherwise discarded. Hrsg is actually a Whrb recovering heat behind a GT. As Hrsgs are large in number, big in size, more complex and quite different from the other types of Whrbs they are classified separately. The differences between the two are listed in Table 22.3. A list of sub-topics covered under this main topic of Whrbs is given below.

1. Whrb versus Hrsg—a comparison
2. Classification of Whrbs
3. Whrbs in chemical plants
4. Whrbs in steel plant
5. Whrbs in oil refineries
6. Whrbs in cement plants
7. Whrbs for exhaust gases in various plants
   a. Whrb in copper plant
   b. Whrb in sponge iron plants
   c. Whrbs in carbon black industry

22.4.1 Whrb versus Hrsg: A Comparison

Whrbs and Hrsgs are the two types of boilers recovering heat from waste gases. In construction, design, operation and in every way, the two are different. Table 23.3 provides the comparison.

22.4.2 Classification of Whrbs

Broadly Whrbs recover heat from process waste gases or engine exhaust gases and produce mostly MP or LP steam for power and process. They can be classified as

1. Process integrated boilers to cool the process gases Such boilers are employed in
   - Chemical plants like H_2SO_4, HNO_3, NH_3, H_2 gas, caprolactum plants
   - Steel plants like sponge iron, coke oven plants
   - Refineries in FCC unit
   - Smelters and converters
2. Exhaust gas boilers for diesel Gen, kiln, furnace, incinerator exhaust
3. Waste gas-fired boilers to burn waste or tail gas in C black, ferroalloy, caustic soda plants and in blast furnaces

Whrbs can be **smoke tube, hybrid** or **water tube** boilers depending on the volume of waste gases and

TABLE 22.3

Differences between Whrb and Hrsg

Parameter	Whrb	Hrsg
Gases	Usually dusty	Always clean
Type of plant	Process	Power and Cogen
Cycle	Steam or Cogen	Combined or Cogen
Upstream plant	Process	GT
Fired/unfired	Both	Both
Designs	Wide variety	Limited
Steam pressure	Usually <70 bar	Up to 150 bar
Steam temperature	<500°C	Up to 570°C
Pressure levels	Single	1, 2 or 3
RH steam	Almost never	In large units
Tube construction	Mostly plain	Mostly finned
Erosion	Yes in many cases	No
Corrosion	Many cases	Only occasionally
Draft	Balanced mostly	Pressurised
Fans	Required	Only in fresh air mode
AH in the backend	Yes mostly	Never
Boiler refractory	Plenty	Negligible
Upstream gas cleaning	Many times	No need
Downstream gas cleaning	Almost always	No need
Boiler population	Less	Plenty

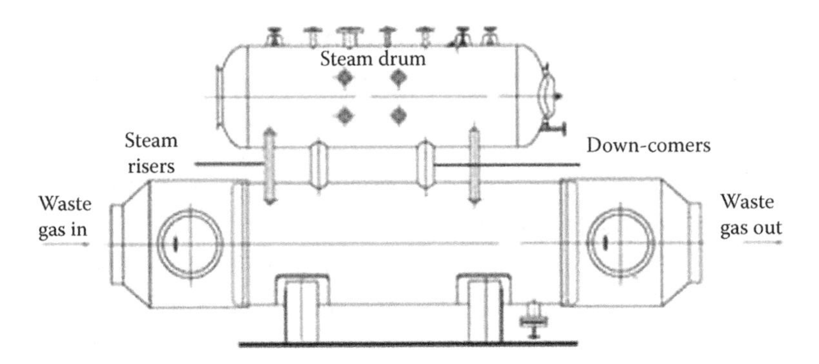

FIGURE 22.2
Whrb in sulphuric acid plant in a fertiliser unit.

the amount of steam generated. **Natural** as well as **forced circulation** is common. Reliability and integration with the main process is seen as more important than fractionally higher use abb. eta for uniformity.

Whrb is a vast subject with many types of applications and several designs. Only a selected common and large Whrbs are described here.

22.4.3 Whrbs in Chemical Plants

In most cases the heat from chemical plants is small, requiring modest WH recovery systems and small boilers. Whrbs in H_2SO_4 plants in fertiliser units and S recovery plants in refineries are described here briefly. The other Whrbs in chemical plants are also of similar construction being usually small units.

a. *Whrb in H_2SO_4 plants*: H_2SO_4 is one of the most important chemicals with wide range of uses. The fertiliser industry is the main user. Steel, rayon, staple fibre and petroleum refining are other users. Liquid S is fired in furnace and hot sulphurous gases are produced which are cooled in the Whrb to produce SH steam for power and process. WH recovery is an integral part of to H_2SO_4 plants. Figure 22.2 shows a typical Whrb in a H_2SO_4 plant in a fertiliser unit. Shell and tube HX construction for these unfired boilers is quite common. Water is at HP. A separate steam drum is needed for including the necessary drum internals. A typical large boiler can produce around 300 tph of steam.
b. *Whrb in S recovery plants in refineries*: S is removed from sour NG in amine scrubbers. The first step is to reduce the acid gas to elemental S at high temperature under controlled conditions. The hot furnace gases are then cooled in Whrb to condense the gaseous S. The cooled gases then go to a reactor for the residual S compounds to get catalytically converted to elemental S for its removal. The gases are further cooled in a tail end Whrb, consisting of Eco and SH. Figure 22.3 shows a typical whrb boiler in S recovery unit of an oil refinery. This is a hybrid boiler with open furnace, BB and shell and tube HX arranged in line to extract WH of the gases. Boilers similar to the configuration shown in Figure 23.2 are also employed.

22.4.4 Whrbs in Steel Plant (*see also* Blast Furnace Gas *and* Coke Oven Gas)

Any steel plant has at least one stream of WH from where heat can be effectively recovered. The following are some examples.

COG and **BFG** are the major by-products in steel plants. In an integrated steel plant their use for heating blast furnace and other equipment and power generation occurs always together. Figure 22.4 shows the schematic of the two gases in a typical steel plant.

Under the topics of BFG and COG the gas properties have been described under gaseous fuels.

The two gases are widely different. BFG is very lean and COG is fairly rich. As the two of them are available in integrated steel plants it is usual to burn them together

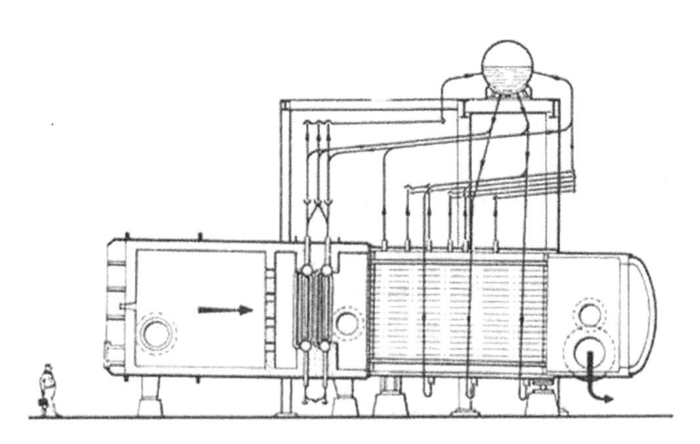

FIGURE 22.3
Whrb in a sulphur recovery plant in an oil refinery.

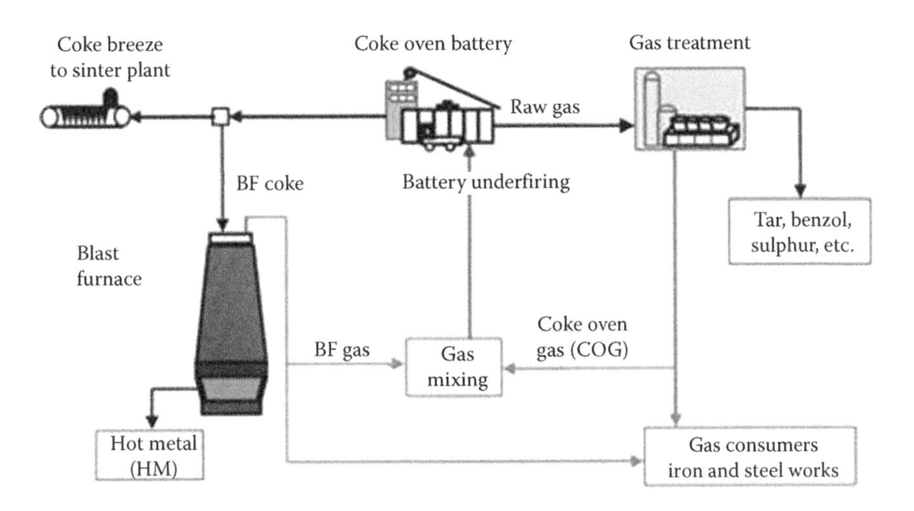

FIGURE 22.4
Whrbs for COG and BFG in an integrated steel complex.

using multi-fuel scroll burners. Such a burner is shown in Figure 22.5. This burner can burn three gaseous fuels in addition to FO in the centre. Such burners can take PF as well. Scroll of the burner provides a swirl to the BFG which helps it to give a longer residence time in the furnace. See also Scroll Burner under Burners.

In pig iron plants, however, no COG would be available and the boilers have to burn only BFG.

Package and **field-erected** oil/gas-fired boilers are suitably modified to burn COG and/or BFG. For COG alone the modifications are minor as it is a high cv gas. For BFG, however

- Furnace has to be enlarged for a low HRR of ~180,000 kcal/m^3 (~20000 Btu/cft).
- Furnace has to be lined with refractory to reduce the cooling surface and increase the radiation for a better flame stability and reduced support oil.
- Clear spacing in the tube banks has to be increased for a lower gas velocity to avoid erosion.
- More SBs may be installed depending on the dust load of the gas.
- ID fan is required to be very big due to large flue gas quantity.

See Figure 15.13 in oil- and gas-fired boilers which shows a typical standard field-erected boiler converted to firing BFG where all such modifications are incorporated.

Coke dry quenching Whrb

Another Whrb in integrated steel plants and coke plants is the coke dry quenching Whrb. Red hot coke from coke ovens is discharged into the quenching chamber at ~1000°C for quenching by gas to ~200°C. Cooling gas enters ~180°C and gets heated to ~800°C and dissipates

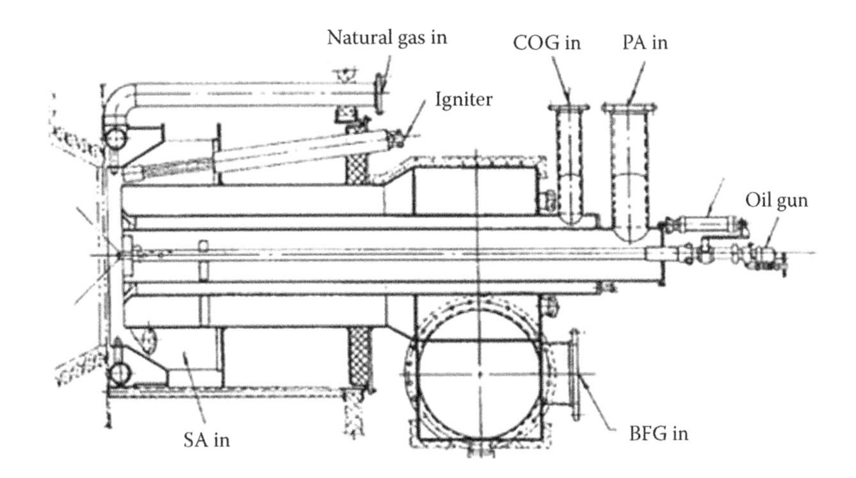

FIGURE 22.5
Multi fuel scroll burner for firing COG, BFG, NG and FO.

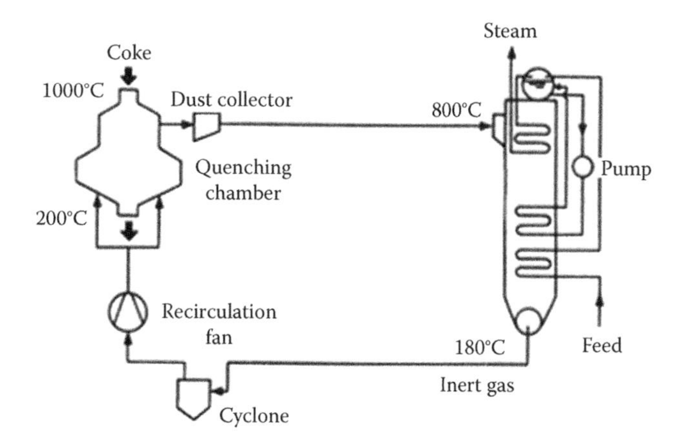

FIGURE 22.6
Schematic diagram of a coke quenching plant with its Whrb.

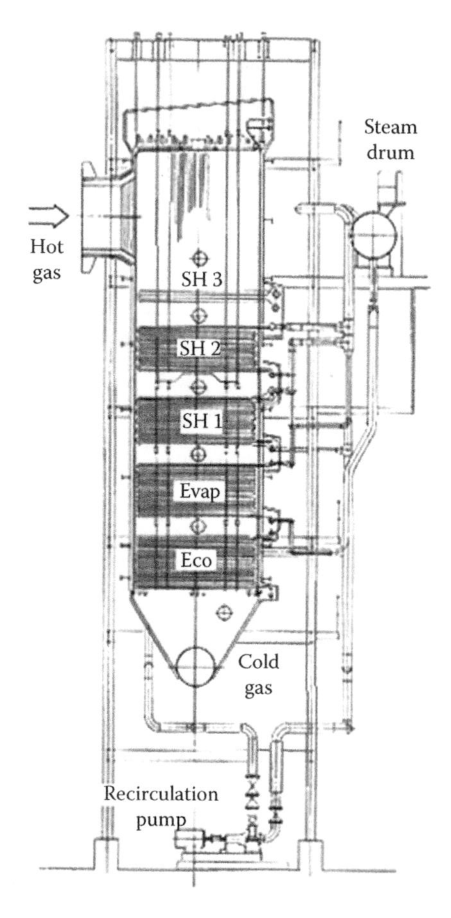

FIGURE 22.7
Whrb in a coke quenching plant.

its sensible heat in the unfired Whrb. A dust collector located between the quencher and Whrb removes the coke fines. Yet the tubes in the boiler are to be spaced generously apart, and gas velocities are to be kept low to prevent tube erosion. Also erosion shields are to be provided on tubes for longer life. Boilers can be made quite compact by adopting horizontal instead of sloping tube layout and forced circulation of bw. Due to the high gas inlet temperatures steam can be generated at high ps & ts >100 bar and 530°C.

Figure 22.6 shows the flow diagram and Figure 22.7 the cross-sectional view of a typical HP boiler.

22.4.5 Whrbs in Oil Refineries

22.4.5.a *Carbon Monoxide Boilers* (see also *Carbon Monoxide Gas* under *Gaseous Fuels*)

Large amounts of HP flue gases containing carbon oxides (largely CO) are available in refineries as by-products of the catalyst regeneration process in

a. Fluid catalytic cracking units (FCCU)
b. Residue cracking units (RCCU)

CO gas properties are given in Table 7.3. This CO gas is utilised for steam production for

- Heat recovery from the flue gases
- Conversion of CO to CO_2 before discharging the flue gases to atmosphere
- Lowering the gas temperature prior to catalyst fine recovery devices.

Depending on the regeneration process two types of heat recovery systems, namely, unfired flue gas coolers or fired WH boilers are available.

1. *Unfired flue gas cooler* is used when little or no CO is present in the flue gases due to full oxidation of CO to CO_2 in the catalyst regeneration process. Here sensible heat contained in the flue gases is utilised to either (a) produce and SH steam or (b) preheat process oil streams for energy conservation purposes. The heat recovery section is composed either by horizontal or vertical bare tubes.
2. *Fired CO boiler* is employed when C deposits on catalyst are removed with low excess air in the regeneration process resulting in flue gases rich in CO. This CO needs to be burnt to CO_2 before its discharge to atmosphere.

   CO boiler must fulfil these requirements. CO boilers act also as incinerators where residence time at high temperature is the key factor for the proper CO oxidation. Two designs of fired boilers are available.

   - Typical industrial natural circulation boilers with water walls where the radiant furnace section is adequately sized to provide the required residence time.

- Adiabatic (refractory) type where oxidation of CO gas takes place in a combustor and the heat is fully recovered in a separate section by convection. This type is used particularly for a minimum but flexible steam production and high turndown capacity. This design can also meet very stringent limitation on CO emission to atmosphere.

Large boilers producing nearly 300 tph have been built.

- CO gas has a very low cv of <~1.5 MJ/Nm^3, ~350 kcal/Nm^3 or ~40 Btu/scft, comprising ~50% each of sensible and combustible heats.
- The gas is delivered at HP of 800–1500 mm wg and high temperature of 600–700°C.
- It has considerable amount of CO to make it sufficiently lethal.
- Auxiliary firing of ~10% capacity is necessary both for ignition of fuel and protection against flame out.
- **Ignition temperatures** are high and vary from 610–655°C (1130–1215°F).
- Furnaces are lined with refractory and maintained at ~1000°C (1830°F) for helping ignition and stabilising combustion.

In addition to 10% continuous auxiliary firing across all burners it is usual to provide 100% supplementary firing and have a full gas bypass to make cracker unit and boiler fully independent of each other. This way the cracker can operate when the boiler is down and more importantly, the boiler can supply full steam with cracker unit down. The power generation can be made fully dependable this way with no interruptions. The fluctuations in the gas quantities and heat values can be met adequately with large supplementary fuel support.

Full refractory construction is possible only in small boilers. Larger units are essentially standard **package** and **field-erected** oil/gas-fired boilers suitably modified to deal with low Gcv gas.

- Furnace has to be large to provide enough residence time. A typical volumetric HRR is ~0.4 MW/cum (~360,000 kcal/cum or 40000 Btu/cft) with minimum auxiliary firing to ensure complete combustion.
- **Excess air** in furnace should be a minimum of 20%.
- All banks should be designed with suitable clear spaces and low **gas velocities** to prevent *erosion* due to the catalytic dust.
- Adequate number of **SBs** should be installed to keep the surfaces clean.
- CO gas entry into furnace should be at high velocity of 45–54 m/s or 150–175 fps to generate adequate turbulence to mix the fuel and air thoroughly.

Figure 22.8 shows how a package boiler is modified to fire CO gas.

22.4.6 Whrbs in Cement Plants

WH is present at two locations in any cement plant—at the inlet and exhaust ends of rotary kilns as depicted in the schematic in Figure 22.9. Both are low-grade heats.

- On the raw material inlet side of the kiln the exhaust gases are cooled to about 300–350°C in a four-stage preheater and 200–300°C in 5–6-stage preheater. These gases are partly used for heating the raw materials and the rest are exhausted to atmosphere.
- On the discharge side of the kiln cold air is used for quenching the hot clinker from 1000 to 100–120°C which heats the air to temperatures of 260–300°C. This hot air is also partly used for drying.

WH recovery in cement plants has not been quite popular for a long time mainly because

- They are low-temperature waste gas and air with limited heat.
- To recover this WH, the boilers have to be quite large.
- Gases are dusty requiring the boilers to have generous tube lanes and cleaning devices making them further more expensive.
- Steam has to be used only for power generation as it is not needed in cement processing, thereby losing ~60% of heat to the condenser. This is in sharp contrast to any cogen application.
- In many cement plants located near the limestone mines water scarcity is a normal feature demanding expensive air cooled condensers.
- Physically, the two sources of heat are usually far apart so that either two separate streams of power generation are needed to be installed at either end of a kiln or some lengthy and intricate piping needs to be done.

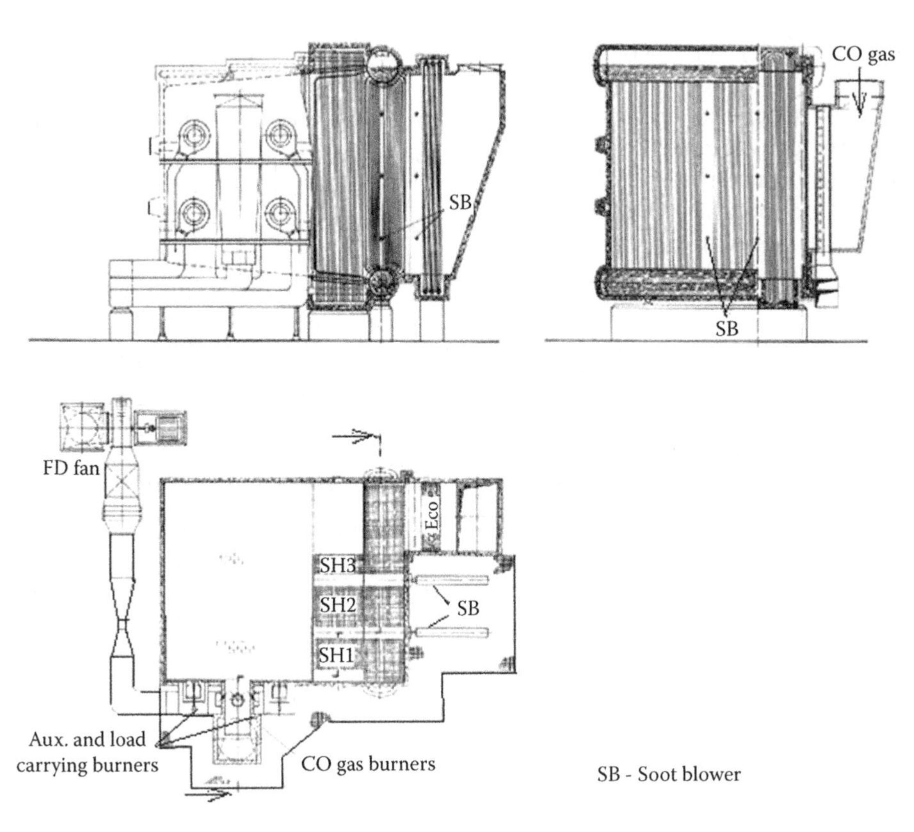

FIGURE 22.8
Bi-drum bottom supported boiler for firing CO gas in refineries.

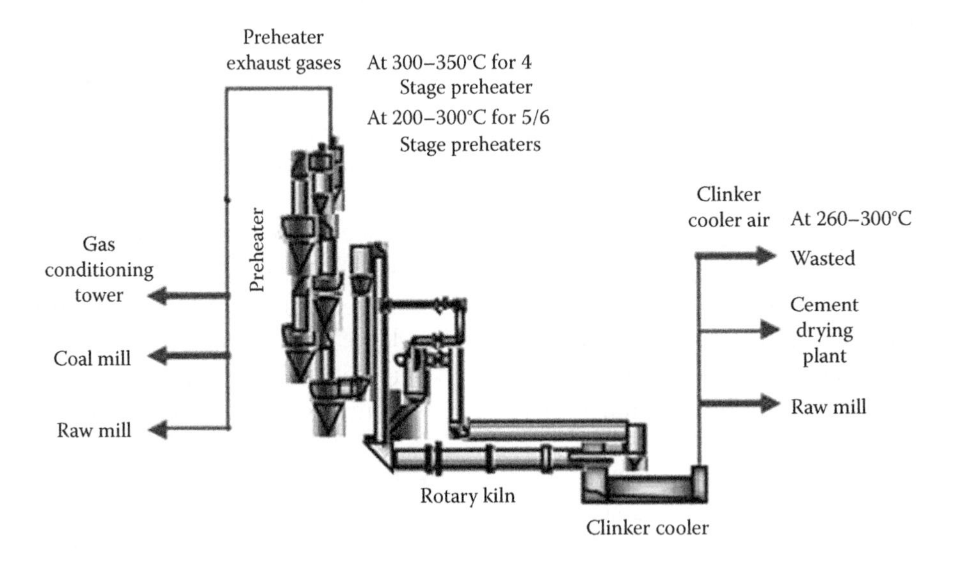

FIGURE 22.9
Waste heat in cement plants—at the inlet and outlet ends of kiln.

With these limitations, WH recovery becomes viable only in larger plants and also where the costs of electricity are high. ~40 kWh of power generation is possible per each tonne of clinker produced. Considering this potential and ever-increasing costs of fuels, large cement plants are planned these days with power generation facility right at the concept stage when adequate space can be made available. Schematic arrangement of such a power generation plant is shown in Figure 22.10.

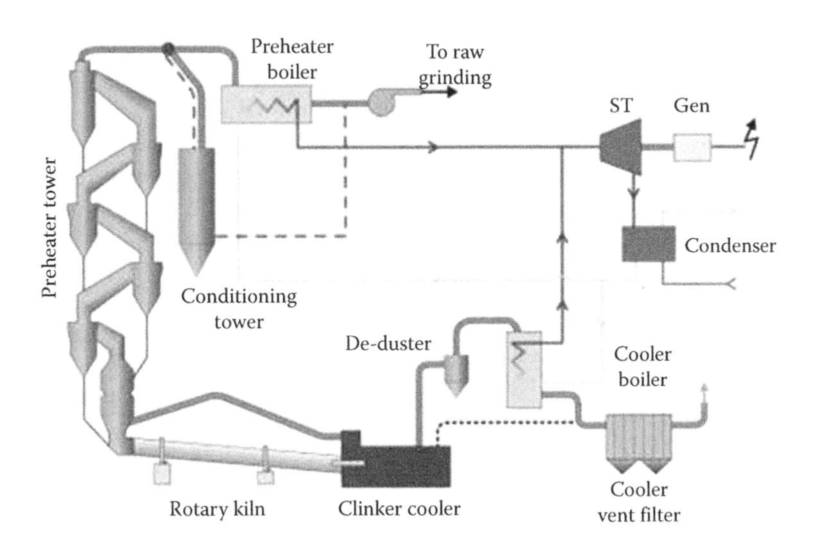

FIGURE 22.10
Power generation in a large cement plant from Whrbs at preheater and clinker cooler.

TABLE 22.4
Indicative Properties of Waste Gases in Cement Plants

Parameter	Units	Preheater Exhaust	Clinker Cooler	Preheater Exhaust	Clinker Cooler
Plant size	MT/day	1500		2000	
Flow	kg/h	85,000	100,000	92,000	155,000
	Nm^3/h	59,000	82,000	63,000	120,000
Temperature	°C	300–350	280–300	300–320	260–280
Dust	g/Nm^3	65	15	65	15
Composition					
CO_2	%	28		30	
N_2	%	69		67	
O_2	%	3		3	

Indicative properties of waste gases are tabulated in Table 22.4 which are taken from operating plants. Accuracy may be low but they provide some indicators.

It is still not usual to plan cement plants with Whrbs as they are not integral to the process like most other plants. Since they are later additions space and head room constrain the size and shape of the Whrb. Hence cement plant Whrbs are very flexible in their configuration. They can have horizontal or vertical layout of tubes, natural or forced circulation of water and low or medium steaming conditions.

Figures 22.11 and 22.12 show typical Whrbs for cooling preheater gases with forced flow vertical tube configuration and clinker air with forced flow horizontal tube arrangement, respectively.

22.4.7 Whrbs for Exhaust Gases in Various Plants

22.4.7.a Whrb in Copper Plant

Every Cu plant has a large Whrb to extract heat from the high temperature tail gases that come out at ~1300°C.

In the most popular Outokumpu smelting process Cu concentrate (crushed, ground and cleaned chalco-pyrite $CuFeS_2$) containing 20–40% Cu is fed, along withO2 enriched air, into a flame in the reaction shaft of the smelter. The settler section of flash smelter contains molten *matte* and *slag* at temperatures of ~1350°C. The matte or *blister Cu* (50–70% Cu) is withdrawn for conversion into metallic Cu. The slag contains almost all the impurities in the feedstock and is discarded.

The dusty tail gas containing 20–60% SO_2 at ~1300°C is fed into a Whrb to cool the gases and produce steam. After cleaning in **ESP** the dust-free SO_2-rich gas is sent to another plant for conversion into H_2SO_4. The dust is periodically removed from the HX tubes within the Whrbs by spring hammers. The remainder of the dust

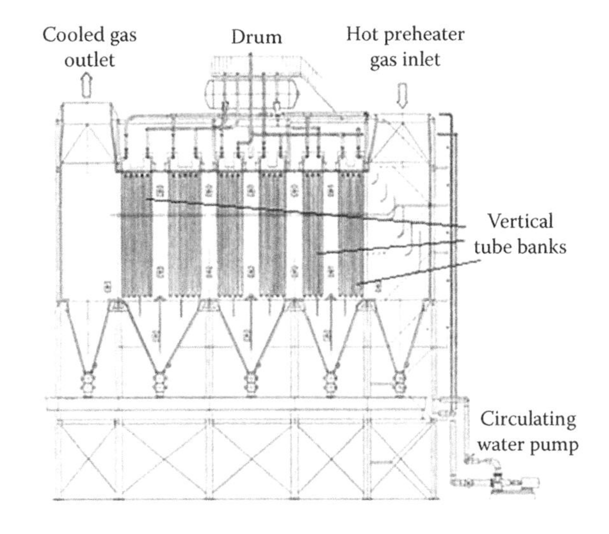

FIGURE 22.11
Typical forced circulation vertical tube Whrb for cooling preheater exhaust gases.

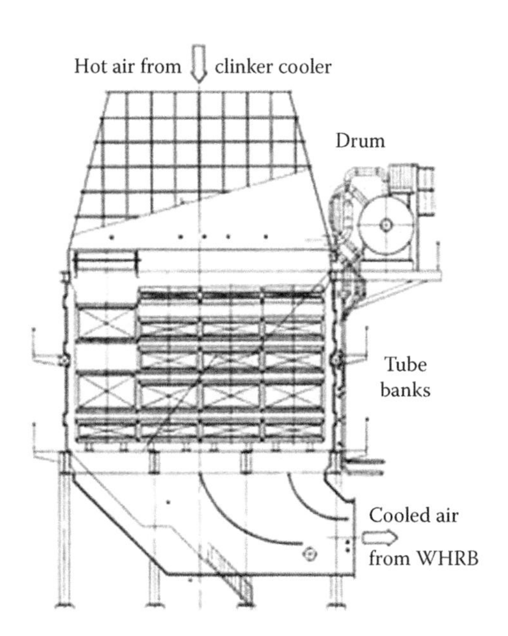

FIGURE 22.12
Typical forced circulation horizontal tube Whrb for cooling gases from clinker cooler.

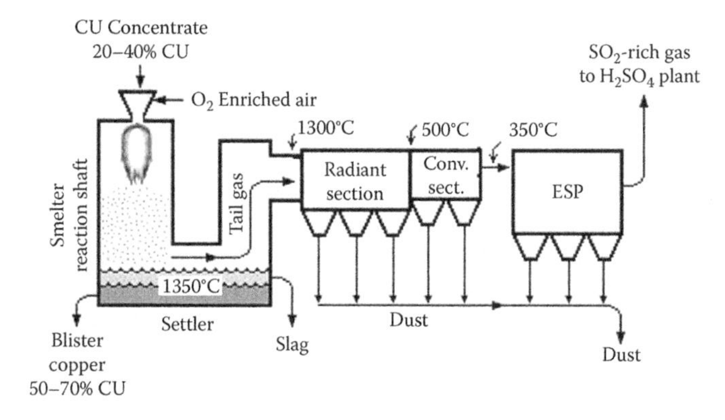

FIGURE 22.13
Schematic arrangement of Whrb section of a copper plant.

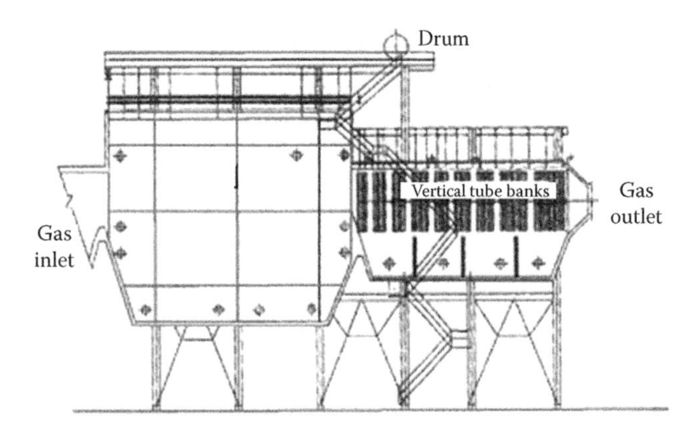

FIGURE 22.14
Typical forced circulation vertical tube Whrb in a copper plant.

is removed continuously by ESP. All the dust is then recycled into the reaction shaft feedstock. Figure 22.13 depicts the scheme.

These Whrbs typically generate saturated steam at a pressure of ~40–60 bar and at low SOT of ~250–285°C. All HSs to operate well above the dew point temperature to avoid corrosion.

The Whrb contains two distinct sections, namely, the **radiant** and **convection** placed one after the other giving it a name of tunnel boiler. Gases are cooled by radiation in the first chamber which has water walls on both sides and roof. Lances are located at all strategic points for the removal of dust to lessen fouling of surfaces as the gas travels in the convection section. It is normal to provide vertical tubes and have spring hammers for the removal of dust to have a reasonably clean HS. Convection tubes are plain and widely spaced. Figure 22.14 shows a typical tunnel-type Whrb.

22.4.7.b Whrb in Sponge Iron Plants

Direct-reduced iron (DRI) or sponge iron is an alternative and easier route of iron making that has become popular in the last three decades. It is made in several parts of the world by either NG or coal-based technology by reducing iron ore in solid state at 800–1050°C (~1470–1920°F) using reducing gas ($H_2 + CO$) or coal. The specific investment and operating costs of direct reduction plants are low compared to integrated steel plants and are more suitable for many developing countries where supplies of coking coal are limited.

Sizable WH is produced in the DRI plants based on coal where Whrbs produce steam for power generation. The heart of the DRI plant is the inclined rotary kiln which is fed with the raw meal (iron ore, coal, and flux) on the rear (upper) end and fired with coal from the front (lower) end. Iron ore reduces to iron and is discharged on to a rotary cooler at the front from where the cooled end product comes out. At the rear end hot gases are treated in an after burning chamber before sending to a Whrb followed by an ESP before discharging cleaned gases to atmosphere. The process of making sponge iron by the coal route is explained in Figure 22.15.

Hot and dusty kiln gas at ~1000°C is admitted into the Whrb where the gas enters a downward shaft which is fully water cooled (as shown in Figure 22.16). This helps to cool the gases and shed its dust as it takes a sharp u-turn at the bottom. The convective tube banks are located in the uptake shaft or in the down take as shown in this arrangement. Forced circulation is normally employed. From the boiler design stand point

- Tube spacing has to be generous in convective banks
- Tube arrangement in-line

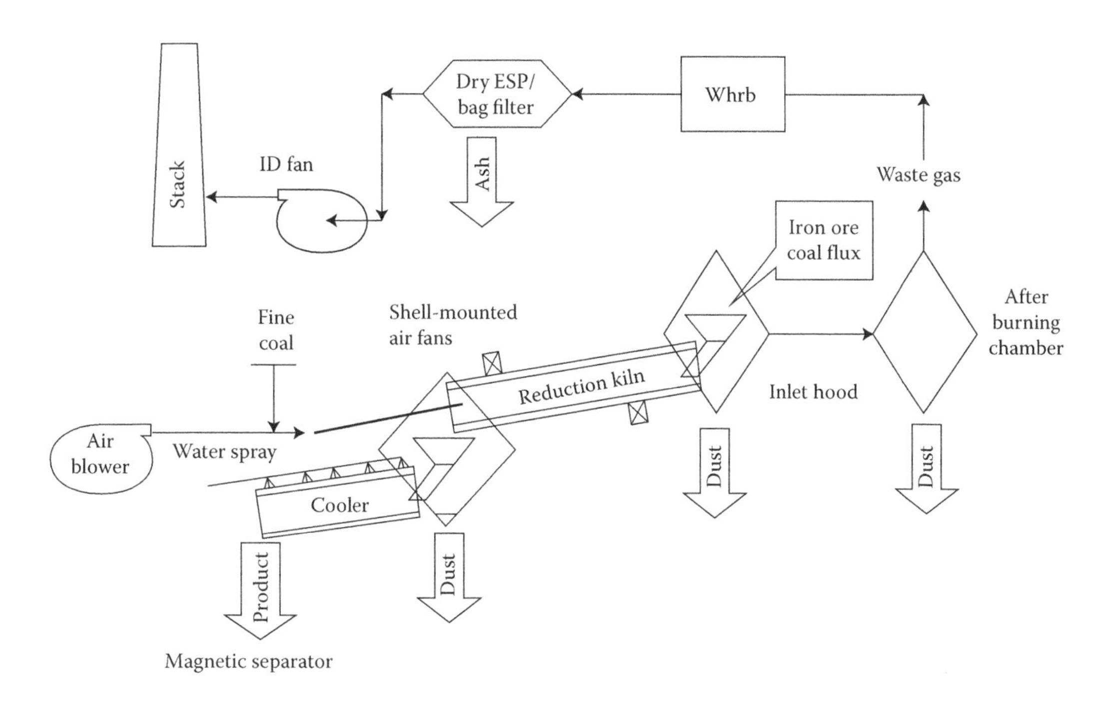

FIGURE 22.15
Process of making sponge/direct reduced iron.

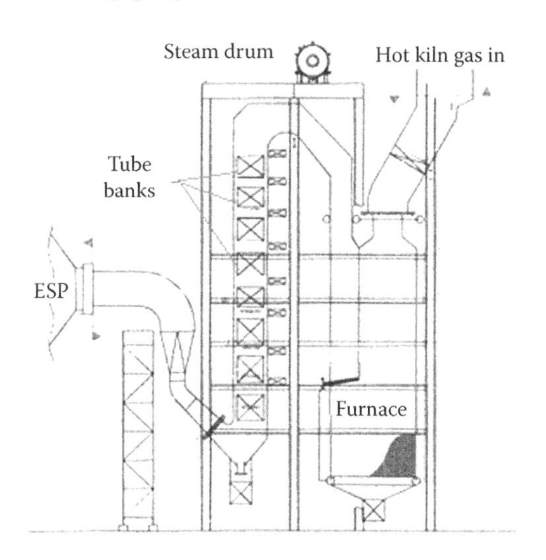

FIGURE 22.16
Whrb in DRI/sponge iron plant.

- Tube shields are to be given at appropriate locations
- Gas velocities should be kept low (<10 m/s) to prevent erosion which can be severe as the dust contains iron particles

As the gas is not corrosive it is usual for steam to be produced at MP and HP and temperatures suitable for captive power generation. Also it is normal to cool the gas to low temperatures of ~140°C. Sponge iron plants also do not need process steam and hence the entire steam is used in CPP.

22.4.7.c Whrbs in Carbon Black Industry

C black is an industrial raw material consisting of 95% or more of amorphous C, with particle sizes in the order of nanometres, produced under a well-controlled manufacturing process. It is the best reinforcing material for rubber and naturally the tyre industry is the largest user consuming ~80% of total C black produced.

Mostly the industrial C black manufacture is based on the process of continuous imperfect combustion of C black oils, such as FCC decant oil and coal tar, using reactors. Pre-heated air and oil are used in the process to obtain higher yield. The imperfect reaction is stopped by injection of quench water at the appropriate place in the reactor.

The manufacturing process generates both sensible heat and chemical energies in tail gas similar to CO gas in refineries. Also in the process a lot of CO_2 is produced. Without WH recovery less than 60% of the sensible and only 50% of the chemical energies are utilized for the C black manufacturing process. It is essential to recover this heat from gases and also reduce CO_2 emissions. Typical composition of tail gas is given in Table 22.5.

TABLE 22.5
Typical Composition of CO/Tail Gas in a Carbon Black Refinery

Constituents	H_2	CO	CH_4	CO_2	N_2	H_2O
% Volume	8.0	10.2	0.2	2.4	36.2	43.0

CO and H_2 are the main constituents providing heat energy in this lean gas that has a lot of M. It is not easy to burn this gas. Special burners and refractory lined furnace of appropriate shape and volume are needed. Being slow burning gas a large furnace volume with low volumetric HRR are required. Continuous fuel support has to be provided as flame outs can be dangerous with such high CO content in gas. Entire boiler has to be pressure tight.

MP steam is generated in Whrb at temperatures suitable for Cogen power. Steam generation mostly ranges from 20 to 80 tph.

22.5 Water

Life on earth is unthinkable without water. Finite amount of water on earth participates in an elaborate recycling scheme (hydrologic cycle) that provides for its reuse. Industry is a minor participant in the global water cycle.

Pure water (H_2O) is colourless, tasteless, and odourless compound of H and O. To some degree, water can dissolve every naturally occurring substance on the earth, which makes it a 'universal solvent'. As water invariably gets contaminated by the substances with which it comes into contact it is not available for use in its pure state.

While generally beneficial, the solvency of water poses a major problem to industrial equipment.

- Corrosion reactions cause slow dissolution of metals by water.
- Deposition reactions produce scale on heat transfer surfaces.

The control of corrosion and scale is the major focus of WT technology.

The following are the sub-topics covered under the main topic of water:

1. Impurities in water
2. Solids in water
3. Terminology of water
4. Hardness of water
5. Alkalinity of water
6. Water types
7. WT for boilers
8. Dealkalisation of water
9. Demineralisation of (DM) water
10. Deaeration
11. Deaerator
12. Oxygen scavenging
13. bw conditioning
14. Drum WT
15. AVT of bw
16. Chemical dosing of boilers
17. Boiler blowdown, continuous and intermittent
18. Effects of water on boilers
19. Water quality for fw and bw
20. Caustic embrittlement
21. Amines

22.5.1 Impurities in Water

Impurities depend on the source of water. They include both suspended and dissolved solids. Suspended solids are not completely soluble in water and are present as floating particles. These particles usually impart a visible turbidity to the water.

- Dissolved and suspended solids are present in most surface waters.
- Seawater is very high in soluble NaCl; suspended sand and silt make it slightly cloudy.
- Ground waters are relatively free from suspended contaminants, because they are filtered as they move through the strata. This filtration also removes most of the biological contamination.
- Groundwater chemistry tends to be very stable over time. A groundwater may contain an undesirable level of scale forming solids, but due to its fairly consistent chemistry it may be treated effectively.

Table 22.6 provides a comprehensive list of impurities, their effects and the methods of removal.

22.5.2 Terminology of Water

The following are some of the commonly used words of the water technology:

- *Hardness:* It is the *soap lather destroying property* of water caused by the presence of soluble salts of Ca and Mg. *Total hardness* is the sum of concentrations of the Ca and Mg salts or hardness, each expressed in terms of $CaCO_3$. Hardness forms scales.
- *Alkaline, temporary* or *carbonate hardness*: If the total alkalinity is greater than the total hardness, the hardness is considered to be alkaline hardness, that is, HCO_3, CO_3 and OH.
- *Non-alkaline, permanent or non-carbonate hardness*: If the total alkalinity is less than the total

TABLE 22.6

Water Impurities, Their Effects and Removal

Impurity	Description	Effects	Method of Removal
1. Hardness	Ca, Mg salts as $CaCO_3$	Scale formation	S, DM, internal treatment, surface agents
2. Alkalinity	HCO_3, CO_3OH as $CaCO_3$	Foaming, carryover, embrittlement, CO_2 in steam causing corrosion in condensate lines	S, DM, HZ softening, dealkalisation by AX
3. Free acids	HCl, H_2SO_4 and so on as $CaCO_3$	Corrosion	Neutralisation with alkalies
4. CO_2		Corrosion in steam and condensate lines	A, Da, neutralisation with alkalies
5. SO_4^{2-}^		Forms $CaSO_4$ scales	DM
6. Cl^-^		Adds to corrosive nature of water	DM
7. Na^+^		Corrosion by on combining with OH	DM
8. SiO_2		Scale in B and insoluble deposits in T	Adsorption in high;y basic AX in DM
9. Fe and Mn	Fe^{2+} (ferrous), Fe^{3+} (ferric)	Deposits in B, water lines	A, F, lime S, CX, surface active agents
10. O_2		Corrosion in B, HX, water lines	Da, $NaSO_3$, corrosion inhibitors
11. DS		Foaming	S,CX by HZ,DM
12. SS		Deposits in B, HX, water lines	F
13. Oil		Excessive foaming and hence carryover	Dual media or activated C filtration
14. Turbidity		Imparts unsightly appearance to water; deposits in water lines, process equipment, and so on; interferes with most process uses	Coagulation, settling and filtration

Notes: S = softener, DM = demineraliser, Z = zeolite, A = aeration, Da = deaeration, F = filtration AX = anion exchanger, CX = cation exchanger, ^ = adds to solids, total solids TS = DS + SS, B = boiler, T = turbine, HX = heat exchangers.

hardness, the hardness equal to alkalinity is considered as non-alkaline hardness, that is, chlorides, sulphates and nitrates.

- *Alkalinity* is the ability of natural water *to neutralise acid.* It is the concentration of alkaline salts present in the water. It is measured by titration. Bicarbonate, carbonate and caustic alkalinity are caused by HCO_3, CO_3 and OH, respectively.
- *Total alkalinity* is the sum of the above expressed in terms of $CaCO_3$ and known as M reading.
- *Dealkalisation* is the *removal of bicarbonates by ion exchange.*
- *Demineralisation or deionisation* is the partial or virtually complete removal of the dissolved solids by treating water in anion and cation exchangers.

22.5.3 Solids in Water

Water carries two types of solids, namely *dissolved* and *suspended* solids, the sum of which is the *total solids.*

- *Dissolved solid* is *matter in solution* state which would be left behind if all water is evaporated. High levels promote foaming in boilers which is objectionable. Demineralisation reduces the dissolved solids.
- *Suspended solids* are the *solids in suspension* which are removed from raw water by sedimentation, coagulation and filtration. Suspended solids cause deposits in boilers and HXs.

22.5.4 Hardness of Water

Hardness is the scale forming tendency of water when heated. It is also *the character of destroying soap-lather.* This is caused by the soluble salts of Ca and Mg picked up from rocks and soils. Total hardness is the sum of concentrates of Ca and Mg salts.

Hardness is measured by chemical titration and results expressed in mg/L or ppm of $CaCO_3$.

$CaCO_3$ hardness is a general term that indicates the total quantity of divalent salts present and does not specifically identify whether Ca, Mg and/or some other divalent salt is causing water hardness. In theory, it is possible to have water with high hardness that contains no Ca.

Hardness is commonly confused with alkalinity as both results are in ppm or mg/L of $CaCO_3$. Alkalinity is a measure of the amount of acid (H ion) water can absorb (buffer) before achieving a designated pH. Just as with hardness (mg/L $CaCO_3$), alkalinity is also a general term used to express the total quantity of alkali present (OH ion acceptors). If $CaCO_3$ is responsible for both hardness and alkalinity, these values will be similar if not identical. However, where $NaHCO_3$ is responsible for high alkalinity, it is possible to have low hardness and low Ca. Ground or well water has little or no alkalinity and can have low or high hardness.

Temporary alkaline or *carbonate hardness*: This hardness is caused by dissolved $Ca(HCO_3)_2$ and can be removed by boiling or by adding $Ca(OH)_2$ (lime) to water.

In alkaline hardness, the total alkalinity is greater then total hardness.

Permanent non-alkaline or *non-carbonate hardness*: This hardness is caused by dissolved $CaSO_4$ and $CaCl_2$ and cannot be removed by boiling, but by certain processes such as ionexchange using polyphosphates (such as 'calgon') or zeolites. The Ca and Mg ions are replaced by Na ions. Hard water forms scales, which is the left-over mineral deposits that are formed after water has evaporated. In non-alkaline hardness, the total alkalinity is greater then total hardness.

22.5.5 Alkalinity of Water (*see also* Hardness)

Alkalinity is the concentration of alkaline salts in water as measured by titration. It is indicated in mg/L or ppm of $CaCO_3$. Bicarbonate, carbonate and caustic alkalinity are caused by HCO_3, CO_3 and OH ions, respectively. *Total alkalinity* is a sum all the three alkalinities and reported in terms of $CaCO_3$. It is also known as M reading or M alkalinity.

Alkalinity is the ability of water to neutralise acid and is an expression of buffering capacity. A buffer is a solution to which an acid can be added without changing the concentration of available H^+ ions (without changing the pH) appreciably. It essentially absorbs the excess H^+ ions and protects the water body from fluctuations in pH.

Alkalinity contributes to foaming in drums leading to carryover. To control foaming frequent bowing down will be needed which means a loss of boiler chemicals. Bicarbonate and carbonate produce CO_2 in steam which gets carried over and causes corrosion in condensate lines. Also it causes caustic embrittlement.

Dealkalisation is the removal of alkalinity in ion exchange process by substituting the CO_3, HCO_3 and SO_4 ions with Cl_2 ions.

22.5.6 Water Types

In power or process plants, water passes through several stages and it is known by different names at each stage which are listed below and depicted in Figure 22.17.

Raw water is the water received at the plant premises, with usually with no previous treatment, from natural sources like river, lake, well, mine and so on.

Treated water is the water that has been treated in the plant premises.

Softened water is the water with hardness substantially removed.

Condensate is the steam cooled to water, which is returned from either the condenser in the power plant or the process plant without mixing with any other water.

Demineralised water is nearly free from total ionisable dissolved solids, rendered so by ion exchange materials of the DM plant.

fw is the treated water suitably softened/demineralised, degasified and deaerated to make it suitable for feeding the boiler. It is the combination of makeup water and return condensate.

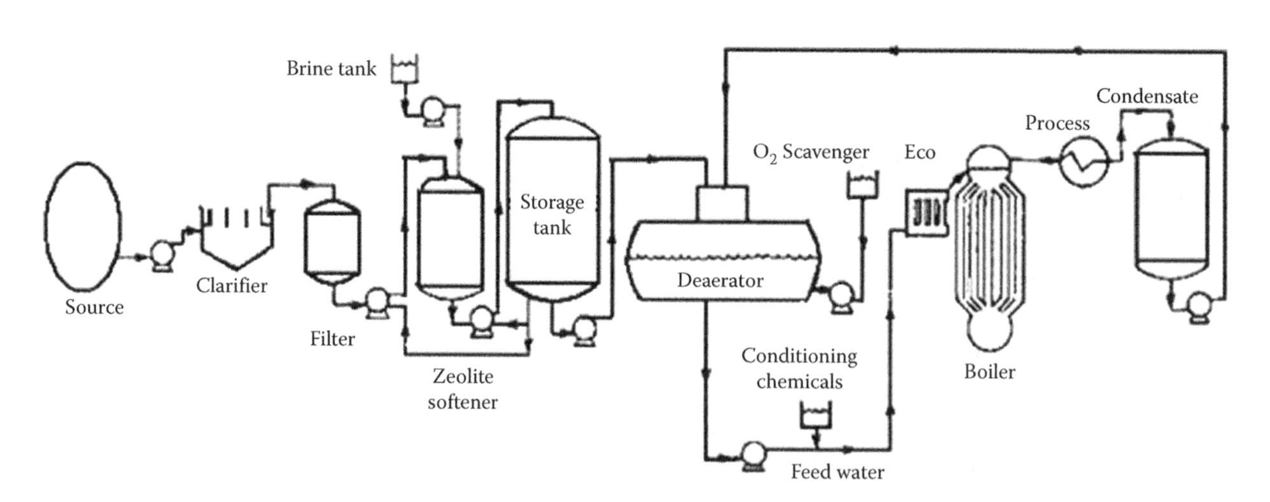

FIGURE 22.17
Stages in water cycle in a process plant.

bw is the water within the Evap circuits of the boiler plant.

Makeup water is the water added to the plant to compensate for the evaporation and other losses.

22.5.7 Water Treatment for Boilers

WT is the procedure of *rendering the available raw water suitable for the required process.* In case of WT for boilers the purpose is to convert raw water to fw.

Boiler WT treatment essentially consists of the following three stages:

1. Clarification to remove the suspended solids—This treatment consists of sedimentation in the large open tanks followed by filtration.
2. Softening or DM to remove hardness and dissolved solids to avoid scale formation.
3. Degasification to remove CO_2 and other dissolved gases to prevent corrosion.

The extent and type of treatment process is dependent on the impurities in raw water present and the purity levels required in fw. Note that *WT* is the process carried out *external* of the boiler island while *water conditioning* is done *within* the limits of boiler island.

The object of WT, external treatment and internal conditioning together, in one word is cleanliness—*cleanliness of the wetted parts.* This, in turn, facilitates the production of clean steam which keeps the boiler, piping and ST—all protected.

22.5.8 Dealkalisation of Water

Dealkalisation of water is the removal of alkaline (carbonate and bicarbonate) ions from water. This is done by chloride cycle anion ion exchange dealkalisers. Dealkalizers are most often used as pre-treatment to a boiler and are usually preceded by a water softener. Dealkalisation is highly needed for the healthy operation. It is mainly helpful for the reducing blowdown.

- Alkalinity often dictates the amount of CBD. Lower alkalinity reduces the CBD.
- High alkalinity promotes boiler foaming and carryover. To reduce both it needs frequent IBD.

Lowering of alkalinity in dealkalisation increases the cycles of concentrations of boiler solids and reduces blowdown and thus the operating costs. The reduction of blowdown keeps the WT chemicals in the boiler longer, thus minimising the amount of chemicals required for efficient, non-corrosive operation.

Unless minimised, CO_3 and HCO_3 alkalinities in bw decompose by heat of bw, releasing CO_2 into the steam. CO_2 combines with the condensed steam in process equipment and return lines to form carbonic acid (H_2CO_3). This depresses the pH value of the condensate returns and results in corrosive attack on the equipment and piping.

22.5.9 Demineralisation of Water

Demineralisation is substantial or complete removal of minerals in water. This is achieved by treating water in both *cation* and *anion* exchanges in an *ion exchange* process. It entails removal of electrically charged (ionised) dissolved substances by binding them to positively or negatively charged sites on resin, as water passes through the resin packed columns. This ion exchange process can be used with different configurations and ways to produce de-ionised water of various qualities.

In a simple system shown in Figure 22.18 a strong acid cation + a strong base anion resins are used in two separate vessels. The cation exchange resin is in the hydrogen (H^+) form and anion resin is in the hydroxyl (OH^-) form. As water flows through cation column all cations of the water are exchanged for H ions. Thereafter, when de-cationised water flows through the anion column, the negatively charged ions of water are exchanged for OH ions. The H and OH ions now combine to form water. This way, all the ions are removed including silica. A degasifier is installed between the two exchangers for the removal of CO_2. Deionisation typically does not remove organics, virus or bacteria except through 'accidental' trapping in the resin. Composition of water as it passes through the various stages is given in Table 22.7.

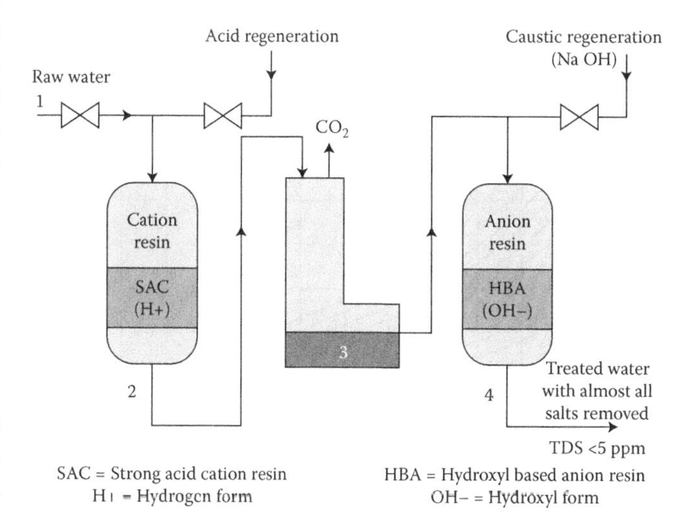

FIGURE 22.18
Scheme of demineralisation of water.

TABLE 22.7

Stage Wise Transformation of Raw Water to Demineralised Water

Raw Water	After Cation Exchanger	After Degasifier	Demineralised Water
1	2	3	4
$Ca(HCO_3)_2$	$2H_2CO_3$	H_2O	H_2O
$Mg\ Cl_2$	2HCl	2HCl	H_2O
Na_2SO_4	H_2SO_4	H_2SO_4	H_2O
Na_2Sio_3	H_2Sio_4	H_2SiO_3	H_2O
pH 7.6	pH 2.0–2.5	pH 2.0–2.5	pH 8.5–9.0

22.5.10 Deaeration

Deaeration is removal of air and gases from water. In boiler plants it is mainly the removal of dissolved O_2 from the fw. In the degasification tower before deaeration, O_2 is not separated because of its strong affinity to water. It is possible to remove O_2 only by heating the water to its saturation temperature when the solubility of O_2 dramatically reduces (as shown in Figure 22.19). About 98% of O_2 can be removed by boiling and the remaining has to be scrubbed out of water, which is done by reducing the water to very fine droplets.

Deaeration is extremely important because even traces of O_2 corrode steel parts in a special manner, that is, pitting. **Pitting**, as the name suggests, is the formation of deep pits by corrosion. In boilers, particularly on the top portions of headers and tubes in Eco, O_2 collects to attack the metal. Disastrous failures have taken place as the boiler parts are under pressure and the incipient weakness often goes unnoticed.

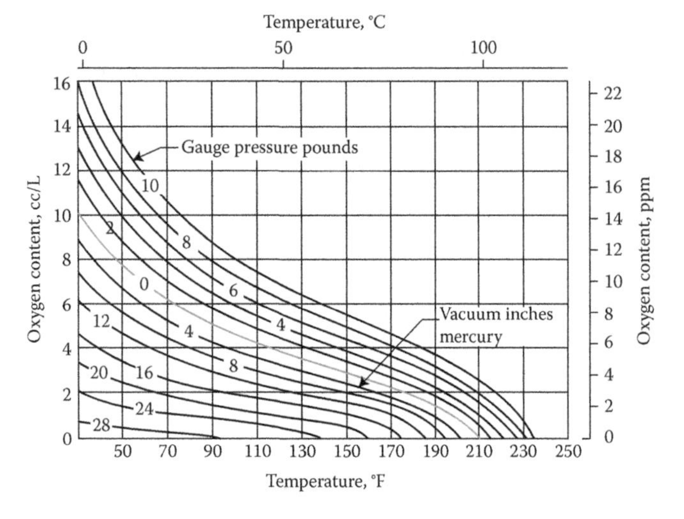

FIGURE 22.19
Solubility of oxygen in water under differing pressures and temperatures.

22.5.11 Deaerator

Deaerators are cylindrical vessels with spray nozzles inside (along with heating trays sometimes) for heating the incoming fw with steam to its saturation temperature and scrubbing to remove the entire dissolved O_2. In a properly designed and operated deaerator O_2 level at exit is <0.005 cc/L by volume or 0.007 ppm (7 ppb) by weight.

The purpose of spray nozzles and trays is to split the incoming water stream into very fine droplets so that the air and gas are released. The deaerators are usually mounted on the storage tanks which are required to provide 10–20 min or more of fw storage in the cycle. The deaerator (singular) design can be either

a. Horizontal
b. Vertical

to suit the available layout and head room. The vertical design is more compact but needs greater head room compared to the horizontal design. In Figure 22.20, a vertical deaerator assembly with spray arrangement mounted on a horizontal storage tank is shown.

Makeup water to replenish the losses in the steam cycle is admitted at the desired pressure into the deaerator tower through the spray nozzles, which atomise incoming water to fine droplets. Water droplets get heated almost instantaneously by the flow of steam in this preheater zone and O_2 is simultaneously liberated. Steam and O_2 then travel upwards and get vented. Heated makeup water enters the main body of deaerated water after it is thoroughly scrubbed by the incoming steam in the scrubber section. Condensate is returned to the deaerator and is not subject to scrubbing action.

Deaerator has controls like level and overflows on water side and pressure and vent flows on steam side. The tank is protected against overpressure by SVs. Anti-vortex baffles at the mouth of water discharge pipe prevent a possible suction of air bubbles which can cause cavitation in BFPs.

In spray- and tray-type deaerator shown in Figure 22.21, in addition to the spray nozzles, the deaerator dome contains a set of ss trays which help in further breakdown of water droplets as they descend over them. This enhances steam water interaction helping the deaeration efficiency and lowering the steam consumption.

The scrubbing section is, in a way, replaced by tray section. Spray- and tray-type of deaerators are very popular in the power stations.

Deaerators are very reliable and rugged equipments. If liberated O_2 is promptly and fully removed from the top of the vessel causing no internal corrosion, deaerators give long trouble-free service. The two limitations of deaerator are that they cannot remove

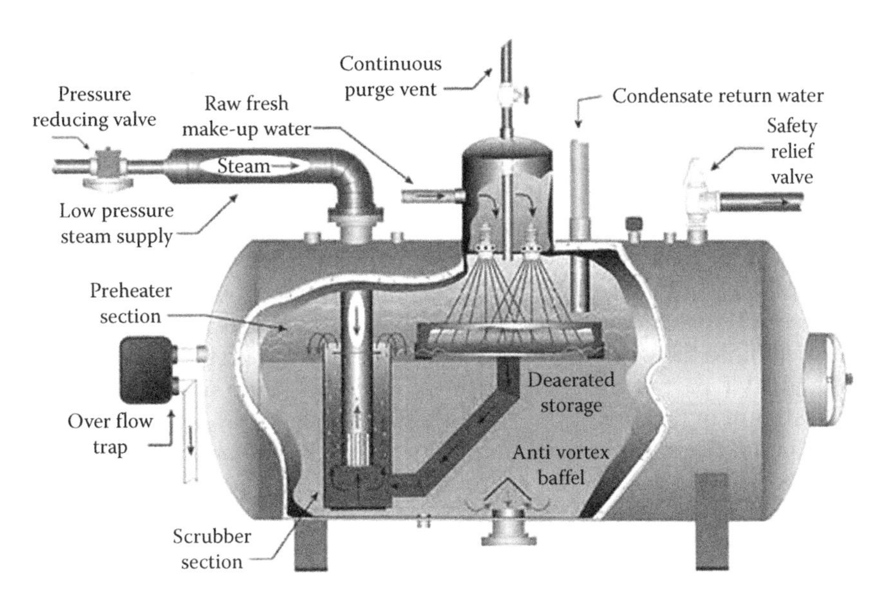

FIGURE 22.20
Vertical spray-type deaerator mounted on a horizontal storage tank.

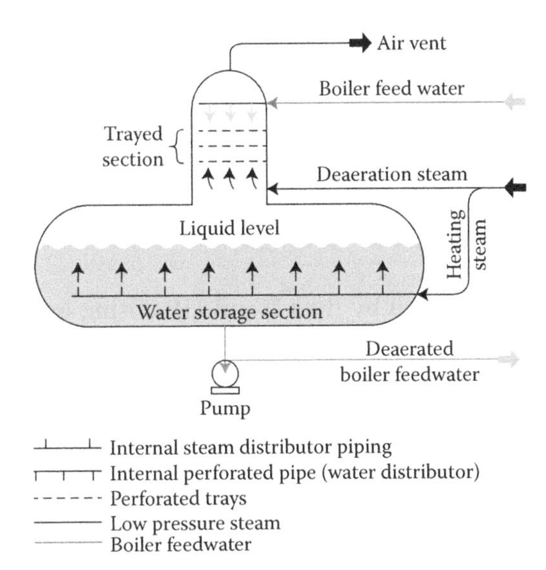

FIGURE 22.21
Spray- and tray-type deaerator.

- The dissolved gases found in the waste streams of many industrial plants
- Bulk of the combined CO_2 in the fw, which has to be controlled by neutralising and filming amines

22.5.12 Oxygen Scavenging

Heating and scrubbing action of water in deaerator leaves it nearly free of dissolved O_2 to <0.005 cc/L by volume or 0.007 ppm by weight. Removal of last traces of O_2 from fw is highly important to prevent pitting corrosion. This is done by adding chemicals called O_2 scavengers.

Depending on the boiler operating pressure, Na_2SO_3 or N_2H_4 are employed in most power plants. Na_2SO_3 is used for boilers pressure up to 70 bar favoured mainly by its low cost, ease of handling and non-scaling properties. N_2H_4 is used for boiler pressures >70 bar as it adds no solids to the bw. Moreover, Na_2SO_3 decomposes at HPs into H_2S and SO_2 both of which cause corrosion of the return condensate systems.

22.5.13 Boiler Water Conditioning

Deaerated fw enters the boiler Evap/circulating system in the steam drum after passing through the Eco. fw absorbs more heat and circulates in the Evap system before progressively turning to steam. The concentration of solids in bw is therefore much higher as steam leaves behind its solids in bw as it leaves the drum. It is necessary that bw should not

- Cause corrosion of the tube internals
- Form scales on the inside of the tubes
- Be carried over to the SH and ST and form scales

Bw conditioning is the combination of

1. Adding suitable chemicals at appropriate places (HP and LP dosing) in the bw circuit (see Chemical Dosing).
2. Carrying out appropriate CBD and IBD (see Boiler Blowdown).
3. Adopting a suitable drum water conditioning treatment (see drum WT) to achieve the above triple objectives of *no scales, no corrosion or no carryover.*

22.5.14 Drum Water Conditioning (*see also* Water Conditioning for SC Boilers *under* Utility Boilers)

Drum water conditioning mainly involves maintaining the correct chemical balance in the bw as demanded by the specific conditioning treatments such as, the conventional phosphate, coordinated phosphate, all volatile, chelant and so on to achieve the *objectives of no corrosion, no scale and no carryover.* HP dosing system injects the chemicals needed directly into the steam drum. For treatments like chelant and AVT part of the chemicals have to be dosed at LP upstream of BFP.

There are mainly four types of drum water conditioning treatments, even though in actual practice, there are variations to these treatments which are practiced by different plants to suit their individual needs.

a. Conventional phosphate or phosphate hydroxide treatment

This treatment is the most popular for LP and MP boilers operating up to 70 bar. Here, the principle is

- To maintain the pH level of 10.5–11.2 in the steam drum with excess OH
- To convert the hardness constituents namely, Ca and Mg salts, as flocculent precipitate which is then blown off

Orthophosphate residuals are maintained between 20 and 60 ppm as PO_4 and hydrate alkalinity between 200 and 400 ppm as OH.

b. Coordinated phosphates treatment

This is for HP boilers operating at >70 bar, where the alkalinity levels of the conventional phosphates treatment are unacceptable as caustic corrosion is of serious concern. The treatment chemicals are a combination of tri- and di-sodium phosphate. PO_4 concentration is maintained as a fixed relation to pH of bw.

c. Chelant treatment

Organic agents react with residual divalent metal ions namely Ca, Mg and Fe in fw to form soluble complexes which are removed by blowdown. Na salts of ethylene diamine-tetra acetic acid (EDTA) and nitrilo acetic acid (NTA) are the two most common water chelating compounds. Chelants are added only to the fw as they tend to degrade at higher temperatures. Chelant treatment is popular up to pressure ~100 bar.

d. AVT

For HP boilers operating at >140 bar, AVT is frequently adopted. Only NH_3 and N_2H_4 are added to the bw to control pH between 9.2 and 9.4. This treatment is suitable with high quality fw and zero solids in bw.

22.5.15 All Volatile Treatment of Boiler Water

AVT is rendering fw contamination-free by employing a *pure water* approach for HP boilers operating at >140 bar, where all control is directed solely to the fw quality. **Corrosion** and deposition are kept under check by the removal of all chemicals likely to cause problems, namely, the hardness, salts and alkalinity and so on. The operation is with 'zero solids' in bw which naturally gives pure steam output.

However, in case there is a contamination of fw there is possibility of excessive tube side corrosion as the buffering ability of the bw is not there anymore. Also the deposits are much harder to remove. Only NH_3 and or N_2H_4 are added to the bw for pH control between 9.2 and 9.4.

22.5.16 Chemical Dosing of Boilers

Thoroughly deaerated and scavenged fw enters the Eco and thereafter the main Evap. In **Eco**, there is practically no concentration of solids in water like in the **Evap**, where the water recirculates several times before converting to steam. The main objectives are to protect

- Eco tubes from internal corrosion principally from the dissolved O_2
- Evaporator surfaces from corrosion, scale formation and carryover

This is achieved by the LP and HP dosing by injecting suitable chemicals in right measure at appropriate places in the water circuit (as shown in Figure 22.22).

1. *LP dosing*: DM makeup water and the return condensate enter the deaerator together to get heated and stripped of the dissolved gases, notably dissolved O2. At the exit of deaerator LP dosing is done for
   - Raising fw pH value by the addition of NH_3 to 9.2 to 9.5 or 8.5 to 9.2 depending upon whether there are steel or cupronickel tubes in the FWHs
   - Removal of traces of O_2 by dosing Na_2SO_3 usually for boiler pressures <70 bar and N_2H_4 for >70 bar
   - Bw conditioning by chelant dosing if chelant treatment is adopted
   - bw conditioning by NH_3 and N_2H_4 if AVT is adopted for after boiler protection
2. *HP dosing*: In HP dosing, the chemicals are injected directly into steam drum for
   - Raising and maintaining pH of bw, by injecting NH_3, for corrosion prevention

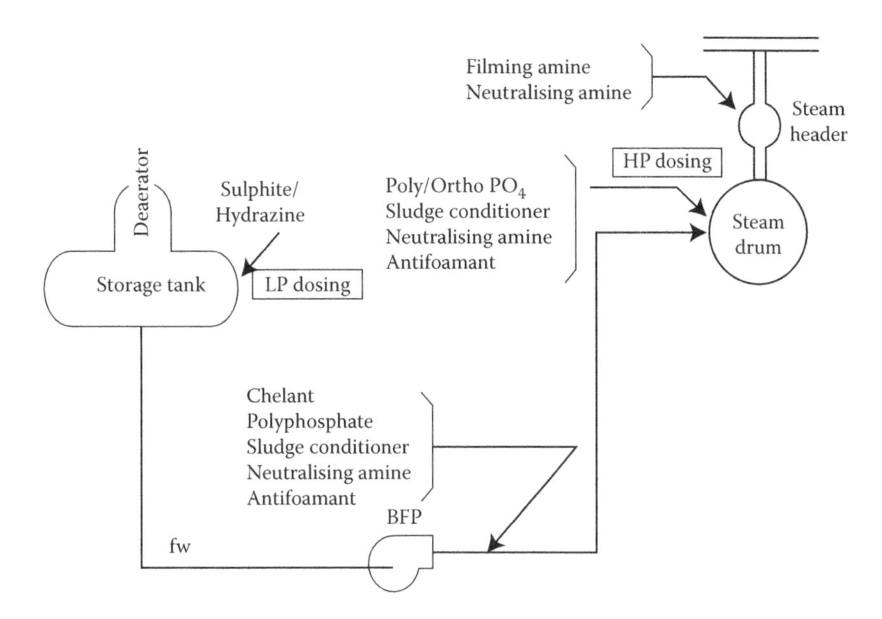

FIGURE 22.22
HP and LP dosing and other chemical injection into boilers.

- Forming soft sludge of the suspended solids, by dosing phosphates, to help its removal by IBD
- Conditioning the sludge by dosing sludge conditioners for its better and easier removal
- Preventing foaming inside the drum by dosing anti-foamants

Blowdown which is expressed as percentage MCR is given by

$$\text{Blowdown} = \frac{\text{Solids in fw}}{\text{Maximum permissible solids in bw} - \text{solids in fw}} \quad (22.1)$$

22.5.17 Boiler Blowdown, Continuous and Intermittent (*see also* Blowdown Tanks *and* Blowdown Valves *under* Valves in Boiler)

In bw conditioning, addition of chemicals to regulate BQ goes hand in hand with blowdown scheme. There are two types of blowdowns:

1. *IBD* also called *dirty blowdown* is for periodically discharging **sludge** from bw. **IBD valve** is a specially designed quick opening valve to discharge high quantities of water from the lowest possible point in the circulating system to enable quick drain out of sludge. No heat recovery is possible as the blowdown is periodic.
2. *CBD* or *clean blowdown* is meant for controlling the TDS at LP and MP (usually <70 bar) and dissolved SiO_2 at HP. Finely regulated micrometer valves withdraw bw from stream drum and discharge to CBD tank on a continuous basis. **CBD valve** opening is decided by the station chemist based on maintaining the desired SiO2 or TDS.

22.5.18 Effects of Water on Boilers

fw duly deaerated, dosed with O_2 scavengers and pH level raised suitably enters economiser or bw circuits where there is no economiser. fw gets heated immediately to saturation temperature on joining the bw in circulation and progressively gets converted into steam, leaving behind its dissolved solids which form scales on the tube and drum internals. The three main effects of bw on the steel tubes and drums are

1. *Corrosion*—It is the process of accelerated oxidation of steel. The high temperature of bw further aids the corrosion mechanism. The formation of metal oxide, besides making the part weak, impedes the heat transfer and restricts the flow. In time, this leads to the overheating and failure of the tube.
2. *Scaling*—It is the deposition of separated impurities of the bw on tubes and drums as it get evaporated. Scaling impedes the heat transfer and causes flow restriction. It can also aid corrosion as it shields the underlying metal from protective chemicals in the bw.

3. *Carryover*—As the steam bubbles disengage from the water surface in the steam drum, minute water particles along with impurities, tend to escape into the steam space, which is called carryover. Carryover reduces the purity of steam and the entrained impurities from bw tend to form fine scales on the inside of SH tubes causing overheating and failure. The impurities carried over can be solid, liquid and gaseous impurities of bw. Carryover arises from the increased concentration of suspended and dissolved solids in drum water and also high disengagement speed of steam and water.

22.5.19 Water Quality for Feed and Boiler Water

For long and corrosion-free life of boiler, monitoring the quality of water on a continuous basis is most important—both for fw and bw.

Boiler manufacturer's duty is to recommend the fw and bw quality based on the materials of construction and the heat flux. There are general norms established by authorities like ASME, ABMA and so on and based on the specific constructional features the boiler manufacturer alters the recommendations to suit.

The guidelines normally prescribe the bw quality only and it is expected that fw is conditioned with chemicals by LP dosing in such manner that the bw conforms to the recommended limits.

Table 22.8 gives ASME guidelines for water quality in subcritical drum-type boilers up to ~140 bar.

The other points are to be understood about fw treatment are:

- The limiting values of bw are to be achieved without excessive blowdown.
- With spray attemperation SiO_2 and solids should minimum. Total solids should be <3 ppm to prevent deposition in SH and <0.1 ppm with austenitic section in SH.
- Spray water silica limit is <0.02 ppm (20 ppb) when the boiler is feeding a ST generating power.
- O_2 scavenging should always supplement physical deaeration in deaerator.
- pH should be 8.5–9.2 with Cu alloy and 9.2–9.5 with steel FWH.

22.5.20 Caustic and Hydrogen Embrittlement

When steels suffer a sharp reduction in their abilities to deform (loss of ductility) or to absorb energy during fracture (loss of toughness) with little change in other mechanical properties, such as strength and hardness, it is termed as embrittlement. Embrittlement of steel leads to sudden and violent failures.

Embrittlement is a form of stress corrosion cracking which takes place along the grain boundaries. This happens in steels when there is a combination of sufficient tensile stress and environment causing corrosion.

a. *Caustic embrittlement or caustic stress corrosion* is an *intra-granular or inter-crystalline* **corrosion** cracking similar to chloride stress corrosion when there is combination of
   - High concentration of caustic (high pH) in the environment
   - High tensile stress in metal

   MS, low CS, low AS, and **ss** are all prone to this type of corrosion crack. Cracks due to caustic embrittlement were first encountered in the operation of riveted steam boilers along the seams under the rivets. The crevices underneath provided places for evaporation of water facilitating concentration of caustic. The modern drums do not have this problem. However, drums and headers where tubes are rolled and

TABLE 22.8

ASME Guidelines for Feed and Drum Water Quality in Subcritical Drum-Type Boilers up to ~140 Bar

Drum Pressure		fw			bw		
~(barg)	(psig)	Iron (ppm Fe)	Copper (ppm Cu)	Total Hardness (ppm $CaCO_3$)	Silica (ppm SiO_2)	Total Alkalinity (ppm $CaCO_3$)	Sp. Cond. (μ mhos/cm) (Unneutralised)
0–20.7	0–300	0.100	0.050	0.300	150	700	7000
20.7–31.0	301–450	0.050	0.025	0.300	90	600	6000
31.0–41.4	451–600	0.030	0.020	0.200	40	500	5000
41.4–51.7	601–750	0.025	0.020	0.200	30	400	4000
51.7–62.0	751–900	0.020	0.015	0.100	20	300	3000
62.0–69.0	901–1000	0.020	0.015	0.050	8	200	2000
69.0–103.5	1001–1500	0.010	0.010	0.0	2	0	150
103.5–138	1501–2000	0.010	0.010	0.0	1	0	100

expanded into place are still susceptible to this type of failure.

In addition to high stress and free NaOH, crevices are needed to provide localised increase of NaOH concentration for this phenomenon to take place.

b. *Hydrogen embrittlement* is due to H_2 and is encountered only in HP boilers of 140 bar and above, particularly with AVT of bw. When pH of bw lowers materially, steel is attacked which generates H_2. If this happens under hard, adherent, non-porous deposits atomic H_2 permits the metal structure to form CH_4. CH_4 being bigger molecule than C atoms, excessive pressure develops within the metal structure leading to rupture along with crystalline borders. Ruptures are sudden, violent and often disastrous.

This phenomenon is not fully understood. H_2 embrittlement does not affect all metals equally but high strength steels are most vulnerable.

22.5.21 Amines

Amines are derivatives of NH_3 wherein one or more H atoms have been replaced by a substituent such as an alkyl or aryl group. They are organic compounds and functional groups that contain basic N atom.

Amines are used mainly in bw treatment of industrial and small utility boiler plants primarily for corrosion protection of piping and internals on the pre and after boiler sections. There are two types of amines—neutralising and filming, depending on their action.

1. *Neutralising amines*: The most commonly used amines in industrial practice for corrosion inhibition in the after-boiler section are
   - Cyclohexylamine
   - Morpholine
   - Diethanolamine

   They act by neutralising the condensate pH.
2. *Filming amines*: A filming amine, normally, octadecyclamine is commonly used either alone or in combination with the neutraliser. Volatilisation of these amines at different points in the system gives a wider protection of the steam and condensate lines.

Filming amines are used to establish a continuous protective film over the after-boiler surfaces which prevents contact of potentially corrosive steam and condensate constituents with the insides of system components. They are added directly into the steam headers.

A filming amine programme is less costly than neutralising amine programme to administer.

22.6 Water Carryover (*see* Steam Purity)

22.7 Water Drum (*see* Pressure Parts)

22.8 Water-Holding Capacity of Boiler (*see also* Steam Drum in Pressure Parts)

Water-holding capacity of boiler is the amount of water that a boiler can hold while in operation. Internal volume of all the tubes, drums and headers excluding that of SH and RH has to be estimated and multiplied by the density to arrive at the weight of water.

Water-holding capacity is an indication of the thermal inertia of the boiler. The higher the volume and the longer the start-up time, the more sluggish the response and the better the tolerance for operational errors.

Single-drum boilers have lesser water holding than bi-drum boilers.

22.9 Water Lancing (*see* Soot Blowers)

22.10 Water Level Indicators (*see also* Direct Water Level Indicators *and* Remote Water Level Indicators)

Water level in drum-type boilers is perhaps the most important indication for the boiler operator. Nothing can give him more comfort than seeing the drum level at its normal.

There are two types of WLIs—direct (DWLI) and remote (RWLI), both of which are installed in the modern-day boilers. As it affects the safety of boiler level indicators are very important boiler auxiliaries. All boiler codes provide elaborate guidelines on the type, selection and installation and so on.

The essential guidelines for WLI extracted from ASME Section 8.5 are

- Every drum-type boiler should have at least one DWLI.

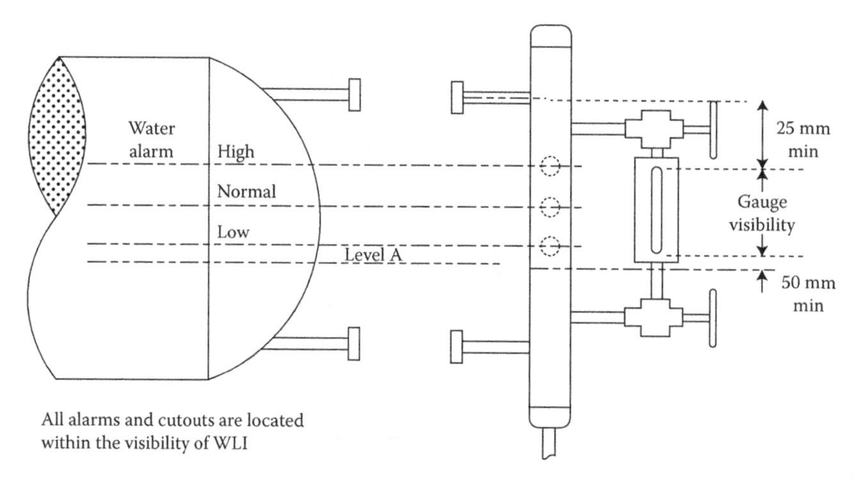

FIGURE 22.23
Aligning trips and alarm levels in drum.

- Two DWLIs are mandated for boilers with working pressure >400 psia (~28 bar).

Alternatively, 1 DWLI + 2 independent RWLIs can also be provided.

- When two independent RWLIs are installed and are operating reliably one DWLI can be shut off but kept in serviceable condition.
- The lowest visible water level in DWLI or RWLI should be at least 50 mm above the lowest permissible water level (50 mm above the upper end of the highest downcomer tube).

Figure 22.23 indicates the way the gauge glass is aligned with various levels in the steam drum.

22.11 Water Tube Boilers (*see* Boilers)

22.12 Water Wall (*see also* Furnace Wall Construction in Heating Surfaces)

Furnace walls can be made of full refractory, tubular or mixed construction. Several decades of boiler development saw brick walls change to tube walls. Furnace walls cooled by tubes are also called water walls.

22.13 Water Wall Circuit (*see* Circulation)

22.14 Wear

Wear is a progressive loss of surface material due to mechanical action of impingement of abrasive particles. Abrasion and erosion are the two types of wear.

a. Abrasion

Abrasion is the *progressive loss of surface material* in which solid particles move in contact in a *parallel* manner. It is the process of wearing away a surface by friction. Abrasion affects the high spots of the surface without any significant effect on the main body. Unlike in erosion, the loss of material is comparatively less in abrasive wear.

Abrasion resistance can be built by a boundary layer of high and preferably hard spots.

Abrasives are used in manufacturing to remove material from a surface to smoothen or create controlled surface roughness. Coatedabrasives, grindingwheels, sandblasting, and polishes are employed.

b. Erosion (see Ash Side Erosion in Ash)

In erosion, hard particles impinge at an *inclination* with great energy and destructive power. Impinging particles destroy both the boundary layer and the matrix and hence, *abrasion resistance materials cannot withstand erosion.* Abrasion, on the other hand, is the sand papering where hard particles move parallel to the surface.

Wear in boilers

Wear in boilers is contributed mainly by ash. Naturally, boilers firing NG and light FOs are practically free from the problems of wear. On the other end of the scale severe wear is experienced in coal-fired boilers, particularly those that have high level of abrasive ash.

'Pure coal', the carbonaceous matter, is relatively soft. It is the impurities like slate, sand and pyrites that make coals abrasive. Quartz content of these minerals is the main contributor to the abrasive nature.

Abrasion, abrasiveness or YGP index

This property is tested in a calibrated mill, containing four blades of known mass. Sized coal of <6 mm is placed in contact with the blades and is agitated in the mill for 12,000 revolutions at a rate of 1500 rpm for 8 min. The abrasion index is determined by measuring the loss of mass of the four metal blades. *Yancey–Geer price* apparatus or equivalent is used for conducting this test. *YGP index* so derived provides valuable insights into the relative abrasive nature of different coals but there are limitations in predicting the actual wear in field conditions.

Coal washing is the only way of removing the impurities and making coal less abrasive. In boilers, there is generally more erosion and not so much abrasion. Ash comes into being only after combustion and thereafter, 60–80% of ash gets gas borne and travels at high speeds impinging on the PPs to cause erosion. This is explained in detail under the topic of ash side erosion in ash.

Coal abrasion

Coal has sliding movement and also gets air borne causing abrasion and erosion, respectively, depending upon the type and level of impurities. Mills and feeders are the equipments subject to abrasion. Wear parts of the coal mill, namely balls, rolls, rings, races and liners, are all susceptible to abrasive wear. Periodic weld overlay on the rolls and tyres is the solution. Coal pipes carrying powdered coal from mill to burners are also prone to erosion, mainly in the bends. But, the inside of coal pipes experience abrasion mainly on the bottom part, due to the sliding of coal streams. Lining of pipes with cast basalt or ceramics is the way to mitigate the problem. Ceramics stand up to wear much better than **basalt**. Alternatively, wear-resistant pipe material can be employed.

22.15 Welding

Welding and related processes are central to boiler making both in manufacturing and erection. A boiler is, after all, an advanced fabrication and most activity revolves around welding. Electric arc welding is the most common type in boiler making.

Electric arc welding is the process of joining one or more metals by the application of localised heat in the area intended for adhesion. A weld pool is created by the heat and the parts melt locally. When the heat is removed the pool solidifies creating a permanent metallurgical bond.

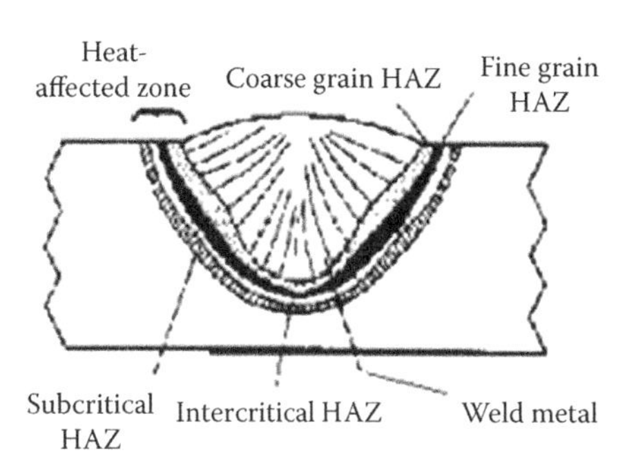

FIGURE 22.24
Heat-affected zone around a weld.

Usually *filler material* is introduced in the molten pool to fill in the gap kept between the parts. Both the parts experience intense heat and when cooled they are hardened up to a certain distance of the weld line. This is known as *heat-affected zone* (HAZ), as shown in Figure 22.24.

Arc welding can employ either ac or dc current depending upon the materials and electrode. There are four types of arc welding that are employed in boiler making. These are

1. Gas tungsten arc welding or tungsten inert gas welding	GTA welding/TIG welding
2. Gas metal arc welding ormetal inert gas welding	GMA welding/metal inert gas welding (MIGwelding)
3. SAW	SA welding
4. Shielded metal arc welding	SMA welding

1. *Gas tungsten-arc welding (GTA welding) or tungsten inert gas welding (TIG welding)*

Heat is generated by using W electrodes as one pole of the arc. The gas is usually Ar, He, or a mixture of the two. A filler wire provides the molten metal if necessary.

The TIG welding process is especially suited to thin materials producing welds of excellent quality and surface finish. Filler wire is similar in composition to the materials being welded.

TIG uses a permanent non-melting W electrode. Filler metal is added separately making the process flexible. It is also possible to weld without filler material.

Usually for TIG welding ac current is used. AC TIG welding usually uses argon as a shielding gas. By changing the diameter of the W electrode, welding may be performed with a wide range of heat inputs at different thicknesses ranging from 0.5 to 5 mm. For >5 mm AC TIG welding is less economical compared to MIG welding due to lower welding speed.

DC TIG welding with electrode negative is used for welding thicknesses above 4 mm. The negative electrode gives a poor oxide cleaning compared to AC-TIG and MIG, and special cleaning of joint surfaces is necessary. The process usually uses He shielding gas. This gives a better penetration in thicker sections. DC TIG welding is applicable for welding thicknesses in the range 0.3–12 mm.

2. Gas metal arc welding (GMA welding) or MIG welding

External gas, such as Ar, He, CO_2, or gas mixtures shield the weld zone. Deoxidizers in the electrode prevent oxidation in the weld puddle, making multiple weld layers possible at the joint.

MIG welding is a relatively simple, versatile, and economical welding apparatus to use. This is due to twice the welding productivity over SMA welding processes. In addition, the temperatures involved in MIG welding are relatively low and are therefore suitable for thin sheet and sections <6 mm.

MIG welding can be easily automated, and lends itself readily to robotic methods. It has virtually replaced SMA welding in present-day welding operations in manufacturing plants.

MIG employs an Al alloy wire as a combined electrode and filler material. The filler metal is added continuously and welding without filler material is therefore not possible. Since all welding parameters are controlled by the welding machine, the process is also called semi-automatic welding.

This process uses a dc power source with the electrode positive (DC, EP). By a positive electrode the oxide layer is efficiently removed from the Al surface, which is essential for avoiding lack of fusion and oxide inclusions. The metal is transferred from the filler wire to the weld bead by magnetic forces as small droplets spray transfer. This gives a deep penetration capability to the process and makes it possible to weld in all positions. It is important for the quality of the weld that the spray transfer is obtained.

SAW

SA welding shields the weld arc using a granular flux fed into the weld zone forming a thick layer that completely covers the molten zone and prevents spatter and sparks. As it acts as a thermal insulator deeper heat penetration is possible.

The process is obviously limited to welding in a horizontal position and is widely used for relatively high speed sheet or plate steel welding in either automatic or semiautomatic configurations. The flux can be recovered, treated and reused.

SA welding provides very high welding productivity—4 –10 times as much as the SMA welding process.

Shielded or manual metal arc welding (MMAW) (SMA welding/MMA welding)

This is one of the oldest, simplest and most versatile arc welding processes. The arc is generated by touching the tip of a coated electrode to the work piece and withdrawing it quickly to an appropriate distance to maintain the arc. The heat generated melts a portion of the electrode tip, its coating, and the base metal in the immediate area. The weld forms out of the alloy of these materials as they solidify in the weld area. Slag formed to protect the weld against forming oxides, nitrides, and inclusions must be removed after each pass to ensure a good weld.

The SMA welding process has the advantage of being relatively simple, only requiring a power supply, power cables, and electrode holder. It is extensively used in construction, shipbuilding, and pipeline work, especially in remote locations.

22.15.1 Comparison of Welding Processes

Table 22.9 compares the popular welding processes in boiler making.

TABLE 22.9

Comparison of Commonly Used Welding Techniques in Boiler Manufacturing

Process		Applications	Advantages	Limitations
Tungsten inert gas welding	TIG	Tube to tube root welding SH, RH and Eco tube welding	High quality and finish Low cost Easy automation	Low productivity Less penetration High skill needed Radiation from arc
Metal inert gas welding	MIG	Tube to welds panel repair and attachments	Average quality Moderate productivity Easy automation radiation from arc	Less penetration Periodic stoppages due to tip wear Radiation from arc
Submerged arc welding	SAW	Membrane panel fabrication Header and drum fabrication Heavy structural fabrication	High quality and finish Very high productivity Radiation from arc shielded	Only horizontal welding
Manual metal arc welding	MMAW	Tube to header weld most repairs	Highly portable Very cheap Any position welding	Low quality Low productivity High skill needed Cannot be automated

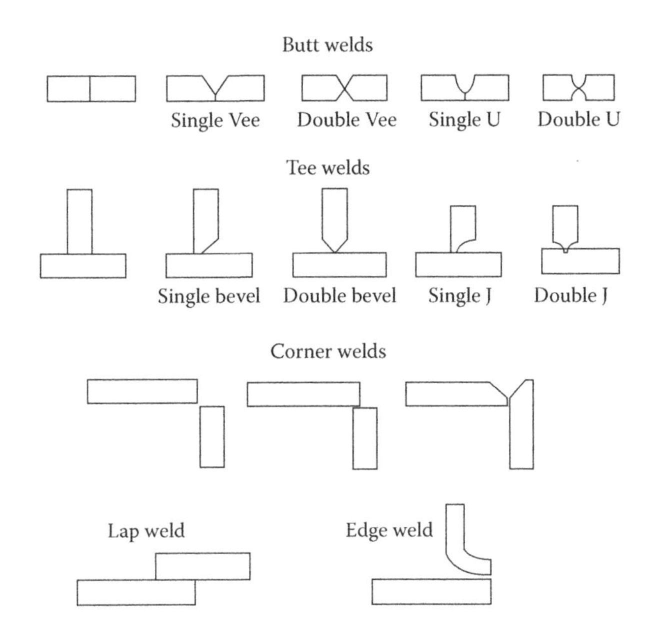

FIGURE 22.25
Different types of welds.

22.15.2 Types of Weld Joints

Figure 22.25 shows various types of possible. In PP fabrication, however, lap, edge and corner joints are not permitted as full penetration is not possible. Boiler codes go into extensive details and specify the welds that are permitted in various parts.

22.15.3 Welding Positions

There are five or six welding positions, with increasing difficulty of welding, as standardised by ASME. Four of these are depicted in Figure 22.26. Suffix F and G denote fillet and groove welds, respectively. In addition to the four positions shown position number

- 5 F and G are for pipe fixed horizontal
- 6 G is for pipe fixed @ 45° upwards or downwards

22.15.4 Welding Electrodes

There is a vast variety of welding electrodes, each differing depending on the task it is meant for and how it affects the welding process. Electrodes are made to deal with a specific kind and range of electrical currents and come with different coatings and in different sizes depending on the welder's requirements.

For arc welding high currents are needed to create large heat that can melt the parent metals. Electrodes pass the current needed for striking the arc. They are shaped either in long rods or stable wires, with a contact point at the tip of the rod, to channel the powerful electrical current into the metal. As the tip of the electrode nears the metal, the current jumps into the metal, creating the dazzling arc creating the heat that melts the metal.

Electrodes are rated according to a system using a letter, such as E, followed by series of four or five numbers. Often, the diameter of the electrode is also expressed in inches before the identification, so the entire code looks like 1/16-inch E6010. This electrode has a diameter of 1/16 inch (the smallest electrodes produced) and is used for arc welding (shown by the 'E'). The number system shows what the tensile strength of the electrode is by the first two numbers in kpsi, while the latter numbers show what type of coating the electrode has and what currents it can be used with. Sometimes other letters or numbers are added to show specific information.

Consumable electrodes slowly burn away as they are used needing periodic replacement. These are purchased in bulk and are much less expensive than the permanent versions. They are easier to use. These electrodes are coated with chemicals known as flux. Most flux coatings are designed to create a smoother, protected weld. As flux burns it produces a small cloud of gas, shielding the weld from oxygen and other contaminants that can ruin its cooling process. The metal of the consumable electrode is also burned and added to the weld to help with stability. There are many different

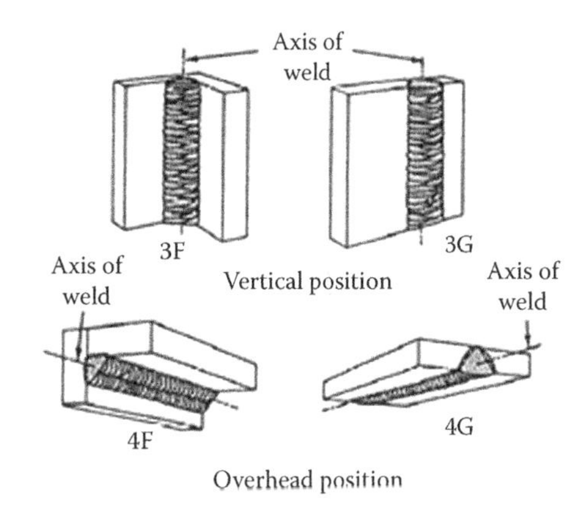

FIGURE 22.26
Welding positions.

kinds of flux coatings with slightly different effects to choose between.

Non-consumable electrodes do not burn away as they channel the electric current and last much longer. These electrodes, such as W rods, are more difficult to use and must have a shielding gas. This gas can be one of several compounds and is spread out at the site of the weld to perform the same protection tasks that the flux does in consumable electrodes. Non-consumable electrodes can be used with a wider variety of metals, especially the heavier metals.

22.15.5 Orbital Welding

This is a welding technique in which the arc from a W electrode is rotated around the tubing/piping weld joint yielding a more precise, reliable and easier method than the manual welding. Modern-day orbital welding systems offer computer control where welding parameters for a variety of applications can be stored in memory and called up when needed for a specific application. The skills of a certified welder are thus built into the welding system, producing enormous numbers of identical welds and leaving significantly less room for error or defects.

The main feature of orbital welding is the ability to make high-quality, consistent welds repeatedly at speeds close to the maximum which offer many benefits like far superior productivity, quality and consistency. It is used in applications where

- A tube or pipe to be welded cannot be rotated or where rotation of the part is not practical.
- Access space restrictions limit the physical size of the welding device.
- It is difficult for a manual welder to use a welding torch or view the weld joint.
- Inspection of the internal weld is not practical for each weld created.

In boiler shops stub to header welding offers a perfect application for orbital welding. Compact orbital weld heads can be clamped in place between rows of stubs where a manual welder would experience severe difficulty for making repeatable welds.

Orbital welding uses the GTA or TIG welding process as the source of the electric arc that melts the base material and forms the weld. An electric arc is established between W electrode and the part to be welded, as shown in Figure 22.27.

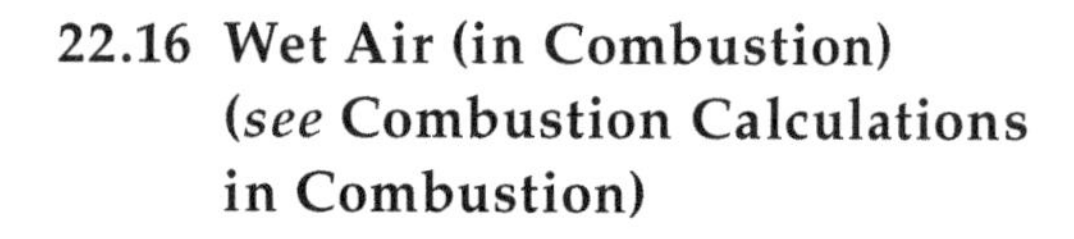

22.16 Wet Air (in Combustion) (*see* Combustion Calculations in Combustion)

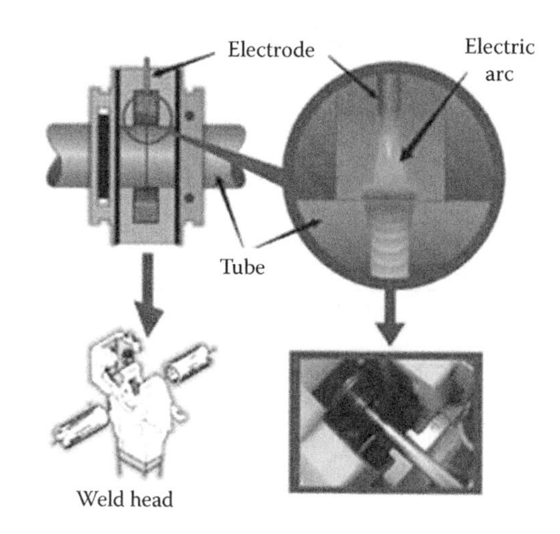

FIGURE 22.27
Orbital welding for welding of mainly tubes and stubs in boilers.

22.17 Wet Bulb Temperature T_{wb} (*see* Psychrometric Chart)

22.18 Wet Flue Gas (*see* Properties of Air and Flue Gas)

22.19 Wet Scrubbers (*see* Dust Collection Equipment)

22.20 Wet Steam (*see* Properties of Steam and Water)

22.21 Wet Wall Flow (*see* Properties of Steam and Water)

22.22 White Liquor (*see* Kraft Pulping Process *and* Recovery Boiler)

22.23 Wire Gauge

Thickness of wires and strips is checked in industry by rectangular or circular strips with slots made in them, called wire gauges. These are very easy to use and accurate enough to decide acceptance. A typical circular wire gauge is shown in Figure 22.28. Tube thicknesses are also measured by wire gauge as the rolling is to the same standards of thicknesses

Higher the gauge number lower is the thickness. Thicknesses higher than 0 are indicated by adding 0s. Each higher number is thinner by ~10%.

Tube size is specified by its od and thickness. Tubes are rolled in sizes from 12.7 to 127 mm od (0.5 to 5″) od and in several thicknesses. Conventional boilers rarely accept tube thickness lower than 3.25 mm (10 swg) from considerations of

- Thinning due to tube bending
- Welding
- Gas side erosion

Hrsgs employ lower thicknesses where both these factors are absent. Also the tubes of Hrsg are strengthened by the fins welded on the outside. SH tubes of utility boilers are the thickest in the whole boiler. Tubular AHs employ thin tubes varying from 10 to 14 swg.

Tube thickness is normally specified by the wire gauge. Standard wire gauge (SWG) is the British and Birmingham wire gauge (BWG) is the American standard. Gauge conversion chart given in Table 22.10 provides the wire gauge numbers and the equivalent thicknesses in both systems for sizes normally used in boilers.

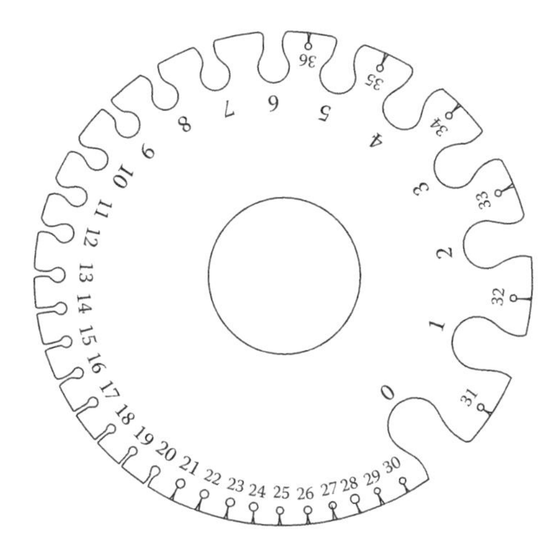

FIGURE 22.28
Circular wire gauge.

TABLE 22.10
Gauge Conversion Chart

	SWG		BWG	
No.	(inch)	(mm)	(inch)	(mm)
0/7	0.5000	12.7000	—	—
0/6	0.4640	11.7856	—	—
0/5	0.4320	10.9728	0.5000	12.7000
0/4	0.4000	10.1600	0.4540	11.5316
0/3	0.3720	9.4488	0.4250	10.7950
0/2	0.3480	8.8392	0.3800	9.6520
0/1	0.3240	8.2296	0.3400	8.6360
1	0.3000	7.6200	0.3000	7.6200
2	0.2760	7.0104	0.2840	7.2136
3	0.2520	6.4008	0.2590	6.5786
4	0.2320	5.8928	0.2380	6.0452
5	0.2120	5.3848	0.2200	5.5880
6	0.1920	4.8768	0.2030	5.1562
7	0.1760	4.4704	0.1800	4.5720
8	0.1600	4.0640	0.1650	4.1910
9	0.1440	3.6576	0.1480	3.7592
10	0.1280	3.2512	0.1340	3.4036
11	0.1160	2.9464	0.1200	3.0480
12	0.1040	2.6416	0.1090	2.7686
13	0.0920	2.3368	0.0950	2.4130
14	0.0800	2.0320	0.0830	2.1082
15	0.0720	1.8288	0.0720	1.8288
16	0.0640	1.6256	0.0650	1.6510
17	0.0560	1.4224	0.0580	1.4732
18	0.0480	1.2192	0.0490	1.2446

22.24 Wobbe Index (*see* Gaseous Fuels)

22.25 Wood and Firing (*see* Agro-Fuels)

23

Y

23.1 Yancey–Geer Price Index for Coal

Abrasion (*see* Wear)

23.2 Yield Point (*see* Tensile Test in Testing *and* Metal Properties)

23.3 Yield Strength (*see* Tensile Test in Testing *and* Metal Properties)

List of Abbreviations

A	Ash
ACC	Automatic combustion control
AFBC	Atmospheric fluidised-bed combustion
AH	Air heater
APH	Air pre-heater
ARC	Automatic re-circulating control valve (BFP piping)
AS	Alloy steel
AVT	All volatile treatment
BB	Boiler bank
BCP	Boiler circulating pump
BEGT	Boiler exit gas temperature
BFBC	Bubbling fluidised-bed boilers
BFG	Blast furnace gas
BFP	Boiler feed pump
BHN	Brinnel hardness number
BL	Black liquor
BLR	Black liquor recovery
BMCR	Boiler maximum continuous rating
BO	Bought outs
BOP	Balance of-lant
BQ	Boiler quality
BRIL	Brick, refractory, insulation and lagging
BS	British standards
bw	Boiler water
CAD	Continuous ash discharge
CBD	Continuous blowdown
CBM	Coal bed methane
CC	Combined cycle
CCS	Carbon sequestration and storage
CCS	Cold crushing strength
CCGT	Combined cycle gas turbine
CCPP	Combined cycle power plant
CDS	Cold drawn seamless
CEMS	Continuous emission monitoring system
CFD	Computational fluid dynamics
CG	Chain grate
CI	Cast iron
CFBC	Circulating fluidised-bed boilers
CMM	Coal mine methane
COG	Coke oven gas
Cogen	Cogeneration
CPP	Captive power plant
CS	Carbon steel
CSDH	Combined spray and desuperheating unit
CSEF	creep strength-enhanced ferritic
cv	Calorific value
CV	Control valve
CW	Cooling water
daf	Dry ash free
DCS	Distributed control system
DDC	Differential damper control (of fans)
DG	Dumping grate
DeSH	Desuperheater
DIN	German standards
DLN	Dry low NO_x
DM	Demineralise
DMMF	Dry mineral matter free
DN	Nominal diameter
DNB	Departure from nucleate boiling
DT	Destructive testing
DWLI	Direct water level indicator
ECO	Economiser
ERW	Electrically resistance welded
ESP	Electrostatic precipitator
ESP	Emergency shutdown procedure (in BLR boilers)
EVAP	Evaporator
FBC	Fluidised-bed combustion
FBHE	Fluid bed heat exchanger
FC	Fixed carbon
FCC	Fluidic catalytic cracker
FCV	Feed control valve
FD	Forced draught
FEGT	Furnace exit gas temperature
FFO	Furnace fuel oil
FG	Fuel gas
FGD	Flue gas desulphurisation
FGR	Flue gas recirculation
FO	Fuel oil
FOT	Furnace outlet temperature
FSSS	Furnace supervisory safety system
fw	Feed water
FWC	Feed water control
FWH	Feed water heater
FWR	Feed water regulator
fwt	Feed water temperature
Gcv	Gross calorific value
GE	Gas–electric
GEN	Generator
GHG	Green house gas
GI	Galvanised iron
GR	Grit re-firing
GR	Gas recirculation
GT	Gas turbine
HAZ	Heat affected zone in welding
HC	Hydrocarbon
Hcv	Higher calorific value
Hhv	Higher heating value
HEA	High energy arc
HFO	Heavy fuel oil

HFS	Hot finished seamless
HGI	Hardgrove index
HP	High pressure
HR&A	Heat released and available
HRR	Heat-release rate
Hrsg	Heat recovery steam generator
HS	Heating surface
HT	Hydraulic test
hv	Heating value
HVT	High-velocity thermocouple
HX	Heat exchange/heat exchanger
I &C	Instrumentation and control
IAD	Intermittent ash discharge
IADT	Initial ash deformation temperature
IBD	Intermittent blowdown
id	Inside dia
ID	Induced draught
IR	Infra red
IVC	Inlet vane control (of fans)
IWCG	Inclined water-cooled grate
JIS	Japanese industrial standards
Kpsi/ksi	Kilo (1000)psi
Lcv	Lower calorific value
LDO	Light diesel oil
LFO	Light fuel oil
Lhv	Lower heating value
LNB	Low NO_x burners
LNG	Liquefied natural gas
LP	Low pressure
LPG	Liquefied petroleum gas
LPT	Liquid penetrant test
LWL	Low water level
Llwl	Low low water level
M	moisture
Mpa	Mega (10^6) Pascals
MB	Mass burning
MCR	Maximum continuous rating
MDC	Mechanical dust collector
MEE	Multiple effect evaporators
MFT	Master fuel trip
MHVT	Multiple-shield high-velocity thermocouple
MIG	Metal inert gas (welding)
MIGI	Magnetic impulse gravity impact (rapping in ESP)
MM	Mineral matter
MM	Manufacturers' margin
MMAW	Manual metal arc welding
MP	Medium pressure
MPT	Magnetic particle test
MS	Mild steel
MSL	Mean sea level
MSSV	Main steam stop valve
MSW	Municipal solid waste
MW	Molecular weight
MWe	Megawatt electrical
MW_{th}	Megawatt thermal
NB/nb	Nominal bore
NCV	Net calorific value
NCR	Normal continuous rating
NDE	Non-destructive examination
NDT	Non-destructive testing
NG	Natural gas
NLPP	No load pump pressure
NPP	Non-pressure part
NPS	Nominal pipe size
NPSH	Net positive suction head
NRV	Non-return valve
NWL	Normal water level
OC	Open cycle (in GT)
Od/OD	Outside diameter
OFA	Over fire air
O&M	Operation and maintenance
OT	Once through
OTSG	Once-through steam generator
PA	Primary air/proximate analysis
p&t	Pressure and temperature
PC	Pulverised coal
PCE	Pyrometric cone equivalent
PCO	Plastic chrome ore
PF	Pulverised fuel
PFBC	Pressurised fluidised-bed combustion
PG	Pulsating grate, pressure gauge, pusher grate
PHG	Pin hole grate
PLC	Permanent linear change
PM	Particulate matter
PP	Pressure part
PRDS	Pressure reducing and desuperheating
PSH	Primary superheater
PTC	Performance test code
PV	Pressure vessel
PWHT	Post-weld heat treatment
RAH	Rotary air heater
RD	Relative density
RDF	Refuse derived fuel
RFO	Residual fuel oil
RG	Refinery gas, reciprocating grate
RH	Re-heat/re-heater
RHOP	Re-heater outlet pressure
RLA	Residual/remnant life assessment
RT	Radiographic testing
RUL	Refractoriness under load
RWLI	Remote water level indicator
SA	Secondary air
SAE	Society of Automotive Engineers
SAW	Submerged arc (welding)
SB	Soot blower
SBV	Steam by volume
SBW	Steam by weight
SC	Supercritical
sub-c	Sub-critical

Scaph Steam coil air pre-heater
SCR Selective catalytic reaction
SDNR Screw down non-return valve
SG Specific gravity
SH Superheat/superheater
SHOP Superheater outlet pressure
SNCR Selective non-catalytic reduction
SOFA Separated overfire air
SOP Superheater outlet pressure
Spec Specification
SR1 Seconds Redwood 1
SS Spreader stoker
ss Stainless steel
SSH Secondary superheater
SSF Seconds Saybolt Furol
SSU Seconds Saybolt Universal
ST Steam turbine/softening temperature of ash
STC Steam temperature control
STIG Steam injected gas turbine
SVLP Safety valve-lifting pressure
SWH Saturated water head
SWAS Steam and water sampling system
TA Tertiary air
TAH Tubular air heater
TC Thermocouple
TDS Total dissolved solids
TEG Turbine exhaust gas
TG Travelling grate/turbo generator
TIG Tungsten inert gas (welding)
TMCR Turbine maximum continuous rating
TP Terminal point
TR Transformer rectifier
UA Ultimate analysis
UP Universal pressure
USC Ultra supercritical
UT Ultrasonic testing
UV Ultra violet
VFD Variable frequency drive
VG Vibrating grate
VM Volatile matter
VOC Volatile organic compounds
VWO Valve wide open
WH Waste heat
Whrb Waste heat recovery boilers
WLI Water level indicator
WT Water treatment
WTE Waste to energy

Symbols of Elements and Compounds

Symbol	Element/Compound	Remarks
Al	Aluminium	
Al_2O_3	Aluminium oxide	Alumina
Ar	Argon	
As	Arsenic	
B	Boron	
C	Carbon	Also graphite
CCl_4	Carbon tetra chloride	
CH_4	Methane	
CH_4S, CH_3SH	Methyl mercaptan	
C_2H_2	Acetylene	
C_2H_4	Ethylene	
C_2H_6	Ethane	
C_3H_8	Propane	
C_4H_{10}	Butane	
C_4H_8	Butylenes	
C_5H_{12}	Pentane	
C_6H_{14}	Hexane	
C_6H_5OH	Phenol	Carbolic acid
$C_{10}H_8$	Naphthalene	
CO	Carbon monoxide	
CO_2	Carbon dioxide	
CS_2	Carbon disulphide	
Ca	Calcium	
$CaCl_2$	Calcium chloride	
$CaCO_3$	Calcium carbonate	Limestone, calcite
$Ca(HCO_3)_2$	Calcium bicarbonate	
CaO	Calcium oxide	Quick lime
$Ca(OH)_2$	Calcium hydroxide	Hydrated lime
Ca_2SiO_4	Calcium silicate	Also $2CaO{\cdot}SiO_2$
$CaSO_3$	Calcium sulphite	
$CaSO_4$	Calcium sulphate	Gypsum ($CaSO_4{\cdot}2H_2O$)
$CaMg(CO_3)_2$	Calcium magnesium carbonate	Dolomite
Co	Cobalt	
Cr	Chromium	
Cr_3C_2	Chromium carbide	
Cu	Copper	
$CuCl_2$	Cuprous chloride	
$CuFeS_2$	Chalco-pyrite	
Fe	Iron	Ferrum
Fe_2O_3	Ferrous oxide	Haematite
Fe_3O_4	Ferric oxide	Magnetite, iron ore
FeS_2	Ferric/iron sulphide	Pyrite
FeS_x		Marcasite
$FeSO_4$	Ferrous sulphate	
$Fe_2(SO_4)_3$	Ferric sulphate	
Fe_3C	Iron carbide	Cementite
H	Hydrogen	
HCl	Hydrochloric acid	
HCN	Hydrogen cyanide	
HFC	Hydrofluorocarbons	
H_2CO_3	Carbonic acid	
HNO_3	Nitric acid	
H_2SO_3	Sulphurous acid	
H_2SO_4	Sulphuric acid	
He	Helium	
Hg	Mercury	
Ir	Iridium	
K	Potassium	
KOH	Potassium hydroxide	
K_2SO_4	Potassium sulphate	
Mg	Magnesium	
$MgCl_2$	Magnesium chloride	
$MgCO_3$	Magnesium carbonate	Magnesite
MgO	Magnesium oxide	
MgOH	Magnesium hydroxide	
Mn	Manganese	
Mo	Molybdenum	
N	Nitrogen	
NH_3	Ammonia	
N_2H_4	Hydrazine	
$(NH_4)_2SO_4$	Ammonium sulphate	
$(NH_4)HSO_4$	Ammonium bisulphate	
$(NH_2)_2CO$		Urea or Carbamide
NO	Nitric oxide	
NO_2	Nitrogen dioxide	
NO_x	Nitrogen oxides	
N_2O	Nitrous oxide	Laughing gas
Na	Sodium	
NaCl	Sodium chloride	Common salt
Na_2CO_3	Sodium carbonate	Soda ash, washing soda
NaOH	Sodium hydroxide	Caustic soda
Na_2S	Sodium sulphide	
Na_2SiO_3	Sodium silicate	
Na_2SO_3	Sodium sulphite	
Na_2SO_4	Sodium sulphate	Salt cake
$Na_3PO_4{\cdot}12H_2O$	Trisodium phosphate	
Nb	Niobium	Also columbium
Ni	Nickel	
O_2	Oxygen	
O_3	Ozone	
P	Phosphorous	
Pb	Lead	
Pt	Platinum	
PFC	Perfluorocarbons	
Rh	Rhodium	

Symbol	Element/Compound	Remarks
S	Sulphur	
SF_6	Sulphur hexafluoride	
SO_2	Sulphur dioxide	
SO_3	Sulphur trioxide	
Si	Silicon	
SiC	Silicon carbide	
SiO_2	Silicon dioxide or silica	Sand, quartz
Ti	Titanium	
TiO_2	Titanium oxide	Rutile

Symbol	Element/Compound	Remarks
V	Vanadium	
VO_3	Vanadium trioxide	
VO_4	Vanadium tetraoxide	
V_2O_5	Vanadium pentaoxide	
W	Tungsten	Wolfram
WC	Tungsten carbide	Carbide
Zn	Zinc	
Zr	Zirconium	
Zro_2	Zirconium dioxide	Zirconia

List of Symbols

A	Area
AC, ac	Alternating current
A_g	Area to gas
A_s	Area to steam
bhp	Brake horse power
Btu	British thermal unit
BWG/bwg	Birmingham wire gauge
°C	Degrees Celsius
Cq, Cd	Coefficient of discharge
C_v	Flow coefficient
c_p	Specific heat at constant pressure
c_v	Specific heat at constant volume
cfm	Cubic feet per minute
cS	Centi stokes
cP	Centi Poise
cu	Cubic
cft, cu ft	Cubic foot
d	Inside diameter
D	Outside diameter
DC, dc	Direct current
dia	Diameter
D_e/d_e	Equivalent diameter
D_i/D_o	Inside and outside diameters
D_h/d_h	Hydraulic diameter
dscf	Dry standard cubic foot
dscm	Dry standard cubic meter
E	Young's modulus
emf	Electromotive force
f	Friction factor
°F	Degree Fahrenheit
fpm	Feet per minute
fps	Feet per second
G_a	Mass velocity of air
G_g	Mass velocity of gas
g	Acceleration due to gravity
g	Gallon
gm	Gram
gpm	Gallons per minute
gr	Grain
h	Enthalpy
H	Head of fluid, enthalpy per mole
h	Hour
hp	Horse power
Igpm	Imperial gallons per minute
J	Joule
°K	Degrees Kelvin (°C + 273)
K_{Re}	Reynolds number factor
k	Thermal and electrical conductivity
kg	Kilogram
km	Kilometer
ksi	1000 psi
L	Litre
L	Length
lb	Pound weight
m	Mass, meter
mg	Milligram
mho	Electrical conductivity
mil	1/1000 of an inch
μmho	Micro mhos
micron, μ	1/1000,000 of meter
N	Number, Newton, rpm
N_{Nu}	Nusselt number
Nvh	Number of velocity heads
Oz	Ounce
p	Pressure
P	Absolute pressure, pressure drop, poise, power
ΔP, Δp	Differential pressure
Pa	Pascal
P_r	Prandtl number
pH	Negative log of hydrogen ion concentration
ppb	Parts per billion
ppm	Parts per million
psf	Pounds per square feet
psi	Pounds per square inch
psia, psig	psi absolute and gauge
Q	Heat, volume
q	Rate of heat flow
R	Universal gas constant
°R	Degrees Rankine (°F + 460)
Re	Reynolds number
rpm	Revolutions per minute
S	Siemens for electrical conductivity
s	Surface area
s	Specific entropy kcal/kg K
SWG/swg	Standard wire gauge
shp	Shaft horse power
sp.gr, SG	Specific gravity
St	Stokes
μS	Micro Siemens
T	Absolute temperature in °K (273 + °C)
ΔTm, Δtm	Log mean temperature difference (LMTD)
TS	Tensile strength
t	Long ton, 1016 kg/h, 2240 lb/h, temperature
t_a	Temperature of air
t_b	Bulk temperature of fluid
t_f	Film temperature
tc	Ton metric, 1000 kg/h, 2205 lb/h
tph	Tonnes per hour
tpd	Tonnes per day

U Film conductance
U_c Convection conductance
U_{cc} Convection conductance in cross flow
U_r Radiation conductance
U_{rg} Radiation conductance due to gas
UTS Ultimate tensile strength
UV Ultraviolet
V Velocity of fluid, volume
v Specific volume, volume
W Watt
wt Weight
wg Water gauge
YP Yield point
YS Yield stress
yr Year
′/ft Foot
″/in Inch
° Degree of angle or arc
μ Absolute or dynamic viscosity
ν Kinematic viscosity
γ Specific weight
ρ Density
η Efficiency
λ Ratio of specific heats

Organisations Related to Boilers

Listed here are some of the organisations that are related to boilers and are mentioned in the book.

Public Institutions

AISC	American Institute of Steel Construction
ANSI	American National Standards Institute
ASME	American Society of Mechanical Engineers
ASHRAE	American Society of Heating, Refrigerating and Air-conditioning Engineers
ASTM	American Society of Testing Materials
API	American Petroleum Institute
AWS	American Welding Society
BLRBAC	Black Liquor Recovery Boiler Advisory Committee
BSI	British Standards Institution
CEN	Commit é Europé en de Normalisation (European Committee for Standardisation)
CRIEPI	Central Research Institute of Electric Power Industry, Japan
DIN	Deutsches Institut Fur Normung E.V. (German National Standard)
DOE	Department of Energy, USA
EN	European Normalisation (European Standard)
EPA	Environmental Protection Agency, USA
EPRI	Electric Power Research Institute, USA
ISO	International Standards Organisation
JIS	Japanese Industrial Standards
NFPA	National Fire Prevention Association, USA
NIST	National Institute of Standards and Technology, USA
OSHA	Occupational Safety and Health Administration, USA

Private Organisations

B&W	The Babcock and Wilcox Company, USA
BWE	Burmeister and Wein, Denmark
DHIC	Doosan Heavy Industries and Construction, Korea
FWC	Foster Wheeler Corporation, USA
GE	General Electric, USA
HPE	Hitachi Power Europe, Germany
IHI	Ishikawajima-Harima Heavy Industries Co., Japan
KHI	Kawasaki Heavy industries, Japan
MHI	Mitsubishi Heavy Industries, Japan
P&W	Pratt and Whitney, USA
RR	Rolls Royce, UK

References and Further Reading

Codes

Topic	S. No	Name	Name	Year	Body	Comments
BOILER	1	B&PV CODE	Boiler and Pressure Vessel Code Section I—Power Boilers	2010	ASME	
	2	BS 1113	Specification for design and manufacture of water-tube steam generating plant (including superheaters, re-heaters and steel tube economizers)	1999	BSI	Replaced by BS EN 12952-1
	3	R101	Guideline for ordering high-capacity boilers	2005	VGB	
COMBUSTION	1	NFPA 85	Boiler and Combustion Systems Hazards Code	2001	NFPA	
	2	NFPA 8506	Standard on Heat Recovery Steam Generator Systems	1998	NFPA	
EFFICIENCY	1	Code PTC 4.1	Steam-generating units	1964 (Reaffirmed in 1991)	ASME	Power Test Codes
	2	Code PTC 4	Steam-generating units—ASME	1998	ASME	Power Test Codes
	3	Code PTC46	Performance test code on overall plant performance		ASME	Power Test Codes
	4	Code DIN 1942	Acceptance test code for steam generators (VDI-rules for steam generators)	1994	VDI	
	5	Code BS 2885	Code for acceptance tests on stationary steam generators of the power station type	1974	BSI	Withdrawn and replaced by BS EN 12952-15 2003
	6	BS 845 Concise Procedure	Assessing Thermal Performance of Boilers for Steam, Hot Water and High Temperature Heat Transfer Fluids	1987	BSI	

Books

Topic	S. No	Name	Author	Year	Publisher
BOILERS	1	Steam its generation and use 41st Edition	J.B. Kitto and S.C. Shultz	2005	The Babcock and Wilcox Co, USA
	2	Combustion Fossil Power	J.G. Singer	1991	Combustion Engineering Inc., USA
	3	Industrial Boilers and Heat Recovery Steam Generators	V. Ganapathy	2003	Marcel Dekker, USA
	4	Boilers and burners	Prabir Basu, Cen Kefa, Louis Jestin	2000	Springer
	5	Boiler operator's handbook	Kenneth Heselton	2005	Marcel Dekker, USA
	6	Boilers for Power and Process	Kumar Rayaprolu	2009	CRC Press, USA
COMBUSTION	1	Combustion and Gasification in Fluidised beds	Prabir Basu	2006	Taylor and Francis, USA
	2	Fluidised-Bed Combustion	Simeon N. Oka	2004	Marcel Dekker
	3	The John Zinc Combustion Handbook	Charles E. Baukal		CRC Press

Topic	S. No	Name	Author	Year	Publisher
FANS	1	Woods Practical Guide to Fan Engineering	B.B. Daly	1978	Woods of Colchester, UK
	2	Fan Engineering 9th edition	R. Jorgenson	1999	Howden Buffalo Inc., USA
	3	Fan Handbook	Frank P. Bleir	1997	McGraw-Hill
FUELS	1	Handbook of Coal Analysis	James G. Speight	2005	Wiley Interscience
FLUID MECHANICS	1	Flow of Fluids	Crane Company staff	1982	Crane Company, USA
GAS TURBINES	1	The Gas Turbine Handbook 2nd edition	Tony Giampaolo	2002	Fairmont Press
	2	Gas Turbine World 2009 GTW Handbook Volume 27		2009	Pequot Publishing Inc., CT, USA
INSTRUMENTATION AND CONTROL	1	Thermal Power Plant Simulation and Control	Damian Flynn	2003	The Institution of Electrical Engineers, UK
	2	Power Plant Control and Instrumentation	David Lindsley	2000	The Institution of Electrical Engineers, UK
	3	Boiler Control Systems Engineering	G.F. (Jerry) Gilman	2005	ISA (The Instrumentation, Systems and Automation Society), USA
MATERIALS	1	Handbook of Comparative World Steel Standards 3rd edition	John E. Bringas	2004	ASTM ES67B
	2	Carbon Steel Handbook		2007	EPRI
POWER PLANTS	1	Steam Plant Operations 6th edition	Everett. B. Woodroff, Herbert B. Lammers, Thomas B. Lammers	2004	McGraw-Hill, USA
	2	Combined Cycle Gas and Steam Turbine Power Plants	Rolf Kehlhofer	1997	Penn–Well Publishing Co, USA
	3	Fossil Fuel Fired Power Generation	International Energy Agency	2007	IEA, France
	4	Power Generation Handbook	Philip Kiameh	2002	McGraw-Hill
	5	Power Generation from Solid Fuels	Hartmut Spliethoff	2010	Springer
	6	Power Plant Engineering	Black and Veatch	1996	Chapman and Hall, USA
PUMPS	1	Pump Handbook 4th edition	Igor Karassik and others	2008	McGraw-Hill, USA
	2	Pump Handbook		2004	Grundfoss
	3	Selecting Centrifugal Pumps 4th edition	KSB staff	2005	KSB AG, Germany
	4	Centrifugal Pump Design	KSB staff		KSB AG, Germany
	5	Pump Handbook	Grundfoss staff	2004	Grundfoss, Sweden
STEAM TUBINES	1	Steam Turbines 2nd edition	Heinz P. Block, Murari P. Singh	2009	McGraw-Hill
VALVES	1	The Safety Relief Valve Handbook	Marc Hellemans	2009	Elsevier, UK
	2	Valve Selection Handbook, 5th edition	Peter Smith and R.W. Zappe	2004	Elsevier, UK
	3	Crosby ® Pressure Relief Valve Engineering Handbook	Staff of Crosby Valve Inc.	1997	Crosby Valve Inc., USA
	4	Control Valve Handbook 4th edition	Staff of Fisher controls	2005	Emerson Process Management
WATER	1	Handbook of Industrial Water Treatment			GE Power and Water, USA
	2	The NALCO Water Handbook 2nd edition	Frank N. Kammer	1988	McGraw-Hill, USA

Technical Articles/Reports

	Technical Paper	Author	Year	Organisation	Publication
Ash Corrosion					
1	Fireside corrosion testing of candidate superheater tube alloys, coatings, and claddings phase ii	J. L. Blough and W. W. Seitz		FW Development Corporation	FWC
2	The role of fireside corrosion on boiler tube failures, part i	Dr. Rama S. Koripelli, Dr. David C. Crowe and Dr. David N. French	2010 April		Coal Power
3	Stress corrosion cracking	Dr. R. A. Cottis	2000	Corrosion and Protection Centre, UMIST	HMSO
Ash Deposition					
1	Fire deposits in coal fired boilers	Roderich M. Hatt	1990	Island creek corp, NY	Pergamon Press
2	Overview of coal ash deposition in boilers	R. W. Borio and A. A. Levasseur		Combustion Engineering, Inc.	
Biomass Firing					
1	Experience of fluidised bed technology for biomass plants in different applications and development towards new applications in RDF burning.	Kari Niemelä and Kim Westerlund	2003	Foster Wheeler Energie GmbH	FWC
2	Biomass fired fluidised bed combustion boiler technology for cogeneration	Report	2007	UNEP-DTIE Energy Branch	
3	Foster wheeler experience with biomass and waste in CFBs	Edgardo Coda Zabetta, Vesna Barišic, Kari Peltola, and Arto Hotta	2008	FW, Varkaus, Finland	FWC
4	Combustion of different types of biomass in CFB boilers	Vesna Barišic and Edgardo Coda Zabetta	2008	FW, Varkaus, Finland	FWC
5	Biomass power generation by CFB boiler	Koji Yamamoto	2001	Solution Engineering Center, NKK, Japan	NKK technical review no.85
6	Principles of burner design for biomass co-firing	Bradley W. Moulton	2009	FW North America	FWC
7	Efficient and low emission stoker fired biomass boiler technology in today's marketplace	Richard F. Abrams and Kevin Toupin	2007 March	Babcock and Riley Power	Babcock Power
8	Poultry litter to energy: technical and economic feasibility	B. R. Bock		TVA Public Power Institute, Alabama	
9	Advanced biopower technology assessment	Jim Easterly and Ajay Kasarabada	2008 January	Black & Veatch, KS	
10	Biomass for power generation and CHP		2007 January	IEA Energy Technology	IEA network
11	Power generation technologies firing biomass	Folke Engström	1999	FW Development Corporation	FW Review
Bagasse Firing					
1	Properties and operating experience with bagasse as a boiler fuel	T. N. Adams, G. D. Whitehouse and d. Maples		The univ. of British Columbia and Louisiana state univ.	
2	Fuels and furnaces	P. R. A. Glennie	1966 March	The SA Sugar Technologists' Association	

continued

	Technical Paper	Author	Year	Organisation	Publication
3	Bagasse-fired boilers with reference to co-generation	Norman Magasiner		Thermal Energy systems, SA	Sa
4	Opportunities for improving the performance and reducing the costs of bagasse-fired boilers	A. P. Mann, T. F. Dixon, F. Plaza and J. A. Joyce		Sugar Research Institute	
Black Liquor Recovery Boilers (BLRB)					
1	Development of recovery boiler technology	D Esa Vaikkilainen	2003	Jaakko Poyry oy	
2	Black liquor recovery: how does it work?	Magnus Marklund and others			
3	Fundamental approach to black liquor combustion improves boiler operation	J.L. Clement, C. Cox and others	1995 April	B&W and Champion International Corp	B&W
4	Black liquor firing presentation	J.L. Clement	1996 April	B&W	B&W
5	Improving recovery boiler furnace reliability with advanced materials and application methods	Joan L. Barna and Keith B. Rivers	1999 January	B&W, USA and Canada	B&W
6	Recovery boiler reheat steam cycle	T.E. Hicks, W.R. Stirgwolt, and J.E. Monacelli	2009 October	B&W, USA	B&W
7	Modern recovery boilers	Roxare	2009 April	Pulp and Paper circle	
8	Recommended good practice safe firing of black liquor in black liquor recovery boilers	The black liquor recovery boiler advisory committee	2010 April		BLRBAC
9	Recent advances in the application of acoustic leak detection to process recovery boilers	J.J. Kovecevich, D.P. Sanders and others	1995 September	B&W	B&W
Burners					
1	Burner development for the reduction of NO_x emissions from coal fire electric utilities	Bireswar Paul and Amitava Datta	2008	Jadavpur University, Kolkata, India	Patents on Mech. Engg
2	Foster Wheeler's low NO_x systems for tangential fired power boilers	John Grusha and Brad Moulton, PE	2006 October	FW North America	FWC
3	Unique union of low nox combustion controls—foster wheeler's tln system	Orest Walchuk and John Grusha Ken Barna	2001 December	F W Energy Corp. Duke Power Co.	FWC
4	Wall fired low NO_x burner evolution for global NO_x compliance	Tom Steitz, John Grusha, and Ross Cole	1998 March	FW	FWC
Bubbling Fluidised-Bed Boilers (BFBC)					
1	Bubbling fluidized bed installation capitalizes on sludge	Steve Charlson, B&W., Brian Taylor Fraser Papers, Inc., USA	1999		B&W
2	Is the future of BFBC technology in distributive power generation?	Simeon Oka	2001		
3	From liquor to sludge-conversion of a recovery boiler to a bubbling fluid bed	J.F. Cronin	1999	Babcock & Wilcox, USA	B&W
4	Construction of the first and largest circulating fluidized bed power plant utilizing Korean low grade anthracite coal	E.K. Kim		Korea Electric Power Corp.	
5	Bubbling fluidized bed or stoker—which is the right choice for your renewable energy project?	J.P. DeFusco, P.A. McKenzie, and M.D. Fick	2007 May	Babcock & Wilcox, USA	B&W

Circulating Fluidised-Bed Boilers (CFBC)

1	Recent ALSTOM POWER large CFB and scale up aspects including steps to supercritical	Jean-Xavier MORIN	2003 October	ALSTOM	IEA Workshop on Large Scale CFB, Zlotnicki, Poland
2	CCT experience of Tong-Hae CFB boiler using Korean anthracite	Jong-Min Lee, Jae-Sung Kim and Jong-Jin Kim		Korea Electric Power Corporation	
3	Status of the 200 MWe Tonghae CFB boiler after cyclone modification	Jong-Min Lee, Jae-Sung Kim, Jong-Jin Kim and Pyung-Sam Ji		Korea Electric Power Corporation	
4	Design considerations of B&W industrial and utility size reheat/non-reheat ir-cfb boilers	S. Kavidass, M.J. Szmania and K.C. Alexander	1996	Babcock & Wilcox	B&W
5	B&W IR-CFB operating experience and new development	M. Maryamchik and D.L. Wietzke	2005	Babcock & Wilcox	B&W
6	B&W IR-CFB operating experience and new development	M. Maryamchik	2008	Babcock & Wilcox	B&W
7	Why build a circulating fluidized bed boiler to generate steam and electric power	S. Kavidass, G.L. Anderson, and G.S. Norton, Jr.	2000 September	Babcock & Wilcox	B&W
8	Next generation CFBC	S. Rajaram	1999	Bharat Heavy Electricals Ltd, India	Chem. engg. science Pergamon
9	Start-up and initial operating experiences at the coke fired petropower cogeneration facility	Bruce C. Studley, and Henry J. Somerville		FW Power Systems, FW Santiago, Inc.	
10	Combustion of pitch/asphalt and related fuels in circulating fluidized beds	Song Wu, Matti Hiltunen and Kumar Sellakumar	2002	FW Development Corporation	7th Int. conf. on CFB, Niagara falls, May 02
11	Foster Wheeler experience in combustion of low-grade high-ash fuels in cfbs	Vesna Barišic, Edgardo Coda Zabetta, and Arto Hotta	2008	FW Energia Oy, Varkaus, Finland	FWC
12	Foster Wheeler compact cfb boilers for utility scale	Stephen J. Goidich, and Timo Hyppänen		FW Energy Intl, USA & FW Energia Oy	
13	Designs of large scale CFB boilers	Ragnar G. Lundqvist		FW Energia Oy Finland	
14	CFB technology—toward zero CO_2 emissions	Timo Jäntti, Timo Eriksson, Arto Hotta, Timo Hyppänen, and Kalle Nuortimo,		FW Energia Oy Finland	
15	New air-staging techniques for co-combustion in fluidized bed combustors	J. Werther and others	2000 October	Technical University Hamburg	VGB conf. on Research for Power Plant Tech. 2000"
16	Large-scale CFB boiler with different cyclone separation efficiencies—operational experiences and analysis	Rafal KOBYLECKI, Marek ANDRZEJCZYK and Zbigniew BIS		Czestochowa Univ. of Tech., Poland	

continued

	Technical Paper	Author	Year	Organisation	Publication
Coal					
1	ACARP report of the Australian coal industry's research program		2008 September		
2	Coal facts		2009 October	World Coal Institute. UK	
3	Panorama 2010—World coal resources	Geneviève Bessereau and Armelle Saniere	2009 November	IFP, France	
4	BP Statistical Review of World energy		2009 June	BP of UK	
Clean Coal Technologies					
1	Clean coal technology and the energy review	Mike Farley and Julie Kelly	2006	Mitsui Babcock	Mitsui Babcock
2	Developments and prospects on the application of clean coal technologies in India	Mr. P. Selva kumaran and others	2006	BHEL, India	Clean Coal Day in Japan
3	Foster Wheeler experience with clean coal technology in China	Scott L. Darling and Sean Li	2007	FW Energy International, USA	FWC
Co-Firing					
1	Design issues for co-firing biomass in wall-fired low NO_x burners	S. Laux, D. Tillman, and A. Seltzer	2003	Foster Wheeler, USA	FWC
2	Co-firing of biomass and opportunity fuels in low NO_x burners	S. Laux, and D. Tillman		Foster Wheeler, USA	FWC
3	Cofiring biomass in coal-fired boilers	David Tillman	1999	F W Development Corporation	FW Review Spring 1999
4	Large scale utilisation of biomass in fossil fired boilers	Erik Gjernes, Hans Henrik Poulsen and Nicholas Kristensen	2007	Burmeister & Wain Energy A/S	BWE
5	Biomass co-firing		2003 March	European Bioenergy Networks	VTT Processes Finland
Fluidised-Bed Firing (FBC)					
1	Development of fluidised bed combustion—an overview of trends, performance and cost	Joris Koornneef, Martin Junginger and Andre' Faaij, Utrecht University	2006	Elsevier	
2	Developments in fluidised bed combustion technology	Zhangfa Wu	2006	IEA clean coal centre	
3	Combustion of various fuels in fluidized bed boilers	Elmar Offenbacher, Helmut Anderl, and Kurt Kaufmann		Austrian Energy & Environment, Austria	
Fuels					
1	Orimulsion ® a fuel from heavy oil, engineered for performance	Cebers O. Gómez-Bueno and others, Venezuela	1998		
2	Reducing power production costs by utilizing petroleum coke	Kevin C. Galbreath	1998 July	Energy & Environ. Research Center Univ. of N. Dakota	
3	Notes on heavy fuel oil		1984	American Bureau of Shipping	
4	Liquefied Nnatural gas worldwide		2008 June	California Energy Commission	
Gas Firing					
1	Blast furnace gas-fired boiler for Eregli Iron & Steel Works (Erdemir), Turkey	J. Green, A. Strickland and others	1996 April	B&W Canada	B&W
Gas Turbines					
1	The theory and operation of Evaporative coolers for industrial gas turbine installation	R.S. Johnson Sr	1988	Solar Turbines Inc., San Diego, USA	ASME

Heat					
1	Two phase flow	Dr. Dennis R. Lilles		Los Alamos Science	
Hrsgs					
1	Vertical natural circulation (VNC) HRSG designed for heavy cycling duty in the Brighton Beach Power Plant/Canada			Austrian Energy	Austrian Energy
2	Vertical natural circulation HRSGs for Brighton Beach power station		1993 March	Austrian Energy	MPS
3	HRSG optimization for cycling duty	Pascal Fontaine and Jean-François Galopin,	2007 November	CMI Energy	Power Engineering
4	When cycling goes up, reliability comes down	Peter S. Jackson	2005 April	Tetra Engineering Group Inc.	Power
5	Fatigue damage to HRSG tube-to-header joints		2004 March		Power
6	Designing HRSGs for cycling	Lewis R. Douglas, PE,	2006 March	Nooter/Eriksen Inc.	Power
7	New pre-commissioning options for controlling corrosion in HRSGs	David Daniels	2003	Power	Power
8	Building on Benson	G. Volpi and G. Silva	2006 March	Ansaldo Caldaie	Modern Power Systems
9	Benson once-through heat recovery steam generators			Siemens AG, Power Generation KWU	Siemens
10	Application of SCR and CO catalyst systems to simple cycle combustion gas turbines	David R. Logeais	2003 June	Deltak, L.L.C.	Proceedings of ASME
11	Evaluate extended surface exchangers carefully	V. Ganapathy	1990 October	ABCO Industries, Inc., Abilene, Texas	Hydrocarbon Processing
Industrial Boilers					
1	Coal-fired industrial boilers	Dept of Trade and Industry, UK	2002		
2	Air pollution control for industrial boiler systems	J.B. Kitto	1996 November	Babcock & Wilcox	B&W
Instrumentation					
1	Boiler furnace pressure excursion and set points	B.S. Tanwar, Manager-Control & Instrumentation	2010	ALSTOM Projects India Limited, Noida	Int. Journal of Computer Applications
2	A guide to boiler drum level equipment and control concepts		2001	Clarke-Reliance Corp.	
3	Chordal and tube thermocouples		2009 October	Storm Technologies Inc.	
4	Boiler control overview		2000 September	Siemens Moore Process Automation	
5	SAMA diagrams for boiler controls		2000 September	Siemens Moore Process Automation	
6	Fossil fuel power plant boiler combustion controls		2004 September	Instrumentation, Systems, and Automation Society	

continued

	Technical Paper	Author	Year	Organisation	Publication
Lignite					
1	Low NO_x combustion technologies for lignite fired boilers	Takanori Yano, Kenji Kiyama, Kazuhito Sakai and others		Babcock-Hitachi K.K., Japan	Hitachi
2	Lignite and brown coal	Kirk-Othmer Encyclopedia of Chemical Technology, 5th edition	2007		John Wiley
Life Extension and Condition Assessment					
1	Creep-rupture assessment of superheater tubes using nondestructive oxide thickness measurements	T. J. Wardle	2000 May	B&W	B&W
2	Regaining lost capacity of aging boiler components by performance evaluation and residual life assessment: an exciting challenge for NDE	R. J. Pardikar, N. Ayodhi and P. S. Subrahmanyam	1996 December	BHEL, India	Conference on NDT
3	Boiler life extension requires accurate condition assessment	Timothy B. DeMoss, Associate Editor		Power engineering	Power engineering
4	Boiler fitness survey for condition assessment of industrial boilers	G. J. Nakoneczny		B&W	B&W
5	Condition assessment of boiler and its components	Dr. P.U. Pathy	2005 November	Middle East NDT Conference	
6	Automated condition assessment of boiler water wall tubes	A. Vajpayee and D. Russel		Russel NDE systems, Canada	
Materials					
1	Materials for boilers in ultra supercritical power plants	R. Viswanathan and W. T. Bakker	2000 July	EPRI	ASME
2	Materials overview and fireside corrosion considerations for advanced steam cycles	Michael Gagliano Gregory Stanko Horst Hack	2008 December	Foster Wheeler North America Corp., USA	FWC
3	U. S. Program on materials technology for USC power plants	R. Viswanathan and others	2005		ASM International
4	COR-TEN®—Weather & Corrosion Resistant Steel	Technical Data		AJ Marshall, UK	
5	Why new U.S. supercritical units should consider T/P92 piping	P. Jason Dobson, PE,	2006 April	Cummins & Barnard	Power
6	Comparison of ASME specifications and European standards for mechanical testing of steels for pressure equipment	Elmar Upitis and Michael Gold	2005 December	ASME standards technology LLC, NY, USA	
7	Materials overview and fireside corrosion considerations for advanced steam cycles	Michael Gagliano Gregory Stanko Horst Hack	2008 December	Foster Wheeler North America Corp., USA	Power-Gen International
8	Best Practice Brochure		2006	Department of Trade and Industry (DTI)	UK Government
Municipal Solid Waste (MSW)					
1	Electricity from municipal solid waste		2000	Pace University, NY	
2	Energy recovery from municipal solid waste			Bureau of Energy Efficiency	
3	Refuse combustion		1996 October	U. S. Environmental Protection Agency	
4	Stoker type municipal solid waste incineration plant			Takuma CO., LTD.	Waste Treatment Tech. in Japan

NO_x Control					
1	Ultra-Low NO_x integrated system for coal fired power plants	Galen H. Richards, and others		ALSTOM Power, Inc.	
2	Low NO_x combustion technologies for lignite fired boilers	Takanori Yano, Kenji Kiyama and Kazuhito Sakai		Babcock-Hitachi K.K., Japan	Hitachi
3	NO_x control for large coal-fired utility boilers: selection of the most appropriate technology	Philip Canning, Allan Jones and Philip Balmbridge	1999 April	PowerGen plc, UK.	The Australian Coal Review
4	NO_x reduction strategy using a SOFA system in tangentially fired boilers at lingan generating station	J. Pham and D. Wasyluk W. Small	2009 December	B&W USA, Nova Scotia Power	B&W
5	Reburning technologies for the control of Nitrogen oxides emissions from coal-fired boilers	TOPICAL REPORT NUMBER 14	1999 May	Clean Coal Technology	DOE
Performance Testing					
1	Better testing	Richard F. Storm, Stephen K. Storm and Sammy Tuzenew	2003 September	Storm Technologies, Inc., USA	World Coal
PFBC					
1	A large capacity pressurized fluidized bed combustion boiler combined cycle power plant	Hideaki Komatsu, Masakatsu Maeda, and Masaru Muramatsu	2001	Hitachi Review Vol. 50 No. 3	
Pulverisers					
1	Coal pulverizer design upgrades to meet the demands of low NO_x burners	Qingsheng Lin and Craig Penterson	2004 April	Riley Power Inc.	Babcock Power Inc.
2	Grinding systems for advanced firing systems	Mechthild Angleys, Wolfgang Rieker, and Peter Stegelitz	2002	ALSTOM Power Boiler GmbH	ALSTOM
3	Dynamic classifiers improve pulverizer performance and more	Robert E. Sommerlad and Kevin L. Dugdale	2007 July	Loesche Energy Systems Ltd	Power
4	Ball mill pulveriser design	A. E. Kukoski	1992	FW Energy corp.	FWC
5	ATRITA pulverizer system upgrade for PRB coal conversion	Steve Stodden and Qingsheng Lin	2006 May	City Utilities of Springfield and Riley Power Inc.	Babcock Power
6	Optimizing pulverizer and riffler performance	R. Brown	2003 December	EPRI	EPRI
7	Pulverizer performance upgrades lower fuel costs	Teresa Hansen	2007 May	Power Engineering	Power Engineering
8	Coal mill and combustion optimization on a once-through, supercritical boiler with multivariable predictive control	Steve Barnoski, Donald Labbe, Jim Graves and William Poe Manager	2005 June	Dayton Power & Light and Invensys	ISA POWID Symposium
Pulverised Fuel (PF) Boilers					
1	Coal characteristics and biomass firing in pulverized coal boilers	David Tillman Richard Conn Dao Duong.	2009 August	Foster Wheeler North America Corp	FWC
2	Anthracite firing in large utility arch fired boilers	Justin P. Winkin and J. Antonio Garcia-Mallold	1997 April	FW Energy International, Inc.	FWC
3	Anthracite firing—largest steam generators	P. Brower/J. Winkin and Ge Changqin		FW and Hebei Elec. Power Corporation	FWC

continued

	Technical Paper	Author	Year	Organisation	Publication
4	B&W'S advanced coal-fired low emission boiler system commercial generating unit and proof-of-concept demonstration	D.K. McDonald and D.A. Madden	1997 November	B&W	B&W
5	New coal-fired steam generator design for high plant efficiency and low emissions	Brian Vitalis and Phillip J. Hunt	2000 November	Riley Power Inc.	Babcock Power
6	Improvements in pulverised coal combustion technology for power generation	Central research institute of electric power institute, Japan (CRIEPI)			CRIEPI
7	Optimum air flow management on coal fired burners	Innovative Combustion Technologies			
Refuse Derived Fuel (RDF)					
1	Design and performance requirements of FB boiler firing municipal RDF in Ravenna, Italy	Michael M. Murphy	1999 May	Energy products of Idaho	EPI
2	Design and performance of EPI fluidized bed RDF-fired power plants worldwide	Michael M. Murphy		Energy products of Idaho	EPI
3	Thermal treatment of waste fuels in CFB boilers—latest design and experience	H. Anderl/E. Offenbacher/T. Ortner	2004	Austrian Energy & Environment AG	
4	Fireside tube corrosion in an industrial RDF-fired boiler—Kodak's experience	Robert D. Blakley		Eastman Kodak Co. Rochester, NY	
5	Considerations for the design of RDF-fired refuse boilers	J.S. Gittinger and W.J. Arvan	1998	B&W, Palm Beach Resource, Florida	B&W
6	Experience of fluidised bed technology for biomass plants in different applications and development towards new applications in RDF burning.	Kari Niemelä and Kim Westerlund	2003 May	Foster Wheeler Energie GmbH	FWC
Refractory and Insulation					
1	CFB refractory improvements for biomass co-firing	Andreas W. Rau		Foster Wheeler	Power Engineering
2	CFB refractory repair	C. Sur, A. Nagar, D.K. Singh and I.N. Chakraborty		ACC Refractories, India	Power, Feb 2006
3	Understanding refractory failures	Gary J. Bases	2006 February	BRIL Inc.	Power
4	Insulation and lagging fundamentals	Gary Bases	2005 February	BRIL Inc.	Power
Super Critical Technology					
1	Design factors and water chemistry practices—supercritical power cycles	Frank Gabrielli and Horst Schwevers	2008 September	Alstom USA and Germany	Alstom
2	Advantages of ultra super critical technology in power generation	Hans H. Poulsen	2005 May	Burmeister & Wain Energy A/S	BWE
3	FWPG Benson VT boiler process and operational description			Foster Wheeler power group	FWC
4	Design and operating experience of supercritical pressure coal fired plant	Paul Armstrong and others		Hitachi USA and Japan	Hitachi
5	Supercritical boiler technology matures	Mark Richardson and others		Hitachi USA and Japan	Hitachi
6	Constant and sliding pressure options For new supercritical plants	Brian P. Vitalis and Phillip J. Hunt	2005 December	Riley Power Inc.	Babcock Power

7	Steam generators for the next generation of power plants aspects of design and operating performance	Dr. J. Franke, R. Kral and E. Wittchow	1999 December	Siemens AG, Power Generation KWU	VGB Power Tech
8	Supercritical boiler technology for future market conditions	Joachim Franke and Rudolf Kral	2003 October	Siemens Power Generation KWU	Siemens
9	Innovative boiler design to reduce capital cost and construction time	Joachim Franke and Rudolf Kral	2000	Siemens Power Generation KWU	Siemens
10	Coal fired plants: horizontal boilers make 700°C steam economic	David Smith	2000 May	Siemens Power Generation KWU	Modern Power Systems
SC-CFB Boilers					
1	The utility cfb boiler—present status, short and long term future with supercritical and ultra-supercritical steam parameters	Stephen J. Goidich, and Ragnar G. Lundqvist		Foster Wheeler	FWC
2	Design aspects of the Uultra-supercritical CFB boiler	Stephen J. Goidich and others	2005	FW North America Corporation	Pittsburgh Coal Conference
3	Supercritical boiler options to match fuel combustion characteristics	Stephen J. Goidich		FW North America Corporation	FWC
4	Integration of ultra-supercritical OTU and CFB boiler technologies	Stephen J. Goidich and others		FW	
5	Lagisza 460 MWe supercritical CFB—design, start-up and initial operation experience	Timo Jäntti and Riku Parkkonen	2009	FW Energia Oy, Varkaus, Finland	FWC
6	World's largest circulating fluidized bed boiler begins commercial operation	Bob Giglio	2009	Foster Wheeler Global Power Group	Power, December 2009
7	Lagisza, world's largest CFB boiler, begins commercial operation	Kalle Nuortimo	2010	Foster Wheeler Varkaus, Finland	Modern Power Systems, April 2010
8	Turów rehabilitation project. The world's largest CFB repowering project	Rafal Psik and Janusz Jablonski Jan Wyszynski		Energia Polska Sp. Elektrownia Turów S.A	FWC
9	Compact circulating fluidized bed boiler operational experiences of a 262 mw_e at Turow power station	Sekret robert, mirek pawel, nowak wojciech, and jablonski janusz, walkowiak roman	2005	Czestochowa univ. of tech. and Power plant Turow, Poland	
10	Design and scale up philosophy of OT CFB boilers with SC parameters	Arto Hotta and Ilkka Venalailen	2006	FW Energia Oy, Varkaus, Finland	VGB Power Tech, April 2006
11	Supercritical boiler options for firing low volatile Chinese coals	Stephen J. Goidich and David E. Wagner		FW North America corp.	FWC
SC PF Boilers					
1	The World's first supercritical FW-BENSON vertical PC boiler—the 750 MWe Longview Power Project	Stephen J. Goidich, Richard J. Docherty, and Kenneth P. Melzer	2009 May	FW North America Corp.	FWC
2	Babcock & Wilcox Company Supercritical (Once Through) boiler technology	J.W. Smith	1998 May	B&W, USA	B&W
3	Developments in pulverized coal-fired boiler technology	J.B. Kitto	1996 April	B&W, USA	B&W
4	Vertical tube variable pressure furnace for supercritical steam boilers	D.K. McDonald and S. S. Kim	2001 December	B&W and US Dept. of Energy	B&W
5	Experiences with coal fired USC boilers in Denmark	Knud Bendixen		Burmeister & Wain energy A/S	BWE

continued

	Technical Paper	Author	Year	Organisation	Publication
Stoker Firing					
1	Considerations for the design of rdf-fired refuse boilers	J. S. Gittinger and W. J. Arvan	1998 June	B&W, Palm Beach Resource Recovery	B&W
2	Fundamentals of stoker fired boiler design and operation	Neil Johnson	2002 July	Detroit Stoker Company	Detroit Stoker Company
3	Renewable fuel grate firing combustion technology—the European experience	Robert S. Morrow		Detroit Stoker Company	Detroit Stoker Company
4	Characterising fuels for biomass—coal fired cogeneration	Norman Magasiner		Thermal Energy systems, SA	Thermal Energy systems
Waste Heat Recovery					
1	Waste heat recovery			Bureau of Energy Efficiency	
Waste to Energy					
1	Technology options for municipal solid waste-to-energy project	Sudhir Kumar	2000 June	TIMES (TERI Information Monitor on Enviro. Science)	
Water					
1	Boiler and feed water treatment: Do the right thing(s)	David Daniels	2004 September	Power-Volume148, Issue7	
2	Boiler feed water—reducing scale and corrosion	C.F. "Chubb" Michaud	2001 April	Water Conditioning & Purification	
3	Boiler systems		1998 June		Gulf Coast Chemical
Wood Firing					
1	Modern wood fired boiler designs—history and technology changes	Kevin Toupin	1995 August	Riley Power Inc.	Babcock Power Inc.

Appendix 1: Combustion Constants

TABLE A1.1

Combustion Constants of American Gas Association Combustion Constants[a,b,c]

							Heat of Combustion				Moles/Mole or ft³/ft³ Combustible						lb/lb Combustible					
							Btu/ft³		Btu/lb		Required for Combustion			Flue Products			Required for Combustion			Flue Products		
No.	Substance	Formula	Molecular Weight	lb/ft³	ft³/lb	Sp gr. Air 1000	Gross	Net	Gross	Net	O_2	N_2	Air	CO_2	H_2O	N_2	O_2	N_2	Air	CO_2	H_2O	N_2
1	Carbon[c]	C	12.01	....	....	....	.....	.....	14.093	14.093	1.0	3.76	4.76	1.0	....	3.76	2.66	8.86	11.53	3.66	....	8.86
2	Hydrogen	H_2	2.016	0.0053	187.723	0.0696	325	275	61.100	51.623	0.5	1.88	2.38	....	1.0	1.88	7.94	26.41	34.34	....	8.94	26.41
3	Oxygen	O_2	32.000	0.0846	11.819	1.1053																
4	Nitrogen (atm)	N_2	28.016	0.0744	13.443	0.9718																
5	Carbon monoxide	CO	28.01	0.0740	13.506	0.9672	322	322	4.347	4.347	0.5	1.88	2.38	1.0	....	1.88	0.57	1.90	2.47	1.57	....	1.90
6	Carbon dioxide	CO_2	44.01	0.1170	8.548	1.5282																
Paraffin Series																						
7	Methane	CH_4	16.041	0.0424	23.565	0.5543	1013	913	23.879	21.520	2.0	7.53	9.53	1.0	2.0	7.53	3.99	13.28	17.27	2.74	2.25	13.28
8	Ethane	C_2H_6	30.067	0.0803	12.455	1.0488	1792	1641	22.320	20.432	3.5	13.18	16.68	2.0	3.0	13.18	3.73	12.39	16.12	2.93	1.80	12.39
9	Propane	C_8H_8	44.092	0.1196	8.365	1.5617	2590	2385	21.661	19.944	5.0	18.82	23.82	3.0	4.0	18.82	3.63	12.07	15.70	2.99	1.63	12.07
10	*n*-Butane	C_4H_{10}	58.118	0.1582	6.321	2.0665	3370	3113	21.308	19.680	6.5	24.47	30.97	4.0	5.0	24.47	3.58	11.91	15.49	3.03	1.55	11.91
11	Isobutane	C_4H_{10}	58.118	0.1582	6.321	2.0665	3363	3105	21.257	19.629	6.5	24.47	30.97	4.0	5.0	24.47	3.58	11.91	15.49	3.03	1.55	11.91
12	*n*-Pentane	C_6H_{12}	72.144	0.1904	5.252	2.4872	4016	3709	21.091	19.517	8.0	30.11	38.11	5.0	6.0	30.11	3.55	11.81	15.35	3.05	1.50	11.81
13	Isopentane	C_6H_{12}	72.144	0.1904	5.252	2.4872	4008	3716	21.052	19.478	8.0	30.11	38.11	5.0	6.0	30.11	3.55	11.81	15.35	3.05	1.50	11.81
14	Neopentane	C_6H_{12}	72.144	0.1904	5.252	2.4872	3993	3693	20.970	19.396	8.0	30.11	38.11	5.0	6.0	30.11	3.55	11.81	15.35	3.05	1.50	11.81
15	*n*-Hexane	C_6H_{14}	86.169	0.2274	4.398	2.9704	4762	4412	20.940	19.403	9.5	35.76	45.26	6.0	7.0	35.76	3.53	11.74	15.27	3.06	1.46	11.74
Olefin Series																						
16	Ethylene	C_2H_4	28.051	0.0746	13.412	0.9740	1614	1513	21.644	20.295	3.0	11.29	14.29	2.0	2.0	11.29	3.42	11.39	14.81	3.14	1.29	11.39
17	Propylene	C_2H_4	42.077	0.1110	9.007	1.4504	2336	2186	21.041	19.691	4.5	16.94	21.44	3.0	3.0	16.94	3.42	11.39	14.81	3.14	1.29	11.39

18	*n*-Butene	C_4H_8	56.102	0.1480	6.756	1.9336	3084	2885	20.840	19.496	6.0	22.59	28.59	4.0	4.0	22.59	3.42	11.39	14.81	3.14	1.29	11.39
19	Isobutene	C_4H_8	56.102	0.1480	6.756	1.9336	3068	2869	20.730	19.382	6.0	22.59	28.59	4.0	4.0	22.59	3.42	11.39	14.81	3.14	1.29	11.39
20	*n*-Pentene	C_6H_{10}	70.128	0.1852	5.400	2.4190	3836	3686	20.712	19.363	7.5	28.23	35.73	5.0	6.0	28.23	3.42	11.39	14.81	3.14	1.29	11.39
Aromatic Series																						
21	Benzene	C_6H_6	78.107	0.2060	4.852	2.6920	3751	3601	18.210	17.480	7.5	28.23	35.73	6.0	3.0	28.23	3.07	10.22	13.30	3.38	0.69	10.22
22	Toluene	C_7H_8	92.132	0.2431	4.113	3.1760	4484	4284	18.440	17.620	9.0	33.88	42.88	7.0	4.0	33.88	3.13	10.40	13.53	3.34	0.78	10.40
23	Xylene	C_8H_{10}	106.158	0.2803	3.567	3.6618	5230	4980	18.650	17.760	10.5	39.52	50.02	8.0	5.0	39.52	3.17	10.53	13.70	3.32	0.85	10.53
Miscellaneous Gases																						
24	Acetylene	C_2H_6	26.036	0.0697	14.344	0.9107	1499	1448	21.500	20.776	2.5	9.41	11.91	2.0	1.0	9.41	3.07	10.22	13.30	3.38	0.69	10.22
25	Naphthalene	$C_{10}H_3$	128.162	0.3384	2.955	4.4208	5584	5654	17.298	16.708	12.0	45.17	57.17	10.0	4.0	45.17	3.00	9.97	12.96	3.43	0.56	9.97
26	Methyl alcohol	CH_3OH	32.041	0.0846	11.820	1.1052	868	768	10.259	9.078	1.5	5.65	7.15	1.0	2.0	5.65	1.50	4.98	6.48	1.37	1.13	4.98
27	Ethyl alcohol	C_2H_3OH	46.067	0.1216	8.221	1.5890	1600	1451	13.161	11.929	3.0	11.29	14.29	2.0	3.0	11.29	2.08	6.93	9.02	1.92	1.17	6.93
28	Ammonia	NH_3	17.031	0.0456	21.914	0.5961	441	365	9.668	8.001	0.75	2.82	3.57	....	1.5	3.32	1.41	4.69	6.10	....	1.59	5.51
														SO						SO		
29	Sulphur[c]	S	32.06	......	......	......	....	....	3.983	3.983	1.0	3.76	4.76	1.0	....	3.76	1.00	3.29	4.29	2.00	....	3.29
30	Hydrogen sulphide	H_2S	34.076	0.0911	10.979	1.1898	647	596	7.100	6.545	1.5	5.65	7.15	1.0	1.0	5.65	1.41	4.69	6.10	1.88	0.53	4.69
31	Sulphur dioxide	SO_2	64.06	0.1733	5.770	2.264																
32	Water vapour	H_2O	18.016	0.0476	21.017	0.6215																
33	Air	......	28.9	0.0766	13.063	1.000																

[a] Based on Fuel Flue Gases, 1941 edition, American Gas Association.
[b] All gas volumes corrected to 60°F and 30 in Hg dry.
[c] Carbon and sulphur are considered as gases for molal calculations only.

Appendix 2: The Periodic Table of Elements

The Perodic Table of Elements (Table A2.1) accompanied by a List of Elements, Atomic Numbers and Masses (Table A2.2) given in this Appendix 2 provide highly useful ready reference material for readers in their combustion and other chemical calculations.

TABLE A2.1

Periodic Table of Elements

Legend: Alkali metals · Alkaline Earth metals · Transition metals · Lanthanide · Actinide · Other metals · Non-metals · Halogens · Inert gases

Columns = groups I–VIII = number of e- in outer shell

Rows = periods 1-7 = outer shell number

←IA	IIA	IIIB	IVB	VB	VIB	VIIB	⌐VIIIB		¬	IB	IIB	IIIA	IVA	VA	VIA	VIIA	VIIIA→
1 H 1.00																	2 He 4.00
3 Li 6.94	4 Be 9.01											5 B 10.8	6 C 12.0	7 N 14.0	8 O 16.0	9 FI 19.0	10 Ne 20.2
11 Na 22.9	12 Mg 24.3											13 Al 27.0	14 Si 28.1	15 P 31.0	16 S 32.1	17 Cl 35.4	18 Ar 39.9
19 K 39.1	20 Ca 40.1	21 Sc 44.9	22 Ti 47.9	23 V 50.9	24 Cr 52.0	25 Mn 54.9	26 Fe 55.8	27 Co 58.9	28 Ni 58.6	29 Cu 63.5	30 Zn 65.3	31 Ga 69.7	32 Ge 72.6	33 As 74.9	34 Se 79.0	35 Br 79.9	36 Kr 83.8
37 Ru 85.5	38 Sr 87.6	39 Y 88.9	40 Zr 91.2	41 Nb 92.9	42 Mo 96.0	43 Tc 98	44 Ru 101	45 Rh 102	46 Pd 106	47 Ag 108	48 Cd 112	49 In 115	50 Sn 119	51 Sb 122	52 Te 128	53 I 127	54 Xe 131
55 Cs 132	56 Ba 137	57 La 139	72 Hf 178	73 Ta 181	74 W 184	75 Re 186	76 Os 190	77 Ir 192	78 Pt 195	79 Au 197	80 Hg 200	81 Tl 204	82 Pb 207	83 Bi 210	84 Po 209	85 At 210	86 Rn 222
87 Er 223	88 Ra 226	89 Ac 277	104 Rf 267	105 Db 268	106 Sg 271	107 Bh 272	108 Hs 270	109 Mt 276	110 Ds 281	111 Rg 280							
	Lanthanide		58 Ce 140	59 Pr 141	60 Nd 144	61 Pm 145	62 Sm 150	63 Eu 152	64 Gd 157	65 Tb 159	66 Dy 162	67 Ho 165	68 Er 167	69 Tm 169	70 Yb 173	71 Lu 175	
	Actinide		90 Th 232	91 Pa 231	92 U 238	93 Np 237	94 Pu 244	95 Am 243	96 Cm 247	97 Bk 247	98 Cf 251	99 Es 252	100 Fm 257	101 Md 258	102 No 259	103 Lr 262	

Group: There are 18 groups in the periodic table that constitute the columns of the table. Lanthanoids and actinoids are numbered as 101 and 102 to separate them in sorting by group.

Atomic number: The number of protons in an atom. Each element is uniquely defined by its atomic number.

Atomic mass: The mass of an atom is primarily determined by the number of protons and neutrons in its nucleus. Atomic mass is measured in Atomic Mass Units (amu) which are scaled relative to carbon, ^{12}C, that is taken as a standard element with an atomic mass of 12. This isotope of carbon has six protons and six neutrons. Thus, each proton and neutron has a mass of ~1 amu.

Isotope: Atoms of the same element with the same atomic number, but different number of neutrons. Isotope of an element is defined by the sum of the number of protons and neutrons in its nucleus. Elements have more than one isotope with varying numbers of neutrons. For example, there are two common isotopes of carbon, ^{12}C and ^{13}C which have six and seven neutrons, respectively. The abundances of different isotopes of elements vary in nature depending on the source of materials. For relative abundances of isotopes in nature see reference on **Isotopic Composition of the Elements**.

Atomic weight: Atomic weight values represent weighted average of the masses of all naturally occurring isotopes of an element. The values shown here are based on the IUPAC Commission determinations.

TABLE A2.2

List of Elements, Atomic Numbers and Masses

Element		Z	Mass	Element		Z	Mass	Element		Z	Mass	Element		Z	Mass	Element		Z	Mass	Element		Z	Mass
Hydrogen	H	1	1.008	Calcium	Ca	20	40.08	Yttrium	Y	39	88.91	Cerium	Ce	58	140.1	Iridium	Ir	77	192.2	Curium	Cm	96	247
Helium	He	2	4.003	Scandium	Sc	21	44.96	Zirconium	Zr	40	91.22	Praesodymium	Pr	59	140.9	Platinum	Pt	78	195.1	Berkelium	Bk	97	247
Lithium	Li	3	6.941	Titanium	ti	22	47.88	Niobium	Nb	41	92.91	Neodymium	Nd	60	144.2	Gold	Au	79	197.0	Californium	Cf	98	249
Beryllium	Be	4	9.012	Vanadium	V	23	50.94	Molybdenum	Mo	42	95.94	Promethium	Pm	61	145	Mercury	Hg	80	200.6	Einsteinium	Es	99	254
Boron	B	5	10.81	Chromium	Cr	24	52.00	Technetium	Tc	43	98	Samarium	Sm	62	150.4	Thallium	Tl	81	204.4	Fermium	Fm	100	253
Carbon	C	6	12.01	Manganese	Mn	25	54.94	Ruthenium	Ru	44	101.1	Europium	Eu	63	152.0	Lead	Pb	82	207.2	Mendelevium	Md	101	256
Nitrogen	N	7	14.01	Iron	Fe	26	55.85	Rhodium	Rh	45	102.9	Gadolinium	Gd	64	157.3	Bismuth	Bi	83	209	Nobelium	No	102	253
Oxygen	O	8	16.00	Cobalt	Co	27	58.93	Palladium	Pd	46	106.4	Terbium	Tb	65	158.9	Polonium	Po	84	209	Lawrencium	Lr	103	257
Fluorine	F	9	19.00	Nickel	Ni	28	58.70	Silver	Ag	47	107.9	Dysporsium	Dy	66	162.5	Astatine	At	85	210	Rutherfordium	Rf	104	263
Neon	Ne	10	20.18	Copper	Cu	29	63.55	Cadmium	Cd	48	112.4	Holmium	Ho	67	164.9	Radon	Ru	86	222	Dubnium	Db	105	262
Sodium	Na	11	22.99	Zinc	Zn	30	65.41	Indium	In	49	114.8	Erbium	Er	68	167.3	Francium	Fr	87	223	Seaborgium	Sg	106	266
Magnesium	Mg	12	24.31	Gallium	Ga	31	69.72	Tin	Sn	50	118.7	Thulium	Tm	69	168.9	Radium	Ra	88	226	Bohrium	Bh	107	267
Aluminum	AI	13	26.98	Germanium	Ge	32	72.61	Antimony	Sb	51	121.8	Ytterbium	Yb	70	173.0	Actinium	AC	89	227	Hassium	Hs	108	277
Silicon	Si	14	28.09	Arsenic	As	33	74.92	Tellurium	Te	52	127.6	Lutetium	Lu	71	175.0	Thorium	Th	90	232	Meitnerium	Mt	109	268
Phosphorous	P	15	30.97	Selenium	Se	34	78.96	Iodine	In	53	126.9	Hafnium	Hf	72	178.5	Protactinium	Pa	91	231	Darmstadium	Ds	110	281
Sulphur	S	16	32.07	Bromine	Br	35	79.9	Xenon	Xe	54	131.3	Tantalum	Ta	73	180.9	Uranium	U	92	238	Roentgenium	Rg	111	272
Chlorine	CI	17	35.45	Krypton	Kr	36	83.80	Cesium	Cs	55	132.9	Tungsten	W	74	183.9	Neptunium	Np	93	244				
Argon	Ar	18	39.95	Rubidium	Rb	37	85.47	Barium	Ba	56	137.3	Rhenium	Re	75	186.2	Plutonium	Pu	94	242				
Potassium	K	19	39.10	Strontium	Sr	38	87.62	Lanthanum	La	57	138.9	Osmium	Os	76	190.2	Americum	Am	95	243				

Appendix 3: Coals of the World

Coal is the most well distributed of all prime fuels across the globe. Coal exists in most countries worldwide and recoverable reserves are in nearly 70 countries. The biggest reserves are in the United States, Russia, China and India. There are sizable assets in Australia, South Africa and Kazakhstan. Brown coal/lignite reserves are also bountiful in the United States, Russia and China. Figure A3.1 shows the reserves of coal and lignite across the world taken from *BP Statistical Review of World Energy* (2009).

With the present rate of use, coal is expected to last for about 120 years. However, at the present rate of mining in China, the coal reserves are not expected to last beyond the next 40 years.

Classification of coals is by their rank. On the basis of their use, coals and lignites are sorted in a slightly different manner, as shown in Figure A3.2 of World Coal Association. Low-rank (lignites and sub-bituminous) coals and large amount of hard (non-metallurgical

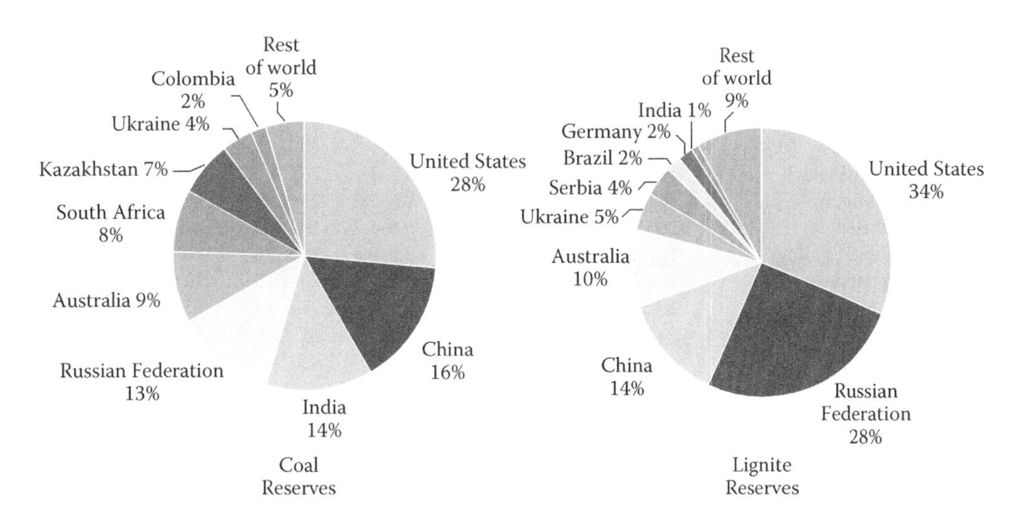

FIGURE A3.1
Reserves of coal and lignite world wide. (BP Statistical review 2009.)

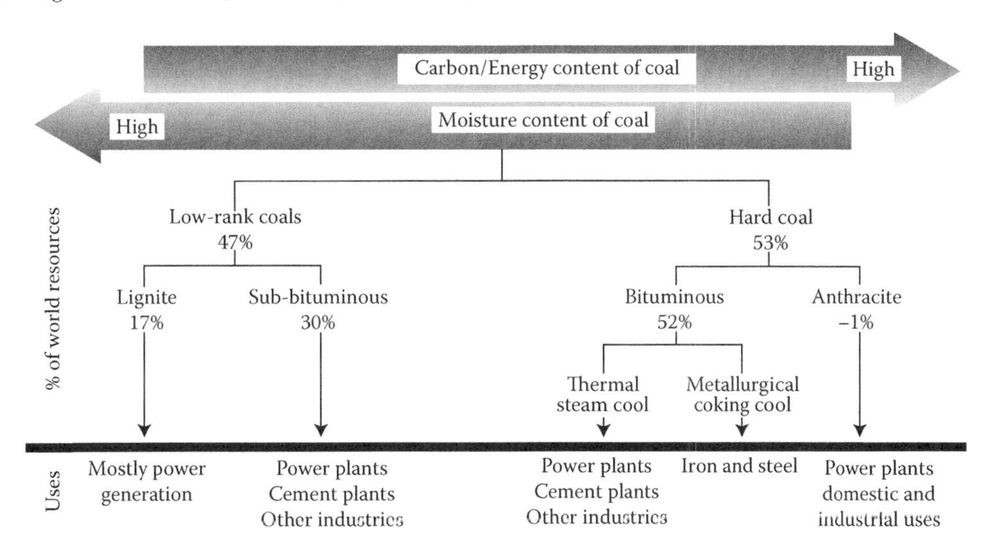

FIGURE A3.2
Coal classification by the end use.

TABLE A3.1

Typical As-Received Analysis of Coals from Various Countries

Country	Mine/Area	Ash%	M%	VM	S%	GCV		
						MJ/kg	kcal/kg	Btu/lb
Austria	Grunbach	5–12	4–6	35		25.1	6000	10,790
Australia	Queensland	11.6	1.5	37	0.7	28.61	6840	12,300
	NSW west	14.6	2.6	30.0	3.7–4.0	27	6450	11,610
	NSW south	11	0.6	23.3		30.5	7290	13,115
Belgium	Low volatile coal	5–7	4–6	14–18		31.8	7550	13,675
	Dry steam coal	7–9	3–6	8–10		31.4	7505	13,500
Czech Republic	Pilsen	9–16	6–13	30–35		20.9–27.2	4995–6500	8990–11,700
	Sadovy	8–15	2–7	27–34		27.2–30.1	6500–7200	11,700–12,940
China	Chihli Kaiping	13.3	0.6	26		31.1	7435	13,370
	Chihli Penchihu	11.2	0.7	234		31.4	7505	13,500
France	Loire	17–25	3–5.5	8–18		23–25.1	5500–6000	9890–10,790
	Nord	5–7	2–4	19–21		29.3–32.2	7000–7700	12,600–13,850
	St.Etienne	12–17	4–5	20–30		26.4–28.1	6310–6715	11,350–12,080
Germany	Aachen (dry steam)	6	4	9–13		31.7	7575	13,630
	Aachen (Med. vol.)	6	4	24–30		31.5	7530	13,545
	Bavaria	11	10	35–40		22.5	5380	9675
	Ruhr	7–9	2–7	10–15		29.7–31.0	7100–7410	12,770–13,330
	Saar	3–7	3–5	38		29.3	7000	12,600
	Saxony	5–6	8	30–36		28.1	6715	12,080
India	Bengal slack	16.2	0.5	9.9	0.3–1	28.3	6760	12,170
	Raniganj	14–17	5–11	34–52	0.6–2.5	23.8–25.9	5690–4900	10,235–11,140
	Orissa	28–31	7–9	27–31	0.6–1.7	19.3–20.5	4610–4900	8300–8815
	Pench Valley	16–32	2–9	17–33	0.5–0.8	20.1–26	4800–6214	8640–11,180
	Chanda	13–15	10–11	31–36		23–25.5	5500–6095	9890–10,965
	Tandur	18–22	5–7	26–30		21.3–24.3	5090–5810	9160–10,450
Indonesia	Palembang Dust	6.6	10.0	28.4		26	6215	11,180
Italy		19	5	39		25.5	6095	10,965
Japan	Kyushu Furukawa	14.4	1.6	34.9		29.3	7000	12,600
Poland	Upper Silesia	4–9	3–7	30–35		28.5–29.7	6810–7100	12,255–12,770
	Katowic	6	4	31.6		31.4	7505	13,500
Russia	Kuznets	5–10	1–7	13–25		28.9–31.4	6910–7505	12,430–13,500
	Ural	21	7	25		33.3	7960	14,320
	Donnetz long flame	21.5	7	40	5.7	23.9	5710	10,280
South Africa	Natal	8.6	4.2	16.7	4.18			
	Transvaal	13.3	2.2	27	0.7	28.18	6735	12,120
Spain	Asturia	8–11	5	7–30		29.3–31.4	7000–7505	12,600–13,500
	Zaragoza	11.3	4–6	28.8		26	6215	11,180
Turkey	Isletme	12	7	30		23	5500	9890
	Zonguldak	15.5	10	27		26	6215	11,180
UK	Cardiff	3–8	1–3	25–35		31.4–33.5	7505–8005	13,500–14,405
	Cumberland	5.5–8.0	2.5–3.5	30–36		31–31.8	7410–7600	13,330–13,680
	Derbyshire	2–7	4.5–12	31–39		27.2–31.4	6500–7505	11,700–13,500
	Durham	3–10	2–8	25–35		28.1–31.8	6715–7600	120,80–13,680
	Midlands	3–7	2–10	21–37		28.1–31.8	6715–7600	120,80–13,680
USA	Alabama	9.2	3.6	30.5		31.6	7550	13,590
	Maryland	6–9	1.5–2.5	20–40		31.8–33.1	7600–7910	13,680–14,230
	Pennsylvania	6–9	1.0–2.5	20–35		32.2–33.1	7695–7910	13,850–14,230
	SW of Virginia	5–7	1.5–2.5	20–35		33.1–34.3	7910–8200	14,230–14,750

bituminous) coals, together contributing to over 70% of coal production, are deployed for power and process steam generation in power stations and industrial plants.

The world is predominantly dependent on coal for electricity generation largely because of the wide geographical distribution and competitive price. More importantly, the price is stable and relatively free from violent gyrations experienced with FO. No doubt the price of coal escalated by an unprecedented factor of nearly 5 between 2003 and 2008, but it remained less

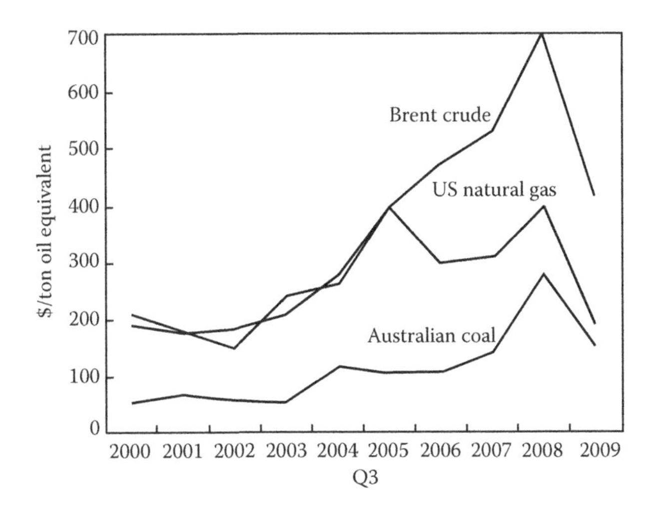

FIGURE A3.3
Variation of fuel prices during 2000–2009. (FR based on Patt's.)

expensive per unit of energy than FO or NG as shown in Figure A3.3 taken from FTP, Paris.

Because of decently high cv, bituminous coals are widely traded, and countries like Japan and South Korea with little energy sources of their own produce substantial electric power from imported coals. Australia, Indonesia, Russia and South Africa are the largest coal exporters.

Coal is the dirtiest of all fossil fuels. It pollutes the environment both on account of the dust and noxious emissions (NO_x and SO_x mainly). Over the years, satisfactory solution to the emissions has been evolved aimed at both reduction and capturing the emissions so that the environment is acceptably clean. However, emission of GHG (CO_2 and CH_4) cannot be avoided. Coal is the worst offender that contributes to more than about 30% and 70% GHG compared with the other fossil fuels such as FO and NG, respectively. Unless CO_2 capture and sequestration (CCS) or oxy-fuel combustion processes are commercialised soon enough, the dominant position of coal may be forfeited in favour of NG which can provide far higher cycle efficiencies and much lower GHG emissions.

Coals are widely available across the globe but their characteristics are also widely variable. Tables A3.1 and A3.2, taken from my book *Boilers for Power and Process* (CRC Press, 2009), provide a brief summary of world coals and lignites to give an overview.

TABLE A3.2
Typical as Received Analysis of Lignites from Various Countries

		Proximate Analysis						
Country	Mine/Area	M%	A%	VM%	FC%	Gcv kcal/kg	S %	HGI
USA	North Dakota	34.8	6.2	28.2	30.8	4005		
	Texas	33.7	7.3	29.3	29.7	4080		
Australia	Victoria	66.3	0.7	17.7	15.3	2055		
Canada	Saskatchewan	35	8.0	23.4	34.6	4050		
Germany	Frimmersdorf	60.7	2.6	20.0	16.7	2355		
	Saxony	53.1	3.5	25.3	18.1	2500		
	Bitterfield	50.0	6.5	57				
Greece	Aliveri	31.0	18.0	30.0	21.0	3130		
India	Neyveli	53.1	4.5	24.2	18.2	2785		
	Surat	24.0	17.0	39.0	30.0	4200		
	Bikaner	51.0	7.3	24.3	17.4	2700	0.8	84
	Kutch	34.4	7.2	32.8	25.6	4090	1.77	102
Russia-Ural	Bogoslovsk	30.0	14.0	22.3	29.7	3385	0.5	
Russia-Far east	Kirdinsky	33.0	12.7	22.3	32.0	3478	0.3	
Spain		47.7	18.8	17.7	15.8	1700		
Ukraine	Aleksandrisk	55.0	10.8	19.5	14.7	2308	3.6	

Note: Conversions from MJ/kg have been rounded up to the nearest 5 to kcal/kg and Btu/lb.

Index

C

G

For Product Safety Concerns and Information please contact our EU representative GPSR@taylorandfrancis.com Taylor & Francis Verlag GmbH, Kaufingerstraße 24, 80331 München, Germany

Printed and bound by CPI Group (UK) Ltd, Croydon, CR0 4YY
01/05/2025
01858592-0001